Under the Heel of Bushido

Other books by Martin Sugarman
published by Vallentine Mitchell

Fighting Back: British Jewry's Military Contribution in The Second World War

We Will Remember Them: A Record of the Jews who died in the Armed Forces of the Crown from 1939 (co-author Henry Morris)

To be published

The Update to the British Jewry Book of Honour of World War One, researched by Harold Pollins, edited and compiled by Martin Sugarman

Jews in the Fire Brigade in the Second World War

Under the Heel of Bushido

Last Voices of Jewish POWs of the Japanese in the Second World War

Martin Sugarman

VALLENTINE MITCHELL
LONDON • PORTLAND, OR

The publication of this book would not have been possible without the generous support of the Pears Foundation.

pears foundation

First published in 2014 by Vallentine Mitchell

Middlesex House,
29/45 High Street, Edgware,
Middlesex, HA8 7UU

920 NE 58th Avenue, Suite 300
Portland, Oregon,
97213-3786 USA

www.vmbooks.com

British Library Cataloguing in Publication Data
A record can ben found on request

ISBN 978 0 85303 877 1 (cloth)
ISBN 978 0 85303 887 0 (Ebook)
ISBN 978 0 85303 908 2 (paperback)

Library of Congress Cataloging-in-Publication Data
An entry can be found on request

Typeset by FiSH Books, Enfield, Middx.
Printed & bound by TJ International Ltd, Padstow, Cornwall.

For all our 3,000 British Jewish veterans and 700 from Israel,
who never made it home after the Second World War

***

שלחני לחבש לנשברי-לב, לקרא לשבוים דרור, ולאסורים פקח-קוח:

He has sent me to bind up the broken-hearted, to proclaim liberty
to the captives, and the opening of the eyes to them that are bound.
(Isaiah 61:1)

***

And for my flowers:

Jane
Gideon, Joel and Leah
Lauryn and Lylah
Polly and Mia
and Eliora

And my mother, Great Grandma Leah

The lizard runs across the sun-drenched earth,
Legs spread and flying, tail erect behind,
Then stops – as only wild things can.
To emulate dead leaves, the earth and fallen trees.

So, too, we ran, the high fierce sun of youth
Beat upon our hearts, we lived and sang.
And so have stopped – and are become
Like dead blown leaves, the earth and fallen trees.

(Untitled, by Signalman Joseph Max Hirsch, RCOC, as a Jewish POW of the Japanese, 'somewhere in Thailand 1942–45'.)

A Prayer

Oh Sleep! Nocturnal genius.
Masking in black velvet day's utter ugliness.
Painting with dreary colours
The loveliness of a hundred happy yesterdays.
Revealing with vivid clarity
The visions of a thousand glorious morrows
Why hast thou forsaken me?

(Written by 5775444 L/Cpl Cyril Wernick, 4th Battalion, Royal Norfolk Regiment, on a sleepless night of pain, in the half-light of Roberts Hospital, Changi gaol, October 1942. Cyril died of diphtheria, contracted in a Japanese POW camp, in hospital in India, on 21 September 1945, on the way back home to England. He was son of Harry and Annie, and husband of Ann, and is buried in Madras, Chennai CWGC cemetery.)

The tears spurted . . . like an irrepressible orgasmic release.
My dam had burst at last and out it all came,
as though all the anguish of the planet had found me,
and out of my bowels all the murders, tortures,
all the Jews burning in Belsen,
all the hellish years,
had gathered in a million voices bursting all bounds.

(Written by Signalman William Allister, RCCOS, on seeing an American aircraft dipping its wings in recognition, as it flew over his prison camp in Japan, 21 August 1945, to announce liberation at last.)

Contents

Illustrations

The illustrations and photographs are credited wherever the copyright holder is known. However, whilst every effort has been made to obtain permission to reproduce the pictures, the author will rectify any omissions if he receives further information as to the copyright owner. All photographs used are published by the kind permission of the family or owner, whether individuals or an organization, or from the Archives of the AJEX Jewish Military Museum.

Abbreviations and Acronyms

AA	Anti-Aircraft
AAOC	Australian Army Ordnance Corps
AASC	Australian Army Service Corps
AB	Able Seaman
ABCIFER	Association of British Civilian Internees in the Far East
AC	Army Corps/Air Controller
ADC	Assistant District Commissioner/Aide de Camp
AGH	Australia General Hospital
AIF	Australian Imperial Force
ARP	Air Raid Precautions
Art.	Artillery
AT	Anti-Tank
ATS	Auxiliary Territorial Service (Women)
Bde	Brigade
Bdr	Bombardier
BEF	British Expeditionary Force
BEM	British Empire Medal
Capt.	Captain
CCS	Casualty Clearing Station
CE	Corps of Engineers
CEF	Canadian Expeditionary Force
CMG	Companion of the Order of St Michael and St George
CMP	Corps of Military Police
CO	Commanding Officer
COFEPOW	Children of Far East Prisoners of War
Coy	Company
Cpl	Corporal
CQMS	Company Quartermaster Sergeant
CSM	Company Sergeant Major

CWGC	Commonwealth War Graves Commission
DFC	Distinguished Flying Cross
Dvr	Driver
ENSA	Entertainments National Service Association
FAA	Fleet Air Arm
FANY	First Aid Nursing Yeomanry (Women)
FEPOW	Far East Prisoners of War
Field Sec. Intel. Corps	Field Security Intelligence Corps
Fl. Off.	Flying Officer
Flt Lt	Flight Lieutenant
FMSVDF	Federated Malay States Volunteer Defence Force
FMSVF	Federated Malay States Volunteer Force
Fus.	Fusilier
GC	George Cross
Gnr	Gunner
GOC	General Officer Commanding
GP	General Practitioner
Gp Capt.	Group Captain
HAA	Heavy Anti-Aircraft
HKVDC	Hong Kong Volunteer Defence Corps
HMAHS	His Majesty's Australian Hospital Ship
HMT	His Majesty's Transport (Ship)
IACC	Indian Army Catering Corps
ICRC	International Committee of the Red Cross
IJA	Imperial Japanese Army
IMS	Indian Medical Service
INA	Indian National Army
Inf.	Infantry
INLA	Indian National Liberation Army
Intel Corps	Intelligence Corps
IWM	Imperial War Museum
JC	*The Jewish Chronicle*
JLB	Jewish Lads Brigade
KIA	Killed in Action
KRRC	King's Royal Rifle Corps
L/Bdr	Lance Bombardier
L/Cpl	Lance Corporal

LAA	Light Anti-Aircraft
LAC	Leading Aircraftman
LCT	Landing Craft Tank
Ldg/Tel	Leading Telegraphist
Lt	Lieutenant
MBE	Member of the the Most Excellent Order of the British Empire
MI	Medical Inspection
MIA	Missing in Action
MiD	Mentioned in Dispatches
(R)MO	(Regimental) Medical Officer
MP	Military Police
MT	Motor Transport
MTB	Motor Torpedo Boat
MU	Motor Unit
NAAFI	Navy, Army and Air Force Institutes
NAM	National Army Museum
NCO	Non-Commissioned Officers
OR	Other Ranks
OTC	Officers' Training Corps
P. Off.	Petty Officer (RN)/Pilot Officer (RAF)
POW	Prisoner of War
Pte	Private
RA	Royal Artillery
RAA	Royal Australian Artillery
RAAF	Royal Australian Air Force
RAC	Royal Armoured Corps
Rad/O	Radio Operator
RADC	Royal Army Dental Corps
RAEC	Royal Army Education Corps
RAFVR	Royal Air Force Volunteer Reserve
RAMC	Royal Army Medical Corps
RAOC	Royal Army Ordnance Corps
RAP	Regimental Aid Post
RAPWI	Rehabilitation of Allied Prisoners of War and Internees
RASC	Royal Army Service Corps
RCAMC	Royal Canadian Army Medical Corps
RCCOS	Royal Canadian Corps of Signals

RCOS	Royal Corps of Signals
RE	Royal Engineers
Regt	Regiment
REME	Royal Electrical and Mechanical Engineers
RIAMC	Royal Indian Army Medical Corps
RIASC	Royal Indian Army Service Corps
Rif.	Rifleman
RMS	Royal Mail Ship or Royal Mail Steamer
RNF	Royal Northumberland Fusiliers
RNZAF	Royal New Zealand Air Force
RSM	Regimental Sergeant Major
RTU	Return to Unit/Returned to Unit
S/Sgt	Staff Sergeant
Sgmn	Signalman
Sgt	Sergeant
SOAS	School of Oriental and African Studies
SOE	Special Operations Executive
Spr	Sapper
SS	Steamship
SSVF	Straits Settlement Volunteer Force
Stkr	Stoker
Sub Lt	Sub Lieutenant
SVC	Singapore Volunteer Corps/Shanghai Volunteer Corps
TD	Territorial Decoration
TNA	The National Archives
Tpr	Trooper
UN	United Nations
USAAF	United States Army Air Forces
Wing Comm.	Wing Commander
WO	Warrant Officer (Army or RAF)
WO/OB	Wireless Operator/Observer (RAF)
WOP/AG	Wireless Operator/Air Gunner (RAF)
WRAF	Women's Royal Air Force
WRENS	Women's Royal Navy Service
YMCA	Young Men's Christian Association
YMHA	Young Men's Hebrew Association

Foreword

Life was divided into two for many people on 1 September 1939 when the Nazis marched into Poland. Before and after. Others, often former Communists, recalled 17 July 1936 as the time that the clocks stopped – when General Franco initiated the Spanish Civil War and aligned himself with Hitler and Mussolini. A few – those who survived Japanese camps and prisons – remember 7 December 1941 when Japan attacked Pearl Harbour.

British Jews of that generation understood that it was only twenty miles of water that saved them from sharing the fate of their brothers and sisters in mainland Europe. They repeated that mantra of salvation incessantly and it was passed down the generations. Others recalled with pride that Jews – premature anti-fascists – were represented disproportionately in the ranks of the International Brigades that stood in the path of Franco's tanks. But who knows about the Jews who bore witness to the extreme brutality which was meted out to many a hapless prisoner in Japanese camps? Forced labour, decapitations, torture, massacres, medical experimentation, starvation rations, death marches, comfort women – all this was part and parcel of the military tradition of the Japanese Imperial forces which looked up to Emperor Hirohito. When his son, Akihito, visited London in 1998 to be invested with the Order of the Garter, Jack Caplan, together with other veterans of the camps, turned their backs on him as a mark of disrespect as he passed by in the Mall. When Jack Caplan set fire to a Japanese flag, the police, the diplomats and the politicians were not amused. While he undoubtedly spoke for a section of popular opinion in Britain, he was also the Jewish dissident who would not forget. And yet he went to Japan a few years later to meet a new generation of Japanese. He returned with an understanding that the sins of the fathers should not be visited upon the sons. Amidst the recurring bouts of malaria, Jack Caplan somehow learned to manage the past and to live in the present. When he died in 2004, the wife of the then Japanese Ambassador to the Court of St James attended his funeral at a Jewish cemetery and brought with her an expression of condolence from the Emperor.

Three centuries ago, the progenitor of Hassidism, the Ba'al Shem Tov, famously said that 'forgetfulness leads to exile while remembrance is the secret of redemption'. This statement rightfully greets visitors to Yad Vashem, the Jerusalem institute which commemorates the Holocaust. To remember what happened in those Japanese camps all those years ago is undoubtedly painful, but to forget is even more agonising – it truly leads to a shameful exile of the intellect. Martin Sugarman has undertaken a mammoth but important task to record the testimonies of those who served and suffered – Jews from London's Hackney and Shoreditch, from Liverpool and Glasgow, from Canada and the United States, from Australia and New Zealand; Sephardi Baghdadi Jews populating the trade route from the Middle East to the Far East; Ashkenazi German Jews escaping Hitler for the Dutch East Indies. This book records how their Jewishness and their Judaism became meaningful and fortified them in dark times; how the communication of a mangled Yiddish was transformed into a defiant allegiance between Jews of different origins.

Religious Jews such as David Arkush and Rabbi Chaim Nussbaum stretched tradition to match the reality they inhabited. Martin Sugarman writes:

> In April 1943, David (Arkush) managed to organize a Passover service with about forty Jewish Allied POWs attending: in lieu of *matzoh* (unleavened bread) and kosher wine, he managed to provide rice cakes made with sago flour, charoset (a condiment symbolizing the cement that the Jews used when making the bricks as slaves in Egypt) from grated coconut rice coffee, *maror* (bitter herbs symbolizing the bitter times inflicted on the Jewish slaves) made from mint leaves, the hard-boiled eggs (symbolizing the circle of life and cycle of spring renewal) made from spinach leaves and rice wine and some bananas.
>
> Those who were unable to walk were brought in on stretchers and those too ill to come were visited afterwards with food and cigarettes. Using appropriate prayers from his British Army Jewish prayer book and a written version of *Adon Olam*, a well-know Jewish hymn, it was held in the Christian chapel.

The traditional Passover cry 'Next Year in Jerusalem' therefore had a special meaning for those who were doing their utmost not to succumb to the black misery of their situation. In Changi prison, a makeshift synagogue was established. Minyanim (a quorum of ten males) were assembled to allow prisoners to say *Kaddish*, the memorial prayer, to remember dead parents. At Bampong the men building the notorious Burma railway held Rosh Hashana and Yom Kippur services in 1942.

The main purpose of all this was more than religious, more than a question of identity. It was to enable survival – to strive towards saying the *shehechayanu* prayer on the day of liberation, to thank God for keeping them alive to see that day. Harry Courts, a veteran of nine camps and the Burma railway, understood that it was not only stamina that would ensure survival. Even though his *siddur* (prayer book), *tefillin* (phylacteries) and *tallit* (prayer shawl) had been stolen, he persevered in praying every day in the belief that he would win through.

Jewish doctors applied their expertise to prison guard and prisoner alike. The Canadian Captain Jacob Markowitz saved the lives of countless hundreds who had contracted gangrene. Japanese doctors refused to hand over medical equipment so ingenuity and imagination produced solutions. Over 4,000 transfusions were carried out using broken bottles as funnels while the blood was prevented from coagulating using a bamboo whisk. For his courage and determination, Markowitz was awarded the MBE after the war. But more importantly he earned the lifelong veneration of those he saved.

In addition to the Japanese, there were Koreans, Indians and other colonized groups who believed that by allying themselves to the cause of Japan, they would achieve independence for their countries. While the vast majority had never met a Jew, European stereotypes of the bloated, wealthy, hook-nosed Jew were being introduced to an Asian audience. Although Jews were occasionally victimized because they were Jews, they were more often humiliated by vengeful guards because they were seen as representing the colonial master.

There are many tales of profound cruelty in this book. Sidney Lawrence of Hove, Brighton was forced by his captors to watch the gang rape of 15 Dutch women in Java by dozens of Japanese soldiers. He wept with anger and helplessness. He later witnessed the dropping of the atomic bomb on Nagasaki.

Around 20,000 European Jews managed to flee before the invading Nazi armies and ended up in Shanghai. After Pearl Harbour, these Jews were moved into a ghetto in the Hongkew district. One of those in control of the ghetto was Kanoh Ghoya of the Japanese Imperial Navy who called himself 'King of the Jews'. At the war crimes trials after the war, Ghoya turned out to have been a spy for the United States.

Keeping a diary in a prison camp was punishable by death. Some risked all to record their terrible experiences. But many privatised their anger and compartmentalised their grief – and moved on to another stage in their lives. Others suffered psychologically and physically – and were unable to experience the post-war sunlight. Most did not talk about what they had gone through. Martin Sugarman has therefore provided them with a collective voice through this important work. He has shone a light on a dark corner of Jewish history and preserved their stories for those yet unborn.

Jacob Markowitz rightly termed the bushido code as 'remorseless villainy, portrayed as chivalry'. Yet somehow these remarkable men and women inexplicably overcame the despair and the humiliation. They survived to dance on Hitler's grave and to forge their own destiny.

But who are these people? They are aunts and uncles, parents, grandparents, friends and neighbours. Old men and women who sit alone in synagogues. Taxi drivers and millionaires. They are all around us, yet up until now, they have not told us their story. Succeeding generations now have the opportunity to enter into their inner world and to learn what happened to them long ago and far from home.

Colin Shindler
Emeritus Professor, School of Oriental and African Studies, University of London
London, 2013

Acknowledgements

This book could not have been written, of course, without the veterans – both military and civilian – and their families who so courageously shared very painful and sometimes horrific memories with me in their homes or through their writing. I will never be able to repay their patient candour, nor indeed their humour and humility. For though many still bear the physical and mental scars of their incarceration and humiliation, this humour also still astonishingly shines through.

I have also included testimonies of some of the civilian internees, for they too suffered so appallingly at the hands of the Japanese; a moving and accurate glimpse of this can be seen in the incredible, partly autobiographical film and book by J.G. Ballard, *Empire of the Sun*, and the television series *Tenko*.

The book is a snapshot only of what the Jewish POWs suffered, because of course so many had passed away before I found the time to carry out this research; it is a sad fact that we are living today, at the beginning of the twenty-first century, on the cusp of the disappearance of that generation who witnessed what is within these pages. Nevertheless, the breadth of these personal testimonies gives an accurate and representative picture of what happened.

Those who served in the Far East theatre of war endured the worst; the extremes of geography, distance, climate, disease and hunger, and the brutality of a fanatical enemy, combined to make the fighting and captivity especially horrendous. And the virtual impossibility of home leave to visit loved ones, due to isolation and distance, was probably the most painful of all burdens. Not for nothing did the British and Commonwealth Forces in Asia describe themselves as 'The Forgotten Army'.

When Stephen Ambrose, the great American historian of the Second World War, realized that he was very ill, he wrote an epilogue in his last book. He said that the privilege he had over many years of meeting and telling the stories of the war veterans was deeply important and moving to him, and he quoted our wonderful Bible, the Torah: 'Truly, my cup runneth over.' I feel exactly the same.

So let this book bear witness and be a tribute to the bravery, fortitude and almost unbelievable 'can-do' attitude of all those who suffered so much under the regime of Japanese captivity, and especially the Jewish veterans who survived, whose story has never really been told. The way in which most POWs adapted to normal life after a living hell, following liberation, is a lesson to us all in the indestructability and determination of the human spirit. These are our people and this is what they did. We will remember them.

Special thanks go to Harvey Kaplan of the Scottish Jewish Archives; Jane Rosen, Richard Hughes, Tony Richards and Rod Suddaby of the Imperial War Museum; researcher Jane Sybylla Flowers; Beverley Amphlett of the *Journal of the Royal Army Medical Corps*; Shannon Maguire of the Sydney Jewish Museum; the Second World War Experience Centre; the staff of the National Army Museum Library in Chelsea; Janice Rosen of the

Canadian Jewish Congress Archives; the staff of the RAMC Museum; Helen Nield and Roger Farbey of the British Dental Association; Nicola Webber, Brenda Goldberg, Roz Currie and Matt Caro of the AJEX (Association of Jewish Ex-Servicemen and Women) Jewish Military Museum; the staff of the Singapore National Archives; the editors and readers of the Australian and Canadian Jewish newspapers; Peter Devitt of the RAF Museum, Hendon; Burke Penny of Canada; Peter Starling of the Army Medical Services Museum; the staff at the National Archives at Kew: Dr Robert Letovsky of Canada; *The Jewish Chronicle* newspaper; Dr Harvey Cohen, Lionel Sharpe and Susan Faine of Australia.

WO Gerhard 'Harry' Heilig (RAF, retired), Holocaust escaper and Jewish refugee from the Nazis, became a Special Operator in 101 Squadron, Bomber Command, in the Second World War, and 'got his own back'. Now living with his family in Vienna, and an author himself, he applied his typical RAF 'eagle eye' to scour the manuscript and make corrections. I am greatly indebted to him.

I also want to thank Heather Marchant of Vallentine Mitchell for her usual firm but gracious guiding hand throughout this project, and Joanna Lawson for her meticulous and professional detailed correcting of my far from ordered first draft of this book. Last but not least, my wife Jane, who so diligently corrected my grammar and other errors, has, as ever, my undying admiration and love.

It should be noted that I was unable to access several personal testimonies of Jewish Community members of Singapore who were interned in Japanese camps on the Island; these may be read only by a private visit to the National Archives in Singapore and with the permission of the families.

Any views expressed in the book, together with any errors or omissions, are mine only.

MARTIN SUGARMAN
London, 2013

A Note on the Place Names of Camps and on References

Many different sources – archival, personal and those used in many history books – spell names of Japanese POW camps slightly differently in English transliteration. The author has tried to use the spellings taken from the map in this book, which is reproduced here with the kind permission of the Little, Brown Book Group, the publishers of *Surviving the Sword* by Brian MacArthur, and the illustrator Duncan Stewart. But the author has also kept spellings the way the witnesses used them in their various written testimonies. Whilst every effort has been made to be consistent, some differences in these spellings may have crept into the text. Any errors are entirely those of the author.

Note: Ubon is in South East Thailand.

References

Few footnotes appear in the book because the content is almost exclusively based on personal interviews with the veterans, carried out over a period of about ten years, or on taped and written interviews in various other public and private archives Where more information was relevant, the author has added some footnotes to assist the reader, but relies on those who wish to investigate further to consult the hundreds of books written about the war in the Far East and the POWs who endured there.

Bushido: This term was originally based on the ancient Japanese samurai code, or 'way of the warrior', loosely analogous to 'chivalry'. Bushido emphasizes frugality, loyalty and self-sacrifice. But in the twentieth century, it became a propaganda tool of the Japanese military which expressed Bushido as 'the purity of war and duty of death'. It thus provided a 'spiritual shield' to permit and justify that soldiers fight to the death and ill-treat prisoners of war.

Kwai: 'Kwai' is Thai for 'river', therefore 'Kwai Noi' is the correct name, rather than 'River Kwai' (which means 'River River').

Map

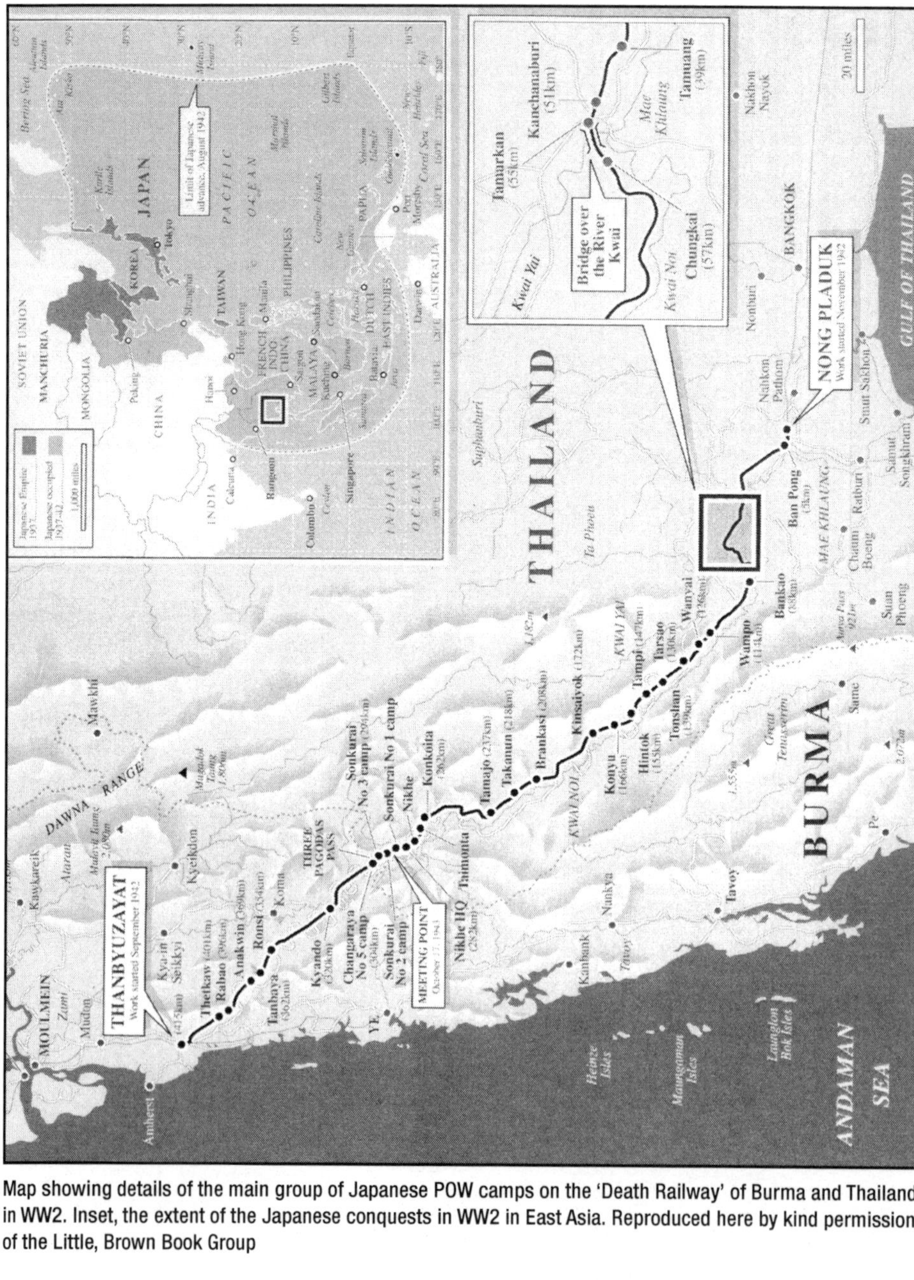

Map showing details of the main group of Japanese POW camps on the 'Death Railway' of Burma and Thailand in WW2. Inset, the extent of the Japanese conquests in WW2 in East Asia. Reproduced here by kind permission of the Little, Brown Book Group

Introduction

This book is not, nor indeed could ever be, an attempt to retell the complete story of Allied POWs of the Japanese; for this the reader can turn to an already existing excellent library of published books as well as archives. Rather, my personal mission has been to gather together in this volume some of the recorded memories, virtually all never published before, of a number of Jewish POWs, from the personal testimony they gave to their families or to the author or to the Imperial War Museum (IWM), and which have been carefully collected and filed in the archives of the Jewish Military Museum of the Association of Jewish Ex-Servicemen and Women of the UK (AJEX), the IWM and the Singapore National Archives. For those veterans I knew, I have recorded their narratives as they were told to me, or written down, with no embellishments or deletions.

In 2005 I read an excellent book, *Surviving the Sword*,[1] by Brian MacArthur. Sadly, and to my intense anger, he wrote gratuitously (on page 226) that Jewish POWs were 'very rare' in the Japanese camps. When I spoke to him and challenged this view with some facts, he accepted what I told him: which was that with even a cursory investigation I had discovered over 540 Jewish British, Dutch and Empire POWs, excluding those from the USA, and that some of these – and I named them – had been distinguished by their courage and decorated for bravery in the camps. Mr MacArthur took note, but in the second edition he would only exclude the phrase 'very rare', making no positive comment at all.

This book sets out to put the record straight, although no words can really convey the true horror of what these men, and women internees too, suffered in body and mind. Among the Jewish POWs were several decorated for their courage and life-saving skills while prisoners: LAC John Fisher, BEM, RAF; Dvr Jack Greenberg, BEM, RASC; Capt. Dr Jacob Markovich, MBE, RAMC; Cpl Samuel Marks, MiD; Capt. Adj. Henry Isidore Phillips, MBE; and war artists such as Sgmn William Allister, RCCOS, CEF, Sgmn Joseph Hirsch, RCOS and Cpl Philip Meninsky, 5th Beds and Herts – these are just some examples.

The Interviews

These were conducted over a number of years, the last two being in 2010 and 2011 respectively, when I discovered Sgmn Aubrey Lichman still alive, well and sprightly in Stanmore, at the time of writing, and Air Mechanic John Herman Franken, Netherlands RAF, in Canada. Memories fade, of course, but the words spoken and written here ring true and in accord with the many other published and written or recorded testimonies that have appeared over the past sixty-five years or so on this subject. The experiences are also very different in some cases, but it should be borne in mind that many survivors of the Japanese atrocities mentally blank out some of the worst of the horrors they witnessed or endured, or simply say outright that they will not speak about them in any form at all. Others have conquered these terrible psychological barriers and are keen to talk, even saying in some cases that it has been therapeutic, as they have not told the stories to anyone else until that moment. The author found this particularly moving, and many was the time I had to stop the tape to allow emotions to subside on both sides, as a veteran openly wept at the telling of a terrible event. Others said I should not hold back on asking about the terror, as they 'had long ago got over it and sleep well these days'. They would then add, however, that this was not always the case, and a wife or child would substantiate this in the course of the interview. Many veterans never spoke at all of their experiences and so their testimonies have gone with them to the grave. Thus we may never really know the true measure of what they lived through.

The testimonies are of different lengths because some wrote or spoke volumes and others would say little. I have thus purposely included some short testimonies, as they are often of those who perished and so have nobody to speak for them, or of those who suffered so greatly or felt the suffering more and so could not bring themselves to speak of it. I can in this book at least provide them with a space to record that they passed this way.

Nor do I apologize for including similar facts in different testimonies, which seem to repeat the same incidents, because within the mass of detail are still many very different anecdotes and incidents that make each story so different, and in any case have been described with a different perspective by different witnesses; some measure of duplication is thus inevitable and unavoidable if each person's story is to be told.

The Background

There is a great deal of curiosity on the subject of the Jewish military and other personnel who fell victim to the suffering inflicted by the Japanese on POWs in the Second World War. Some of this curiosity is stirred by the paradox of the general knowledge of the Japanese atrocities against European military prisoners, including Jews, of course, as opposed to the generally 'proper' behaviour by the Nazis – with some notable lapses – who at the same time were murdering millions of Jewish civilians in the Holocaust in Europe.

It is a bizarre and seemingly contrary fact that the chance of dying in a German POW camp was only about 5 per cent, but in a Japanese POW camp it was 25 per cent.[2] I have done further analysis to show that the proportion of deaths of POWs of the Japanese from the UK, the Commonwealth and the Dutch Forces was 36,000 out of 132,000 captured – about 27 per cent in round figures. Among the Jewish prisoners it was 136 out of 541 – about 25 per cent – much the same, but the figures for the Dutch Jewish deaths are a large underscore,[3] as I was unable to discover the true figures for this group.

Nazi war crimes were largely – but by no means only – confined to civilians, Soviet POWs mostly excepted. But the Japanese were equally brutal to those they considered inferior, whether military or non-combatants. For example, it is estimated that more than 300,000 'coolie' labourers – Malays, Tamil, Chinese and others – were worked to death on the Burma Railway by the Japanese. The book by Fergal Keane, *Road of Bones* (Harper Press, 2010), with its explanation of the Japanese military code of Bushido, is one of many that will give the reader an introductory and awful insight into the reasons for this.

Other readers will be curious about how Jewish soldiers fared, fighting in the distant Far East tropics, in countries where Jews were hardly known and were certainly a tiny minority, confronting an enemy that originated in a country where almost no Jews had ever lived and where anti-Semitism was virtually unknown. Yet others will be moved and amazed by the fact that images and true stories set in the terrible jungle conditions in which European soldiers fought against a rather bloodthirsty and fanatical enemy – images popularized by books and films such as *The Bridge on the River Kwai*, *Knights of the Bushido*, *Camp on Blood Island*, *King Rat* and many others – should also be experienced by Jewish military personnel who came from Stepney or Leeds, Brooklyn or Sydney, Auckland or Ottawa.

The Roll and Record of Honour

The roll of Jewish POWs of the Japanese came from several major sources.[4] I first trawled through Henry Morris's wonderful books, *We Will Remember Them* and *The Addendum*,[5] which list all Jewish personnel killed in the Second World War. Henry used the Jewish chaplains' cards at the AJEX Jewish Military Museum in London,[6] and also an AJEX survey which was conducted via synagogues and other Jewish institutions, and which captured information not noted on the chaplaincy cards. Information also came from numerous cuttings kept in the archives at the museum about the Japanese POW experience of Jewish servicemen and women.

I then searched – over a period of a year – through all the chaplains' cards myself to seek out those who were noted as POWs and survived, and also those who had died as POWs but been missed by Henry because not all the Jewish community took part in the AJEX survey. However, an immediate problem is that researchers such as Henry Morris and I know that not all the cards are there;[7] this is not surprising, given that not all Jewish personnel enlisted as Jews, and not all saw a Jewish chaplain in order to have a card completed in their name. Furthermore, the information on the cards is only a snapshot: it may record several meetings with a soldier, for example, up until 1941 – then nothing. There is thus no indication that the soldier may have gone on to serve in the Far East – or anywhere else after the war with Japan began – and perhaps been made a POW; this is stated only on those cards where such information came to light via families or newspaper reports, or if it was communicated to the Jewish chaplains via regimental or corps clerical staff. Indeed, many were posted as missing or believed killed when they in fact were POWs, and this was either not known until after the war, and possibly never recorded on the cards, or even never discovered at all. Prisoner information given to the Red Cross and neutral governments by the Japanese – who had not ratified the Geneva Convention on POWs – was minimal and often inaccurate. Add to this the desire of the Japanese to hide the names and the numbers of POWs who died in their hands, to obstruct the War Crimes investigators, and then it becomes clear that we shall never truly know how many missing Jewish personnel died as POWs in the Far East.

Further limitations in using the cards include the fact that many describe a death after the beginning of the fighting with the Japanese, but not specifying whether this was as a POW or not; where it is not so specified, that name has been omitted. Yet it may well be that the persons concerned did die as POWs; we may never know. In a section of the cards labelled 'Missing in Action' (MIA), it is equally unclear if this means missing as a POW or missing as 'Killed in Action' (KIA). For example, 'Missing in Malaya' could mean either. Even when cards are cross-checked with the CWGC (Commonwealth War Graves Commission) Registers website, it is unclear if the serviceman died as a POW or in fighting.

Rabbi Chaim Nussbaum's book[8] revealed many names of Dutch Jewish civilian internees as well as Allied soldiers, and some were named in Jack Caplan's book.[9] The beautiful synagogue that the Japanese allowed at Changi gaol is also described but, unlike the Christian chapel, it did not survive the war.

I have tried to include civilian Jewish internees, but according to Joseph Kennedy[10] there were almost 500 Jews in the Sime Road camp alone in Singapore, and many German Jews not interned at all in Bangkok. The thousands of German and other Jewish refugees interned in Shanghai and other cities, and the members of the Jewish company of the Shanghai Volunteer Force,[11] are also mostly outside the scope of the Roll.

The indomitable David Arkush, retired captain of the RADC (Royal Army Dental Corps) provided me with his personal list of Jewish POWs whom he had known, from his own saved archive; these came from his secret notes kept at the time, at great risk to his life.[12] Similarly, Reuben Kandler of the Royal Corps of Signals (RCOS) kept a secret list which has survived intact in a battered notebook, now in the hands of his family. This also gave further names of not only all the men with him in various camps, with their rank, number, home address and so on, and where and when they were buried if they died, but the Jewish personnel in particular were noted by him – a kind of 'Schindler's List' of POWs of the Japanese. His son Richard now owns this remarkable document and has written a touching book about his father's experiences,[13] as well as filming an interview with Reuben before he died. Because of Jewish POWs such as Arkush and Kandler, all relatives and the Commonwealth War Graves Registration service had a better idea of how to commemorate these men and inform their families of the how, when and where, thus at least somewhat relieving their terrible grief, in that they knew how loved ones had died and where their remains were. We owe those POWs a great debt; discovery by the Japanese of such papers meant severe punishment and often death.

In addition, the unfinished recorded testimonies (obtained by the late Jack Lennard of Hull AJEX) of those who were then surviving Jewish POWs, who have since passed away, also figure in this book, and are kept in the AJEX Jewish Military Museum archives, donated by Jack's daughter, Phillipa Behari.

Researcher Jonathan Moffatt kindly supplied me with information from huge files he had viewed at the Imperial War Museum (IWM) in London and the National Archives (TNA) at Kew on Britons and other nationalities, including the resident Jewish community, in Malaya and Singapore at the time of the Japanese invasion, and who were interned. He also kindly gave me lists of regimental personnel who were present, so enabling me to pick out further Jewish POWs. However, his research is far from complete and many more names of Jewish personnel are likely to emerge. Similar information came from FEPOW (Far East Prisoners of War), COFEPOW (Children of Far East Prisoners of War) and Taiwan POW websites.[14]

Behind every name there is a story, and volumes could be written. Monty (Manasseh) (Nat) Rakusen (of a family in Leeds famous for the matzo factory) was POW after fighting in the Hong Kong Volunteers and incarcerated in Sham Shui Po camp and later sent to a coal mine in Japan in 1943. His wife claimed neutrality, as she had a German father and Portuguese mother, and she and her son Ronald were interned in Shanghai. Reunited after the war, the family lived in Hong Kong until Monty died in 1966 and he was buried there. Mrs Rakusen died in the UK in 1981 and Ronald, now retired, lives in France.[15] Hyman Weisberg, a Cambridge graduate, was a senior civil servant in Malaya; his signature was on the bank notes. He always claimed he survived because he was short and the Japanese could thus relate to him! After the war he lived in Wembley and taught himself Braille so

1. The author at Kuala Lumpur railway station in 2010, where many POWs stopped en route to the Railway of Death from Singapore.

he could teach mathematics to blind people.[16] At Changi, Dutch Jewish internee Loet/Lou Velmans assisted Rabbi Nussbaum with the production of *The Changi Jewish Chronicle*.

From Tamarkan camp comes the remarkable story of Australian Jewish POW Mark Hayman organizing a 'Barmitzvah at a distance' for his son Leon Jacob, in Melbourne, on 27 September 1944/5704. It has become known as 'The Barmitzvah on the Kwai' and an actual bar mitzvah certificate was drawn up and signed by about sixty POWs of all Allied nationalities who attended the ceremony, including several non-Jews and the Japanese interpreter, Matsuyama. The document is now in the archives of the Australian Jewish Historical Society in Victoria. The number of stories is endless.

For many months I went through all the *Jewish Chronicle* newspapers from December 1941 to December 1946, with the help of Harold Pollins in Oxford, and appropriate names on the CWGC registers on their website, to confirm or discover more. Here arose another problem, for it is impossible to tell from CWGC registers if a grave or a name on a memorial (no known grave) indicates a serviceman who died as a POW or fighting the Japanese in battle – the latter being beyond of the scope of this book.

Anti-Semitism

A very few of the testimonies mention incidents of anti-Semitism among the Japanese, though they were certainly influenced by their pact with the Nazis; the most outlandish of examples is the Fugu plan, described in detail in the book *The Fugu Plan* by Marvin Tokayer and Mary

Swartz (Paddington Press, 1979). Put briefly, the Japanese had a passé and stereotypical view of Jews, which included their belief that Jews are all good at business and trade; thus they believed it would be beneficial to force Jews to populate areas of Manchuko/Manchuria, and other locations the Japanese had conquered, to establish a wealthy middle class that would help the economic development of Japanese 'colonies' in Asia. This understated anti-Semitic view was underscored by the name given to this plan, for the fugu is a poisonous fish, eaten as a delicacy in Japanese cuisine, but if not cooked properly it will poison the consumer. In other words, Jews were not to be trusted, but just cynically manipulated.

A Dutch Jewish teenager internee, David Mussry, recalled as a 13-year-old boy the day on 31 August 1943 when Japanese soldiers with fixed bayonets burst into his home at 6 a.m. in Surubaya, arresting his father and tearing down all the Jewish symbols in their home, screaming 'Yudaya bakayroo' (damned Jews). Later that day the soldiers returned for the women and children, who were taken by truck to Surubaya gaol where they met up with the rest of the arrested Jews who hailed from Iraq, Poland and Holland; they were kept segregated from the men.

Harry Benda testified that many anti-Semitic articles appeared in the occupation-era newspapers – especially the *Asia Raya* – written by local Moslems such as Umar Said, inspired by Nazi propaganda against Jews, and that there were anti-Semitic radio presentations by Kempeitai officer Murase. Vera de Vries, born in Surubaya in June 1943, recalls her mother telling her, years later, how German liaison officers were in Indonesia, urging the Japanese to separate Dutch Jews from non-Jews. On their prison camp wall was the notice 'Banksa Jehudi' (Jewish camp); the inmates included a lot of local Jewish doctors who appeared to have given the Japanese the names of local Jewish families in return for extra food.[17]

According to researcher and historian Jan Sybylla Flower, Alexander Oppenheim was ordered to lecture on the Jewish faith in the camps on the Burma Railway, by his senior officers, to counter Japanese anti-Jewish propaganda. Captain M.W. Drower, an interpreter, was asked by the Japanese to identify Jewish officers in 1944.

On the front page of the *Jewish Chronicle* for 11 May 1945 there is also a report from the United States War Department, giving the testimony of a German Jewish refugee surgeon, 55-year-old Dr Walter K. Frankel, lecturer at the University of the Philippines. It describes how on 12 February 1945 at 4.15 p.m. the Japanese came to his house and asked for the passes of him and his wife and sister. The soldier said 'Judasi' and arrested them; their hands were tied behind their backs and they joined other Jews and some thirteen Filipinos, also arrested, in the street. They were taken to the salon of a neighbouring house and Dr Frankel stated that he knew they were going to be killed. He kissed his wife farewell as they were forced to kneel facing the wall, and furniture and cushions were piled around them and petrol was poured over them. The Japanese then set fire to them and one, a sergeant, threw a grenade. His wife was shot dead as she moved. Three others also died but as the fire drove the Japanese out of the building, a maid rushed in and released the others, who escaped.

Many other Jews were murdered in this way, but it should be remembered that the Japanese carried out many massacres in the Philippines as the Americans moved in to liberate the country; whilst Jews were singled out, so were many others.

By contrast, we should remember Japanese such as Chiune Sugihara, the vice consul in Kovno/Kaunas (Lithuania), who freely gave passes to over 6,000 Jews fleeing the Nazis in

1941, to allow them to proceed across Russia to 'neutral' Japanese territory; he continued to do this even when instructed by Tokyo to desist and as a result was sacked from his job after the war and never re-employed in the Japanese Diplomatic Service. When Jewish survivors living in Israel (which made Chiune Sugihara a 'Righteous Among the Nations' at Yad Vashem in 1985, the year before he died) and elsewhere discovered his poverty-stricken status, they sent him, until he and his wife died, a regular allowance to help him live decently in Japan. (For more on the complex and often benign attitude of the Japanese to Jews during this period, see the article 'Japan and the Jews' at www.jewishvirtual-library.com.)

The Legacy

Reliable information on how POWs died can only come from personal testimony from family archives or surviving comrades – rapidly declining sources – or by consulting personal military records, which is a very expensive and sometimes impossible task, as only next of kin are permitted access to such information at present. At some time in the future, such documents may be released into the public domain, as with First World War records. For now, the best that can be done is to attempt to capture as many names as possible from available sources, for the final result, as a proud but sad tribute to our men and women who were there.

Notes

1. Brian MacArthur, *Surviving the Sword: Prisoners of the Japanese in the Far East, 1942–45* (Little, Brown, 2005).
2. Estimates vary, but this is from the IWM *Official History of Casualty and Medical Statistics*, HMSO.
3. Jewish POWs who died are as follows: UK and Dutch – 104 out of 425; Australian – 29 out of 101; Israeli/Palestinian Jews – 1 out of 7; Canadian – 0 out of 7; New Zealand – 0 out of (?); there was one South African who died, included in the UK total. A total of 136 died, out of 541 POWs, or 25 per cent.
4. It was first published in the hugely enlarged second edition of *We Will Remember Them* by Henry Morris and Martin Sugarman (London and Portland, OR: Vallentine Mitchell, 2011).
5. Published by Brassey's, 1989, and AJEX, 1994. The Addendum was written in response to the clamour that arose from the community that relatives and comrades had been missed in Henry's first book.
6. Over 70,000 of these, which were kept by the chaplains, mostly during the Second World War, are now stored at the museum.
7. Museum staff are adding cards frequently when research reveals that personnel are missing.
8. *Chaplain on the River Kwai* (New York: Shapolsky, 1988).
9. *Gorbals to Jungle*, published privately, in the AJEX archives.
10. *British Civilians and the Japanese War in Malaya & Singapore* (London: Macmillan, 1987). The *Straits Times* for 8 September 1945, p.2, confirms this and states further that 170 of the Jewish internees were women and 90 were children. It also noted that two Jewish internees had made successful full escapes from Sime Road camp.
11. See chapter in Martin Sugarman, *Fighting Back: British Jewry's Contribution in WW2* (London and Portland, OR: Vallentine Mitchell, 2010).
12. His records will be kept at the AJEX Museum.
13. Richard Kandler, *The Prisoner List* (Marsworth, 2010).
14. Thanks to George Money of the Essex and Herts FEPOW Association.
15. Letter from Ronald to the author, November 2003.
16. Letter from Ernest Shenton, who knew him, to the author, November 2003.
17. These excerpts are from Jeffrey Hadler's study in the journal, *Indonesia and the Malay World*, 32 (November 2004), available online.

The Tamarkan Barmitzvah, the Changi Synagogue and the Chungkai Kol Nidrei

This article is printed by the kind permission of the FEPOW webmaster.

Australian Jewry's Book of Honour

World War II

Edited by Gerald Pynt, with the co-operation of NAJEX historian Jack Epstein Australian Federation of Jewish Ex-Servicemen & Women

Throughout centuries of persecution, in peace and in war, Jews have clung tenaciously and stubbornly to their religion, finding solace in its wisdom and comfort in its prayer.

How true this was of a unique Barmitzvah service by proxy held in a prisoner of war camp at Tamarkan, Thailand, on September 27, 1944.

The service marked the thirteenth birthday of the son of Victorian Signaller Mark Hayman (8th Division). Jewish parents hold dear the service of a 13-year-old boy, since it signifies the lad has reached the age of manhood. Because Signaller Hayman could not be present at the Barmitzvah of his son Leon, he determined to hold a special service by proxy at Tarmarkan.

With difficulty, a sceptical Japanese commandant was persuaded to allow the service to be held. Negotiations extended for four days before his approval was obtained. He was apparently highly suspicious that the proposed gathering would be simply to hold a religious service, so there was the spectacle of colonels and other officers of field rank on opposing sides seriously conferring about a service for a boy of 13 living thousands of miles away. It must surely have been one of the most unique conferences of the war.

Approval for the service finally was granted on condition that the Japanese colonel's interpreter attended. Because the service proper was to be followed by a social gathering, the date chosen was not a Sabbath. Guests included Australian, British, American, Canadian and Dutch servicemen, as well as the Japanese interpreter.

Despite the restrictions placed on the service, it was deeply impressive. Before leaving, the Japanese interpreter made a speech in which he expressed his best wishes for Signaller Hayman's son and the hope that father and son would be speedily united.

The service was conducted with the assistance of a Singer's Daily Prayer Book belonging to Signalman Neville Milston (NSW), who was a POW with Mark Hayman. Signaller

Milston, who was suffering from a severe bout of malaria, could not attend the service. He said later that on occasions the temptation to sell the prayer book for food was great indeed (its pages would have made excellent cigarette paper). But Signaller Milston's faith and the book survived the horrors of three and a half years of prisoner-of-war existence in Malaya, Burma and Thailand. He brought it safely back to Australia when he was repatriated. Signaller Hayman also returned safely to Melbourne. The boy in whose honour the Barmitzvah service was held became a barrister.

This is the synagogue built within the confines of the infamous Japanese prisoner-of-war camp at Changi. The induction took place in August, 1944. Jews from Holland, England, India, America and Australia participated in the induction service. Regular services were held each Friday evening and Saturday morning and lectures given on Sunday afternoons. The Synagogue, which could hold a little more than fifty men, was the largest religious hall or church in Changi prison.

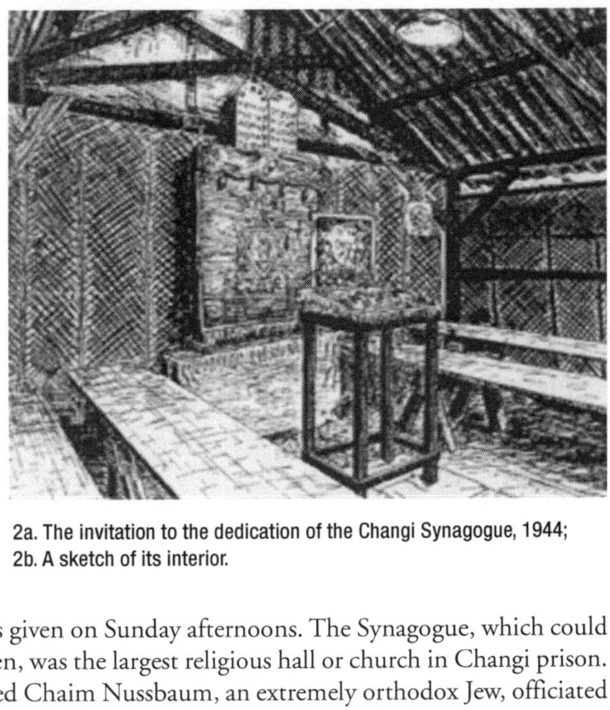

2a. The invitation to the dedication of the Changi Synagogue, 1944; 2b. A sketch of its interior.

A Dutch army Rabbi, named Chaim Nussbaum, an extremely orthodox Jew, officiated at the services. These pictures were provided for this War History by Sgt Sam Biner, who enlisted in the AIF in Victoria. The prayer book he used was given to him by the Japanese after it had been sent out from Jewish sources in America. Its opening page bore the Japanese Censor's stamp No. 320.

The Tamarkan Barmitzvah Document

Reproduced here by kind permission of Dr Harvey Cohen of Melbourne, Australia, this document has about sixty signatures, comprising those of Allied Jewish, and some non Jewish POW's – and even the Japanese interpreter, Matsuyama (see below).

3. The Tamarkan Barmitzvah document.

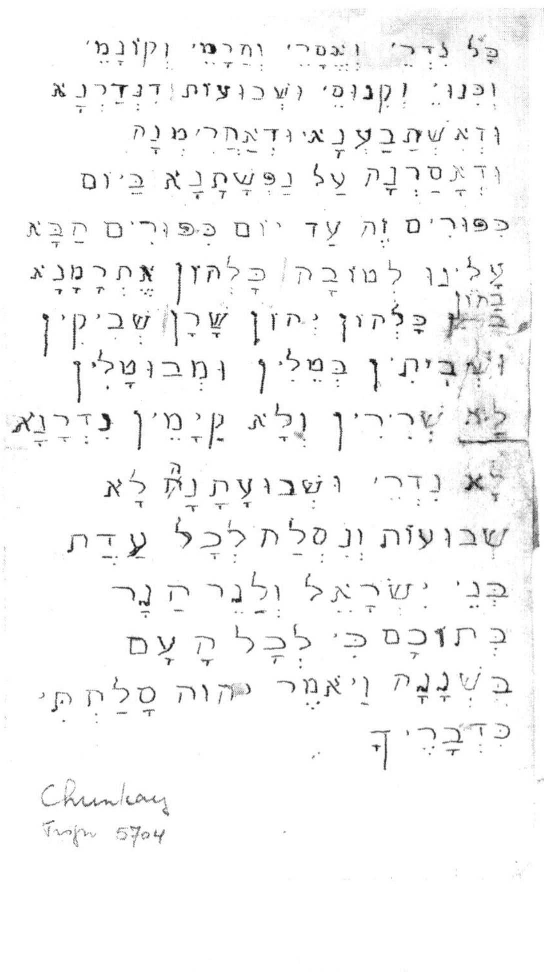

4. The Chungkai Kol Nidrei prayer. This was written from memory by Pte Davids at Chungkai (see David Arkush testimony, p. 55).

The Testimonies

111769 Pte Albert Abel, RASC, 54th Company

The author thanks Albert's son Paul for the information used in this testimony.

Albert was born on 6 July 1920 in Bradford, son of Harry and Rose Abel – Jewish refugees from East Europe – who initially settled in Bradford and ran a furniture shop and coffin-making business. The family then lived at 4 Moseley Street, Blackburn, where Albert was bar mitzvah and also attended the Queen Elizabeth Grammar school.

5. 111769 Pte Albert Abel, RASC.

Albert volunteered in 1939 and was still in France six weeks after Dunkirk, as his unit had been stationed in Vichy and were evacuated from there. On returning to the UK he declined to attend officer training when it was offered.

Trained for desert warfare in North Africa, he was en route to North Africa around the Cape of Good Hope but his troopship was diverted to Singapore when Pearl Harbor was attacked, even though they had no training in jungle warfare. He was told by officers that as the Japanese have slit eyes they can't see very well, using the then pejorative stereotypes which were in common use. His troopship arrived a few days before the surrender. One day they were paraded at Changi airport when an Australian general arrived to give a speech; he said that they must 'fight to the last man', then got on a plane and flew away. Albert always said they were a sacrifice to the slaughter, as before his regiment arrived, most of the troops in Singapore were Commonwealth, and for political reasons it had been felt necessary to have a UK contingent, so the British could argue that troops other than colonials were to be captured. If asked whether he would have preferred to have made it to the desert, as trained, Albert would say that with the way things actually turned out, he lived; had he gone to the desert, who knows what would have happened.

Following the surrender, Albert was in Changi until 20 August 1942, then went by rail in the notorious cattle trucks, taking several days to reach Thailand, where he slaved to help build the railway to Burma. He was at Bampong until 10 August 1944. He was lucky in that he was never sent far upcountry and was apparently mostly near the start of the railway, so did not suffer all the hardships and diseases of those who went into virgin territory – but he did suffer from beri-beri, malaria and other illnesses. He volunteered to learn local languages (Malay, Thai, Cantonese and Japanese) and was used by his officers as a translator, which left him open to being beaten by the Japanese whenever he mistranslated. He also spent time in the cookhouse. However, he very rarely spoke of any of his experiences. One anecdote he told his family was about when guards taunted them by tying poisonous snakes to their ankles and making them run along a track; they later found that the snakes were already dead, but the terror of the men was great.

In one incident, the camp guards had got wind of a planned strike by the POWs. The POWs were assembled and the Japanese asked any who were planning to strike to take a step forward. There had been strong support for striking but, when push came to shove, only Albert and a very few others actually made the step forward. This group was then made to stand in the sun all day, at risk of being shot at any time. A similar incident appears in the film *The Bridge on the River Kwai* (directed by David Lean, 1957).

The Japanese hierarchy meant that the Korean guards were 'bottom of the pile' and often meted out far more brutal punishments than the Japanese. It was considered shameful to be a camp commander/prison guard rather than a fighting soldier, so the Japanese would often take their frustration out on the Koreans, who in turn would beat the POWs. The main preoccupation of the POWs was food – they were always hungry and often dreamt about food.

From August 1944 until 10 February 1945, Albert was at Non Pladuk, and then moved to Bangkok until 11 May 1945. He ended up at Ubon airport in Bangkok, where he heard the same old rumours that war was over (15 August) – and of course didn't believe it. On finding out it was true, he got drunk and shaved his head – and then actually slept through the days immediately following. They had been forced to dig large trenches about that time (August 1945), presumably to be buried in. Although they were not told this at the time, they did suspect it was true, and apparently there were plans to kill them all in September 1945, should the war go further against Japan. When asked if the atom bombs were justified, Albert never answered directly, but said that if not, none of his family would be here because he would not have survived.

The prisoners were not returned to the UK for about six months after VJ (Victory over Japan) Day, when the war ended. The authorities were apparently keen that the UK population should not see just what a wretched condition they were in. On Albert's boat home, he discovered by coincidence that a cousin was one of the doctors on board (Dr Walden – shortened from Waldenberg), and he used to get little 'treats' now and again. He was finally demobbed in 1946. Their arrival into the UK was unnoticed – they felt they had been 'hidden' away in marked contrast to the fighting men returning at the end of the European theatre of war. Many felt their sacrifices were downplayed and this seemed to make their suffering worse in consequence.

Albert studied Mandarin post-war at the School of Oriental and African Studies (SOAS) in London but did not settle into academic life – he apparently found his teachers and academia petty when considering the 'reality' he had lived through. He left after four years without a degree and went to Malaya to work for the civilian government during the 'Emergency', from about 1951. His son and daughter were born there. He was initially a resettlement officer in Perak, organizing and supporting the physical separation of local communities from the communists through building 'protected settlements', and was later appointed development officer when employed in the building of Petaling Jaya, Malaya's first new town a few miles away from Kuala Lumpur. He left in 1961 after Malayan independence and was replaced by local staff, as was appropriate. He could not find employment when he returned to the UK and ended up running a confectionary shop – El-bourne – in West Hampstead from the early 1960s until the late 1970s. During this time, he maintained his interest in all things oriental, talking to visitors from overseas in their own languages (often to their astonishment, as he was 'only' a shopkeeper) and informally arranging English lessons and lessons in oriental languages.

Albert married Lilian Abel (née Lovatt; she was born in 1919 in Cork, Eire) in 1950. She was in the WRAF, where she worked with his sister Dinah (still alive and about 93 years old), who introduced them. The family had moved to Hillfield Road, London, NW6, by then.

Lilian talked about Albert waking up shaking and screaming – dreaming, for example, about running around during Allied bombing raids, trying to catch the bombs before they exploded. This was the nature of his post-traumatic stress disorder. Apparently, POW camps were deliberately positioned around strategically important areas and some POWs died from Allied bombs. Albert used to get regular bouts of malaria in the early days after his return to the UK, and Lillian told of a time when he 'jumped the queue' at a pharmacy, to get quinine because of his shaking. Howls of protest followed from the waiting queue, even though he was obviously very ill. Nobody seemed to be bothered about his condition. Albert attended many of the POW reunions at the Festival Hall in London after the war, and took his family with him.

After Lillian died in 1979, he married again (in 1983). Perhaps ironically, his second wife, Yaeko Kawai, is Japanese; she is still alive, and lives mostly in Japan. She was one of those who belonged to the group to whom he taught English and in turn he was taught Japanese by them when they came to his shop. They were married for about twenty-five years. Regrettably, after the marriage, some of his previous friends no longer spoke to him. Albert used to say during wartime, the Japanese behaved only in a manner that was consistent with their own culture, which explained, and made more acceptable, their cruelty – but that this was in sharp contrast to the manner in which the Germans behaved, which was far removed from their cultural norms. His son Paul states that this is not really correct, as the Japanese had been very good to their prisoners during the First World War, many of whom stayed behind afterwards and married into the local population. It was the policies of the Japanese fascist regime in the 1930s that were responsible for the changes in behaviour and brutality that were demonstrated during the Second World War.

A further irony was that Paul's daughter (Albert's granddaughter) also went to SOAS where she studied Japanese and also lived in Japan for over four years, and now works for Fuji TV in London.

Albert died in September 2009.

D116327 Sgmn William Allister/Alzitzer, RCCOS, CEF and War Artist

Most of this testimony is taken from William's amazing book, Where Life and Death Hold Hands *(Stoddart, 1989), with additional material supplied by his widow, Mona.*

William (Will/Bill) was born to Ukrainian Jewish immigrants in the tiny town of Benito, Manitoba, on 5 October 1919, son of Mr Samuel Allister (originally Solomon Alzitzer) and Luba (née Narovlansky) of 1267 Bernard Avenue West. His family later moved to the Mile End area of Montreal where William lived much of his life. He was one of six children. He graduated from Baron Byng High School and became an actor with the Young Men's Hebrew Association (YMHA), often performing in the Catskills Jewish holiday camps and in Montreal, and a writer and artist before and after the war, working for some time as an actor in Hollywood and New York as well in the post-war period.

He joined the army in March 1941 and underwent basic training at Huntington (Quebec) and Debert (Nova Scotia) where he learnt Morse code. With him was his good friend Hank Greenberg, from Winnipeg; when visiting Hank's mother before going abroad, William promised her that he would take care of Hank. William wanted to go overseas but had to pass the eyesight test – and he had to memorize the chart to do this. He and Hank were determined to get to Europe to fight and 'make a difference'.

Hong Kong

William was embarked on a foggy night on 27 October 1941 on the New Zealand vessel, SS *Awatea*, shadowed by the gunboat *Prince Robert* as protection from submarine attack. Leaving Vancouver with 1,800 men, including the thirty-three from HQ Signals Unit, of which William was a member, they zigzagged across the Pacific in a packed troopship. Conditions were

6. D116327 Sgmn William Allister/Alziztzer, RCCOS, CEF and war artist (standing, right) with another Jewish soldier, Lou Gottheil, taken before being posted overseas.

very basic and tempers frayed, and William said he lived on bread and jam. His unit took turns on duty on the bridge, as Aldis Lamp signallers, staying in contact with neighbouring ships in the convoy. This was a blessing as it got them out of the overcrowded holds, especially as they reached the tropics. Via Hawaii, they arrived in Hong Kong on 16 November after three weeks at sea. From the ship they marched past crowds down Nathan Road, rifles at the slope, boots and buttons gleaming, amidst the terrible poverty of the colony. They were billeted at Sham Shui Barracks, where ironically William would later be incarcerated by the Japanese. With servants to do their laundry and shave them, they often watched as hundreds of poor gathered at the camp wire to beg for food and money. They could have had no idea that soon the roles would be reversed, as locals would throw them food and cigarettes when they became prisoners. Meanwhile, for three weeks they lived wildly, enjoying the bars, food, sights, colour and women of the East.

Early on 8 December, as William and his comrades were waking, bombs suddenly rained down on them. Most of the men were badly trained, some not even having fired their rifles even in training. William was in Bowen Road hospital with malaria, but was sent back to his unit on the front line at Waterloo Road, as serious casualties flowed in and beds were needed. Although massive casualties were inflicted on the Japanese suicidal attacks, the Allies were overwhelmed, and William and his unit were ordered to retreat to Hong Kong Island after four days of intense fighting. Destroying anything that could not be carried, they made for the docks, where there was a mad scramble to get into boats. William saw military police and officers shooting Chinese looters trying to take food from the godowns (warehouses). It was absolute chaos. Finally getting to Hong Kong Island, he and his comrades were posted to Victoria barracks, where they ate a hasty meal of bully beef and biscuits and tea, and then taken on trucks to sleep in local abandoned houses. Next morning they were sent to man a pillbox with their radio equipment at the Wong Nei Chong Gap.

Fifth columnists were everywhere, spying for the Japanese, dressed as local people, and orders were given to shoot them if seen. William witnessed a young and nervous lieutenant interrogating one such suspect near to where he was sitting; with no interpreter, chaos reigned and the officer decided he would have to shoot the suspect; two young soldiers were ordered to take him behind a wall, and shots rang out.

The heavy bombardment continued as the Japanese poured onto the island from boats, and William and his men were ordered to evacuate and retreat further inland, taking only their rifles. It was dark, and he and some friends, hopelessly lost, found an abandoned car and slept in it until daylight. They were woken by gunfire and found themselves surrounded by Allied troops – including Chinese whom they first suspected of being Japanese; William joined in and fired away at the enemy, killing his first men in the war. Again they had to retreat, but a British major ordered a halt along a road and said they would make a stand, but as dozens of stragglers and wounded joined them, they broke through the line and would not stop. Small skirmishes were fought all around William; men died trapped in pillboxes as Japanese threw in grenades; some surrendered, the wounded were bayoneted to death and other small groups fought to the death. As the retreat continued, William and his friends found themselves in the courtyard of the Aberdeen Village hospital. His friend Hank was nowhere to be seen. Nurses ran about tending the wounded, and then a counter-attack was ordered as men boarded two trucks. William's truck headed for the trenches at Sushin Hill, where at dusk the exhausted troops fell asleep.

William and his group then returned to the hospital, which was now surrounded, and all armed men were ordered out by the frantic RAMC (Royal Army Medical Corps) officers to avoid Japanese reprisals. Wandering aimlessly, William and others bumped into a British truck which drove them up into the hills to Signals HQ in the Wanchai Gap. Here he discovered many of his old mates, including Jack Rose (whose testimony is included in this book), but also learned that many had been killed by shelling, and William had the gruesome job of identifying the mangled remains of Hank Greenberg, whose sweater and respirator had his name on; his face was blown away. William was distraught with grief and horror, thinking too of his promise to Hank's mother to 'take care of him'. He got very drunk on liquor he found in an abandoned house and then wandered off into the hills; he was picked up by some mates who dragged him off to sleep in a nearby house.

Next day they were ordered back to Victoria barracks. It was a beautiful morning; they found a bombed-out NAAFI (Navy, Army and Air Force Institutes) canteen and stocked up on bully beef and cigarettes. Striding along, they passed the bunker of a huge cannon; the British crew had decorated it out in a luxurious fashion with looted furniture from the many bombed villas nearby, with lamps and a record player and lavish foods stocks, and were sitting in bathrobes and silk shirts. They said they were not going to fight as they had been spotted by the enemy and would be taken in a day or two. They all sat and drank some liqueurs; William asked what they would do if the Japs arrived. They replied that they would shoot themselves first. William never knew if they ever did.

POW

At the barracks, hair-raising stories abounded of last stands, escapes and massacres. Enemy snipers shot anyone who was out in the open; there was no water (William shaved in beer) and very little food. On Christmas Day came the order to surrender, and men ran through the barracks shouting out orders to lay down their arms, which were collected. There was shock and anger, many saying they should fight on and make a stand; some even committed suicide, fearful of surrendering to a vicious enemy. The men watched the triumphant Japanese parade in their thousands through the British-built streets. But days passed and no enemy guards appeared.

William went down with bad dysentery and was ordered to march several miles to the Bowen Road hospital and take a box of medical supplies as well as his pack; he barely made it on the uphill walk and found the hospital more like a prison than a haven. There was virtually no food; he was put on a tea diet and watched RAMC orderlies pilfer the food of the patients; some left the building to rummage for food in nearby abandoned villas. William was told that Canadians were being sent to a camp at North Point, so he and many others opted to go, feebly marching though the street with their kit as passers-by watched the bedraggled, unshaven Allied soldiers, no longer the masters.

The camp was near a sea wall, a cluster of wooden huts – some with filthy, lice-ridden bunks, others with bare concrete floors – with guards and a high wire fence; twice a day they received a cup of mouldy rice with thin vegetable soup. They kept away from the guards; they watched as one well-dressed Chinese man was seized upon for passing on the wrong side of the street. He was beaten senseless, dragged into the camp in front of the men, hung by his ankles and bayoneted, screaming, to death. Another day, a woman was gang-raped in the Japanese guardhouse and left to die, naked all night on the street outside.

They used tins cans with wicks and peanut oil for light, and slept huddled together for warmth in the cold winter nights. William's precious diary was stolen from his coat at the latrines, so his friend Blacky made one for him to write in, by sanding two strips of wood as covers and boring holes to take a shoelace to bind paper, supplied by the YMCA (Young Men's Christian Association) officer, Porteous. It survived the war. In his diary he described all the men around him; some refused to eat the appalling food and just smoked, and wasted away; some became silent and lay on their bunks day after day, staring and dying slowly; some became insane; yet others showed great optimism, one man setting up as a barber. William found himself acting as a kind of moral advisor, somehow attracting those who wished to pour out their hearts.

Some foraging parties were allowed out, and brought in books; William acted as an informal librarian, encouraging the men to read, to lift them out of apathy. Slowly, new shoots of vitality appeared; William persuaded men to help set up classes – these started in Chinese and Japanese (among the students was fellow Canadian Jewish POW, also from Vancouver, Jack Rose). Concerts were put on, debates and lectures began. But as the weather warmed, blocked latrines became fly-infested and dysentery spread rapidly. Fly-catching campaigns ensued but, meanwhile, resentment against the better-fed officers grew intensely; they began to eat indoors to avoid showing what they ate; men shadowed them to pick up their cigarette butts.

Boards were rigged up over the sea wall to provide 'clean' latrines, with a post to hold onto to keep balance in high winds; often bodies were seen passing by in the sea below, many with hands bound and a rope around the neck. Dysentery began to kill; the sick-hut gave out an appalling stench as men relieved themselves where they lay; doctors could do nothing. Some men refused to go to the 'hospital' and their friends carried them to the latrine and cleaned them up.

Organizing the concerts, taking part in them, encouraging men to read – all this gave William great purpose and a will to live. Then, having studied art in Montreal, William began leading art classes and this in turn inspired other men and awakened unknown talents in them. Blacky suggested that William should paint properly himself. Blacky 'liberated' some tent canvas for William; oil from a Japanese truck engine (taken while William distracted the Japanese driver with idle chat) provided black and brown paint, and finely whittled sticks and old shoe brush hairs became paintbrushes. William sold paintings to officers and obtained extra food for him and Blacky.

William realized that conversation and humour helped survival, and he took every opportunity to initiate and engage in it; with friends he would re-enact scenes from famous movies. But remembering names was a problem, as beri-beri affected the brain. Once he saw a friend weeping because he could not recall his son's name.

Meanwhile the first work parties were ordered – mainly at Kai Tak airfield – and the gruelling work wore out men who were already starving and sick, many with beri-beri. William laboured, shovelling earth into baskets attached to bamboo poles that rested on bony shoulders, and carrying it away; he and his friends made jokes to keep up their spirits against the fatigue and monotony. Sergeant Johnny Payne, an artist and poet, decided to escape rather than be starved and beaten to death; he persuaded a friend, Tommy Marsh, to join him. With three others they tried to escape, but were caught and executed, except for Tommy who made it home eventually to Vancouver. There was uproar in the camp and guards ran about beating everyone. After that, each group of men was divided into seven and the decree was that if one escaped, the rest would be shot. It also caused people to spy on each other, as the lives of all were at stake if one tried to get away.

Blacky made a bunk bed in the hut rafters, free of bugs and people; he would swing up and down on a rope like Tarzan. One day another POW ran foul of a guard who backed him down the centre of the hut at bayonet point, screaming; Blacky swung down from the heavens, landing right between them and roared 'Nanda?' (What's going on?). Startled at this apparition, the guard stared and then ran off.

Starving, the men began to die of diseases – pellagra caused running sores of pus, and swollen testicles; beri-beri caused 'electric' feet – which William got, the pain causing sleeplessness and loss of appetite; he recovered, but many died in the hut ward set aside for the worst cases. Men wept and went insane before the eyes of their friends. Others were bent double in agony, weaving and bobbing like penitents at the Western Wall. Such sights put great fear into the men, as they knew this fate might await them too at some time.

One day William had the possibly suicidal idea of approaching the brutal Japanese Commander Tokunaga, a fat hog of a man, with the idea that the brown rice fed to the pigs should be given to the men instead of white rice; the appeal was that Tokunaga would then be able to make money by selling the useless white rice, but the brown rice contained the vitamins the men needed. William thought of first asking the POW officers to approach the Japanese, but he knew that they were moral cowards and, as they had a much better life and salary than the men, would not want to jeopardize that by seeming 'ungrateful' to the Japanese. So, having carefully rehearsed what he would say, on the next inspection, and literally shaking with fear, William bravely – and against all the advice of his friends – dared to approach Tokunaga and all his entourage and, bowing slavishly, quickly expressed his idea to the commandant through the interpreter, as they were approaching the camp gate. The guards immediately levelled their bayonets, but William said his piece. After the interpreter had finished, and Tokunaga had listened irritably, William was, astonishingly, brutally pushed aside, but left alone. The answer had been no, but at least he had tried.

Many of the Hong Kong Volunteers (HKV) in the camp had local skills and were from professional backgrounds; they ran a lot of the lectures in the camp; some were of Russian Jewish origin (like William's parents) and spoke Chinese and Russian; he befriended some of them (see the testimony of Solomon Bard). But it could not stop the dying; when William approached the officers for some extra money to buy some food for dying starving men, his platoon commander refused.

Weeks passed; the Indian POWs were moved and the HKV were sent to Sham Shui Po; classes stopped and free time ended. The Canadians sank into a gloom, and cholera and diphtheria appeared. Then, in September 1942, William and his comrades were also moved to Sham Shui Po – once a great barracks, it was now a decrepit slum, but it was spacious and had cement huts. Overcrowding led to disease and more deaths. At first, very rough coffins were made from boards, then these ran out and corpses were buried in the nearby cemetery with no protection; the doctors had no serum and so the Japanese beat them for their so-called negligence. William had had severe diphtheria as a child and was probably immune; men sleeping all around him died of it. The gruelling work at Kai Tak also raised the death toll. They rose at 4.30 a.m. in the dark and ate rice for breakfast; then there was a confused, endless head count, the crowded ferry to work, emaciated men suffering interminable slavery in the burning sun before returning to the camp in the dark at 8 p.m: nothing but exhaustion and death. Some men were carried to work on stretchers to fill the manic labour quota. One morning, William heard singing on the walk from the ferry to the airport; it was the Russians, sick and thin, but marching in time to their music, morale high. William asked his Russian Jewish friends about these men and they said yes, they always sang, and had a proper choir. William and some Canadian friends joined their choir lessons; it reinvigorated them, both the singers and the men, to whom they began to give many concerts; for the sick, it was life-affirming. But they could not bring themselves to enter the ward where the dying lay. Some volunteered to work in the appalling stench and filth and William admired them so, but could not bring himself to do it. He remembered well the amazing Anglo-Indian doctor, Ashton-Rose, praised by Dr Solomon Bard in his testimony (in this book), saying how he stood up to the Japanese and saved many, many lives through his untiring, committed work with the sick.

In November 1942 the first post arrived, though sadly many letters were addressed to those who were already dead. It was almost a year before William's family finally heard he was alive and a POW, and an announcement went into the *Canadian Jewish Review*. The letter he received told him that the family were 'overjoyed' he was alive – in fact, he was ill, lice-ridden, starved and a walking skeleton, but he tried to see it from their point of view. William and his comrades cherished this contact with a normal world, and he read and reread his letter many times in the following months.

Slowly some serum was smuggled into the camp and doctors had to decide who would live. Lieutenant Saito refused to move sick men to the Bowen Road hospital, and 130 Canadians died waiting. Only better food would save men from the terrible diseases, and there was no hope of that. For alleged insubordination to his cowardly platoon officer William was sentenced to three days 'honey bucket' (latrine) duty, a most nauseating and horrific job; buckets of diseased faeces, dragged on a cart to the sea wall, splattering the men with vile waste at every bump and jog.

One dark day a new Japanese interpreter appeared: the 'Kamloops Kid' – a notorious tormentor whose childhood experience of racism in British Columbia before the war had left him with evil intent. He was mad with revenge and screamed at the men that the Japanese flag would soon fly over Ottawa and their women would be raped and their men enslaved. His punishments were murderous and for the flimsiest reasons, because he had suffered such racism as a child in Canada.[1] One day a truck load of new sports equipment was brought, bizarrely, into the camp, followed by trucks carrying meat; the men were all ordered to shave and clean up and form games teams; the hospital area was wired off to appear to be outside the camp. The reason: a Red Cross inspection. The emaciated men were ordered to play games and not to say a word to the officials, or they would suffer dire consequences. As the officials were about to leave, Captain Norris, a Grenadier Guard officer, could not keep silent; he walked up to the head official and said that this was all a fake; the men were dying of starvation, and the hospital was there to see. Kamloops was in a fury but the Swiss official then insisted on seeing the hospital. The sight before the inspectors, of dying, starving men lying in their own filth, was appalling. The official fumed and said it was scandalous and would be reported to the ICRC (International Committee of the Red Cross) and a complaint made to the Japanese government. There was great satisfaction among the POWs that now their plight would be known, but then they waited for what would be Kamloops's revenge.

At the next roll call, the courageous Norris was on duty. When he reported that all were present, Kamloops said he was lying and, after working himself into a rage, flew at Norris and proceeded to beat him almost to death as the men were forced to watch. Norris did not die, but he was a physically broken man for the rest of his time as a POW.

In December, work at the airport ended but the death toll still went up. William's close friend Blacky was wasting away in an agony of starvation and beri-beri; once so strong and William's protector, from training to Hong Kong, he was literally a shadow of his former self. One meal time, Blacky picked a fight with William, raging and trying to prove he could still hold his own; William backed away, knowing this was Blacky's way of saying 'I'm not dead yet'. Eventually, out of sight of the others, William took him aside and calmed him down, and helped him back to the hut.

Because the HKV had families in the city, they were able to receive food parcels; on Christmas Day 1943, those of them in the choir invited William and a friend to eat with them. They drank saki and ate themselves senseless, singing Russian songs. For a short while the horrors of the camp receded.

Japan

In January 1943, 700 of William's comrades who were deemed fit enough were herded with him into the hold of a rusty freighter 'Hell Ship' (the *Tatuta Maru*) and transported to Japan. Rumour had spread that a previous transport of men had been drowned when their ship was sunk by Allied submarines (this was later found to be true). As they went aboard, they noted that large red crosses had been painted on the boat, but it was also known that the Allies knew this was often a ruse used by the enemy; William noticed that lots of 'pretend' Japanese wounded, with bandages covered in red paint, were crowded onto the deck to make it look like a hospital ship. Conditions in the crowded holds quickly became unbearable and the food was deplorable, but at least there was some allowance to go topside to use the toilets. They arrived safely in Nagasaki after four days. Nobody had died; it was a miracle, as so many 'Hell Ships' lost hundreds to death from starvation, disease, suffocation or Allied attacks, on horrendously long journeys lasting weeks.

It was a wet and freezing February morning, and a huge shock to the body, after the tropics; morale plummeted. Suddenly a tall Japanese Imperial Guard officer arrived, speaking perfect English, and asked for the senior POW officer, Dr Reid. As they saluted each other the two men recognized one another as former university students in Canada: a second miracle. The Japanese asked if there was anything he could do and Reid asked for clothing and food. At the railway station it arrived – trucks of greatcoats and huge bread rolls, as many as could be eaten. Feeling so happy, the choir members began singing in the Nagasaki sun as the rain stopped.

That night they were crowded onto a passenger train with friendly, tall, well-dressed Imperial Guards who gave them cigarettes and chatted. They were warned that the regular army men they would soon meet were not so nice; William and his friends thanked them for the tip. Through the edges of the blinds, as day broke, they could see picturesque fields with farmers working the land. They arrived at Yokohama in the rush hour and commuters gaped at these emaciated, ragged, dirty, sick, slow-walking, fair-haired giants, as the Japanese rushed past, clean, smart and full of energy. Beautiful kimono-clad women reminded William of his theatre days; *The Mikado* and *Madam Butterfly*. The regular army took over, shouting and prodding the POWs; the public showed fear of these soldiers, something the men were to notice more and more in the days ahead. Another train took them to Kawasaki and then, led by an officer on horseback, they had to march to their new quarters, in an ugly industrial area – long, barn-sized bamboo huts and a compound with a cinder floor. It was called Camp 3D; there was not a tree or blade of grass in sight.

Camp 3D and the Shipyards

Through the interpreter, Kondo, the commandant paraded the prisoners and – hand on sword-hilt – said they must work hard or be punished. The huts had rickety tables alternating with tatami-covered sleeping platforms about a foot off the ground (each one for fourteen men), a shelf slightly higher, and each man was given a bowl, four thin fibre blankets, a hard pillow and a cotton sheet. For four days they cleaned the huts and learnt Japanese army commands – no English was allowed, on pain of punishment – and were given Japanese numbers as their new identities (these were sewn onto their tunics and on ribbons attached to their hats).

Food was one cup of soup and one of rice three times per day, far too little for these grown men, but actually the rice was mostly unpolished barley and it reduced the 'electric feet' agony of beri-beri vitamin B deficiency. They even received tooth powder.

Then on a cold, rainy February day they were marched four abreast two miles to the Nippon Kokan shipyard to work as slaves. Their flimsy plimsolls were totally inadequate in the icy mud. In a large shed which served as the mess hall, they were divided into work groups led by older civilian guards. When asked his occupation, William wrote librarian and painter; so he ended up dangling precariously on thin planks on a rickety bamboo scaffolding at a huge height above the Pacific Ocean, painting ships. The men froze in thin cotton jackets and trousers, far too big for them, so as they marched in fours to work they resembled a ragged freak show, adorned front and back with large Japanese numbers and numbered ribbons hanging from their hats like banners in the wind. Their foreman, Kondo, took them to the paint store shed where the paint master bowed to them and welcomed them; these were older men, polite and friendly, who invited the POWs to approach the stove and stay warm, and spoke respectfully to them. They seemed to disapprove of the bad treatment. One day William was ill with malaria and Kondo sent him to rest in the shed, but the guards there disapproved and slapped him back to work. The paint master hid him in a small storeroom and told him to lie on a bench and sleep. Being treated as equals came as a shock, and William at first resented this because he had hardened himself to hate the Japanese; now he saw a different side of them. William and Kondo slowly became friends and it transpired that this Japanese was an intellectual who had read many western authors in translation; often he read the world news to the men on the way to work, to give them a tenuous link with the outside.

The work on the scaffolding was very dangerous; it was made of rickety bamboo and ladders and loose planks, and they had to carry the heavy tins of red lead paint and the brushes up four floors, while the whole structure swayed in the wind; with slippery army boots, they had to lean over in the icy breeze to paint the ship's sides. William's private effort at sabotage was to work as slowly as possible and conserve energy on the starvation diet. It was freezing in their huts and outdoors, but it was healthier than indoor work where temperature changes brought pneumonia to many men and the death rate went up. Where they could, painters would stand near the riveters' braziers to keep warm; men in the pipe shop were lucky as they had a furnace burning all day. Bathing was impossible so sand fleas found their way onto the men's bodies and caused unbearable itching. At night, men with beri-beri feet plunged them into ice-cold buckets while they shivered in their blankets.

William discovered that the more educated guards believed even the most outrageous propaganda, including the reports of thrown rice balls being used to bring down American planes. Some Japanese teenage workers, trained in martial arts at school, often goaded the POWs to fight; one youngster irritated a Canadian Guardsman so much that he lunged at him and the boy ran a mile to avoid his wrath; another myth was the size of European penises. Often William would be in the toilet and guards would sneak up to look at his; they were disappointed, of course, but William would joke in 'Japenglish' that it was due to the bad food; in Canada, much food meant big organs – and he would make a vulgar gesture with his arm to illustrate his meaning.

Some guards were kind; one student who spoke English would buy ten cigarettes each day and give half to each of his POW gang before work. When Red Cross parcels arrived, the men showered him with fat American cigarettes in return.[2] Some guards entered into trading deals with POWs; one soldier made soap from stolen ingredients, made into turnip shapes, and the guard would smuggle this out to sell, with both of them sharing the profits.

Kondo loved to watch William – whom he called 'Arisuta' (Allister) – do his acting impressions and often joined in mock performances with him in front of the other POWs; the Marx brothers were a favourite. It was funny and raised morale; for Kondo it may have brought some glamour to a drab existence.

Jobs varied for William: tarring huge anchor chains on the shore, or the bottoms of railway carriages; scraping rust off steel before painting; loading sacks of cement. Their work clothes were stiff with paint and dust. Breakfast at 6 a.m. was a tiny amount of rice, beans and watery miso soup; by 10 a.m. the men were starving and shook with hunger until lunch in the large mess hall. Lunch was more rice and watery soup, and afterwards they sat and watched local workers eat from their prepared lunch boxes, shovelling their food in at great speed, and reform-school teenagers in tattered uniforms and smelly split-toed rubber shoes, forced to work as punishment in the yards. Smoking was carried out secretly (it was not allowed at work) as it staved off hunger pangs; all smokers carried little cut-down bully beef tins in which they hid the butts they collected from the floor, heads down (called sniping). At the day's end they rolled their tobacco thus gathered in brown paper, which held together well when moistened.

On one occasion a teenage worker offered William the luxury of a whole cigarette and said he could smoke it. William took the cigarette but said he would wait until later, but the boy said it was OK as he gave permission, so William lit up. A soldier guard passing by saw him and bellowed at William; he looked to the teenager who, cowardly, denied any knowledge. William received a huge blow to his head, then another from this bull of a man, then several others as William tried to stay upright. But he learned that the numbness dulled the pain, a lesson that made him less fearful of future beatings. The teenager smirked at William and eventually the guard, having worked out his anger on William, walked away. Months later he saw the teenager in a punishment work detail, so he had some kind of revenge.

Sometimes they toiled near where female Japanese workers were in view and they flirted with them, joked and whistled; it made them feel like men again, more human once more.

Kondo overlooked this, but the work quota had to be filled and when one POW began taking advantage of Kondo's goodwill and disobeyed several orders to return to work after a break, Kondo went berserk and went after him with a metal scraper. William stood between them and caught the blow squarely on the chest. For days they did not speak, but were finally reconciled. It made little difference, as some months later Kondo was moved away after some trading scandal and William's group lost their benefactor.

In the spring, William was badly beaten by a local worker for singing a popular song in English; he refused to be told what to do, but hunger and humiliation were turning him and his comrades into labouring drones, performing routine jobs by instinct alone. Day followed day in endless dullness and there was little news of the war to raise morale. As they marched through the shipyard gates each morning with all the other local workers, they heard the blaring factory march sounded on loud speakers. It was meant to lift the spirits but to the POWs it was a funeral dirge. In the summer the men could wash with cold water after work; in the winter they could not wash at all. In the bedlam of deafening riveters, sparks, hammering and shouting in the bowels of a half-built ship, William would concentrate hard and drift into a dream world in moments of contemplation, imagining he was besides a bubbling stream, in the sunshine somewhere in Canada. This helped him remain sane in the madness of his prison.

The Artist

One day the paint master asked William if he was really an artist and asked to see his work; William took him three paintings done in Hong Kong (he had kept them in his kit). The Japanese asked if he would do a painting for him and William asked for paints and materials. Several days later the paint master took him to a storeroom where a corner had been cleared among the cement bags, and a studio set up. William was flabbergasted. The Japanese gave William a photo of a vase of flowers and asked him to paint it; they bowed and shook hands and William set to work.

He worked at a snail pace to make it all last; other Japanese came to see him at work, standing respectfully and silently; there was something about a deep artistic strand in the Japanese character, no matter what their class, which amazed William. He played on their awe and made every brushstroke into a theatrical performance, thinking deeply, with long pauses between brush work. The only drawback was that the sitting alone for long hours made him think of food, and his hunger worsened. But he was resting, clean, unregulated, warm and enjoying some respect as each evening Japanese workers came to admire his progress.

Finally the painting was finished and the paint master accepted it; they bowed to each other and then came the gift: two measly cigarettes! But the Japanese redeemed himself by ordering another painting of another vase of flowers, and a week later the paint master brought him a photograph of the painting which had been set in a huge, ornate frame; however, no extra food appeared.

For several weeks William continued to be the genius at work; then there would be another ceremony and two more cigarettes. Suddenly the paint master presented William with the famous photograph of the Allied surrender at Singapore. William was shocked; was this meant to humiliate him? But he had no wish to end this idyllic work and so he played for time, suggesting that each figure needed a series of separate sketches in preliminary preparation. This went on for several more weeks.

Each day the Japanese inspected the slow progress, but his mood worsened as William made the faces of the Japanese officers demonic and those of the Allies angelic. Suddenly one day the paint master grabbed the painting, smashed it on the ground and sent William back to the drudgery of the scaffolding.

Slavery Again

At supper times, mess tins, or tin cans with steel wires, were placed over the POWs place number on the long rickety tables and the volunteer server and helpers brought the large bucket of steaming rice to one end while helpers passed the tins along. The men closed in like ravenous wolves for this nightly ritual. Hungry distrustful eyes watched hypnotically as the servers' ladle was sunk into the rice, lifted, and a heaped portion patted into a bowl and smoothed level; the contents were shovelled into the mess tins, which were then returned to their places, cooling, as the men waited for the completion of the ceremony. When there was more left, it was distributed as 'seconds' in careful sequence, often as many as six getting an extra mouthful. If portions were too large, there would be nothing left for the end prisoners, so a recall of the mess tins was required, the helpers returning them to the server; he removed a dab from each tin to make up the difference. It was subtle torture to watch your precious portion shrink. Then at last it was time to eat.

Servers were often changed because of poor judgement or bribery. For soup it was a matter of how deep the ladle was dipped, as the few vegetables settled at the bottom.[3] As stomachs filled, conversations began with the cheery clatter of spoons on mess tins. Some filled water bottles and drank to feel full, but this stretched the stomach and made you feel empty soon afterwards; the trick was to let the stomach shrink. Diarrhoea further weakened the already starved and exhausted body; men had to squat in the outdoor toilets in the freezing cold several times per night, and then, weak and empty, had to march to work next day, with little to sustain their wasted bodies, unable to absorb anything from their food. The medics tried to keep sick men back from work but it meant the man did not get his ration of food.

The doctors saw to it that the black and burnt rice at the bottom of the cooking pots was given to the sick to eat. William also realized that talking to fellow prisoners about their homes and their future plans kept minds off the horror of the camp and shipyard, and kept despair and sadness at bay for long periods; this was the psychological war of survival, for bad moods could end in loss of the will to go on, and death could follow shortly.

As the winters were cold, so the summers were excessively hot, and scaffolding and steel surfaces were like branding irons to the touch as the men worked stripped to the waist, bathed in sweat and with no shade. One of the Japanese foremen, a particularly comic-looking, short and ugly man, nicknamed Harpo,[4] had sexual designs on William and kept trying to kiss his arms and hands, slobbering like a dog on heat. The men joked with him about it but he was a hopeless works manager and soon disappeared.

One day in late autumn the first post arrived – a huge psychological boost. Some men got something and William was lucky enough to receive four letters, all from his sister Rhea. Each rest day – once per fortnight – he read one page of a letter, dragging out the joy for two months before he had to begin again.

One of the more sadistic guards was nicknamed 'Little Napoleon', so brutal that he was even detested by his peers. He delighted in keeping the POWs away from the stove in cold weather, and ordering his green tea to be made for him while he puffed on a cigarette in a holder and read his newspaper in a superior fashion. In revenge, William and his friends surreptitiously urinated in his kettle just before he returned to the paint shop for a break, and sat as it was boiled for his drink. It was placed before him as William and his comrades stood watching by the open door, and he poured the hot water into his dainty porcelain cup, enjoying his cigarette and newspaper as he waited for the tea to cool; he lifted the cup and sipped and swallowed luxuriously, several times, setting down the cup with great contentment. The hilarity of William and his fellows was hard to contain.

On rest days the commandant ordered the prisoners to enjoy themselves, and William was forced to organize the same old concerts and songs and musicians, month in and month out. One guard became obsessed by the tap-dance routine of one Canadian soldier and frequently came to the hut demanding an impromptu private performance. Once, at 2 a.m., at bayonet point, he came into William's hut and forced the man to dance in the freezing cold, exhausted and half asleep, in his underwear, on his bed. One Newfoundlander feigned madness and the Japanese – always superstitious about such illness – never forced him to work; he became an expert black market dealer and survived the war unscathed.

Often, on the way to work, William and his friends would find discarded tangerine peel at the roadside; it was avidly collected, washed and later chewed on for crucial vitamin C. The men noticed one day that the shipyard kitchen staff often dumped unwanted, spoilt noodles into the garbage bins. But it was difficult to approach the bins unseen by the guards and cooks. However, on one occasion William was left unsupervised to finish a tarring job underneath a railway truck, near the kitchen, and saw his chance to approach the bins; he found heaps of steamy coagulated noodles. He scooped great globules and stuffed them under his shirt, wet and sloppy against his chest. He then dashed into the toilets to hide. The stench was dreadful and the squats were overflowing with faeces, with great white maggots infesting the floor and walls. Despite the overpowering smell, he shut himself into a cubicle and, watching the moving wall of maggots, and squashing others underfoot, he gurgled his way, ravenous, through the spoilt noodles. And to him it was a feast. He felt full for the first time in years.

Another highlight was the arrival of Red Cross parcels – one per week was sent but they only saw one per year; nevertheless it was an unbelievable treasure. Best was the slab of chocolate which dissolved lingeringly in the mouth. Then the powdered milk that could be stirred into cream. Tins of meat and vegetable soup, jam, cheese and butter, and finally fat, smooth American cigarettes – gold on the black market. All this was accompanied by the unfamiliar sound of laughter and yelps of ecstasy. Surreal concoctions were invented – William saw one man cut pieces of garlic into a dish of chocolate shavings. Singing broke out and continued into the small hours. The following evening the emaciated inmates began the trading – what they did not want for what they craved; for some it was cigarettes, for others it was jam. After a crescendo, within days the trading died down as the items were consumed, and then there would be an intense period when the last things were swapped and deals were done at ridiculously low or high prices. In the one clothes parcel received, in 1943, William got a pair of leather shoes, which lasted for over two years.

William described how men with similar post-war plans would meet and discuss in fine detail what they would do when they got home; this was a therapy to take their minds off their present horrors and to stay in touch with what they knew was the real world out there. Two men planned to open a restaurant, and drew detailed diagrams of the kitchen and dining areas; they made up menus and devised costs. Then they began to argue about the details, so real was it to them, and to ask the opinions of fellow POWs, who thought the whole performance bizarre. They almost came to blows.

William himself carefully traded his parcel contents for things that would give him protein and help him stay alive, eking them out for the longest possible period of time, driving as hard a bargain as possible to get what he could. It kept him alive and well but he was aware that it was changing his character – he was becoming hard and selfish, a survival freak – and insidiously eroding his spirit, ethics and humour. It was a far cry from the days at North Point camp where he had been a shining light of the library, classes and discussions. Even his friends said his expression changed during this period – he looked grim and angry.

Winter 1943

The winter of 1943 arrived; freezing oceanic winds shot through the huts and the emaciated bodies; pneumonia returned and hope dried up. Trying to stay warm at night under their thinnest of blankets, the sand fleas rose in their thousands to torture the men's bodies and keep them from sleep; they faced a murderous day of labour on only three hours of slumber. Then one day William found a man who had some mothballs and he swapped half a rice ration for them, ground them to a dust and spread it in his underwear; the fleas died and he slept the sleep of the blessed.

The men were freezing at work and unable to wash; the pores of their skin were clogged and they were ridden with lice; then one day a shipment of coal arrived. It was a miracle. Hot water was prepared. Two huge concrete baths were filled in the washroom, one for washing and the other for soaking – but they were to be for 500 men. Ten sections of fifty men were formed, meaning fifty men in a tub meant for twelve. So a lot of men would end up in the dirty water of others – but it was better than nothing. It transpired that the water was extremely hot and men were scalded but happy. One man – to the hilarity of the others – shouted that he felt like Sam McGee, from a ballad about a Klondike miner who froze to death but, when he was cremated, came alive and said that he was happy at last.

William saw many friends die of illness, starvation and beatings; he witnessed huge, muscular, life-affirming bodies reduced to skeletons, and strong men weep at the sure knowledge that they were soon going to suffer a painful and slow death as they lay unable to move on their beds. Yet there was little room for mourning, as all human energy went obsessively into just staying alive. For years this was to make William angry and hostile. William thought that even if he survived, would there be anything left worth saving? What would happen to his humanity after such horror?

One of the great acts of defiance was their cockney, loud-voiced regimental sergeant major (RSM), the amazing Paddy Keenan, a First World War veteran who twice each day lead *tenko* (roll call), conducted in Japanese, with the Japanese commandant, interpreter and NCOs all inspecting. All the men lined up before their bunks, each section leader reported those present or sick, and then they numbered off in Japanese; getting it wrong meant a wallop. The RSM was always in good voice, his clothes pressed and spotless, brass insignia shining, boots polished, puttees perfect. On such occasions, the act of defiance took the form of spontaneous and intense farting, as there was no regulation against it, and the men saved all their gas for this event; it was used as the commandant and his party passed by, the fanfare following them as they walked by and inspected the men.

One day before *tenko*, however, the RSM made an announcement, clearing his throat: 'It 'as been brought to the hattenion of the authorities that certain practices of the prisoners, 'ave been found to be hoffensive to the hexisting powers. Heffective as of this date, all forced farting will be strictly prohibited.' When the RSM, oozing confidence, spoke and moved about, the dingy hut became a sun-kissed British parade ground, Japan and hunger vanishing; it was magic; they were in full uniform, at home, with rifles sloping, bellies full, the present a triviality.[5]

Other kinds of escapism involved William and his friends re-enacting scenes from famous movies they had seen, and music. The Japanese commandant had allowed a POW officer to leave camp and buy some records. In the old bathhouse, unusable in winter, and nicknamed 'the Hong Kong ballroom', the men would gather on a rest day or in the evenings to listen on the old gramophone. They squatted on tiny footstools – scarecrows with filthy grey skin covered in sores, humped and bony-kneed – for no prison could confine the music of freedom. One of the more bizarre Japanese guards, Yamanaka, had a penchant for disguising himself and trying to fool prisoners into thinking that he was someone else, to test their reaction. To tease him they would always stand to attention when he appeared in a childish outfit, and address him as Corporal Yamanaka, showing immediate recognition; he would scamper away in anger. He especially took a dislike to William.

On one occasion, when William was on night fire-guard duty, he visited the kitchen to watch the cooks and keep warm, stealing two potatoes on the way. Yamanaka, on his way to the toilet, saw them and screamed at them. William was more terrified that he might find the stolen potatoes, which would mean severe punishment. Raving and ranting, he smashed William across the head. Yamanaka had him down as a marked man.

Tokyo decided to broadcast a prisoner show on the radio, to give the propaganda impression of how well the Japanese were treating POWs. The Allied officers agreed, as it meant they could slip in information to the outside world about the names of the POWs and the conditions they were suffering – such as 'conditions are good, tell it to the Marines' – US Army slang for 'do not believe it'. William was named MC, and when the Tokyo technicians arrived to record the show he was given star treatment. Yamanaka hated this and looked for a chance for revenge. The show went well and William was able to pass on much information on names, which were picked up by US West Coast radio hams and forwarded to families; the government, incidentally, covered this up – just the radio amateurs passed the news.

Later William had a malaria relapse and went to work next day wearing a fur hat supplied in a Red Cross parcel, to help keep the shivers away. Yamanaka saw this and screamed at William, who could not hear because of the malaria tablets the doctor had given him (which caused partial deafness) and the fur earmuffs of the hat. He descended on William, ripping off the hat – which he said was worn without permission – and swiping him repeatedly round the face with it, the buckle especially causing him a lot of pain. In a rising frenzy he hit William on the head with his rifle butt, driving him into the large dining hall so all 400 men could see what was going on. He shrieked, demanding to know why William had not answered his orders; in intense pain, William said he could not hear properly, and Yamanaka responded by hitting William with the rifle barrel so his head became covered with cuts and blood. After many blows, the Japanese then finally drew his sword and, through the interpreter, asked if William would like to have his head cut off. William replied: 'Tell him my boss does not want his men working without their heads.' The interpreter refused, telling William this would make Yamanaka even more insane, and he gave a more polite response. Yamanaka walked away. This saved William's life. He had not given in, though, and this gave him a feeling of exhilaration, while his comrades looked on with a mixture of relief that it was not them and guilt that William had taken the blows. So, as a result of this and other beatings, William became eaten up with a hatred that stayed with him for decades, but he was determined that his revenge would be to survive.

The illegal black market flourished; prisoners, guards and civilians were all involved in their enclosed, surreal prison world; being caught meant a term in the notorious Tokyo main gaol. One Canadian became such a true 'King Rat' that the powerful cooks even brought him warm bricks at night to heat his bed.

For smoking a furtive cigarette while at work, on one occasion, William and friends were lined up and severely beaten with a leather belt by one civilian guard.

A parcel arrived for William out of the blue, from his parents; an angora sweater to keep him warm, socks, two shirts – riches beyond belief. But one night the sweater and socks were stolen. William was angry – stealing from a fellow prisoner was the lowest of deeds. He complained bitterly to the commanding officer, Dr Reid, who ordered all known black-marketeers to his room and demanded that the items should be returned with no questions asked; meanwhile they would go on half rations. Theft, he said, was becoming too common. Within days the stolen items were handed back to Reid and William collected them. Other men found the thief and took him aside for a beating, inviting William to have first punch, but he understood the temptation the man had faced and could not bring himself to take part. After cursing the man, he walked away.

One horrible story that William heard at first-hand was from a friend who had to bury his best buddy in a coffin that was far too small; they tried to fold back his legs, but the head protruded in a grotesque manner. By forcing the head, the neck snapped and the lid was closed; the sheer horror of it was that it was like killing the man a second time. The trauma never left this man until the day he died, many years after the war.

Spring was coming; William could feel the warm breeze of the sea. More brown rice appeared and illnesses were slightly reduced as a result; the permanently sick were given light jobs. One civilian interpreter, Koyanagi, who had lived for many years in Detroit in the USA, and in Windsor, Ontario, secretly smuggled in drugs for the hospital at risk to his life; forty years later, William was to meet his widow and son, who testified that he was ashamed of the treatment given to the POWs.[6] William heard of other acts of kindness by civilians. Another older Japanese foreman, seeing POWs struggling, weak from hunger, with the heavy work, led his group of a dozen prisoners to a distant railway car, broke into it, took out a sack full of potatoes, started a fire and boiled them for the men to eat there and then – an ordinary man who risked his job and prison to help enemy strangers.

Rest days were usually one in fourteen; William just used this to rest and lie back; his personal ritual was to reread his letters – all carefully arranged in date order – and look at his family photos and write in his diary. Awareness of the massive divide between their lives and his sometimes sent him into tears of despair at perhaps never seeing his family again. Men would gossip and argue on these rest days, their frayed nerves often leading to arguments as over the years they became hypersensitive to people's annoying habits. A scratched record player, passed from hut to hut in rotation, played its music, and rumours of Allied victories would circulate. Chat and joking kept spirits up, and the photos and letters kept hope alive, but would they all be executed before the Allies arrived, as some rumours had it?

Summer 1944

Then the bombing raids began; they would gaze up at the shimmering, silver specks of the B29s, only a mile away: the well-fed airmen, their friends, and revenge at last! It was August 1944 and it was easy to read between the lines in the English language propaganda newspaper they were allowed to read, and to see that the war was going against the Japanese. The workers' rations got worse; the POWs grew hopeful and the Japanese more morose. Together they all built shelters, the POWs being in just as much danger from bombs. According to the Japanese, Churchill and Roosevelt were both Jews, part of the world conspiracy against Japan, but although the Japanese became more hostile, their hostility was tinged with respect. They knew what was coming, especially when news of D-Day came through, and it was the beginning of the end. Among the prisoners, hopes soared.

A library was permitted by the camp commandant, and the spirits rose of those who, like William, craved the written word; he devoured the books and found ways of getting those who did not want to read to get books for him, above the ration of only one per month – a petty limit set arbitrarily by a sour-faced American naval officer called Quinn, whom everyone tried to ignore. The books sustained William through the winter.

On another occasion the Japanese installed a regular, sit-down toilet. It was like a shrine to western culture, clean and private, and boosted morale enormously.

Christmas

During the winter of 1944–45, many died in the cold weather, but Christmas brought a slight respite: a Red Cross parcel for each man. It seemed unbelievable. The men decided to decorate the huge dining hall to celebrate, and a competition was suggested between the work sections. William was asked to take charge of his group's effort, but he quipped that it was against his religion, as he was Jewish. 'So was Jesus', quipped his comrades, and so he gave way graciously. The whole section was sent to scavenge for anything colourful or bendable. Hidden, creative talents were discovered and all kinds of paper flowers and shapes were manufactured. Vases were soldered from scrap metal and paint was liberated to coat them; cotton wool became snow stuck on windows; rusty pipes soldered to a tin base became candle holders with red paper flames. One man brought a three-foot-diameter tar drum lid, on which William painted nativity scenes; it hung like a giant mobile in the centre of the hall, which began to take on the glow of a manger. It was then decided that the circular nativity scene needed a wreath surround. But from where? They called in one of the big black market prisoners, Charley Clarke, and arranged that William would meet him next day to do the deed.

Next day the two men met at the lunch break, and with a cart and Charlie's brazen air of authority they began searching the shipyard, feigning important work. They were searching for straw Japanese rain-capes, and came upon a pile after some time, stacked near a group of Japanese workers. Charley asked how many he wanted. William said five, but also that he must be mad; they would be caught and beaten to death. Charley told William to stand between the workers and the cart, and he slowly loaded them up. William was dumbfounded and shaking with fear as Clarke slowly loaded the capes, calmly whistling, and coolly left the yard for the camp. William asked how they would get them into the camp; after work, he found them on his bunk.

The danger was that all this material being gathered was difficult to hide from the *tenko* guards who periodically searched the huts. The best places were the upper bunks and shelves which the inspecting guards rarely looked at, as it meant climbing ladders; most of the material was on William's bunk, which looked like a junkyard instead of the neat bed required for inspection. Suddenly a full inspection was announced. A guard began the climb; on reaching the bunk he let out a scream and then the commandant, with swords, boots and medals, also climbed up – an unprecedented sight – screaming and demanding to know, as he descended, what this mess was. Questions and answers started flying, and William felt faint. Captain Reid and Paddy Keenan explained, at high speed, the importance of Christmas as a religious festival, to raise morale, increase output and keep prisoners busy. Finally the commandant was mollified and just demanded that things were tidied up. It was a real Christmas miracle – he walked away.

Next was how to prepare the table – twenty yards of old, drab, bare, warped boards on spindly, gnarled legs. At a conference, sheets were suggested as tablecloths – but they were filthy and there was no soap. Then someone remembered the globules of soap used on the skids to launch the ships, and handfuls were smuggled into the camp in mess tins. Tooth powder was used to make the wreath on the nativity mobile white in colour. Now they just needed the Red Cross parcels; hopes had been dashed before. They were certainly given the day off. By judicious trading, someone had smuggled in some garlic to garnish the rice, extra portions of which had been bought on the black market and stored for the big day. And on 24 December the parcels arrived – one per man; it was astonishing. That evening there was far too much rice but it was mixed with butter, chocolate, cream, sugar; next day was Christmas Day and the eating began, amidst much vomiting and diarrhoea from the previous night's gorging. Breakfast was light as the men saved their appetites for the main meal and put up the decorations. By early afternoon the tablecloths were on and the food laid out: a remarkable sight, with coloured paper flowers in vases and even some oranges, a gift from the shipyard owners. And above, the suspended paper icicles and nativity scene, forming a canopy which shut out the dreary reality of their imprisonment. Even the commandant came to see and pay homage.

Hung above the table was also an imitation scroll that William had made, and he had written in old English calligraphy upon it:

> We said by Xmas '42, we know the war will all be through,
> We said by Xmas '43 we know that we will all be free,
> We said by Xmas '44 we know there will be no more war,
> But now? By Xmas '45 I hope to hell we're still alive.

The last line was closer to the truth than they knew, for orders from Japanese High Command had by then been prepared, that all POWs should be killed if the Japanese homeland was invaded. Their commandant was no doubt aware of this and it may have explained his generosity in the recent months; we shall never know. Whatever the truth, he issued yet another Red Cross parcel on New Year's Day. Gloomy post-Christmas faces burst into laughter.

Spring 1945

By March 1945 the Allied bombing became an almost daily event. It brought work in the shipyard to a crawl because each time the siren went off the men were rushed to the dining hall and shoved under tables and into air raid shelters. Then they doubled back to the camp and were ordered into the prepared slit trenches. Finally it was decided to break up the camp, and William was one of fifty men ordered to pack; the wrench of leaving friends of two years' standing, to be taken into the unknown, was great indeed. Knowing they would be searched, William gave most of his diary to a friend, to return to him after the war.

Another Camp

Hardened to emotional goodbyes, the men said farewell and William's group were put onto two open trucks and driven off. They passed many bombed, ruined towns and eventually arrived in Tokyo, which was a scene of devastation. Soon they were taken to a decrepit coal yard on the edge of the city, called Sumidagawa, a canal and railway junction where barges and railway trucks were loaded and unloaded with rice, beans, iron, coal and other materials. Here they were to work, housed in a shabby two-storey shack lying between two huge coal heaps. There were already 250 POWs in the camp, of varying nationalities, dressed in all kinds of strange bits of uniform, with odd hats and long beards and hair; it all looked unsettlingly disordered – but they did appear to be well fed.

They were searched and numbered and led into the hut. The resident prisoners were somewhat surly, but one approached William and so he asked about the food; it was bad, came the answer, but they supplemented it by stealing copra (coconut oil) and beans, whereupon he gave William a generous handful of soya; the handshake was just a bit too lingering and William suspected he was gay. Indeed, masturbation was much more common in this camp, as the men obtained more food than in other places William had been.

A tall, jovial Japanese sergeant then showed them their beds – typical thirty-seven-inch-wide rattan mats, seven feet long, on a shelf from aisle to wall, soiled and full of lice. The guard carried a large poll, used to crack heads if you did not move quite fast enough. He was known as One-Eye. William witnessed One-Eye's handiwork within minutes: he had caught a Canadian smoking. Ordering him to attention, One Eye removed his gloves, finger by finger, staring with a smirk at the soldier. The he began beating him. He would not lie down and the Japanese beat harder; the veterans shouted that the trick was to fall. He did so, and One-Eye soon walked away. Another veteran warned them of a particularly crazy guard called Piston Fists. Piston Fists sported white gloves and polished riding boots. He liked to practise his punching on the men.

At first, William noticed that the newcomers (Canadian and Dutch) always worked on the coal platform, where the opportunity to pilfer food was zero. It became clear this was deliberate on the part of the American POW foreman who organized the work schedule. In addition, the coal work was back-breaking and filthy, and for already emaciated men this was not good for survival. Later, work was given to them only because on some days more men were needed on the food platform, but the problem then was smuggling the food into the camp; searches were thorough (hiding rice in one's shoes was the best method) and the veterans protested that newcomers being caught would expose their racket; but it was life and death to get extra food and the Canadians refused to stop. Thus resentment between the two groups developed.

Rice, however, was plentiful and the Americans had built a secret oven to cook it, under a bed space, which could deal with thirty tin cans at a time. The steam was dissipated by the sick men during roll call when the guards and POWs were all outside the hut. William teamed up with his buddy Grant – like a married couple, they helped each other and shared all; this was how POWs would survive. Grant began collecting the material to make their own two-tin oven; he spent weeks carving concentric rings in a brick and then put coiled, old wire in the grooves, for an element. Then secretly one night he wired this, by a cable hidden among the beams, to the main electrical supply in the roof , and they were able to boil two tins of rice. But others were doing the same and so to avoid overloading the power, and attracting Japanese electricians, time slots were allocated to each group with a cooker. As one group finished, another group went on line. The Americans did not much like this free enterprise because it lowered the market price of their rice from the big thirty-tin oven.

Some time later the Americans spread the rumour that the Japanese were going to check the electricity supplies. Rapidly, and in panic, the other POWs dismantled the wires and stoves at huge inconvenience, but not the Americans, who said they would take the risk; rice prices rose, but no electrician appeared. Clearly the Americans had made up the story in order to raise their profits again. This game of bluff continued on and off for months.

The coal yard work was exhausting and filthy, with William and his friends shovelling loads; the worst task was pushing the heavy trucks up the railway incline to get to the loading platform. In the evening, tired and starved, they would have a meagre meal and then would go to the wash house alongside the hut, with their thin, small towels, dowsing each other with cold water to clean their bodies. The food smuggling meant the men began to put on a little weight and so the more predatory gay men came on the prowl at this time. William was propositioned on a number of occasions; food was often used as a bribe.

The experienced Americans were very efficient in their work of moving the heavy food sacks, rhythmically picking them up with hooks, bending the knees like coolies, so conserving effort, compared to the awkward Canadians; they resented the newcomers even more for slowing their work pace, which if good would bring extra food as a reward. On the bean platform, William was astonished to witness an American openly roasting huge amounts of beans in full view of the guards and foremen – because the guards were in on the racket, taking a share home to their families. The Americans ran the scam and the Japanese often complied.

Unlike Camp 3D, this place seemed to William to have a devilish nature; in the midst of death and desolation, there was a kind of perverted vitality which fed on itself. Men put on weight but exploited each other; guards used fists instead of palms to beat men. There was a brutal internal camp hierarchy and self-created slavery/capitalist system unlike any camp he had been in. What would all this do to their minds by the time it was all over?

As his rice store was depleted, William was tempted to go with two friends on a 'raid' in the rail yards to replenish his stock. Individuals often did this with great courage and brazenness and got away with it, but risked terrible beatings if caught. Finding the storage area, having covered a distance of about half a mile in all, and in the process of loading up with some rice, they heard a voice suddenly ring out, telling them to stop; it was the sadistic Piston Fists – the very worst person to have found them. He stood them to attention, circling then menacingly. He screamed out: 'Who is going to win the war?' Grant, William's friend, replied that he did not know; this sent the guard into a rage and he slapped his face. He then ordered each man to pick up a fifty-kilo sack of rice and said he would beat only the first man to drop the sack, as a kind of cruel game. He then sat down to watch as other prisoners gathered around. One Englishman whispered that they should all drop the sacks together and so spoil the Japanese fun. So on the count of three, they did just that. Piston Fists went berserk and punched each man viciously in the throat. For months afterwards, William had problems swallowing.

On occasions William went with the clean-up squads into the nearby towns to clear rubble from the Allied bombing; he witnessed the stoicism and real hunger of the civilians struggling to live in the ruins. At least he and his friends could steal and get enough food. Just then, in this chaos, the slightly mad Japanese commandant came to address them: a tiny, wizened man standing on a box, hopping and screaming. He told them to collect all metals and put them in little piles, then made a crazy patriotic speech; in this insanity and bombed-out carnage, the POWs did their work.

The men were allowed to bring kettles of tea to work for drinking during breaks; they made false bottoms so that they could hide stolen rice in them and get it back to camp. William slit the lining in his hat and, by filling it with rice and spreading it around, when asked to lift his hat during the inward searches could do so without the rice leaking out. By tying his cuffs tight he could also hide rice in his shirt sleeves, not detectable under his jacket, as they returned to camp.

Despite increasingly severe punishments by the Japanese, ordered by One-Eye, consisting of standing out on the parade ground all night, for twelve hours, if they were caught, and being beaten each hour and having no food or water, the stealing went on and on, simply because it was life or death for the men. They simply had to take the risks; they also had to compete with the American POWs who hogged the rice platform work in this camp. Inevitably a POW was caught; William watched as the man was stood to attention and his friends witnessed the punishment administered. One walked past the lonely, swollen but erect figure and whispered to him to ask to go the toilet; it was allowed, and a friend was hidden inside, and gave him tea and rice.

So it was that on the rice platform, soon afterwards, William prepared to steal again; he and a friend prepared a long stocking, sewn tight at one end. At work, fifty-kilo sacks were swung onto his shoulders in the railway truck and hour after hour he staggered along the shaking planks, with quivering legs, to load them onto the waiting barge. At an opportune moment he and his friend slipped behind a pile of sacks and slit one open. Whilst William was lookout, his mate filled the stocking and stashed it between some sacks to be picked up on the way back to camp, and the rice would be put into their hats, shoes and sleeves, in the toilet, before being marched off. At the end of the work day they successfully carried out the transfer, but, as he joined the waiting men, William looked odd, with a misshapen hat and rice leaking from his shoes. His friends said he should dump the rice, but it was too late as the guards gathered round them, and he started the march, a mile to the camp. On the journey, the men pushed him and his friend forward so that the rice visibly leaking from his shoes could be trodden into the mud by the others, who, cursing, were frantically dancing about, treading the rice down so the guards would not see it. William was charged with fear, imagining the beating looming before him and remembering the man on the parade ground a few days before. Then his sleeves began to leak rice – they were so threadbare, that they could not hold the weight of the grains.

They entered the gates and lined up in the usual threes in the yard, facing the guard-house; the coal-dust surface made the white, leaking rice look even more conspicuous on the ground. They were ordered to open ranks for inspection. William loosened his belt so his trousers sank a little lower to hide the rice on the floor; he folded his arms to stop rice falling from his sleeves, whilst the other men had their arms at their sides. A guard searched the man next to him, and then moved to William, who was quivering with fear and the effort to remain at attention. His fellow prisoners were staring at him. The guard patted his sleeves, the hands passing to his legs, back to his waist; William gently shook his hat as demanded: nothing. The guard moved on. The order came to close ranks and stand to attention. Because William delayed lowering his arms by pretending to adjust his buttons, so as not to loosen any rice, the corporal of the guard rushed over and slapped his face.

On dismissal, the other men gathered round William in a knot to hide any rice leaking on the way to the hut. Inside, ignoring the protest of their mates, William and his friend hugged each other and unloaded their loot. But there were repercussions – the Americans were angry and were going to give William a beating, but the Canadians closed ranks round William as the Yanks entered their hut. Confronted with this opposition, the powerful American leader thought the better of it and just asked to talk to William. Allowed to pass, he went to William's bunk and sat down and said that they were being greedy. William countered that it was because the Americans grabbed all the food-stealing jobs and opportunities and left them to work the coal; they wanted a fair share too. The American said he would think about it. He then asked to see some of William's artwork and asked if he would do a portrait of him there and then. William did so; the 'Rice Baron' suddenly became a meek model and the roles were reversed. For a brief moment he dropped his guard; the 'tough American' persona vanished and he became an ordinary guy. William saw there was hope for them all yet.

Not so funny was a Japanese guard who took a fancy to William; girl-like, his hands would linger rather longer than necessary on William's body during the searches. William took to hiding from him whenever he was in the vicinity, to avoid him, but one night on the way to the toilet, wearing only his loincloth, William walked straight into him. The guard barred his way and began prodding him and jabbed a bayonet at his throat. Suddenly the American block leader appeared, also on the way to the toilet, and asked the guard what was going on. The guard ordered him away but the American said they must all go to speak to the sergeant; this did the trick. The guard backed off and that was that.

VE Day

The news arrived that Germany had surrendered. But there was not too much celebration, as nothing changed for the POWs in Japanese hands, who had to work like slaves. The guards grew meaner as they knew they were losing the war.

One day a fully armed guard came into the hut and all the men jumped to attention at the end of their beds as usual. He made for William's bunk and confirmed his number, 325. William's stomach churned; what was this about? But the soldier laid down his rifle and its vicious fixed bayonet and made signs to ask if William was the artist. Relieved, he said yes. The guard asked to see William's sketchbook, and sat down, gently turning the pages with approval. Through the prisoner interpreter, he asked if they could exchange work, then opened a little kit of ink, brush and water and painted a quick but beautiful scene on a blank card. He offered it as a gift. William was very moved and thanked him, then he did a quick pencil sketch and gave it to the guard. He smiled and thanked William; he sighed as though he did not want to leave; he folded his artist's kit, took his rifle and bowed. William bowed back. He bowed to all the other men and they in astonishment all bowed back; he bowed again to William. Then he marched off. Whoever he was, he had shown a vision of the possible.

Some time later, William was on a working party to plant potatoes, though many got eaten even before they got into the ground. But among them this day was a new POW, a very young US Air Force officer who gave them all the latest news: rationing; women workers; the newest hit songs, whose words he had to sing to fulfil their craving for being in touch with the real, happy world; the German defeats in Russia and the attempted assassination of Hitler; the new pin-ups they had never heard of – Betty Grable, Rita Hayworth, Jane Russell. It was both satisfying that the world had moved on, and shocking that it had not remained the same. Many of the men had been POWs since 1943, but William had been POW since 1941. Would they ever fit in again? Would they ever return?

William had other worries – he had wet beri-beri and his legs were giving out; no work meant no food – you were totally dependent then on handouts from friends. A deal was struck; the American foreman said he would keep William off work in the hut if he would create pin-up drawings for the men. Crayons and card were given to him and he drew shapely hula girls with huge breasts, and worked in wonderful, peaceful solitude for days on end.

Roused one day from his art by pounding feet, he saw a prisoner with a fifty-kilo sack on his shoulders run through the hut in broad daylight, and watched the sack being seized by waiting hands and hidden away, all in thirty seconds, as a guard patrolled past the window. William was astonished. What had happened was that the sick POWs had been sent to sweep fallen rice from the platforms to supply the pigpen; secretly, near the rice-loading platform, a whole sack would be dropped into the cart and covered in sacking, covered in the loose sweepings, and then brought back to the camp without being searched. From the pigpen it was smuggled to a hut. William was both impressed and angry; no wonder the American racketeers could cook so much rice for sale – and yet they criticized others trying to smuggle their own rice from the stores. It was American enterprise at its most brazen.

William's pin-up drawings were a huge success – one hung near the hut entrance and all the work parties hailed it as they passed, blowing kisses and more basic endearments; christened Lulu, she was carnal escape. The Japanese commandant demanded an explanation and the senior Allied POW explained that it raised the men's morale. Encased in his rigid life of codes, the commandant stared myopically back, then barked out orders that if the POWs worked harder, he would allow removal of the dress!

The Bombings and Liberation

US Air Force raids increased and became a nightly occurrence. But there were no air raid shelters and the men remained in the hut, from whose top bunks it was easy to see the conflagrations, the searchlights, the bombers and the 'ack-ack' (anti-aircraft gunfire). William watched, horrified at the devastation and also at the sad deaths of their young airmen, as hits brought down many bombers; some exploded, some trailed smoke and fell, others exuded white parachutes as men escaped. They saw two airmen come down near the camp and watched angry crowds race towards them, seeking bloody revenge for the deaths of their loved ones. They heard the sounds of the mob, a hum growing to a roar, then the screaming as the crowd converged with pent-up rage on the target. William shivered; they were lost; what a way to die; what were they doing to them? Torn to pieces. This went on night after night.

On 15 August 1945 it seemed a normal, hot, boring day. But something was odd, as the men were sent back to camp at noon, an event that had never occurred before. Rumours spread that the war had ended. William could not believe it, as such stories had been heard before, months ago. To test the rumours, William went outside and lit a strictly forbidden cigarette to provoke a guard. The guard blinked and walked past. It was true! But certain proof was required. The men went to their senior officer, who agreed to visit the Japanese commandant. The men waited and paced outside the hut for half an hour, and then their officer emerged with the Japanese; there was an instant hush. The commandant stood, legs akimbo, arms on hips: 'The rumour is true!' The hut exploded in joy, the men leaping and cheering as the Japanese staff raced away for safety. But William could not get excited about it, and others were so paranoid that they would not believe it. For their own safety, everyone was confined to the camp. Food improved hugely, smoking was allowed anytime, and men slept outdoors to avoid the lice-ridden bunks. William abandoned rice and ate vegetables. The war criminals fled, rightly fearing the prisoners' revenge. The men staged a concert and different men sang songs; addresses were exchanged and future plans talked about. But most, like William, remained numb at what was happening; how would they cope, being released from this narrow, terrible world, into freedom? There was joy mixed with fear. And, on top of this, the men fully expected to be massacred, as the rumour had suggested they would be.

During the following six days they kept seeing aircraft far away but they were not spotted. Then on 21 August a single Allied aircraft approached. Cheering and shouting, the men poured into the yard, waved sheets and jackets and shone mirrors; the pilot acknowledged them and dived low over the camp, his engine roaring, dipping his wings. William said that 'it was the most magnificent symbolic salute ever received'. The men were hysterical, shrieking with joy, and tears flowed. William said: 'My dam had burst at last and out it all came, as though all the anguish of the planet had found me, and out of my bowels, all the murders, tortures, all the Jews burning in Belsen, all the hellish years, had gathered a million voices bursting all bounds.'

This pilot scribbled a note and wrapped it round a packet of cigarettes and dropped it into the camp. The men rushed forward to see it; it said: 'Greetings to men of a brave and free world.' Never did words carry greater power. It was truly over.

In the following days, more aircraft dropped notes asking what the men wanted, and they painted 'CHOW' (food) on the roof in huge letters; but bombers dropped huge canisters far too heavy for the parachute and one smashed their roof, and another killed a woman in a Japanese house. There were Hershey bars (chocolate), tinned fruit and stew, cheese, butter and so forth; years of fantasies came to life. Men ate and vomited; others opened cans and tasted, then cast them aside to taste the next delight. A warning system was set up against the falling parachutes, and men sheltered under the concrete rail platforms when the planes approached. In desperation they painted 'STOP' on the roof.

One container landed near William and sent a shower of cinders painfully into his back as it struck the ground.

Slowly the eating moderated and food cravings were appeased, but now they became very impatient for liberty. William and some friends decided to sneak out to find a brothel. And they were intent on some kind of payback after the years of bitter and outrageous treatment and inhumanity. Outside the camp they followed a road and soon came to a building – some kind of institution – and barged in, confronted by one man who bravely questioned them. They made it clear what they wanted and the women inside saw them; when news of their intent spread, the women began to scamper through windows to escape. The POWs pushed the man aside and saw the women running across the fields and away; except for some old women and children, the building then appeared empty. They were enchanted by the children – they had seen virtually none in four years – and offered them chocolate. At first they refused, afraid to approach these foreign giants, but the old people took it and the children joined in.

The POWs began to leave but William was determined on his revenge, and to get a woman – revenge for being robbed of his manhood for years. He noticed a second floor and went up alone. On the third try, in one room, he came upon a beautiful woman with black flowing hair and sky-blue kimono, sitting at a table; he was sure the fear in her eyes would subside when he took her in his arms. But she rose in terror, fixed his stare with dignity, and noticed William's hesitation; and he could not bring himself to stop her as she glided out of the room past him, gone forever.

William walked to the table and saw that the woman had been painting. Her tubes of colour and brushes and palette were a rebuke. He sat down and continued her work and in this way felt he had consummated their love in colour instead. He had contemplated, but not committed, a crime.

US Marines soon arrived in PT (torpedo-armed) boats along the canal, armed to the teeth, crouching and edging round corners into the camp. It was 27 August. William and his friends beckoned them on, as no Japanese were there. The POWs were in awe, wanting to touch, hear, smell and question these young gods; the Marines were shocked by the conditions in the hut, and told the men about the atomic bombs. Looking for souvenirs, William took one Marine to the Japanese guard hut where some of the younger guards had stayed behind, awaiting orders. The young American was ready to kill them at once but William calmed him down. As the Japanese cowered, the two Marines took rifles and helmets.

Soon the POWs were taken by the PT boats to a US hospital ship to be scrubbed, fed, bedded and pampered, in a world where everything was just perfect, throbbing with life: well-fed, normal human beings, smartly dressed, and beautiful nurses. Thorough medical checks were given and William stood naked as a doctor probed and poked. He frowned and William asked what was wrong. He did not reply. He called his colleague. 'See this?', he said. The other doctor squinted. Alarmed that he had some terrible tropical illness, William demanded an answer. 'Lice!', said the doctor. 'Oh that! We've got plenty of those!', said William. He had also lost one-and-a-half inches in height. This was the first of the many after effects he would suffer over the following years.

Then came the voluptuous showers, with foaming suds, thick white towels, laundered clothes; William still waited for someone to wake him from the dream and tell him to get to the coal tip! After shaving he went to the mess; the choice of food was astonishing, the men piling their plates with something of everything. Many put food in their pockets but were told to put it back as it would be available all day. They queued again and again for more helpings, the bulky cooks gawking at them. When asked questions, some still answered in Japanese; there would be much to learn and unlearn.

On billowing mattresses and in crisp, white sheets and clean pillowcases, with a small bedside radio and full stomach of fine food, William lay in bed, hurled into the future at breakneck speed; it was paradise. Later there was a movie in the ship's auditorium, and the men luxuriated in chairs and were transported to the world they once knew.

It was here William learned that had the atomic bombs not been dropped, their camp and railway yards were marked for obliteration by the air force.

Next a British warship took them on; on this vessel, discipline was more strict, accommodation more cramped and the food plainer. William understood that the British had been through a longer war. At Formosa they picked up some British POWs who were in a terrible state. Then they went to a US rest camp near Manila to recuperate and fatten up. Here William met some comrades whom they had left in Hong Kong. Others went to hospital with a whole range of physical and mental ailments. William and a friend met up with some GIs who showed them round and joked and chatted with them; but when William mentioned they had been POWs they fell silent and gave them a look of awe and pity, with a tinge of guilt and distaste. The jollity had gone, and William resolved not to talk about the past too much after that.

After two weeks they had passes to go to Manila itself and spent an evening dancing and drinking. They paid to dance with club hostesses with frozen faces; what horrors of Japanese occupation had they seen, William wondered? Meanwhile, lots of mail arrived for him, but he began to ask himself if he could ever be part of the normal world again.

They went on to San Francisco on the USS *Joseph T. Dyckman*. Here they ate in a camp which housed German POWs, and they could not believe how well fed they were and the excess of food of all kinds on tap for these enemies. What kind of a war had this been? Even the British Army did not eat this well. There were barrels full of oranges for the Germans to take, and William thought of the orange peel he scavenged from the street in Japan.

William was taken by rail to Seattle, where he was able to wander the streets that evening and have all his dreams in the camps now come true – eat, drink, dressed smartly, no guards, window shopping – it was a struggle to take it all in. He sat in a nightclub and ordered a drink while a beautiful woman sang a sad song in a spotlight. William's thoughts turned to the sadness he had witnessed and suffered – and a terrible melancholy flowed over him.

From Seattle they sailed to Victoria, British Columbia; they were pursued on board by reporters wanting gruesome stories of their captivity. One insisted that William tell him of the beatings he had received, and William obliged for few minutes. When the man insisted it was not gruesome enough, William almost threw him overboard. In Victoria they partied for five days, as moods swung from ecstasy to depression to anxiety to unreality, as they wandered the normal streets of normal people in their own country at last. There were joyous reunions, letters and telephone calls and telegrams. But many were still ill; one of William's friends was in a mental hospital; some faced divorce; and many others would die young from the conditions they suffered. William and the others found themselves back in the same old military barracks they had left in 1941; there was no awareness that these men might act or feel different, no counselling or advice – just the long, nervous train journey home, east across Canada, with four years back pay stuffed in their army battledress.

William could not face seeing Hank Greenberg's mother or the welcome party organized by a wealthy cousin in Winnipeg, and he went on directly to Montreal. At the station, his family waited in line; shaking with emotion, William watched their tears flow and he fell into their arms. But he was dry-eyed, his heart – like the other POWs – secured against emotional onslaught. Only his father's keen, horseman's eyes detected the beri-beri in William's swollen face as he kissed his son, straw moustache and all, Russian-style, on the lips; then his father stood back and let his mother claim him.

Squeezed into the car, they rode home, where his mother raged at the Japanese; William could hear her thinking: 'They did this to my child, my baby whom I breast fed, nurtured, cooked for, loved...'. William later learned that she slept for years with his photo under her pillow, weeping and praying for his deliverance. William walked from room to room, as the family followed; everything was in place as he had left it. They sat and had tea and honey-cake. The family spoke of rationing and trivial difficulties; the distance between them was frightening; they seemed so unaware. He felt they should count their myriad blessings. Thinking of his experiences and of the men whom he had watched die, William swore, 'If I forget thee Jerusalem ...'.

He felt the guarded behaviour of his family. 'Pass the sugar', he said. It was passed silently and swiftly, glances exchanged. 'Oh for Chrissake!', he snapped, 'There's no hidden meaning!' Silence. His brother said gently, 'It's just that ... we feel so sad ... that you have not had any for so many years.' William sighed. 'I know', he said quietly, 'Forget it.'

William knew the barrier between them would grow; they meant well but he felt the need to get away from this unreality. 'But you just got home', they said. 'I have to go', he insisted, and off he went to the hospital to see a comrade. Immediately they talked about the camp, all barriers down, no guarded comments, a storm of their most intimate thoughts, naked honesty of Camp 3D; then they laughed about how they were going to enjoy their new freedom.

William met and married Mona Gurland in June 1946,[7] in a private ceremony officiated by Rabbi Denburgh in Montreal, and they had two daughters, Dorrie and Ada. They eventually moved to Vancouver.

April and May 1983: Japan

William had returned, but could he forgive and forget? Never forget – that would make his experiences all seem meaningless, and anyway he could not, ever. It all came back on this visit to the Nippon Kokan shipyards. Even though the landmarks had gone, he could see where it all happened.

Koyanagi's (see p. 29) son, Ichirou, acted as his guide for the first week in Tokyo. On meeting, they embraced emotionally. His official guide was very helpful at the shipyard, showing William all the places he wished to see – the site of the old paint shop and the prison camp three kilometres away. They even found an old Japanese worker who was in the shipyard at the time, to point places out to William.

He and Mona were taken in a limousine and welcomed by a large committee; in a scrum of managers, journalists and TV reporters, he suddenly broke away and walked – blind and deaf to those about him – to the same gateway and track he remembered so well: a portal to

7. William Allister as an older man.

hell. He heard again the scratchy company song blaring out – even starting to hum it – as he was carried back across time. He turned and stared at Mona and in a daze could not believe that she and the shipyard staff were there. In a reception room he was shown wartime photos of the shipyards, and his mind dissolved in and out of them – the anger, hunger and boredom. Given a helmet and company jacket, he toured the yard with his entourage, conscious that once he skulked behind men like these, waiting to pick up a cigarette butt.

Then they were driven to the site of the old camp and it was, of course, unrecognizable. Locals stared at the group of visitors. Mona could not believe that this was where William, and those of his friends whom she knew, had actually lived.

Later, after a snack, William and Mona were driven back to their hotel and then taken for meal at the home of Ichirou – a great honour in Japanese culture – to meet all the family at a sumptuous meal. The ghost of his friend Koyanagi was present.

For the next seven weeks, William and Mona toured Japan. One might forgive, perhaps, as long as it included being able to express disapproval. William was able to view thousands of years of Japanese culture and to see that the Second World War was a time of sorrow, a blemish on which to rebuild, as after an earthquake. People had been friendly and felt his pain. He heard of a Japanese officer who was so ashamed of the atrocities at Sham Shui Po that he had committed suicide.

As an artist, William would paint his way to peace without hatred.

Epilogue

William kept a diary while a POW, even though detection would have meant death; but – as described above – some of it was lost when he left a jacket in a latrine, and the remainder when a friend threw the pages into the ocean on the journey home.

His book *A Handful of Rice* (Hodder, 1961), though a fictionalized description of his experiences, has gone through several editions and translations and won a literary award. 'What happened to me as a POW aroused anger and racism towards the Japanese people... but after the war I became a lover of Japanese art, culture and history', he wrote. His return in 1983 was 'to exorcise my demons'. He did, and said that 'I got the poisons out of me'; it became an exercise in forgiveness, he wrote; hatred was to him self destructive.

William wrote of this visit in a twenty-two-page article in September 1985 in *Reader's Digest*, called 'Samurai and Cherry Blossoms: An Ex-POW returns'. It won a Canadian Literary Award and is an astonishing description of how an ex-POW felt on returning to the scenes of the crimes. This author urges all to read it.

The article was developed into a book, *When Life and Death Hold Hands*, in 1986. This book tells William's whole POW story in detail.

William became active in the Hong Kong Veterans Commemorative Association. Some of his war paintings survived; two works – one depicting a Japanese sentry, and another a sunken ship in Hong Kong harbour – were sewn inside the trouser leg of fellow POW, John Burton of Toronto. Burton's daughter took ownership after his death and they now hang in a living room in Prince Edward Island. William fully supported the dropping of the bomb on Japan as he says it saved the lives of millions of Japanese civilians and American troops who would have otherwise invaded. After the war, William was a prize-winning documentary film-maker and writer, and an internationally acclaimed artist.

He died aged 89 on 2nd November 2008 at his home in Delta, British Colombia, Canada.

Notes

1. He was hanged for war crimes after the war
2. Forty years later, William met him in Tokyo; the guard had read of William's visit to Japan and they met. But there was no further contact after William returned home.
3. This identical mealtime situation for soup distribution was often described in the same way by those who survived the European concentration camps
4. The Marx Brothers' member with a penchant for blonds!
5. Around 1970, William had an art exhibition in Winnipeg. While greeting the many guests, the gallery owner said a man wished to speak to him. William walked over to the small, grey-haired old soldier who said, 'You may not remember me, I am Paddy Keenan and... '. William shot back, 'Remember you?' and grabbed him in a bear hug. Not only was he not forgotten, said William, but never repaid!
6. They maintained links with the POWs for decades after the war and even after Koyanagi died in 1981.
7. Daughter of Ukrainian Jewish immigrants Max (Mottel) Gurohovsky and Bella (née Spivak).

918211 Cpl Lewis Altman, RAF

8. 918211 Cpl Lewis Altman, RAF.

Lewis was born on 21 November 1915 or 1917 in Stepney to poor Jewish immigrants from Russia, his father being a warehouseman. He left school and joined the stockbroking firm of Raymond & Bekhor (see his obituary in *The Guardian*, 15 August 2008). He joined the RAF in 1940 and eventually became medical secretary to an Australian medical officer (MO), Col Edward 'Weary' Dunlop (he is mentioned several times in Dunlop's wartime diaries). When Lewis arrived in Singapore he thought it was heaven after the bombings, blackout and rationing of London. While at RAF HQ he saw the first Japanese attacks on Singapore, in December 1941. With many other RAF personnel he was evacuated by ship in time, but only to Java, where he became a POW when the Dutch surrendered at Bandung, where he had been working in the military hospital. His debrief sheet at TNA shows that he was in a large number of camps from Changi to Burma, on the notorious Burma Railway.

The Japanese were brutal; besides the starvation rations and nil health care, the Japanese were fond of beating prisoners with rifle butts, sticks or fists, at the slightest excuse. Lewis carried scars all over his legs and feet from such beatings. Guards were totally unpredictable – one moment giving bits of food, the next beating you. As Dr Dunlop's medical secretary and aid, Lewis was as close to the Japanese as one could be in the camps. He and others watched as thousands died building the terrible railway to Burma; so often, there was nothing they could do to save malnourished, beaten, overworked and sick men. He was responsible for keeping statistics and would be sent by Dunlop from camp to camp keeping records of the deaths and illnesses. His work meant that the families of those who died at least knew when and how, and also assisted in the war crimes trials at the end of the war. With no footwear, in the often trackless, muddy jungle, and with the real possibility of being lost in the darkness (which comes suddenly in the tropics), this was a harsh, dangerous and tiring task. 'Sixty years on it was the mud', said Lewis, 'that was the abiding memory; evil-smelling, it sapped morale; we worked in the rain, ate in it, slept in it in leaky tents at night. The camps

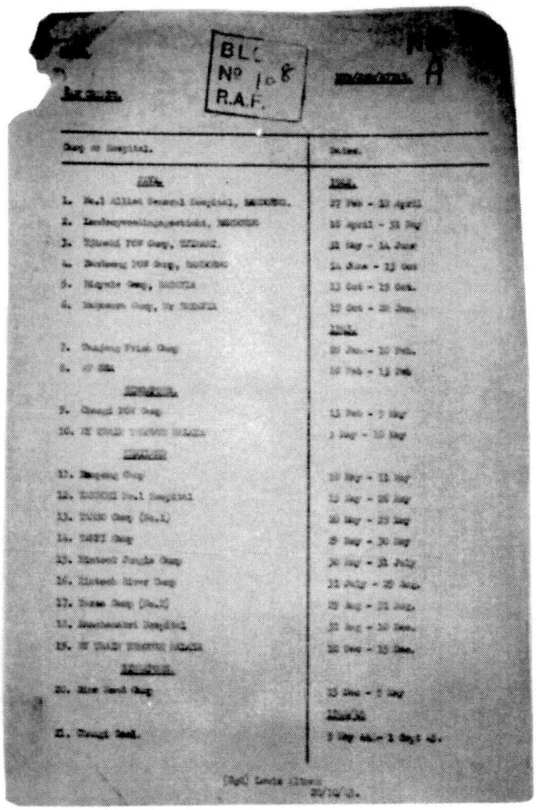

9. **Lewis Altman's original deposition, typed on 20 October 1945 for his debriefing interrogation, from the National Archives at Kew (hereafter TNA), file no. WO344/361. Note his name at the bottom.**

became slippery as ice rinks, a journey to the cookhouse or latrine, especially at night, often meant falling into pools of water and returning to tents with two inches of mud on boots and shorts; roads were a rutted morass of glutinous mud.'

Sometimes amputations were carried out without anaesthetic; often bribery and theft were the only ways of obtaining medicines. In Singapore Lewis was often sent to Afran Road, a Kempeitai Gestapo-type prison where torturing by the Japanese secret police, of alleged miscreant POWs, was commonplace; he had to go in a van and collect the POWs, often half dead, and treat them back at the camp.

One amusing incident he remembered was when Dunlop was approached by a Japanese who said he had VD and would be executed if his officers found out. Dunlop said he would treat him if he could steal the necessary antibiotics with a good supply left over for the prisoners. This was done and Dunlop gave the man a very painful injection in the penis. After he left, Dunlop told Altman that the soldier did not have VD but it was a good way to get medicines. Lewis was often known to entertain his fellow POWs with stories of his exploits in the stock market.

On liberation, Lewis weighed just five stone He later returned to his old firm and then established his own company in 1954, dealing in Far Eastern tin and rubber using links he had set up with local Chinese businessmen immediately after the war ended. He was extremely bright, but secretive and very anti-establishment, often engaging in transactions which tested the regulators, until he was expelled from the Stock Exchange in 1978, following a scandal. In 1945 he married Florence Camp (the marriage was later dissolved), with whom he had a daughter. He died in June 2008.

116381 Capt. Dr David 'Arky' Arkush, RADC

The author interviewed David in January 2001. Other sources for this testimony include various articles written by David kept in the AJEX Jewish Military Museum and in the Arkush family archive where most original material is, and a one-hour DVD made by the Medical Services Museum at Aldershot. The author thanks them for their cooperation.

10. 116381 Capt. Dr David Arkush, RADC.

David was born in Glasgow on 8 June 1914, youngest son of the Revd Samuel Arkush and Bella (née Perle). Samuel was a cantor (a Jewish minister who assists a rabbi in the taking of services, particularly with the chanting and singing) born in Poland, and his first post in Britain was with the tiny Jewish community in Inverness. Later the family moved to Edinburgh, Glasgow, Cardiff and Blackpool. David was educated at Blackpool Grammar School for Boys and the Dental School of Liverpool University.

In 1939, David was 25 and working as a dental surgeon in Edgware, London. He volunteered on Monday 4 September 1939, the day after war was declared; he would have done so on 3 September but it was a Sunday and all offices were shut. David says he thought the Jews should be the first to go, as it was as much their war against the Nazis as anyone else's. Indeed, when he went to Millbank, the RAMC (Royal Army Medical Corps) HQ in Pimlico a few days later, the room was full of volunteer Jewish doctors and dentists waiting to be examined for recruitment.

From Millbank, David returned home to await further instructions, but because his parents were Polish nationals (although his father was naturalized in 1926) and despite the Allies going to war to defend Poland, his call-up was delayed. Such are the paradoxes of politics and war.

His patience wearing thin, in January 1940 he called the War Office to demand why he had not been contacted; they said all was well and he was commissioned as lieutenant on 1 February 1940 into the Army Dental Corps. Posted to Catterick, he was confronted with the problem of what to eat, as he was orthodox, so he lived on food parcels from home. His colonel was very understanding and agreed to post David to a town where there was a Jewish community. He was sent to Hull, where he was billeted with the Shenker family, off Anlaby Road, between March and June. Later he was posted to Aldershot (June

1940), where he did his 'square bashing', and then to No. 1 Field Ambulance, First Guards Brigade in Lincolnshire; here his commanding officer (CO) told him that he had two pet hates: dental officers and Jews. David was both!

David was posted to Beverley in Yorkshire in December 1940, to be in charge of the dental centre. In May 1941 he was informed of his impending overseas posting and went to the RAMC centre in Beckett Park, Leeds, to await orders. He travelled to King George V docks at Glasgow and sailed via the Cape, on the SS *Cape Town Castle*, on a nine-week voyage to Singapore with RAF personnel, arriving on 8 August 1941. He was part of a draft of eight dental officers. On the voyage he placed a notice on ship's orders for all Jewish servicemen to meet him at his cabin for a service, and eight airmen of other ranks appeared; some were related to people he knew. They held regular Sabbath services throughout the voyage and also planned to march to a synagogue service in Cape Town on the first Saturday (Shabbat/Sabbath) of their arrival. But on the Friday night, at dinner with the Chief Rabbi of Cape Town, who was from Manchester and knew David's father, he was informed that he would be sailing next day. He never discovered how the Chief Rabbi was aware of his impending journey.

Life in Singapore was very pleasant. David was posted to the Royal Army Dental Corps (RADC) centre at Gillman barracks and had lodgings in River Valley Road. There was no blackout as in Britain, as war had not begun, and the Jewish community gave tennis and swimming parties for the Jewish troops every Sunday. David even bought a car. With his friend Airman Harry Courts (who was to become another POW), he had a standing invitation to go every Sabbath eve (Friday) to the home of a very wealthy local Jewish businessman, Reuben Meyer, son of Sir Manasseh Meyer. Reuben's widowed sister, Mrs Moselle Nissim, kept house for him and organized the events. However, there was a strong division between officers and men elsewhere, and when David innocently invited Harry for drinks one Sunday morning at the Seaview Hotel, he was acutely embarrassed by the attitude of the other officers towards Harry. David had not quite understood the implications of military etiquette: Other Ranks not allowed. The prevalent 'colour bar' was equally a shock for him

War

As rumours of war became rife, all troops were confined to barracks. David was living in digs, and on the night of 7 December – which he remembers was a Saturday/Sunday – he was awoken by a noise that he had known in Britain: the whistle of falling bombs. He fell flat to the ground as a salvo landed nearby, killing some Chinese families in the houses opposite. Confusion reigned; he recalls that Singapore Radio kept playing, every hour, a cracked record of a middle-aged woman soprano singing 'Keep the Home Fires Burning'. He was later told that this was a secret coded signal for the British in Malaya and Thailand to leave.

As disaster followed disaster, David continued his duties as dental officer, moving to Nee Soon barracks near Singapore City, servicing two anti-aircraft batteries. They were regularly bombed by the Japanese and he helped to deal with many casualties, including the first dead. As the Japanese got closer, David was forced into the city, around 1 February 1942, as the retreat began in his area. He reported to the Military ADMS (Assistant Director Medical Services) but was given no orders, so he attached himself to a general hospital near the docks and began helping the stream of wounded that was flowing in. As a dental surgeon, he knew enough to give anaesthetics and bind wounds.

David specifically remembers Friday 13 February as the worst day, as this was when the Alexandra Central Military Hospital was overrun and terrible massacres were carried out by the Japanese against both staff and patients. David knew a Private Rose in particular, who was a Jewish medical orderly, and four of the dental officers murdered had been on his ship from Britain; he knew them all.

On the day of the surrender, David and three other dental officers moved into an empty flat in the city, in the only skyscraper that then existed. The Japanese shelled it, so they left. An order came through to destroy all liquor stocks because of what they knew had happened after drunken Japanese had captured Hong Kong – the more drunk they got, the more atrocities they carried out. In a rare comical moment, David explained the intense pain he felt as they poured four bottles of Scotch into the drains.

Without a billet and leaderless, the four men walked to try to find the nearest hospital, to seek guidance. They saw a Japanese plane shot down, and what must have been the last British plane leaving was also shot down overhead. David later discovered that Harry Courts got away with the RAF to Sumatra, but his ship broke down near Java; taking to the boats, they rowed to Sumatra, where Harry was taken prisoner when they missed the last ship to Australia. His wife was informed that he was missing, believed killed, for almost four years. David's family did not know he himself was alive until April 1943.

David and his colleagues eventually made their way to the Singapore Cricket Clubhouse, right in the centre of the city, and after two days the Japanese entered the city to round up the POWs after a 'cooling off' period following the ceasefire.

David asked himself how the British could have been defeated by these small, bandy-legged men with scruffy and patched uniforms. On the Padang cricket ground just outside the Clubhouse, dozens of British Artillery guns fired off their ammunition so as to deny the Japanese any supplies; the noise was horrendous. They then carried out the saddest duty a gunner can perform, and blew up their guns.

On the following evening a whole group of heavily armed Japanese soldiers arrived at the Clubhouse, waving their guns in a threatening manner. David and friends had filled bottles with water on the bar as the city water supply had been cut off; the Japanese ordered one of the surgeons to drink some, and then they tried it. But they spat it out angrily, not realizing it was just water. Seeing that the situation could get nasty, David and his friends made their way surreptitiously out of the building with some civilians who had attached themselves to his group, and went to the old Victoria theatre nearby, which was being used as a hospital, where they slept on stretchers that night. Returning next day to the Clubhouse to retrieve his personal possessions, David found that they had been looted by the Japanese, and his wallet and his family photographs were strewn across the floor. He began to pick them up and, amazingly, one of the Japanese helped him. One photograph, however, was missing: an attractive snap of his pretty sister. To this day, David is sure that one of the Japanese took it for a pin-up.

David began to realize what they were in for when he watched from across the road outside the main post office as a Chinese man in a lorry got into a discussion with a Japanese sentry. The sentry smashed the butt of his rifle through the windscreen and dragged the man out and round the corner; he was never seen again, and the vehicle remained there for several days. David also witnessed many times the atrocious sight of numerous decapitated heads of Chinese appearing all over the city on poles, as the Japanese carried out war crimes on those whom they saw as their enemies. He also saw a British Medical Officer, a major, badly beaten for not saluting a Japanese sentry – Japanese military etiquette which simply had not been communicated to the British soldiers at all. David said they learnt quite fast after that.

After about eight days of treating wounded in the hospital, they were told to make their own way to Changi gaol in the north-east of the island. David kept the papers with the names of the doctors detailed to go, together with a diary and many other papers of his time as a POW, most of which are now in the RADC museum. Some original materials, as well as copies, are still with him, put in order by the author, and these will be presented to the AJEX Jewish Military Museum in due course.

With their Red Cross armbands on, the officers (RAMC, RADC and padres) began to march east. En route they passed a Japanese officer seated at the wheel of a parked Hillman Minx car; he saluted them, but David does not know why – perhaps he was a fellow doctor, or just respected the Red Cross. Soon a British truck appeared; a POW whom David had treated saw him, picked them all up and drove them the rest of the way. The roads were choked with prisoners, looking totally demoralized, on their way to the gaol; there was no saluting, discipline had broken down and officers were jeered. It was, David says, a truly awful spectacle of a defeated army. On the other hand, David heard later that when the European women were walking to Changi, led by the matron of one of the hospitals, the matron called out: 'Come on girls, let us show them we are British', and they marched proudly, backs straight, into the gaol, singing 'There'll always be an England'. The Japanese were apparently very impressed, but when local people watching began to cheer them, the soldiers beat them up.

Changi

David arrived at Changi on 20 February 1942. Changi prison was actually two separate sites – a civilian gaol and the military barracks, the latter having been largely bombed, with no water, electricity or proper shelter and little food; here tens of thousands of Allied POWs were crowded in and the prisoners were ordered to look after themselves. They wired themselves in and really did not see much of the Japanese at all. Officers took control and slowly, after initial chaos, some order returned. Army Drill was introduced to help restore discipline; the area was divided up, with the hospital in the centre. David worked here as a dentist. They had managed to get a lorry to bring as much equipment as possible from the city hospitals.

The INA (Indian National Army – Indian soldier traitors who had gone over to the Japanese) were given their own area, and they and the Japanese ordered that they had to be saluted at all times by the POWs – something which went down very badly with the troops. Some British Indian Army men refused to do this, apparently without incident at that time, though there are many recorded eyewitness accounts where some Allied POWs were beaten and even killed by Indians for not saluting as ordered (see Aubrey Lichman's testimony, in this book). On one occasion, David took a group of men to the hospital and they passed an INA sentry. David called out 'Eyes down!' but he saluted and the sentry was satisfied and did not understand the English command anyway; after passing, David called 'Eyes up!' and that little victory brought great amusement to his men.

Nobody was allowed to move from section to section without an authorized armband bearing written Japanese. One story David heard was about a British soldier allegedly finding one of these armbands and putting it on. A Japanese soldier saw it and began laughing; it said 'Clean Prostitute', but of course nobody knew this until later. Apocryphal or not, it raised a much-needed laugh around the camp. Another incident involved a British doctor mistaking a Chinese British doctor for a Japanese and demanding what he was doing in their camp.

Food quickly became scarce, and hunger began to set in after a few days. Personal food carried in kitbags began to run out. David saw one doctor who had a large captured lizard on the end of a rope; he never discovered what happened to it. He saw another man with a large tin of dog biscuits, who became the object of great jealousy.

On one occasion a full colonel in the Indian Medical Corps was caught stealing a spoonful of jam from someone's food supply, and the other officers wanted to court-martial him; it was very unpleasant and a lynch-mob atmosphere developed, but fortunately died down. A swimming pool in the camp made life slightly more bearable, and an old lorry chassis was used as a trolley by David and his staff to go out on foraging parties to get food from traders in Changi village. The Japanese allowed this, and the POWs had money they had brought with them from their pre-hostilities life in Singapore.

Eventually the Japanese began to supply some rations, mostly rice, but how one fared depended on what officers' mess one was in. Some ate everything in one huge meal; others carefully rationed it out. But the British cooks did not know how to cook rice at first and the results were awful – either runny like water or sticky like glue. Also, it only partially satisfied hunger and after a while made one urinate a lot. David described the food situation as a perpetual Yom Kippur fast.

David set up his clinic in a Changi village in an old laundry near Dalby Court (?). He had a Field dental chair and an orderly to assist him. In one incident, a Japanese soldier who had been in car accident was brought to him for treatment and the Japanese were very grateful, saluting and bowing politely to David for the help he gave.

One atrocity David witnessed was when a Japanese soldier machine-gunned to death a Chinese man in his boat who had brought across items to sell from Johore on the mainland.

After a short while, David met two Jewish soldiers – one from Hendon – who were from the 18th East Anglian Division section of the camp; there were apparently a few Jewish POWs among them and they told him that they held a regular Friday night service. David made his way there one Friday and found a service being conducted in a long hut by RAMC Captain Dr Harry Silman from Leeds; there were over forty in the congregation that evening, and more than that on other occasions. David knew the Assistant Chaplain General who was in the camp, and he agreed to allow a Jewish service in the hospital area, which was more central, using the old cinema building. At the beginning of June 1942, the first formal full Jewish service in Changi gaol took place, conducted by David, assisted by Corporal Lou Brodie, with about fifty present; David used his own Jewish prayer book which he had held onto after capture. The synagogue later became permanent, but this was the only time David took the service as he was posted away a week later. Nevertheless, this began David's other main role as a POW, as the leader of the Jewish Community wherever he was sent; he made it his duty to keep the Jewish lads together, under any circumstances, whatever may be.

11. Sketch of David's dental chair, drawn by Fl. Lt I. Smith, RAF, 6 May 1943, made to David's design by Capt. R. Spaulding, Indian Artillery.

In June, David was ordered to depart as dental officer with a group of men being sent north to Thailand. David's CO gave him a minimal amount of medical instruments (forceps, six hand instruments (excavators and scalers), a few ounces of novutox drugs and a small supply of zinc oxide, oil of cloves, and other odds and ends) and told him to do the best he could in his new posting, wherever that was to be. It was 26 June 1942. David was taken by truck to Singapore railway station with 600 other POWs. He took with him a cushion, thinking they would at least be in second-class carriages with wooden seats, but they were faced with cattle trucks, each of which was to hold thirty-one men. Kit was put in the centre and the men sat around the truck walls. Inside it was very hot; near the doors was cooler but you were liable to be splashed with urine and faeces because men came to relieve themselves by the door and the wind blew the material back in. The thin veneer of civilization disappears quickly, David said – at each stop men got out and relieved themselves on the platform; there was no alternative. While attending to his call of nature, the rather pukkah colonel of the Malaya Regiment was heard to point out rather loudly that the last time he was here they had laid out a red carpet for him – now this! Keeping a sense of humour was crucial to morale. At Kuala Lumpur there was, of course, a proper city railway station and the men tried to do the right thing, but one toilet for 600 men left David sickened when he got to his turn, somewhere around number 599.[1]

Each day a bucket of rice-like swill was put in each carriage, and a bucket of leaves floating in water, said to be 'tea'; this was their food and they ate whatever was slopped into their mess tins and did not ask too many questions about the contents. Just after Johore, locals were throwing pineapples for them into the carriages, but these are quite acidic and caused awful diarrhoea (the result needs no elaboration here). At some stops they were allowed to strip off and stand under the railway water tower where they took on supplies, and so could manage to clean themselves a little.

After four days and nights they crossed into Thailand.

The River Kwai

At the first stop (Haji?), a native Thai woman approached them carrying trays of fried eggs which she dished out to the men. They dubbed her an 'angel'. It was only the second egg David had eaten since February, the first being a hard-boiled chicken's egg he had for his birthday in Changi, which he ate secretly and alone so as not to annoy his comrades.

Filthy and hungry, they continued their journey, arriving in Bampong on 30 June, where they were told to detrain and to leave their kit to be taken by lorry to a camp. On arrival they met up with troops who had left a few days before them, and these were all the worst-behaved men from Changi, whom the officers wanted to get rid of. As a result, David and others found that their kit, which arrived before them, had been looted by their British comrades; they had lost virtually everything.

Just as bad was the accommodation – long bamboo huts with atap (palm-leaf) roofs, and long sleeping platforms along the walls, with a central path. David had his mattress stolen and so he laid on the ground under the atap eave and was so miserable and depressed that he said if he had had a razor he would have slit his own throat. The one piece of good news was that the food was better; chicken stew was not uncommon and there was a small native market, which the Japanese permitted the POWs to visit, which was selling biscuits, duck eggs and other items. The sentries, however, did think it sickeningly amusing to search prisoners on their way into the camp and deliberately pat the pockets hard, so breaking any much-needed eggs that were being smuggled in. One good ruse was to put them inside your hat. Generally the Japanese guards behaved reasonably well, and David had a good relationship with one Japanese sergeant who used to tease David about the beard he had grown; the problems began when the Korean guards arrived.

David and all the other medical staff were left to do their jobs but, although the men did not know it yet, the troops were the advanced party for the building of the notorious Burma Railway; meanwhile the men were put to hard labour at unloading trains, storing materials and so forth. Soon the camp swelled to 1,800 men.

Among the Jewish officers at Bampong was Major Everard Woods of the RIASC (Royal Indian Army Service Corps) from Leeds, whom David described as 'an interesting character'. He never owned up to being Jewish until quite late on. Nicknamed 'the City Slicker', he was determined not to kowtow to the Japanese and took every opportunity to defy them. When the officers were ordered by the Japanese at Changi to remove all badges of rank except one pip to be worn on the breast pocket, Woods refused and put up his major's crowns – and everyone else followed him. The CO, Colonel W—, a very tall, regular gunner, was the very opposite, and the Japanese told him that if anyone escaped they would shoot him as a reprisal. After this W— lost his nerve and showed poor leadership. The men lost all respect for him and at night used to sing a song about him: 'Balls to Colonel Q——,[2] Q——, Q——, balls to Colonel Q——, lousy old sod.' Woods detested him. W— threatened to court-martial him but nothing came of it. David described Woods proudly as 'one of our boys, a mensch'.[3] Sadly, David was unable to trace him after the war; there appears to be no record of him in Leeds and none of the Jewish community there knew of him.

Fed up with being stuck in the camp, David offered to take out one of the many British POW working parties that the Japanese regularly demanded, which had to have an officer in charge. Out they marched, and ended up at the Japanese army brothel down by the river Kwai;[4] here they were given the job of digging drainage ditches to help lower the river's level after the monsoon flooding. Meanwhile the Japanese sergeant asked the prostitutes in no uncertain terms, obvious in any language, to show David some free 'affection', but David remembers clearly that they fancied him about as much as he fancied them – which was not at all.

There were regular bathing parties down to the river, the troops always marching to the tune of 'Amapola';[5] whenever David hears that song he remembers the Kwai River camp. The bathing was good, save for the small biting fish which could cause nasty cuts in delicate parts of the male anatomy, and when the river flooded it would reach the latrines in the huts and bring the contents into the living area, which was not only unpleasant but also extremely unhealthy. Generally the camp was a squalid one, but David discovered eleven other Jewish POWs, including Lieutenant Martin Finegold. They started Friday night services and also held services on Rosh Hashana (New Year) and Yom Kippur of 1942, with David managing to get the men time off to attend, with the permission of the CO.

Soon, on 8 October 1942, 1,500 POWs were on the march again to a new camp, setting off for three days on a thirty-mile trek, following the Kwai, and marching to the tune of 'Keep Right On to the End of the Road' (a British music-hall song sung by Sir Harry Lauder). At one stop, at Kanchanaburi, where the exhausted men flopped to the ground on the river bank, a local came along with tins of fish which he opened and shared with all the men. He was an English-speaking Thai who was actually a British agent, and he continued throughout the war to provide extra food to the many thousands of POWs in the many camps in Thailand, saving countless lives. His name was Boon Pong and he became a hero to the British and other POWs in the region, being awarded the MBE in 1947.[6] David got to know him very well, as did many of the POW doctors and officers along the notorious Burma Railway of Death.

Chungkai

From 'Buri' the men boarded barges and were taken upriver to Chungkai, a base camp for the Burma railway project, which only had one huge completed hut at that time. It was designated as a hospital camp, but in name only, as there were no facilities except those that the men themselves provided and smuggled in. David's tiny supply of drugs had now run out. Barges arrived full of desperately ill men (sick, starved, beaten) who had been working on the railway, many already dead on the boats. Lying on split-bamboo platforms in the hut there was virtually no room to lie down and when one turned over, everyone had to follow. However, there was comparative freedom, and men received a small amount of pay, from which officers deducted some for buying medicines and food for the sick on a communal basis. There were no fences and there was a small Thai market (run by local women) where it was possible to obtain extra food. The men called it Piccadilly Circus or Bond Street and would exchange personal items such as watches and pens for food; banana fritters were a favourite purchase. David explored the bush on one occasion, and once found a cave with a recumbent Buddha.

It was here he encountered the Korean guards who were especially brutal; the men quickly learned that if you were beaten, you stayed down and covered your vital parts as best you could, and this would save face and allow the guards to walk away once satisfied they had inflicted enough beatings.

After first working in the open air, David was soon given a dental surgery in the hospital hut. He obtained some Novocain from the Japanese, and some could be bought from local traders; at one point the Swiss consul in Bangkok obtained a supply for the prisoners and got it into the camp. Improvisation was the thing, and broken dentures were repaired either by smuggling them into and out of Bangkok to local dentists, or David used silver wire to bind parts together, or used aluminium rivets made from old mess tins. When the rubber on his syringe plunger perished, it was replaced with leather made from the uppers of army boots.

One day, a Jewish soldier, Charles Simon of the SSVF (Straits Settlement Volunteer Force), who lived in Singapore, approached David and said he had to say prayers for the anniversary of a death in his family at home (*yartzeit*), and so David rounded up a minyan (quorum of ten Jewish men required by tradition for prayers) and they prayed and said the Kaddish (prayer for the dead) in a jungle clearing. This was the beginning of the Chungkai synagogue group which held Sabbath and Festival services and lasted continuously for over three years. David, as the son of a Jewish religious minister, was the nearest they would have to a rabbi.

But soon things changed. Masses of troops began arriving on packed trains from Singapore and Changi, in September–October 1942, to start the building of the notorious 'Death Railway'. Up to 10,000 POWs were in the camp.

Lieutenant Colonel Yanagada was the commandant, an elderly soldier of the old school who tried to deal fairly with the POWs; the really vicious types were the railway engineers, who harassed the men to death in the building of the railway. Yanagada was quite pro-British and even wore his British First World War medal ribbon on his chest. He urged the men to do PT to keep fit, even competently demonstrating himself what to do. He often assembled them for well-intentioned pep talks, shouting loudly and so enthusiastically that the first rows of listeners had the benefit of his splutterings of spittle, which soaked them.

On one occasion, a Japanese holiday, he presented the British CO, in front of all the men, with a fifty-cents gift (practically worthless); another officer commented in very posh tones: 'How utterly humiliating; a British Field Officer presented with 50 cents by a Japanese; should have been at least a dollar!' (Japanese wage tokens were given in cents and were virtually valueless.)

At first, officers did not work, but after some weeks the Japanese ordered that they must because the railway had to be completed. David watched as all the officers were ordered to parade and tools were placed before them. The officers said that the Geneva Convention allowed officers to be exempt, but the Japanese said that they had not ratified this agreement and all were to take up tools. The officers refused, so a lorry was brought with a mounted machine gun (an incident exactly like this was re-enacted in the film *The Bridge on the River Kwai*) and the Japanese threatened to open fire after a count of ten if the order was not obeyed. Discretion is the better part of valour, and in real life, as David said, you do not argue with machine guns, so the officers agreed. David also believes that it helped the men's morale and physical safety to have the officers alongside them as they could, at least in this camp, protect many of the men from some of the worst excesses of the guards. A Japanese striking a prisoner officer was not done lightly.

One of David's regular Friday night minyan at Changi was Harry Hertzberg of the 1st battalion, the Cambridgeshire Regiment. In David's archive is a sheet of paper with the names of many Jewish POWs typed by Harry in red for some unknown reason– possibly when they were in Changi – and dated 2 April 1943 in pencil. Harry suddenly left the camp on a draft (as men recovered in the hospital they were sent to work camps elsewhere) and he wrote a moving letter in pencil on the back of this sheet, thanking David for all his help.

In mid-1943 the death occurred of the first Jewish soldier known personally by David: 1558132 Gunner Lionel Basco, RA.[7] It was the early days of the railway, and the Japanese staff attended and a wooden coffin was used. David led the procession, with the coffin following on a wheeled carrier, and the Japanese officers next with the mourners behind. David recited the Jewish burial service using his British Army Jewish prayer book[8] and as the coffin was lowered the Japanese colonel saluted and said (in Japanese, later translated) that Gunner Basco had died for his country and he sent condolences to his family. The commandant asked why the dental officer was taking the funeral and it was explained that the soldier and officer were Jewish and different prayers were required. This was fully understood and accepted. The grave was marked by a wooden Star of David (Magen David). It is important to point out that, in David's experience, no Jewish POW suffered solely because he was Jewish.

Only once did David experience anti-Semitism, when a Korean guard shouted at him: 'I know you; you Yuday man; you very bad man; you take everything.' David swore back at him. His name was Osahu (nicknamed Harold Lloyd[9] by the men, due to his thick horn-rimmed glasses) and he was unpopular even with the other guards. He used to annoy the British quartermaster, Major Pycock, who was a tough Olympic swimmer. One night he saw Osahu about the camp, shone a torch in his eyes and hit him. Next day Osahu appeared with two black eyes, much to the huge amusement of the men. Nothing was ever done about it as nobody could be identified. Brave Major Pycock!

The prisoners put on the usual shows and sports days, to keep up morale and occupy minds,[10] which the Japanese also thoroughly enjoyed, but as casualties from exhaustion, beatings, blast and work accidents and disease increased with the slave work on the railway construction, the character of Chungkai camp changed and it became a primitive hospital camp. David describes it as just dreadful, with every imaginable tropical disease and more, made worse, of course, by the appalling malnutrition and hopeless lack of medical supplies. Twenty men a day were dying, and David and his colleagues could do little to help.

Whilst at Chungkai, David designed a proper dental chair, which was then made of bamboo and home-made twine, with a cut-down oil drum for a spittoon, by an Indian Army Artillery Officer, Captain Spaulding. It was based on a Charpoy or Indian bed, with a rope seat and moving parts to adapt to any size of patient. It performed wonderfully with several moving parts, and after the war he wrote a paper about it, which was illustrated with drawings, for the *British Dental Journal*.[11] The artist who drew the chair (on 6 May 1943) was Flight Lieutenant I. 'Smudger' Smith, RAF, who died on one of the Hell Ships when being transported to Japan, but the original drawing, which hung in David's surgery

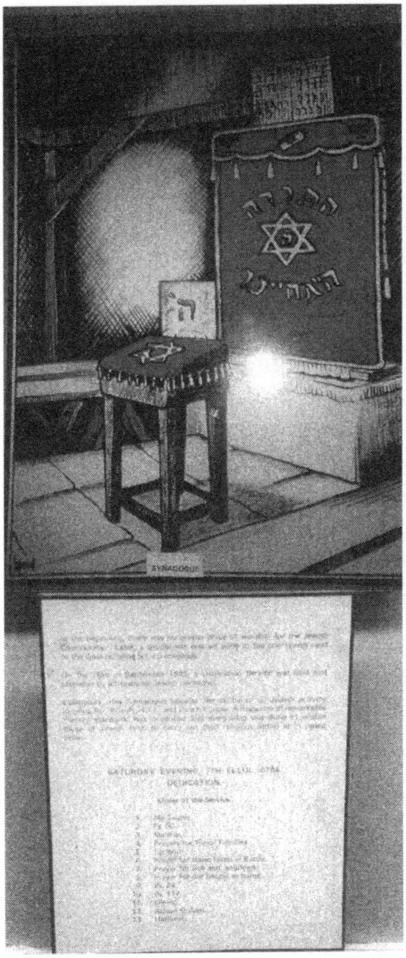

12. Drawing by Australian George Sprod (see p. 86) of the Changi Synagogue interior, showing the *bimah* (reading desk) and Ark curtain. This drawing is currently on permanent display at the Changi museum, Singapore.

for many years, is now at the Royal Army Dental Corps Museum,[12] together with a replica of the famous chair.[13]

Another Jewish character whom David met was Surgeon Commander Epstein of the US Navy, an elderly officer who was on his way home to the USA from the Philippines when his destroyer was caught up in the Battle of the Coral Sea. His ship was sunk and he was captured, and he ended up at Chungkai. As a gynaecologist he did not have much to do, but was fond of saying to patients: 'Take two aspirins, and everything will be all right.' Others included the Welsh Jewish soldier Lance Corporal Kransky and the Australian Major Sam Harris and his two soldier sons.

The great amputation expert was the Canadian Jewish doctor, Captain Jacob 'Marko' Markovitch,[14] whose surgical skills saved hundreds from death from gangrene. He slept next to David in his hut and he was a man of great humour which shone through the misery and helped to keep up morale. David often persuaded Marko to go to the synagogue with him to help present and preserve a measure of normality to the men, in the prison hell, and he would say jokingly to David: 'What time do we storm the gates of heaven?' To everyone's amusement he often wore a Scots Gordon Highlander's kilt and Aussie bush hat on his 40-something, short, bald frame – but was greatly loved and admired. He took an active part in the education programme of the men and lectured mostly on sexual topics. Another character was the middle-aged Captain Willi 'Rosie' Rosenthal, Croix de Guerre; he was originally in the French Army in both world wars, but he joined the British Army (RAOC – Royal Army Ordnance Corps) after getting away at Dunkirk. He amused one and all with French ditties. He always carried his house keys with him and told everyone that one day he would get back to his home in Paris and let himself in – and he did.[15] David also met Jack 'Scotty' Caplan at Chungkai, and his amazing story is told later in this book.

Conditions rapidly worsened, and often the synagogue hut, which was used as a church as well, had blanket-covered bodies on stretchers waiting to be buried, whilst David held his services.

In early 1943 a large draft of Dutch army POWs were brought from Java to Chungkai and among them were several Jewish personnel. One became David's cantor at services,

Private Davids (he went to live in Israel after the war). It was he who tore the flyleaf page from an old book and wrote the holy and crucial Kol Nidre prayer in Hebrew from memory for use at the Eve of Yom Kippur service, as it was not in any of the prayer books the men had brought with them. David still has this page and it is almost word-perfect, dated 5704[16]/1944, Chungkai

Another POW was a Dutch rabbi, Chaim Nussbaum,[17] who enlisted as an ordinary soldier and was at a neighbouring camp. He and David never met, but Nussbaum drew up a Jewish calendar (*luach*) from memory, which he sent for David to use at Chungkai after his own Boots calendar, which he had kept and in which he had noted the Jewish festivals, had run out in 1943. David remembers that Nussbaum told him that there was a Dutch soldier who was a cantor in the camp with him, but he died. His name was Lisser.

In April 1943, David managed to organize a Passover service with about forty Jewish Allied POWs attending; in lieu of matzoh (unleavened bread) and kosher wine, he managed to provide rice cakes made with sago flour, *charoset* (a condiment symbolizing the cement that the Jews used when making the bricks as slaves in Egypt) from grated coconut rice coffee, *maror* (bitter herbs symbolizing the bitter times inflicted on the Jewish slaves) made from mint leaves, the hard-boiled eggs (symbolizing the circle of life and cycle of Spring renewal) made from spinach leaves, and rice wine with some bananas. Those who were unable to walk were brought in on stretchers and those too ill to come were visited afterwards with food and cigarettes. Using appropriate prayers from his British Army Jewish prayer book, and a written version from memory of 'Adon Olam' (a famous Jewish hymn), it was as usual held in the Christian chapel – perhaps the world's only working synagogue with a cross on top of it. David felt emotionally unable to give a sermon, but he reminded everyone that as in days of old, the Jews were slaves to Pharaoh, and they would also win their freedom from the Japanese slave masters. The traditional final song, 'Next Year in Jerusalem', was sung with particular gusto and meaning. On that occasion, David remembers that one Jewish officer, Captain Beck (from Brighton), had kept a bottle of vitamins as his lifeline should he fall ill, but he gave these to Private Yankel Perlman (from Sunderland) who was very sick, and this probably saved Perlman's life. David explained how crucial it was that caring for each other and sticking together kept men alive; lapsing into hopelessness could kill, as much as any illness or beating.

In November, David received a message that a signalman Shapiro[18] had died on the railway and had been buried, with a cross on his grave, by the Christian chaplain. David wrote at once to the chaplain with a sketch of a Star of David marker, asking that the cross be replaced, and this was duly done. Many years after the war, David had a call from a Dr Shapiro of Leeds, asking about his brother and how he died. David was only able to tell him about the incident with the cross.

Like most other observers of the POW conditions, David knew that giving up or giving in to despair was a death sentence and that psychological will was as important as physical strength in staying alive. One of many examples was that of a Jewish POW in his thirties, Leo Lavender, considered quite old for the average soldier. He was small, ill and thin, and covered in sores and the appalling tropical ulcers. David organized a rota so that Leo's comrades constantly badgered him, never leaving him alone, and told him he was going to get better and go home to his wife and family, and that the war would be won. He was to buck up, and David told him that if he died, 'I will have to bloody bury you and I do not want to do that'. Sure enough, he pulled through, and after the war wrote David a very moving letter saying how David had saved his life.[19]

In the nearby camp at Kanchanaburi, David heard that there was a dental officer prisoner, Colonel Jimmy James, who had acquired a foot-driven dental drill and a proper dental technician. David told his boss, the Japanese doctor Nobosawa (nicknamed the 'Horse Doctor' for his lack of skill), that he had a number of patients who needed treatment and, as a sweetener, said this number included one Japanese guard too. The Japanese winked at David as if to say he knew he just wanted to get out of the camp for a while, and he agreed, so long as nothing interfered with Nobosawa's fishing trips in the river. David got permission to take them all to the other camp, accompanied by a guard. On these trips David was able to buy drugs with money from the fund set up by the officers for the sick. In the camp, old Dutch mess tins and plaster of Paris were melted down and used to make dentures, and old dentures taken from dead POWs had to be recycled, for those who had lost or damaged their own. Candle wax was used to make denture impressions, and David was able to treat men whose jaws had been broken during beatings from Japanese guards.

David pointed out that he never sought out Japanese patients but if they came he treated them, as he could always exact some kind of payment to help obtain drugs and food for the hospital. When hacking at the teeth of one soldier, the man drew his bayonet as a reaction to the pain; David decided discretion was the better part of valour and proceeded more gently. On one occasion a Japanese officer came for treatment, bringing an offering of excellent, European-style soap as a gift. David took the hint and inflicted as little pain as possible.

In Chungkai they had a secret radio, run by two officer brothers called Donald and Max Webber. It was hidden in the bottom half of an army water bottle, so designed that the top half was sealed and could hold water and thus fool the guards, when it was shaken during a search, that it was just what it seemed to be. Keeping a radio was strictly forbidden under pain of death. But David was thus able to pass the war news to men in the other camp on his visits. He recalled that when Italy surrendered in 1943, the news reached them via local Thai workers. That day on the way to work, the men sang from a popular song – 'Oh, Antonio you have left and gone away, and left me on my own-io' – to signify that they knew. On one occasion it was news of the huge morale boost of D-Day in June 1944, but the rule was not to tell anyone until one week had passed so that the news could not be associated with David's visit, in order that it might not spark a search of the camp for a radio by the Japanese.[20] Another method was to wrap messages into tobacco leaves made into cigars, and pass these as gifts to men they met from other camps while working with different groups on the Railway. After the war, David and Jimmy James often met at reunions, and Jimmy said that he would never forget David's visits with news from the outside world.

In late June 1944, during one morning *tenko* (counting of prisoners, or roll call), there was a huge noise and as the men looked up a squadron of giant American Liberator bombers passed overhead, the first Allied planes seen since the surrender in February 1942. There was a riot and the padre rushed to get the Union Jack – used for funerals – to wave, but had to be restrained. Sadly, David learnt later that the bombers, in trying to strike at the Kwai bridge, dropped some bombs on Allied prisoners in the camp nearby, who were killed. David said that in another camp, the name of which he cannot recall, Allied aircraft accidentally killed ninety-eight prisoners.

On one occasion in June, Allied planes dropped leaflets over the camp saying 'Hold Tight we're coming!' But it was to be well over a year more before the nightmare was to end. However, at least by this time the Japanese had allowed some Red Cross medical supplies into the camp, including more Novocaine.

But from that day onwards the railway was rarely used during daylight by the Japanese, due to the ever-present bombers, and David remembers one night watching the massive bombing of the fuel dump at nearby Kanchanaburi, to the cheers of the prisoners, calling 'Go on boys!' as they took a grandstand view.

In addition, with the help of the amazing Boon Pong, David was allowed to go to buy drugs from the chemist in the nearby village which Pong had organized in his great work for the prisoners. Under cover of needing batteries for medical torches, as there was no electricity in Chungkai, David was also able to buy the crucial batteries for the secret radio. One day the Webbers asked for a set of headphones to better hear the radio news, and so David asked Pong, who said he would bring them from Bangkok on his next visit. There was a secret and well-established routine: David would enter the shop with his little rucksack, and Pong's wife – who also spoke perfect English – would take the Japanese guard to the back of the shop for coffee. David would buy a tin of jam or some other item, and take off his rucksack and leave it so that Pong could place the required goods into the bag while he went to the chemist's shop, and then he would return. A nod from Boon Pong signified that the deed was done, and David would nonchalantly pick up his rucksack and go. The headphones were duly smuggled by David into Chungkai camp, where he handed them over to a very nervous prisoner friend to pass on. David said the risk was enormous, but it was worth it in order to get information from the outside world to give the much-needed boost to morale among them all.

One evening, Donald Webber was listening in and something fell on his shoulders; it was a cat but he shook with fear because he thought it was a snooping Japanese guard.

David recalls seeing Red Cross parcels only twice in his whole imprisonment. Despite the British and American Red Cross regularly sending them, the Japanese stole them; the POWs always knew they had arrived because the Korean guards began smoking Chesterfield and Lucky Strike cigarettes. On one occasion, David shared one twelfth of an American parcel, meant for one man, but it contained lots of ham and spam – strictly non-kosher. However, amongst the items was one tiny tin of salmon, and David said he would give up his share of everything else if he could have the tin; it was so agreed.

On completion of the railway, many men were sent back to Singapore from the various camps, but David was kept at Chungkai. Drugs for the sick could be smuggled in by the Chinese camp canteen manager, Lee Sung/Soon, who like Boon Pong very bravely risked his life to help save many sick men. David dealt with him a great deal. Soon after liberation in Bangkok, David was walking to a Rosh Hashana service which was to be held at the home of a refugee German Jewish doctor, Jakobsohn, when a car suddenly pulled up. A voice called out 'Doctor, doctor' – it was Lee Sung, who immediately drove him to the house of Jakobsohn, whom he called 'the Jew Doctor' and who is mentioned again later in this testimony.

Around this time, many of the Jewish POWs whom David knew were sent by ship from Singapore to Manila, and Dr Mark Braham (from Leeds)[21] died en route and was buried at sea. When the ship docked, several sick men, believed to be dying, were sent to a hospital by the Japanese; however, soon after this, American bombers accidentally sank the ship, killing many on board, including members of David's Jewish congregation from Chungkai. Jack Berner from Salford was among those taken off the ship, and soon afterwards the Americans liberated him and he was returned to the UK in April 1945. Here he was interviewed by the *Jewish Chronicle,* and he told them of a captain, son of a Blackpool rabbi, who was safe at Chungkai. For security reasons this man could not be named, but it was David, and Jack went to see David's parents to assure them that he had left David in good health and that they were not to worry. David still has the cutting from the *Jewish Chronicle.*

A second Red Cross parcel (this time a British one) was distributed by the Japanese about this time, but the batch of parcels had been diverted from the German POW camps (now liberated) to Asia. Having travelled so long and so far, most of the tins were 'blown' by the time they arrived, and the contents were inedible. To add insult to injury, some of the tins contained rice pudding – David said that the reaction of the men to that are not printable! Three men shared one parcel on that occasion. One other donation came from the Swiss consul in Bangkok, who sent an old gramophone and some records; he clearly had a warped sense of humour as one of the records was the Andrews Sisters[22] singing 'Don't Fence Me In!'

By March of 1945 the Japanese began to separate and move away the officers from the men at Chungkai (to forestall possible revolts)[23] but David was left behind with some other doctors and padres to care for the men. The radio went with the officers and the last war news David heard was of the Allied crossing of the Rhine in late March 1945.

One day in April 1945, the camp interpreter came to David to announce that 'Hitrer is dead!' To confirm this, a few days later a POW on a working party met some British officers and one gave the man a cigarette and told him to give it to MO Captain Mackintosh at Chungkai. When they opened it at the camp, on the paper was written 'GOUT' – code for 'Germany out'. That is how they found out about VE Day.

Liberation

On 26 June, Chungkai was evacuated and all the men were moved south to Taimuang/ Tammuan (this was where the dental chair was finally left and nothing more was seen of it). David confirms that there was a genuine belief among the officer and NCOs (non-commissioned officers) that the Japanese would kill all the POWs rather than give them up, and we now know, from archives and from interviews after the war, that detailed plans were indeed made by the Japanese to do precisely this.

After a few more weeks, David was moved to a subsidiary camp called Kaoren, where the men worked in railway workshops. Here they heard rumours about a peace conference – probably Potsdam – but one day the men returned early from work and said that the Thai workers had all got drunk at the workshops and were saying 'England No. 1, Nippon No. 20, war over.' David and all the others agreed, then and until they died, that the atomic bombs were a miracle that saved them all.

Soon after this, word came that the main camp at Kanchanaburi was flying Allied flags. Discussions went on all evening among the prisoners, and later the senior officer, an RAF doctor squadron leader, marched into the Japanese commandant's office and said that they wished to fly the union flag, as they had heard the war was over. The commandant refused, saying he would not let them do so until he had orders from the main camp. The squadron leader said, 'To hell with you, tomorrow we fly our flags at 9 a.m.' Meanwhile, it was noticed that the Korean guards had disappeared.

Nobody could sleep, and next morning long bamboos were found and the Union flag (forbidden until then, except at funerals) was prepared, together with the Australian and Dutch flags. David polished his officer's pips for the first time in years and put them on. He commanded the British soldiers as they paraded, as he was the only army officer, and as a POW violinist played 'God Save the King', the squadron leader gave the order and the flag was hoisted. David gave the salute. It was, he said, 'very, very moving'. The Japanese remained in their barracks from then on. The prisoners' first meal on being freed, he remembered, was roast duck – an unbelievably good taste after more than three years of starvation.

David realized that Rosh Hashana was approaching, and he got a lift by crowded truck into Bangkok; as they approached the city they saw the first white woman in nearly four years. The men all cheered and she waved back. At the Oriental Hotel a Christian padre told David about a service being held for Jewish POWs at the house of a Dr Jakobsohn, the German refugee eye specialist mentioned above. Other refugees and some other POWs were present, and David was called up to read from the scroll – a great honour. He was invited to stay for lunch and marvelled at the normality of a fine meal in nice company – the first time for a very long time. As the festival finished that evening, the doctor's 18-year-old daughter brought out a gramophone and asked David to dance – truly a magical experience for him.[24]

Two weeks later a Captain Ross arrived by parachute[25] with a radio man, and he explained what was to happen next. It was that day that David first learned from Ross about the Holocaust, and it was for David a terrible moment; he, of course, knew absolutely nothing about it until then. He said that he retired to his room and wept bitterly.

Soon after this, two doctors parachuted in and one recognized David as a former fellow dental student with him at Liverpool University.

As the doctors had to stay with the sick, who were sent on as they recovered, David was one of the last to leave the camp in September 1945. Eventually he was sent by train to Bangkok. En route they stopped to stretch their legs at a small railway station, and a Japanese officer dressed in a very elaborate uniform approached him, gave a very smart salute and said that he was the station master, and asked if the British officers would like a cup of coffee. It was agreed, and David said this was the moment when he really understood that it was over and that they were free.

At Bangkok airport they slept rough on the tarmac for a couple of nights and then flew by Dakota to Rangoon in Burma. Here David looked for the Jewish chaplain, Rev. Cecil Bloch, whom he met and who gave him four recent copies of the *Jewish Chronicle* newspaper to read; David declared this was the best thing he could have done for him. Relaxing in his bed next day, he nearly fell out, because in one issue he read in the announcements section that his sister-in-law had had a baby – which of course he knew nothing about.

After ten days they went by a P&O liner to Southampton, arriving after about a month, in late October. As David's father was ill, he asked to be quickly processed and he was soon allowed home, quite recovered, and later demobbed.

In 1981, David returned to the Far East for a visit with his wife, Shirley. They travelled to Singapore and Malaysia, then Bangkok, and on to Kanchanaburi and the bridge, and then took a boat to Chungkai, to visit the graves of three Jewish soldiers whom David had buried there – including Abraham Aaron,[26] an orthodox Jew from Liverpool, whose Jewish prayer book he kept and later gave to Abraham's family. It was very emotional, and in trying to say the Kaddish prayer over the graves he broke down in tears as all the awful memories came back – his lost youth and the futility of it all, given that even the railway virtually no longer exists. Ironically, David mused, Japan remains one of the world's richest nations. David said that as he stood at the cemetery, he could hear in his mind the strains of the Last Post, that had so often been played there those many years ago. It was hard to believe, he thought, that such terrible things had happened in what was now so beautiful and peaceful a place. The huts of the camp had long ago disintegrated and the jungle had reclaimed the site.

He noticed that the head gardener was tending the graves, and he later came over and laid some cut flowers where David was praying. In broken English he explained that he had been 16 years old when the camp existed and he had seen what happened there. David told him that he had known a Thai boy called Roong who had helped the prisoners, and the man said that he was that boy! After one final look, David and his wife returned to the boat and Bangkok, and he felt he had exorcised some of the past.

David regularly attended the FEPOW (Far East Prisoners of War) reunions after the war, as well as the Malayan Campaign Officers Reunions. In 1995, on the fiftieth anniversary of the war's end, David was given the honour of leading the Far East veterans group which led off the Annual AJEX Remembrance Parade in Whitehall. In 2010 the only other surviving British dental officer POW died, and David at 96 was the last one remaining. Despite all the work he did and the risks he took, David received not one honour or mention. The very keeping of his diary and other POW lists was in itself an act of defiance and potentially could have quickly led to his execution if they had been discovered. Much of this material was written on looted and distinctive naval notepaper he picked up at Singapore while waiting to be sent to Changi in 1942.

At Kanchanaburi camp museum, a sketch of both David and the famous dental chair are on view to the many tourists who go there, but without any labels of explanation. David kept and presented to the AJEX Jewish Military Museum one of the Death Railway sleeper spikes, which is proudly on display.

David has the Pacific Star, the 1939–45 Star and the Defence and War Medals.

In the diary that David kept while he was a POW, the entries for some days are very brief, whereas others are quite long. I have paraphrased some of the more significant parts below. As mentioned above, the diary was written on naval message notepaper that David managed to salvage before the surrender. Comments in square brackets are the author's.

Changi

25/5/1942
Writing by the light of a diesel oil lamp with a pen I bought today for $5.

26/5/1942
Woke to someone playing the piano in the mess ... do they know I am alive at home? Too tired to go to the show tonight, 'Arms and the Man' ... housey housey [bingo] in the mess ... passes the time ... reading 'Gone With the Wind'.

1/6/1942
Met Capt. Silman and CSM Blashki, AIF.

3/6/1942
Went on scrounging party tonight [buying food from locals outside the camp].

5/6/1942
Going to 18th Div. Schule [synagogue] tonight.

6/6/1942
We had a minyan [quorum] at the schule. I acted as Chazan [Cantor]; arranged again for next week.

8/6/1942
My 28th birthday; had a treat – two hardboiled eggs for breakfast [a gift] and were they good! Had a tin of spaghetti and rice and coconut cake as a birthday cake for supper.

10/6/1942
Regimental concert in evening – very good, especially Gunner Goldberg.

12/6/1942
Chazan at schule again – 14 men altogether; arranging to have one central service in hospital area for all Changi. Great excitement as [all men] received first pay today.

13/6/1942
Officers scrounging party not too good today – natives scared as one was taken away today.

16/6/1942
Guest night – Jack Redhill [FMSVF Jewish bandmaster – see 'The Record and Roll of Honour', UK and Dutch Army list, in this book] played selection of Yiddish tunes at my request [– he later composed the Malayan National Anthem].

18/6/1942
Depressing news that am to go up-country with a group; hoping for the best. Went to concert with Charlie Stolarski [Jewish AIF].

19/6/1942
Able to send my first postcard home. Schule services now arranged for every Saturday; up to 30 British and Aussies turn up, including Cpl Brody.

[On 26 June, David left Singapore for Thailand.]

Bampong

3/7/1942
Left Singapore 12.30 p.m. on 26th June ... buying food at every station ...resembled a

Sunday School party ... but crowded 25 to a zinc wagon ... lying on luggage ... doors open, not so bad. At 8a.m. next day arrived Kuala Lumpur (KL) ... fed quite well on way ... stopped at Ipoh, Sungei Patani, Prai by Sunday morning ... used platforms as latrines all the way along ... Haji, Thailand by Monday morning ... Tuesday 30th June arrived Bampong and walked 2 miles to camp ... discovered all my personal bed rolls, etc stolen. Huts primitive, latrines appalling, thick mud, two roll calls each day – really feel we are POWs now ... Japanese with fixed bayonets everywhere ... rice and stew for every meal ... sometimes meat and veg added ... have to supplement with tinned food brought from Changi. Have done two extractions with patient seated on a box ... no dental care or hospital yet ... were marched to a river where we could bathe and wash clothes ... drinking water heavily chlorinated and rationed ... difficult to keep clean ... met some of my Changi Minyan today and hope to hold services ... Japs gave 10 cigs each to us today.

4/7/1942
My orderly Fisher, washed some clothes for me; Jap came and taught me *Ha Isha* (dentist) and *Taiyee* (Captain).

8/7/1942
Getting a lot of uneatable limed rice, usually for pigs...protests brought no result ... reading Pride and Prejudice.

11/7/1942
Three men gone missing ... I wish them luck. No cigs given today ... punishment? Appalling flies in cookhouse but they do their best.

13/7/1942
The three men recaptured today ... held with hands behind their backs, then taken away.

15/7/1942
Moved to officers hut today ... more comfortable ... nearly fainted at roll call ... bad diarrhoea ... starved today ... better this evening thank God.

22/7/1942
Dental chair now made by Sgt Oliver ... very good indeed ... my toothache treated by Major Peacock ... bought some tinned fish ... growing a beard and look a villain!

31/7/1942
Now constant diarrhoea ... possibly too many bananas ... have extracted tooth from a Jap ... gave me his thanks ... I would cheerfully kick them all in the crutch ... they have stopped our black market with Thais when we go to wash at river ... holding our first Jewish service in Thailand tonight. One new man Simon at service tonight ... two boiled eggs a day keeps me going.

8/8/1942
On working party to a Jap brothel ... to put soil on a flooded garden.

15/8/1942
Our Jap rations are rotten and lots of pork so I must supplement with my private store [*buying*].

17/8/1942
Went in car with 3 guards to Bampong to buy dental filling cement ... emergency operation last night ... surgeon Smythe (survived massacre at Alexandra hospital as his cigarette case deflected a Jap bayonet) removed 20 inches of small intestine from a soldier ... I watched; sadly soldier died later. Jap Colonel read us a message at roll call, interpreted by a Japanese, who said, 'This order will be printed on paper. Please get it to the bottom of all men.' A lot of us laughed.

21/8/1942
Received [Jewish] prayerbook from Pte B. Barnett at another camp ... able to copy out several prayers [later returned to Barnett].

30/8/1942
40 miles trip to Kan(chana)buri on a working party.

11/9/1942
Evening of Rosh Hashana (Jewish New Year); services organized and got men off work for that day.

14/9/1942
Had a minyan for service but one man went to work even though I got him time off. Wore my best and only good shirt for first time since leaving Changi and polished my pip – officers only allowed one pip by the Japanese no matter what rank! Food worse – may be moving. We prayed we would not be here much longer as POWs.

20/9/1942
Preparing for Kol Nidre (Eve of Yom Kippur/Day of Atonement service) tonight.

22/9/1942
Fasted on Yom Kippur. [A massive sacrifice, given that they were so hungry.]

23/9/1942
Three Loyal Regiment men escaped last night but gave themselves up next day [David does not mention what happened to them]. I will not try till our troops are much nearer. Malaria on the increase ... beard fully grown now.

8/10/1942
We are moving; cut off beard.

Chungkai

21.10.1942
Left Bampong Oct 9th to march 17 miles to next camp but got a lift in Jap medical lorry half way – Yeggi the Jap interpreter even bought us an iced coffee en route. Next day marched again 10 kilometres ... accommodation bad and most slept that night in the rain ... Oct 11th again marching in the rain and after 13 kms reached Kanchanaburi ... utterly exhausted, Drs Black, Mackintosh and myself held each other up singing 'Keep right on to the end of the road'. Eventually embarked on a barge to the next camp on the Kwai river but virtually no huts prepared and we slept in open or in crowded huts. No fence and lots of space and natives selling food; our food is awful. Had two months pay but most taken by Japs for food and 'compulsory saving' – both a racket of course. New troops arriving from Singapore and so food prices locally soar as demand increases ...

2/11/1942
... among new arrivals are Smith and Morris and another Jewish officer, Capt. C. Beck. Hoping to arrange services in the next week or so.

14/12/1942
Now have about 15 regular worshippers, but are lots more [who do not come]; Morris, Beck and Gaisman do not come ... good news from North Africa ... also allied planes bombing Bangkok ... Bampong POWs heard the bombs from their camp ... borrowed a dental engine from a neighbouring camp ... am going to hopefully retrieve my dental chair from Bampong.

21/12/1942
Sent my second card home.

27/12/1942
Lionel Basco died at 9 a.m. of dysentery and TB – a Jewish soldier 125 A T Regiment. I took the service at 3 p.m. with a minyan of 10 and the Japanese Colonel. Joe Morris said Kaddish [ancient Jewish prayer for the dead].

15/2/1943
Lots of movement in and out of the camp – 1,000 Dutch arrived for a week – very nice fellows, five Jews among them who came to Friday night service. Lose lots of friends as they move on.

30/3/1943
Good war news, but 4 men (one of them RAMC) escaped today, were recaptured and shot; I heard the volley. It was sickening and I shan't ever forget.

18/4/1943
Thousands of sick from other camps area arriving as we had been warned, including many Dutch and Australians. Cpl Bluestone, one of my minyan, badly beaten by Japs today with broken mandible and is in hospital. Many funerals as sick die.

10/5/1943
Some mail arrived from June/July last year; wonderful day.

24/5/1943
Last November a Signalman Shapirio died at Tarsau and was buried under a cross. I wrote to Padre and this has been changed to Star of David; Pte Aaron is ill; Capt. Markowitz has joined the hospital team with us. Pte Aaron (from Liverpool – Recce Corps) has died of dysentery and malaria at 11.30 a.m. [date not given here] – tried to jolly him up but illnesses too much for him. Buried at noon; Dutch and British Jewish soldiers attended. Joe Morris again said Kaddish.

11/10/1943
Buried Pte Mutsemakker, Dutch army; death is common but always terrible.

[Next diary entry written at Kaoren camp]

17/8/1945
The war is over at last.

Notes

1. Many years later, in 1981, David returned to Malaya and on reaching Kuala Lumpur railway station he suddenly had a terrible stomach ache; he believes this was a psychosomatic reaction to seeing the building where he had such a terrible last visit.
2. A play on the name of the notorious wartime Norwegian collaborator, Quisling.
3. A term of endearment in Yiddish, meaning a true man.
4. Actually the Kwai Noi.
5. A popular Spanish song of the period made famous in English by Deanna Durbin in the 1930s.
6. Boon Pong Sirivejjabhand was a Thai merchant and a member of the Free Thai Army and the British V Force, who used his position as a supplier of food to Japanese POW camps as a cover to smuggle medicines, money, messages and food into the camps along the Kwai River. He is credited with saving thousands of lives, at great danger to himself and his family.
7. A member of 125 Anti-Tank Regiment, son of Hyman and Matilda of Hackney. Died 1 May 1943, buried at Chungkai CWGC cemetery, Thailand.
8. These were issued to all British and Commonwealth Jewish troops in both world wars. The Americans had their own version as well. David also had his Blackpool United Hebrew Congregation siddur (prayer book) with him and he still has this now battered but treasured volume. It bears the Japanese censor mark, as it was first taken from him by the camp guards, then later returned. On the inside cover are written the dates and names of the camps he was in. He also kept lists of every Jewish POW he met, which survive intact in his archive.
9. Famous American silent comedy film star of the early twentieth century.
10. Astonishingly, David has photographs taken by a POW of some of these activities, showing Japanese soldiers enjoying the events and posing with the POWs.
11. *British Dental Journal*, 7 October 1995, pp. 268–9.
12. This was presented to the RADC Museum by David in 1984, and an article about this appeared in the *Journal of Dental Practice* that year. The general to whom David made the presentation was the nephew of his wartime colonel. An earlier article by David was published in the *British Dental Journal*, 16 August 1946, pp.128–9.
13. Another touching item is one given by a grateful Dutch patient to David: a piece of soap into which he had carved his head-and-shoulders profile – now on display at the Army Medical Services Museum, Aldershot.
14. Awarded the MBE for gallantry as a POW – see his testimony in this book.
15. Soon after the war, David and his wife Shirley spent part of their honeymoon in Paris and met up with Rosie.

16. The Hebrew/Jewish calendar year.
17. His testimony is included in this book.
18. Israel Shapiro, 137 Field Regiment, RCOS (Royal Corps of Signals), son of Harris and Annie of Leeds. Died 19 November 1942, buried at Kanchanaburi.
19. Many years later, David gave a talk about his experiences to a Jewish society in Stanmore, and Leo Lavender was there when David told this story. At the end, David told Leo to stand up and Leo declared that it was all true. Everyone cheered the two old comrades. Sadly, Leo died suddenly in 1984 on a railway platform, on his way home, but David knew that he had helped to give him forty years of life. Leo's son later journeyed to Thailand, with David's guidance, and visited the places where his father had been. He wrote later to David to say how deeply moved he had been by the visit.
20. In fact, at that camp a radio had been discovered through careless security, and the prisoners responsible were beaten to death; this is related in Eric Lomax's book *The Railway Man* (CCV Books, 1998).
21. RAMC, son of Peter and Ada of Leeds, husband of Sonia. Died 21 September 1944, named on Singapore memorial.
22. A famous American singing group.
23. For the Japanese, separating the officers also meant not needing to waste time at work dealing with officer POWs protesting the poor working conditions and holding up the work schedule.
24. Finchley Synagogue magazine, October 1997.
25. The Allied air forces had taken aerial photographs of the camps and knew exactly where they were located. Some examples are in David's archive.
26. Reconnaissance Corps, Loyal Regiment, died 2 June 1943, buried Chungkai. David also noticed the grave of a Dutch Jewish soldier he had buried, Private Mutsemaker, with a cross on the headstone; he had this changed by the CWGC.

Jacob Ballas, Singapore

13. Jacob Ballas, Singapore.

Jacob was born in Amara, Iraq, on 21 January 1921, only child of Shua and Grace, a poor orthodox family, and was brought first to Labuan and then to Singapore as a child, and educated at St Andrew's School. His parents belonged to the majority, poorer Jews on the island, living at 19 Wilkie Road and later at Short Street. As a small boy, he helped his mother sell her home-made flat bread (roti) to the local Jewish community to make ends meet, and attended cheder (religious school) at Bencoolen Street synagogue every afternoon. He started the first Jewish Scout Group on the island – Deft Patrol – with ten other Jewish boys from St Andrew's School. Even though he passed his scholarship exams, he was not offered a university place because he did not have British nationality.

When war broke out he was running a car sales business and his expertise meant he was made head of the St Andrew's unit of the Medical Auxiliary Service, deploying vehicles to bombing incidents to evacuate the wounded.

His car sales business was closed down as the Japanese invaded, and he lived by selling his personal possessions and second-hand watches until interned. He remembered being forced, as a Jew, to wear a red armband by the Japanese. Jacob was interned by the Japanese in the last six months of the war in Sime Road camp, with his mother; his father died just before this. Jacob said the camp was well organized and there were work groups and a fair distribution of food; hut leaders organized the work and cooking. There were about 600 Jews, and the women's huts were separate from the men's. For a while he was a hut leader, but his main work was hauling wood on lorries from the reservoir area. Once per week the families were allowed to meet and that was when he met up with his mother. Every evening after work, people were very tired and ready to sleep, and always hungry, despite some extra food that came from the gardens.

Generally, life was better than on the outside, as you only had the guards to worry about; people could meet and talk, and food was regular, though sparse; outside you lived in fear of who may come and break into your house.

After the war, Jacob went into business as a stockbroker and stayed in Singapore, becoming a millionaire. He often visited family in Canada, Iraq and Israel. He was a leading member of the Jewish community and gave large sums to Jewish, Israeli and Singapore charities. He never married, and he died in June 2000.

Capt. Dr Solomon Mathew Bard, HKVDC Field Ambulance

The information in this testimony was taken, with permission, from Solomon's astonishing book, Light and Shade *(Hong Kong University Press, 2009), and from personal correspondence with the author.*

Solomon was born in Russia on 13 (Russian 26) June 1916 at Chita in Siberia. He was one of three brothers. His father was a Polish-born watchmaker and his mother was from the Crimea. His grandfather had been a soldier in the Czar's army and fought at the battle of Plevna in 1877. While living in Chita during the First World War, his mother had hidden an escaped Austrian-Hungarian POW in their house, at great risk, as she felt so sorry for his predicament and condition. This brave and humane act, as will be seen below, greatly affected Solomon's life later on. In 1919, three White Russian officers were billeted with them during the civil war between Whites and Reds, and Solomon remembers well how they spoilt him and took him riding when they went to water their horses at the river. When the White Russian General Semionov occupied Chita, he arrested the local merchants, including Solomon's father, but his mother personally appealed to Semionov and he was released, probably for a ransom.

Harbin and Shanghai

In 1920, aged 4 years, Solomon began Hebrew school, though he also learnt Russian and Polish (and some Chinese later on). In 1924, as Soviet influence reached into Siberia, the family decided to leave for Harbin in Manchuria – one of the last families to do so officially with Soviet passports. Harbin was a city to which a lot of Russians and Jews had fled and lived happily among the local Chinese. Here Solomon was bar mitzvah and learned Jewish history at his primary school, where special lessons were provided for the Jewish children. He also joined the Zionist youth movement, Hashomer Hatzair, and later the more militant Betar. His his love of music began, and he learnt the violin as well taking English lessons.

But storm clouds were gathering and in 1932, when the Japanese occupied Harbin, his family decided that he and his older brother Leo, an optician (their brother David had died in Chita years before), were to go to Shanghai to learn better English so that Solomon could prepare to go to study medicine at Hong Kong University. His uncle David, already in Shanghai, found them rooms in the French Concession area. There then occurred an amazing coincidence. Solomon could not gain admittance to the best English-speaking school (Public and Thomas Hanbury for Boys) because he did not have a resident parent or guardian in the city. To help keep himself, he got a job with his uncle David, playing in an orchestra at a club. His mother knew that the escaped Austrian soldier she had saved in the First World War was a successful pharmacist in Shanghai, and he offered to help her with anything he could, if ever she needed his assistance. Solomon found him; he readily agreed to be his guardian and Solomon was able to enter the school. Completely cosmopolitan but run on English public school lines, Solomon fitted in well in the sixth form and completed his examinations to obtain entry to Hong Kong University

He arrived in Hong Kong in January 1934, to study medicine at the university. His parents, meanwhile, had emigrated to Argentina. He graduated in January 1940, but he had enlisted in the HKVDC on 3 September 1939, on the day war broke out, and was commissioned in June 1940; he also became a British citizen. In September 1941 he travelled to Shanghai to marry Sophie Patushinsky (she had been born in Harbin, also of Russian Jewish parents, but brought up in Shanghai), who was also a medical student at Hong Kong University, and then hurried back to Hong Kong again.

War

On 7 December the HKVDC and all Allied units were on high alert and at action stations. Solomon reported for duty next morning at 8 a.m. and said goodbye to his wife, whom he would not see for three years and eight months, except during a brief lull in the fighting. Issued with his revolver and given six men, he was posted to his position – an ADS (Advanced Dressing Station) in the bunker at the Pok Fu Lam reservoir near Victoria Peak on Hong Kong Island itself.

The Japanese advance was swift; they outnumbered the defenders by at least two to one and had artillery and air cover. By 10 December, Solomon's area was being heavily bombed and he was ordered to report to the west of the island to Mount Davis Fort, leaving his sergeant in charge at his bunker. As the mainland was totally overrun by 12 December, the Japanese now concentrated on the island. The Japanese sent over a launch to demand unconditional surrender but this was rejected. During the several-hours-long truce, Solomon went home, to Pok Fu Lam Road, not far from Mount Davis, to take a shower and see Sophie, but she was working at St Stephen's College, which had become a hospital. On his way back to his post, as it was getting dark, he walked in the middle of the road making loud footsteps with his boots so no trigger-happy sentry would shoot him.

On 13 December the enemy bombardment was intensified and one gunner ran into the fort to say an anti-aircraft position nearby had received a direct hit. The men formed a corridor to let Solomon through as, roused into action at last, he left the bunker, the men encouraging him with pats on the back, to reach the gun position. When he got there it was carnage, with dead and wounded lying everywhere; he gave what emergency treatment he could and then went back to get a party to bring in the dead and wounded men to the medical room, from where they were taken by ambulance to hospital; that was the last he saw of them.

On Sunday 14 December, a nine-and-a-half-inch armour-piercing shell penetrated the bunker and embedded itself in the concrete floor, but failed to explode; it was five feet from where Solomon stood and sixty men narrowly escaped being blown to bits. The CO ordered evacuation, as another shell nearby could cause a sympathetic detonation. Small parties of men were sent out to escape between shell bursts, and Solomon was ordered to bring up the rear. Waiting his turn was terrifying because the unexploded shell could go off at any time. The air became thick with sulphur, possibly given off by the shell, and the men had to use respirators for the next half-hour or so. When his turn came to leave, Solomon ran up the slope and down the lee side away from the shelling, and then rested under a tree for a smoke. After a while a voice shouted that he was to move to a new HQ in a basement in a nearby building, named Felix Villas, where his medical post was to be in a concrete guardroom near the Fort.

Shelling continued for the next two days, an attempted Japanese landing was bravely driven off, and a second truce was called to demand surrender; during this time, Solomon went again to see Sophie and this time found her, staying for a few hours at St Stephen's. On 18 December the battle resumed and some hits were scored on enemy artillery emplacements; the Hong Kong garrison was doing well. Solomon's post was now moved to a flimsy wooden hut and it was frequently attacked by air and artillery but with no hits. That night the Japanese landed undetected and advanced inland; there was fierce fighting but they had consolidated and were unstoppable. They took no prisoners, massacring wounded men and, when capturing gun emplacements or dressing stations, bayoneting everyone to death, including civilian and military medical personnel. Fierce fighting continued by land and sea to no avail. However, Mount Davis still held out. Solomon took shelter from the shelling in a house, when on his way to a meeting at HQ, and befriended a stray dog hiding in the building, but a strange impulse compelled him to move on, grabbing the dog, and as he reached the next building, the first house received a direct hit – another brush with death. On reaching HQ, a mug of hot, sweet army tea brought relief. The dog stayed with him until the surrender.

The battle continued, with both sides taking heavy casualties as defender and attacker fought to the death, with the Japanese murdering any prisoners they took. By 25 December the battle was ferocious, the Japanese enraged at the fantastic defence being put up by the Allies; at St Stephen's another Japanese hospital massacre took place. Finally the Governor of Hong Kong decided that the lack of water and ammunition, and the huge civilian and military casualties, meant that surrender was the only way, and so the fighting ended at 3.15 p.m. on Christmas Day. At Mount Davis the order to surrender reached the men that evening, and Solomon's comrades destroyed all their personal weapons and artillery.

Solomon reported to the HKVDC rendezvous point at the university and the men all then spent 28 and 29 December at Murray Barracks on Garden Road. Then they made their way to Sham Shui Po Camp. Solomon did not know where his wife was. They were humiliated and depressed, all soldierly appearance gone; with their few belongings, they were jeered at by a few onlookers. The Japanese guards were properly behaved but it was the most degrading experience of Solomon's life.

POW

The camp was a former British Army barracks on the western shore of the Kowloon peninsula – a parade ground, two large four-storey buildings and wooden huts – and more than 5,000 POWs were crammed into a space designed for about 1,500. This was to be home for the next three years and eight months. There were no beds, bedding, toilets or wash-houses – it had all been looted. It was chaos. But after two to three weeks, these things did emerge and the major problem became lice, due to the overcrowding and poor hygiene. Food was totally inadequate – mostly rice and vegetables, but only half what normal men needed to stay alive and well. However, relatives not interned were allowed to send food parcels – despite being short of food themselves – and these literally kept people alive; they were always shared with the less fortunate POWs. The Japanese,

14. Capt. Dr Solomon Bard, MO, HKVDF (standing, far left), inside Sham Shui Po Camp. This is one of very few photos taken in the camp and was snapped by the Japanese interpreter Tsutada in 1943.

meanwhile, stole the Red Cross parcel deliveries. Only three per person were given out for the whole three years or more, and these were given only towards the end when the Japanese knew the war was lost. The canteen was useless, and often closed as a punishment for minor infringements. Men on working parties received a tiny amount of pay which could buy very little indeed. After the war it emerged that even this pay was from Britain, via the Swiss; it was not even Japanese money.

The camp had a number of army doctors, orderlies and dentists from the Allied forces (UK, Canada and HKVDF) and a makeshift hospital was set up. Serious cases were sent to a proper hospital run by POW doctors at Bowen Road hospital on the island, but there were long delays and many died waiting to be transferred. The senior Japanese doctor in the camp, a Captain Saito,[1] was completely uncaring and rarely seen. When men were sick he demanded that they work, which brought confrontation with the Allied doctors.

The Japanese medical sergeant, however, risked his life by giving to the POW doctors the drugs that he smuggled into the camp for the Allied sick.

Around April 1942, a Major Ashton-Rose of the RIAMC (Royal Indian Army Medical Corps) was appointed senior Allied doctor; he not only stood up to the Japanese but was a very capable administrator. But there were no medicines to speak of, and poor hygiene facilities, and he could not hold back the terrible illnesses that came with malnutrition: diphtheria, beri-beri and malaria – in 1942 about 35 per cent of the men died from a diphtheria epidemic in particular. In 1943 a distressing illness called 'electric' feet, or burning feet, was rampant. Probably caused by vitamin deficiency, it was exceedingly painful and treated with morphia, of which there was a reasonable supply; Solomon luckily escaped this but did have malaria and dysentery during much of his internment. By the end of 1944, deaths and transport to Japan reduced the camp population to 1,500, with inmates just about functioning on minimum food.

Punishment, Administration, Morale and Recreation

As elsewhere in the 'Co-Prosperity Sphere' prison camps, face-slapping and other punishments were frequent and severe for the most minor infringements (such as not saluting a guard quickly enough) and were humiliating (Solomon suffered slapping twice). Escaping was a capital offence. There were two roll calls a day, called by a POW bugler. There were about twenty Formosan guards under a Japanese commandant, with interpreters, and men went on outside working parties, mostly extending the airport; officers did not work. Those left in the camp did fatigues such as cleaning and cooking, and the doctors worked in the 'hospital'. The British liaison officer picked by the Japanese was a weak and cowardly man – an RASC (Royal Army Service Corps) major – who together with his minions did nothing to help the POWs and obeyed all Japanese orders without question; this group was hated by the rest of the internees. Some interpreters were sadistic but one, nicknamed 'Holy Joe' (real name Watanbe), was a Christian pastor and risked his life smuggling medicines and messages into the camp. Another, Tsutada, although hating the British because of the treatment he received as a schoolboy in Singapore, obtained essential building materials for the camp and took internees' spectacles or dentures for repair into the city.[2] The POWs were allowed to send and receive only one card a month.

The year 1942 was a bad one for the Allied war effort, and many died in the camp; Solomon tried to think positive, keep busy and not think too much about the future. He observed how 'giving up' could hasten a man's death. Some news got in through messages and some via smuggled Chinese newspapers. When news of the German defeat at Stalingrad reached them this was a huge boost. Good military discipline also helped, with senior officers and NCOs enforcing military behaviour at all times when possible – saluting, wearing uniform, showing rank. Religious services also helped; as well as church services, Jewish services were held, using prayer books donated by the YMCA (Young Men's Christian Association) and stamped as 'Passed' by the Japanese censors.

Cigarettes were a currency but also a danger to health (men exchanged their food to get them) and a danger to discipline, as it was common to see senior officers demean themselves by picking up the cigarette butts of Japanese guards.

Besides the usual sedentary pursuits such as chess, many POWs discovered hidden talents in carving, knitting or painting (sports declined as physical deterioration set in) – all encouraged by the Japanese, as such activities assisted in the good order of the camp. Academic lectures were given and a library evolved; a concert party was formed to put on shows, and sometime in 1943 Solomon decided – as he had some past musical training – to form an orchestra. There were Canadian army bandsmen and amateur players in the camp, and old piano scores were adapted by Solomon; two Canadians made a cello from an old oil drum; programmes were written by hand and beautifully inscribed. In all, Solomon produced three 'Bard concerts', but the Japanese ordered that 'Rule Britannia', 'God Save the King' and 'Home Sweet Home', were forbidden. In an act of defiance, Solomon had the orchestra play 'Plymouth Hoe', an anthology of naval themes in which was hidden the strains of 'Rule Britannia'. The Japanese, in the front row, did not notice and applauded enthusiastically along with everyone else. In the evenings, men often talked well into the night, the favourite topics being family, home and food. Solomon's little group acquired a pet cat, which ably fed itself by catching sparrows.

Throughout 1943, sickness, death and punishments remained at a high level. Some POWs cultivated little gardens or the communal vegetable garden; the latrine provided manure. The local Red Cross representative was a Swiss who lived in Hong Kong; he visited occasionally but was always with senior Japanese officers and he was not allowed to speak to the inmates, or they to him. Blankets in the hospital were drawn up to the men's necks so he could not see their emaciated bodies. Nevertheless, one brave HKVDC officer once said to him, in French, 'We are starving'. There was no response from the visitor and nothing else was said; after the visit, the officer was hauled off and badly beaten. Solomon supposed that the Red Cross man at least was achieving something by making a restricted visit, rather than no visit at all.

The families were allowed to bring food parcels (often selling their possessions in order to acquire this food) and see the POWs, but only from a distance away from the wire fence. Children were sometimes allowed closer. Solomon's wife was not interned and so could send him food. Many walked long distances to do this as there was no public transport. The parcels were examined by the guards and then passed on, but if messages were found, the recipient was beaten; messages soon stopped.

As in other camps there were inmates and guards who traded illicitly across the wire. Personal possessions (pens, watches, gold teeth, etc.) went out; food, cigarettes and medicine came in. Some of these traders were unscrupulous but others, when asked, brought in food and medicines to help save the lives of sick POWs in Solomon's care.

Notable Events

Four officers made an escape in April 1942 but, as a result, about fifteen men were arrested, taken out of the camp and badly beaten and tortured; at least one is thought to have died; the others were returned about one month later. Solomon had assisted in the escape but someone may have betrayed him; instead, a man called Sergeant Beard was arrested in place of Solomon, whose name (Bard) had probably been misheard by the Japanese as 'Beard'. In addition, the canteen was closed and the delivery of parcels to the POWs was stopped for several months; the fence was electrified (resulting in two deaths later on) and many officers were transferred out to the Argyll Street camp in Kowloon. About three weeks later, all the POWs were paraded for a roll call, and several lorries drew up and a number of Japanese soldiers jumped out. A Japanese officer then ordered the prisoners to sign a paper which was a promise that they would not try to escape. Their colonel (Royal Scots) approached the officer and pointed to his copy of the Geneva Convention, declaring that they could not sign; the enraged Japanese snatched the book and slapped the colonel in the face with it (just like the scene portrayed in the film *The Bridge on the River Kwai*), threw it to the ground and stamped on it. He ordered that all would sign and that they would be called out individually to do so. Solomon's turn came and he was taken into a building where a Japanese officer sat at a table, pointing to the paper. Solomon calmly stated that it was his duty as an officer to try to escape. The Japanese shouted back that he had surrendered and so had no rights. Solomon still quietly insisted he would not sign. A sentry was called, Solomon's hands were tied behind his back with wire and he was pushed to a wall and ordered to stand there, and left for three hours with the sentry. After some commotion, a shout went up outside that the POWs had been ordered to sign by the GOC (General Officer Commanding), but that it would be invalid as it was done under duress. Some twenty men still did refuse and were taken away but reappeared after two weeks unharmed.

But escapes continued and Captain Douglas Ford, Royal Scots, and two other officers were caught in the planning stages, probably betrayed by an informer, and shot on 18 December 1943.[3] A Catholic chaplain demanded better conditions for the men but was badly beaten.

In September 1942 Solomon remembers that about 120 Chinese POWs, mostly HKVDC, were released in line with propaganda that the Japanese had no quarrel with Asian people, but it later emerged that many had escaped from Hong Kong and rejoined Allied forces in China and India.

From late 1942, air raids by the Allies became common; Solomon recalled one in October 1943 by the US Air Force that lasted eight hours. In this and other raids, Japanese ships were sunk and oil stores destroyed. Cheering meant a face-slap, but many did it anyway.

Near the end of 1943, Solomon and another RAMC officer, Captain Lancaster, were summoned to the commandant's house, to play bridge with the overall commandant of all POW camps, Colonel Tokunaga.[4] The atmosphere was frosty and the players spoke only about bridge; they were served tea but no food. Solomon and Lancaster were ill at ease but played the game. After two hours they were sent back to the camp with two packs of cigarettes each. Their fellow POWs found the story amusing, and the cigarettes were shared out. During the following year there were three more of these sessions, always ending with tea and two packs of cigarettes.

By 1944, the death rate was considerably lower; burial parties to the plot opposite Argyll Street camp consisted of a chaplain, one officer (they took turns to do this), four men and the Japanese guards. On one occasion Solomon acted as the Jewish chaplain for a Jewish soldier who had died; he did not remember his name.

One ruse used to help sick men was when Solomon and his fellow officers insisted on having batmen (servants) to assist them in their duties; the Japanese surprisingly agreed and so several sick men were saved from having to go on working parties.

Liberation

In August 1945, liberation came peacefully; news about the war reached the camp, and people on working parties were getting 'thumbs up' from locals. Then news arrived of a powerful bomb dropped on Japan. On 15 August the POWs saw the guards lined up and listening to what turned out to be the emperor announcing the surrender over the radio. The next day the Allied officers took over the camp and ordered all Japanese to leave, which they did. A tattered British flag was found and raised over the camp in an emotional ceremony. Solomon was shaky with dysentery but left his bed to take part. The infamous British liaison officer (mentioned earlier in this testimony) and his gang were arrested and later tried for collaboration, but unfortunately acquitted.

Allied planes began dropping supplies, and leaflets advised the POWs to stay in the camp; the first Allied person to arrive in the camp was a Canadian war correspondent, armed only with a camera! But it was difficult to enforce the rule and so passes were issued and Solomon went to see Sophie. Armed Japanese were still about in the streets, but when passing some soldiers and later some Japanese nurses they stopped and bowed and Solomon saluted back. Two weeks later the Royal Navy arrived in the harbour.

Post-War Life

Solomon remained in Hong Kong for several months afterwards to help care for the sick members of the HKVDC. Then he and Sophie went to Shanghai to see Sophie's parents, and also Solomon's – they had been in Shanghai for the wedding in 1941 and were trapped there for the whole war. In February 1946 he and Sophie were repatriated to the UK; Solomon flew in stages (in a Liberator bomber, then a Sunderland, taking two weeks) and Sophie followed by ship. Everywhere en route he was greeted as an ex-POW hero. In London he was allowed to stay in officers' clubs until he found accommodation, and at the War Office he was given his back pay, travel warrants, petrol vouchers and – not least – a bottle of gin and a bottle of whisky. He got a job at St James's Hospital, Balham, where he did a medical refresher course, and later at St George's in the East Hospital at Wapping. Sophie arrived and resumed her medical studies at the West London Hospital; they lived in Holland Park. He was finally demobbed in Guildford in July 1946. But the hospital pay was always poor and so they decided to return to Hong Kong in 1947, where Solomon was invited to join a friend in general practice. Their daughter Monica was born in April 1948 in Hong Kong, followed by sons David and

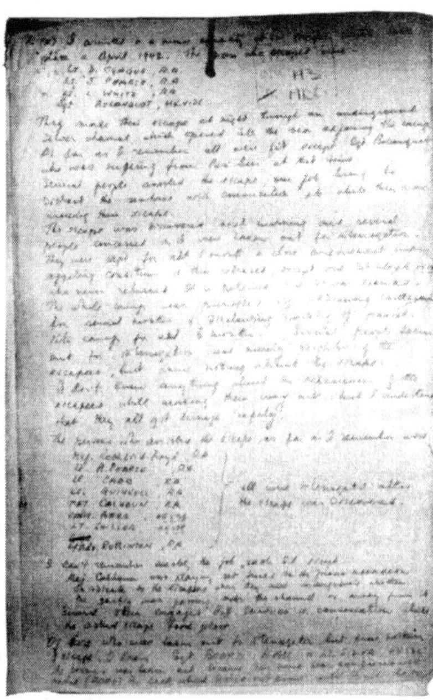

15. Capt. Bard's debriefing transcript from TNA WO 344/363.

Paul. Solomon became the university health officer in 1955, until he retired in 1976. Sophie was a GP. Solomon was also prominent in the Hong Kong music scene, and an archaeologist. On visiting family in Israel, a country he adored, he met many former Jewish refugee residents of Harbin from his schooldays. Sadly, son David died in a car accident in Australia in 1979, a country to which his family moved by the 1990s, and Sophie died in 2006.

Solomon remained in the Hong Kong Volunteers as an MO and in 1953 received the Efficiency Decoration; in 1982 he became Honorary Colonel of the Royal Hong Kong Regiment, still doing his duty. At time of writing in 2011, he is alive and well in Australia.

Edited Transcript of Capt. Dr Solomon Bard's Testimony at Liberation, Concerning an Incident at Sham Shui Po Camp, Hong Kong

I assisted in a minor capacity at one escape which took place in April 1942; the escapers were Lt D. Clague RA, Lt J. Pearco RA, Lt L. White RA and Sgt Bosanquet HKVDC.

They made their escape at night through an underground sewer which opened into the sea adjoining the camp ... all were fit except Sgt Bosanquet who had beri-beri ... several of us distracted the sentries with conversation. The escape was discovered next morning and several were taken for interrogation for about a month in close confinement under appalling conditions, then released except Sgt Lloyd HKVDF, who we believe was executed.

The whole camp was punished by the withdrawal of the canteen for several months and refusal to allow parcels into the camp for six months; most taken for interrogation knew nothing of the escape but were just neighbours of those who had escaped. It is thought all the escapers got away.

Among those who helped were Lt Pearce RA, Lt Carr RA, Lt Guinness RA, Maj Calhoun RA, Capt Bard HKVDF, Lt Shiller HKVDF, Bmdr Robinson RA.

Maj Calhoun played the accordion to signal the escapers when the sentry had passed the channel from which they would emerge. Others distracted sentries with conversation. Those taken for interrogation included Lt Silva HKVDF and Sgt Beard – whose name was mistaken for mine [– something he did not find out about until 3 years later].

(From the National Archives, Kew, file WO344.)

Notes

1. Later convicted of war crimes, sentenced to death, commuted to life, but probably released soon after the war.
2. Convicted after the war of ill-treatment, and given two years imprisonment.
3. See the CWGC website – Ford was awarded the George Cross posthumously for helping organize escapes; tortured and starved, he gave nothing away to the Japanese; also Lt Col Lanceray Newnham, GC.
4. Like Saito, he was later sentenced to death for war crimes, but spent only a few years in prison.

T/80091/EC2464 Capt. Clifford Moss Beck, RIASC

16. T/80091/EC2464 Capt. Clifford Beck, RIASC.

Clifford was born 11 November 1915 and lived at 141 Anson Road, London NW2, and 43 Dunstan Rd, Golders Green, London. After posting overseas he was captured at Singapore in February 1942. He worked on the Death Railway for two years and then was sent to Chungkai. In April 1944 he was approached by a tall, thin Dutch POW, Stockfisch, who said they should organize a Seder (ritual meal) for Passover, for the Jewish POWs.

Stockfisch said he had worked out the date by asking the Catholic padre when it was Good Friday. The following evening, several Jewish POWs met between one of the sleeping huts and the bamboo fence; among them were David Arkush, the red-haired David Wince and two Jewish Gordon Highlanders from Glasgow. They sat in a circle on the ground; it was a warm, dry, star-filled night. Stockfisch led the service; he had obtained burnt scrapings of congealed rice to serve as matzoh and tea-coloured water as wine.

Clifford wrote movingly: 'The order of service I can't remember, but we did thank God, with every reason, that "he had preserved us and kept us till this season" (the words of the Shechechayanu blessing, said on festivals and other joyful occasions). There was no need for bitter herbs; the Angel of Death had visited us all too often.'[1] They repeated the traditional 'Next Year in Jerusalem' with true meaning.

In an article in the *AJEX Journal* in the 1980s, he wrote about the pile-driving job he did, among others, on the notorious Burma Railway. A tree trunk would be held perpendicular in a tripod, and a heavy weight was suspended over it on a thick rope round a pulley; the POWs would pull on the rope, lift the weight and then drop it on the trunk to drive it into the ground as a base for the bridges carrying the railway over rivers and low-lying swamp. It was a twelve-hour day in the blistering heat, with two short fifteen-minute breaks and very little food and water. In September 1945, Clifford was flown home after liberation via Israel (Palestine). As they landed, he remembered the other potent words of the Passover service: 'This year we are slaves; next year we will be free in the Land of Israel.' On that Passover in Thailand, Clifford said, never did he think that he would be in Israel, a free man, seventeen months later.

Note

1. Referring to the bitter herbs partaken to remember the 'bitter' times suffered by the Jews as slaves in Egypt; and the Angel of Death who slayed the first born of the Egyptians.

195194 Capt. Dr Allan Berkeley/Berkovitch, RAMC, attached RA

17. 195194 Capt. Dr Allan Berkeley/Berkovitch, RAMC/RA, on his wedding day in 1945 with his wife, Ida.

Allan was born Abraham Berkovitch on 28 August 1912, in Govan Street, the Gorbals, Glasgow, to Jewish immigrant parents from Russia, Harris Berkovitch (a cabinetmaker) and his wife. By coincidence, the family of Ida, who was soon to be his wife, lived in a flat in the same building. Allan's family later moved to 9 Dixon Avenue, Glasgow. Allan attended Strathbungo primary school and then Queen's Park secondary, from where he went to Glasgow University to study medicine, becoming the first Jewish student to win a blue in soccer.

Allan was a GP in Sunderland when he volunteered for the army and was sent on a tropical medicine course in London. From there he was attached to the 48th Light Anti-Aircraft Regiment in Southend. Soon they were sent via Glasgow to embark for Singapore on 8 December 1941, the day after the attack on Pearl Harbor. On the journey out, Allan was approached by the senior padre on the troopship and reminded of the church service on the coming Sunday. Allan said he would not be coming and the padre asked if he was a non-believer. Allan said no, he was Jewish. The padre said, 'You must be joking; there are no Jews in Scotland!' Allan reminded him that not only were there hundreds of Jews and several synagogues, but that many were prominent members of the wider community in all walks of life.

Java

While at sea, the ship was diverted to Java, as Singapore had fallen. Once in Java – stationed at Garout/Gardet – they had the humiliation of the Dutch High Command surrendering; after three days of waiting, on 8 March 1942, the Japanese appeared, and Allan became a POW and was reported missing to his parents.

Allan was taken to a camp in Batavia, and put to work – officers and men together – on road building and in the docks, after which came *tenko* and counting off in Japanese, and the appalling rice and watery soup diet, at starvation levels. Often a hot communal bath was provided – without soap – but those who came late had to put up with the filthy, cold result. Allan played the banjo, and to help keep up morale he often organized concert parties for the men, in which he played a prominent role, despite the terrible conditions.

In July 1942, he was sent to be the medical officer at a camp at Bandoeing, where he was delighted to meet men of his own regiment. After six weeks he was returned to Batavia, and then later ordered to Borneo; this was cancelled at the last moment, and turned out to be good luck, as most of those sent to Borneo died in the notorious Japanese death marches there. Allan reported the usual atrocities – no drugs and the Japanese stealing of Red Cross parcels resulting in men in his care needlessly dying. He remembered always telling his men that when on work parties they should seek out wild tulip or other flower bulbs and eat them for nourishment to help supplement their poor rations.

Japan

In November they were told they were being transported to Japan; this was the nightmare journey on the Hell Ships, packed like sardines into sealed holds with no facilities and appalling food for days on end. In the evening they were allowed briefly on deck in the cold to be hosed down with seawater. Many died of dysentery and were buried at sea.

On 29 November 1942, they docked at Moji in Japan, where they spent three years in Fukuoka 4B camp, an old YMCA building. Of Allan's 250 men, 160 died within six months, mainly because the Japanese deliberately refused to supply simple remedies such as saline or glucose for the many dysentery cases. Eventually some drugs were given from the Red Cross boxes. One of his Jewish orderlies, Walter Levy of Glasgow, stole drugs from the Japanese MI room, at great risk to his own life, but saving many other lives.

At first food was not bad, but meagre – rice three times per day, some fish and meat pieces occasionally in the watery cabbage soup; when Allan requested some milk for the sick, it was always refused. Then in August 1943 the rations were reduced and the men began to starve. The new commandant, Captain Saito, refused to act when illness began to take hold. In March 1944, Allan approached the Japanese doctor, Lieutenant Hegashi (about 34 years old) and his orderly Sergeant Major Tarniguchi (about 30 years old, and a drunkard), for more food. Nothing occurred, but a few days later Tarniguchi gave Allan a severe beating. Allan testified that the Japanese were actually stealing a large portion of the official supplies of food sent to the camp for the POWs. Allan put those POWs with dysentery on a special diet. When Allan asked for rice instead of the beans which irritated the alimentary tract, Saito refused.

The POWs were given some Japanese clothes on arrival, but the Red Cross clothes were refused them; when boots wore out the Japanese did not renew them but stole the Red Cross boots and socks meant for the prisoners, many of whom had to go to work with totally worn shoes or none at all, even in the cold winters.

Most men had dysentery on arrival, and at first the Japanese helped to keep infection down by supplying antiseptic for the wooden-lined latrines, and using locals to empty the cesspits. But this all soon stopped and the men had to clean the latrines and had no antiseptic supplies. Soap was almost unknown and baths were allowed only once per week, with forty to fifty men sharing one tub of water which soon became black and cold. Lice were rampant, and boiled water for washing was not supplied because the Japanese said there was no fuel. So Allan asked if he could purchase a hot iron, to iron and kill the lice; this was refused. To keep spirits up, and so help keep up morale and hope and humour, Allan had the men bet on how many lice they could find and kill on a bed. In the summer there were flea epidemics, causing total sleeplessness, and the Japanese did virtually nothing to ease this. Bed space was so cramped that the fleas spread easily from man to man.

In December 1944 there was a mumps epidemic but men were forced to work, so Allan devised ways of deceiving the Japanese and giving the men false, very high temperatures

– and they were allowed to stay in bed by the Japanese medics. Allan was then told by the Japanese that the water should not be drunk for a while, as it was infected, but they provided no alternative and beat men whom they saw drinking; this was outrageous, given the men's loss of fluids from sweating after a heavy day's work. The Japanese refused Allan permission to test the water himself so that he could confirm it was definitely polluted.

Failing to salute a guard meant a severe beating. Even if the men were taking much-needed rest, they had to jump to attention and bow when a guard, often deliberately, came into a room while they were resting. No music or singing was allowed without express permission, except on the rare rest days; these were on Sunday, then fortnightly, and then there were virtually none at all. Smoking was only permitted after meals for a few minutes and at bedtime, but not in the bed space; at any other time it meant a beating, and the men always had to stand near a brazier to flick their ash into it. If ever news from the Pacific War was bad, the Japanese medical orderly frequently refused the necessary drugs to keep men alive, as a revenge; Allan named a Sergeant Terada as particularly guilty of this murderous attitude. Sergeant Terada also often refused antiseptic and dressings for men returning from work with sometimes severe cuts and wounds. Even though the Red Cross sent bottles of plasma, this was rarely given to Allan when he needed it to save lives. Hegashi also stole all the anti-syphilis drugs that Allan had brought from Java, and when Allan asked for some blood tests and drugs for six of his men who had the disease, he was refused. Allan estimated that up to 40 per cent of all Red Cross medicines, food and clothes were stolen by the Japanese and many POWs died as a direct result of this.

On Christmas Day 1944, Allan asked Saito for Red Cross parcels for the men, as he knew that many were in their stores. After a long argument Saito agreed, but said that all the parcels would have to be opened and consumed that day – thus denying the men the chance of eating the items slowly over a period of three or four weeks, which was much more satisfying and healthy. The result was that the poor men ate and ate, and then vomited. This was deliberate waste and torture by the Japanese commandant.

Working conditions in the docks, factories and railways were horrendous: sick and weak, starving men carrying huge rice sacks; loads of iron which often dropped and broke limbs; falling and swinging loads from ships cranes; working in all weathers, soaked and cold, or burnt by the sun, and working fourteen-hour days with virtually no breaks and low rations; brutal guards dealing out blows. As a result, health suffered and accidents – many of them fatal – were common. One RAF warrant officer had his back broken by a swinging load while in a ship's hold. Sergeant Joseph 'Danny' Maloney, RA, was severely injured in June 1943 when a heavy load fell on his back, while he was on a working party, and paralysed his legs; the Japanese did try to help but he died of his injuries in February 1945 because a useless Japanese orderly gave him the wrong treatment, refusing Allan's advice. If a sick man was thought not to be working hard enough, he was beaten.

American Sergeant Carpenter, who had been badly wounded in the Philippines, died of his battle wounds after a load fell on him at work, and caused great loss of blood. Allan said that plasma from the store would save him, but Saito refused and Carpenter died as a direct result.

When the men returned from working parties they were searched and anything found that was deemed contraband was confiscated and a severe beating was given, to boot; even a man with a button missing would be beaten, or placed in the so-called guardroom – a tiny, darkened, cold cubicle – with no food or blanket, and a can for excretions; it was known as 'the condemned cell'. One man spent ten days in the cubicle, with only a little water and no emptying of his can. Frequently he would be brought out and held while a guard punched him hard in the face with a clenched fist; if he fell he was kicked until he got up; if he fainted he would be aroused with cold water; if he dodged a blow he would be beaten tenfold.

Another punishment for the usual minor infringements was to make a man lie on his stomach and then raise his body on his arms with palms flat on the ground, for one hour. If he moved he was hit with a large stick on the back; other methods of punishment were standing hatless in the hot sun for long periods; walking about with arms tied round the shoulder blades; being tossed over a guard's back; hit on the face with a leather strap; hit on the back of the legs or back with a flat shovel – and many more that Allan witnessed.

Often locals, sometimes as young as 14 years, would hit or jeer at them on the way to workplaces, but the women were invariably kind and slipped them food, often getting a slap from a husband in return. Sometimes work involved moving ammunition or bombs – this was illegal, as it was war-related. Often guards crept into rooms and stole the belongings of POWs at night or when they were at work. Any possessions, such as watches, had to be handed in when they arrived, but were never given back at the end of the war.

Saito was a short, swarthy, stocky man about 45 years old; he was married and had a teenage daughter. He had short-cropped hair, a deliberate walk and an angry, loud voice; he closed his eyes when ranting. He wore black-and-tan boots and in summer often wore a short-sleeved singlet. He never helped Allan with requests, no matter what they were, and stole Red Cross parcels for his guards' or his own use, including clothing and medicines. He often instructed guards to beat the POWs whenever they wished.

There was a civilian Japanese medical orderly called Inouye, in his late twenties and a little highly strung. He was married with a baby; he was very efficient and had been a chemist and in the Home Guard. At first he had been very kind but after some months this changed totally and he became deranged at the slightest provocation, often beating prisoners, throwing men downstairs, once trying to throw one man from a window thirty feet up, and often hitting men's heads against walls. Inouye also badly beat Americans Private Spencer and Sergeant Powell for having food on them, and Sergeant William Bray for going to the air raid shelter during a raid when he had been told not to.

On another occasion, Inouye had seen a British officer with severe wounds dressed by Allan after he had been beaten by a guard. But Inouye said a POW beaten for misbehaviour must not be cared for; he then hit Allan in the face and his head hit a wall as he fell back; Allan was unconscious for three days.

On 10 March 1945, Tarniguchi sent for Allan in his quarters; he was very drunk. He struck Allan repeatedly with his fist on his face and body, then made him kneel with hands up and hit him repeatedly with his rifle butt in the ribs; he said he would decapitate him and produced his sword, swinging it in a threatening manner. Some guards went to fetch Tarniguchi's wife, as they thought he was going too far, but the beating continued for about half-an-hour, before she arrived. Afterwards Allan was unable to walk properly for several weeks.

Another guard called Sergeant Terada, aged about 30, who spoke some English, was also quite mild at first. Then he developed the habit of snooping on the working parties and beating men for the slightest slacking in work, especially beating the sick ones who had to stop working and return to the camp. Allan noted that he viciously attacked American

Corporal Joseph Cusick and also David Silvie, RA (who were sick men), for allegedly not working hard enough; he fractured Silvie's arm and then disallowed an X-ray, which meant that Allan could not treat him properly.

Allan's fiancée Ida (née Schuster) was in the WAAF, and she finally heard he was a POW via the Red Cross and wrote allotted cards to him.

Liberation

Eventually US aircraft came and dropped leaflets telling the POWs to draw large circles into which they would aim to drop food canisters; it was manna from heaven. Among the supplies was penicillin, but there were no instructions and Allan did not know how to use it. Lives would have been saved had he or someone else known. Finally, when liberation came, they were exposed to more danger, as they had to pass through Nagasaki just a week after the bomb was dropped. Allan says he will always be haunted by the terrible sights he saw and the sick and wounded civilians he was simply unable to treat.

Allan arrived home in November 1945; he lived at 20 Cathkin Road, Glasgow, and married Ida on 22 December. Later they had a family. He became a prominent and popular GP in Glasgow in a mainly Muslim area of the Gorbals, helping local people adapt to their new life, and at the same time defending Israel against Muslim assertions. He was an active member of the Jewish community and AJEX until he died in 1990. At time of writing, his wife is 92 years old and still in Glasgow, with a son in Israel (also a doctor), a country Allan loved and frequently visited.

The War Crimes Affidavit

Immediately after his liberation, Allan gave written testimony for the Japanese War Crimes Tribunal – in his case headed 'Fukuoka, 4B POW Camp, 29th November 1942, until September 1945' – to a magistrate, James F. Langmuir, at the Central Police Station in Glasgow. It has been included in this narrative and gives a formidable and shocking picture of what he experienced and saw.

On 12 February 1946, Allan's former CO, John (M? F?) wrote to him (the original letter is at the Scottish Jewish Archives), making the following comments:

> One thing I would like to tell you, and I hope you will pass it on to your wife, because wives like to hear this sort of thing. It is this. Many, many of the men of the regiment who have written to me since they got home, said how much they owe to you, and what a great job of work you did. They were not just 'blathering', they wrote with real sincerity and tremendous gratitude and several of them have said they hope to meet you again and tell you what they owe to you. Remembering as I do the splendid way you looked after us all in Glodok, especially during the dengue epidemic, it is easy for me to visualise the whole hearted manner in which you must have taken care of the men all the way through and I am glad to know that they really appreciated it. There were times when it was infernally hard to keep on one's toes, and make the effort which had to be made, and I congratulate you on a great job of work. I hope we may meet again one day.

T179739 Dvr Joseph (Joe) Bernstein, RASC

The source of this testimonial is a series of telephone interviews with Joe in October 2004, by the author.

18. T179739 Dvr Joseph Bernstein, RASC.

Joe was born on 12 March 1915 in Dumbarton, son of Sarah and Samuel, and attended Adelphi Terrace School in Glasgow, Queen's Park Higher Grade School, Battlefield School and the Gertrude Jacobson orphanage. He went to work at Goldberg's department store but in 1940 he volunteered and was sent to the RASC at Altringham. He married Rita Fearnleigh on a forty-eight-hour pass in October 1941 at Garnethill Synagogue, and they lived at 61 Cecil Street, Glasgow; Joe was then shipped abroad from Greenock, on the HMT *Duchess of Argyll*, via Canada, Cape Town and the Maldives to Singapore, a journey of eleven weeks.

Attached to the infantry in Singapore City, at the time of the surrender he came across a disused Bren gun carrier. He put his hand in the back of the carrier and pulled out a chitty signed several weeks before by his brother in Liverpool from where the carrier had been shipped ages before. The chitty was stained with blood and Joe presumed his brother had been killed. However, he discovered later that his brother had only cut his finger while loading the vehicle in Liverpool and he survived the war. Soon after, a sergeant major asked him why he was not in Liverpool, mistaking him for his brother – an amazing coincidence.

The British were ordered by the Japanese to march in order to Changi prison, where Joe found himself incarcerated in terrible conditions, as described in all the literature. On one occasion he had to sell his gold wedding ring to get 140 duck eggs on the black market which was run by the British officers. He was allowed to get them piecemeal over a period of time and they saved his life. To this day, he says, 'duck eggs are not my favourite food'.

A padre, Foster-Haigh, set up a seventy-man choir and orchestra (using instruments taken at the surrender from the army band) in Changi and offered proper training to six men who could sing. Joe could sing and was already well known in the concert party as 'Scotland's golden-voiced tenor'; as a result, if he sang for one hour on a working party, the Japanese would give a one-hour break to the whole working party. This enabled them to conserve energy and strength in the dreadful heat and conditions.

One working party (F Force, consisting of 600 men) consisted of recovering Ford, Chevrolet and Leyland engines in a huge open lorry park from abandoned Allied vehicles for the Japanese, and crating them up to be sent to Japan. At every opportunity, Joe and his comrades put sand in the sumps, plugs and other sensitive parts of the engines in order to sabotage Japanese efforts. This was very dangerous and, if they had been caught, would have resulted in execution. On one occasion Joe smashed up a Rolls Royce so that the enemy would not have it.

The guards varied. The educated Japanese officers were the most brutal and on one occasion Joe was struck across the face by an officer for no reason, with a hard, rubber, engine fan belt, breaking a tooth (at the age of 89 years, Joe still has all his own teeth, except the one knocked out by the Japanese guard sixty-two years ago), but, unusually, Joe testifies that the best one was a Korean guard called Kawahara San, who called Joe 'Joe San'. He would, at the risk of his own life, take Joe and his friends to villages to barter for food, warn them when Japanese were approaching if they were taking illicit breaks, and secretly give them war news – all punishable by immediate execution if caught. Joe would pretend to him that Joe's black hair meant he too was descended from Far Eastern ancestors. At the end of the war, about twenty POWs, including Joe, signed an affidavit on behalf of Kawahara to say he had protected them, in order that he should not be punished in war crimes trials.

After some time, Joe and some comrades were given the choice of going upcountry to work, being fooled by the Japanese that conditions would be better than at Changi. He accepted – this was, of course, a mistake, as thousands of Allied POWs soon discovered.

Several hundred were sent by train for about three days in terrible conditions, towards Thailand. Food was appalling and water came from the engine on the infrequent stops. The steel carriages were so hot that severe burns resulted if anyone touched the sides. Somewhere near Kanchanaburi they were marched in brutal conditions for 100 miles through the monsoon to the Death Railway and had to build their own huts with tools supplied by the Japanese. The guards and officers were mainly engineers and very brutal. Joe testified that some of the British officers were awful and sought only to look after themselves rather than take the lead in caring for their men; some documentaries on this issue have been made to support this allegation, which was largely hushed up at the time.

One great man, however, was the Australian MO, whose name Joe forgets. With a badly infected calf and unable to walk, Joe went to see him. He was laid on a table and told to look away for three seconds; the MO took a scalpel and, without anaesthetic, made the cuts which cured Joe, whilst others – who were waiting nearby to be dealt with – fainted at the sight (at this point Joe laughed heartily and said, in his lovely Scots accent, 'I can laugh now, but then the pain was awful').

Joe and his friends often sabotaged the railway work, but it became a problem because if a train was derailed or damage was found they were simply called out in the middle of the night to repair it. Every day was a terrible struggle, and Joe said that he survived by waking each day and saying to himself that someone was waiting for him and that he would go home, come what may. He was twice offered promotion in the camp to lance corporal, but refused because of the responsibility.

A good friend of Joe's was Canadian Frank Mills. One day they met in Frank's hut, and Frank said that he had had enough and was going to die. To shock Frank out of his depression, Joe got two friends with chunkles (a type of spade used in the Far East) and stood over him and said they were going now to dig his grave and would come back, and they went outside and started. Frank said: 'You are not going to bloody-well bury me!' He survived. Soon after the war, when Joe returned and worked as a major buyer for Goldberg's store in Glasgow, he noticed a man and woman in the queue outside his office. He went and asked what they wanted, and the man revealed himself as Frank Mills; the woman was his wife. They had come from Toronto where Frank had gone to live, and Frank remembered that Goldberg's was Joe's place of work. They spent the day together and were in touch until Frank and his wife died in 1998.

A constant feature of camp life was the propaganda talks, usually given to them by a Japanese officer, about how badly the war was going for the Allies. Like most of the men, Joe suffered constantly from dysentery, pellagra and malaria, but no work meant no food, so everyone had to struggle on; he seems not to have suffered any long-term effects. Evenings for the POWs were often spent reminiscing, to keep spirits up.

On one visit to the MO, Joe returned to find that all his kit had been stolen. When other POWs found out, they all contributed something; he remembers that one man gave him the all-important half-shell of a coconut as a dish – probably one of the most important and life-sustaining items a POW of the Japanese could have, as it was the receptacle for storing and eating food.

Meanwhile, Joe's wife did not know his fate and had only received the usual War Office letter saying he was 'missing presumed POW'. After some eighteen months she did get a half-burnt postcard saying he was alive in enemy hands.

After about a year-and-a-half, the men were shipped back to Changi, probably in summer 1944. The journey was again horrendous, over about three days – many of those who had been present on the journey there were now missing, having died on the railway as they worked alongside Joe. At Changi they were set to work building the airfield, accompanied by the usual brutal treatment and poor food. On only one occasion does he remember Red Cross parcels being given out and even then one between several men – the Japanese just stole them.

It was at Changi, however, where Joe became involved with the POW synagogue and Rabbi Chaim Nussbaum, a Dutch Jewish POW officer (see his testimony later in this book). The Japanese allowed the construction of the synagogue, and of various churches, and Joe was the obvious choice as *chazan* (cantor) because of his voice. Many non-Jews attended services too – there was only one prayer book, which Joe remembers Nussbaum had acquired – as a source of solace and inspiration, and even the Japanese came to seek advice from Nussbaum, whom Joe remembers as a great and spiritual man, who kept up the morale of all who knew him (see his book *Chaplain on the River Kwai: Memoirs of a Rabbi POW* [Ulverscroft, 1987]). Nussbaum was responsible for the idea of men giving 10 per cent of their rations for the sick in the hospital. Joe laughingly remembers now that 'It was not difficult to fast on Yom Kippur – we had nothing to eat anyway!' – but at the time it was horrific. The choir had non-Jews who would learn the Hebrew words in transliteration, and sing at the services with Joe – an Aussie (George Sprod, who made a drawing of Changi Synagogue which is now on display at the Changi Museum – see Figure 12), a Brit, an American and Joe the Scots Jew.

Once, Joe was in Changi gaol hospital, seriously injured, after a Japanese guard had hit him on the back of the head with a rifle butt for no reason; the British MO ordered him to hospital. He recalls Rabbi Nussbaum visiting him, covered in sweat, and giving him two priceless bananas. It was a life-saver, and typical of the man. They kept in touch for years after the war when Rabbi Nussbaum emigrated to Toronto. Joe recalls that all the Jewish POWs gave 10 per cent of their meagre pay from the Japanese to Nussbaum for the sick men. The Christian chaplains told Nussbaum that they could not get the non-Jewish men to do the same.

At Christmas, Padre Duckworth asked Joe if he would sing the solo at Mass. Joe asked if the padre knew that he was Jewish, and Duckworth said: 'Well, we don't mind if you don't!' Joe sang, to the portable organ played by Bill Williamson, 'Come All Ye Faithful' in Latin (he still remembered the words, and when I interviewed him he sang on the telephone to me at this point) and 'Silent Night', among others, and received a rapturous welcome; he said it was indeed a very moving occasion. It was held in the open air on a beautiful moonlit night and hundreds attended. The Japanese sat in the front row with their rifles between their knees.

Joe was one of the first to know of the liberation. A radio was hidden in a broomstick with two screws acting as dials. A Jewish Australian POW, Alan Rabinow – he was a friend of Joe's from the synagogue – was involved in this, and he woke Joe, after coming down three floors in the cell block to the huge crowded dormitory where Joe slept, and told him that the Japanese had surrendered; he had heard it from Delhi radio. But nobody was to tell the Japanese for fear of reprisals, and for a whole week the POWs knew and the Japanese did not. Work continued normally for a week until one day the Japanese disappeared – Joe says they knew that the POWs would have killed them if they had stayed. After the war, Joe received regular parcels of food from his Australian friend for many years.

Soon British and Chinese troops arrived, and on one occasion Joe remembers Lord and Lady Mountbatten arriving to see them. The thing on most men's minds was sex, and the opportunities in Singapore City were huge; but Joe and a Jewish POW friend, Jimmy Bines, had vowed not to indulge; all Joe could think of was food! After some days, Joe was shipped home in October 1945 on the Polish ship *Sobiesky* and arrived in Liverpool to be met by his brother-in-law, who was in the RAF, on 16 November 1945. Joe was the first man off the boat. He was taken to an MO at Maryhill barracks who accused him of having 'been with a woman' (that is, that he had a sexually transmitted disease); Joe knew he had not, and a second opinion revealed scabies contracted on the ship home. 'Some doctor!' said Joe.

Joe had a joyous reunion with his dear wife, later in Glasgow, after four years apart. He and she agreed that he had not changed at all and he had no need to spend any time in hospital, as many others had.

Joe had no knowledge of pension claims and never asked about a pension until the mid-1990s; he then received it, but they would not backdate the payments.

Until only three or four years ago, Joe could never talk about his experiences. He had terrible nightmares for many years afterwards. He would never go back to visit, as it is too painful for him, though Jewish and non-Jewish ex-POWs have gone, of course, including some in Glasgow. He only ever attended one reunion in Blackpool, many years ago, but found it an uninspiring drinking-fest, but he is loyal member of the FEPOW Association.

In the mid-1990s he was visiting his niece Sylvia Carter in Palmers Green, London. Talking about the war, she mentioned that her neighbour's father and father-in-law had both been Jewish POWs of the Japanese. They were Louis Weiner (aka Winton), who had been in the RAF, and Morris Tiller, a gunner. Joe agreed to meet Louis, and Louis asked him if he remembered him. Joe did not, but Louis told him that in Changi, Joe had given him two hard-boiled eggs and a roll and saved his life.

After the war Joe founded the Glasgow Jewish male voice choir, which continued for thirty years and raised huge sums for charity all over the UK; the choir's highlight was going to perform in Israel after the Six Day War in 1967. Having been *chazan* in Changi, Joe was also *chazan* to the Giffnock Synagogue choir for many years.

Joe had the 1939–45 and Pacific Stars and the War Medal; he claimed them only in 2000 after his family told him that he should do so. As for the Japanese, Joe always said that you simply cannot go on hating all your life.

Freddy (Elfrieden) Bloom, Civilian Internee No. 2669, Changi Women and Children's Camp, and 115204 Maj. Dr Philip Maurice Bloom, RAMC, Imprisoned in Changi Military Camp

Much of this account is taken from Freddy's book, written in the form of letters to her husband; the letters were written like a diary and never read again by her and Philip until published in 1980.

19. Mrs Freddy (Elfrieden) Bloom and 115204 Maj. Dr Philip Bloom, RAMC.

Freddy Bloom was born on 6 February 1914 in New York as Elfrieden Wenzel, the only child of Robert (a German consul) and Emmy (a journalist). Freddy graduated in psychology from Barnard College, Columbia University, in New York and then took a further degree in languages at Trinity College, Dublin, after her parents moved to Ireland.[1] She first married a medical student (named Gilbert) who became MO to a regiment in the Indian Army, and in 1939 he was posted to Malaya, where he died from pleurisy in Penang in 1940. Freddy suffered a miscarriage after a bout of dysentery. She then got a job as a journalist in Singapore, where she met Jewish army doctor Major Philip Bloom, RAMC, in December 1940 at a dance.

Philip Bloom was son of the president of the Durban Hebrew Congregation in South Africa, born 7 November 1904. He qualified as a doctor in South Africa but had been a gynaecologist in London. He volunteered as soon as war was declared in September 1939 and was sent in December 1939 to Singapore, and was later MO (major, RAMC) commanding all casualties in the General Hospital when war with Japan broke out. Freddy was working on the *Malaya Tribune* newspaper when she met Philip.[2] When the fighting worsened in January 1942, she was working as a nursing assistant in the General Hospital alongside Philip, who had become her close friend.

During one air raid, Philip heard that a bomb had hit the X-ray department, where he knew Freddy was working; he ran to the room and found her underneath a large cupboard that had been blown onto her by the blast, but she was protected by the bulk of a brave RAF orderly who had thrown himself over her to save her, and both were unharmed.

Philip told Freddy that she had to either leave or marry him. She chose to stay (much to the annoyance of the US consul, who told her to leave) and Philip went to order the marriage licence. A few days later, on 6 February, Freddy changed from her soiled nurse's uniform into a clean one borrowed from a colleague, and with two friends as witnesses they ran in between their work shifts to the registrar between air raids, Philip buying a sapphire ring on the way from an Indian street trader he knew. They married at the Registry Office; Registrar North-Hunt hurried through the service as bombs shook the building during the actual ceremony (he was killed three days later), and they returned at once to continue work at the hospital. Among those who congratulated them on the wards were half-naked survivors of the *Empress of Asia*, which had been sunk that day, who clutched towels around themselves and toasted the couple with mugs of tea. That night they honeymooned in Freddy's apartment. It was the height of the siege, and nine days after their wedding day they were both captured, and spent the next three-and-a-half years apart in separate but adjacent camps.[3]

As MO commanding the hospital, Philip had to make heart-rending decisions as to which wounded were to be evacuated by a designated ship, the *Wu Sueh*, and who to be left behind; the ship did get away to Colombo, one of the few. Among the many volunteers working with Freddy was the Jewish jeweller Mr Grand, whom she especially remembers carrying stretcher after stretcher, never once sheltering during the bombings.[4] Both Freddy and Philip saw many of their friends and colleagues die in the hospital.

After the surrender the Japanese expelled all Allied service wounded from the General Hospital and moved them to makeshift locations in the cricket club, the theatre and the Fullerton building, where Freddy worked and Philip was again in command – over 1,500 patients were moved. Hundreds of wounded were crammed into offices, lounges, kitchens, bars and so on; bodies of the dead were put on landings. There was so much to do; staff worked until they were exhausted and then slept. There was no water and the conditions resembled those of the Crimean War. Eventually, by the end of February 1942, the wounded were moved to the POW hospitals and Freddy went into the Katong women's camp. They were told to take clothes and food for ten days, and as many of those women who had been evacuated from the island had stayed at the Singapore Club, where Freddy and others were sleeping, there were lots of abandoned clothes and tinned foods to take with them, which had been left lying around the rooms. On 2 March they were taken in trucks, but not before Philip and other doctors had thrown mattresses into the lorries; they would need them. That was the last time she saw Philip for many months.

Prison

Freddy kept a diary in a linen hardback notebook throughout her captivity; it was more detailed at the beginning because as the pages began to run out her comments had to be rationed. At first there were 400 women from all kinds of backgrounds in the camp – wives of colonial administrators, planters, factory owners, nurses, teachers, missionaries and those who were simply too old to have left; most were European but some were Malays, Indians, Chinese or Eurasians, and there were also Americans and two Japanese wives. Although they were confined, the Japanese just laid down some basic rules and then left the women to run their own camp; how and if they survived was going to be up to them.

On arrival, the Japanese confiscated their cameras, radios and other contraband; there were many soldiers' wives and children whom Freddy recognized. There were only three large houses for 400 people, so overcrowding was appalling. But there were gardens, and some plants and vegetables already growing. Freddy and three friends found an old garage with a chauffeur's quarters and moved in together and fitted it out with furniture that they had found. The houses were near the sea and they were allowed to bathe; there were six latrines in the camp and daily rotated work parties to empty them; there was also running water.

Freddy's first work party was wood-chopping for the kitchen; breakfast was tea and a biscuit; lunch was rice and watery vegetable soup; then they had tea and a biscuit at 4 p.m. and a piece of bread with butter or jam at 7 p.m. Some had brought small stashes of tinned food. Most of the women wore men's clothes to work, especially in the vegetable gardens where night soil was used. Then suddenly on 8 March they were ordered to prepare to move to Changi. It was a 14-kilometre hike and many began to eat their tinned food stores as they were too heavy to carry. With their pitiful baggage, some with prams and carts piled with bedding and pots, they formed up in fives, with Freddy in the front row. The big worry was protection from the sun and needing good walking shoes. Freddy had a large black umbrella for use as a shade or walking aid, and a haversack with a bottle of water. She had trousers, a nurse's veil as a turban and a long white blouse. Five abreast and dressed in all kinds of haphazard clothes, the 400 marched off. However, the old and ill were driven and the guards allowed a stop each mile or so and helped with water. By noon, towels, newspaper, lampshades and anything else were used to shelter from the cruel sun. Locals watched in silence Though utterly exhausted as they approached the gates of the gaol, the women suddenly got into step and began to sing at the tops of their voices: 'There'll Always be an England'. The men in the neighbouring prison cheered them until they were hoarse.

Changi

Changi was all grey walls and iron bars and no greenery at all. Freddy and four friends went in search of accommodation (the cells they were offered were tiny – nine feet by twelve feet – and filthy) and a guard said that if they looked after the stores they could live in the large, shelved tool room. They soon put up their beds and covered the mattresses with various coloured spreads they had brought. Freddy placed a picture of Philip on her little shelf. A prison kitchen was soon set up but most food was cooked in the men's gaol and brought in for the women and children by the stronger men. After being served from huge drums, the prisoners had to sit where they could to eat, there being no dining area. The main meal was boiled rice, day in and day out, sometimes with some vegetables, for the weak and everlastingly hungry stomachs.

Meanwhile a little fox terrier dog named Judy had been adopted by the group. She had been at the hospital and taken by lorry to Katong and, despite orders to shoot all animals, had followed them all the way to Changi. Now she was the camp pet and loved by all, especially the children, but she always returned to the tool room billet to sleep.

One of the doctors examined everyone to see how the poor diet was affecting them; they were getting half of what they needed and a quarter of what they were accustomed to. Anxiety was high, as most did not know where their men folk were. There was little general news except the English-language Japanese propaganda sheet, the *Syonan Times*. On 21 March it was decided to hold elections for various committees in the camp; heads of buildings, floors and, of course, liaison with the Japanese. Bickering was inevitable because being on a committee meant that you allocated work to others. There was also discord about the Eurasian women who claimed to be wives of British soldiers and received extra rations for their children, whilst others were thus forced to accept less (there were about sixty children in the camp). Having children they were also given bigger rooms, but even though these had running water they did not keep them clean, and bed bugs became rife. There were fights and one woman was hospitalized. Freddy was proposed for office but refused and was happy being an official food server, spending her spare time sewing, nursing babies, cleaning lavatories, and on carpentry and writing. For four hours each day she helped in the hospital, where the main complaints were exhaustion and dysentery. It was known as the Infirm Ladies Annexe (ILA). Freddy applied her background in psychology to help people. She mainly cleaned and bathed the elderly patients. Women would bring a baby to her and say it was a 'Bloom' baby, delivered by her husband of course; this always brought a lump to her throat. On one occasion she was feeding a sick baby, and the next day was measuring it for its coffin; she said it was the grimmest experience of her life.

Women of like mind formed little self-help groups. Concrete walls and floors and iron stairways were swept clean with little brooms made of twigs; wood ash was used as a toilet cleaner and bleach. In the cells a variety of tins were used to store water for use for bathing, ensuring eyes were kept open for prowling Sikh guards. There was no toilet paper and there were no facilities at all for menstruation, though malnutrition meant that most women ceased their cycle during imprisonment. Relaxation and proper sleep were impossible as there were virtually no beds or chairs, just the floor; only after many months were some made by the men and allowed into the women's camp. As a result, fatigue was common. Five more women wormed their way into the crowded tool store and tempers became frayed. Some of the original group became depressed about their internment.

The Japanese asked for volunteers to work in a canteen in a hotel in the city. There followed an intense debate about 'working with the enemy', but some said it would be psychologically good to get outside the camp. It was decided that the camp liaison officer, Lady Heath, would ask the Japanese a series of questions about the work and she would also ask her husband, General Heath (in Changi POW camp), for advice. But the scheme fell through. The guards could sometimes be humane; on one occasion one proudly showed Freddy how to correctly wear her kimono; another helped them fold their laundry; another brought flowers to the hospital.

After a short while, Freddy established contact with Philip via a policeman named Hockenhull whom she had got to know when she was a reporter; he brought the food into the women's camp. He would leave a matchbox on a stairway and she would retrieve it; in the first one was a note from Philip saying 'We know you are here. Keep your chin up.' She would reply and leave the matchbox for Hockenhull to pick up on his way out.

Freddy soon became an indomitable leader of part of the camp, as there were factions which sometimes argued.[5] In March 1942 she was asked to set up a camp newspaper which she called *Pow Wow*, and the first issue appeared on 1 April; for eighteen months it became a major source of news, advice and a morale booster for the internees living in the most terrible conditions under Japanese rule. The paper varied in length from four to eight pages and appeared every Wednesday without fail, though later it became fortnightly. It had to be hand-typed by her team – usually ten copies – and was passed from block to block and seen by all the internees. Some copies even reached the men's camp. The women came from all races and classes but the newspaper had something for everyone and was a unifying factor among them, as they tried to keep in touch with some kind of normality in those terrible days. Essential contents included items on camp behaviour and rules, entertainment events, a birthday column and competitions. The birthday column especially was an inspiration as it gave internees the excuse to get together and celebrate with others and exchange small gifts. Barely did a few days go by without somebody having a small party. One edition, the ninth, was devoted just to the children, with nursery rhymes and puzzles.

The Japanese insisted on prisoners standing and bowing in their presence. One Welsh nurse was in the habit of bowing and then, as the Japanese turned away, making a rude gesture, which cheered those who saw it. When nurses were asked to work outside the camp for the Japanese they refused unless they went as an allied unit. Bring-and-buy sales were organized. Men from the Changi POW prison – which they could see from the roofs of their prison – were brought in to do the cooking and other heavier chores and the food improved somewhat. Husbands and wives got to know who was coming in the working parties and furtively contrived to meet and chat and secretly pass food to each other, as well as copies of *Pow Wow*.

Philip

In Changi Military Camp, Philip worked at Roberts Barracks, which was the chosen site for the hospital; equipment and drugs had been smuggled in after the surrender, from the outlying hospitals, by hiding them under the stretchers of wounded men in ambulances and lorries. Until water and electricity were back on line, conditions were appalling, but within weeks order was established. Except for occasional inspections, the Japanese left the Allies to run things themselves, with Sikh guards commanded by Japanese sergeants; the Sikhs took a sadistic delight in their new-found authority over the British, whom they deserted in return for promises from the Japanese of help to get the British out of India.

Philip's priority was to set up communications with Freddy; in this he was far from alone as there were many others in the camp with wives in the women's section. The arrangements began with police officer Hockenhull and others getting on working parties to the women's camp and dropping and retrieving messages to and from the wives. They felt great elation at outwitting the Japanese, and although many were caught and flogged, the communications never stopped, even though many attempts were aborted by messages being discarded in the bushes, if the carriers suspected they would be found out. With regard to Philip's own family, it was not until Christmas 1942 that his parents in Durban heard he was officially a POW and alive.

Soon Philip was commanding a dysentery ward in the mornings and reading and gardening in his spare time. At first some officers fared better because of their secreted supplies of tinned food; but in due course hunger became a great equalizer and all were malnourished, ill and miserable. Anyone with valuables could trade 'under the wire' with locals on the black market. On occasions Philip accompanied a Canadian doctor who crept out of the camp at night and climbed palm trees to retrieve coconuts, which were much sought after. Often Philip would climb the stairs to the top of the hospital block and gaze over the three miles to the camp where Freddy was, longing to be with her.

Freddy

On 6 April 1942 there was an Easter service and the hymns were sung with great sadness. The Japanese sent flowers and the Red Cross group organized a meal. Later, Freddy's group celebrated by emptying a tin of soup onto their bland rice meal. Some days later a department store in Singapore was allowed to bring in items for sale and there was much bickering over who could buy what; shoes were in particular demand.

People began to die of malnutrition and associated illnesses, but morale remained high. Noise could be intolerable: echoing walls and corridors, the gramophone, children, carpenters hammering, dishes clattering, and constant chatter. On 24 April, Freddy was in hospital with mild dengue fever; she was fussed over and ordered to rest in the tranquillity of the hospital for some weeks, and here she celebrated St George's day with some marmite spread, and friends brought her little treats, especially cigarettes which were now running out. It was painful to listen to the evening bugle in the nearby soldiers' camp, filling the women with both sadness and pride. The emperor's birthday was on 29 April and the prisoners were forced to observe a two-minute silence, standing at attention and facing Tokyo. The women had strategically placed themselves around the prison, and in the split-second stillness that followed the silence they all flushed the toilets, in a coordinated snub. A tin of pineapple between three people and a packet of washing powder between two were given as gifts from the Japanese. But during the day, probably in answer to the toilet trick, several women were slapped in the face by guards. Later a Japanese guard was caught trying to sell a bottle of whisky to some men, but they stole it from him and he slashed a man's face in revenge; he was given two years in prison by the commandant.

When somebody died, the women would try to attend the funeral, not only to get out of the camp, but also to make clandestine purchases from locals. On 9 May Freddy was taken by ambulance for an X-ray on her back to Kadung Kanban (Myako) hospital in the city. Never one to miss an opportunity, she secreted about twenty letters in her bra, hoping to give them to friendly people to pass on to the husbands in the men's camp – including one for Philip, of course. She rushed from room to room in the hospital until she found an RAMC officer and thrust her hand into her bra and gave him the letters to deliver. It worked, as she discovered later. At the hospital, she and the other patients from the camp were greeted warmly by local staff and given a lovely meal with white tablecloths in the nurses' home. She made friends there, who later sent food to her via the ambulance to the camp, and in one case hid ten cigarettes in a hollowed-out loaf of bread. Food in the camp contained virtually no meat or fish; even vegetables were rare.

Philip

Philip received Freddy's letter with great excitement but at a time when he was laid low with dysentery. Rats now infested the camp and were running about the atap hut in which he was living. But Philip's camp by now had a canteen, gardens and a piggery (urine was an important source of irrigation), and a university in which he took part both as teacher and student. Then, in July, all officers above the rank of colonel were ordered to camps in Japan and this was followed in August by the Selerang episode, whereby the Japanese ordered all Allied POWs to sign an undertaking not to escape – an act that was a breach of the Hague Convention. The British refused and so on 1 September the Japanese GOC of POWs in Malaya (Fukuye) ordered all 20,000 Allied prisoners into the confines of the very small Selerang Barracks, designed to hold about 1,000 men.

Philip was ordered to stay with the hospital patients as the mass of POWs were moved to Selerang; eventually the Allies had to give way and sign, because conditions in Selerang became so appalling that men began to die of disease (other testimonies in this book give more details on this incident – it ended after four days). That day, Philip was called by the Roberts Hospital commander, as the officers had been ordered to witness the execution of four British soldiers who had tried to escape; one of them had been taken from the hospital to be shot. Philip was angered and sickened by what he was forced to watch, as Sikh soldiers carried out the act in such a way that it took twenty-one shots for the men to die slow and painful deaths.

The Japanese had meanwhile agreed to pay the Allied officers Occupation Dollars for use in the camp to buy essentials for all the prisoners; it also meant they could transfer some to the women's camp, where it often made the difference between hunger and starvation. Great achievements occurred in Changi Hospital and many lives were saved by the ingenuity of the medical staff. Philip's part was played in the dreaded dysentery ward. He had only some salts and his bedside manner – which psychologically became hugely important to the patients' survival and made the difference between life and death. He was greatly angered when a man died for want of the most simple and basic drugs.

With food short in Changi, Philip worried that Freddy too was suffering, but her notes always assured him she was well. He also witnessed the severe beating of two officers by Sikhs for not just refusing to salute them, but making a rather rude gesture at them. The other doctors knew that Philip had a keen interest and background in psychology and they sent men with mental health problems – and there were many suffering such – to him for patient treatment. Many were worried about their womenfolk at home and the decline in their own sexual appetites and abilities; he explained that the constant problems of malnutrition explained this and that a return to normal life would resolve all such issues. Many were grateful to him for easing their burden.

Philip's officer group acquired some ducks and chickens via the canteen and began producing eggs in a makeshift coop, feeding the birds on snails and scraps of garden produce. Most eggs were sent to the hospital but they kept some back for themselves, which meant roughly one egg every ten days or so – something to look forward to. Occasionally one animal would be killed and roasted, which was a great and rare treat. With three others, Philip formed a supper club, so that at 10 p.m. each evening, by pooling resources, they could enjoy one small cup of coffee sweetened with gulam and one piece of bread (the Changi cooks could make it with maize, tapioca or rice flour) with a banana spread on it.

It was in Changi that Philip met an artist whose work he so liked that he bought a sketch from him and promised to pay him five guineas after the war; this turned out to be Ronald Searle, who became famous.

In August the hospital was ordered to Selerang Barracks; Philip, like others, was suffering from all kinds of nutritional illnesses which were just kept under control with various herbal and other remedies – yeast, a vile fluid made from lalang grass, hibiscus leaves, and so on.

Freddy

Soon Freddy and her friends had to leave their roomy tool shed to make way for the batches of new prisoners now finally being brought in from the city, having escaped the first imprisonments. Her new accommodation had to be one of the many old cells – number 30 A IV on the second floor. It was seven feet by ten feet, and there was a three-by-six-foot concrete slab, eighteen inches high, in the centre, with an Asian squatter (hole in the floor toilet) in the corner, and a tiny window, one foot by four feet, along the top of the wall – and all this had to be shared with someone! Whilst officially accepting this cell, she actually slept on the verandah nearby, refusing on principle to live in such a hellhole (the verandah also became her office for writing *Pow Wow*) but with some pilfered planks they made some shelves and covered all with coloured towels to give a more homely look and then put their beds in. She later realized that the cell was actually more private. Freddy continued her work helping the elderly for four hours per day.

Articles in *Pow Wow* often included tips on how the women could keep clean and neat; this was regarded as essential to morale and the fight against self-pity and was an act of defiance against the Japanese – the women refused to lose their dignity and pride. Brown leather shoes could be polished with the inside of an old banana skin; for white canvas, a little rice water with talcum powder could be used; hair should be plaited, or pinned up, and always with a ribbon. As a result, surviving photos of the women in the camp often show them as well turned out, hiding the reality of the psychological trauma, illness and hunger that they were actually suffering.

Bad feeling developed over one woman who was in charge of the Red Cross section in the camp. She was incompetent, failing to organize the hospital properly and making large profits in the 'bring-and-buy' corner, with nobody sure what she was actually doing with the proceeds, and she was always ingratiating herself with the Japanese.

Lectures began, as the blackout meant that less entertainment was possible by way of outdoor shows. Freddy led a talk about the post-war social problems they would face after victory. Discussions got heated. She also started taking art lessons. One woman gave a talk on how she developed her bust! As expected, there was a lot of talent among the prisoners and there were recitals and concerts. One singing group Freddy joined would gather within earshot of the men's camp and sing messages to them, but the Japanese suspected something and they were ordered to stop.

On 19 June, Freddy received a smuggled letter from Philip. Others received communications too. What everyone missed now were the simple things; a chair to sit in (backs ached for support of some kind, other than a concrete wall); forks to eat with instead of only spoons; seeing trees; having a bath. On 4 July, Freddy and five other Americans celebrated Independence Day with a small feast, deliberately beneath the flagpole of the Japanese Rising Sun.

On 15 July the civilian men were allowed to come and give a concert at the camp, on condition that communication with them was forbidden. It did not work, as the women clung to the prison bars and called to their husbands. As a result, one of the chief camp guards, Tokuda, ordered the bars to be boarded up. Freddy noted that lesbian feelings among some of the women had begun to develop.

On 10 August, Freddy was suddenly visited in her cell by Commandant Okasaki and two unidentified men who were obviously Kempeitai (Japanese Gestapo-type police).

They wanted to see all the *Pow Wow* files and search her cell. They took away the typewriter and paper; apparently someone had been seen in the hospital, by a Japanese guard, with a copy of *Pow Wow* that somebody else had illicitly sent in. As it was non-political, however, no action was taken. A few days later, Sikh policemen (part of the turncoat Indian National Liberation Army – INLA) replaced the Japanese guards. Freddy spent another spell of almost two months in hospital with an attack of beri-beri.

Meanwhile, the women could hear the men play cricket across the walls, Freddy did a series on 'Birds of Changi' in *Pow Wow,* and an arts and crafts fair was held, where the men and women entered items for judging and a good time was had by all, helping to take their minds off the awful conditions in the camps.

On 4 September Asahi became the new commandant. He was a former embassy official in London and he appeared to behave well and spoke excellent English. Each day ten women were delegated to take the camp dustbins to the main square and ten husbands met them, emptied the bins and returned them to the women; each day ten wives waited eagerly for their turn to meet their men and maybe secretly say hello or touch, while under guard. They dressed as well as they could for this event. One day, the Sikh guard was ill and fainted and there was pandemonium as the couples gathered round him but only in order to exchange news and kiss and pass messages. The women were also given permission to occasionally exchange parcels with their men, once they had been thoroughly searched each way.

On 29 September the women were allowed to go on a bathing party in the sea nearby, and on 7 October Red Cross parcels arrived and there was much bickering about who would get what, as the contents of each had to be shared between more than one person. Civilian husbands and wives were now allowed to write to each other in the adjacent camp, but not to military personnel. Later some meetings were allowed too. At the same time, at the end of October, all prisoners had to sign a promise to the Japanese that they would not try to escape, and air raid practices were introduced; some prisoners received letters via South Africa. A mock circus was held for the children in the camp and was a great success, with Freddy playing a clown. Hundreds of hours of work went into its organization. On 3 November they were allowed to send postcards home, but an order came that all water now had to be boiled for fear of disease breaking out.

News came of someone in the men's camp being badly beaten for trying to pass a message to the military camp, and four men were flogged because they had stayed out overnight from a working party to find some local women and were caught getting back in.

On 21 November an outbreak of diphtheria in the camp was announced and at the same time the Japanese made a search for a radio, as they were convinced by rumours that one existed. Freddy noted that the Sikh guards kept bursting into the women's cells, usually when they are washing; it became 'safe' to wash only after dark when many of the guards went off duty.

On 8 December the rainy season began and the corridors were awash and the roofs and walls all leaked. Umbrellas, where available, were kept up indoors while people slept. The green mackintosh sheet Freddy had was thus a godsend at this time, and ranked among her most prized possessions, along with a lavatory brush and a chair. People survived on the food only by being able to supplement it with Red Cross bits and pieces and private purchasing with hoarded money.

On 15 December it was announced that civilian husbands and wives could meet for two hours on Christmas day, and military husbands and wives for half an hour. Men and women in both camps began preparing gifts for each other and the children; the men were allowed to perform a concert in the women's camp. The Scottish women had a St Andrew's Day party. On the day, Freddy and Philip had thirty-seven minutes in the sun in each other's arms outside the walls. He looked gaunt but was well and gave her a gift of pyjamas; she gave him slippers. It was all a huge escape from reality.

1943

Over 250 new prisoners arrived in January and now it had to be three to a cell, but at least the camp had running water and lights and the concrete meant it was easy to keep clean, save for the constant plague of cockroaches. At last the corridors became officially permitted as living spaces. In Freddy's block the corridor was 120 feet long and seven feet wide and each person got a space five feet wide by eight feet long. Freddy and her companion had their bed and a small office space for *Pow Wow* in the corridor, which became known as 'Hurricane Alley' because of the way it funnelled the wind.

There was serious concern over the amount and quality of food reaching the children. By a supreme sacrifice by virtually all the camp women, the children received all the protein and vitamin food they needed, by the women denying it to themselves.

On 6 February it was Freddy's birthday and her first wedding anniversary. Friends brought Freddy lots of gifts – a housecoat made from a bedspread, hankies, a coffee strainer, tins of food, and so on. Philip later told Freddy that he celebrated by opening a tin of precious bully and beans and sharing it with his mess friends. Time for the women was becoming more regulated – keeping living spaces and bodies clean, cooking, sewing, cleaning common areas, for Freddy the *Pow Wow* preparation – and for free time, there was writing and sketching, bridge, practising French conversation with a group of others, reading and knitting and learning Braille. Often small groups got together for tea and coffee mornings, obtaining wild flowers and tablecloths to decorate the rooms for a brief respite of normality in the hellhole of the prison. Heated arguments were endless among some groups, often ending in physical fights, and rumours were rife about the end of the war.

On 12 February the civilian men were allowed in again to give an excellent classical concert; the ladies dressed up, and husbands and wives were allowed to meet for one hour. Freddy watched in envy. One nasty incident involved the stealing of jewellery and its sale outside. The culprit was found, tried and found guilty of what in the prison was a heinous crime (Freddy does not say what actually happened to her). One amusing story, however, came out of the men's camp. One prisoner received a letter from a friend saying 'I hope these Japanese bastards are not treating you too badly.' The Japanese censor, thinking he was being very clever, crossed out 'Japanese' and inserted 'Nipponese' instead.

By this time, the camp had split socially and politically into two – the Red Cross faction versus the Rest. But to counter this unpleasantness the Japanese gave permission for the inmates to keep ducks and chickens as a food source, allowed the men to give concerts and lectures once per week in the women's camp and gave permission for prisoners to walk within a restricted distance outside the walls three times per week. They were also allowed to bathe in the sea once a month, heavily guarded by Sikhs, and there they often saw and waved to Australian POWs, but were not allowed to speak to them. In April, Philip was able to send $10 to Freddy. Meanwhile, internees were planting shrubs and bushes all over the courtyards to brighten the greyness of the walls, and pictures were hung everywhere. Sadly, despite many receiving letters from home, Freddy had none and was quite depressed about what her family in the USA were doing on this score. She was, however, able to receive and send cards to Philip occasionally. She wore his pyjamas as a way of imagining he had his arms around her.

On 10 April, Freddy noted that over 100 Jews from Singapore were suddenly thrust into the men's camp without notice, and various privileges in the women's camp were removed – no more art classes or sea bathing or lectures from the men, an edition of *Pow Wow* stopped because of jokes about poor food and clothing, and everyone had to speak Japanese to the authorities – no English. Beatings and solitary confinement for minor infractions increased. On 6 May, Freddy and all the Americans were called to the commandant's office (now a Lieutenant Tominaga was in charge). A telegram from the American Secretary of State, Cordell Hull, was ceremoniously read to them, telling them they 'were not forgotten'. She also heard at last from her mother.

In May a suggestion was made that the Japanese should be asked if husbands and wives could live together in some of the larger houses on a rota basis of perhaps a week at a time. Amusingly, the older wives objected vociferously and said it was a kind of prostitution. Freddy and friends were absolutely flabbergasted at such a weird view and nothing came of it. Meanwhile, in June, Freddy had her typewriter returned to her by the commandant and food improved, but an edict arrived that all dogs in the camp were to be brought to the commandant's office to be shot. There was a huge outcry and a petition organized, with the children mobilized to sign it too. A minority of the internees supported the killing of the dogs, and in the end Judy the dog and one other were reprieved but two more were killed. The same bizarre minority tried to have smoking banned; this too was firmly voted down. In prison, every facet of life becomes hugely exaggerated and intolerance had to be challenged.

A week later the internees were allowed to send their third postcard home and food improved somewhat as Red Cross parcels were allowed in (the military POWs saw virtually none of these), but all electrical appliances were suddenly confiscated from the camp. The children in the camp had become very wild and naughty, so Freddy offered to care for them after their morning school lessons. Sixty-three children were a handful; Freddy does not elaborate on how she managed them. An unannounced search was carried out by the Japanese and various items such as cigarette lighters and candles were taken away. The women were allowed, however, to perform the circus show again, but this time for the men's camp. Freddy said that all the activity helped to keep people busy and keep away from the constant squabbling that went on.

On 1 July, Tominaga summoned Freddy to his office concerning something she had written in *Pow Wow*; thinking that she was going to be punished, she stuffed food about her person in case she was going to be locked up. In fact, he was smiling about an item she had written and congratulated her on her wit. Such was the mindset of the Japanese.

In August there was a cholera scare and everyone was inoculated in the women's camp; then there was a suicide, to make life more grim. In September the internees held another art exhibition but, significantly, any paintings of the prison were immediately confiscated by the Japanese (it is known now that they were very sensitive to the outside world knowing the squalid conditions in which they kept prisoners vis-à-vis possible war crimes). One boost came in August; a large number of Jewish internees arrived but they, with great foresight, smuggled $50,000 into the camp, which they promptly handed over to the general funds; this generous and civic-minded act saved a lot of lives in the remaining years of internment.[6] The Japanese allowed an issue of fabrics to the women and this led to a huge outburst of frock-making – excellent for morale; in addition, all the married couples were permitted a meeting.

The Double Tenth

On 10 October 1943 (known as the 'Double Tenth'), the newspaper stopped appearing because Freddy as its editor was arrested by the Kempeitai, together with others, suspected of fomenting revolt, of being spies[7] and of transmitting messages by hand and by a secret radio, and organizing sabotage in Singapore from the camp.

Buildings were picketed by Japanese with fixed bayonets, and Kempeitai officers in civilian clothes and carrying pistols raided the camp and began prowling round, led by a Lieutenant Kodama. Every woman was body-searched, all possessions were searched and an all-day roll call was carried out, with no food distributed; no explanations were given. This went on from 10 a.m. until 6 p.m. – many fainted from exposure in the sun. All recreation was stopped and fires for cooking were forbidden. Similar incidents occurred in the men's camp. Within a few days two radios were indeed found, but a third one never was.

Freddy was held for five months in prison and subjected, with her fellow prisoners, to many brutal interrogations in appalling conditions. There were beatings and electrical and water tortures while they were kept in filthy degradation (see the story, 'The Christmas Present', below). The aim was to get confessions to the spurious charges.

Freddy was arrested with about fifty other men and women – anyone who had any post of responsibility in the camps was a target. At first they arrested her cellmate, Katherine De Moubray, in error but then saw their mistake. Freddy was taken in an elegant car with an interpreter to the YMCA HQ of the Kempeitai and then driven onwards to Smith Street lock-up. Any questions she asked were waived aside; her small bag, her hairpins, her rings and the elastic in her knickers were all taken. This was done in disturbing silence. From other parts of the building she heard shrieks and groans, the like of which she had never heard before. She was not to know that at least one man who preceded her had already been beaten to death, and another badly injured. She was then taken down a narrow passage, twenty-four feet long and four feet wide, with cells either side with thick wooden bars along the whole length. The cell floors were seventeen by ten feet in size and about one foot from the ground; on the outer wall of each cell was a barred window, ten feet up, and about three feet by two feet in size. Outside each cell, rows of shoes were neatly ranged; Freddy counted fifteen pairs, as a sentry opened a low gate into the cell through which she had to duck as she was pushed in and told where to sit.

Two rows of silent, cross-legged men sat facing the bars, hands folded on their laps. She took a place near the wall. Three or four of the men were European – one she recognized was the charming Walter Curtis, a camp interpreter – and the others were Indian, Chinese or Eurasian. They all sat still, looking straight ahead as an armed guard patrolled outside every ninety seconds. Across the passage she saw another fifteen men sitting quietly facing them in another cell. One was a young blond-haired businessman named Perry whom she remembered from Penang; he winked in recognition and gave the thumbs up. Elsewhere in the building, the shrieks continued.

The cell was bare, with one European-style toilet in the corner, with no seat, and a running-water tap for flushing and drinking; from the ceiling hung a large, bare electric bulb. After some time her back began to hurt so she leaned against the wall but the guard prodded her with his gun through the bars and shouted. Later she heard a shuffling and two Japanese, naked from the waist up, dragged in an unconscious man. He was put in the cell opposite and two fellow prisoners gently placed him in the corner and then returned to their places.

Freddy told the guard, with Curtis as interpreter, that she wanted to use the toilet, and he asked a guard in the outer office; this man shouted back in English that she should use the toilet like everyone else. She rose to go and Freddy felt the men withdraw into themselves to give her the privacy that was not really there. Those opposite bowed their heads to avert their eyes. This chivalry never varied during the whole of her incarceration.

Freddy and Perry found a way to communicate by spelling out letters on their knees when the guards were not watching; this was especially useful when those who had been questioned wanted others to know what they had been asked and what they had said; it was also a symbol of resistance.

That evening there was a great clatter as a teenage Chinese conscript brought in a huge bin of rice and filled sixteen round, rusty tin dishes (five inches across by two inches deep) and slipped them into the cell. They ate with their fingers and afterwards washed the tins in the running water over the toilet. Later, hot tea was served and the conscript poured it into their tins as they held them out; Freddy's had rust-holes which had to be covered with her fingers while she drank, to save the tea; it was very hot so her fingers blistered. Without a word, a very ugly, bearded Chinese finished his tea, took her tin, covered the holes with his thick, dirty hands and held it for her to drink, like a mother feeding a child; he did this gallant thing for her for the next five months. He was a local communist leader and also a Catholic, and very tough, but very kind.

Later the tins were taken away and as it got dark the light was turned on and the window blacked out; the guard patrolled day and night. The light stayed on and they were ordered to sleep and lie down on the bare wooden floor, body next to body. Agonized cries continued from elsewhere in the building. Freddy noted that not all the men behaved well all the time, but those who broke had every reason to in the infernal prison in which they were incarcerated; the miracle was that so many were strong and also thoughtful towards her. It became clear to her that cruelty debases those who inflict it, and that dignity lies in self-discipline and not status.

Freddy described how she was aware of all the detail of what was going on around her and happening to her, but there was no fear as one normally knows it; it was beyond reality; the inmates all felt for each other rather than feeling independently. For weeks they sat immobile, the same routine day after day, relieved by her and Perry signing to each other. All kinds of bugs swarmed on the cell floor; the air stank of lavatories, unwashed bodies and foul breath. Only the visits of the interrogators broke the monotony, as a name was called and a prisoner would leave at any time of the day or night, and nobody would know when or if they would return. Sleep was fitful as the comings and goings and the screams of beatings from far-flung rooms kept everyone awake. This was the awful routine, day after day. Freddy herself, already ill when she entered the prison, became even more emaciated, not least due to the shock of the terrible things she heard and saw, as strong men were taken from their cells and returned broken and barely recognizable. She was convulsed by apprehension; would this happen to her and would she stand up to it?

By the end of December, the pressure slackened. Freddy got to know men under circumstances few other women have shared. It was always extremely difficult was for her to recount this period, and the sadism of the Japanese, because it was so painful.[8]

The Prisoners

Walter Curtis, the camp interpreter, spoke Japanese and used it to help them all. A devout Catholic, always reading his Bible, he was selfless and received many a beating when trying to improve the lot of his fellow prisoners. His family had got away to South Africa. His favourite Grace Before Meals was 'Praise God we do not have to eat this lot again.'

Dr Stanley was a tall, handsome man; he was beaten to death in the first few days following the most atrocious series of tortures. Another young doctor was finding it all too hard but was helped by a charmless but psychologically supportive young Scottish engineer. A cultured Chinese, Doctor Lim – very fearful and astonished at the constant humour and levity of the dogged Europeans – tried with futility to ignore the rest of them. He only laughed once, when the toilet overflowed its contents onto the floor.

Dr Fowey Clark, a Eurasian, had no idea why he had been taken from his home; he later died peacefully in his sleep. A Eurasian 18-year-old, Guy Webb, was an undertaker; he felt the hunger acutely and so filled his stomach with lots of water; he was a Larry Adler fan and mimicked his hand movements on an imaginary harmonica. Francis was very pro-British; he was a south Indian plumbing expert and hated the Japanese. He had helped smuggle radio parts into the camp as he drove the Japanese commandant's car. Everyone respected him.

Mahinder Singh was a big, bearded Sikh (who is mentioned in Freddy's story, below) and became a particular friend of Freddy's. At first they thought he was a stool pigeon (an informer); the guard forced Freddy to sleep next to him, to try to degrade them both. But he became her protector. One night she felt a movement inside her blouse; terrified, she woke, waking Singh too; it was a mouse. It also amused them to pretend to be deadly enemies when the Japanese guards were about, to make them think that their forcing the two together had worked.

He gallantly allowed Freddy to use his sacred Hair Comb which kept his topknot in place; only she was given this privilege, despite requests from the others to use it. They had long, bizarre conversations as a result of his poor English; he often mispronounced words, to Freddy's great amusement. In fact she later remembered who he was (his turban had been taken from him) – a young guard in the gaol who had helped her referee a children's hockey game some months before. When she was ill he took off his shirt and wrapped it around her; when he was ill she washed his clothes.

Another inmate was an elderly, emaciated Chinese drug addict with a tubercular hip who also became Freddy's friend; dressed in black and racked by a hollow cough, he had a wicked sense of humour (after the war he was the first to come to the camp and visit Freddy with a basket of fruit and eggs).

Others came and went, but one of the most memorable was Norman Coulson, a tall, craggy Geordie who was a water engineer from Durham, who with Francis (mentioned above) kept the water supply going in the camp. The Kempeitai accused him of all kinds of sabotage and he was badly beaten, tortured and flogged on many occasions; on return to the cell, bloodied and cut, he always winked at Freddy and gave the thumbs up. For several weeks he was taken every day and sometimes at night too. Once he was brought back unconscious and Freddy washed his bloody face, despite the bullying shouts of the guard. Norman never broke and was a deeply religious man with a wife and child. Freddy often read passages from the Bible to him, and he talked about his family.

When Freddy once menstruated, probably due to shock (as mentioned above, most – if not all – women stopped menstruating when in the camps as their bodies reacted to the awful food, disease and stress), Curtis spent hours explaining to the guards that she needed sanitary towels; eventually a roll of toilet paper was thrown in. Two months later, sanitary towels were thrown in, but she did not menstruate again for over a year; instead she used the towels as dressings for Norman's wounds. Brought back from questioning once, more dead than alive, he quipped that he never thought he would ever need those things! But in his hand he gripped a grubby aspirin which he had asked for because he knew Freddy had a toothache. Two weeks before Christmas, this gallant man was taken and she never saw him again.[9]

Most of the prisoners brought in were military, emaciated and had long beards; it was hard to tell them apart as they were hustled along the corridor or dumped unconscious into cells.[10] Philip, meanwhile, had been told that Freddy was on a working party in Singapore and that is why she could not sign his receipts when he sent money to her through intermediaries.

When Freddy was first taken for questioning, she froze as her name was called out – 'Belooem!' She put on her shoes and was taken to the YMCA interrogation centre. She was made to stand before an interpreter and interrogator seated at a table. She was shown a shopping list she had kept for two hospital patients, which the Japanese were certain was coded radio wavelengths, and also a Braille nursery rhyme that they were sure was a code. They did not believe her explanations; over and over they asked the same questions, to which she gave the same answers, never deviating from her story, but she was not tortured and was sent back after half-an-hour. Day after day she lived with the constant fear that at any time, as she herself saw, a prisoner could be called for many times and return unscathed, but then on the last occasion be sent back bloodied and beaten.

Dysentery broke out in the cells and everyone had to constantly and urgently use the toilet; there was neither time nor inclination to observe any discrimination of gender or timidity or fastidiousness – no cloth, rag or paper – for the lavatory was equally toilet, wash basin and drinking water. Bare hands with no soap had to suffice as the primitive essentials of hygiene. It was abominable. In all this foulness, Freddy's resistance to disease naturally diminished; skin disease took hold and spread, hunger grew and the beatings continued.

On 9 December she was called again; a sergeant smashed her on the head with his fist and she flew across the room into a wall; she stood back up, was questioned again and then sent back.

Philip

Philip was in a bad state, with tropical ulcers on his legs and bad swelling of his hands and feet; his medical colleagues ordered him to hospital where he stayed for seven weeks. Slit trenches were ordered to be dug and the Japanese stopped newspapers; this was understood by the officers as a sign that the war was going against the enemy. Friends joined Philip for his fortieth birthday, celebrated with coffee and coconut milk. He later watched an astonishing cricket game from his verandah. The wicket keeper had one leg; one batsman had an artificial leg and his substitute had one arm. When the one-legged batsman hit the ball, the other one ran for him. Among the fielders, one-armed men ran and the one-legged men threw the ball to wicket. Philip was moved with pride at the cold courage and spirit of these brave hearts.

A few days later the Japanese arranged a propaganda broadcast concert where some men were allowed to send messages home on the radio programme; balancing collaboration with need, at least some families would know their men were still alive. More worrying was that Philip had not seen Freddy's signature for some weeks on the money allowance he sent to her. Was she well, he wondered?

At Christmas he received the green pullover Freddy had made him, via Katherine de Moubray; he still did not know of Freddy's plight. The pullover fitted him perfectly and he was both seething with anger and choked with emotion; he had sent her some face cream and soap wrapped in a handkerchief. Meanwhile, as thousands of prisoners were being returned from the Railway to Changi, Philip was working intensely with the medical teams to help the diseased and starving men survive and to return to some kind of reasonable health.

Finally, in February 1944, after much badgering of the Japanese commandant, he received news that Freddy was in the YMCA lock-up with some other women and was 'well'. In May, Philip and the POWs were moved into the now-vacated Changi gaol where Freddy had been incarcerated (the civilians had been moved to Sime Road camp, as described later in this testimony) and by amazing coincidence discovered on the walls a crayon sketch by Freddy. He left it there to brighten the surroundings.

The prison was grossly overcrowded so immediately the engineers rigged up proper sanitation and soon things were as good as could be expected, to prevent the outbreak of the many dreaded diseases which would have decimated the men, who were by now emaciated and sick. The best buildings were used by Philip and his medical teams as a hospital for the sick. Philip also gave medical and historical lectures to the men at the 'university'.

Freddy

Sometimes a brutal guard would call a prisoner to the cell bars and then for no reason strike him with a rifle butt; another abuse was to force everyone, old and sick, to stand at any time of day or night and do exercises in the cell. Freddy would often have her hair roughly pulled and be searched, as guards were convinced she was hiding something. Her clothing was repugnant, her feet infected; emaciation meant it was painful to sit or lie down; sleep was impossible.

The prison routine continued until Christmas; their bug bites became septic and the wounds of men stank. On Christmas Day 1943, Walter Curtis gave Freddy a small, coloured, holy picture which he kept in his Bible. Perry and Dr Choo were taken and beaten black and blue. On return, as he grasped the cell bars to take off his shoes, Perry dropped a small piece of toilet paper wrapped round a sliver of soap, for Freddy. That night she washed her hair; Guy Webb the undertaker lent her his shirt to dry her hair, and the Sikh lent her his comb.

1944

In January the questioning slowed down, the guard stayed in the outer office and the inmates were allowed to sit where they wished in the cell, but the security officer continued his 'raids' into the cells, searching their possessions for so-called contraband. Cicely, an older woman and a doctor, became Freddy's friend and they protected each other and chatted, determined to remain strong for the sake of the men. Everyone looked ill: pale, dirty, covered in bites and scabs, emaciated, and some puffed with beri-beri. But Freddy kept her nails manicured using two slivers of fashioned wood pulled from the floorboards by Mahinder, and Guy Webb gave her a metal fly button for use on her toenails. These tools she kept in a piece of red cloth from Francis's shirt that he had given her; she kept all these items for years after the war. Occasionally a new Japanese sergeant would allow Cicely and Freddy to wash in the Japanese guards' washroom, which was dirty but seemed heavenly to them.

One day Freddy noticed that a Japanese had used a copy of the *Syonan Times* as toilet paper and not flushed the toilet; instinctively she and Cicely pounced on it and washed it clean of excrement to read about the battle at Cassino. On return to the cells, the news was passed round silently.

On another occasion, a particularly ugly sentry kept stopping outside the cell and stared at Freddy hour after hour during his duty. Suddenly he stopped, put down his gun and gave her some Red Cross chocolate. The other encounter was when a barrel-chested Samurai type was thrown into the cell opposite, wearing only a loincloth, for several days – for some military misdemeanour no doubt. Two or three times a day he would rise and go through a mimed sword fight in silence, save for appropriate grunting; it was unforgettable. He was beaten by his escorts before being taken away.

On another day, the sun shining on the glass cell window produced a brief camera obscura effect and reflected in colour, upside down, onto the cell wall, the view of the bustling street outside, giving them a momentary glimpse of the normal world. Then it disappeared as suddenly as it had come.

Freddy was by now very swollen all over her body with beri-beri; Walter and Cicely asked the Japanese in vain for medical help. Then, in March 1944, Freddy collapsed. Japanese guards appeared and brought her hot, sweet coffee. Her trousers were removed and she was wrapped in a sarong; the Japanese then brought her vitamin B pills, which were administered by Cicely, and within a day she felt better and the swelling receded somewhat. On 24 March she was called to the office and was told that she could return to Changi. 'You are not well', the officer said. 'This is not a healthy place', she replied. Next day she and about a dozen others were called out; those left and those leaving smiled at each other and all said they would meet soon and have a party. In the office they were given back all the things taken from them five months before, and driven in silence in an open lorry to Changi.

She and Cicely received a wonderful welcome. Children shouted: 'Mrs Bloom's back!' Katherine de Moubray wept as they hugged. Freddy was sent straight to the camp hospital and slowly recovered; she had lost several stone in weight.

Her delight was that Philip was well. Sometimes military POWs would be seen working in fields outside the gaol, and stones and sticks would be laid out to spell names and the wife or girlfriend would rush to see their man from a top-floor cell, and wave through the bars. News got to Philip in Changi that Freddy was trying to contact him. So some days later Philip arranged to be on a working party near the women's gaol. Freddy saw 'Bloom' spelt out on the ground. She was helped to climb the stairs and she saw Philip some seventy-five yards away.

In May she was transferred by lorry with all the civilians to Sime Road gaol in the city, a former RAF base but now very run-down – but here they at least had a chance to garden. There were trees and flowers and open vistas. The camp was on a hill and separated from the men's huts by a fence. Freddy's feet remained infected, but food was better and there were weekly concerts. On 18 July, Norman Coulson's body was brought for burial. The imprisoned Bishop of Singapore was able to hold a special memorial service for him, which Freddy was able to attend. Terrible news came of others who had been with her in the cells, who had died since, including the brave Perry.

Freddy and Katherine were in No. 1 hut, with the usual atap sides and tin roof, with a tiny kitchen attached for all the inmates who were sharing. The men prepared the food which was brought for distribution by a trolley three times per day. Food was not as plentiful, and the big problem was rat infestation.

From about August 1944, Freddy began taking classes with the children again; she focused on the older girls to help prepare them for outside life. Tension in the camp mounted in November as flights of Allied bombers were seen.

1945 and Liberation

By January 1945, Freddy had received cards from her mother, and small amounts of money came from Philip. But inmates continued to die of malnutrition-related diseases and there was an increase in mental illness as the strain began to tell; some indulged in alcohol production using fermented rice and sugar. In April, the wives and children of the Jewish community were suddenly sent into the camp and later a malaria and dysentery epidemic broke out. On 17 April the Japanese killed Judy the dog, alleging that food shortages meant that keeping pets was no longer viable; this wanton act caused much upset in the camp. On St George's Day they received the second Red Cross parcel in three years, and on 11 May they heard about the German surrender and received yet another Red Cross parcel. Clearly something was in the air, and in June the number of prisoner roll calls was increased.

Meanwhile, in Changi, Philip noted the start of a rabbit farm to supplement food and also that one soldier had invented a way of using wood ash and coconut oil to make soap and toothpaste; rubber trees were tapped to repair boots and make bandages. In April he was made senior medical officer at a new camp known as No. 2 X5, just outside Changi, with a hospital of 150 beds. The death rate was high as only the sickest were sent here. More and more prisoners were transferred to No. 2 X5, and dogs, birds and rats were added to the cooking pots in desperation. Philip and other doctors, at the end of their strength, were all ill, but worked on. They were hanging tenaciously to life and some patients even ate their skin ointment as it contained palm oil.

At Sime Road, by August, illness had increased, especially among the elderly, as the food crisis worsened. By 21 August there was still nothing official about the war being over but news had been trickling in. Some guards disappeared and others began ingratiating themselves with the prisoners. Red Cross parcels and held-back letters suddenly appeared.

20. Women internees digging trenches at Sime Road Camp.

Finally some British Paras with red berets arrived at the camp; they warned everyone to stay in the camp as there were still thousands of Japanese soldiers on the island. One was so shocked at the state of the prisoners that he seized a Japanese guard and broke his arm in front of Freddy.

But nothing would stop Freddy seeing Philip, and so she and Katherine put on their best dresses (especially kept for this day) and crept under the wire. They brazenly hailed a cab and told the astonished but amused Chinese driver to take them to Changi gaol. The dilapidated vehicle bumped along and on arrival they paid him in good old Straits Dollars (also carefully hoarded for this day). From the gate they started to walk towards the hospital; POWs in loincloths and shorts approached them as if they were from outer space, staring and shaking their hands. 'First white women in 3.5 years!' they all said. This happened repeatedly as they made their way along the pathways, much to their amusement.

On reaching the hospital, Freddy found that Philip was away and he was sent for; meanwhile she was shown to his bed in his quarters which he shared with six other doctors; she looked at his bed with its clean but tattered sheets and began to sob; then he appeared and took her in his arms. Knowing they were both safe, Freddy returned to Sime Road in the rickety taxi. Some days later they met again and took a taxi to Freddy's old flat in Chatsworth Road, but the flat had been stripped; they then went to see Goh Sai Poh and her family; she had risked her life in smuggling food to Freddy when she was interned. They gave them money and the two families stayed in touch and visited each other for years after the war.

Soon they were on the SS *Monowai* and on the way home with other emaciated service personnel, via Colombo; here they were welcomed by hundreds of small boats with bunting flying and sirens sounding; onshore, church bells rang and car horns hooted. The Red Cross and other groups welcomed them with food and clothes and toiletries. Then on to Tewfik, Egypt, and finally Liverpool where they spent their first night together in an attic room, separated from another army couple by twenty beds! But they were more or less alone.

Epilogue

Freddy always said that she never held any grudge against the Japanese, despite her near-death, and in 1994 accepted an invitation to visit the country as a gesture of reconciliation.

Penniless, she and Philip moved in 1945 to London, where Philip became an obstetrician and psychiatrist, while Freddy set up the Youth Book Club to encourage teenagers to read good literature. She had a daughter, Virginia (Ginny), who was born deaf due to the terrible effects of malnutrition on Freddy's body during her incarceration. But Freddy worked with her daughter against all the conventional wisdom, using her energy, wisdom and experience to set up the National Deaf Children's Society which changed beyond recognition the attitudes of the public to deaf people and enhanced the lives of thousands born with the disability. With her unstinting devotion to Ginny, the child became fully articulate and this work has benefited tens of thousands ever since through Freddy's internationally recognized work as a broadcaster and author on the subject of teaching deaf people, until the day she died. She was awarded the OBE for this work in 1966. Freddy and Philip also had a son, William.

In 1980, Freddy published a book on the story of her life as a POW, told in the form of letters to Philip (which she was never able to post, of course). It was called *Dear Philip* and it became famous.[11] They had never looked at it for thirty-five years after the war. They stayed in touch with former POW colleagues always. Philip died in 1984; Freddy died in May 2000.

The Christmas Present, by Freddy Bloom (1949)

Singapore, Christmas, 1943

I woke up and watched them drawing back the last of the blackout curtains from the outside. This meant that soon the glaring lights that had been on all night would soon go out. I tried to stretch my legs and kicked Reddy. He did not stir, and vaguely I wondered whether he was unconscious or just sleeping the deep sleep of exhaustion. It did not really matter; in either case it would do him good.

I turned my head and looked at the Sikh who was lying next to me. Poor Mahinder Singh. When they had brought him in he was tall and strong and magnificent. Now he was tall and skinny and just very young. His beard and long, coiled hair were no longer sleek and shiny. His smooth brown skin had a greenish hue. Perhaps dark skins always turned green when fair skins turned pale.

I looked at my own hands. They were quite beautiful. Slender, smooth, and white, with nails that I had carefully bitten into a good shape.

Just as well that I could not see the rest of me – except my feet. I had seen so much of them recently. They had made us all take off our shoes as we entered the cell. None of us had stockings or socks. Who would think of wearing such things after 22 months of internment? Anyway, most of the women in Singapore had gone about bare-legged even before the Japanese came.

I thought of the silly line of shoes outside the cell. Sixteen pairs, all shapes and sizes, but mostly trompahs, native wooden-soled, one-strapped sandals. My own were real shoes, white leather with crêpe soles and they would have to last me until the war was over. I certainly was not wearing them out at the moment. Perhaps they would not fit the next time I was called out for questioning. The last time there had been difficulty putting them on. I looked at my feet again. They were white and fat and dimpled, like a baby's. That was beri-beri. The shoes would have to fit. Everybody always made a proper business of putting on shoes when called out. It gave you a chance to do something outside the daily routine, and a chance to collect your wits and fight the blue funk that filled every inch as you thought of the questioning to come.

'We Would Show Them – but What?'

This was Christmas Day. Perhaps nobody would be taken out for questioning. What a hope! The Nips would pile it on thick just to show us. This was Christmas day and we would show them. This was the Kempeitai, the Japanese Gestapo, and we were their prisoners. Oh well, it was Christmas Day and I was going to wash my hair.

I looked across to the WC in the corner. Dr Choo was washing. He was always washing. Washing had become an obsession with him. Other things had become an obsession with him, but it was hard to know what they were for he talked so little. Of course, the Nips did not allow us to talk, but he was the only one who did not disobey them. He was scared. So were we all, but it affected each in a different way. He seemed to sink more and more into himself. The other day, when Mac (Macintosh) had accidentally bumped into him, he had almost shrieked, 'Don't touch me. Don't touch me. I'm superstitious.' Mac had barked back, 'I'm no devil, fool.'

Mac was a dour Scotsman. Imprisonment had made him more dour, more Scottish, and somehow, more of a man. Mac was hard. Dr Choo was not hard. He was just a hard-working Chinese doctor who had suddenly been torn from his work, his wife, and his children. He swore he did not know why. I looked at the WC again. Dr Choo was using it. I looked away.

Mahinder Singh woke up. He rolled his head on his stiff neck and started to massage his body. He turned to me and smiled. 'May Christmas, Memsahib.' 'Merry Christmas, Mahinder.' 'Today Kismas. Today good day. Today all go home. Catch big eating.'

After 75 days of the Kempeitai, Mahinder was still the complete optimist. He was my friend. We had played a great joke on the Japanese. This joke was our bond; it was our strength.

Mahinder and the Mouse

They had brought the young Sikh policeman in a few days after me and they had insisted that he sit next to me during the day and sleep next to me at night. This, they reckoned, would be most degrading to the proud Sikh and to me, a white woman, the only woman in a cell with fifteen men of different races. Degradation meant nothing to me, but at the start I was a bit frightened. I tried to stay as far away from him as possible, but that was not easy, with sixteen people in a cell 10 feet by 18.

The first night we lay next to each other I awoke because a mouse had slid up between the warped floorboards and slipped up my shirt. My femininity had never included a horror of mice, but to have one scuttling round my ribs under my shirt was another thing and I sat up with a start. We had to lie so close, that this woke Mahinder, who sat up too. Just then the mouse leaped out of my bosom and scurried away. The young Sikh put his hand on my shoulder, smiled a lovely paternal smile and whispered, 'Thik hai, Memsahib. It is only a little mice.'

At that point the Nip sentry on night duty screeched at us and we both lay down quickly.

From that moment we were friends. We never let the Japs know. When they were near, Mahinder and I would play at being enemies, and this was the best game we had. When they beat him up, I nursed him. When I was ill, he took off his shirt at night and put it over me.

Mahinder's greatest gesture was when he lent me his comb. The Sikh religion forbids men cutting their hair, so they all have beards and silky tresses, which often reach below their waists. The hair is then combed up, twisted, and coiled into a knot on the crown of the head and held in place with an adroitly inserted small wooden comb. Ordinarily a turban is worn. They had taken away Mahinder's turban in case he tried to hang himself with it (they had removed the elastic from my knickers for the same reason), but had left him his comb. After we had been together for some days, he watched me trying to comb my hair with my fingers, and then, to my utter astonishment, he handed me his comb. His gesture meant a great deal.

Thereafter he lent it to me every day, but he would never lend it to anyone else. Mahinder had proved himself, as mentioned, my friend and we were wishing each other a merry Christmas. Then we both looked towards the WC. That WC! How our life revolved about it. It was no ordinary toilet. It had no wooden seat, and instead of the ordinary flush arrangement, it had a tap that could be turned on fast or slow. When the bowl was full, it automatically drained itself. This was a very practical arrangement, and for a time I thought of having a similar fixture in our home in London, until it occurred to me that in England we do not use toilets for such varied purposes. In the Smith Street lockup in Singapore the WC was used for washing, for drinking, as well as for our basic needs.

Initiation

When they first flung me into the cell with all these strange men, I had sat for some hours until it had become imperative to catch the attention of the sentry who marched up and down outside the bars that formed one wall of the cell. Since he spoke no English, I pointed to the WC and then to myself. He obviously understood, for he went out and spoke to the corporal on duty, and within a few minutes a Japanese voice bellowed, 'If the woman wishes to wash, drink, or pass water, there is a toilet in the corner of the cell.'

The embarrassment of the men about me was so strong that it broke through my own. When I eventually got up, each man seemed to try to disappear within himself. Now, after so many weeks, we all took each other very much for granted. Across the cell Walter was reading his Bible. Walter was, in some respects, a privileged prisoner. He was an Englishman who had lived for years in Nagasaki, spoke fluent Japanese, and understood Japanese customs and manners. He had been our camp interpreter and done a wonderful job until he had been arrested for running a strictly forbidden radio. Now he still acted as interpreter in our urgent demands for medicines, foods, and the barest essentials of life. Most of the time his efforts were in vain, but any vaguely human touches that were ever shown us by the Nips were almost always due to his efforts. He was a deeply religious man and had somehow obtained permission to bring his Bible into the cell. As I watched Walter, he looked up and smiled.

Though there seemed to be a rule that a sentry should be on constant guard, marching up and down, passing each cell about once a minute, this had slackened considerably during the past few weeks and was only in force when higher officers were expected at our lockup. Ironically enough, the headquarters of the Singapore Kempeitai were housed in what had been the YMCA. This morning the guards were all busy with their own affairs and paid no attention to us.

The First Present

Walter got up and came to me. 'Merry Christmas, and here is something the postman left for you.' With care he took a little picture out of the Bible and gave it to me. It was the type of highly coloured little holy card that is often given to children in Sunday schools. Perhaps one of Walter's children had put that particular one in that particular Bible. Now he was giving it to me as a Christmas token. I held it carefully; it was very beautiful.

And then our first meal of the day arrived. Choy, the young Chinese conscript, clanked down the corridor with a high stack of tin dishes, the size of small cake tins, and a bucket of rice. He stopped outside our cell, counted the number of inmates, slung a dollop of rice into that many tins, unlocked the small door in the barred wall, and slid all the dishes on the floor. The men sat down in the places they had occupied all day. I got up and handed each man a tin. This had been my duty ever since the beginning, when there had been squabbles. There never was very much rice.

We all ate meals with our fingers; there were no utensils. When we had finished, we waited for Choy to come back with the tea bucket. Then we lined up at the door while he poured tea (no milk or sugar, of course) into the tins. This led to difficulties and contortions, for many of them were rusted and full of little holes, and we had to try to hold the tins so that the fingers plugged all the offending holes. Since the tea was often boiling hot, my fingers used to blister, and often one of the men would hold the tin for me while I drank.

Greetings to Cicely

One of my cell-mates was a Chinese Communist called Tang, who was the head of one of the up-country resistance movements. He was short and squat and, unlike most Chinese, grew a thick stubbly beard. He was the toughest man I ever knew.

No matter what the Nips did to him, he never changed his expression and never made a sound. When anybody tried to nurse him, he just shook his head and said, 'Tidapa', Malay for 'Never mind.' He spoke no English. Sometimes I thought that I would rather be a white woman in Japanese hands than a Japanese woman in Tang's hands. He was tough ... and he was always the first one up to hold my tin when the tea was hot and he would tilt it carefully like a mother feeding a child.

Remembering Christmas, 1943, there is a great deal of sentiment but no sentimentality. We were a group of extremely mixed people sharing a most unpleasant experience. Some showed up better than others. As far as I personally was concerned, there was not a man, European, Indian, Chinese, Eurasian or Malay, who was not thoughtful and kind and they had a great deal more than my presence to think about.

That particular breakfast differed from the others in one respect. Three or four cells down was Dr Cicely Williams, another woman and a good friend. They had brought her in the day after me. We had seen each other on one or two occasions since then. As far as we knew, we were the only women who had been taken from Changi Camp. While Choy poured out my tea, I whispered, 'Christmas greetings to the other lady.' He did not bat an eyelid. Later, when he collected the empty tins, he growled, 'She say you too.'

I was just going to the WC to wash my hair when there were heavy footsteps in the outer office. Quickly, we all sat down in a straight row facing the bars, knees bent, feet tucked in, and waited. We spent at least 14 hours a day in that position. To straighten your legs was considered very bad form. And so we sat on Christmas Day as we had sat for weeks before silently, all in a neat row, looking through the bars into the corridor, and then through another set of bars into the cell opposite, where a similar row of miserable beings faced us. The thing to do was to wiggle into a position where the gap between one's bars coincided with the gap of the person opposite and one had a clear, if not large, view for signalling.

Sign Conversations

Opposite me sat Perry. We had known each other before the war in Penang, where we had played games at the same club, danced at the same hotels, and been invited to the same cocktail parties. The next time we saw each other was through the gap in the bars. It did not take us long to work out an inconspicuous sign language and we spent the tedious hours having long conversations. We also passed on information about what had happened in various interrogations.

Christmas morning we wished each other all the best and while the sentries marched up and down we made rude remarks about them. Suddenly the noise in the outer office increased. Three of the Nip big-shots stamped down the corridor, followed by a drip of slouching, arrogant interpreters. They looked at us the way a person looks at a harmless beetle before stepping on it. They called out a few names in front of cells farther down, then turned round and stopped at our cell. We looked into space, our hearts pounded, there did not seem to be enough air to go round. They called Dr Choo's name and turned to the cell opposite, where they called for Perry and two Chinese. The rest of us relaxed.

Cell doors were opened and those who were due for questioning got out and put on their shoes. Perry held onto the bar in front of me while he put someone else's trompahs on his swollen feet. He wiggled his fingers at me and before he left he winked.

We continued to sit, looking straight ahead. Only those who had been badly beaten and tortured were allowed the luxury of lying down. When the Japs were working on a man, he never got much rest.

I wondered about Perry, about Dr Choo. We had heard some cars leave. That meant some of the prisoners were being taken to the YMCA for questioning. That was bad for they were very thorough. Others were being questioned here. The noises that went with questioning were too familiar by now. It is almost impossible to identify voices under such conditions, and yet one cannot stop trying.

We sat until the second meal, just like the first, was brought round. We put a tin of rice aside for Dr Choo, and it was eyed greedily, for when a man has been questioned he has either been given food or he is in no condition to eat. In either case the cell may share his ration.

Perry Comes Back

After 'lunch', we sat down again. I wanted to wash my hair, but thought it better to do nothing until the big-shots had left. After all, it had not been washed for ten weeks; Boxing Day would really do as well as Christmas. We sat. A couple of people from cells farther down were brought back. They did not look too bad. We sat some more. A scuffle outside and two interpreters dragged an unconscious figure down the corridor. We could not be sure, but he seemed to be Chinese. We sat some more.

Hard to say what time it was when Perry came back. He seemed a bit stiff and his face was bruised, but not too bad. He did not look at me, but as he bent to take off his shoes, he held the bar nearest me. As he bent, his hand slowly came down the bar. When he eventually let go and turned to enter his cell, there was a tiny parcel on the floor in front of me.

It was not until a good deal later, when most of the prisoners, including Dr Choo, had returned and the Japanese officers from headquarters had left, that I could examine the parcel. It was a single sheet of toilet paper, and inside was a sliver of real soap. They had allowed Perry to wash up after his interview, and he had stolen a Christmas present for me. Before the third and last rice meal of the day, I took my precious gift and, with great ceremony, washed my hair, with soap in the WC, and a Eurasian lad lent me his shirt to dry it.

And then, of course, there was Mahinder's comb ... Many years have passed since then. Most of the people who were in that prison died. I was lucky. We are back in London and since then have had two wonderful babies. Looking back to Christmas, 1943, I remember that was the day I washed my hair and Walter gave me a holy picture.

[The original illustrations were by the legendary artist and cartoonist Ronald Searle, as published in Leader Magazine. *This moving story contains pejorative language directed at the Japanese, who are today respected allies – please remember that it was wartime, and that the persons involved suffered greatly.]*

Notes

1. Obituary in *The Guardian*, 22 July 2000.
2. Freddy also edited a women's magazine which was used for Allied propaganda, and this may have had something to do with her later arrest by the Japanese on the Double Tenth.
3. She was in Katong women's camp and later Changi women's camp, and he was in Changi POW camp.
4. Leslie Bell, *Destined Meeting* (Odhams, 1959), p.39.
5. Dorothy Dixon, granddaughter of an English internee, whom she does not name, wrote that her grandmother and Freddy were rivals (see www.tighsolas.ca/page757). There were accusations of hoarding Red Cross parcels and breaking censorship rules, and of being hidebound by too many committees; indeed, researcher Sibylla Flowers, who knew Freddy, said that when she was released in 1945, she made a vow 'that her lips were sealed as regards Changi and its occupants', as some people – not surprisingly, she implied – behaved badly (email from S. Flowers to the author, 19 May 2011).
6. See *Destined Meeting*, p.138.
7. The arrests followed the successful raid by Australian Commandos on the harbour in Singapore which sank six Japanese ships.
8. *Destined Meeting*, describes, using extracts from Freddy's own account, and in somewhat idealistic terms, the war experiences of Freddy and Philip.
9. His body was bought for burial to Sime Road gaol, where Freddy was by then incarcerated, on 18 July 1944; he had been beaten and starved to death.
10. Chapter 15 in *Destined Meeting* contains the most horrendous descriptions of the tortures inflicted by the Kempeitai on many of Freddy's male cellmates; it is necessary but gruesome reading for those wishing to have an insight into the Japanese methods.
11. Published by Bodley Head in 1980.

217704 Pte Jack Bloom, 55th Infantry Brigade, RASC

21. 217704 Pte Jack Bloom, RASC.

Jack was born on 9 December 1909, son of A. Bloom of 18 Tudor Road, Hackney, London. After postings to Bulford, Galashiels and Lichfield, he was sent to the Far East. He wrote the following piece for the AJEX journal, some time before he died.

Deep in the Thai jungle, 110 miles north of Bangkok, a cold monsoon dawn breaks over a POW camp. I open my eyes and listen to the heavy rainfall beating down onto the atap roof of our bamboo hut. Above the din I hear the Buddhist priests chanting from a nearby temple. Maybe they are praying for us.

Inside the hut 250 men lay side by side on long bamboo beds with barely an inch to spare, some covered with well worn blankets, others making do with rice sacks. My thoughts turn to the day we surrendered in Singapore:

'I buried my gun in the evening sun
Thought of mum, her only son
Grabbed my chum, stared to run
Faster, faster to the docks
Cockneys and Jocks, all were there
Alas in despair, what no ships?
Goodbye fish and chips!

The jungle begins to stir, monkeys dominate the early risers, flocks of geese braving the monsoon rains, swoop low over the camp, skimming the swollen Mai Khiong (Kwai) river. Dozens of rats invade the hut, scorpions sleep under the beds, bullfrogs croak, mosquitoes retreat to the drains, unloading their deathly offspring. A King cobra slithers into the hut, rears its head. I have never seen men move so fast, shivering, swearing at the unexpected cold shower. The snake paid for its folly!

Breakfast time, we make a dash for the cookhouse and on a small bamboo table stand three 4 gallon petrol cans. First I dip my rusty mess tin and mug into boiling water, then the head cook fills a half pint cup of rice, levels it off with a small bamboo cane and tips it into my mess tin. Tea or coffee optional, sugar and milk out of the question – for three and half years!

By the time I get back to the hut the rice is a wash out. Fortunately I have some Thai dollars that enable me to buy duck eggs, bananas and local tobacco. The black market in these operates at night. We leave camp at the risk of a severe beating if caught, make contact with the Thai traders, buy stuff and re-sell at profit in the camp; this is to survive.

At sick parade a long queue waits patiently to see the MO – complaints include dysentery, malaria, jaundice, beri-beri, typhoid, leg ulcers – anything to get out of working on the horrific railway, but with no real drugs except some quinine, only the most serious cases are considered.

At the cookhouse we collect our packed lunches – rice, dried fish, tea in the waterbottle, starvation portions. We parade, are counted, given tools and march out with two armed Korean guards. Heavy rain lashes my half naked body on the seven kilometre walk, with vultures circling above and monkeys screaming. Snakes and lizards are few; we eat them. Unsmiling peasants watch us as they work their fields. Forest everywhere and the treacherous bamboo, which once it scratches you can cost you a limb in these conditions.

Four Buddhist priests share two umbrellas and make way as we pass, as we arrive at the work site. I sit on a damp log and roll a cigarette. Our sergeant, a tall Yorkshireman, towers over our guard, a short fat engineer with prominent gold teeth. Orders are given; 'Pay attention! The Imperial Japanese Army expects every man to do his duty, work hard, we promise good food, good accommodation, and good pay, anyone breaking the rules will be punished! Your task today is to clear 100 meters of jungle!' Raspberries are blown, another says cut the Nelson stuff.

So another work day begins, the shouts of 'Timber!' echo everywhere and hordes of angry black ants dig into my flesh as a yellow centipede crawls up my leg; one bite can be fatal.

Several times I slip away for a ben-jo (relieve oneself) and smoke, but just before lunch a guard catches me; he beats me up and drags me to the taskmaster in charge; he goes into a rage and strikes my face with his fist and then makes me stand to attention with a jungle fruit stuck into my mouth. I think of my mum and dad, and my girlfriend who promised to wait for me; I must stay strong and have the spirit to survive. I eat my lunch.

At work I obey, to stay alive. Back at camp afterwards we rush for the evening meal – fried rice, soya sauce, chilli, tea. In the hut crude oil lamps flicker, the end of another day; I stretch out on my bamboo bed and close my eyes, listening to the rain.

Jack became a taxi driver after the war when he was sacked from his job as a chauffeur for eating too many cakes at work!

VX39085 Capt. Dr Victor 'Vic' Brand, MC, RAAMC, AIF

22. VX39085 Capt. Dr Victor Brand, MC, RAAMC, AIF.

Victor Brand was born on 16 July 1914, second of five children of Isaac and Toba Brand, who came to Melbourne, Australia, from Poland via Israel in 1913. Although they were very poor, Victor got to Caulfield Grammar School, and went on to Melbourne University to study medicine in 1931, graduating in 1937.

When war broke out, Victor was working as a locum in Melbourne's poor working-class districts and other parts of Victoria. He enlisted in November 1940 and became MO to the 2/29th Infantry Battalion training near Bathurst. He married Emilie Georgette (Fifi) Fry, a French immigrant from Marseilles, in June 1941, before sailing for Singapore in July 1941 with the 27th Brigade.

Victor's battalion landed on 15 August and by 15 September were at Segamat in central Malaya. When the Japanese invaded in December, Victor's battalion were engaged in heavy fighting, holding up the advancing Japanese Imperial Guards for six days while Allied forces withdrew. Sixty per cent of his battalion were killed and the 150 seriously wounded left in trucks were massacred by the Japanese at Parit Sulong. Victor never got over this incident, as he had fixed Red Cross signs on the lorries before leaving,

Victor was posted Missing in Action in mid-1942 and then, in January 1943, Fifi heard that he was a POW in Malaya; occasional reports in Japanese radio broadcasts mentioned a Dr Bran or a Captain Grand.

In September 1943 he was awarded the MC for heroism at the battle of Muar in Malaya on 19 January 1942. The citation read:

> Capt Brand was with his battalion near Bakri. The RAP (Regimental Aid Post) was in the centre of the perimeter, and constantly under heavy shell and mortar fire. With complete disregard for his own safety he left the slit trenches in which all other personnel were forced to shelter and remained out in the open during the heavy and continuous bombardment attending to the wounded, both Australian and Indian. Shortly afterwards the battalion withdrew leaving only a small party of carriers. Capt Brand, however, remained behind and refused to leave until he had attended to all the wounded and placed them on vehicles. He moved them to the road but owing to a road block was unable to proceed. He then went through the

> vehicles collecting the waiting wounded, whom he personally escorted back to the battalion.

Victor's detailed account of his part in the fighting can be found in *A History of the 2/19th Battalion AIF* (AIF Association, 1976).

In April 1943, Victor was transported with 'F' Force to Thailand, with 7,000 other POWs, to work for a year on the Death Railway, operating a primitive hospital for 200 men at Tamuronpat, 240 kilometres north of Bampong in Thailand, between Takanoon and Konkoita: 'I got there by truck with two Korean guards, 3 medical personnel and a British officer, Capt. White and his batman ... at one point we rounded a corner and there was [a] huge elephant screaming with outstretched trunk ... probably his first sight of a truck.' They came to a small clearing about 500 metres west of the railway and were offloaded. His only supplies were a wok, a machete and some quinine, to treat the many diseases and infections. He occupied a tent with the medics and then built bamboo huts for the patients. He was soon laid out by dengue fever but recovered after some days. Then came the cholera.

One night a party of POWs arrived carrying sick men on bamboo stretchers, knee deep in mud. Victor took in the sick men and realized that they had cholera. The first case was a New Zealander of the SSVF. For treating cholera, Victor wrote, 'I cut down on a vein in the ankle and inserted a length of bamboo, with half the bore of a drinking straw, attached to lengths of stethoscope tubing fixed to the bottom of a pint sized container. The idea was to replace fluids into the blood and pour about six pints of blood and boiled and strained river water with added rock salts into the patient's blood stream in 24 hours.' Still there were many deaths; a man could be very well in the morning and dead by evening. Even if a man recovered he might be carried off by a secondary illness such as dysentery or malaria. Victor always said that his suffering, awful though it was, was nothing compared to that of the men on the railway in Thailand.

The men were able to supplement their diet with peanuts and sometimes meat from stray yaks which wandered away from local herds. The tethered animal would be despatched with an axe, but with great difficulty. Victor would catch the blood and fry it up with palm oil for the sick. In early May the monsoon began and so they were constantly repairing roofs and using split bamboo to cover the mud to form usable pathways. Victor was amazed to see Japanese arrive on the railway on their way to fight in Burma; they struggled with heavy equipment in the mud and were badly and mercilessly beaten by their own NCOs and officers

One day after the monsoon, the Japanese 'Jungle Cinema' arrived. The film was a propaganda history about Japanese warriors fighting Caucasian villains named Goldberg, Cohen and Rosenberg. Victor was astonished and could only assume it was cheap anti-Semitism inspired by the Japanese alliance with Nazi Germany.

By maintaining strict hygiene (boiling mess tins in water and keeping latrines clean) Victor was able to keep infection down. Eventually, by September, he was ordered to send the very sick by boat to Kanchanaburi. He was left with twenty lightly sick who had to be paraded each morning by him, and he had to send whom he could on working parties for the Japanese. He often resisted and was beaten for his stubbornness. By December he was back in Kanchanaburi with 'F' Force remnants and then sent back to Singapore.

Roy Mills describes Victor in his book, *Doctor's Diary and Memoirs* (1995):

> 'He gave wonderful assistance.' And Captain E.C. 'Robbie' Robbins remarked: 'Victor may be small in stature, but in my mind he was a giant of a man.'

On liberation, in August 1945, Victor went down to the docks in Singapore and had a meal aboard a British battleship. He returned to Changi:

> waving a slice of white bread, the first seen in over 3 years! One evening a hospital ship was moored at the wharf with large openings on the side. There were rows of beds and nurses in uniform, all brilliantly illuminated. Outside in the dusk was a long line of emaciated men dressed in a few rags. I could see them enter into the light and be divested of their rags and mess tins and taken away to be washed. They returned dressed in pyjamas and were put to bed between white sheets, so tenderly.

Back in Melbourne he practised as a GP and graduated as an anaesthetist in the 1950s, and worked at Alfred Hospital until 1971, finally retiring in 2003. Although he was against Australia's involvement in Vietnam, he joined a voluntary group of doctors and nurses and served in a civilian hospital at Bien Hoa, just north of Saigon, in 1966. A passionate supporter of Israel, he worked in a military hospital just after the Six Day War. He and Fifi had a large family. Years later, Victor visited the CWGC cemetery at Kranji near Changi. On seeing the hundreds of white headstones, he wept uncontrollably.

In April 2006, on Anzac Day, just before his 92nd birthday, he proudly led the 8th Division AIF as one of the last surviving decorated officers. He died weeks later on 9 June.

2591725 Sgmn Israel Jack 'Scottie'/ 'Jock' Caplan, RCOS

Jack's story has been told in his superb book, Gorbals to Jungle, *first published privately in 1960, but it is so extraordinary I have taken the liberty of including extracts of it here. His audio-recorded memoirs can be heard at the Imperial War Museum and information from them has been incorporated here too, together with additional material from the AJEX Jewish Military Museum archives and my own conversations with him.*

23. 2591725 Sgmn Israel 'Jack' Caplan, RCOS, on embarkation leave with his mother and brother Gershon, 7 August 1941.

Jack was born on 3 August 1915, one of seven children, and lived in the Jewish area of the Gorbals, Glasgow (South Portland Street), known as 'Little Jerusalem'. His parents, Morris and Fanny, were Lithuanian Jewish immigrants from Kovno and Vilna, and his father was a tailor; his maternal grandfather had been a rabbi. Jack survived polio and attended Gorbals Primary and Adelphi Secondary school, leaving at the age of 14 to work in his brother's printing shop. Growing up in a tough neighbourhood, he learnt to defend himself through boxing. He was involved in fighting fascism in Glasgow in the 1930s and, like many young Jews, he joined the Young Communist League in order to take action against Mosley's men. He planned to go to fight in Spain against Franco, but by the time he was ready to travel, the government had clamped down on those leaving the UK to do this.

On 10 March 1940, Jack volunteered and joined the RCOS and was posted to Enfield, London, where they became part of the newly formed 18th Division. He was a motorbike despatch rider. By October 1941 they were in Macclesfield and then on their way to the Middle East (all the equipment was prepared for that theatre of war – khaki uniforms, shorts, topee helmets) on the SS *Sobieski*, via King George V docks in Glasgow, to Halifax, Nova Scotia (where they were re-embarked to US naval ships, in this case the USS *Mount Vernon*), to Trinidad, Cape Town and Mombassa.

On the way, Jack was involved in two incidents on the *Mount Vernon*. In one, during a deck-top talk about the alleged superiority of the Tommy over the Japanese soldier, by an Intelligence Officer, Lieutenant Simmonds, Jack shouted out 'Bull-s—t!' and was put on a charge. The second involved a rather more nasty anti-Semitic incident. Jack was an army boxer, nicknamed 'Benny, the fighting Jock', after the famous fighter Benny Caplan. One day while trying to sleep on the crowded, hot deck as they entered the tropics, Jack overheard a conversation among a group of Army Regulars who had served in Palestine/Israel with British forces there. One red-haired loudmouth was boasting to the others about the Jewish women he had 'had', using obscene gestures to make his point. Hot, and in a bad mood owing to the impending charge, Jack got up and approached the group. He tapped the ginger-haired man on the shoulder and told him that no Jewish girl would go with a windbag like him. Startled and nervous – because he knew about Jack's boxing skills – the man replied that he was just telling his mates about life in Tel-Aviv and the 'Jew women'. 'Right', said Jack, 'and I'm a Jew too!' 'You can't be', said the man, 'You're a Jock!' He then muttered and walked away, saving Jack further trouble for being on a charge for fighting.

En route to South Africa, Pearl Harbor was attacked and so the convoy was diverted to Singapore, where they arrived on 13 January 1941. Chaos reigned, as the Japanese were bombing the city as they arrived in the pouring rain. Jack was batman and driver to a Signals officer and was billeted in tents near the docks. For many days and nights he endured bombing, shelling and Imperial Japanese Army (IJA) infantry attacks as the Japanese approached the city. The Japanese planes bombed with impunity, as there were virtually no anti-aircraft guns available and the infantry could only shoot back with small arms as the Japanese came in very low to dive-bomb the vulnerable Allied forces. Jack never saw one shot down, and they flew over at 10 a.m. and 3 p.m. precisely each day, so confident were they of no Allied aircraft or AA opposition. On one occasion Jack went to the ablutions hut to wash and noticed he had left his razor in his tent. As he left his tent to return, he was blown in the air by a terrific explosion; the hut had been blown to pieces and he had just missed being inside.

Nevertheless there was some time for light relief, even during the bombing; one evening Jack enjoyed an intimate liaison with a beautiful and cultured Chinese girl called Lilly Yen whom he met in Lulu's dance hall in the Lavender Street district, famous for its night life. Afterwards, on his way back to barracks, he had the sickening experience of helping to pull a severely injured civilian from the wreckage of a badly bombed house, and passed bodies blown up by the Japanese air attacks, and screaming women and children; such were the incongruities of events in a city under siege and about to fall to the enemy. He never saw Lilly again.

On 20 January, four of his closest friends were killed while on a wire-repairing mission, when the truck was blown up by a dive-bomber;[1] he had just missed going on the truck, as he was delayed at his tent – otherwise he would have been killed too. Jack openly wept for the first time since he was a boy. On another occasion Jack was on guard duty when a lorry load of Gurkhas arrived to go on jungle patrol. As he watched, they stripped down to their underwear and, with their kukris (a type of curved knife), made their way in single file into the bush. Later they returned with huge smiles, and showed the British troops their little pouches, in which were the bloodied ears that they had cut off many ambushed Japanese soldiers. Like a North American Indian scalp, this was proof of a kill.

Towards the end of the fighting on 15 February, the Japanese bombing intensified. Jack was by now attached to 135 Field Artillery, helping fire twenty-five-pounders at the causeway as the Japanese streamed over; the enemy took huge casualties but wave after wave of charging Japanese infantry was impossible to stop. Suddenly came a message that at 4 p.m. there would be a ceasefire.

When the surrender was announced there was a terrible and sudden silence, and Jack's initial feeling was one of relief at the abatement of the dreadful noise and shelling and killing. Jack watched as the gunners spiked their guns; some men committed suicide, including a senior NCO known to Jack. Jack was given the job of immediately destroying all stocks of the officer's mess whisky (he noted that, as a Scotsman, this was something that broke his heart) – but it did not stop several officers getting drunk.

Changi

Disarmed, on 17 February, Jack joined the thousands of others on the fourteen-mile march to Changi gaol, passing the terrible debris of battle: craters, ruins, bodies. Few guards accompanied them, but there was nowhere to run. There was no brutality, and stops were allowed, some even being permitted to buy coffee en route. At the gates they were greeted by the wounded, the cooks, and others who had gone on ahead in trucks. News filtered in that some British officers were being summarily executed.

Their guards at first were not Japanese but embittered Sikhs who had gone over to the enemy and used every opportunity to humiliate the British and other POWs, constantly demanding that the British should salute them at all times and frequently using their boots and rifle butts to beat Allied troops. Soon afterwards, the Sikhs disappeared. Food was short and tempers became frayed, but discipline was strong under Jack's famous Australian CO, Colonel Toosey; as a punishment for arguing with an NCO, Jack was given three weeks' fatigue duty on reduced rations, thereby becoming a prisoner of the prisoners.

In an article written for *Wartime News* in February 2007, Jack described how he witnessed whole families of Chinese – men, women and children – brought to Changi in covered lorries, their thumbs tied behind their backs, being pulled and beaten off the vehicles. Then they were forced to walk into the sea and were massacred in front of everyone. The men watched in horror as the sea turned red with blood; many wept. It happened every day for about ten days, twice a day.

Jack said that at first they were receiving food only from old army dumps, not found by the Japanese, and this consisted of two army biscuits and a tiny piece of cheese, three times a day, and water was brought in a water-wagon. But they were allowed to go sea-bathing and move around the camp freely, and religious services were held. However, men began to die daily owing to the lack of medicines, their unattended wounds and the growing incidence of deadly tropical diseases. More and more prisoners arrived; latrines overflowed as more and more had to be dug. Soon dysentery broke out. Clothing began to wear out and men took to wearing a tiny piece of cotton around their middle, known as a G-string or 'ball-bag'.

Slowly the Japanese began to send in rations, using POW drivers (mostly Australians) as 'trusties' to bring in supplies from the city; here they also gathered news of the war from local Chinese and Malays (who gave them medicines and food for the POWs), but they reported the atrocities against Chinese civilians, who were decapitated and their heads were displayed on poles around the island. News arrived of especial ill-treatment of Gurkha POWs by the Kempeitai (Japanese Gestapo-type police), for they had inflicted heavy casualties on the Japanese during the fighting for Malaya due to their jungle fighting abilities. Apparently there were mass executions of Gurkhas who had become POWs.

Jack remembers well the camp library, lectures, debates and craft classes, all set up to raise morale, among all the bad news and pessimistic rumours. Suddenly one day, the Japanese ordered a parade of all POWs, who were then marched to line the Changi road on both sides, evenly spread out. They were given new shirts and shorts. After they had waited for an hour in the hot sun, a motorized Japanese convoy began to drive past, with armoured lorries, posh cars carrying high ranking officers, and a car with a cine camera filming it all for Japanese propaganda purposes. Failure to wave and cheer was met with immediate punishment by Japanese guards. This notorious film is well known and often shown in historical documentaries. Afterwards the new clothes were taken back by the Japanese. Jack also witnessed the Selarang incident when all POWs were confined for days in the small barracks square, in growing filth and squalor, at Changi, until they signed the illegal 'promise not to escape'. As working parties were formed in a haphazard way (as the Japanese demanded slave labour for various tasks in Singapore), Jack became separated from old friends. At first, working parties had a British officer attached as an overseer, but later the officers also had to work. Jack was taken by truck to help build a shrine at Bukit Timah hill, where fierce fighting had raged, and the Japanese wanted to show 'respect' to both sides who had lost lives. Day after day Jack toiled, using monotonous 'chain gang' methods of moving spoil, the men ensuring that the sick and weak men were in the section where the empty spoil baskets were passed back up the line to the building site. There was a ten-minute break each hour, and one hour for a cold rice midday meal; the working hours were 8 a.m. to 8 p.m. in the burning sun. If a tool went missing, everyone stayed until it was found. For ten days' work they got the equivalent of 2.5 pence in today's money. The guards would boast that many prisoners were going to die working for the emperor, and as the Japanese advanced in the war front, conditions and treatment and rations got worse. Beatings were given on the slightest pretext; on one day Jack was beaten twice and then at the meal break he accidentally trod on the mess tin of a guard, when he had gone to ask him for his leftovers. The guard threw Jack in the air in a judo move, and then went to kick him on the ground; as Jack lifted his leg to shield himself, a rain of blows descended on him and eventually a large stone was used on his head which knocked him out completely. When he woke up, he saw two bloodied comrades with arms raised above their heads holding heavy stones; they had tried to protest Jack's beating and were being punished with this method, having already been beaten with sticks. If the stone was dropped, beatings followed until the POW collapsed.

Slapping was commonplace, even for Japanese soldiers who broke a rule, but the blows to POWs were usually to the ears, and not the cheek, and this is why many surviving POWs had – and still have – ear problems in old age.

Jack was allowed to send his first card home in June; he later discovered it reached his mother two full years later in 1944.

Work in the docks was slightly better in comparison, loading scrap metal onto ships bound for Japan. Some men would secretly creep on board such ships and steal and eat sugar and eggs from the galley, providing themselves with essential life-saving vitamins. Because of this, the POWs would often bribe the Japanese guards with fountain pens, watches, rings or money to get placed on a docks working party. Quite often the Japanese sailors aboard these merchant ships simply would not notice the POWs taking eggs, cigarettes or chocolate from the ship stores – and would even cook the eggs for the men. Jack speculated that the Japanese sailors were more friendly than the soldiers, perhaps because they had travelled and seen western countries.

Jack's first brush with the Kempeitai came in Singapore. Jack testified that these men seemed to wear no badges of rank, although they wore the same uniforms, but they were much feared by the Japanese soldiers themselves and Jack saw that the soldiers always bowed to them when they visited camps. Jack had a friend, Bill Duncan, who was a pianist before the war in one of the big bands. One day he was stopped with another friend by a Japanese officer who demanded to know where the secret radio was. Bill had no idea, but the other man did. They were taken and beaten and starved for three days without result, and then told they would be sent to the Kempeitai HQ for further questioning. The Kemeptai beat the men again for three days but again they said nothing. They knew Bill was a musician and they warned him that they would pull out his fingernails if he did not talk. Again he told them he knew nothing. So they stabbed his hand to a bench and pulled out the nails of one hand, and again on the next day did the same to the other hand. Jack later saw what had been done. They also broke Bill's right arm. But he survived, and Jack saw him play again after the war. Jack also witnessed a British escaper who had been recaptured; he was taken by the Kemeptai to a tree on the edge of the camp, and openly shot.

The Railway

Some time in November 1942, Jack found himself on the march to Singapore railway station. The men were told they were off to better camps in Thailand. They were lined up on the platforms amid the usual screams, pushing and shouts from the Japanese, and now the new Korean guards, who were especially brutal. While they waited, Chinese civilians secretly bargained with the POWs for their shirts or shorts, when the guards' backs were turned, and Jack sold his for extra food. One Australian stole a guard's rifle when he went to the toilet, and sold it.

Then a train approached, made up of open and closed cattle trucks, and thirty men were squeezed into each truck, which could normally only hold ten in comfort. For five days and nights they suffered appalling conditions on a journey north through Malaya: suffocating heat by day, freezing by night. At night you could not turn without asking six others to do the same. Three times each day a scheduled stop would occur and buckets of rice were pushed into each truck and a nearby tap could be used to fill water bottles. Toilet stops were also made, but those with dysentery simply had to relieve themselves in the truck, thus causing horrific conditions for others; this brought the flies, which brought further disease. At night stops, despite Japanese threats, locals would secretly approach the cars and bring food in exchange for clothing, watches, pens, rings and so on from the POWs. Sadly, Jack actually witnessed what happened when one Thai woman was caught; the Japanese dragged her by her hair to a tree, tied her by the hair to the tree, and one guard clubbed her to death with his rifle butt.

After three days Jack felt he had developed malaria. A gunner friend, Lofty, helped nurse Jack and lent him his blanket, a great sacrifice. Determined to get extra food, Jack asked another trusted friend, Bill Forrest, to sell another pair of shorts for him, as he was too ill to move, but Bill was seen and beaten by the Korean guards and lost the valuable shorts. However, for fear of having too many sick men to hand over to the new camp commandant on arrival – and thus be punished – the guards allowed the POWs to trade with locals on the last day and so obtain some extra food. Jack obtained and ate twelve eggs and six bananas to build up his strength.

His total possessions now consisted of one G-string, one small army knapsack, a rice-bag blanket, one mess tin and spoon, one water bottle and an illegal penknife.[2]

Tamarkan

They arrived at Tamarkan Camp near Kanchanaburi, taken by truck the last few miles from the railstop. This was more of a transit camp during the two months that Jack was there. There was a hospital but it was really just a hut were men went to die of old wounds, disease and exhaustion – there were virtually no medicines. On arrival, they were ordered to march in and salute the Japanese guards with an 'eyes right' and then were ordered to parade to have all their kit searched for illicit items. But they were saved from this when the Korean guards were so astonished by the red hair of a Scots soldier, whom they all closely inspected by ruffling his scalp, that they forgot the inspection and waved the men away, so Jack's penknife – possession of which was a serious offence in the eyes of the Japanese – was safe. In the camp was an advanced contingent of Argylls and Gordons, and Jack felt immediately at home. The men were to build the bridge over the Kwai Noi. The commandant was a Captain Suzuki.

The newcomers had to build their own huts first – about 150 to a hut – but although work was hard, the organization and discipline of Colonel Toosey meant that everything ran smoothly in the camp, and hygiene was a top priority; morale among the men remained high as a result. The officers had slightly better conditions, with just ten men to a hut. Other heroes of Jack's were Colonel E.A. Smyth, and Captain J. Markovitch, RAMC, whose medical skills, compassion and inventiveness saved thousands of lives in the camp. There were many other men who were heroes.

Jack's friend Joe carved some beautiful chessmen using Jack's precious but illegal penknife; the two men played a great deal after work. One Japanese guard often came into the huts to check what the men were doing in their spare time and he asked to be taught the game; Jack and Joe obliged, and within about forty-five minutes he was quite good at it and often returned to play. One day he failed to turn up, and then suddenly appeared again five days later, in a bad mood for some reason; he went along the rows of beds and demanded in a frenzy if the men were American or English. Everyone said they were English and got slapped. Jack said he was Scottish and the guard demanded that Jack should teach him a Scottish phrase. Jack knew no Gaelic and told the guard this but the Japanese insisted and slapped Jack's face; again Jack explained, so the guard drew his bayonet and pressed it into Jack's navel, drawing blood, as he pushed him against the wall of the hut. Thinking on his feet, Jack said, 'say "Ich bin ein galischen mamzer" ' (Yiddish for 'I am a yellow bastard', although in another version of the story Jack said 'Schechter mamzer' – terrible bastard). For several days afterwards the guard would always repeat this phrase, shouting it out after inspection, but of course none of the POWs understood what he was saying. Then one day a Jewish POW who had recently arrived in the camp (Lewis, RAF[3]) heard the guard repeat this sentence and could not believe his ears, wondering how a Japanese could speak Yiddish. He eventually discovered that Jack was responsible and he advised Jack that this was bit foolhardy, because the guard might one day find out what it meant. Fortunately, a few days later, Jack was transferred to another camp.

Jack also witnessed at least one other public execution when two POWs were recaptured after an escape attempt and ceremoniously decapitated in front of the assembled men by an officer wielding a samurai sword. Jack said he did not remember what camp this was in and it is not mentioned in his book. Nor is another sad story, when he brought food to a dying comrade in the hospital; as he lay there he told Jack that when he passed away, Jack must have his bush hat, a piece of essential equipment against sunstroke. Jack kept that hat for over fifty years and it is now an exhibit in the AJEX Jewish Military Museum.

The men often helped each other. By virtue of their skills, some got easier jobs; Jack's friend Joe was a tailor and so worked in the Japanese compound, repairing and making clothes for the guards, and was given Japanese rations and good accommodation. As a result, he was able to smuggle extra food to Jack on many occasions. One racket that developed was selling the clothes of sick men in the hospital, to obtain money for them to buy extra food. For doing this the man would obtain a commission and both parties were satisfied. Others who made money would lend it to officers, who were asked to sign much larger IOUs to be redeemed at the end of the war if either survived. Jack knew at least one man who obtained £5,000 from officers in 1945, which was at that time a huge sum of money.

In one Thai camp, on the emperor's birthday, Japanese prostitutes were brought in for the amusement of the guards. Jack noticed they were extremely ugly and ungainly women, but one POW known to Jack, who worked in the Japanese cookhouse and was friendly with the Japanese sergeant, was asked if he would like to go with one of the women. He said yes and some days later developed syphilis. When he went to the medical officer about this, he was told not only that there was no medicine, but also that if the Japanese commandant found out, both guards and POWs would be punished. A short while later he threw himself under a Japanese train near the camp.

A canteen and library were opened, and when the tiny Japanese wages came through for the slave labour, a percentage was always taken for the sick men by the officers and these monies also helped to subsidize the canteen for all the men. Officers were allowed into nearby towns under escort to purchase extras for the men, especially a sweet peanut toffee the men craved, local cigarettes (called 'Sheaves of Rice', Jack remembers – comparable to the British Woodbines) and eggs – all helping to keep the men alive on their starvation diets. The indefatigable Toosey insisted that the Japanese should give one day's rest in every ten and allow the building of a concert platform for the men. Jack found that in all camps where books were available, he was able to escape the real-life horrors of being a POW by immersing himself in reading, and he consumed hundreds of books in his years as a prisoner.

One problem was that the Japanese engineers who worked alongside the men worked very hard but were well fed; nevertheless they expected the POWs – who were on a starvation diet and sick with disease – to do the same, and nothing would persuade them otherwise. The result was brutal beatings and punishment that resulted in a downward spiral of more illness and exhaustion, especially stoning and face-slapping on a huge scale.

Jack remembers receiving about one Thai dollar per day – about two old shillings per week (ten pence today) – which could buy one dozen duck eggs and a hand of twelve bananas. But beatings and punishments grew worse and Jack received three such beatings in one week. An order came through that all POWs had to bow and not just salute the guards. Jack failed to see a guard on one occasion, and he slapped Jack for not saluting him; Jack held up his arm to defend himself, and for this he was made to run the gauntlet between twelve Japanese soldiers who beat him senseless with sticks and fists.

When the wooden bridge was completed, the Japanese found it was built on poor foundations and began a second concrete one alongside it – but Jack had moved on before it was finished. He began keeping a diary, using blank pages torn from library books, toilet paper and general waste. He kept it, folded carefully, in the wallet section of his small paybook. One hot free day, lorries drew up outside the camp and out came hundreds of well-dressed Dutch and Javanese soldier POWs, complete with suitcases filled with clothes, kit and food. The British were astonished. Among the Dutch were three Jewish soldiers with whom he became good friends, including George Sammes, who survived the war.

The Javanese especially knew which plants were edible and so were able to teach the POWs how to supplement their food rations. Some British and Australian POWs began a coffee-making business; they would go under the wire at night and trade coffee and sugar with locals, then sell it to the POWs by the cup. Jack became a partner with one man, Owen Burke, whereby Owen would make the coffee and Jack would sell it; by agreement with the British camp adjutant, the business was made 'legal', as sale of cheap drinks was good for morale. The 'wealthier' Javanese/Dutch POWs proved good clients. Jack remembers one moving incident when a very tall, long-bearded Dutch prisoner known only as Big Maxie brought his accordion out one day and, standing under a tree, serenaded a spellbound camp with a concert of popular and classical music.

Suddenly the dream was broken by the shouting of 'The Undertaker' – a notorious, six-foot-tall Korean guard called Kaneshiro, who boasted that he had killed seven POWs[4] – and everyone expected Big Maxie to fall victim to his brutality; instead he requested an 'Englando songo' – 'Annie Laurie'. Concerts in general, said Jack, were a fantastic boost to the men as it reminded them of the normal life they hoped to return to soon. He had strong views about the environment, pointing out that only those who had never lived in the tropics romanticized them; in fact, the savage ants, poisonous animals of all kinds, and nerve-wracking chirping of crickets twenty-four hours a day were all psychological irritants that would wear people down. Only the sunsets might be described as events of great beauty.

Guards and Propaganda

Among the worst guards were 'Bastard 1' and 'Bastard 2', who specialized in hiding when POWs went to the latrines and then suddenly appearing and beating them for not saluting quickly enough. 'Pig' would accuse a man of insolence and then practise his vicious judo on the unfortunate victim. 'The Yank', however, whom Jack came across at Nong Pladuk, was a Japanese who had lived in Hawaii and astonished everyone by speaking perfect English with an American accent (he had gone to visit his sick mother in Japan and was caught there when Pearl Harbor was attacked; he was made an interpreter). He befriended many POWs, played chess with them, warned of approaching searches (for radios or hidden weapons) and had no love for his Japanese masters; he was never known to have punished a prisoner, but the men could never be quite sure of him and had to be guarded about anything they said to him.

The Japanese gave English propaganda sheets to the POWs, and they were the same as the Japanese-language ones read by the guards. They raised morale because they were so grotesquely untrue – the POWs knew that the Japanese were experiencing losses – and such lies convinced them that they could never lose the war against people with such a bizarre mentality. One lurid story told of how a Japanese pilot had run out of ammunition and so he flew his aircraft alongside an American plane and used his Samurai sword to kill the American pilot in mid-air. Another told how a Japanese motorcycle despatch rider delivered a message to his CO two hours after he had died, because his samurai spirit had kept him alive until he reached the officer.

Kensayo

Around January 1943, orders came to leave Tamarkan by truck; Jack's Dutch friends were tearful, and gave him food for the journey. When a truck stuck in the mud, the guards went berserk with their beatings when the starved and sick prisoners could not budge the vehicles; in the heavy, cold rain, the men struggled to manhandle the lorries into the camp, under a rain of Japanese blows. The men approached Kensayo camp, a quagmire of filth, with swarms of flies the like of which Jack had never seen before,[5] and dilapidated bamboo huts. When the sick and wounded were hustled out of the camp to help them, the newly arrived men were shocked at their horrific condition and were fearful of what must lie ahead for them. Yet these men did what they could to assist and even made jokes about this being a Butlin's holiday camp; Jack was forever amazed at the spirit of the Tommies.

At Kensayo they met 'Rat-Face', a notorious Korean brute who vowed to kill many 'Englando solja' because he alleged that his brother had been killed in the fighting in Singapore. Within four months, most of the men in Jack's transport were dead, mostly of the dreaded cholera, despite the amazing lifesaving efforts of Captain Christison and Captain Butterfield, RAMC. Butterfield segregated the cholera patients, organized cremation to stop the spread of the disease and was beaten for defying the guards by keeping the sick back from work, but he refused to be cowed. The Japanese tended to keep their distance for fear of catching cholera. Jack had malaria but as a POW he inexplicably escaped both dysentery and tropical ulcers, of which he saw so many others die. He witnessed horrific scenes of dead and dying men in the so-called hospital huts, lying in their own filth, knowing their terrible fate and yet remaining cheerful. The smells were unspeakable. For years afterwards he had terrible memories of the open cremations – many of them of his friends – and he never forgot the stink of burning flesh.

Leaving the sick to do light work in the camp, Jack and comrades went off to build the notorious railway, either jungle-clearing or cutting through rock and building embankments – very heavy work even for the fittest of men. The chain-basket method was used, and if anyone even slightly faltered, a severe blow or a stone thrown at them was the result. Tea breaks were given, of ten minutes in the hour – not from compassion but because the Japanese knew the men would die if they did not drink something and rest. Jack described an incident which was a typical occurrence in the camp, when a POW was struck down by a stone, then kicked up, then dragged away for a slapping and beating in sight of everyone; finally he was made to stand hatless in the burning sun with a stone raised above his head, without a break, or food or water. With bare feet on hot, sun-baked rocks, and naked except for their G-strings, all the men suffered horrendous body burns, heatstroke and excruciating pain when urinating, as they were so dehydrated. As a result, many drank from stagnant pools and thus died later of disease. Any lost tool at the day's end resulted in a severe beating. Jack said that the heavy rains were even worse than the beatings, because they caused a chill which penetrated a man to the bone.

Later Jack was transferred to rock-boring; holes were made with a twelve-pound or fourteen-pound hammer and chisels of varying lengths, then filled with dynamite and blown up. One day Jack was working with a sick Glaswegian named Legget and Jack could see they were going too slowly for the liking of the guards. This would mean a severe beating. Other men saw what was happening and came to assist Jack to finish the work while Legget surreptitiously rested. But even then the guards would give the men another task if they managed to complete the first one, thus extending the misery and fatigue the men were enduring. Men lost their minds under this terrible stress; Jack witnessed this horrible phenomenon many times. He also witnessed many sick men saying they wanted to die, and he and others often had to try and inspire them to stay with it, to return to their families and spite the Japanese by surviving. The number of crosses in the cemetery grew. Men resorted to catching and killing cats and dogs whenever possible. On one occasion Jack asked some Aussies if he could share some meat they were cooking and they gladly offered him some. It was delicious, and only afterwards did they tell him it was snake. En route to camp, each man had to bring some wood for the cookhouse fires, or a beating ensued.

Another food source was fly maggots, so long as they were collected from dysentery-free areas and lightly cooked. Imitating the Aussies and some of the Argylls, Jack – who was by now desperate – decided to get men to agree to sell their shirts or other items to locals, and he would go under the wire and trade; in return for the risk, he would keep half the money. Friends warned Jack not to try, and instead to use others who took the commission but also the risk. He decided to go ahead anyway.

On a moonlit night he broke out and made for the barges along the river, only a short distance away; he traded a shirt for five dollars after some very hard bargaining and made his way back. Suddenly a Japanese shouted at him and he stopped in terror. Jack was taken to the Japanese compound and an NCO appeared; Jack was shaking with fear and sweating, as he was slapped across the face several times and fell down; then another soldier urinated on him. Jack was furious and screamed: 'Nippon soldiers dirty bastards!' The first soldier produced a pick and shovel and told Jack to dig a grave; he heard the bugler sound lights out from the camp and thought of where he could have been. After some considerable time, Jack was ordered to stop digging and squat down, and he was offered a cigarette. The soldier engaged Jack in a conversation about masturbation, of all subjects. Suddenly he stood and told Jack to go back to the camp and never trade with Thais again. Sadly, Jack had lost the five dollars; what would he tell the soldier whose shirt he had used? And he faced the humiliation of explaining his injuries to those POWs who told him not to go in the first place.

Jack was by now ill and depressed, with chronic malaria and debility as diagnosed by Dr Christison. He had collapsed on the way back from a working party one evening. Christison told Jack that he could be saved while others could not; the Japanese had given permission to send thirteen men south to a hospital camp, and Jack was going to be one of them. After four months, Jack was sent south to recover, to Chungkai base hospital camp, and this undoubtedly saved his life. The journey by barge down the river is a hazy memory but Jack remembers that one man went mad and others near him died on the journey.[6] This was around May 1943.

Chungkai

Jack remembers Chungkai as the best of the camps he was in; it was a hospital and transit camp where food and medical treatment was slightly better, the guards kept a lower profile, the men received tiny but regular amounts of pay, and there was a canteen, a library, and a POW camp police force to keep order and an air of peace. The only work parties were for the upkeep of the camp to benefit the prisoners, such as wood cutting, vegetable growing or latrine digging. Fit men made regular contributions to the sick fund for extra food for the hospital, and officers were allowed to go to local shops and supplement the food rations using camp money. Jack began to make a recovery. Jack observed that the bigger men died first – perhaps because they needed more food.

Jack was often on burial detail and, being small, was designated to go into the grave with the body to remove the one Union Jack that was used for all funerals. The horror of looking into the wasted faces of the dead stayed with him always.

Then, in July 1943, the first letters from home arrived – but there were none for Jack. Sadly, as the names were called out, there were many who had already died. On 17 August, Jack wrote in his diary: 'Still no news from home.'[7]

One day Jack's old friend George Sammes arrived; the two men fell into each other's arms; each had thought the other dead, as often happened. Sent to Chungkai with him was the big Dutchman who played the accordion, but he was very ill; they went to see him and he died while they were at his bedside.

Jack recalls meeting David Arkush, Clifford Beck and Joe Cohen at the Passover (Seder) meal organized in the camp, but cannot be sure if it was here or elsewhere, as he was only at Chungkai for six weeks.

Nong Pladuk

About July 1943, Jack was moved to Nong Pladuk, where the main work was loading and unloading railway trucks; there was also a large petrol dump and railway workshops. The food was slightly better. Jack said that were no killings, as far as he remembered – or none that he knew about – but Suzuki and the Korean thugs ('the Undertaker' and others) were still with them. One new experience was that as each working party left the camp, the men were accompanied by music from the camp orchestra. On one amusing occasion they played 'There'll Always Be an England'. The men sang it with gusto and the Japanese guards joined in, but had no idea what they were singing and did not realize its importance to the POWs. The men grinned with delight at getting 'one back' at the Japanese.

The camp was kept relatively clean, with military discipline; each man had to kill 100 flies a day; the latrines were well maintained, and cholera and typhus vaccinations were given.

There were incidents of homosexuality among the POWs; one disused hut, No. 19, was used as a rendezvous where attractive Eurasian men especially were sought after by gay men among the POWs; Jack witnessed one very brutal fight between a Dutchman and an Argyll over competition for a young Eurasian man. It became such an issue that Colonel Toosey ordered that it must stop. On the matter of the lack of women, Jack declared that hunger and the struggle to survive were such prevalent problems that men rarely if ever thought of sex, but those with easier jobs or who were better fed were sometimes able to indulge.

Jack had by now lost many friends, and it was mentally difficult for him and others to reconcile this with the quiet, beautiful tropical evenings spent after work admiring the landscape and amazing sunsets. Friendship with another Jewish POW, Joe Cohen from Manchester, helped; they often played chess together.

In early September 1944, the camp woke to the noise of an aircraft. The Japanese ordered the digging of shelters for the guards. Days later they heard bombing in the distance and the Japanese became nervous. Then around 7 September the camp was bombed at 2 a.m. by two Allied planes, as searchlights and anti-aircraft guns opened up; almost 100 Dutch POWs were killed. Probably the camp had been mistaken for barracks for the nearby petrol dump and workshops. A further raid killed more POWs, including a British officer. But the men now realized that this was the beginning of the end for the Japanese, as the Allies must be not too far away. News was also reaching them of D-Day and the fall of Paris. The attitudes of the guards changed somewhat for the better. A Red Cross parcel was given out by the Japanese, but it was one between ten men.

News came in February 1945 that Jack was to be shipped, with Cohen and others, to Saigon. By lorry and railway truck they were sent to Singapore and embarked on a filthy tramp ship, the *Awa Maru*, part of a small convoy; the guards shouted and beat them as they were herded aboard like cattle.

The *Awa Maru*

Approximately 4,000 men were crammed aboard the Hell Ship in the usual appalling conditions for five days and nights in what Jack described as purgatory. The holds were deep and dark, and wooden shelves lining the hull were their beds. The holds stank of animal faeces and rubber cargos that the ship normally carried. The men were allowed to climb on deck only to relieve themselves or to get food.[8] The lucky overflow were camped on the deck. On one occasion an Allied plane flew nearby and they dreaded the submarine attack that might follow; the POWs were forced into the hold at gunpoint. The men planned to break out and take over the ship if they were attacked, and not drown like rats in the hold. Jack kept his small penknife handy, and the men on deck grabbed rubber or wooden items in case of being torpedoed; one Australian even owned a small pearl-handled revolver which he showed to Jack. The men even decided on which guards each one would kill. Then Suzuki gathered all the men on deck and told them that as they were being shadowed by American submarines, all men were to stay topside and there was to be absolutely no smoking to reveal their position. Suzuki had become more mild-mannered – did they all realize that defeat was near? Or fear a POW revolt on board? That night the men hummed and sang quietly – then suddenly the engines stopped. A nearby Japanese merchant ship blew up, hit by a torpedo and sank immediately, but nothing else happened and, despite the terror of what might happen to them, the night passed without further incident.

On the fifth day they arrived at the mouth of the Saigon River; not only had two of their escort destroyers been sunk en route, but the prisoners were glad to see that Saigon itself had been heavily attacked by the Allies, and the airfield, harbour and town had been heavily damaged. The POWs were roughly herded onto the quay with shouts and punches, but it appeared that not one of them had died. Jack never discovered, however, how many of the coolies, in a neighbouring hold, had died.

Saigon (Vietnam)

The men were shoved into empty quayside godowns (warehouses), which with concrete floors and corrugated roofs were at least not infested by the usual jungle vermin and insects. On the first work parties they were amazed to see French white men and women freely living in the city; they cheered the POWs as they drove by in the lorries but were not allowed to speak to them; Jack saw pity in their eyes. Most of the work involved loading and unloading lorries and trains, and repairing the damaged airfield. There were many Allied air raids, and on the fourth day Jack's group was strafed by a fighter plane. Two of Jack's mates who had been with him since Kensayo were killed, having survived all this time. This was to happen dozens of times in the coming six months until liberation. The food was better, and meat and new kinds of vegetables appeared in their diet. The work was less strenuous than on the Railway.

News arrived of the fall of Mandalay to the British and this caused great optimism among the POWs. Then after a few weeks the group was moved again, this time thirty miles away to a large estate where they were put under canvas. One day pork was served, and Joe Cohen would not eat it, of course, as an orthodox Jew. Jack offered to eat his share. But the smell was so delicious, that Joe succumbed and ate his fill. On another occasion an Australian had made an escape attempt and Jack witnessed his public torture. After being spreadeagled, in front of the whole camp, water was poured down his throat and then a guard jumped on his stomach; this was the water torture, and the man died.

Yet again the men were moved – this time without their officers, who were sent to Malaya. Jack and his friends were sent in March to mountain country where they were to build an airfield. Machine guns were placed all around the camp and it was felt – with some truth, it turned out – that the Japanese had a plan to kill all prisoners before they surrendered. Here, while bathing in a stream, Jack had his first experience of being attacked by leeches, which revolted him; colleagues rushed to help with burning cigarettes to remove them.

Extra food was obtained by men when on working parties; they would ask the guards to go to dig a latrine and then make for a path which natives used. On meeting the locals, trade would take place and once contact was made the locals turned up regularly each day. Men could also make money by repairing clothes, knitting, and even making handicrafts to sell to locals; the ingenuity was endless. In the evenings, games of cards, chess and shove-halfpenny helped while away the hours after work. Sometimes there would be sing-alongs and occasionally a proper concert party.

On one occasion their new airfield was bombed by the Allies. A few days later they were moved again.

Phnom Penh (Cambodia)

The French and locals here were much more anti-Japanese and shouted encouragement to the prisoners when they saw them in the streets. The Japanese did not react – clearly the end was in sight. Morale soared as news of Japanese defeats secretly flowed in. The men did only light work and after a short stay were moved yet again, taken in proper railway carriages into nearby mountains to build a road, then after a few days again moved in open trucks – Jack did not know the names of the places where they went. On one occasion they helped load ammunition onto railway trucks, for sending to the front, and travelled a while with Japanese soldiers, but they did not know where. One train they loaded up was attacked by Allied aircraft and again over fifty POWs were killed, many of them known to Jack.

By May they were back in Phnom Penh, where they heard of Germany's surrender, but men with Jack were still dying of malnutrition and illness. How much longer could they last? Then on they went to Saigon again, where what was to be Jack's final work party was being organized; the Japanese had decided to search for a so-called fabulously wealthy lost city in the jungle.

On 8 August they left by boat downriver, but then after a few hours turned back to Saigon. They were greeted by happy, shouting crowds; later they discovered that this was all about the dropping of the atomic bomb on Hiroshima. Some days later leaflets were dropped by an Allied plane – Japan had surrendered. Back at camp, men gathered in excited groups whilst the Japanese became obviously sullen and subdued.

Australian soldiers were the first under the wire to collect the leaflets, which were written in several languages, telling the POWs to remain calm and in the camps, and that help was on the way. The men were jumping and shouting, and many – including Jack – openly wept. Japanese guards watched as the Rising Sun flag was hauled down by the prisoners in the camp and the Union Jack, used for funerals, was raised, to great jubilation; the British and Dutch national anthems were sung and hidden alcohol appeared.

Suddenly a smartly dressed lone British officer in RAF uniform marched through the gates of the camp, armed with a revolver. The crowd of silent, amazed prisoners opened before him like the Red Sea before the Children of Israel, as he entered the camp. 'Who is the Japanese officer in charge?' he asked Jack in a broad Scots accent; Jack was amazed and pleased and said, 'Suzuki; he is a dangerous man, though.' The officer smiled and said, 'A Glasgow man, eh? I am from Edinburgh. Soon other officers will arrive by parachute like me and we will get you home.' The officer then bellowed in Japanese and a guard came running and bowed to him. Another Japanese order followed and the guard ran at the double to the commandant's office. Suzuki came out of his hut dressed in white breeches and puttees and open-necked shirt, casually picking his teeth; he had dined well. The officer raged at him and then slapped his face. The POWs were stunned. Suzuki bowed and ran back to his hut. The officer explained to the assembled POWs that he had told Suzuki he was improperly dressed and had insulted his captors, and he ordered him to return properly dressed. Suzuki soon returned with full uniform, buttoning his top button on his tunic, his cap and sword in hand, servile, respectful and standing to attention. After a short lecture, the officer did something Jack had only ever seen in a movie – he tore off Suzuki's badges of rank from his shoulders and collar, took out his samurai sword and broke it over his knee and threw it aside. To the assembled Allied prisoners it was like an incredible dream.

Lorries soon arrived with food and clothing and the men were free to leave the camp. Jack said that words cannot describe how they felt. Local people welcomed them but they noticed the feeling of enmity from the local Vietnamese against the French, whose rule they hoped soon to be rid of.

Jack and Joe wandered the town and had their first professional shave and haircut – with a manicure – since 1942. They warily entered the Club Colonial Français and were treated to drinks and food by hordes of French men and women. As they were about to leave, a tall, fair-haired young man approached them and spoke in Yiddish, asking if they were Jewish. Jack, who spoke Yiddish, was astounded. The man, Monsieur Herzlich, took them to his home and they met Madame Herzlich and the couple's baby daughter, Monique. Monsieur and Madame Herzlich allowed them to take a hot bath, gave them fresh clothes and then a huge slap-up meal. The couple showed the men a huge bed and said they could stay as long as they wanted, gave them a wad of money and said they should go out for the rest of the evening and enjoy themselves. Monsieur Herzlich warned them away from prostitutes and young local nationalists – in other words, to stay out of trouble. When they got home, they could not sleep in the soft bed and ended up sleeping on the floor.

They woke startled, asking why there was no reveille or *tenko*? Then they realized where they were. After a hearty breakfast, they went to have their photograph taken and bought some small gifts for the family. They stayed one more night and then heard the sound of many aircraft landing on the airfield nearby. They bid goodbye to the Herzlich family and made for the camp; here they discovered mail from home (the first since February 1942) and were given an aircraft seat for evacuation for the next day. Jack's happiest news was that his three brothers had survived the war in the forces and all was well at home.

On this last night, Jack and friends went back to say farewell to the Herzlich family, who told him they too would be leaving soon, for France. Whilst his friends went back to camp, Jack stayed on and went to a club; on the way back that evening he was kidnapped in the street and beaten by local nationalists, but freed by a policeman. Next day, 12 September, 1,400 men marched out of the camp, having been preceded by fifty amputees who were on the first flight out. Dakota aircraft landed, bringing in troops, and the POWs boarded in their place. That night they were in Bangkok, Thailand.

Home

Next day they flew to Rangoon in Burma and the pilots flew low over the Death Railway to show the men how it looked from the air. In Rangoon they were cared for by the RAPWI (Rehabilitation of Allied Prisoners of War and Internees). They were taken to long huts with tablecloths and flowers and were fed a huge European meal. Each man was handed a letter from the King, welcoming them home. But sad news reached them: the first aircraft, with twenty-five amputees, had been lost at sea after leaving Saigon, with no survivors.

While spending two weeks at the British General Hospital, where fourteen of them were being helped to recuperate by British nurses, an RAMC officer, Captain Saffer, visited Jack and asked him if he would like to go with him to Rangoon synagogue for the Jewish New Year (Rosh Hashana) services. The building was beautiful and the Jewish congregation were local Burmese in appearance. Soon afterwards, Jack and his colleagues were taken to the docks and they embarked on the SS *Ormonde* for Southampton. Here they were debriefed for several days in a transit camp, and one day in late November Jack arrived in Glasgow.

In 1946, Jack went to Paris to find the Herzlich family to thank them again, and here he met his French wife-to-be, Claudia Lange, with whom he later had four children. He eventually went back into the printing business in Canterbury. Jack became a founder of the FEPOW association, lobbying for years for welfare payments for veteran POWs, and for compensation from the British and Japanese governments, which he did live to see just before he died. He also joined both AJEX and the Burma Star Association and was a

24. **Jack Caplan burns the Japanese flag in protest at the emperor's visit to the UK in May 1998 near Buckingham Palace (from the *Sun* newspaper).**

member of the more militant Japanese Labour Camp Survivors Association. He had a deep hatred of the Japanese, and when the emperor made a state visit to the UK on 26 May 1998, Jack was with a large group of ex-POWs in the Mall. As the emperor passed, the men turned their backs as Jack, in the front row of the crowd, set light to a Japanese flag in protest at the visit. A group of policemen came forward to arrest him, but a phalanx of veterans with medals and green Burma Star berets defiantly faced the police and stood around him, and the police backed away. A colour photograph of this incident made the front pages of many national newspapers. Yet, in 1983, Jack received at his home a former Japanese officer, Lieutenant Masao Hirakubo, as a gesture of recon-ciliation.

There is one final anecdote that does not appear in Jack's book, which he related to this author. It appears that they made a stop on the way home in Singapore – though this incident may have occurred in Saigon. With a fellow Scotsman, Jack went out on the town one evening and, on reaching a street of nightclubs, they began to investigate some upstairs rooms. Singapore and other cities were still in some turmoil and Japanese troops were being used as local police all over parts of Southeast Asia until normal governance could be reinstated. On bursting into one room, the two men came upon a Japanese officer in bed with a local woman; before anything was said, the other POW grabbed the sword, unsheathed it and plunged it into the officer's chest, killing him at once. Jack was horrified and fled the building; he never discovered what happened afterwards.

Jack received the 1939–45 and Pacific Stars and War Medal.

In 1967, Jack was still living in Scotland, in East Kilbride, and set up a group of volunteers to go to fight in the Six Day War in Israel – a country he supported enormously and often visited for the World Jewish War Veterans Conventions and as a tourist; the embassy politely turned them down but thanked them for the offer. Just before he died, Jack received an invitation via AGATE, a reconciliation organization, to go to Japan on a visit of reconciliation in October 2002 and meet young Japanese and ex-servicemen who wished to apologize for what had happened to the POWs. At first he refused but his wife and friends persuaded him to go. When he returned he telephoned me and said that he had changed his mind a good deal and found the people he met charming and genuine. He said he had lectured to schools and appeared on Japanese TV, laid certain demons to rest and was no longer full of the prejudice he had.

Jack always said that his survival was based on his firm belief that his mother was watching over him and willing him to survive and get home; he never told his comrades this for fear of being mocked, but he maintained she was his guardian angel. For many years until he died he suffered from bouts of malaria. Soon after the war he often had severe outbreaks of a rash, which was diagnosed at the Queen Elizabeth Military Hospital at Woolwich as strongyloidiasis – fortunately treatable.

I met Jack on several occasions. He was always cheerful and smiling, looking like a mischievous gnome with his ruddy face and grey hair and broad Scots accent; he was enormously proud to be Jewish and a passionate Zionist, and never let his terrible experiences get the better of him. He came to the AJEX Jewish Military Museum and presented his .303 rifle and bayonet (almost taller than he was) – that he had somehow, miraculously, kept all those years – and his bush hat, mess tin and implements and his paybook. I also had the honour of seeing him several times with his wife at the annual AJEX parade in Whitehall, which he always attended, and where he always greeted me with a hug. He was a major player in the national and international fight for compensation for the POWs from the Japanese and British governments, and lived to see the final payout from Britain, rather than from the perpetrators themselves.

When he died in 2004, the wife of the Japanese ambassador, Masako Orita, attended the Jewish funeral with a delegation, bringing condolences from the Emperor.

For my part, I still miss him.

Notes

1. Driver Bill Emmett, Signalman Reg Niblow, Corporal Wood and Lance Corporal Hill.
2. The mess tin, spoon and knife are now in the AJEX Jewish Military Museum.
3. The author has been unable to find a Jewish RAF POW of this name, and this may be an error of memory on Jack's part
4. Together with Captain Suzuki, later executed for war crimes; he, like most guards, was totally unpredictable and could be jovial one moment, even giving POWs cigarettes, and brutal the next. He had a heavy stick with notches on for each POW he had killed – twenty-five by the end of the war; when Jack knew him, he had seven notches.
5. The camp site had previously been occupied by Tamils and other coolies who had no idea of hygiene and turned it into an open sewer; disease was rampant and the first job was to build on a new, adjacent site. However, it hardly helped to prevent the spread of diseases.
6. Jack returned to Kensayo in 1977 with a group of veterans; the site was then a luxury hotel and the owners allowed the men to stay with no payment for two nights when they told them that they had been POWs there. Jack described it as a most bizarre feeling, comparing the luxury to what he had experienced thirty-five years before. The owners invited Jack to meet them and to describe what had happened there, and he pointed out that where the swimming pool was had been the hospital hut; that the gardens had been the Japanese quarters; and so on. He said that they could hardly believe it.
7. Jack received nothing until August 1945, at the end of the war, when the Japanese decided suddenly to hand over thousands of letters they had cruelly kept back from POWs.
8. On some ships even this was not allowed.

The Story of Rob Cassuto's Dutch Jewish Family

This testimony has been taken from the van Rob Cassuto website: www.robcassuto.com.

My grandparents were the Cassuto and van Zuiden families, who in the early twentieth century emigrated to the then Dutch East Indies, where both my parents and myself were born.

Jews in the Netherlands Indies until the Second World War

Jews have never been in the Dutch Indies in great numbers. In the heyday of the Dutch colonial presence in 1932, there were around 2,000 out of 300,000 Dutch colonials, who lived among the then 60 million natives. One of the first Jews was Leendert Miero, who at the end of the 18th century was [a] soldier on Java and then began a trade. In Jarkarta his grave is still to be seen. Near Banda Aceh there is a Jewish cemetery at Poncut, which indicates [a] Jewish presence there at some time. A traveller from Jerusalem, Rabbi Jacob Saphir (1822–1886), who in the 1850s on his journey to collect funds for the Jewish community in Jerusalem also visited Batavia (Jakarta) ... mentions in his report 20 Jewish families of Dutch or German origin there.

As the 19th century progressed, there were gradually more Jews, mainly traders. Well known is the residence of the journalist Alexander Cohen, the rebel and anarchist who moved to the east and for a number of years was a soldier in the Dutch East Indies Army.

Around the turn of the century and beyond, the Indies was advertised extensively in the Netherlands as a place to serve in the colonial ranks. Many Jews responded. During the late 19th century and the first half of the twentieth century, more and more Jews emigrated to the colonies. They started a trade, served in the domestic administration of the colony or in the colonial army, were active in [the] teaching and medical professions, or were 'planters' in the 'ondernemingen' (extensive plantations of coffee, tea or rubber).

My two grandfathers were examples of Dutch Jews who went to the Indies for this development.

My grandfather Cassuto emigrated in 1915 to the Dutch East Indies as a young lawyer and teacher (later director of schools), educating so-called 'natives' for administrative and managerial positions. My grandfather on my mother's side, van Zuiden, after his education at the Royal Military Academy in the Netherlands, enlisted in the Royal Dutch Colonial Army, the KNIL. That was in 1905 or 1906. He spent many years in various outposts of the vast colonial empire. My assumption is that my grandfathers were also motivated by the greater freedom in the colonies. There were no heavy ties to the Jewish environment and there was less discriminatory prejudice and more career opportunities.

More romantically expressed: the adventure that lay ahead was less bounded by barriers for those of Jewish origin. In addition, there was a migration of Jews from the Ottoman

Empire to Southeast Asia and some ended up in Sumatra and Java. In Surabaya, there was even a small community of Sephardic Jews, from Iraq. They were called Baghdadis and had a modest synagogue, the only one in the then Dutch East Indies. Later in this story we will encounter them again.

To what extent was there a Jewish life in the 'emerald green belt' (literary nickname of the archipelago)?

Rabbi Jacob Saphir[,] who in the 19th century visited the Dutch East Indies[,] sighed [lamented?] in his report that the Jews there were hardly practising their religion anymore. They no more [i.e. no longer] circumcised their sons and did little to observe the Jewish holidays. In the twentieth century that still held true.

There was no rabbi in the whole of the Dutch Indies. If someone insisted on a circumcision of his/her son (brit mila) ... [that person] had to hire the rabbi of Singapore.

There was no synagogue, except in Surabaya, where the few hundred Baghdadis still were practising the Jewish way of life. There were Jews in larger numbers (about one hundred) in Batavia, Semarang and Bandung, but they ... [did not form] a real religious community.

One may ask whether these numbers are correct, because I suspect that many Jews didn't profess themselves as such. In that case, the estimation of 2,000 Jews in the 1930s may be on the low side.

My grandparents lived in the same way as all the other Dutch colonials.

They were members of 'Society' (the Club of the Dutch community, nicknamed 'the soos'). They were active in drama clubs. A journal article preserved by my grandmother reviews a successful stage performance, which my grandmother appears to have co-produced.

And a hilarious picture of my otherwise very modest Cassuto grandfather in his young Indian years shows him dressed as a woman in the play 'Charlie's Aunt', showing a glimpse of their participation in the pleasures of a carefree life.

They celebrated Christmas and Santa Claus, and were involved in charities and in a variety of celebrations around [connected with?] the royal family.

My grandfather van Zuiden is to be seen on [i.e. in] photos sitting at long tables on festive occasions, surrounded by many more ... companions [whose names are not known].

The younger brother of my father ... [who was also] born on Java ... wrote in his memoirs about his experience as a Jewish boy:

> Though being a Jewish boy my Jewish background meant little to me. The idea that once [sic] others would attach great interest to it didn't in the least occur to me ... The fact that I was Jewish no more impressed me than the fact that I had two hands, two legs and a nose. It was just a small part, not worth noticing.

Later he wrote about the holiday time of his family spent in the Netherlands (it was 1929 and he was ten years old): 'In Holland, I met my Jewish family and I discovered that they were quite different than the Dutch colonials. Intuitively I began to realize that being a Jew meant "being different".'

Partly from sources, partly from hearsay, I assume that the Jews in the Dutch Indies were looking for friends in the Jewish circle. In the time that my grandfather Cassuto and his family lived in Bandung ... [they] made many trips to the plantations at Cigombong, near Bandung on the plateau of Preanger, which was run by their good friend, the Jewish 'planter' Albert Zeehandelaar. Other suspected Jewish families figure in photo albums.

So my grandfather Cassuto met with my other grandfather van Zuiden, when the latter[,] after many wanderings on military posts in the outer regions[,] was stationed in Bandung in an administrative function. They were good friends.

Both were members of the Masonic lodge of Bandung. Many Jewish residents of larger cities in the Dutch Indies were members of a Masonic lodge.

Although the vast majority of Jews were no longer practising Judaism, they had a need ... to discuss the deeper things of life, to philosophize and to celebrate certain events together. The lodge offered[,] of course not only to Jews, but also to others[,] this opportunity to meet kindred spirits in a free atmosphere. In those meetings many contacts were made.

Sometimes Jewish celebrations were held. I met a man at the lodge who like me had lived as a Jewish boy in Bandung. He was a few years older than me and he could remember that in the Freemasons Lodge Christmas as well as Chanukah was celebrated. Without doubt, the lodge also offered many opportunities for what we now call 'networking'.

Also Zionism attracted the attention of the Jewish residents of the East. There seems to have existed a periodical[,] from 1926 until the Japanese occupation[,] called 'Erets Israel'.

Jews in the Dutch Indies during and after the Second World War

The Jewish colonists in the Dutch Indies in the first decades of the last century lived a relatively carefree and luxurious life. They lived in beautiful villas with many servants around them. My parents had a wonderful childhood. 'Oh, Indonesia! How safe and beautiful you were – so many miles and years away', sighs my father's brother later in his autobiography. The two brothers had a wonderful childhood.

Dark clouds loomed over the tropical paradise but most seemed successful in ignoring them. An alarming happening, of course, was the invasion of the Germans into the Netherlands. A letter from my father from 1945 contains a compelling testimony of the shocking news.

My father had[,] after his studies in the Netherlands[,] just arrived in Batavia (Jakarta) and had a job as an administration official; he was working at the telex, when ... in May 1940 [came] the report of the German invasion; he describes it:

> On May 10, 1940 I was in Batavia in the Office to the Assistant Resident (Provincial Governor). [On] Wednesday, May 8, when I asked if I could spend the weekend in Bandoeng, the assistant governor said that given the situation in Europe he could not allow this. [On] Friday May 10 about 11 o'clock we got the message of the invasion of Holland. Immediately, all suspect elements were arrested, and then in about 2 hours the message was to be made public.

We worked, in shifts, day and night at the office. All kinds of people gathered at our office offering cooperation and financial gifts. Some Arabs donated about 5,000 Dutch guilders as a contribution to the war effort. Small old Indonesian men asked to be sent to Holland to join the fight. Many who actually could not afford it brought gold and silver objects. Great was the anger against every one who claimed themselves 'NSB-ers' (supporters of the National-Socialist party). The Deutsche Klub in Bandung and the NSB clubhouse on Naripan Road were destroyed by the Bandung schoolchildren.

In the Dutch Indies life returned to the old routine. However, the young Dutch men were called into military service and were given military training.

My parents – both born and raised in Java – had as a young married couple returned to the Dutch Indies at the end of 1939 after a Dutch interlude of several years of study; my father began as a young civil servant in the colonial administration, but soon after his arrival he too had to comply with being called into military service for some months of training as an officer.

The colonial community liked to believe that the new world war would pass by the 'Emerald Belt'. But at the end of 1941, this unbridled optimism finally was dashed. Instead came the sudden bombing of Pearl Harbour [sic] on 7th December 1941. At the same time the Japanese armies attacked Singapore. On December 8 the Dutch government declared war on Japan. The Japanese army invaded the Dutch Indies on January 10 1942. Singapore fell. After a short, courageous and hopeless endeavour to defend the vast empire of thousands of isles, the Dutch Colonial Army surrendered on March 8 and the Japanese occupation began. Military prisoners of war were largely employed as slave labourers. A large group, including my father, was transported under inhuman conditions to Burma and Thailand to construct a railway through the jungle – the infamous Burma railway. Other men, women and children of European descent were interned in civilian camps across Indonesia.

What about the Jews in this turbulent time?

Initially it was no issue for the Japanese, whether people were Jewish or non-Jewish. In the beginning it was even an advantage [to be Jewish]. The Japanese regarded the Jew as 'Asian' and they let him in alone [i.e. and they left him alone]. Particularly the Baghdadi in Surabaya will have profited from this view. To my opinion [i.e. to my knowledge] many assimilated Jews didn't disclose themselves as Jews to the Japanese, if only because they no longer felt identified with Judaism. They felt themselves firstly as all Europeans, mere Dutch nationals [i.e. as primarily European, as Duch nationals], and they behaved as such.

So did my mother and her parents, who in the beginning of 1940[,] after years of retirement in the Netherlands[,] had decided to visit their newly married daughter in the Dutch Indies. By this action, my now since long retired grandfather and his wife escaped the Germans hunting for the Jews and my grandparents certainly ... saved themselves from the gas chambers ... [only] to fall prey to starvation by the Japanese. On their identity card issued by the Japanese occupiers was to be read: 'Bangsa' (race, nationality): 'totok Blanda' (born in the Netherlands, pure Dutch). Although their journey to the then Dutch East Indies [was] in my opinion actually a flight[,] they never defined it as such, but I presume this escape motive will surely have been going through the head of my sober and realistic grandfather.

Other Jews who decided to escape from the Germans in the Netherlands – and also from Germany, Austria and other European countries – fled to the Dutch Indies, be it ... [in] a more leisurely way during the late [nineteen] thirties or in panic during the outbreak of war in Europe. An example is Lydia Chagoll, who describes how in 1940 as a young girl she and her sister came with her parents from Brussels via a long escape route through France, Portugal, Mozambique [and] South Africa, eventually arriving in Batavia (Jakarta) and finally ending up in a Japanese camp. We will deal with her book in a while.

In the course of 1942 the Japanese sealed off areas with barbed wire (kawat) and plaited bamboo walls (bilik, gedek)[,] transforming them to internment camps[,] and drove the white women and children there together. Before being summoned to go in the camps, many went voluntarily, like my mother. She was born in the Dutch East Indies, and initially there was no need [for her to go] but she nevertheless preferred to go into ... the camp – in this case the Bandung quarter Cihapit – where her mother, friends and acquaintances were already and where it was safer than outside the camp.

As for Japanese attitudes towards the Jews, the average Japanese soldier had no idea of what a Jew was. Perhaps some officers were more aware of it [i.e. what being a Jew meant]. Initially, there was no official policy implying a discriminatory or special treatment of the Jews. The Japanese actually had nothing against the Jews. Japan itself absorbed even a large group of Jewish refugees. Over two thousand Jews in possession of papers, provided by the Dutch consul in Lithuania[,] Jan Zwartendijk[,] and Japanese diplomat Chiune Sugihara, consul general in Kovno, Lithuania, had travelled through Siberia and the Far East aiming to go to America. They were stranded in Japan.

The Japanese were both hospitable ... [and] intrigued by the refugees. In particular, the Rabbis and Yeshiva (Judaic college) students looked strange to them. Members of the Photography Club Tanpei made photographs of the picturesque refugees; the photos were exhibited under the title: The Wandering Jew. One of the photographers in a photo magazine wrote in his portrait of a Yeshiva student: 'How the eyebrows [sic] of the displaced man express not only grief and misery ...but also the tenacity of a people desperate and scattered about the world. However, they can not hide their sorrow. They fight in order not to be beaten.'

After the war, most refugees remembered how interested the Japanese were and how they were not possessed with [i.e. did not display] anti-Semitism and didn't subject them to anti-Semitic treatment, which in pre-war Poland too many had experienced.

Some, about a thousand of them, before the attack on Pearl Harbour [sic], did reach the US and Canada. In the autumn of 1941 the others were deported to Shang Hai, where more than twenty thousand other Jewish refugees had sought refuge. There they formed a ghetto, where life was difficult, but despite that there were more than 50 newspapers and magazines in Polish, German and Yiddish. The history is available on the website of the US Holocaust Memorial Museum.

Jews Must Report

The relatively mild attitude towards the Jews by the Japanese occupiers of the Dutch Indies underwent[,] in the course of 1943[,] a change. As a gesture to the German allies the Japanese authorities called upon the Jews to report to the police. According to Jacques Presser in his Dutch book about the persecution of the Jews in the Netherlands and the colonies[,] this attitude change was put in motion after a visit by a German consultant[,] Dr Wohltat[,] to Java. This consultant had to point out to the Japanese allies how dangerous the Jews were and that they should no longer be left alone. Those who obviously were Jews (Iraqi Jews for example) and those who went and registered disappeared into the camps (as did the Freemasons), though some were there already. The Japanese lacked the German ruthlessness on this subject; many 'stateless persons' remained undisturbed.

The Japanese were probably wondering why the Germans were so obsessed by the Jews. Presser reported that in one camp – it is not stated which – some ten Jews had been compared to the caricatural images in the 'Der Stürmer' Nazi journal – but it did not work out as the Japanese appeared largely to ignore such anti-Jewish propaganda.

When a Jew died, his funeral was performed with the required ritual. It proved particularly difficult to convince the Japanese that not all Americans were Jews and that Roosevelt, whose mother was called Sarah, was not [a Jew].

In general, the treatment of Jews in the camps seems to have deviated little from the treatment of the others in the Cimahi camp.

My mother ... [was meant] to report as a Jewess but she did not. She has told me that she felt that the Japanese attached little value to it and therefore nobody noticed her disobedience. Many assimilated European Jewish women (maybe some men at Cimahi camp) will have followed the same line.

As for the camp experience of my grandmother, my mother and I, we were in the same boat as the other European women and children.

Those Jewish women who had reported as Jewish were sent to camp Tangerang, just west of Batavia (Jakarta). Part of the camp was intended for Jews. Memories of camp Tangerang and Adek (the camp where[,] at the end of the war[,] residents of camp Tangerang were to go) are to be found in the book 'Camp Memories' of [i.e. by] Miep Bakker. She mentions in her preface, that 'Jewish women were brought in.' She does not mention a separate department. But she describes a number of events featuring Jewish women such as Ms Cohen, who got beautiful curls in her hair after an attack of typhus; and then there was the young Jewess who had lived for years in Japan and who was now acting as an interpreter[,] whispering in the ear of Miep Bakker not to get into a discussion with the angry Japanese commander.

For a description of the camp from the Jewish experience [from the Jewish point of view?] I make grateful use of the book 'Six Years and Six Months' of [i.e. by] Lydia Chagoll. She came with her family from Brussels, [and] arrived after many wanderings in the Dutch Indies and her mother had reported as a Jewess after the summons of the Japanese. From Tjideng camp, where they first stayed, she was transferred with her sister to Tangerang. There was, Lydia recalled, a Christian section and a Jewish section, where Jews [of all nationalities] lived ... religious or not, in various big rooms, with women who were married to Jews as well. For example, there were also a Javanese with her two daughters, and a Chinese with her child.

There were also European Freemason women.

The only man besides the doctor who stayed as a prisoner for a long time in the camp was the famous violinist Szymon Goldberg. He fled Nazi Germany in 1933. He was on tour in the Dutch Indies with his wife, also a pianist, and the equally famous piano virtuoso Lily Kraus. All three were interned in Tangerang. Goldberg and his wife had a separate room assigned [to them], stirring the jealousy of other women. Occasionally they gave a concert for the Japanese commander.

After a few months Szymon Goldberg was transferred to another camp. His wife was moved from the separate room to the public sleeping room, [where] her bed was opposite that of Lily Kraus. Lily Kraus lived completely isolated from other women. The other women considered her completely crazy, because ... [she] did yoga, reports Lydia Chagoll.

Miep Bakker tells how at a certain moment in the hangar that had served as a gym, a big black piano was brought in. Lily Kraus was required to study every day. Once a month she was [meant] to give a concert, which actually happened a few times. Miep Bakker tells how during [her spare time] she would sit in the corner of the entrance to the hangar[,] listening ... [with complete] delight to the piano sonatas of Mozart, which Kraus was world famous for.

Szymon Goldberg and Lily Kraus have survived the war and have continued their career[s], Kraus despite the damage to her hands.

The 'Irakers'

A shed in the Jewish section was populated only by Iraqi Jews in Surabaya, the Baghdadis that we already ran into [i.e. that I mentioned earlier]. I would like to quote Lydia Chagoll's vivid description of these very special camp residents:

> Irakers! Temperamental people, some dressed as gypsies. No one ventured a step in the Iraqi barracks, which also during daytime remained shrouded in mysterious darkness. It was always noisy. There was always something going on. In loud voices they gave expression to their moods. They came regularly to blows. With an infernal noise, crying and shrieking, they fervently beat one another with whatever was to hand. When there was nothing they went to fight with bare fists, even with the teeth. If one of them, however, was attacked by another camp resident then they stood as one man, a strong power, a united group. At that moment it was wiser not to interfere, it was wiser not to look, quite probably you would catch an unintended blow. No, it was best to stay as far as possible.
>
> The Irakers preferred to deal with only their own group members. A few tried to be their friend, but everybody avoided to be [i.e. avoided becoming] their enemies.

The Japanese did not know very well how to cope with the Iraqi Jews. Chagoll [writes]:

> It was on the advice of the Germans that the Irakers had been transferred from Surabaya to the Tangerang camp with its Jewish section. But the Japanese were puzzled. The Irakers looked physically so different and behaved so differently in comparison with the other camp residents. The Irakers were not Westerners, not Europeans, not whites. The white skinned Europeans were worthy of their contempt, but not the Irakers not so!

Notwithstanding, the Iraqi division was likely the most religiously observant place throughout the archipelago. There was even a kosher kitchen! [This is the] ... striking description of Chagoll:

> Maybe because the Japanese accepted Irakers as non Europeans, or in order to bully the whites, or because there were not enough guards to contain the Iraqi furies, or in order to frustrate the Germans who were in spite of all still white skinned, or simply not to stir 'soesa' (troubles), the Japanese gave in to urgent requests of the

> Irakers and provided them with a kosher kitchen. Only the Iraqi community was officially observant of the strict Jewish religious laws. To live accordingly was something else. That had to be taken with a grain of salt and a pinch of salt of very rough quality. The other Jewish population in the camp, with a few exceptions, was quite liberally oriented. So Irakers wanted a kosher kitchen. The Irakers were so convinced of their right that they could move mountains to reach their goal. There came a kosher kitchen. A totally separate kitchen service, only controlled by them from the outset till distribution.

Lydia suspected, probably not wrongly, that the kosher kitchen also was a good opportunity for processing smuggled food.

Adek Camp and Capitulation

In the spring of 1945, the whole camp moved to Adek, a building in Batavia (Jakarta). Lydia Chagoll painted, in a brief paragraph, a vision of the inhabitants of her shack:

> Part of the Jewish Diaspora was united in our Adek shed. A small Palestine without men. There were nine countries represented: Netherlands-Belgium-Austria-Germany-France-England-Romania-Iraq-China and of course the Dutch Indies. All together some fifty women, with or without children. Housewives, lawyers, nurses, beauticians, prostitutes, office clerks, salesmen, seamstresses, business women. Together we shared a continuous [communal?] bed. Everyone was entitled to fifty centimetres. It was a small hut, about 9 to [i.e. by] 5 meters. The group did well. I cannot remember quarrels, only small frictions. Together we tried to make the best of it, interfering as little as possible with each other.

Chagoll proceeds with her personal story of exhaustion, hunger, disease and apathy, and finally the liberation in August 1945, which was not liberation. Repatriation in April 1946 and the difficult adjustment to European life came later. T[his is t]he story of many children, including me, who suffered in this way, which is also reported elsewhere.

In summary we can say that a small ... [number] of the Jews in the Dutch Indies had been interned separately as Jews, and they ... [had] been treated as bad[ly] as – not worse [than] – ... the non-Jewish internees in other camps[,] including the assimilated Jewish women, children and some men (like my grandfather held captive in the Cimahi camp).

Among the soldiers made prisoners of war and forced to perform slave labour, such as at the Burma railway, there ... [were] also a number of Jewish men, my father was one of them. To my knowledge, the Japanese ... [did not make] a case of special selection of Jews.

Indonesia

After the war most Jews of European stock had to live through the post-war turbulence (called the Bersiap period) such as the independence struggle of the Indonesians[,] and this anarchic period with its robber gangs resulted in many victims.

Many Jewish families had lost their homes and possessions and returned to Europe and the Netherlands.

Some stayed in Indonesia. In the [nineteen] fifties there still seemed room for a Jewish presence. According to Beth Ha-Tefoetsot (The Diaspora Museum in Tel Aviv): 'there were 450 Jews in 1957 in Indonesia, in Jakarta and Surabaya, Ashkenazi Jews and Sephardic Jews. The Sephardic community in Surabaya, especially, started to bloom. Oral witnesses gave an optimistic picture of the 1950s.

'They say that in this period thousands of Jews lived in Surabaya, and that the centre of the city was dominated by them ... [as] by the Chinese today. The community had acquired a new synagogue and had made a badminton court behind it. The young people did well in sports, studied the holy writings and learned Hebrew, they celebrated the holidays and celebrations of life with their family. But this glory would not last long.'

When in the beginning of the [19]60s the New Guinea affair stirred nationalist and anti-Dutch sentiments, many left Indonesia to [go to] the United States, Australia and Israel. In 1969 there were 20 Jews in Jakarta and [in] Surabaya there were] 25. Some went to Israel and formed a … [close-knit] group [as they were so few in number]. 'The Dutch-Jewish immigrant community doesn't understand the immigrants from Indonesia and has no desire to', says Shoshanna Lehrer, the initiator of the association of Israelis with an Indonesian past, 'Tempo Dulu' ('ancient times'). 4 to 5 times per year there is a gathering with a lecture or so [i.e. or two] about Indonesia and each member brings something for the Indonesian food dinner. The members fall roughly into three groups: Dutch Jews, Jews from the Baghdad community of Surabaya and former refugees mainly from Germany and Austria.

Now there are only a few Jews: 20 in Surabaya, the remainder of the Iraqi community, and perhaps a few individuals in Jakarta.

The synagogue in Surabaya is still there.

In the colonial era a Dutch doctor lived there. His house was purchased in 1950 and rebuilt. The outside of the synagogue is white. Inside it is an undeniable orthodox Sephardic synagogue. The simple Holy Ark is empty now and the Torah scroll has been moved to the larger community in Singapore. There is no rabbi or teacher anymore. In the case of the desk, a clutter of books is to be found, including pre-war Dutch books, prayer books of the army from the Second World War, and shiny new prayer books, sent by a Sephardic institution in New York.

1282616 LAC Harry Alec Courts, No. 13 Group, RAF

25. 1282616 LAC Harry Courts, RAF.

A devout Jew, Harry Courts had the dubious distinction of surviving twenty-three different Japanese POW camps. He was from 58 Linthorpe Rd, London N16, and was the husband of Rita. His parents were Morris (born in Poland) and Ada, and Harry was born in Bow. He attended the Coopers Secondary School and joined the RAF in Newcastle in July 1939. He was posted to Singapore in 1941, based at Kallang, when on 13 February 1942, just before the surrender, he was ordered to be evacuated with other RAF personnel on the SS *Trang*, to Java. The boat ran aground and all but thirty-five men were taken off by other vessels, including survivors of the *Repulse* and *Prince of Wales* which had recently been sunk by the Japanese.

In the lifeboats, Harry and sixteen comrades started a fourteen-day journey, rowing to Java, island hopping. At Banca Island they heard Java had been captured and so they began to make for Sumatra, spending another fourteen days rowing at sea. Arriving exhausted in Sumatra, they found that they had just missed a gunboat that would have got them away, and were told to continue to the town of Padang, where they arrived on 14 March, only to be told that the last naval vessels had just left for Australia. On 16 March, Sumatra fell and Harry was taken prisoner. At about this time his wife was told he had been lost at sea, and it was two years before she heard that he was a POW.

Harry was transported on the Japanese ship *England Maru* with 500 others to Mergui in Burma; he was the only Jewish POW in that group. Here he was hospitalized with dysentery and had all his kit stolen from him, including his prized Hebrew prayer book, tefillin (phylacteries) and tallit (prayer shawl). By January 1943 he was working on the Death Railway; his only kit was a wheel hub that he used as plate; a ragged pair of long trousers, much patched; and a rubber sheet for a bed. By December 1943 his family had heard that he had survived, and they threw a big party to celebrate that at least he was alive.

He was in nine different camps from Thambuzayat to Kanchanaburi, labouring on the terrible Death Railway, 'digging, laying rails, living on rice and water' (see *JC* supplement on Jewish POWs, undated, p.31 in AJEX Jewish Military Museum Archives). Beatings and hunger were their constant companions, but Harry was convinced that it was not only stamina but attitude and hope that kept people alive; those who gave up, whatever their state of health, soon died. Harry prayed every day and earned the respect of his non-Jewish comrades. He was small of stature and they looked after him. He had an excellent knowledge of the Jewish calendar and was able to calculate the Jewish festivals and Holy Days so he could pray at the correct and appropriate times; by the war's end he had discovered that he was only one day out on the calendar.[1] In one camp he met a Dutch Jewish medical orderly who arranged to hold a Passover Seder meal for the Jewish POWs, but Harry was moved on before it took place. He said that one day's pay could buy you a banana from a local trader; two days' pay got you an egg.

From the last railway camp at Tamarkan, Harry was sent by train to Phnom Penh in Cambodia and then by barge to Saigon. Of the original 500, there were now only 150 left. Here he was for the first time in proper huts in an old French army camp, instead of huts made of jungle bamboo and atap. They were sent to build an airstrip under the supervision of brutal Korean guards. When one guard struck him for no reason and knocked off his glasses, a Japanese remonstrated with him and gave Harry his spectacles back. It was in Saigon that he heard of the Japanese surrender; he went into town and made contact with a Dutch Jewish family called Schindler, with whom he spent Yom Kippur 1945.

Next day he was flown to recuperate at Rangoon, then to Calcutta where he crossed India by train, then via Tel-Aviv in Israel, to London, where he arrived in October 1945. He rejoined the family textile business and in 1975 he made a pilgrimage with his wife to the scene of his ordeals on the Railway. He died in February 1993.

Note

1. My interview with his son Leonard in August 2012.

270544 Pilot Officer/Flt Lt Oswald 'Ossie' Nathan Diamond, DFC, Pilot, No. 1 Squadron, RAAF

The following account is based on a recorded interview made in Brisbane on 3 July 1980, and other information comes from family sources. Ossie was the first Allied serviceman in the Second World War to strike at the Japanese (this is explained below) and the first bomber pilot to shoot down a Japanese Zero using fixed machine guns – Browning .303.

26. 270544 P. Off. Oswald Diamond, DFC, RAAF.

The grandson of a rabbi from Kiev, Ossie was born on 27 October 1916 in Brisbane, Australia, son of Jewish immigrants Joseph Dashevski from Kiev, and Rose Mendoza from London.[1] He was bar mitzvah at Margaret Street Synagogue and attended every Saturday with his father. He went as a boarder to Brisbane Boys School and always kept his Jewish background in his heart (though his wife, Norma Epton, was not Jewish and his children were brought up as Anglicans).

Ossie was in the militia from 1936 to 1938, and a chemistry student when war broke out. He flew at the Brisbane Aero Club before the war and then, having been commissioned at Point Cook air base, he and his squadron (No. 1) were kitted out with tropical gear (shorts, pith helmets, and so on), went through various medical checks and were posted to Malaya by ship in 1940, with ground crews and aircraft. The Lockheed Hudson bombers he was to fly were waiting for them in Singapore, flown previously via Alice Springs and Darwin.

Ossie said that the RAAF (Royal Australian Air Force) took a dim view of the lax methods of the RAF, especially when the RAAF crews needed spare parts; all the RAF bases seemed to close for the weekend, despite the approaching war clouds. The British seemed to regard Malaya as a soft posting and this was reflected in the lack of preparedness for what was to come.

Ossie trained intensely, though restricted by fuel and ammunition conservation; runs included bombing practice (using mock bombs) at high and low level, air-to-air and air-to-ground gunnery, and navigating over water by day and night. In between times, however, life was good; he, like most officers, was able to buy a small car, and he visited the pleasant Cameron Highlands for leave, stayed with wealthy planters in their homes, and drank at Raffles Hotel in Singapore.

No. 1 Squadron – the first Allies to engage the Japanese in the Second World War – was at Kota Bharu base on the Thai border (east coast) when the war broke out in December 1941.[2] Intelligence was very bad and they were to be taken by surprise later on by the fast Zero planes used by the Japanese; they had seen high-flying Japanese aircraft, clearly on reconnaissance, for some weeks before. When obsolete Brewster Buffalos were sent to intercept, they could not reach the altitude required and had to turn back. Their own reconnaissance planes were often fired on by Japanese ships.

In the early hours of the morning of 8 December, an alarm came through from Indian HQ on the beach that there was a lot of movement of small boats off the coast. The Indian troops engaged them, and even more boats carrying Japanese troops appeared. Ossie and six other crews were ordered to 'bomb up' and attack immediately,[3] taking off with some difficulty from the soggy grass airstrip in the heavy rain. It was soon apparent that it was an invasion fleet of warships and landing craft.

Ossie was the third or fourth aircraft off (no. A1632) and they bombed the Japanese ships at mast level. They were only a few kilometres from the invasion beaches. On the second run, they hit one ship at 3.30 a.m., with a stick of two 250lb bombs, followed by a terrific explosion,[4] and then hit it again with two more; it sank a few hours later. He also strafed the ship and enemy tanks and infantry on the beaches, receiving rifle fire from the Japanese in return.[5] But either the blast or an anti-aircraft gun hit them, and Ossie lost the port engine, so he made straight for base, which was quite nearby. Two other RAAF aircraft were shot down. Changing aircraft at once, they continued bombing and strafing the enemy all morning and caused a lot of damage and casualties, as far as they could see, especially on the troop-carrying barges, which fired back with mounted machine guns. The Japanese later confirmed that these were some of their highest casualties in the campaign – between 8,000 and 10,000 men killed or wounded.

On a second sortie that day, Ossie bombed a crucial Thai railway bridge at Pasir Mas, causing severe disruption to the Japanese advance. Ossie also witnessed, on one of nine reconnaissance flights he made, the attack on the *Prince of Wales* and the *Repulse*. On night flights he saw large burning arrows in the jungle, pointing towards Singapore, to guide Japanese bombers to the city; this was the work of the many Japanese spies and Fifth Columnists who had been working in Malaya for some years.

In a further incident, Ossie's Hudson acted as a fighter escort to accompany another Hudson, with bombs, to attack the Japanese landing at Endau, also on the Malaya east coast. They were attacked by Zeros, just as they were radioing the report of the landings to Singapore. Four Zeros surrounded them so they made for a nearby large cloud to avoid the enemy fighters. Each time they emerged, a Zero was waiting, and one got on the tail of Ossie's fellow pilot Dave Colquhoun, hitting him but fortunately doing little damage, so Ossie turned in and blew the Zero straight out of the sky with his machine gun. A few minutes later, Ossie's co-pilot Flying Officer George Evans shot down another Zero with one of the two side Lewis guns. But sadly Ossie also witnessed how Zeros easily shot down the gallant swordfish pilots in their obsolete planes, watching many men go to their deaths in this way.

At Kulai Krai they had to evacuate the aerodrome at the last moment, by railway, as the Japanese overran it. Remaining fuel, ammunition and any damaged aircraft were blown up, but one managed to take off with some troops to Singapore, to fight another day. When Ossie ran back to his quarters to collect his small pack, he saw Indian soldiers fighting hand to hand with the Japanese in his hut as they retreated. On the two-day train journey they were strafed and bombed, and when fuel got low they had to tear up the floorboards and fittings to burn as fuel in the engine to get them to Singapore.

At Sembawang, Ossie's squadron were constantly bombed by the Japanese, who broadcast that they would drop Geisha girls on the Australians. But the Aussie aircraft were well hidden in pens. Nevertheless, Ossie was flown with three others on a Quantas DH86 out to Darwin to bring back more Hudson aircraft. On the way back they refuelled at Timor and landed them safely at Singapore and flew them in the remainder of the battle.

When Singapore fell, the squadron was evacuated to Surabaya in Sumatra to a secret airfield called simply P2, from where they bombed a lot of Japanese shipping and again were in combat with a lot of Zeros trying to protect the Japanese invasion of the Dutch Indies. The Japanese never found P2 because it was so well camouflaged. Though they were unescorted as usual, the Australian element of surprise was good as they would come in very low out of the sun, to bomb the Japanese ships. Ossie also bombed the railway yards and docks at Singapore, by then occupied by the Japanese, by day and by night, causing severe damage. On one run near Sumatra (one source says over Banka Island) in early March, they hit a troopship and came over again to machine-gun it; suddenly four Zeros jumped them so they made a run for it. Two gave up, but two stayed to chase them. Ossie flew so low to avoid the Zeros that the rear gunner told him they were causing a wash on the water like a speed boat, at just sixteen feet above the sea! He wove from side to side and saw the Zero's bullets stream past into the water each time. Then one Zero came in from the side and, for some unexplained reason, Ossie left his seat and sat on the control panel, guiding the aircraft with his hand as he felt the Japanese pilot would aim for his seat. Sure enough, a few seconds later, bullets came streaming through the seat where he had sat, and down through the floor. When he sat down again, the seat was hot; he had lost his left undercarriage and the starboard engine and tailplane were shot away. After this sixty-kilometre chase, weaving and diving, Ossie shouted to Evans, asking which way to go, and they made for home, streaming smoke; the Zero broke away.

They made it to Palembang RAF/Dutch aerodrome by following the river at low level but high enough to clear the trees. Here they crash landed on the one wheel and saw that the whole aircraft was covered in holes like a sieve. Ossie then saw a Japanese paratrooper caught in a tree – he began to bring his machine gun round to shoot, so Ossie had to shoot him dead: the first man he ever shot. He felt quite ill afterwards.

Ossie tried to take off in another Hudson which was damaged. But, having tried, he had to abort and crashed-landed again; this time they let out the fuel and, using a flare gun, set it alight and destroyed the plane.[6] There was a lot of gunfire nearby and it was clear the Japanese were about. For a while Ossie and the crew took part as infantry against the encroaching Japanese troops; they had one Lewis gun from the plane, one had a rifle, and Ossie and Evans had a revolver each; they killed two Japanese on a machine gun, but it was manned again, so they decided to try and skirt round it to escape. Hiding in the thick grass, Oscar's co-pilot asked, 'Do you know it is Friday the 13th today?' and Oscar said, 'Well, thirteen is my lucky number, mate.'

They got to a road where they met some Ambonese (Dutch native) troops, who were evacuating the aerodrome towards Palembang town. Ossie took command as there were no officers and his adrenalin was up. As they went along, the Ambonese suddenly disappeared into the forest, leaving Ossie and his three crew. Knowing they would probably be shot if captured, they continued alone through a swamp and paddy fields, waist-high in water in most places, keeping their cigarettes alight to burn off the leeches, and came to a native village as dark approached. Ossie put his gun into the back of a local and demanded to be taken to Palembang, reaching it after a ten-hour trek. They found RAF HQ and were told that Palembang was being evacuated too, and they were ordered, as air crew, to escape to Java. They had been very lucky and they managed to get through the Japanese lines to the town. They got onto a truck to the aerodrome and there found various personnel trying to get away; about nine of them got into a Hudson and Ossie, with great difficulty and having not slept for forty-eight hours, took off to Java.

Several Hudsons greeted them and escorted them to Bandung airfield. Ossie had a near escape when again Japanese captured the aerodrome and he got away in a jeep. Communications and Command was now in chaos. In one incident, he saw a Dutch pilot come to the aerodrome, put his wife and child in a slit trench, take off in his Brewster Buffalo, engage two Zeros, shoot one down, land with his head bleeding profusely, pick up his family and go home. He had done his bit. It appeared to be every man for himself, as initiative dictated.

Ossie led a convoy of cars to the nearby port of Telakap to get RAAF men away from Java by ship. At the port, they saw that all ships had been sunk so they returned to Bandung. On the way a Japanese tank blocked the road. Ossie by now was riding an abandoned British Aerial motorbike; he could not stop and hit the tank full on. He flew through the air, hit a hedge, passed out, and when he woke up the convoy was there and the tank gone; he had a broken foot and collar bone. Patched up by Wing Commander Ramsey-Rae, they continued in a car to a rendezvous on the coast that they had been told about, by radio, where some RAAF Catalinas from Darwin would pick them all up. Sadly, these aircraft were shot down. Then local natives gave them away and the Japanese sent a message, telling them to come and surrender or the Japanese would come themselves, and take no prisoners. So the CO of the convoy decided to go to a nearby hill station by car to see exactly what the situation was. Here there were hundreds of Japanese about but nobody took any notice of the RAAF men, who were armed and had more guns in the boot of their car. Ossie thinks that the Japanese mistook them for German advisers who were known to be serving with the Japanese.

Ossie was taken to hospital in Bandung, commanded by Colonel 'Weary' Dunlop, and it was here he was taken prisoner with about 120 other RAAF men, on 8 March 1942, as the Japanese arrived. Meanwhile he was posted as Missing in Action, presumed killed, his mother receiving a letter to that effect on 12 May 1942. For his courage during this period, Ossie was awarded the DFC (Distinguished Flying Cross – part of his DFC citation is given at the end of this testimony).

He was taken to the famous Cycle Camp in Java with about 3,000 other POWs, and after six months was taken to Changi in Singapore, then on to the Burma Railway, then to Formosa by ship. This was one of the notorious Hell Ships, and the conditions – overcrowding, poor food and water, and filthy hygiene – made it a journey of horror. With two mates, they used a groundsheet to gather condensation from the sweat of the men as it fell from the roof of the hold, and rationed it strictly from a bottle between them. It saved their lives.

Oscar was then sent to Kyushu in Japan, where he worked as a slave in the mines at Shirakawa; he was there at Christmas 1943, and until March 1944. Then on to Pusan in Korea, and then to Mukden in Manchuria; having been very ill, his health improved here because the cream of the Japanese Army were stationed in the area and the camp cooks were able to make bread and porridge from soya beans. The camp here was more orderly than many he had been in and officers had to work or get lower rations. It was here that one of the incidents occurred which were often described by ex-POWs as having happened only in their camp. The men often smuggled bits of food into the camp, which they had traded with locals while out on working parties. When searched at the camp gates they always managed to secretly pass items to men already searched; frustrated, a Japanese officer, believing he knew good English, screamed out: 'You bloody Australians; you think I know bugger nothing, but in fact I know bugger all!' Often Oscar witnessed damaged Japanese aircraft making their way back from the Russian front, and convoys mangled by Russian shelling. On their secret radio they heard about the atomic bombs.

Russian tanks broke through to their camp and liberated them; they killed a lot of the Japanese and Korean guards, then drove straight on; this was followed by American aircraft dropping cases of Spam, coffee and cigarettes by parachute. One huge canister knocked out the cookhouse and water tower. Some of the men broke into a nearby brewery and got drunk on Japanese beer. Mukden contained many nationalities and included many senior American officers captured at Corregidor and Bataan, including the CO, General Wainwright.

Oscar told his wife and children many years later that during his imprisonment he had his teeth knocked out with a rifle butt for answering back to a Japanese camp commandant, and that in another punishment he had bamboo slits pushed under his fingernails. He also said that on one occasion he was put in an underwater cage (probably in Thailand) in a river, and survived by sucking air in through a hollow reed.[7]

Ossie was taken to 113 Indian General Hospital in Calcutta to recover from his POW experience, as he had severe amoebic dysentery. He had managed to 'liberate' a DC3 on release and with permission from his CO, Wing Commander Curly Davis, flew it himself with several other POWs, including an American navigator, to Kunming in China, where he blew a tyre on landing. Here they were kitted out with whisky and new uniforms by local USAAF (United States Army Air Forces) crews, then went back to Calcutta. From there he begged a lift on a flight to Sri Lanka (Ceylon) and stayed in an expensive hotel, charging it to the RAF. Finally a DC-4 took him to Perth via the Cocos Islands. Here a mate gave him a lift by air to Brisbane. He reached home on 4 September 1945, one of the first Australian POWs to arrive.

Sadly, Ossie had nothing to say in the tape-recorded interview about his time as a POW. He was awarded the DFC, the 1939–45 and Pacific Stars and the Defence Medal. He died on 25 August 2003, and is buried in the Jewish section of Toowong cemetery in Brisbane, next to his parents.

The DFC citation includes the following:

> Flying a Hudson aircraft he carried out two night attacks on enemy landing forces, setting fire to a 10,000 ton transport which turned over and sank. The same day on a third sortie he blew up a railway line above Pasir Mas near the Thailand border. During 14 hours action at Kota Bharu, he completed four flights in all, and during

> one of the attacks, one of his engines was put out of action and the aircraft severely damaged, by pom-pom fire, but he flew safely back to base.
>
> Later this officer attacked an armoured wireless transmission ship approximately 250 miles off Singapore and obtained two near misses and carried out nine reconnaissance and escort flights in the South China Sea. It was on one of these flights that the Japanese landing forces approaching east Malaya was sighted.
>
> From Sumatra, Flight Lieutenant Diamond completed two further sea reconnaissances and one bombing raid on Singapore, when damage was done to the marshalling yards at that base.
>
> On 14th February 1942 he was engaged in attacking the large concentration of Japanese sea forces approaching Palembang, Sumatra. During this attack, his aircraft was badly damaged but eventually he reached Palembang, where he crashed landed after the aerodrome had been surrounded by parachute troops.
>
> He then ordered his crew into another Hudson which had 6 inches of the propeller tips broken off. And, under tommy gun fire from the Japanese, he attempted to take off but could not get airborne. He abandoned that aircraft and after an unsuccessful search for a serviceable aircraft, led his crew back into the paddy fields, still under hand grenade and small arms fire. After 10 hours during which he had assisted a party of the Dutch army, he got his crew safely back to Palembang town. During all that night they had travelled through swamps full of leeches in order to avoid the Japanese.

After the war, Ossie married Norma (who at the time of writing is still alive in Brisbane, aged 92), had three children and ran a dry cleaning business in Brisbane, where he was well known among the Jewish community and the Australian War Veterans (RSL – Returned Services League). He often flew his own Auster to Fraser beach, where he was a fisherman in his retirement days.

Notes

1. Rose was directly descended from the famous eighteenth-century Jewish English boxing champion Daniel Mendoza, and a second cousin to actor/comedian Peter Sellars, also a Mendoza.
2. Due to the International Dateline differential, in real time Oscar and his crew hit the Japanese a full one-and-a-half hours before Pearl Harbor was struck on 7 December.
3. His crew were Flt Lt O.N. Diamond, P. Off. G.B. Evans, Sgt R.C. McMillan and Sgt C.A.V. Wells.
4. This was the Japanese 10,000-tonne transport *Awasaki Maru*; the action is depicted in an oil painting in the book *They Flew for the King* by Frank Harding (a limited edition published by Aviation Art Australia and Folklore Gallery, date unknown).
5. *Australian Aircraft Magazine*, May 1990, pp.14–15.
6. A photograph of Japanese paratroopers shouting triumphantly by Ossie's burnt-out aircraft was discovered after the war in the Japanese archives, by a researcher of the RAAF Historical Branch.
7. Email from daughter Karin, 12 August 2011.

1441067 Gnr Robert Dykes, RA

The source of this testimony is my telephone interview with Robert Dykes in August 2004.

27a. 1441067 Gnr Robert Dykes RA, lying in the foreground, taken in the first days at Changi.

Robert was born on 20 January 1916, son of Nathan and Jean, and lived at 28 Newark Drive, Glasgow. He was in the TA and was called up in 1939 to the 52nd Searchlight Regiment, RA. He was sent to Singapore and arrived on 13 January 1942, being attached to the 6th RA Regiment.

Robert remembers clearly the fighting around Bukit Timah and the retreat to Singapore City. On being ordered to surrender, he remembers the huge crowds of Commonwealth POWs in the great square in Changi gaol. Here the soldiers were forced to sign the notorious 'no attempt to escape' agreement, mentioned in previous testimonies, after senior officers gave in to the Japanese in order to avoid bloodshed and starvation. Conditions, while they waited in this area, were appalling, with the heat, no water or food, and no sanitary facilities.

Robert was among those imprisoned first in the wooden barracks at Changi – they proceeded to tear up part of these to get fuel for cooking. There were many abandoned uniforms and much personal equipment lying about and they used this to replace their worn-out clothing as best they could.

27b. 1441067 Gnr Robert Dykes, RA.

In March, he suffered the hellish train journey north to Bampong on the River Kwai in Thailand, which was to be the beginning of his life as a POW on the 'Railway of Death'.

The journey of about five days was with about 500 other men with no food, sanitation or water; sometimes at a stop they could get water from the engine. Natives who tried to give them food at stops were punished by the Japanese and Korean guards (the latter were especially vicious). Individual Japanese guards could be more reasonable; in groups they were brutish and intimidating.

It was at Bampong that he met Captain David Arkush of the RADC (whose testimony is in this book). They had to build and live in bamboo and atap huts in filthy, muddy, disease-ridden conditions with appalling food and sanitation. Among the other Jewish POWs he met were Charles Simon from Singapore (SSVF) and Willy Bluestone (see 'The Record and Roll of Honour', UK and Dutch Army list, in this book). Later he was imprisoned at Chungkai and Tamarkan. On liberation by American paratroopers he weighed only six stone. He married Helen after the war and had three children: Anthony, Geoffrey and Shirley. He made one return pilgrimage to the Far East in 2005.

After telling me all this, Robert refused to say more, and so sadly this is all we and his family know of his terrible ordeal. He died in 2010, aged 94 years.

Sion Elias, Interned in Singapore by the Japanese, aged 12 years; and his Older Sister, Flossie

The information below was taken from an audio-taped interview with Sion, recorded in London in 2002, now held in the sound archives of the Singapore Museum.

Sion's grandparents were originally from Iraq, where they spoke English and Arabic; they then moved to Bombay. Their parents came as children from India to Singapore, and married there. As a result, Sion and his generation were brought up speaking Hindustani and understanding Arabic, but they learnt English in Singapore as well. Most of the Jewish children went to the church schools in order to have the best education.

Sion's family lived in Burma Road, and he and his sister testified that about 100 Jewish men were sent to Changi, three months after the surrender in February 1942. Freddy Isaac was the first to be taken after a few weeks by the Kempeitai and, faced with torture for a month, was ordered to give 100 names of Jewish men he knew who were to be interned. But Elias Ezekiel and Elias Ezekiel (two different people: cousins with the same name), were taken from their shop in Beach Road by the Japanese, who were looking for carpenters, and never seen again. Ivor Nathan was similarly taken, and it was known that a number of Jews were murdered in the sea on Changi Beach in those early days. Some time later a Malay man came to Sion's house and told him that he knew that two Jewish civilian men with blue eyes were among the soldiers murdered in this incident: the Japanese counted them as enemy soldiers because of their blue eyes. This is how Sion and the others knew that it was the Ezekiel cousins who were shot.

In addition, many in the Jewish community were killed in the bombing at the harbour as ships taking off civilians were sunk by Japanese aircraft; Sion said they were making for British India but were never seen again and perished at sea. Among them was an elderly man named Solomon; a woman named Leila was rescued at sea and brought back to Singapore. Sion said the synagogue had an air raid shelter in the basement and it was therefore important to be near the building when raids occurred.

Before internment, the Jews were forbidden to work by the Japanese, who called them 'Judai-jin' and forced them to wear a Star of David badge. This meant they lived in extreme poverty and so, to earn some money, Sion and his brother used to make buckets of tea and sell it in the streets. Sion's mother Flossie was the matriarch of the family and organized the selling of their possessions to buy food; all kinds of different people bought their jewellery, furniture and various household items.

The Japanese arrested the hundred Jewish men (whose names were forced out of Freddy Isaac) suddenly, in the middle of the night, and took them to Changi. Next day, ten brave women went to the Kempeitai to ask to be allowed to take the men some food, clothes and bedding; they were very fearful and asked the other women to care for their children if they did not come back. On arrival the Kempeitai were shouting at them and drew their swords in a very threatening manner. But the women told them they were just the wives of the new prisoners and wanted to know where they were. The Japanese agreed to a date when the women could visit the men and the word was passed round to all the Jewish women concerned, to meet with their supplies and store them, clearly labelled for each man, in a house in Nankoolen (?) Street, ready to go to the prison. They organized a lorry and on the appointed day went to Changi.

Ten Jewish prisoners were picked from among the families of the women who were in the truck, and ordered to unload the lorry in front of the women sitting in the back of the truck in the pouring rain, but they were not allowed to talk to them, only glimpse them through the barbed-wire fence. Then they had to leave. A Japanese official, Shinozaki, gave the women all the assistance he could and the families were very grateful to him. Two months later they were allowed to organize another visit, but two Jewish youngsters paid an unofficial visit to deliver things to their father beforehand, and the Japanese became suspicious that they might be spies. They were arrested and threatened with shooting, but later released. Other families complained that they would like to go to visit their men too. Sion's sister reminded them that her group had been the first and risked their lives going to the Kempeitai when others refused to accompany them. So the visits needed to be organized properly in future, so that everyone had a chance to see their men folk. At the next visit, however, the group were refused entry, owing to the incident with the young men, and all the items for the men had to be brought back. This was a great blow to the women and the two young men were ostracized for this.

Eventually the 100 arrested Jews were sent to Sime Road Camp and built huts there for the rest of the Jews who were soon to join them at some unknown future date.

One kind Japanese guard was nicknamed Shorty; he was a Catholic and when the first Jewish inmate died in Changi (Baruch?) he came to the funeral and prayed alongside the Jewish mourners; the funeral (at Thomson Road Jewish cemetery) was used as an excuse by the Jewish prisoners to get outside. This was because they could meet their families who were still free in their homes, as the 'grapevine' always buzzed when a death occurred and so everyone would gather at the cemetery and exchange news, gifts and so on with the prisoners. It was more of a social occasion for those not from the family of the deceased.

Another kind Japanese, who was American-educated, was approached in the camp by Sion's brother-in-law to ask about the birth of his baby; his wife Mary (Sion's sister-in-law) had become pregnant just before he was arrested. The Japanese visited the house where Sion's sister lived, and saw the baby and passed the good news back to the man in the prison.

Once a week some of the Jewish prisoners were allowed out of the camp to work emptying the garbage in the city; when word of this reached the women, they would gather nearby for a chance to see them and wave. A Catholic Indian named Francis worked as a driver for the Japanese and used to smuggle letters in and out of the Changi civilian gaol so that families could communicate; this brave man was later tortured to death by the Japanese for smuggling radio parts to the military POWs at Changi (see the testimony of Freddy Bloom, above).

Some months later, all the rest of the male Jewish community were arrested and sent to Sime Road camp in open lorries. The Jewish section was known as 'Aldgate' by the British prisoners (Aldgate, in London near Petticoat Lane, was a very Jewish area). This was a very frightening time for the women as there were no menfolk to protect them and there was not only the fear of Japanese harassment of women but also a certain amount of lawlessness in general in the Singapore City, from local gangs. When the rest of the Jewish community were eventually sent to Sime Road, there was a feeling that they were in fact safer in the camp than in the city.

The Jewish prisoners smuggled valuables into the camp when they were arrested, as they knew they would have to use them to obtain money to give to their families outside who were living at below subsistence level. One of the Jewish inmates, named Kalif Moredchai, was a man in his thirties who had come from India as a young man and had been a hawker in the roads near the synagogue in Waterloo Road and Middle Road. He had been one of the first arrested and spent time in Changi too (he survived, and lived for a while in London after the war). He was known as the 'gulam molucca king' (gulam molacca is a sweet syrup commonly eaten as a snack in the Far East, and much sought after in the camps). He had got to know many Japanese soldiers before internment, who bought street food from him and one, nicknamed 'Puss in Boots'[1] – because of the huge boots he wore – was his particular friend and later became commandant at Sime Road where Mordechai was later sent. If the Japanese wanted to buy and sell valuables they would go to Mordechai and he would continue his deals with Puss in Boots in Sime Road camp, although not to the benefit of the other prisoners. Many thought he did far too little to help other people with the gains he accumulated. He paid two men to guard his trunk in the camp, like a kind of 'King Rat' character (the name of a Hollywood film starring George Segal, made in 1965, based on a true story by James Clavell). These guards were from wealthy families and it amused the others to see them working avidly for the poor hawker who had come up in the world and had some power in the camp because of his business acumen. They also enjoyed better food, of course, and so stuck with Mordechai for that reason too. Mordechai was often seen at the Green House (the Japanese main office in the camp), such were his trading links with the corrupt Japanese officials. The British prisoners organized a different system, whereby they had an agreement with Shenton Thomas (the High Commissioner) to sign IOUs which would be paid after the war to people who loaned money, but most of these were never honoured, according to Sion.

Sion made the point that some German Jewish refugees were not caught up in this because, as they were German, the Japanese considered them as Allies. Such was the obtuse thinking of the Japanese, as of course their German friends regarded all Jews as enemies.

Sion also noted that the Japanese kept the cinemas open and free; in this way they achieved two things. They could always show their propaganda films to a captive audience who craved evening entertainment, and these films featured lots of alleged Allied defeats. Also, when they decided to round up people, they would open only one exit door and as people came out they were pushed into waiting lorries and captured, saving the Japanese huge amounts of energy and manpower.

When the Jewish women and children were interned at Sime Road at the start of 1945,[2] Puss in Boots beat two of the Jewish women, in one of many incidents of brutality that took place; he kicked one and beat the other with the side of his sword (her name was Juliette, but Sion could not remember the family name). Sion heard this story from one of the women when he visited her in Israel in the 1980s, before she died. Bowing to the Japanese was compulsory, and failure to do so properly would result in kicking and slapping. A Hungarian Jewish wrestler called Combulases, who used to work in the Happy

World amusement park, was caught stealing chillies from an Englishman's garden; he was reported to the Japanese, who gave him a public flogging of thirty lashes with a cane and rubbed chilli into his wounds. Among the Jewish internees were three Chinese women who had married Jewish men, and some Chinese men who had Jewish wives. The Jews spoke Arabic when the Japanese were around and nicknamed them 'shorties'; among themselves they spoke English and Malay.

The gardens were managed by an Englishman and each day he allotted food to the kitchen, to supplement the rations. The camp barber was a British Merchant Navy Jewish man named Wimbush.

Sion himself was of borderline age and taken to the men's side of the camp with his father and his older brother Joe, and separated from his mother. He thus missed his bar mitzvah, and on the tape he joked that maybe he should sue the Japanese for this, and that it was a good job it was not his circumcision. He also pointed out that there were a lot of military prisoners in Sime Road who had discarded their uniforms in order not to be sent to the much worse military POW camp at Changi.

Every so often the families were allowed to meet each other, and mothers and fathers and children got together. One Eurasian woman in Sion's sister's hut was the girlfriend of Puss in Boots. He built a little room for her at the end of the hut and would come to her each evening for sex or to take her out in the town; in return she would get fruit, chocolate and eggs from him. She was very kind in sharing such gifts with the other prisoners. After the war she went to live in America, and married there.

Sion witnessed the former High Commissioner, Sir Shenton Thomas, being humiliated by being made to carry sewage for the Japanese; this was part of their degradation programme. He was in hut 127 and Sion was in hut 128. Others, such as bankers and the owners of big plantations, were made to do the same thing.

The Chinese assisted the prisoners by secretly selling them food under the barbed wire. The Indian Sikh guards had sticks with bayonets on but were not really interested in stopping smugglers and often took bribes to look away; the Japanese did not actually guard the Camp at Sime Road. Sion and his sister testified that the Japanese gave double rations to those helping to build two large tunnels in Sime Road, which they said were for storing ammunition, but the prisoners knew that they were intended for use as execution pits for the prisoners in the event of an Allied victory.

The Jews at Sime Road organized a kosher kitchen and one of the cooks was Yosef Cohen. Sion joked that it was not exactly kosher, as he often found snails in his food, and in any case there was virtually no meat to eat, but it satisfied the religious needs and commitment of the Jewish internees to be able to say that they attempted to keep kashrut where they could. Often the Jewish prisoners ate tins of Spam from the Red Cross parcels they occasionally received, and only found out after the war that it was made from pork (this made Sion laugh hugely on the tape). Sion remembers creeping up on his brother Joe, on one occasion – he was frying an egg over a candle in an old tin, and was afraid someone might come and take it from him. On another occasion the brothers obtained some gulam molucca from Mordechai to sell to the Europeans in the other huts. It came in two-ounce balls and cost about five Straits Dollars per ball. Sion took his share and tried to sell it but found nobody was interested, so he ate the whole lot, as he was hungry; he was told off by his two brothers for eating the profits. He called this incident 'Death of a Salesman', as his brother Joe had to pay Mordechai[3] from his earnings, on behalf of

Sion, and he was not trusted to sell the stuff again. On one occasion an Englishman from Liverpool (possibly Jimmy Jigs) took Sion's brother onto his shoulders and said, as they nonchalantly passed under the papaya tree which someone else had cultivated, that he would slow down, and he was to pick a papaya secretly and share it with him later; such were the tricks people got up to, in order to stay alive and obtain food. The British sailors were particularly mischievous and would pilfer lots of food from the Japanese kitchens – and they were never caught.

There were quite a lot of snakes in the camp but Sion knew of nobody who died from a snake bite. It was not unusual for snakes to be caught and eaten.

Wealthy British prisoners who had brought gold and diamonds into the camp used these to trade for food. As a result, some civilian internees with outside contacts were able to make a lot of money, which made them rich after the war; they saw this as a payback for the colonialist behaviour of the British moneyed classes over many years of occupation in Malaya.

Each hut (men and women separate) had a leader or commander; Sion's was Jacob Balus (?) and their hut was christened 'Commercial Street' by the English (another predominantly Jewish area in the East End of London). His sister's hut leader was Rachel Menashe. The Aldgate leader was Max Menachem. Sion attended school in the camp and they learned Japanese as well as other languages. Toilets consisted of deep open boreholes with planks over them; Sion said it was possible to slip into the hole and so he always used a local field instead. He recalled with much amusement one old Englishman who always went to the same hole, despite the fact that there were dozens of them all over the camp. When Sion asked him about it, he said that if he fell in he at least knew it was his poo in the hole and nobody else's!

The Jewish internees took full part in the concerts that were put on to keep up morale – many were good singers and musicians – and there was even a Jewish gambling club, held every Saturday night in the Aldgate hut, called the Casino, where even the Japanese came to play mah-jong or cards; the British played bridge and gambled too. Prayers were held by the more religious Jews but there were no organized synagogue meetings.

After the war the British put the Jewish and other civilian internees into the Empress Hotel (now demolished) and here they were given large numbers of Red Cross parcels to live on; as nobody had any money or possessions, and some had lost their homes, much of what was in the parcels made its way onto the black market and made a lot of money locally for former prisoners, to help them start their lives again. Each British internee received £75 compensation from the government in the 1940s – then a considerable sum – but the Jewish internees, although British passport holders, received nothing because their parents were not born in the UK.[4]

Just before liberation, leaflets were dropped telling prisoners not to eat too much rich food from the Red Cross parcels, as it would make them ill or even kill them, and also to give any cigarettes to the Japanese as gifts to 'soften' them in case there were plans to massacre the civilian internees. Then one day British paratroopers actually dropped into the camp; Sion said he will never forget that joyful sight.

Mountbatten came to visit them and signed passes for them to visit the harbour and go to see HMS *Sussex*; Sion kept the pass for years after the war. The sailors gave them wonderful white French stick loaves, freshly cooked on board, and dark chocolate, neither of which they had ever eaten before.

After the war, several Jewish lads joined the British Army to do their bit but in 1948 they deserted to go to Israel via South Africa, to serve in the War of Independence. In South Africa, local Jews hid them and helped them on their way. Until the 1980s these men ran the boat service on the Sea of Galilee in Israel.

Notes

1. Identified as Tominaga – he was tried for war crimes after the war and sentenced to death but it is not known if he was hanged or if the sentence was commuted.
2. There were approximately 600 in the camp in three huts.
3. Mordechai later moved to the UK and married, but when his wife passed away he went to live in Israel, where he lived until he died.
4. After many years of campaigning, surviving British POWs and widows were given £10,000 compensation by the Blair government in 2000. Diane Elias, who was interned as a child in Stanley Camp, Hong Kong, brought a test case for civilians; it was finally won in 2006 and so civilians also received this compensation payment.

914053, LAC John Fisher, BEM, RAF

The sources for this testimonial are my telephone interviews with John Fisher in June 2004, when John was 88, and a DVD interview John made with his nephew Malcolm Fisher in 2011, when John was aged 95.

28a. 914053 LAC John Fisher, RAF, BEM, when first in the RAF.

John was born on 20 December 1916 at 89 Evering Road, Hackney, London, and attended the Bethnal Road and Upton House schools. He was bar mitzvah at Phillips Lane, a synagogue, now demolished, in South Tottenham near Seven Sisters Road, after the family moved to Wood Green. He had brothers, Samuel and Simon, and sisters Yetta and Betty. His father was Joseph (a cabinetmaker in Brick Lane) and his mother was Anne.

John volunteered on 4 April 1940 for the RAFVR (Royal Air Force Volunteer Reserve) and served in Kent airfields in the Margate area and then was sent with Bomber Command to train as WOP/AG with 144 Squadron in Hemswell, Lincolnshire, on Hampdens. Here he was trained to send bogus signals to German bomber and fighter aircraft approaching the UK, to a dummy airfield nearby, so the Germans would bomb the dummy aerodrome and leave the real base unharmed.

While stationed in Lincolnshire, he befriended a girl called Enid, but before further training began he was sent by ship with his friend Bill Ogburn via Glasgow and the Cape, during the Rashid Ali fascist riots, to the large RAF base at Habbaniyah in Iraq to join the Blenheims of 84 Squadron. After landing in Aden they went by rail to Iraq. However, his squadron had moved on and so they followed them north to Kirkuk and Mosul, and thence by rail to Alexandria via Haifa (Israel) to join Auchinlek's army in the big push of 1941 in North Africa. In North Africa he took part in the bombing of the retreating Italian forces as far as Derna, from where he and Bill were sent back to Heliopolis (Cairo) and then were shipped to Singapore with 250 men of the ground crews (half the squadron).

They were diverted to Sumatra (in February 1942) because Singapore had fallen. From Sumatra, they fled the Japanese by ship to an airfield near Cheriban in Java, about forty miles inland, where they worked as ground crew helping bomb the Japanese, while under Dutch command. They were with about forty other British Army men with Bofors guns. After some time only about three Blenheim bombers were left and they were used to evacuate the last few VIPs to Australia. The men were all in huts dispersed around the airfield and Bill was on duty one night as a machine-gunner while John slept. Early in the morning, Japanese paratroops landed and attacked, and while the army tried to hold the enemy back, Bill was killed at his machine gun post; the RAF were told to flee in trucks as they were only lightly armed.

They were ordered to the coast at Tillijap where they were told that the Royal Australian Navy would rescue them. But there were 50,000 mixed Allied troops there and no rescue came. John and some of his friends searched for some boats to try to get away but were not successful. The senior RAF officer (a group captain) told them to stay, and in return the Japanese would not harm them. After three days, the Japanese arrived and they were taken POW. John was given POW number 141.

POW

The POWs were marched many miles under brutal conditions; arbitrary shootings, bayoneting and beatings occurred and virtually no food or water was provided by the Japanese guards. Many men died. At Boeeglodok, near Batavia, they were put in a camp in filthy condition with no hygiene, hordes of lice, poor food and more beatings. Here they were set to work to clear the old Dutch airfield for use by the Japanese air force.

After a few weeks, they were shipped to Singapore together with US, ANZAC, Dutch and Javanese POWs on a two- to three-day journey, which John described as tolerable, and sent to Changi gaol where conditions, though better than in Batavia, were still bad. In December 1942, John and 500 men were sent on a gruelling, overcrowded 3,000-tonne tramp ship on a journey which lasted more than a month, to Hokkaido in Japan. Conditions on board were appalling; there were no lights in the hold; there were open buckets for latrines, which overflowed and fell everywhere in stormy seas; dysentery was rife; and with only tropical clothing it felt freezing as they went north. Many died of illness, disease, hunger and beatings. Conditions in these type of Hell Ships are well documented elsewhere.

John remembers one evening talking to RAF Flight Lieutenant Harper Bongard, who slept in the next bed; the next morning Bongard was dead. This incident is engraved on John's memory.[1]

The men were allowed on deck only for one hour a day. John fell ill and does not remember much else until they reached Tokyo in the snow. From here they were transferred by ferry to Hokkaido Island and John was somehow carried to the camp which was servicing a mine. It was not until August 1943 that John's family discovered, via the Vatican, that he was alive as a POW.

The camp was called Hakodate and was the most northerly in Japan, in north Hokkaido Island, only 300 miles from Vladivostok across the sea in Russia. It was extremely cold and the POWs wore padded/quilted suits like the Japanese. For footwear, they used straw bound around their feet or kept their boots, issued at Changi gaol before they left for Japan.

There were no medical personnel or facilities, but one hut was put aside for the sick and this is where John was kept for a while with a severe chest infection. One day, one of the dying men in the hut kept asking for another blanket. An RAF sergeant, Phillip Gibbs, pushed the man, in his delirium and illness, outside the hut into the snow. The man banged the door to be let in and eventually he was allowed back, but died the next day. After the war, in 1946 Gibbs was court-martialled in the UK and John was called as a witness because he had been lying ill and saw the incident. He argued that Gibbs was unwell like everyone else in the hut and not aware of his actions and had no medical knowledge of how to deal with the man, just like everyone else in the hut. While waiting for the verdict, Gibbs's attorney came and thanked John for his evidence. Gibbs was acquitted.

One day a Japanese doctor came into the hut and spoke German (the Japanese learnt their medical studies in German). A Scots soldier had dysentery and John was able to translate for the doctor because he knew Yiddish from his upbringing in Hackney, and a bit of German from school. From then on John became the camp interpreter and medical orderly and so did not have to go down the mine. As the room was the only one with a fire, guards also came in to keep warm and through one guard having a Japanese-English dictionary, John learnt good Japanese too.

The camp population was around 400; as men died, more were brought in to replace them. Guards were also changed.

To begin with, the POWs were given little workbooks in which a seal was stamped for every day they worked, and for which they received ten sen or one tenth of a yen per day; this could then be exchanged for cigarettes with the Japanese, and these became a currency within the camp. However, this system fell into disuse after a few months.

The conditions of camp life were terrible, and have been described in the many terrible books about Japanese POW camps. Food was mainly cold boiled rice with green tea now and again, but usually only water, for breakfast; for those in the mine, a lunch ration was wrapped in leaves, sometimes with a piece of cucumber-like vegetable, and in the evenings the same again, sometimes with a small piece of fish. Each hut had a rice bucket from which was measured a portion of rice per man from a small aluminium bowl which everyone had to use to eat from; this caused many infectious diseases once one man fell ill and spread it to everyone else. As a result of this starvation diet, and mine work, men died constantly of malnutrition, exhaustion and the consequent illnesses. Several died every week, but as they did new men were brought in – John remembers many were Americans (from Wake Island and Guam) and some Australians. Those who died were cremated in Japanese style in barrels full of sawdust. Sometimes John had to do this and often limbs had to be broken to fit the bodies in; this was a horrific job and everyone had to take a turn to do it. There were few official prayers but sometimes friends of a deceased POW would secretly hold a service at some stage.

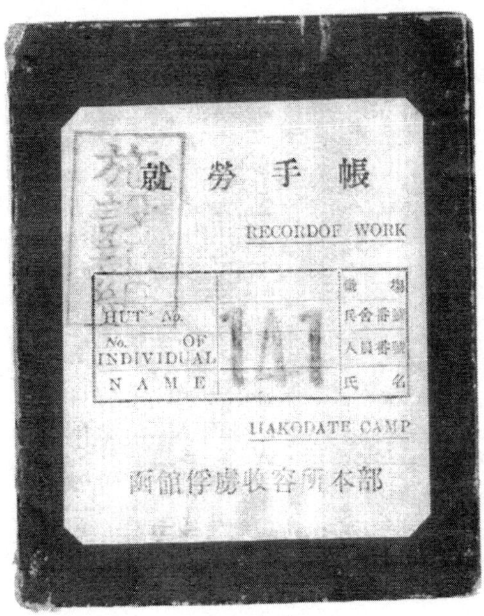

28b. John's Japanese POW 'pay book'.

Each morning, at about 6 a.m., a Japanese NCO would oversee roll call and it was here it was decided who would work and who was too sick. John argued for as many as possible not to work if they were ill, but as rations were only given to those who worked, and then had to be shared, it was a terrible choice to have to make, as to how much food the camp as a whole could afford not to receive. If the NCO was in a bad mood, John would often suffer beating with a rifle butt if he was too persistent; on other occasions he might get away with it.

After about a year of coping alone, an Australian Army doctor, Captain George Bristow, arrived in camp. John became his assistant and they maintained a close and warm partnership, looking after the POWs, for the rest of the war. Each evening after work they would both decide on which men would go sick next day. Every case was argued by John with the Japanese NCO responsible, with Bristow looking on as John translated. In this way John saved many lives of men who otherwise would have died in the mines, and always at the risk of being beaten by the Japanese NCO. His BEM (British Empire Medal) was given for this courageous work.

Dr Bristow sometimes suggested improvements to the Japanese which John had to pass on. For nine months they asked for a septic tank to reduce disease from the latrines; eventually they got it through sheer persistence.

In addition to the soldier guards, there were some older Home Guard Japanese. They had sticks which they used to beat men in the mines or whenever there was an alleged infringement of rules. One very cold night, a man with dysentery soiled his blanket just before the evening inspection of the little ward. John knew that this was classed as 'damage to the Emperor's property' and so hid the blanket in the snow just outside. The NCO spotted it. As punishment, the guard took a pair of hair clippers and cut a swathe through the centre of John's hair. He felt deeply humiliated, and he was then taken to an open-sided windblown cage on the clifftop overlooking the freezing sea, and left for the night. This was a death sentence and John, at what he described as probably his lowest point when a POW, was about to give up. But another guard came and told him he must keep walking around the cage to keep warm and survive; he found the strength to do this with the kinder guard's encouragement, and so he lived

John explained, as other POWs have done, that there was a hierarchy of punishment among the Japanese themselves. If an officer hit an NCO, the NCO beat the prisoners and so on.

On only one occasion was a man sent to hospital in the nearby town – a Greek merchant seaman with appendicitis. They forced a young American student doctor to do the operation without anaesthetic, but it was a botched job and eventually a Japanese doctor intervened and completed it. The man lived to tell the tale. John was present and recalled that the operating theatre floor was covered in water containing antiseptic and all those present wore rubber boots – this was the Japanese way of cutting infection.

The POWs never received Red Cross parcels or post – the Japanese stole them but claimed it was because they were too far north for the parcels to be delivered.

Towards the summer of 1945, food almost ran out; the Japanese were starving too and each day John and others, with guards, would go out foraging for roots and other wild food, as many Japanese civilians also did at that time.

Near the camp baths was a wall which separated the POWs from the guards' bath house; the wall did not reach up to the ceiling so John could hear the Japanese talking. One day John heard the Japanese talk about the German surrender in May 1945; they then knew that the war was beginning to end.

Liberation

One morning, the Japanese said that the men would not go to the mine as there had been an accident; then the afternoon shift was kept back because the Japanese said there was cholera. The men began to suspect something was wrong. Next morning the guards had gone. John and Dr Bristow went to see the commandant, Colonel Hutakiyama, to demand a radio. This was a risk, as he could have killed them. But he gave way and the men learned that the war was over. They picked up messages telling all POWs to stay put until help arrived. Next day the commandant disappeared too.

Then a message came via the radio for help from a nearby camp of Chinese POWs and John went with Bristow and a sergeant, to do what they could. Conditions were unbelievably bad – resembling those in German concentration camps – and John realized that compared to the Chinese camp, their camp had been bearable. They asked the Chinese what they wanted most and they said guns – but John told them that the Japanese had already fled.

After another day or so, US aircraft from the carrier *Hogarth Bay* dropped chocolate, and bread still warm from the ship's ovens. The men were overjoyed. Included in the package was a message asking if a certain American POW was there, as he had a relative aboard the ship.

After about ten days an aircraft landed and a US lieutenant with two NCOs accompanied John and Dr Bristow to the commandant's office – he had been ordered to return to the camp. John will never forget that when they entered his office, the American said 'Get up you son of a bitch when a US officer enters the room!' He did!

The POWs were taken to Tokyo and then flown to Manila by Dakotas; here they were given full medical check-ups. John's overwhelming feeling was a numbness that it was all over, mixed with real fear of the freedom of a normal life ahead and how he was going to cope with it after three-and-a-half years of hell. From here they were flown to Hawaii, then to San Francisco, then they travelled north by train to Canada and across to New York, and finally sailed on the *Queen Mary* to Britain, arriving in December 1945.

In London, John saw a specialist because his eyesight was bad. He was told that lack of vitamins had caused partial blindness and that his sight would improve as his health improved – but as he got older, his eyes would deteriorate as a result of his ordeal. This proved to be the case and John became partially sighted in old age and consequently receives a war pension. He returned home to his family, then living at 67 Chardmore Road, Clapton, London.

After the war, John went to live in the USA (Los Angeles and Santa Monica), doing various jobs. His brother and sister also went there to marry and settle. In 1984, unmarried and in his sixties, John returned to Lincolnshire for a visit to see if his old girlfriend Enid from those early days was still alive. She had thought him dead during the war, and had married a local farmer. The farmer had died and so she married John, and returned to live in the USA. In 2001 they came back to the UK, to live in Budleigh Salterton, Devon, and then moved to Lincoln where, in their eighties, they split up. John had nephews and nieces and stepchildren in the USA and UK, and returned to live in the USA, and died there in 2011.

John kept hold of a list of the POWs which had come into his possession near the end of the war. It was typed in English by the Japanese or someone among the POWs at the behest of the Japanese, perhaps to try to mitigate their guilt by somehow showing they kept a tally of the POWs. It is annotated in Japanese and John did not know why; he kept this list until he died.

In 1990, when working in his store in Santa Monica, John met an airline pilot and asked him to see if he could find Dr Bristow's address next time he was in Australia. He did so, and the following year the doctor and his wife came to a medical convention in Santa Monica and the old comrades spent time together; this was a most moving time for John, as both men admired each other so much. Except for one squadron reunion in the UK in the 1940s, John has never met any wartime friends again.

John has the BEM, awarded for courage (proposed by Captain Britow of the RAAMC) and for his work as a POW, and it is on display in the AJEX Jewish Military Museum. He also has the 1939–45, Africa and Pacific Stars and the War Medal. John's BEM (Military) was promulgated in the *London Gazette* on 1 October 1946:

> For gallant and distinguished services rendered whilst a POW in Japanese hands.
>
> Whilst a prisoner of war in Japanese hands, this airman was a volunteer medical assistant at the prisoner of war camp at Hakedate. He worked hard, at all times, to serve the interests of the sick allied prisoners. LAC Fisher's knowledge of Japanese made him invaluable as an interpreter, enabling the medical officer to appeal directly to the Japanese doctor. The medical officer was greatly assisted thereby in protecting allied prisoners who were unable to perform the heavy work allotted to them. As LAC Fisher had been a combatant, he was liable for the heaviest of manual labour. Despite this he was always willing to jeopardise his comparatively safe position in the camp by assisting the medical officer in his arguments with the Japanese authorities. Many allied Prisoners of War owe the preservation of their health to his efforts on their behalf.

Note

1. John's recollection was that Bongard, from Christchurch, New Zealand, was thrown overboard, but in fact he is buried at Yokohama and so he must have died just before arrival in Tokyo harbour.

3780682 Pte Leon Frank, 13th Platoon, 13th King's Liverpool Regiment, 77th Indian Infantry Brigade – Chindits

29. 3780682 Pte Leon Frank, in 1943. He was captured on the first Chindit raid into Burma.

Leon Frank was born on 21 April 1920 in Green Dragon Yard, Shoreditch, London, son of a Russian Jewish tailor, Nathan, and a British-born mother, Sarah. He learnt tailoring alongside his father while he was at Canterry Road (?) School, and became his assistant after leaving secondary school in Leyton, aged 14 years.

After joining the ARP in 1939, Leon joined the army in June 1940 and was sent to Glasgow for training at Jordanhill College barracks with the King's Liverpool Regiment. Army life appealed to him and he enjoyed meeting people from varied backgrounds and the camaraderie of the forces. It was at a dance in Glasgow that he met his wife.

Later he was moved to coastal defence duties, but on 7 December 1941 he sailed from Liverpool on the *Duchess of Oronsay* for South Africa, the same day that the Japanese attacked Pearl Harbor. They travelled by convoy via Durban, South Africa, where they stayed for five days in Clarewood transit camp, in relative luxury; he could not believe the bright lights and excellent food after the rationing and blackout of Britain and Blitz. All the local people were extremely friendly to the British troops. Later he learned that the other half of the convoy had sailed on to Singapore, straight into Japanese captivity.

From here he went to Bombay, on the RMS *Andes*, arriving in February 1942, where at the dock stores they swapped pith helmets for bush hats, and then journeyed by train for three days to a base at Meadows barracks in Secunderabad, for garrison duties. With him were other Jewish lads; he can remember Private Sid Kaplan and Private Len Berkovitch. In all his service, Leon never met any anti-Semitism, except – ironically – when he was a POW in Rangoon (see below). At Secunderabad, he became a regimental policeman (RP).

In one incident he had to deal with a very large Cameronian soldier who got very drunk at the local canteen (called the Coffee Shop) one Friday evening and assaulted a local Anglo-Indian woman who had refused to dance with him. His Welsh police sergeant, Evans, ordered Leon to arrest him, even though he was half the Cameronian's size. So Leon took off his RP armband, put his arm around the man's shoulders and said he knew a great place where he could really enjoy himself; he led him gently to the police cells and locked him up for the night.

Later in 1942 they left Secunderabad; the local servants who did odd jobs for them were very sad to see them go, and one boy ran alongside the column near Leon, calling out

'Do not go Frank Sahib! Do not go!' Leon adored the Indian people and thought them all kind and polite.

The men went east by train to take part in intense jungle training on the northeast border of India. They now became the 77th Indian Infantry Brigade, comprising men of many regiments, including Gurkhas, as well as mules, horses, elephants and bullocks – affectionately known as 'Wingate's Circus'. They did not know it then but they were to become Chindits, special forces to be inserted behind enemy lines in Burma. They trained very hard in mud and downpours , practising river crossings, stealth and fieldcraft, ambush-laying, mule driving, hand-to-hand fighting, jungle route marches, navigating by the stars, and many other skills they would need in the jungle war ahead. Many could not make the grade and were sent back to their units. Quite a few were killed during training. One sad incident was the death in training of a Jewish officer, Lieutenant Neville Nathan Saffer. Standing in a circle for rifle inspection, Saffer's 14th platoon were just alongside Leon's. For this procedure, the officer inspected the rifle by looking down the barrel to see if it had been properly cleaned, after any bullets had been ejected. Suddenly a shot rang out: a soldier had accidentally fired a remaining bullet, which passed straight through Saffer's brain, and he died at once. Everyone was horrified, especially the soldier concerned. Lieutenant Saffer was buried with a wooden Star of David on his grave.

After training they were taken by steamer to Manipur up the Brahmaputra river, and then walked 300 miles over ten days, through mountains to the Burmese border, using their newly adopted single-file style (snake column) that they would use in the jungle. It was now early 1943. On this journey they passed the remains of men and equipment left from the terrible Allied retreat from Burma, of the year before, including skeletons sitting in abandoned trucks near Kohima.

Burma: Operation Loincloth

At Imphal – their springboard base before going into Burma – they were given 'God Speed' speeches by General Wavell, General Stillwell (the American commander), and Brigadier Wingate himself, who said that they were to take part in intense activity behind enemy lines in Burma, and that any man who did not want to go should fall out now, 'as we want no rotten apples on the tree'. He also warned them that sick and wounded would be left behind, but supply would be regular and by air. Leon thought he was at least with people he had known for years. Nobody stepped forward. Leon could see the bearded Wingate quite clearly and later in the campaign saw him leading his horse with a sick private soldier in the saddle.

Seven columns then marched off east (Leon was in Column 7 under Major Jelkes; his platoon had the specific job of guarding the explosives section), about 2,500 men altogether, and traversed the Naga Hills and then crossed the Chindwin, by boat or swimming. Leon remembers swimming over with men with rifles ready in the boats; he thought it was in case Japanese appeared but later was told the riflemen were ready in case of crocodile attack! Their rations were quite good, being American-supplied. Drops in the hills were relatively easy to retrieve, but on the plains they were exposed to Japanese mortars and Leon took part in the many skirmishes which took place when trying to bring supplies in from under the nose of Japanese patrols. Leon also witnessed times when wounded had to be left behind, usually near water, in the shade, either to be cared for by Burmese or to fall into captivity; in fact, it is known that many were never seen again.

Although there were seven columns, they were numbered one to eight, as Column 6 had been so decimated by illness that they were dispersed among the other seven. For weeks they marauded through the jungle, attacking Japanese camps in independent columns, destroying supply dumps and blowing up the Burma Railway which was so essential to Japanese supplies. On several occasions, Leon's platoon would stand guard while the explosives teams did their work on the railway, small dams, bridges and so on. But conditions were appalling: impenetrable jungle that had to be hacked through with machetes; cold nights, hot days; and being dependent on air drops for food and other supplies. Packs were heavy and the monsoon raged, and many fell ill and had to be left behind, as Wingate warned they would be. Exhaustion was setting in.

One night, Leon's platoon was ordered with others from HQ Company to set up a road block – sixty men in all – and then return to rendezvous with the other columns next day. The road block was uneventful, but they missed the rendezvous and got lost and so were not fed that day. They later found a trail of elephant droppings and food wrappers and realized they must be from No. 3 (Gurkha) column, lead by Major Mike ('Mad') Calvert, the only column to use those animals. For three days the sixty men with Leon tracked the droppings, then heard gunshots. Calvert was attacking a railway station and blowing up bridges and was delighted to see them, so they joined the attack. Then the whole column, using deliberate hit-and-run methods, retreated into the jungle.

Then radio orders came to Calvert to send the sixty men back to the main force, meaning a difficult march of twenty miles to join the other columns. They were just roasting a bullock, so they grabbed some meat and all marched off for the next three days. Different columns joined and split , but eventually Leon continued east with his column and crossed the Irrawaddy, which was even further behind the lines. Their sabotage work continued but then orders arrived to return to India. The operation was over.

On reaching the Irrawaddy on the return journey, they tried to cross in some boats provided by local Burmese, not all of whom could be trusted. The first boat, caught in midstream, was shot at from the far side; the Japanese had found them, probably due to the air drops giving away their position. The men in the boat were all killed. Wingate ordered all columns to split into small groups and make their own way home; Leon's column split into thirty-man units (about one platoon). Their officer, a Dane called Lieutenant Petersen, decided to go north to Allied-held Fort Hertz, using local guides, and leapfrog to safety from village to village.

For a while this worked, when one day two Burmese guides took them to a crossroads and disappeared – but they had led them right into a Japanese ambush. Heavy fire was directed at them when one of the Dennet twins screamed out, 'Japs!' and gave away their position. Leon and his comrades literally rolled down the hillsides to get away and crossed and recrossed rivers to put their pursuers off the trail. But they had tracker dogs and so their officer decided that everyone would do better if they split up, and it was every man for himself. Leon and five others, including Lance Corporal Jordan (who removed his stripe in case of capture), decided to go east to China.

Starving, they would go to villages and demand food at gunpoint; bearded and armed, they looked like bandits. They also found wild fruit. One day they entered a valley and saw a hut on stilts above a stream with wild fruit growing nearby and rice in the hut. They stopped and ate. Exhausted, they fell asleep and failed to post a guard; the next second, Japanese burst in and they were prisoners.[1] It was Leon's twenty-third birthday.

Prisoners

Jordan was bayoneted in the hand but an officer stopped any further violence. The atmosphere was quite calm. All six were ordered outside and lined up, and facing them was a half-circle of heavily armed enemy. The first thing they did was steal the British medical supplies. One also stole Leon's ring that his fiancée had given him before he left the UK. Leon remembered saying to his mates, 'The bastards are going to shoot us in cold blood.' They were sure they would be killed. But the soldiers were Imperial Guard and well disciplined. The men's hands were tied and they were also tethered together in a chain. The officer asked if they realized their position and so Leon took it upon himself to say, 'Yes sir we are POWs and ready to obey orders!' The officer replied, 'That eez good!' One Japanese soldier even told Leon he looked like Robert Taylor! Then they were all marched off to the Japanese jungle camp. All their possessions were taken and they were interrogated, but of course the Japanese knew all about the Chindits, and Leon and his comrades had seen the propaganda leaflets dropped by the Japanese earlier in the campaign, telling them to surrender and they would be treated well. They showed one to the officer.

They were lucky and were fed and made to do jobs around the camp. Leon was made an officer's batman and shared a hut with other batmen, who always saw that he was fed properly. One Japanese private called Tagata spoke some English; he was a barber from Tokyo. Obtaining permission from his sergeant, he invited Leon – under armed guard – to eat at his hut, serving him delicious tea, with rice and kidneys. He said to Leon that after the war he should visit Tokyo and meet his parents. On another evening the Japanese cook asked Leon if he knew American cowboys songs and so he sang 'Home on the Range'. The cook presented him with a huge bowl of tomatoes which he took to share immediately with his fellow prisoners, having literally sung for his supper.

After two weeks they were sent by truck to Maymyo to a so-called 'Discipline Camp', but on the way they made an overnight stop at another jungle camp .They were taken to a hut and locked in while their guards went to eat. Later some other Japanese entered and tied their hands and feet behind their back and ordered them to kneel down in the execution position. A big officer swaggered in and drew his sword, and began swishing it around. Leon remembered thinking, 'Please God let it be quick!' But either it was a sick joke or their own guards returned in time, and ultimately nothing happened. Next day they continued to Maymyo, which had a pleasant climate and had been a leave town for British soldiers in peacetime.

As they entered the prison they were beaten up very badly by mainly Korean guards; Leon did not, at this point, want to go into details of what had happened to him. They were forced to learn the Japanese words of command for *tenko* (roll call), how and when to bow, and especially how inferior they were to the Japanese; they were drilled a lot and forced to pay homage to the emperor at all times. As Leon's eyes became affected by malnutrition, he would bow to every Japanese, even though it was not necessary for a private soldier; as he could not see clearly the badges of rank, he took no chances.

After about three weeks they were moved to Rangoon civilian gaol with other prisoners, in steel cattle trucks on the railway. The conditions were typically appalling, both with regard to food and latrines. At a stop in Mandalay, the doors were opened to give them some fresh air, and so that the bodies of those who had died en route could be taken out. A Japanese soldier called Hianossi (one of the Imperial Guards who captured them) recognized Leon and told them to wait a moment; he returned with a huge bunch of bananas and threw them to the POWs.

By the time they reached Rangoon, even more of the POWs had died. Someone used Leon's hat to defecate in, and as he had no hat his eyes were affected by the sun and troubled him for the rest of his life.

Rangoon

Rangoon gaol was a massive building full of hundreds of POWs who slept in large barrack-room-type cells. It was divided into national compounds – Chinese, British, Indian and so on. The SBO[2] was Brigadier Hobson, who was a liaison officer with the Chinese army and was in Rangoon when Leon arrived. A very distinguished man, he was well respected by the men. Leon was housed in Block 6 and then Block 3. The conditions were horrific: starvation diet, gruelling climate, appalling latrines (ammunition boxes, which were emptied for use as night soil in the Japanese gardens), no medical attention and, worst of all, disease. Cockroaches, lice and scorpions were everywhere. Each day Leon would see the bodies of POWs who had died stacked like logs under the stairways, waiting to be taken away. He noted that RAF and USAAF POWs were treated much worse and were in solitary confinement at first, and only later allowed to join the others.

They were sent on work parties – ditch digging, road building, loading barges, and so on. Quite suddenly, Lance Corporal Talbot, just 19 years old, developed a huge swollen stomach and then died – their first casualty. Lieutenant Roberts explained that it was a disease of malnutrition called beri-beri, a severe lack of vitamin B, which begins in the feet, spreading all over the body – including, in men, hugely swollen sexual organs especially. It is extremely painful, and this was the wet variety of beri-beri. Leon caught dry beri-beri, which gives excruciating pain in the tips of the fingers and toes, leading to deadly gangrene if not treated. He and his friends would have to massage each other's feet and hands to stop infection and give pain relief, especially at night when they lay down. Leon's friend Harry Kinson, a fellow tailor, died of diphtheria. Every day men were dying. Another man (a signaller) had a burst appendix and took two days to die in terrible agony of peritonitis. On one occasion when he was ill, Leon worked in the POW hospital as a fly-swatter. One day two American B17s were shot down nearby and in one aircraft two survived but were very badly burnt and brought to the gaol. Leon watched in horror as the British MO, Colonel MacKenzie, begged the Japanese to give him a revolver with one bullet to shoot one of them as he was in such agony, screaming out 'Jesus let me die!' Leon described the American airman as resembling a joint of burnt meat with his bones protruding, and he could not understand how someone in that condition could still be alive. The Japanese refused the mercy killing and the airman died soon after. Leon was so distressed on recounting this incident, which he had witnessed, that his voice trailed away as he wept. The other airman lived but gangrene set in and, with treatment refused by the Japanese, he died too.

There was a propaganda English-language newspaper that the POWs were given, which of course spoke only of enemy victories, but one day on work detail they stole a Japanese newspaper which, though biased, gave more of the truth. A Japanese speaker translated it for them, and they were able to guess that things were not going well for the Japanese by this time.

Leon discovered that the Japanese were carrying out war crimes by experimenting with injections to the brain on POWs. One, Lance Corporal Hardy, was executed by bayonet stabbing after escaping and being found wandering about the streets in Rangoon. His officer, Lieutenant Hardcastle, was deemed responsible for not reporting his absence and was beaten to death in front of Leon and his comrades. On many occasions Leon saw the bodies of British soldiers who had been bayoneted to death for minor infringements; the bodies were left deliberately in courtyards for the POWs to see, as a warning.

A Chinese general POW, Chi, was asked by the Japanese to join them. He spoke good English; he refused and they had him murdered in a set-up brawl between rival factions in the Chinese compound of the prison. While working in the cookhouse – where they sometimes got extra food by scraping the bottoms of the cauldrons in which rice was cooked – Leon scalded his right ankle. He was given copper sulphate to put on the wound but it gave him lockjaw and his legs temporarily seized up. To get around afterwards he had to haul himself along the walls. While he was ill with the lockjaw, a soldier called Joe Boyle came round with some extra rice and asked if anyone wanted any. Leon, speaking with difficulty, said he would like some. He replied: 'You would do, you F—g Jew-faced bastard.' Leon could do nothing, lying crippled and ill on his mat. On another occasion, while loading a rice barge, Leon overheard another POW called Sharp making anti-Jewish remarks; Leon went for him but his mates pulled them apart, shouting that they were all suffering and must stick together at all costs. These are the only two instances of anti-Semitism that Leon recalls during his whole army career.

Leon's good friend was Dennis Walmsley, who was a Lancashire lad and very handsome, like the actor Ray Milland; it was important to have a 'buddy' and men would care for each other through thick and thin. Leon explained: 'Your friend was your life; brother, father, mother sister; when I couldn't eat properly they minced my food and fed me like a baby.' He recovered, by some miracle. Dennis and Leon had the same birthday and were born in the same year; they were also captured on their birthdays! When Dennis died, Leon was greatly upset. It was common for such a death usually to result in the death of the partner, but Leon was determined to stay alive.

Bodies were covered in sacking and taken to a nearby cemetery for burial, but Leon said that dogs often dug up the bodies to eat very soon after burial. When he had a very painful boil under his armpit, he went to see the MO, who took a sharp knife and called out: 'Private Frank, hands on head!' 'Yes Sir!' 'Attention! The bands are playing!' – swish! And the boil was lanced and out poured the pus.

One of Leon's Jewish Chindit friends from the 13th platoon, 378037 Private Len Berkovitch, who was very religious, had kept his British Army issue Jewish prayer book and prayed every night in the prison. The other men all sat quietly and respectfully as Len put on his bush hat and faced Jerusalem. Whenever he did this, an RAAF Flight Sergeant would always call out, 'Quiet everybody, a man's saying his prayers.' Berkovitch was a very kind man and on Yom Kippur, amazingly, he fasted and gave his food to Leon, who was ill at the time.

One day when out on a working party, Leon managed to pinch a pair of Japanese gym shoes and white socks from a warehouse. Later a Japanese guard saw him wearing them. The guard demanded to know where he got them (Leon noticed the guard's own shoes were in tatters) and Leon said that an officer had given them to him for his hard work. 'You presento me!' shouted the guard and took them from him – but gave him a small lacquered cigarette box in exchange; this was a small act of kindness, for had the guard given him over as a thief, Leon could have been executed. He clearly remembers that he neither received nor was allowed to send any mail home in the whole two-and-a-half years he was a POW.

One POW, always wearing a bowler hat, used an old toilet potty as his mess tin, and this was a source of great amusement, Leon remembers. At first the POWs were allowed to receive small amounts of pay via their officers to buy from local Burmese – eggs, fruit and so on. But prices soon became exorbitant and that supply fell away.

There was an outbreak of cholera but they were fortunate in that only twelve POWs died. The first was a Private Ashcroft, who had survived a short death march of POWs to get to Rangoon, only to die of illness.[3] The bodies of cholera victims were cremated by the men and lime was spread around the prison to kill off infection. The thousands of flies round the latrines were thought to be the source. In another incident, they were working at the docks loading rice sacks into a warehouse. One of the men, an Inniskilling, tried to rest and lay down on the mound of sacks – out of sight, he thought. But he was spotted, and the Japanese beat him so mercilessly that he later lost his mind. In another work detail, a Japanese officer approached them and screamed 'No move!' Everyone stood as if frozen, knowing he was looking for an excuse to punish someone. Indeed, one POW (he told Leon later) moved his eyes and the officer struck him so hard that he flew through the air, into a brick wall. Such was the daily brutality Leon saw and suffered himself in similar ways.

Once Leon witnessed Brigadier Hobson in only a loin cloth, and made to stand all day in the sun because he had stood up for prisoners in an argument with the Japanese about their ignoring of the Geneva Convention.

By the spring of 1945, the POWs noticed that the Japanese were getting edgy and there was a tension in the air; there were lots of comings and goings in the prison. On 25 April, between 400 and 500 POWs, including Leon, were given some cast-off Japanese clothing; Leon had only a loincloth and no shoes by this time. They were given some rice in buckets and taken on a five-day march north over very rough terrain to a village near a river, called Pegu, a distance of about fifty miles. They were meant to be taken to Thailand. They were warned that any attempted escape would result in execution by decapitation. They noticed Japanese sappers laying mines on the road and bridges and they knew this was because the Allies were closing in. On the way they were attacked by British RAF Mosquito aircraft on several occasions. At Pegu they worked in fields near the village, the Burmese giving them extra food sometimes, probably because they could see that the British troops were on the way.

Suddenly one morning the Japanese disappeared. Using some stolen Japanese sheets and blankets the men made a Union Jack flag and four-feet-high letters saying '400 prisoners', and laid it in the paddy fields so the Allied pilots could see who they were. Next day, Allied aircraft flew very low over them and circled and flew on; but one later flight machine-gunned them, tragically and accidentally killing their CO of three years, Brigadier Hobson, as well as several villagers. The POWs were deeply saddened by Hobson's death, especially in such pointless circumstances. The experience was the closest Leon had come to death and he was screaming for his mother, he said, as he fell to the ground to hide from the strafing aircraft.

The British officers brought out their badges of rank and put them up, and the Burmese began feeding them chicken and rice; it was an astonishing feeling, said Leon.

Liberation

That evening they did not know what to do, but a Burmese girl suddenly appeared and they asked her for water and to be taken to a village. She led Leon and an American fellow POW, Frankie Hubbard, to another village and fed them. The headman told them that the Japanese had gone. He had heard American tank tracks nearby (Shermans with British crews) but Leon was afraid that if they wandered to the road, the Allied troops might shoot first if they saw men wandering out of the jungle. The next day they were taken to another village where the headman spoke perfect English; he was worried that the Allies would shell his village, thinking the Japanese were still there. An American soldier, Major Lutz, dressed as a Burmese, had crossed the lines to warn the British that the villages nearby were friendly, that they should hold their fire and there were a lot of POWs wanting to come in.

At dawn next day, the POWs received orders from two Burmese runners to march in column of three towards the road; they helped each other along, propping up the walking wounded and sick, like a scene from *The Bridge on the River Kwai*, many half-naked and emaciated. As they left the jungle, they came to paddy fields but there was still no sign of troops. Suddenly, dozens of armed Allied soldiers appeared from behind walls and ditches – they had been lying prone. They were free! Of Leon's ten-man team captured as Chindits, he was the only survivor from Rangoon prison. Of the 220 Chindits captured in Longcloth (1st Chindit Expedition, Operation Longcloth, 1943), he was one of only twenty-eight who survived captivity.

Stretcher bearers were called forward. Leon was lying on the ground, like others, feet badly cut from the walking, and an Indian soldier picked him up like a baby and took him to the ambulances waiting on the roadside. He weighed only six-and-a-half stone. Soldiers walking by gave them items of clothing and food – they could not do enough for them. The feeling, said Leon, was indescribable. They noticed dead Japanese snipers who had tied themselves into tree tops, killed by Allied troops. By sunset, the men arrived at a huge field surrounded by a great square of massive tanks and guns, of the 17th Armoured Division, and sat down on groundsheets, each with a blanket, mug, spoon and mess tin laid out in regimental order, waiting for them. They were given mugs of cocoa with rum and hot stew. Above them the moon and stars shone in the evening sky; if heaven existed, said Leon, this was it. Unbelievably, seven Japanese tried to attack the camp that night, but were all killed in a suicide raid.

The next day bulldozers flattened an area and Dakotas flew in, one after the other, each taking twenty prisoners; they were flown for about an hour to Akyab and fed with boiled eggs and bread and butter, and from there they were flown to Kamilla Forward Base hospital, where they stayed for some weeks to recover. Here Leon met the Jewish chaplain, Major Jaffee, who told him that he had nothing wrong with him that good Jewish chicken soup would not cure, and that he should write to his mother at once. He said he had to leave because he was taking matzohs to the front line for the Jewish troops to celebrate Passover. Major Jaffee also wrote to Leon's family, and he kept that letter all his life. General Slim (Commander of all the 14th Army troops in Burma; later Field Marshall) also came and visited them.

From here, Leon went by hospital train to Chittagong and then on a hospital ship to Madras, and another hospital train to Secunderbad barracks, now a hospital, where his journey had really begun. Here an Indian fruit-seller came round the wards and asked

Leon if he wanted to buy, but he said he had no money and so the Indian walked away. Then he came back and said: 'You King's Liverpool Regiment? Regimental police? Frank Sahib?' 'Yes', said Leon, and the Indian exclaimed: 'You used to be big wallah, now small wallah – what happened?' Leon explained that he had been a prisoner. The Indian embraced him, almost in tears; he recognized Leon, and the Indian told him to take all the fruit, then left.

From here Leon was flown home via Poona, Karachi, the Persian Gulf, Israel and Libya, to Taunton in Somerset; when the men got off, they knelt and kissed the ground. Then Leon went home to London, where flags were outside his home to welcome him. Six weeks later he married his girlfriend Sylvia, who was waiting for him, and they had three children. He was at home when the atomic bombs were dropped and he remembers being saddened when he thought of all the innocents who died in those attacks. But he knew it had to be.

After the war, Leon always belonged to the FEPOWs and AJEX (the Jewish war veterans of the UK) but he was never keen on reunions. He never really suffered from nightmares after the war, as he had a loving and understanding wife. For the last twenty years of his working life, he was a security officer at the British Museum.

[*Interviewed by the author.*]

Notes

1. Earlier they had passed a Burmese on the track and they think he must have betrayed them to the Japanese.
2. Senior British Officer in a POW camp.
3. Ashcroft had told Leon that on this march a mutual friend, Turner, had been wounded earlier and fell while marching in the POW column. A Japanese soldier asked Ashcroft to take Turner's pack off; then he bayoneted him to death before his eyes.

Cpl Jacob 'John' Herman Franken, Royal Netherlands Navy Air Force, Aircraft Mechanic

The author thanks Jacob and his son, Ezra, for this personal interview.

30a. Cpl Jacob 'John' Herman Franken, Royal Netherlands Navy Air Force: as a young man after liberation in 1948.

Jacob Franken, known as John because in the Far East it was too hard to pronounce his full name, was born in Semarang in Indonesia on 10 April 1922 and grew up in Purworejo. He was the son of a Jewish regular soldier, Leopold Meyer Isaac Franken, from Almelo in Holland, who was posted to serve in Indonesia. On leave in Holland he married Roset Kass, a Jewish orphan from Amsterdam, and they returned to live in Indonesia. John was bar mitzvah in the Jewish community in Purworejo. After high school he went to technical school in Jakarta.

He was conscripted in July 1941 and the war broke out on 7 December. In March 1942, they had orders to be evacuated to Australia, to finish the mechanics course. He was put on an aircraft with eighteen other men, in the middle of a dogfight above the airport, and flown 600 kilometres away to a little harbour in south Java, from where they boarded a small merchant ship to Australia with over 300 other Dutch forces personnel. They were spotted by a Japanese seaplane and soon two big Japanese warships intercepted them. John and three of his friends went to the other side of the boat to prepare to jump overboard, but they were surrounded by other Japanese ships and the sea was full of sharks; they surrendered as the Japanese boarded their ship. The Japanese commander ordered some alcohol to be brought to the deck and he made them drink, making a toast to the fall of the Dutch East Indies. As he made the toast, the Dutch spat out the drink.

Unknown to John until after the war, his mother died in a Japanese internment camp; also three of his brothers were POWs of the Japanese – soldiers Albert, Paul and Joob (or Joseph); this must be a POW record – four brothers all POWs of the Japanese. All survived the war but died young. Albert was left for dead by the Japanese in a swamp, but a native girl found him and nursed him back to health, and they later married and went to live in Holland. John's fourth brother, Louis, was deported by the Nazis in Holland and died in Mauthausen death camp.

John described some of his experiences:

> We were working building schools for the sex slaves, we were ordered to work in schools to make partitions so that the girls have a little privacy. The girls were picked up from the street in trucks by the Japanese. They were brought to the hospital for a check-up for venereal disease and then they were put in those schools,

> waiting for the officers first then the soldiers to come and use them. And my job was to...put up a curtain about 15 feet long, made of very soft material, across the room and after every soldier, the woman or the girl then has to walk across it to wipe off the semen and waiting for the next soldier. I had to provide clean water for them, clean the rooms and so on. They were crying for help, you cannot describe those things. It's beyond imagination. This is what people do to other people.

Sometimes a girl was abused up to forty times a day, but the Dutch POWs could not help them, on pain of death.

Sometime in late 1942 John was shipped to Japan. He travelled on one of the Hell Ships in the most appalling conditions, made worse by the fact that the ship also carried live animals for a zoo in Japan; the smell in the overcrowded airless holds was atrocious. There were about 600 POWs on board, half Dutch and half Indonesian soldiers. The journey took about two weeks, but John cannot be sure as the memory is a blur. When they arrived in Japan, for a while the main job was clearing rubble from buildings bombed by the Americans. Then John was sent to work in the shipyards at Kawanami, near Nagasaki:

> ...day in and day out and on punishment if you don't work, really work. If you don't, they come down and beat you up; we were allowed only one bath a week. If you go in the bath, the first people get the clean water and the rest of them get the rest. We had about 1,000 people in the camp so you know, we shift every week so another room will get the first to be in the bath.

As a welder, John was lucky in having a slightly more comfortable job, as working with welding material warmed him a little, and the required heavy protective clothing meant he could keep warm in the very cold winters.

Soon the prisoners heard volunteers were needed for a coal mine, 'where they can have a bath every day because you're completely black from the coal, we said, well, why not, we'll work in the coal mine. So I volunteered to go to the coal mine. Just for the bath, and that saved my life.' It was while in the coal mines that the atomic bomb was dropped on nearby Nagasaki

> We know it was bombs, they were bombing day and night and we know the planes are flying over. But when the bombs hit Nagasaki, I was on the nightshift. And I think it was 11:00 a.m. or something like that. So when we came up, we didn't see much, we were too far away from the atomic bomb. But we saw the smoke in the distance and we said, my gosh, so they must have hit an ammunition dump and we were enjoying it. You know, every time the Americans did something very drastic, we're celebrating, we knew that freedom was coming soon.

John was liberated by the Americans and taken to recuperate in the Philippines. When he returned to Japan many years later to speak to schoolchildren about his terrible experiences, he saw that a school had been built on the site of his old camp. He married Sonya (an Auschwitz survivor) after the war and went to live in Canada and had a family. He worked

30b. John as an older man.

30c. John's wartime medals, left to right: Dutch Indies Star, Java Cross, Devoted Service Medal.

as an aircraft mechanic for Air Canada all his working life. He was involved in trying to get apologies and compensation from the Japanese government for the horrific treatment the POWs received, and in 2007 was awarded the Medal of Orange Nassau from the Dutch Queen for his work. He also has the Dutch Indies Star, the Java Cross and the Devoted Service Medal.

The following narrative is John's story in his own words, published for the first time; I have excluded some items already mentioned in the above testimony. Anything in square brackets has been added by me for clarification purposes only, otherwise John's story appears here as it was written.

WWII Memories of John Franken

One

The memories [of Japanese atrocities] have been pushed back by most of the victims but they are still there, appearing in recurring nightmares. The Queen of Holland, when in exile in England [in WW2], made all kinds of promises in her radio speech in London. She told about compensation after the war. They turned out to be only promises and they left a bitter after-taste. The stupidity of the Allies to release the Japanese criminals from any further trials in 1958 and the lack of punishment of their crimes, is unbelievable. To top it all ... the Prime Minister of Canada accepted the apology from the Japanese for their treatment of the Canadian POWs in Hong Kong. He knew that the War Amputations Organization was trying to get some compensation and that they had no co-operation from the Canadian Government. Once in a while I have visited the reunions of ex-POWs and ex-civilian camp-survivors, some of whom were used as child slave labourers. We talked then about our past and realized that we all have the same experiences. The trouble is, our groups are getting smaller every year.

I would also like to dedicate this story to the many friends who never made it and I hope that they did not die in vain; that the Dutch Government and the Allies show some appreciation for their War Veterans who are still suffering from the mental and physical wounds of that war. The response, the understanding and the knowledge of the politicians is deplorable.

This story is not a happy one. There were moments of sorrow and sometimes laughter, but there was always hope in the camps. The Japanese 'humour' is very hard to understand. If we feel sad, they laugh. I can't forgive and forget what they did to the women and children in the name of their so-called God, the Emperor. Their treatment of their women and minorities are still of the dark ages. The Japanese expansion theory is only put on the backburner and it is only a matter of time or history will repeat itself, if we let it happen... This story is about my experiences of growing up in Indonesia, the former Netherlands East Indies, and my time as a Prisoner of War.

I was born on April 10 1922, in Semarang the capital of Central Java, the heaviest populated island of this chain of islands. I was the fourth of five sons. My father was born in Holland on March 2 1877 and my mother Roset Kaas, was born on July 28 1881. She grew up in a Jewish orphanage in Amsterdam. My father joined the army in 1898 and was shipped to Indonesia. He fought in the Atjeh war in the Northern part of the island of Sumatra. After a few years he went for leave to Holland and got married. He went back to Indonesia and settled in Batavia, now called Djakarta, the capital of Indonesia.

After twenty-five years of service, he retired and received a small pension. When I was one year old, my parents moved to Purworedjo, a little village in the Southern part of Mid-Java. My father started a small store and for many years, the income was very small. He also became a travelling salesman ... and he was away several weeks at the time. My mother and my oldest brother ran the store, but it turned out to be too much for my mother. Then my father closed the shop and rebuilt it into a hotel as there was a big demand for accommodation for the travelling salesmen coming through. Most of them were selling imports from Holland. This way I learned more about my Jewish background as many of the salesmen were Jewish. There were only a few Dutch people in the village, mostly teachers, priests and officers of the army. We were the only Jewish people and learned at an early age about anti-semitism. My younger brother Albert [born 18 September 1923], was of a more explosive character than I and there were many fights.

My mother ran the hotel with a firm hand and was always fair to everybody and she was always there when needed. She had many friends among the natives and was very much loved. She was up very early in the morning to make breakfast for the guests. At 7 a.m. everything started, also the schools. At 1 p.m. everything was closed until 4 p.m. till the hottest part of the day was passed [35°C]. And at 8 p.m. the day was over. After seven years of elementary school, I went to high school for one year, but I did not like it. In 1937, when I was 15 years old, I went to Djakarta together with my younger brother Albert. My oldest brother Joop took over the hotel.

Djakarta is a city about 60 km to the East of Purworedjo. My father was seldom home as he was still doing the salesman route. He passed away May 3, 1941 from blood poisoning of an infected tooth and was buried in Magelang with full military honours. He left my mother a small pension to survive. Before he died he called my younger brother and I to his bedside and told us to learn a trade and never to try the business so that we would never be hungry.

In July 1941 my younger brother Albert and I were mobilized. I went into the Navy Air Force and my younger brother into the regular army as a male nurse. My brother Louis, no.3, went to the Forester school in Malang, a city in East Java and he did very well. He won a prize in the lottery and went to Holland to continue his studies at the forestry university. When the Germans rounded up the Jews in Holland, he joined the resistance groups, but he was caught. He died in the concentration camp Mouthausen [Mauthausen] May 5, 1944. He was 27 years old!

My younger brother Albert was caught by the Japanese and so were my other two brothers. They were put to work as slave labourers at the Burma railroad, which is described in the book 'Chaplain on the River Quai' by Rabbi Chaim Nussbaum. My brother Paul, who still lives in Toronto, Canada, is also mentioned in it. His wife Ester was in the Japanese concentration camp Adek in Batavia, with their son Louis who was then only 4 years old. When Albert was too sick to work due to the high fevers of malaria attacks, he was thrown into the swamps. A Thailand girl found him and nursed him back to life. He married her after the war and took her to Holland. He worked in the army hospital as a male nurse for emergency care. He died on April 16, 1970 of a heart attack. I was at his side, when I was on a holiday in Holland from Canada. My brother Paul, who was already an optician in Indonesia, died in Montreal May 30 1970, also from a heart attack. Joop, my oldest brother, became a guard at a military base in Holland and died on October 27 1965, of a stroke.

In the forces, I went to a Navy base in Surabaja, the capital of East Java, in July 1941, to sign a contract for ten years. The course would take eight years, two years of which was for training for the age group of 18 years and over. And, it took three years of training for the ones from 14 to 17. I mention those two groups as it was an important factor of my fate when I became a Prisoner of War.

Two

The Japanese attacked Pearl Harbor on the eighth of December in 1941. We listened to the speeches of the Governor General and the Queen, who declared war on Japan. We were transferred to Malang, a small city on the slopes of a mountain. In an old sugar factory we used the old machine shop to make guide-fins for deep sea bombs, which were attached under the wings of the Catalinas water/land planes, to combat Japanese submarines. Sometimes the sirens went off and we saw some dogfights between our Brewsters and the Japanese Navy Zeros. We stayed in Malang until the end of January 1942 and then we were sent back to Surabaja to prepare for evacuation. From Tjilatjap, a harbour town on the South coast in Mid-Java, we would be sent to an unknown destination.

In the meantime, at the navy base in Surabaja we had to make use of air-raid shelters which looked like the form of the end of a big cigar. The walls were three feet thick. The air pumps had to be turned by hand to get fresh air inside. One day I was sitting on the toilet in the quarters where I was staying, when the sirens went off. In a hurry I ran, with the others, to the airs-raid shelters and as soon as I saw the heavy door close behind me, I heard the bombs dropping all over the area. One of them hit the toilet where I had been on just minutes before the air raid. The airport was built on a swamp and when the bombs hit, the landing strips looked like a rice field had just been ploughed. The damage at the airport was terrible. Some of the hangars, filled with fully loaded planes ready for combat, exploded.

There were also machine guns installed along the waterfront. One day, the commanding officer on an inspection tour, accidentally pulled the trigger of one of those machine guns and of course, the bullets came flying out. They went through the walls of the main building. Some of the fragments hit a student mechanic by the name of

Smethorst, in his behind. On the next air raid one of these machine guns shot down a Japanese navy Zero.

Then the order came to actually evacuate to Tjilatjap, a distance of about 600 km from Surabaja. Fourteen of us were assigned to an airport bus with an Indonesian driver. Each of us received a gun (model 95) with ammunition, just in case we would meet some Japanese soldiers – what a joke! While we were loading up the bus with food and drinks, the sirens went off again for another air raid. We were at the outskirts of the airport and we saw the dogfight above the navy base. We took shelter under the bus and when the raid was over, the driver had disappeared. Only one of us could drive the bus. His name was Theo Snellen van Vollenhoven. (He survived the war and is now a retired bus driver in Los Angeles.) More about this later.

We discovered when we were on the road, that all the road signs were removed over the whole island of Java. Every so often, another man had to sit in front of the bus near the driver to show him the way, depending on which area they had been familiar with. And also, if he had relatives living there, he was allowed to say goodbye. When we passed through Djakarta, I said goodbye to my mother. That was the last time I saw her alive. This stayed engraved into my mind. My mother died on November 2, 1944 in a Japanese prison camp in Ambarawa.

In that same camp was a woman Mrs Van der Velde with two children, a boy Benno and a girl Elise. Benno today is an accountant in Amsterdam, Holland. His sister is married and a grandmother in Jerusalem, Israel. Their father, who passed away in a Japanese prison camp, performed the marriage ceremony of my brother Paul in 1937. As he (Paul) was a pious Jew, the condition was that he would be married by a rabbi if he ever got to Holland. Mrs Van der Velde went to Holland after the war where she found out that she had TB. I had then married a sister of my brother Paul's wife. When Mrs Van der Velde went to Switzerland for treatment, I adopted her daughter Elizabeth until her mother returned all healed from her illness. Then her daughter went back to her. I emigrated to Canada in 1951 and divorced in 1955, but I kept in contact with Mrs Van der Velde. She was the one who brought me in contact with my second wife, Sonja, through correspondence. I married her in 1960.

Three

The war continued in 1942 and our planes were on a mission. The Navy took heavy losses. The coasts of the island of Java were now wide open for landings by the Japanese invasion forces. All the other islands were already taken. We were stopped with our bus at that time, in the middle of the night near a little village, Ngawi, for rest, refreshments and fuel. The Japanese were very close. We did not stay long and continued our trip to Tjilatjap. We heard that the airbase of Morokrembangan in Surabaja was completely demolished. A long time before this war broke out, the Japanese had sent many spies to Indonesia. The Navy-Air Force had put old planes on the water with fake bamboo guns and when the bombings started, not one of the old fake planes were hit, only the real ones.

The Japanese had landed at Rembang and were trying to cut off the evacuation from Tjilatjap. Only a few of our side were thinking of forming a guerrilla group, but nothing was organized at this point in time. When we arrived in Tjilatjap, we had to return our guns and ammunition and there was a lot of commotion and disarray. The harbour was full of ships of the Dutch Merchant Marine. In the meantime, a lot of air raids took place as the Japanese knew that evacuation was in progress. The boys in the three-year course were on the boat the 'Kota Baruh' and next to it was the Java-China-Japan line ship, the 'Tjisarua', the biggest of them all. Many of the corporals and sergeants of our Navy training school were there and they asked us to join them. The other ships were full of women and children. The most important thing was to get on board as soon as possible and try to leave this chaotic place. Many ships were leaving, also planes loaded with officers and their families. We stayed another day, without any explanation. All fourteen of us boarded the 'Tjisarua' going to an unknown destination. We had one 7 cm gun on board! We heard that some of the ships that had left ahead of us, had been torpedoed after all the effort of getting out of this mousetrap on the South coast. There must have been a circle of submarines waiting outside the harbour. This was a rumour and nobody knew for sure! Everything was upside down as the world was collapsing around us.

About 650 men were on board the Tjisarua and it was very crowded. We slept on the upper deck side by side, with our navy bag as pillow. We were very happy that we were on aboard and leaving, not knowing what lay ahead of us in the future. We left as it became dark and we went in the east, keeping close to the coast line trying to avoid any waiting subs. We left behind a quay full of newly arrived war armaments and complete planes, still in their crates, soon to be burnt so that nothing would be left for the enemy. The next day we came to the end of the South coast of Java and we changed direction to the south-east. We then understood that we were on the way to Australia. The sea was very calm. Nobody complained as it was just like camping. It was the best under these circumstances. As food we got porridge with jam and we did not complain. (There was much better food for the officers; they got meat and potatoes with onions.)

We were apparently escaping the ring of subs and were starting to feel safe. With a calm sea, the night sky full of stars, the atmosphere became more relaxed as we were almost there, but ... not for long. It was around eleven o'clock in the morning when the ship's siren went off, the sign of alarm like there was an air raid in progress. We were roughly awakened out of our dreams about Australia. Three long blasts and one short one, which means the letter 'V' for aircraft alarm. One single plane with a red ball under its wings, came flying over. It circled once and disappeared in a Southern direction. We were hoping that the main fleet was too far away to block our escape before night time. Maybe, maybe, we were hoping, praying in silence. But it was in vain! At three o'clock in the afternoon, we saw a smoke plume at the horizon. A Japanese warship! We would not see Australia after all. Shortly afterwards another one appeared, a corvette. They were signing with their flags and we stopped our ship. The two warships were just circling, about a dozen torpedoes pointed at us and also their cannons. Gabel and Van Veen told me, 'Let's go to the other side of the ship, then we can jump off if they start firing.' How naive we were ... On the other side of the ship we saw more torpedoes pointed at us and they kept circling like sharks. They lowered a sloop with a crew of Japanese marines on board and they came to our ship. We lowered a rope ladder to let them come on board. Then another three warships appeared on the horizon, three battle ships with double towers and now we discovered the enormous war power of the Japanese. We were always told that they could not fight with their little crooked legs. That they all wore glasses and were cross-eyed, that their ships were top heavy with the armament, but this was something different from what we were told.

These were beautiful ships and very modern, much better equipped than ours. We were now experiencing the painful price of socialistic politics. On the battleship decks we saw hundreds of Japanese Navy men all lined up in their yellow-brown tropical uniforms. They looked at us from under their caps. We felt humiliated and powerless with our 'only 7 cm' gun. And to top it all off, one of our officers went to the gun to fire it! He was hastily told to get lost as he would put all our lives in jeopardy. Was this the first time that they looked at some white people? It was our first sight of the Japanese enemies and we felt powerless. The cruisers were circling so close that we could see their faces and their dirty smiles of triumph. We were just a little stopover on their mission to entertain their crew. The three battleships stayed for a while and then left the scene. We started to realize that this was the beginning of a period of becoming Prisoners of War. The voices of the marines who came on board were very loud and mad sounding, but maybe they were just asking, 'How are you?' We were all told to come on deck to listen to a speech. I will never forget this! The Japanese officer jumped on the railing of our ship and walked to the front. He made a good impression doing this. He made a speech in perfect English and told us that they would win the war; that we were now prisoners of war without any honour. We were very lucky that the weather was good and that there were hardly any waves so that nobody was seasick. The officers told us that we had to follow all the commands or we would be punished. The cruisers kept close watch. The engines of the 'Tjisarua' started again and we went in the North-East direction. We were now on the way between the Sunda Islands under escort of one warship. We could do nothing with the Japanese crew on board. The next morning the other warships came alongside again. I did not know if they were the same ones. After waking up I thought that I was waking up from a nightmare, but the truth was that the Japanese flag was on top of our mast. Yesterday it was Red, White and Blue and now it was a red ball. We were now in the Strait of Lombok and some men were talking about escaping. But with the strong current and sharks around, it would be just suicide. On March 8 we were all called on deck and given a tin cup which was filled with wine from a wooden barrel. We did not know what was behind this but we soon found out. We had to drink the wine and when this was finished, the Japanese officer told us that we had just drunk to the downfall of Java. We felt like throwing up and he was laughing his head off. For him this was a big joke. This was their kind of humour! I never got used to it and we got more of it during the next three and a half years. While on board we were counted and counted a million times over and over again.

When we entered the harbour of Makassar, we passed beautiful little islands. Was this the opportunity for escaping? It was noon, small boats were around us from the local inhabitants and we did not know if they were trustworthy. They were not really known for their loyalty to the Dutch Government. Everything looked so beautiful and it was hard to believe that a war was going on. With a lot of screaming and pushing we disembarked. On the way out, I passed the ship's kitchen where a nice steak was just sizzling. I looked around and found a piece of cloth, took the steak out of the frying pan and threw it at Gys Flammang as it became too hot and he threw it back at me. I hid it behind my shirt, not knowing that it would take another three and a half years to get my next steak. We could not take anything with us but a toothbrush and a towel and were told that the rest of our clothing which were in our blue navy bags, would soon follow. Many did not believe this and took most of their belongings with them.

We boarded a landing craft and went ashore with a lot of 'kurahs', and kicks with the butts of their guns, we had to climb on the quay. On a small space in a corner, surrounded by fences, we had to stand to attention. We were counted and counted and counted, a thousand times. In the meantime, it became completely dark. The burnt out and empty buildings and sheds we saw across the streets, looked very depressing. We saw ordnances and motorcycles passing by and more luxury cars with red headlights loaded with stolen merchandise of the Dutch civilians who had already been put in separate camps. We had to wait a very long time while we were sitting around on the ground.

Four

Suddenly there were the loud 'kurahs', at the top of their lungs they screamed and we learned the meaning very fast. In this case we had to stand up very fast and be counted again. We had to walk in a long row. At first we just walked, then we had to walk faster and then in a slow run. As I was then 20 years old, it did not bother me very much, but my thoughts were with the much older ones who were carrying also heavy luggage. It did not take very long or they had to throw it to the side of the streets. It was too heavy to carry and even younger ones had to do the same. The natives, watching us from the sides of the roads, had never seen the Dutch running like that, all along pushed and shoved by these little people. They were enjoying this scene very much, they were shouting at us while fighting for the luggage which we were not able to carry any more as is was thrown to the side of the road ... There and then I realized that this was the end of the Dutch East Indies being a Dutch colony. Now we had surrendered to the Japanese and now we also had the hate of the local natives upon us. We stopped in the darkness of the night in front of a big ugly building, which was a women's prison... In groups of fifty we had to enter after being counted a thousand times and one more, before we entered the cells pushed by an animal of a Jap who used a heavy piece of wood for counting. Everybody received a hit when he passed by this bully.

The cell was made for 18 women and we were 130 persons in it. Many were sitting, some were laying down. The ones at a wall had a little back support. The prisoners of this prison, thieves and murderers, were released and we took their places. The boys in the next cell had come in a day earlier. We found out that they were the ones who survived the Battle of the Java Sea and were of the battle ships 'De Ruyter', the 'Piet Hein', 'Van Amstel', 'Dubois', 'Kortenaar', 'De Java', the American destroyer 'Pope', the English cruiser 'Exeter' and the 'Encounter'. Their cells were also very overloaded and we were trying to get information from each other about friends and about the results of the battles. No paper headlines, but careful information through the prison walls by morse code.

We were very hungry and very tired from all the emotions and were trying to sleep on the hard cement floor. The cell had only one window with steel bars and one wooden toilet which had to be emptied many times. The smell was unbearable. At that time we were still a mixed bunch of sailors and officers together. The officers were thinking that they were the only ones there in the cell and took possession of the only window for fresh air. We made it clear that we also had some rights for fresh air. The next morning we received our first breakfast, a slice of bread. After that we had to wait till the evening for a little rice and a piece of dried (flying) fish. Plates were not available. We had to use anything, a handkerchief, a hat, or your bare hands. We were let out the next day on an open cement platform for a so-called bath. We were sprayed with a hose by a small native boy. He most likely washed his buffaloes like that as he enjoyed this spraying very much.

We washed ourselves very well and even held our underwear in the stream of the hose. It would not be cleaned, but it had the fresh feeling of cleanliness. That bath was followed by some quick exercises to loosen our muscles as they were very stiff from sitting in those crammed quarters. We had to run three times around the enclosed area. This was followed by counting and counting and we had to make rows of five on our knees.

The Japanese officers and ships crew came to see us in our cells like it was a zoo. They arrived especially around meal times as this had a special attraction for them, just like feeding the monkeys ... One day, on one of these visits, one of the English 'Java Sea Battle' survivors made a remark, 'Bloody bastards!' One of the officers who happened to understand English overheard his remark. He was taken out of the cell, fastened to a pole with his hands tightened above his head and then he got a beating with a ships' rope which was soaked in water first. It was the first time of many tortures I had to witness.

But, slowly the treatment was getting better. The food arrived twice a day in two big galvanized washing tubs. They were filled with rice and still, with the dried flying fish. The fellow prisoner who distributed the rice was responsible that everybody received the same quantity. Drinking water was in an earthen-baked tub where we could drink it with a half coconut shell tied to a handle. The dry cooked rice was not so healthy as it made us constipated. One day my friend Gys Flammang was called to the guard commander to give him a massage. When he was finished with it he had to pass by a storage room full of cans, among which were five pound cans of dried prunes. He managed to steal one can which he smuggled under his towel into the cell. We could not open it as we had no can opener. I found a rusty nail and punched small holes next to each other till it was enough to pull the can open. The four of us finished the whole can and felt much better the next day. I had not gone for the last seven days. The others got laxatives from our own doctors who had taken over the health care in the prison camp. There was no treatment for other illnesses though as there was no medicine available. The time for exercise also became longer, so did the time outside the cell, sometimes a whole day. We were moved into a bigger cell and were now sleeping shoulder to shoulder on a bamboo platform instead of on the cement floor.

On March 18, 1942, the day started as usual by counting, then breakfast of rice and fish and then counting again, when I heard a commotion outside our window to the courtyard. I heard a lot of 'kurahs'. This meant, 'Fast, fast'. I saw a young Ambonese boy of about 18 years old, ready to be shot by an officer of the guard house. The people of Ambon are more in love with the Dutch Queen as many Dutch should be. The young man was called to the guardhouse by the Japanese officer where he was told to take down the picture of the Queen of Holland, which he did. The officer threw the picture on the floor and started to dance on it, broke the glass and then tore the picture to shreds. The young Ambonese then hit the Jap officer in the face. The officer took the boy to the courtyard in the back of the guardhouse and asked if he would like to be blindfolded, but the boy refused. He was then told that he would be shot. They tied him up to a pole, the officer walked backwards, took out his pistol and got ready to shoot. We were watching with horror through the window. At that moment the boy managed to release one arm and started to yell at the top of his lungs, 'Long live the Queen!' The officer kept on shooting and the boy kept on yelling. But the boy was not dead. The officer walked over to him and put the gun to his temple. Now we knew what to expect in the future!

We were told to do some work outside the prison walls, in small groups, to clean up the debris of the bomb raids by the Americans. We also had to load coal into the ships or clean the streets. In bygone days it was unknown for Dutch people to do this kind of work, but now we had to do it for the new masters. The natives soon saw, that this was very humiliating for us and realized that the reign of the Dutch was over. Sometimes when we were outside working, they watched us and sung a song to us that went like this, 'The rain came, the lamb ran away, The Japanese came, and the Dutch ran for cover' (in their own language in rhyme form). They were really rubbing it in.

In the evening when we were together after mealtime, we were telling the stories of our experiences of the day and about the situation in town. We talked about home where ever this might be or, about our families. We were speculating and made up fantasies of how the war should end as this imprisonment could not last forever. In our expectations and fantasies we sometimes imagined that the Australian troops would soon liberate us, maybe in a matter of weeks. Of course there were the ones among us who were more realistic and they would like to know where we based these rumours on. These men were the negatives and the pessimists who we did not like to listen to. They were saying that it would take at least a year. We believed in only predictions of 'weeks' and were on the positive side. Soon we had bets going about when the war would end ... We started telling jokes and sang songs to keep our spirits high, especially in the early days of capture. From different cells songs were coming loud and clear in Dutch. The English and Americans were joining in with their own repertoires. Many of us liked to go out and work outside the prison, even if some jobs were very dirty and depressing. But when we were outside, we were able to smuggle in books, which we found when we had to clean out abandoned houses and schools. Besides that we could bring in bananas, eggs, soya sauce etc.

To keep ourselves occupied, we made our own games. When after a few weeks the food started to improve, our own cooks were assigned to prepare and buy the food at the market. We were not spoiled, but it tasted much better, especially the rice with a little bit of vegetable soup and small piece of dried fish. We were also trying to keep the hygiene as pure as possible applying strict discipline. Our cell was mopped every day and the only toilet, a drum with cover, was emptied daily and more times if possible, cleaned out and dusted with calcium powder to eliminate the stench. Slowly we acquired more utensils from the abandoned houses, which we had to clean out from debris, so that the Jap officers could live there.

Then there was that part of changing clothing and underwear which was not always possible. The so-called room service which the Japanese had promised by sending our luggage from the ship, never materialized. The ones who had to throw their luggage to the side of the road when they had not been able to carry it any more, had practically nothing to wear. It was very hard to keep our limited wardrobe clean, especially in the tropics where it is so hot. Finally, when it was about time to leave the prison to go to other quarters, our luggage was brought in by truckloads and dumped right in the middle of the courtyard. Everybody could look for his own property. This proved impossible and then we decided to distribute it to the ones who had nothing. It was now in the beginning of April. We had been about three weeks at this 'hotel' and we were slowly getting fed up with it. We were told that we were going to another location, but nobody knew where it was going to be. There were no trucks and we had to walk so it could not be far. We were counted in rows of five, over and over again and then we were on our way. We carried everything with us and we looked like a mass of landed immigrants. The natives were again lining up along the side of the road like before, but this time they got nothing and even a group of POWs walking on the street did not look so special any more.

Five

When we came to the new place, we had to wait in front of it. And then, after being counted again, we had to go back where we came from. Somebody had the instructions wrong and we had been sent to the wrong place. We would have to try again the next day. It was a big chaos and we were buying food from the natives with the little money we had. The next day it was a repeat performance, only this time we were going to the ex-Dutch army barracks. It was a big improvement from what we had before. The Dutch army POWs who were already there, were walking freely all over the grounds in the shadow of big trees. The campsite was made up of four quarters with a centre point which was the mess hall. After being counted, John Koch, Joop Brinkert and I were in barrack 16, altogether 116 men. The floor was made of stone tiles and each of us got the width of three tiles, a total of 24 inches. I managed to get a hold of a straw mat so that I was not laying directly on the cold floor. Soon we were surrounded by the Dutch army POWs and treated with candy as a matter of sharing. They had bought it from the inhabitants when they went out on cleaning jobs.

My monetary possessions were only 15 guilders which I had hidden in my shoe. The rest was taken as they searched me on the ship when I was captured. In room 16 I was lucky with a place right under a window where I could climb in and out onto a little grass field. We could walk around and see other boys and exchange stories about their experiences. We were pleased not to see the Japanese guards very often, only when we went to the toilets. When we had to pass by the guard we had to make a proper bow. In the beginning we did not do this right. Then we were called by the soldier to stand to full attention in order to receive a blow to the side of our head with the inside of his fist. With your eyes closed you would not know when the fist would come. The idea was to fall immediately to the ground, otherwise the guard would call another prisoner and he had to hit you as hard as he was able to. If this was not done to the guards liking, he would do it for you, until it was done right. We practised how to hit and make a bad fall! If the guard laughed loudly it meant that you did a good job.

Every day the rumours were different, from landings to fighting on the island, but it was only wishful thinking. In the evening we were singing all kinds of songs or were listening to the experiences of other prisoners. Many were playing self-made games like chess and checkers and many were writing diaries. The food was much improved as we did our own cooking in the camp kitchen. Everything in the barrack was kept very clean and we had our own inspection team. Our officers were kept in different and better quarters.

We had a lot of trees in the grounds and also coconut trees. Around the camp was a barbed wire fence and about ten feet further out another one, which formed a piece of no-man's land in between where nobody was allowed to be. Lots of smuggling was done by throwing food and other items over; then we would throw the money over fastened to a stone, until the guards put an end to it by replacing the outer barbed wire with sheets of tin plates and that did stop the smuggling. We were quite a mixture in the camp, Australians, Dutch, Indonesians, English, Americans etc. My friend Joop Brinkert survived by swimming for twenty hours in the Java Sea. (He now lives in Delft, Holland and is coping with health problems which are a result from his POW time.)

The toilets were set a little further away from the barracks. They were made of stone and there were no doors in them. There was a trench about 10 inches wide and a foot deep over the length of the floor, through which a small stream floated the faeces away. At the end was a cesspool. Once a week, a large truck with a big container on a rig, sucked the faeces out of the cesspool and drove it out of the camp to the seaside where it was dumped. These weekly cleanings were very important to us, as this was our only source to order food and other things from the outside. When the faeces were dumped into the sea, the six inch diameter hoses were flushed with sea water and then filled with the goodies we had ordered. They passed easily by the guards who never looked into the hoses. We did not have seats on the toilets so we had to squat. Also paper was not available. Everybody carried a bottle of water with him. In Indonesia today, this is still the custom. One hand to hold the bottle, the other to wash! The Indonesian crews on the cruisers today always wear a glove on their left hand according to their custom.

Then came the announcement again that work crews would have to clean the debris from the bombings. We were divided in groups of ten to report for work in the morning, after of course, first being counted. Washing was a big problem as soap was hard to obtain. When you were counted, you had to turn your head sideways and you called out the number. After that you had to turn your head forward again. If you did not do this fast enough, you got hit on the head with a stick of about 3/4 inch diameter or you were called in front of the group and hit in the face.

One day the guard asked for volunteer telegraphists and the Navy boys who were familiar with the Morse code stepped forward in the hope to get news from the outside world. They marched to the main gate and when they returned in the late afternoon, we were eager to hear some news. But they told us that they had to carry telegraph poles the whole day. The poles were 40 feet long and one foot in diameter. This was the Japs' kind of humour. On a later date they asked for volunteers for a swim and there were many who went, escorted by two guards. When they returned they were full of red blotches all over their whole body. They had to bring in a small stranded boat by dragging it to shore. It sat in water full of very poisonous jelly fish, the Portuguese Man of War.

Many buildings, stores and harbour warehouses were destroyed. The cleaning had to be done by us 'Horios' or POWs. The ones at the harbour were very special to us as there were still supplies of clothing and canned food under the debris. Much was spoiled by fire and water. But some was still in good condition and we ate it on the spot as it was too risky to bring it into the camp. One day we found a burnt out wine cellar under a store. The guard had no notion of it. There were three big wine barrels and many bottles. Some of us kept a watch and kept the guard busy to distract him. One by one we went into the cellar to have a taste of the wine. Some of the men could not stop and over-indulged themselves. Some of them were quite drunk and on the way back to the camp we were the happiest prisoners ever! The guard did not know that we were drunk, he thought that we were giving a good show. But at the camp when we were being counted, we had a hard time standing straight. We managed somehow and we also had to hold our breath so the guards would not smell the wine. One prisoner was caught who tried to smuggle a bottle of wine in. He was severely punished and beaten with the 'famous' rope soaked in water which was followed with a lot of screaming and yelling. When the man became unconscious, he was revived by a couple of buckets of water and then received a second beating with the same rope. But this time he was tied to a pole, so he could not fall over. In the olden days this must have been the cat of nine tails. At this point the man was too drunk to react as he was completely numb. He started singing at the top of his lungs. He was then taken off the pole and with a rope he was tied to a bicycle and dragged with his face scraping over the ground. Then our own camp commander, by the name of Gortmans, who also witnessed this torture, walked slowly over to the Japanese bully, by the name of Yoshida. He snapped at him in the Malay language, 'Tjoekoep', which means 'That is enough.' The poor victim was now unconscious and Yoshida acted surprised. He got off his bicycle and released the Navy man to us.

Who was this man Gortmans? He was a man who with a small group of soldiers kept fighting a guerrilla war, even after the official capitulation. They fought in the mountains but weakened by disaster and exhaustion, the group became smaller and smaller. One way or other, the Japanese let him know that if he did not stop fighting, action would be taken to destroy the families of the group. So he had to capitulate. The Japanese thought that they had been fighting a bigger force and were surprised that Gortmans only had a handful of men. This was the reason that Gortmans could ask for a few concessions before he surrendered. The officers were allowed to keep their sabres and there were no reprisals taken against the soldiers' families. Due to this behaviour, Gortmans automatically became the camp commander of the POW camp. Everybody had the highest respect for him, so did this bully Yoshida.

But some other day this same bully Yoshida saw a prisoner not bowing properly. It was an Englishman. He was tied to a pole of the barbed wire near the guardhouse, with his face towards the street so that the city inhabitants could see him. Soon a small crowd had gathered. The Englishman was hit with a billiard cue. His hands were tied behind his back and were also hit, so that both his wrists were broken. Yoshida kept on hitting and hitting till his pants were wet with semen. When the ordeal was over and the Englishman was untied, he looked Yoshida straight in the eyes and walked away ... that was sheer willpower! When I saw this happen, I asked myself how people can do this to others, sometimes in the name of their Gods, to inflict such pain to their fellow man. This man must have had parents or a family at home. I did not know how all this fits together.

The smuggling of food became more intense and sometimes we had eggs for breakfast. We dug long trenches of about a foot deep and covered them with branches and grass, so it would not be noticed. At the end of the dug-out we made a small fire where we cooked the eggs and other goodies. We were rather free to do what we liked except when we found out that there would be an inspection or counting. There was one good Japanese soldier, who risked his life to give us, usually, a day's advance notice so that we could prepare ourselves. Then it was time to hide all our food treasures. Usually we tied it to the tree branches, out of sight.

There was a day that we ran out of wood and we had to find something to make a fire. We decided to take a piano apart and strip it of the wood. This piano was standing in a little cabin on the campgrounds near the guardhouse. Some of the soldiers sometimes used it to practise on. When we were finished with it, only the strings were left. When the Japs found out, all hell broke loose. Our barrack, which was closest to this cabin, was suspected. But nobody came forward. We all had to stand in line and each of us received four hits with the famous rope on our behind. Mine had all the colours of the rainbow, from purple to black. Yoshida was a specialist in torture. To name some things he did, for example, if you were caught smuggling bananas, you had to run a couple of rounds around the camp with a banana sticking out of your mouth. Or, you had to kneel with a stick behind your kneecap, which stopped your blood from circulating. When you got up, you could not stand on your feet and the pain was unbearable. And also, if you had a moustache, he pulled it out, sometimes the skin came with it! My behind stayed black for weeks from the beating.

We were still sent outside the camp to do slave labour, especially cleaning up the debris. On one of those occasions, I had found a small bag of finely ground hot dried peppers, the hottest chilly [*sic*] you could find. It was called 'lombok rawit'. I had hid it inside my shoes which I carried over my shoulder as I walked on bare feet. I had used this method many times before, but this time the guard called me back. He stuck his hand into my shoe and found the bag with chilly. As punishment I had to eat the whole bag. My mouth and stomach were burning, but as it did not go fast enough to his liking, he pushed it down my throat. For weeks afterwards I was very sick and could not eat any food. I am still on medication today. The doctors found many scar tissues and fissures which may bleed at any time.

Many things were happening and we still heard all kinds of rumours. If you were caught with a sliver of newspaper in Malay or Japanese, the punishment was severe. Sometimes they put iodine into your nose. Or, you were put into a push-up position and if you were sagging, you would come down on hot coals which they had put there underneath your stomach, so you kept yourself straight. Your stomach muscles were painfully tight for weeks afterwards

But the rumours became stronger about a landing by the Americans on the other side of the island. There were three men, an army sergeant, an officer Mr De Haan and Entrop, a sailor who slept in one corner of my room, who planned their escape in deep secrecy. We all got along very well. One of them, a big man about six feet tall with a beard and very kind blue eyes, was well educated. He told us many stories about his adventures in Holland. The escape was well planned but not the results which followed. The next morning when we were counted, they were missed. The whole camp was called together and the camp commander gave a speech. The Japanese commander was furious. Right away ten hostages were taken from the corner where Entrop was sleeping. The Jap camp commander let us know that they definitely would be killed if the escapees did not return in three days and that this would continue until they all would be back. This as a threat to ward off other escapes.

After three days the escapees were back in camp. The natives told the Japanese search team where they could be found. One of my friends, Plasse, was asked what he would like as his last meal. He asked for eggshells. When the Jap guard did not understand and asked why he wanted eggshells, he said that he needed all the calcium he could get for his bones when he was dead. The three were severely beaten while they were still in their green army uniforms and had their puttees around their legs. The Japanese camp commander ordered the death sentence. He was also going to show us that he was not entirely without feeling so that he released the ten men back to our room. They were quite shaken up by the ordeal and it did make a mark on their minds forever! The warning was given that at the next escape these ten were the first ones to be executed.

On the day of the sentencing of the three escapers, they were given their last meals. Then they were transported to an open field outside the campsite where they had to dig their own grave. They had to kneel in front of the grave and were decapitated. Jan Oorlog tried to have the sentence changed, to no avail, as the Japanese did not respect the Geneva Convention. All my friends and I were trying to cope with these experiences even till today.

Note 1: There were no doctors waiting for us as it happened in the case of hostages today. After the war we spent three months in Manilla in the Philippines to recuperate. Then we were sent back again to Indonesia by the Dutch Navy to fight there. We almost had a mutiny on board of the ship, which was then diverted to Balikpapan when we had time to calm down.

Note 2: On August 13, 1993 I went back to Japan as moral support for my friend Marius Van Dyk van Nooten who gave a speech at the Convention for the International Mental Health in Tokyo. I did not realize that after almost fifty years the hidden feelings of my experiences in the prison camps had never been taken care of. When I listened to my friend's speech I broke down!

Six

One day a Japanese guard found a rusted gun under a heap of firewood and the Japs made a real big thing about it. We all had to come together on the big square in the middle of the camp. There we got a big speech and were told to reveal how the gun got there or we would all be shot. We had to line up in a long row and again we were hit on the behind. This was my second time for a black and purple behind!

After a while I had a more permanent job with a couple of other friends working around the houses where the Japanese officers were staying. I saw lots of trucks passing by with native girls on it between 14 and 17 years old and some were even younger ... They were transported to the hospital for a check-up prior to sexual contact and rape by Japanese officers and soldiers. After the check-up, they were transported to the schools where the classrooms were transformed to small rooms. In each room was a bamboo bed with a tikar (a straw mat) and a small pillow. I had to work in those schools to clean up. When we arrived at the school, we were first counted and then we were told that we were not allowed to talk to the girls. We would arrive around eight a.m. when it was still quite peaceful. We had to clean the grounds around the school first and then we would see the girls coming out of the bathhouse, clean and fresh for the coming events. Slowly the schoolyard would get busier with officers and soldiers.

The more beautiful girls were kept in a different part of the school for the officers. At the entrance of the yard a soldier would stand whose job it was to punch holes in a card so that the soldiers would know when it was their turn to have intercourse with the girls. They had to stand in line waiting for their turn and they seemed very happy about these weekly sexual activities. As soon as five or six soldiers had gone in, my friend and I were called in. Our job was to remove a long towel, stretched across the room, about four feet off the floor and we had to install a clean one. After each intercourse, the girls had to walk across the room with the towel between their legs to clean off the semen of the soldiers and then to be ready for the next one. Some of the girls were accepting it, others were just like zombies.

I felt very uncomfortable and very sorry for them. They had been picked up in the villages to be raped. The officers would get the virgins first. I can still hear their screams for help, 'Toeloeng, toeloeng!' (which means 'help' in Malay). This would also leave a mark for the rest of their lives. As there was soap around that place, I sometimes managed to smuggle a piece back into the camp. I had sewn a special pocket inside my pants between my legs and there I hid the soap. It was such a delight to wash yourself with soap.

The UN right now is studying the wartime brothel issue, about the systematic rape and secular slavery during the wartime. There were over 150,000 women, mostly from the Korean Peninsula, not counting the ones from the Asiatic countries, who were sent to the front lines during World War II, to provide sex for the Japanese soldiers. It took twenty years for the Korean women to come out with their experiences, another twenty-five years to get justice, when finally after fifty years, the Japanese admitted their crimes. But, the women still did not receive any compensation for their suffering.

When time went on, the food in the camp was not so good any more and the supplies from the natives were not getting better either. But we invented other means. We waited for a full moon and in groups of six, we went outside. We waited till the changing of the guards was over. Then one of us would be the look-out. Another one climbed in a coconut tree which was more than 40 feet high. The other four took a blanket at each corner and stretched it out until it was completely flat. After the signal was given, a coconut would be on the way down to be caught on the blanket. The coconut meat and the milk were delicious, especially the young ones were very nutritious. We buried the husks and coconut shells in holes in the ground.

Sometimes the men played practical jokes on a friend and even on Japanese soldiers, although with them there was always the danger of backfiring. First I tell you the joke we played on a fellow prisoner. This guy was a bachelor and very sexy, and he liked himself a lot. Also, he was very fond of sex stories. We would usually talk about sex and we would make sure that he would hear us. Then he would move up closer and closer to listen in. So, we made up a story that sometimes there were native girls near the barbed wire waiting to have sex with the prisoners. There was a marine corporal who was an expert in dressing up as a native girl. After we had talked about the fact that the girl was waiting, the bachelor had come near us again to listen in. And when it was dark, he went to the barbed wire where the corporal stood waiting, dressed as a native girl. While everybody was watching, the corporal was trying to have sex through the first barbed wire, which was not as closely watched by the guards. I thought it was quite mean to do this to a fellow prisoner. But, another time we played a joke on the Jap. Every morning a big bag of bananas was delivered to the camp and put under a tree near the guardhouse to be picked up by the Japanese cook to be used for the Japanese soldiers at lunchtime. Under this tree was also a bench where we could sit. So every time a prisoner went to the bench to sit down, he took a few bananas out of the bag and replaced them with twigs and small pieces of wood so that on the outside the bag remained the same size. After the bag had been emptied of the bananas, we were waiting to see what would happen when the Jap came to pick up the bag. He almost flipped over when he lifted the bag up as he had expected the bag to be heavy with bananas. The 'kurahs' were heard almost a mile away, but luckily he walked away, and we realized what a dangerous game we had been playing.

The evenings were very boring and we went to sleep early because it was lights-out by eight p.m. and we were looking forward to the weekends when it was clean-up time. We visited friends, talked about the future or visited the doctor for scrapes and pains at this time. There was still medicine available from the Red Cross, but for infections the only medicine was a solution of permanganate which we used for flushing out infections. I still use it today for gargling when I have a sore throat. We also had exercises and lectures from civilians who were assigned to teach on many subjects. So we also had maths classes, chess games, made with camp-made chess pieces. We had sing-songs, all this in order to try and keep our morale high. We even had a dentist for whom we had made a dentist drill from an old bicycle wheel turned by hand. There was no anaesthetic to kill the pain though, if a tooth had to be pulled.

Seven

In August rumours went through camp that 1,000 POWs were to be chosen to be shipped to a different country. I was to be one of them! With other prisoners we were told that we were going to a colder climate. Then we found out that we were going to Japan on a freighter, the 'Asamah Maruh'. We started to say our good-byes to the ones who stayed behind and exchanged addresses in case we would meet again after the war or, to notify family in case we would not survive the war. The 'Asamah Maruh' was a part passenger ship for soldiers and geishas who were already on board. We had to board the freighter on the side and had to go below deck. There were some cows there and a terrible smell of animal waste. On October the 14th, 1942 we sailed out of the harbour of Makassar. The portholes remained shut for the whole voyage and washing yourself was not included. Lots of us became very seasick and were throwing up constantly all over the place. There was not very much space to begin with and no air circulation. The only air that could come in was to come from the opening at the top of the steel stairs to the deck. We were allowed to go on deck once a day going up the steel stairs one at a time. Many thoughts were going through our minds thinking about being trapped down below in case we would be torpedoed by an American submarine. Some of my friends were so affected by the heat and lack of air that they were fainting. The smell of the urine and vomit was overwhelming. Boiling hot air and darkness, it was like the times of the slave transports from Africa to America, now that I think back. We found a water hose and connected it to one of the outlets and so we tried to somewhat spray ourselves clean ...

We knew that our next stop would be the Philippines and we were zig-zagging for a while to avoid the submarines. After ten days I was ready to give up. Nothing could be worse. When we arrived in the harbour of Manila, we saw the carnage the Japanese had done to the American fleet ... The warships were laying on the bottom of the harbour with the smokestacks sticking up above the water and I said a prayer for the ones who must have drowned with their ships. After our ship had taken in food, drinking water and fuel, we sailed out of the harbour again in the middle of the night.

Late in the afternoon on October 23, we arrived in the harbour of Nagasaki, Japan. The temperature was about 14 degrees C. and many of us caught pneumonia when we got on deck. The difference in temperature from below deck and above was so extreme that it was unavoidable to catch something. The harbour looked beautiful though, with the hills in the background and in clear blue sky. Flat boats came alongside the 'Asamah Maruh' who took about fifty POWs and their belongings on board. An old man was navigating the boat I was on, with a long paddle at the rear and there was one guard at the front. After we landed we were counted again and again. Then we had to march to the famous Fukuoka II (aka Nagasaki, actually about 5 miles from Nagasaki on the island of Koyagi Shima – now joined to the mainland), a wooden camp with a guardhouse and a big barbed wire fence all around it. Again we were counted and counted and then searched. Every sharp thing was taken away.

The camp was built in a U-form with 18 rooms on each side and in the connecting rounded part were the washrooms and the kitchen. Each room was for fifty-two POWs.

There were bunk beds, two rows of thirteen, on each side of the room. The lower bed was about two and a half feet off the ground and the space underneath had sliding doors where we kept part of our belongings. Over the whole length in the centre of the room were narrow tables and benches. At the end of the room was one big glass window wall. I was going to be in this room for the next few years, room 14. My number was printed on a band of cloth which I had to sew on all my clothes; it was no. 620. My friend Harry Fryling, no. 621 also survived the war and lives today in Amstelveen, Holland. He and I still correspond together. He still suffers from the after-effects of the POW-period today. (So do I, as after fifty years my emotions still are raw about that time. As I mentioned before, I experienced this at my visit to Tokyo, when I attended the Convention of International Mental Health with my friend Marius Van Nooten. These feelings were not taken care of after the war as they do today when hostages are released. This should have been done by the Navy Air Force at the time when the war was over. Instead, they put us right away back into the war zone in Indonesia, which at that time was fighting for their independence.)

After a few days of rest, dysentery broke out among the men and many dehydrated very quickly. They got opium drops to stop the outflow. Many did not make it to the toilets as the waiting lines became longer and longer every day. Many died in this first period in this camp. Also because many suffered from the pneumonia which they caught when they were exposed to the drastic temperature changes by coming out of the boiling heat of the hold of the ship, into the cold fall air above on deck. There had been nothing to cover up, no warm clothing or anything...

On the first of November 1942, we had to report for work. The only thing we heard again was, 'Kurah, kurah!' Slowly we learned the meaning of some words. And we even could choose some of the work we liked best. We were put to work at the Kawanami shipyard owned by Mitsubishi.

I had a suspicion, as I remembered the volunteering for the famous telegraph jobs in Makassar which turned out to be carrying telephone posts. All kinds of work was available, such as drillers, cleaners, welders etc. Then we got the speeches ... about working hard and being punished if caught loafing, about violating rules...I did not like to volunteer for the heavy work so I chose to be a cleaner. This meant I had to sit in front of a big heap of dirt that had been collected off a ship that was being built. The main thing was to take out all the nuts and bolts and other metal objects for cleaning and re-use. There was little control and the guards were far away. We were now under the control of some wharf foreman. It was a very cold job without gloves and my fingers became badly swollen. The noise around us was unbearable so that I put a piece of wet newspaper in my ears. The officers did not have to work. I was jealous of the riveters who had a little coal stove where they had to heat up the rivets before they were installed on the steel plates to be hammered in. The ship being built was resting on big wooden blocks which we also had to place when the keel was laid.

What I did was a very dirty job and it was very dark under the ship. There were lots of pools of water and I always had wet feet. One day while we were resting, we made a small fire to warm ourselves when a guard happened to spot us and then all hell broke loose! We had to make push-ups in complete straight position for twenty minutes. Every time we sagged down, we were hit on the behind with the butt of his gun. My stomach muscles were hurting for weeks afterwards. I went to see our doctor. There were four of them, Dr Niewenhuis, Dr Walvis, Dr Huisman and Dr Syred. Dr Huisman told me to use hot towels to relieve the pain, for all twelve of us.

Our lunch consisted of an amount which could fit in two sardine cans. One small one with some seaweed and the other with rice. Some of our boys filled these cans up prior to going to work, but once there, we had no chance to heat it up. We also could not leave the cans out of our sight for a moment or they would be stolen in no time. When we were passing an office we were looking for scraps of food and especially the smokers were after every single cigarette butt they could find. After a day's work we were starving when we came back to the camp. Then the Japanese were promising extra food for better production.

There I got a break. They were asking for electrical welders. I had learned that at the Technical School, so I volunteered and was told to make a test piece. Just before the foreman came for inspection, I had heated up the piece with an acetylene burner. The foreman was very pleased and the next day I got my welding outfit. It was made up of a canvas jacket and trousers plus a welding cap. Now I could sabotage by so-called 'cold welding'. It looked good on the outside but it was not melted together on the inside. After a while I was promoted to weld the big masts of the ships and I started to do a better job. The trick was trying to stay healthy as you would get only half a ration of food if you became sick. The weather became colder and colder and I was glad I had my canvas outfit. It happened one day while I was hammering off the slack, I got a sliver in my eye and it became seriously infected.

I was transferred to the wood mill shop where I had to assist an elderly Japanese at the smith's fire. I had to hammer with a five pound hammer. I had to make the parts for the wood mill shop under the guidance of the old man. I made a roaster of steel rods inside the chimney to heat up the lunches for the boys who were working outside and this was greatly appreciated, especially in the winter. Then I had to get the coal for the smith's fire. I had to travel through a little village nearby where the dried fish was hanging outside and so sometimes I had dried fish for lunch. I had trouble walking in my rubber shoes as there had only been a size 11 available. I filled the front end up with straw which kept my toes warm, but in the meantime the shoes were flapping around. Sometimes I had to help outside and one time a big steel plate fell on my foot but luckily enough, it landed on the tip of my shoe which was filled with straw and that saved my toes.

On another occasion I had to help with the direction of steel plates which were to be hoisted to the side of a ship. I had to signal the crane operator by hand movements so that the large plates could be pulled out of the big piles. He could not see what he was doing, so that made me the eyes of the operator. The cables of the crane were one inch thick and there was a large metal block which kept the cables taut. It happened one time that the cable kept rolling off the main roller uncontrollably as the operator had not paid attention. The cable block kept on coming down! It grazed my head and it stopped just above my hand which was just above the steel plate wall of the ship. I had to force the block away and only the inside of my hand was bloodstained.

One afternoon the steel cable slipped off the main pulley at the end of the beam attached to the main tower, about 150 feet high. I was ordered to go up there and put the cable back on the pulley roller. It was quite an ordeal. I also had to take all the equipment with me up there to attach to the cable, in order to hoist it back in place. Only when I was back on the ground again I got a reaction to what I had done. I needed a rest and went under the ship out of sight or so I thought. I was spotted by a soldier and received the same punishment as before. This time I only had to make push-ups without sagging my behind. He let me off for a change with fifteen push-ups but my muscles were sore, just the same.

Thank God that I was not a smoker. I could trade my ration of cigarettes for warm clothing and food. One of my friends was quite a smoker. He pretended to be nuts and performed very well. The Japanese made him do all kinds of tricks. When they threw away their cigarette butts, he crawled to it and from about three feet distance, he made a jump for it. He got a sign sewn on his jacket that read, 'Bakka', which means 'crazy'. After the war we had some good laughs about it.

My friend Van Veen lost his life. He was one of many. He was working about 50 feet above ground transporting concrete in a wheelbarrow in order to fill up a pillar that would support a rail track for the new electrical crane. Because of his rubber shoes being too big, he tripped over a bolt which was there to support the new track. He fell to his death by breaking his neck. Of course the Japs said that it was his own fault. Four of us who were his closest friends, paid our last respects and brought him by boat to Nagasaki for cremation. We had to pick some bones and some ashes and took this back in an urn to the camp. This urn was placed in a special room with others who had died due to sickness and accidents.

Eight

Especially the first winter was terrible as we were not dressed for it. I managed to get a hold of empty cement bags in which I cut holes for my arms and head to keep the wind out. In the morning after counting, we had to stand naked in a row in front of the benches in the middle of our room and rub ourselves with a little straw brush attached to a rope to kill lice. You had to rub your back till you were red all over. The next thing to do was to run to the washrooms and after that to get ready for the next counting. Then it was time to go to the wharf. The Army released us to the Navy and they released us to the civilians and in the evening this went in reverse order. That first winter was also the first time I saw snow. I was shivering all the time and my hands and feet were always cold. On the weekends we could take our bath in the community bathhouse. The water was so hot that you had to let yourself slide into the pool very slowly. Your blood would rush to your skin and you would become beet-red and it would make you dizzy. On one side were the women and girls and the men on the other side. While we were waiting for our turn, we were standing there with our pants and shirts in our hands looking for clothes lice and at night the bedbugs were eating our blood. The cavities of the woodwork in our rooms were full of them.

In the beginning of December of 1942 about 300 POWs came in from Singapore and they were quite warmly dressed. I traded with cigarettes for some of their airforce shirts. They had lots of news and told us about the sinking of the 'Prince of Wales' and the 'Repulse'. Then Christmas came along and the Japanese surprised us when we each received an apple. We were very happy with this small token. We also got four days off just like them. We deserved this rest and we used part of it to fix our clothing and did our laundry. In room 18 was a soldier who got a violin from the Japanese to play on special occasions and weekends. He was a master. For the New Year we received our first winter outfit of the same quality as the soldiers. It made the winter more bearable. Only our shoes in the rain were always wet. We felt miserable when it rained as we got soaked and had a hard time keeping warm. The road to the wharf was a real country road and when it rained it was very muddy. I used to slide instead of walking and there was always that guard with the stick.

Then it was my turn to get dysentery. The old Japanese man, my boss, hid me under the driving belt of the main power supply and every morning, as soon as I was at work, he brought me some soft rice. On the way to the wharf I had to break out of the marching ranks to go on the side of the road and then I had to run to catch up with them all. I lost a lot of weight and I was able to count my ribs. From Dr Niewenhuis I got opium drops to stop the outflow, but I surely had to cut some more holes in my belt. I count myself very lucky to have pulled through as many died from this disease. Our resistance was so low and there were many trips to the crematorium.

Then the Red Cross parcels arrived. With them were also very good leather shoes, but we were not allowed to wear them to work. They were water-tight and had heavy leather soles. (I think that it would have made the other Japanese labourers who worked alongside us jealous, if we were wearing new shoes.)

That first winter made me feel real miserable. The days were very short and I was wondering how long this prison time would last and if I could survive it now that my health had deteriorated. Sometimes I was thinking of hitting a brick wall, but then I went to see my friends John Koch, Joop Brinkert, Harry Fryling and David Brandon. Then we talked and talked about willpower and trust in God as He would take care of everything. That was also the time that I started to go back to my roots. David was more educated in the Jewish religion and I wished that I had known more about it. I was brought up in no-where-land in the middle of Java in Indonesia where Jewish life was non-existent. We talked about home, what the future would bring and about our plans when we would get out of prison.

The hunger got worse. We started to talk about food all the time. The Red Cross boxes had to be divided. We received a two pound box between the four of us but we were keeping it for special occasions. Our water retention got slowly worse due to lack of salt in our food. Sometimes we had to run four times during the night and the line-ups before the toilets also did not get any better. Then the four of us found a solution by bringing a tin can with seawater from the shipyard to camp. We boiled and boiled this water on the kitchen fires till it was all evaporated. A black salt powder was left in the bottom of the tin can which we collected and added to our food. This took care of that problem and many followed our example and did the same thing. Then our feet started to swell up due to a lack of vitamin B. The sawmill where I was working at the time, had a machine at the smith's fire to separate the chaff from the rice. I managed to gather a couple of cups per day which I ate after I had cooked it in some water above the smith's fire. And it worked, slowly my legs got back to normal.

At work we were always on the lookout for the guards. This was not only necessary for us POWs but also for the civilian workers who also had to put in the necessary war effort. They also could be reported and punished. We knew our way on the ships that were being built, very well. The guards would not come to the places where there was a chance of accidents. For example, the riveters were throwing the red hot metal rivets way up to the riveters who would catch them in a metal cone. Where this was going on, we were quite safe and were never caught for resting or committing other infractions. The noise was unbearable and the cranes overhead kept on running with the beams and steel plates for the ships. One day an accident did happen. A steel cable carrying maybe a ton of beams broke just about 20 feet away from me after it had just past over my head. Nobody was hurt this time. But there were many accidents. A friend of mine tripped and fell over the side but luckily landed on a coil of riveting air hoses which saved his life.

Our sleep at night was so often interrupted due to the attacks of the bedbugs so we tried to catch up a little during the day when we were at work. Some of us would be on the look-out for the guards although they would not come to the dirty places as they were wearing white uniforms. Most of us would sleep during lunchtime when at that time everything came to a standstill. It would be over in no time when everything came to life again. We were happy when the days had gone by without any mishaps or beatings. And then we were counted and counted and again when we entered the camp, just in case someone might have escaped on the way back. After everybody was in, it would be time for a smoke. But only when the guard came around to supply the match. If anyone tried to light up before that, if was considered a crime and you would be severely punished as this meant that you were in possession of matches. We would exchange news, if there was any. At one time I got two cigarettes which I traded for food.

Around six o'clock it was feeding time and every room in turn picked up the barrels with rice and the soup with sometimes a small piece of meat in it. Then came the tea water as it looked only slightly yellow. But it was hot which was important to me. We would talk about what happened that day but mainly about food. Palte, a friend of mine could imitate very well. He could imitate eating a banana and the whole room was eating with him. When the horn sounded it was the sign that it was rubbing time again and search time for lice. And then it was time to hit the deck, at 9.30 pm it was lights out.

On the weekends we had cabaret, but in such a way that the Japanese would not notice it. If a guard was on the way we called 'Rood voor' which meant 'Red is on the way'. The players would make up their own tunes and songs in their own words. These light points on the weekends kept us going and kept our spirits high. It would give us a laugh now and then. We received Red Cross parcels four times a year and we divided them among ourselves. I am sure that there were enough parcels for each and everyone of us, but the guards must have taken part of them because their food was also not so good due to war rationing. Later on we got our bath once every ten days in a cement pool about 12 by 15 feet. The idea was to get in there first and wash yourself. Even if the water was boiling hot, it was clean only, for the first groups. By the time it was the turn of the last rooms of men, the water would be black but still nice and hot. So then they rotated the turns and every time another room was first. At this time when everyone was naked, you could see how skinny we were, ribs and bones sticking out. You could see the special bites of bedbugs and lice as there were welts and marks all over our bodies. But at least we felt clean every ten days.

Then there was a big catastrophe. Where we were working there were four ships in the docks. Two near the sea, separated by a movable hollow pontoon door. The other two were separated again by another movable wall. When the ships near the seaside were finished and ready to go, the doors of the locks were opened and the seawater let in. The ships would rise till they were even with the sea level and then they were pushed out for further completion. But on this day, January 24, 1944, there was a spring tide which happens only when there is a full moon. The first door collapsed and the whole dock flooded in no time. The two ships in the dock were lifted up and pushed against the second hollow movable wall which cracked. Then the water was seeping into the second dock next to the first one. We saw that an accident was ready to happen and every POW got out of there as fast as possible. But the Japanese labourers just kept on working which is still beyond my comprehension. Then the dam burst and the other two ships rammed the ones which were not in floating condition. The whole thing happened in a flash. We pulled lots of

Japanese workers to safety. But many were drowned under the ships. When we were counted, nobody of the POWs lost their lives.

On both sides of the dock were many concrete floor work levels which were used to supply the ships below with the steel components. There were ten floors on each side. While we were being counted on the side of the dock, one of the Japanese workers fell down from the tenth floor right in front of us. His brains were smashed all over the pavement but the guards apparently found it a big joke. They were laughing their heads off when this worker had his accident. Until today I try to understand the Japanese kind of humour! There was another unbelievable incident where the Japanese workers had to hammer on a big steel plate which just came out of the furnace. One of them had found a kitten and everyone was stroking it. For some unknown reason the man threw the kitten on the red hot steel plate where it gave a shrieking yell which was bone chilling. Then it slowly burnt to a small heap of ashes. All the Japs had the laugh of their life like it was a big joke. These heavy muscled men against a small poor kitten. Hard to understand!

Nine

Some of my friends received mail through the Red Cross. I was worried about my mother and my brothers. We were allowed to send a card with three sentences, all standard and I was praying that they would receive it. After the war was over I found out that my mother had died on November 3, 1944 in a Japanese prison camp in Ambarawa on Java, Indonesia. We lived day by day and trusted that one day an end would come to this nightmare. The rumours were getting wilder and especially when we were at work, we sometimes managed to smuggle in a newspaper.

We would occasionally see badly damaged ships in the harbour which were to be repaired. The Japanese did not like the rumours we heard. We were threatened with severe punishment if we would spread them. In June we heard about the invasion of France and a possible end of the war in Europe. A little earlier in In March 1944 we heard the air raid alarm going off daily and we were overjoyed to see the first B29 bombers flying over. It was not possible to have a sign on top of the roofs of our camp buildings to tell them up there that we were a prison camp. Up there they were free and down below we were POWs. We were put to work by building big tunnels into the rocky hills as air raid shelters. The air raid sirens were now going off several times a day. On 14th May 1944 we saw the first jet fighters flying over which meant that aircraft carriers were not too far off the coast of Japan or, that the Americans had captured an island not too far off.

We were not sure if the Japanese would continue to fight if the Americans would land on the main island.

On the 27th of May we were told to get ready to be transferred to another location. Only a few hundred would have to stay behind. On the 17th of June the names were made known and I was one of the men to be moved to a coal mine. We packed our few belongings in a bag and took it with us. We left around two o'clock a.m. to go to the train station and then we did not leave until seven in the early morning. The windows were tightly shut and we sat like sardines in a can. There was hardly any room to walk over to the toilets. We arrived in the evening after an almost 12 hour ordeal. We had to walk a good 40 minutes to the coal mine camp while a truck took our bags with our possessions. The camp looked very bad from the outside and we did not know what to expect. Later on we found out that there were improvements compared with our previous camp Fukuoka II. As soon as we had arrived we were told to gather together to listen to a speech from the camp commander. We were told to work hard, not to try to escape ... the same old routine. Then we were searched and counted and counted. Anything of value was taken, also any writing material.

In every camp we had been, there was a sadist in command. In Makassar it was Yoshida, in Nagasaki it was Bokubo, and here it was 'Crazywind'. You had to be sure that you had all your buttons closed at all times. If you had missed one, you would get the punishment of your life. But here the coal miners could take a bath every day and we also received more food. The rooms were small, six men to a room but it was a little more private. We were being trained to work in the mines. It was very dangerous and there were many accidents. I became skinnier by the day but tried to remain healthy. We had to walk down the shaft while the Japanese used a little train. I did not see the daylight until the weekend as we left very early in the morning when it was still dark and it was evening when the work day was over.

Then we heard that the planes had been flying over to bomb Tokyo or many other places many times a day. We felt that freedom could not be far off but I did not want to expect too much.

Ten

In Spring 1945, we heard that Germany had surrendered. Then on August 9 my friends saw a flash of B29s in Northern direction and we knew something had happened. In the evening we saw a red glow in the distance. Another city had been bombed. On August 15 we saw the Japanese soldiers listening to the radio with long faces. We did not hear any air raid sirens any more and the night shift did not have to go to work in the mine. The tension in the camp was growing. We felt that this was the end of the war!

Our officers were invited to come to the main office and were offered chairs to sit on. There they heard that the Emperor had decided to end the war. It was hard to make the change-over to freedom and to this day, I thank the Lord that I made it. There were no psychiatrists waiting to debrief us as happens nowadays after hostages are released from capture. As I mentioned before, I am still suffering today from the effects of all the years in the prison camps.

That whole following night, nobody slept a wink. The next day was not like any other 'next' day. It was just like a dream. I took a hot bath and washed all my clothes, whatever few I still had. So did everybody else. We wanted to look our best. No more working in the mine. Everything else was stopped, the mine was even closed. Slowly our officers took charge. We received the Red Cross parcels out of the storage rooms and the food improved. Thanks goes to Gys Flamming, our friend who was a male nurse. He advised us to take just the right amounts of food as our bodies had to adjust to the richer consistency of it. I saw many men walking around all bloated up after they had been overeating all that better food, which did not agree with their starved bodies.

Then we did paint the roofs of the buildings with the letters 'POW' and not long after we were attacked by B29s who this time dropped barrels with food attached to parachutes. Many parachutes did not open and the drums fell to pieces while the food was scattered over a wide area. The Japanese even enjoyed the food. Soon our stockrooms were overflowing with food. Then our camp commander asked us to sign with mirrors to the B29s to stop dropping down more food. But they continued to do it. Apparently the telegraphist had just asked for more food while in the meantime he pretended that the B29s must have misunderstood his messages. In one of the next days, a B29 came so low in a dive that it seemed that we could almost touch it. We could even see the pilot. He dropped pamphlets telling us that the war was over, that we would soon be freed and transported.

Then came the first Americans and there were lots of handshakes. We all made small trips outside the camp. The population also looked very poor because they also had suffered. Three of us had our picture taken and we paid with a blanket. Then came the 18th of September when we were taken out of the camp and went by train back to Nagasaki. Now every window was open. We yelled at the farmers in the fields. Some of us were chewing gum and smoking the best of cigarettes. The train arrived in Nagasaki where a month before the atom bomb had been dropped. We were all very quiet when we saw the damage. The train took about 40 minutes to get to the main station where the American Red Cross was waiting for us. We could only see some streets and the remnants of some concrete buildings. Everything was burnt. The town was no more! The burning smell was still there and it seemed the shadow of the dead was everywhere. We heard only the clicking of the wheels of the train as we went through this mass grave!

How can you put freedom in writing? We were treated like kings. There were nice girls and women who served us tea and coffee. We were taken to a bathhouse to clean ourselves, to be de-loused and disinfected. All our possessions were burnt, except for a few souvenirs which were also disinfected. I was flown to Okinawa. And from there, I went by ship to Manila, where in a huge tent camp I was to recuperate from the ordeals in prison.

I was sad that it had to take the atom bomb on Hiroshima and Nagasaki to convince the Emperor to end the war which he had started!

2365549 Sgmn Benjamin 'Bennie' Gold, RCOS

I interviewed Ben in September 2001. Special thanks to Ben's son Maurice and grandson Adam, who have compiled an album of Ben's wartime memorabilia, and especially to Adam, who published a book about how his wife and he traced Ben's footsteps in the Far East, soon after Ben died.

31. 2365549 Sgmn Benjamin Gold, RCOS.

Ben was born on 3 August 1920 in Chicheley Road, Hampstead, London, son of Morris and Rachel (née Kronick). In 1924 the family moved back to Brick Lane where his parents had previously lived, first to number 167 and later to number 4, which was a well-known salt-beef shop, Bloom's. Ben's parents ran a knitting and wool shop (N. Gold and Son, after his grandfather Nathan). Ben attended Virginia Road primary school and was bar mitzvah at Duke's Place synagogue. When war began he was apprenticed to his father.

Ben was involved in the 1930s anti-fascist movement through Habonim, the Zionist youth group and wanted to get involved in the war against the Nazis. He enlisted in late 1940 into the RCOS and was initially sent from his home (then at Station Road, Harrow, as the family had moved again) to Glasgow for training in Morse and radio work; he was based at Maryhill barracks. He was then sent to St John's Wood barracks in London, from where he was posted to Malaya in November 1941 (via Iceland, Nova Scotia, Sierra Leone, South Africa, Colombo, Singapore and finally to Kuala Lumpur) attached to the 9th Indian Army, smartly attired in white uniform and topee with brass buttons. They arrived on 28 November.

Singapore

Each radio unit consisted of 4 men – a driver of their vehicle, two radio operators and a runner – and Ben and friends were posted to the Cameron Highlands, mountains which run down the spine of the peninsula; his job was to maintain radio links with Singapore. Forbidden contact with both the white planter officer class and the Indians, he befriended local Chinese people.

When war broke out his unit was sent north to oppose the Japanese invasion from Thailand, but at every location they were bombed and strafed, and slowly but surely the retreat south began. Ben remembers that they moved from plantation to plantation towards Singapore, constantly being chased by Japanese Zeros, and on the front line of the retreat sending code messages back and forth. Ben was able to converse with many local people because they knew Dutch and they could understand his Yiddish; this made him very useful to the other men. They lived on canned peaches, local Tiger beer and tinned sardines. Finally they crossed the Causeway to Singapore and arrived at their HQ. After a meal, they were paraded next morning and Ben was put on a charge for wearing a dirty uniform and not keeping his vehicle clean; meanwhile, the Japanese were shelling the parade ground using observation balloons from across the Straits. Such was the pomposity of some of the chair-bound British officers.

The next thing Ben knew, he was made up to a Military Policeman (MP), as they were needed to help organize the evacuation of nurses and civilians from the harbour in Singapore City. He helped carry the children on board as families made their way up gangplanks to waiting ships of all shapes and sizes; in some instances they could not get aboard, as bitter Australian troops, believing they had been betrayed by the British government in being sent into the death trap of Singapore, had taken over some ships for themselves and were actually shooting at those trying to board, whilst the Japanese were also shelling the harbour at the same time – it was utter chaos. Then Ben was horrified to see ships that he had helped to load with women and children being strafed and bombed and sunk by Japanese fighters as they left the harbour; it was a nightmare that stayed with him all his life.[1]

On the night of 12 February 1942, a lone yacht glided into the harbour and was designated to pick up skilled personnel needed for the war effort – pilots, engineers and so on. As it was unarmed, Ben and his team of about eight MPs, who were armed, were ordered aboard as protection for the 300 mainly RAF evacuees. They were strafed the next day and two men near Ben were killed as the MPs fired back with rifles and one Tommy gun, but they made it to Batavia in Java, where they were all billeted in a school. Ben, meanwhile, was posted as missing, believed killed in action, and his parents received a telegram to that effect. However, Ben always kept, on his father's advice, a £5 note folded in his pay book, so he went to the Hong Kong and Shanghai bank in Batavia and sent a cable via Panama to his father's bank (Midland Bank) in Commercial Road, Whitechapel, saying he was alive and well in Java. When his sister took the cable to the War Office they would not believe it.

Ben caught malaria and was delirious for several days. When he woke up, the whole school and the streets around it were deserted: the men had moved on and left him. Ben began walking east and hitched a lift on a water-tanker with Indonesian soldiers. He left them, but much of what happened in the following period is a complete blank in his memory, probably because he was still ill.[2] He linked up with an American Texan regiment for a few days and then met a British soldier with whom he teamed up, and they decided to make their way to Australia somehow. While walking by a ravine, they saw in the road below an abandoned convoy of vehicles with boxes of Dutch guilders from the banks. Ben's friend disappeared into the ravine after the money and left Ben alone. Ben never saw him again. Ben only remembers blacking out, and the next memory he has is being in a large pit guarded by a Japanese soldier, at Garoet, Java, on 18 March 1942.

POW

Soon, other British prisoners arrived, and together with some Dutch civilians they were all marched to a railway station some miles away and sent on a long journey to a POW camp at Tanjok Pryok, where the senior British officer was the famous author Laurens van der Post. It was the summer of 1942.

Ben and others were subjected to an enquiry by the British officers as to how they had been captured. Ben was sent on working parties to the nearby docks, loading and unloading Japanese merchant ships. He took part in sabotaging Japanese food stocks while loading them on ships. On 26 September 1942 he was put aboard one of the Hell Ships and sent back to Singapore; the conditions on board, in locked-down holds, can be imagined. The journey was horrendous and Ben, like most of the men, was suffering from dysentery and was very malnourished. On arrival, the men were counted off the ship and then paraded half-naked through Singapore to Changi gaol so that the locals could see the degradation of the European soldiers. Some days later (on 9 October) they were put aboard yet another boat to be moved with 3,000 POWs to Kuching in west Borneo (now Sarawak). Ben remembers this journey as even worse, if that were possible. They arrived on 13 October 1942.

Borneo

This camp (Batu Lintang) was about twenty miles upstream along the Kuching River and held about 1,800 soldiers (British, Dutch, Indian and Australian) and about 500 civilians – men, women and children – who were mainly planters, civil servants and missionaries, and their families. There were also some Americans. The British CO, a colonel, decided Ben would be best placed to work as a clerk with the Japanese commandant, who was in the Kempeitai (Japanese Gestapo-type police), perhaps because the CO knew that Ben could somehow communicate with the Japanese better than others; as the Japanese knew some Dutch, they could understand Ben's Yiddish, with its Germanic content. His instructions were to keep the Japanese busy and prevent their brutal forays into the camp, and to keep the British officers informed of what the Japanese were up to. This included secretly making a note of any reports coming into the main camp office from Sandakan camp (north Borneo) and other satellite camps, about which men had died there. Ben reported back that the numbers at Sandakan had drastically been reduced, until virtually all the record cards kept by the Japanese were withdrawn. This was a clear indication that some kind of catastrophe had occurred, and after the war it was discovered that this had indeed been the case. Ben was given an armband with the Japanese word for 'Office' on it; thin and tall (six feet), with spectacles stuck together with latex gum straight from a rubber tree, he cut an eccentric figure, but he was doing important work.

There was a secret radio in the camp and one day the Kempeitai made a dawn raid to confiscate diaries, photos, letters and so on, looking for clues to help them find the radio that they suspected existed. Ben, who was known to the Kempeitai as 'Goldo' (they could not pronounce the letter 'd' at the end of an English word), was shown some of the captured diaries by his boss, Lieutenant Kuba, and told that the phrase 'The gen is good today', found in the diary of a Lieutenant Burberry, meant the radio must be a reality. But Ben persuaded the Japanese that it meant 'gin', and that the officer had obtained alcohol. For this offence, Burberry was badly beaten up, but had the proper interpretation been given, he and others would have been tortured and executed for hiding a radio. Torture and humiliation were routine; on one occasion, Kuba, a chain-smoking drug addict, who boasted that he had eleven children in Nagasaki, simply stubbed out his cigarette on Ben's neck during a conversation, and then offered him the stub to smoke. Ben said that the deep scar had never properly healed. On another occasion, another Kempeitai officer, who had taken a dislike to Ben, kicked him with hobnail boots, on his foot, for not bowing low enough, and this caused a deep cut which did not heal until after the war. Ben discovered after the war that all Kuba's family had been killed by the atom bomb that was dropped on Nagasaki. Kuba himself was hung as a war criminal.

The Bishop Affair

One day, Ben's Japanese boss, who expected him to know everything about western civilisation, religion and the prisoners, engaged Ben in conversation as he sat cross-legged on a chair while Ben stood in his loincloth, and his wooden clogs, made from old floorboards. The Japanese told Ben there was a Church of England bishop (of Sarawak) in the guardroom who had been arrested for smuggling food into the hospital under his white cassock. He was to be tried and perhaps beheaded. Ben explained that a bishop would think it his Christian duty to help the sick and also was appointed by the 'King Emperor' in Britain, which meant that only an emperor could take his life. The Japanese officer stopped him in mid-sentence – he was enormously impressed by this explanation and praised his 'friend Goldo' for describing this great and royal tradition of the British Empire to him, and agreed with Ben's analysis. The next day the bishop was back in the men's compound and no further action was taken against him. Ben had saved his life.[3]

Sometime in 1944, Ben thought of escaping; he could see the river from the camp and thought if he could get to a sampan (boat), escape might be possible. One day he made his way into the jungle, but within five minutes was lost; it was so dense and the sky was so hidden by tall trees that he had no idea where he was. He heard Japanese being spoken and saw some soldiers making their way back to the camp; he came across some oil drums filled with aviation spirit and filled his water bottle with it for the sick bay (it was used to clean wounds and helped to stop infections). He had not been missed, and he gave the precious water bottle to the doctors; he never saw it again.

About a year later he tried again; he was working on the roof of the museum in Kuchin on a working party. The Japanese were looting the slates to send back to Japan. Behind the museum were garden allotments, and in the distance were some boats tied up along a quay. Ben made his way through the allotments but he lost sight of the quay because of the thick vegetation; he also thought that if he did get away, the rest of the working party would be severely punished. Instead, not one to miss an opportunity, he grabbed a load of vegetables and smuggled them back to the cookhouse; he was very popular for that offering and allowed to work for three days in the cookhouse as a small reward.

One day Kuba called Ben to get into the back of a van with a guard and another British prisoner, and they were driven to a brothel outside Kuchin. The aim was to obtain a piano for the camp. Ben and the other soldier were then ordered to lift the piano onto the van, but this was impossible in their condition. In the end, the prostitutes, the guard and the driver were pressed into lifting the piano into the van while Ben and friend watched, with some amusement. It was used in all the concert parties in the camp.

Around the beginning of 1944, an edict came down from Japanese HQ that all POWs were to write postcards home saying that the treatment the POWs were receiving was wonderful and that the food and accommodation were excellent. Bowing low and wearing only his 'Jap-happy' G-string, wooden clogs and straw hat, Ben said he must first ask his officers in the British section of the camp. The British SBO refused and so the Japanese brought all the sick out of the huts – about 400 men, Ben said – and left them in the intense heat on the white clay parade ground, all day without food or water. The British had to give in or see men die. Ben was sacked from his job by the Japanese in retaliation for the initial British resistance to the order, and joined the work parties next day building an aerodrome nearby; some of his British fellow prisoners, not knowing he was working for the camp's intelligence group, sneered at him for having to return to manual work. But he still managed to take part in the sabotage of the petrol dumps at the airport, by adding sand to the fuel. His malaria worsened, and he also developed beri-beri and dengue fever. At the time of liberation, he weighed just over six stone.

Ben said that generally speaking, the British officers behaved badly in the camp; the Dutch and Australian officers lived with and cared for their men, whereas the British demanded a separate compound, grew a lot of extra food which was only for themselves, and ran a mini-university. They showed no leadership; it was a scandalous business and was investigated after the war, but it is not known if any action was taken.

Some time later, Ben was ordered to be a batman for the British officers, and he was doing this job when liberation day arrived on 13 September 1945. Australian soldiers marched into the camp and a general demanded the Japanese records. Ben was sent for, as the Japanese claimed that he knew where all the records were, even though it had been over a year since he was dismissed from his post there. Ben was able to lead them to the space beneath the Kempeitai huts where the Sandakan records had been dumped by him on the orders of the Japanese, and these records were used in the war crimes trials concerning the massacres of Allied POWs that had taken place in the many small satellite camps around the island. Ben was interviewed by the Allied military police to give evidence in some of these trials, as well being asked how he got away from Singapore, why he was working in the Kempeitai offices, and what he had heard and witnessed in the camp. He regarded some of this as insulting, because there was an indication, in the manner of the questioning, that they were suspicious about how he had got to work in the Japanese offices and precisely what his role was there.

Eventually Ben was taken by an American MTB (motor torpedo boat) from Kuching harbour to a waiting hospital ship, on which he was taken to Labuan (Sabah), arriving on 15 September, and then on to Bangalore in India in November 1945 on the hospital ship HMAS *Wanganella*. At a rest camp in Poona, Ben was presented with a watch and carved walking stick by the Punjab regiment, as they knew he had suffered as a POW. He arrived in the UK on 20 December 1945 at Portsmouth, where they were met by the mayor and brass bands. He went on by train to Waterloo, in sealed carriages in case men absconded, and on arrival they found that all the kitbags of the men had been stolen. His sister met him at Waterloo, having been telephoned by a relative of the family in Portsmouth.[4] Ben was sent to Mount Vernon hospital in Northwood for medical treatment for all the diseases he was carrying. The doctors gave him a year to live. He was kitted out in hospital blues and an Australian bush hat and boots, and he was given crutches that he had to use.

For five months they received no pay and had to rely on pocket money from their families. Countess Mountbatten visited and insisted all the men should be given a travel warrant to go anywhere in the UK, as a gift. Ben told his family, and his brother paid for him to go to the Kosher Green Park Hotel in Bournemouth. On arrival, he was offered a huge chicken meal, but because he was a vegetarian he could not eat it, and he said he was going upstairs to lie down in his room. When he woke up, he was back in Mount Vernon hospital. He has no idea how he got there, but thinks that he had a malaria attack and was taken by ambulance to London; the staff had found him in his bed.

Ben attended many FEPOW reunions after the war, often with his wife Pearl; in one at the Festival Hall, he was mobbed by emotional men who had counted him as their friend in those dark days; his wife was astonished at how they received him. He never missed an AJEX annual parade and marched proudly with the Jewish FEPOWs, among others. Ben died in September 2006 aged 85.

Postscript

In a letter written from Labunan on 20 September 1945 to his parents and family at 229 Station Road in Harrow, Ben wrote:

> Once more sitting up taking notice of the world, content, happy, at peace on white sheets, and with the best, the kindest care that anyone could wish for. It's a grand feeling, clean shaven, with a good breakfast inside me, well I am just fine ... the last card [that I received from you] was dated July 1944 ... I'm not impatient, all the good things seem to be coming my way. I've always hoped and I've always believed that all of us would come out OK ... I've everything to live for and look forward to...I've had my share of the incidents in the POW camp. Luck has been on my side ...it could have been worse...most of the time we had a secret radio ... on August 16th I was on a 15 man working party as cook ... the camp relaxed ... we had new kit and soap, rice and inner camp freedom ... on the 31st the very first bread and butter, dropped with tinned food and medical supplies, by parachute. The Aussies arrived and on 13th September, three years less one month to the very day after landing at Kuching, I was raced down river in a USA torpedo boat to the RANS Hospital ship Wanganella. Fever knocked me out ... still a bit shaky with kind but strict orders to take things easy ...

Below is a letter to me from Ben's son Maurice, dated 12 July 2011, which is a final and touching envoi for his father, a delightful man:

> Thanks for sending your article about my father. You certainly got more facts out of him than any of us in the family ever did, which is a tribute to your talents in getting him to open up to you.
>
> With me he tended to ruminate on some of the big existential issues. And I also picked up his attitudes from how he lived his life, how he behaved and what others, his friends, said. I'm thinking of things like why did he survive while others, often outwardly tougher than he was, perished? What should he think of the Japanese and how should he behave if he met a Japanese person? What should he think of the atomic bombs dropped on Hiroshima and Nagasaki? And what is life about anyway?
>
> Clearly he was quite a robust young man at the time of the fall of Singapore. But he'd been a somewhat sickly child and came from quite a cloistered, over-protective and fussy family. His mother was a particularly timid lady. He was an avid reader from a young age, a bookworm. He didn't smoke or drink, and was a virtual

vegetarian. I didn't know that he was given the job in the Kempeitai office by his British CO. But he had a number of qualifications for the role. He'd taken a course at Pitmans after leaving school at 14 and was a quick typist, a skill that I suspect not many of his comrades shared. Add to this his reading and his ability to communicate with the locals based on his Yiddish; he was probably an ideal candidate. (To say nothing of his rather bookish appearance.) Anyway, it seems to me that having the office job was a major reason that he survived. Other factors included: not smoking, being able to 'turn off' through sleep when things became traumatic and being liked and somewhat protected by his mates.

His great friend of his early married life Ben Cronin used to tease that he was like 'bloody Jesus Christ'. And the reason for this jibe was that he refused to bear a grudge against the Japanese. His attitude was that life's too short to waste energy on hate and bitterness. If he met a Japanese person he would react in what was clearly a visceral way, which he wouldn't if he met, say, a Ghanaian. He would be rather overly deferential in an almost mocking way. But intellectually he never gave way to a sarcastic comment or bad word. His mind ruled his heart.

As for the atomic bombs he was never triumphant. He was aware that, by shortening the war, the bombs may well have saved his life. But he took it as a sad and ironic reflection that he was spared at the cost of so many others.

He often asked 'do we work to eat or eat to work?' He had a great work ethic and was very committed to his shop. It became a great frustration at times but the idea of not working was anathema to him. But at the same time what he really wanted to do was to help others. I think this tendency was already there when he was in the camp. But it does seem that his terrible experiences reinforced for him the idea that life is about helping others. Family, the Jewish religion, pottering in the garden were all very important. But he always seemed most in his element taking a group of disabled people on holiday in a minibus!

Notes

1. Years later in 1975, Ben went for a check-up at Edgware Hospital and while a nurse was asking about his medical history, she suddenly ran off; later she returned, clearly having been crying, to explain that Ben must have been one of the men who helped her board the Empire Star in Singapore. She had survived the war but would not tell him her name or what had happened to her; they never met again.
2. Long after the war, Ben's doctor told him that as so many terrible things had happened to him, his memory had blocked out certain incidents, as a form of mental self-defence.
3. The Church of England Archives reveal that the bishop was Francis Hollis (1884–1955) – with thanks to Mme Sophia Ang, secretary to the Bishop of Kuching, 2011.
4. Ben's sister Golda wrote to him in her immaculate handwriting throughout his captivity, and many of these letters are preserved in Ben's album; he received most of them only at the end of the war.

5775488 Pte Leonard Goldfarb, 'B' Company, 4th Royal Norfolks

Leonard spoke to me, but he found it very difficult to recount the horrific experiences he had been through and had to stop; hence the story is poignant but brief.

32. 5775488 Pte Leonard Goldfarb, 4th Norfolks.

Leonard was born on 26 March 1918 in Vallance Road, Stepney, London, son of A. Goldfarb, and later moved to 34 Knightland Road, Clapton, then to Islington. He joined the army on 13 January 1940 and became a regimental PT instructor. He was sent overseas to Singapore and on arrival his battalion went straight into action. He was wounded at the Battle of Bukit Timah; knocked unconscious, he awoke on the battlefield early next day, surrounded by dead and wounded comrades, his own wounds being all down his back. Leonard was sent to Queen Alexandra Hospital and here he was placed in the basement, semi-conscious, as wards were overflowing with wounded. When the terrible massacre of staff and wounded by the Japanese took place upstairs, those in the basement escaped attention and were saved.

FEPOW records show Leonard was in Changi from February to September 1942, and then Chungkai from September 1942 to June 1943.

In one camp he found that most Japanese officers spoke good English and had been educated in the USA; they were quite friendly. He was ill with malaria, but one officer wanted a tailor to make two little sailor suits for his children in Japan. Leonard and a Jewish friend, Davidson, offered to do this and as a result were allowed to sleep in the officers' huts to recover from the malaria. Later he was able to work in the cookhouse on light duties for a while.

Bampong was a transit camp and as a result there were opportunities for bartering and trading, at which Len excelled; he used to go 'under the wire' and trade with local Chinese, mainly for food and medicines to help the sick. He claims that the film *King Rat* was based on his story! However, Corporal Bluestone is another candidate for that role.

One night Leonard and his mate got a little tipsy, drinking alcohol with the Chinese, and were caught getting back into the camp with food supplies for the hospital. Len managed to escape but his friend was tortured. He was made to stand in the centre of a ring of guards who wielded large bamboo sticks, and he was consistently beaten until his body sank into the sand up to his waist. He was then hung by his feet from a tree for two days without food or water. Leonard did not say whether he survived.

At Chungkai and Changi camps, Leonard organized extra food from the Japanese officers for a Seder night meal at Passover 1942, for all the Jewish POWs, assisting David

Arkush. While at Chungkai he met the Australian Harris family who became famous because they were a father and two sons, all taken POW together. He helped them with extra food when they were starving and ill, and after the war they kept in touch. When Leonard's brother wanted to emigrate to Australia after the war, the Harris family assisted him as a small payback for what Leonard had done for them.

At one stage, Jewish POW David Wince arrived in Leonard's camp (probably Tarso, where Leonard was from June 1943 to June 1945) in a very emaciated condition from upcountry. Leonard was able to bring him extra food and helped to nurse him back to health. They remained friends from then on. In 2007 I was able to put them in touch with each other for the first time in sixty years (at time of writing, David still lives in Canada). It was not until June 1944 that his family discovered through the Red Cross that he was alive.

Other camps he was in were Kanchanaburi, Tamarkan and many others, up as far north as Niki Niki, all on the Death Railway. His debrief notes mention Wunjea (April 1945), Wunpo and Brankasi (? June to August 1945) and Kori (August 1945).

In mid-1945 the Allies bombed the Kwai bridge, where he was at the time, and hit some of the POW huts near which Japanese anti-aircraft guns had been deliberately located. Leonard was buried alive under a hut and dug out, wounded by shrapnel; the Jewish Canadian doctor, Captain Markovich, pulled out thirty-four pieces of shrapnel from his body, with no anaesthetic. He witnessed Markovich operate many times, using surgical tools he had made himself from bits of scrap metal.

On liberation, Leonard spent several months in Rangoon hospital because he was so ill, and indeed suffered mentally and physically since the war from the ordeals he went through and witnessed. He accompanied a Jewish POW, Cyril Wernick, by ship to the UK via India, where they recuperated for some weeks, but Wernick sadly died en route, in India, as a result of his terrible illnesses. At New Year 1946, Leonard met up with three Jewish POWs: Davidson, Sid Shulman and Moss 'Morris' Orchard/Orchant to have a meal. Crossing the road in London, they were knocked down by a car and all ended up in hospital, but survived.

On the fiftieth anniversary of VJ Day, Leonard was taken with other veterans to meet the Queen at Buckingham Palace.

1549822 Gnr Harold Buchanan Goulding, 89th LAA Battery, 35th LAA Regiment, RA

The information in this testimony is taken from an audio-tape recording held at the IWM, and an unpublished extended essay written by Harold, titled 'Yasme'.

Harold was born on 1 February 1917. He volunteered in January 1940 and went into the RA in Wotton near Woodstock, Oxfordshire. His parents were living at 39 Cholmley Gardens, London NW6. His first training was on Lewis guns. He was posted overseas on 13 November 1941 and sailed from Gourock in Scotland on the SS *Empress of Japan*; they were due to sail to the Middle East and all their kit was prepared for that theatre of war. In Durban, South Africa, they heard that they were switched to go to Singapore on the SS *Narkurda*, arriving there on Friday 13 January 1942. There was a huge rainstorm in progress and the harbour was being bombed by the Japanese; Harold saw two Allied Brewster Buffalo aircraft shot out of the sky by Japanese fighters. The men on board disembarked at the double and made their way to shelter. Their uniforms provided neither protection nor camouflage

Harold said that one of the ships in his convoy carried specially adapted Hurricanes with air-cooled engines for tropical use; but, if they did get into service, they made no difference to the outcome of the final battle. There were also many rumours about a big landing by the Australians which was going to push back the Japanese.

There was a lot of complaining by the men that no Allied aircraft were to be seen opposing the deadly Japanese bombers, and that they flew at heights beyond the range of the LAA (Light Anti-Aircraft artillery) – what was needed was HAA (Heavy Anti-Aircraft) guns – and there were very few, if any. Consequently the men felt vulnerable and impotent, and morale was thus low.

After a few days, Harold was posted with some of his unit upcountry to the last mainland aerodrome in Allied hands, just north of Kuala Lumpur, called Kalang, which they held for only a few days. It was very simple for the Japanese simply to dress in singlets and shorts and infiltrate the Allied lines, because to most Europeans they looked like Malays; using these tactics, many men were killed and the retreat continued en masse. Another common tactic the Japanese used was to climb trees and lob grenades into passing Allied vehicles travelling along the roads. This happened to one vehicle Harold was in, but the grenade bounced away. In any case, his unit could not stand and fight as they had too few rifles between them.

He experienced the retreat down the peninsula as the Japanese constantly used boats to get behind the Allied lines, and witnessed the efficient use of tanks and bicycles by the Japanese forces as they surged towards Singapore. The British did not believe tanks could be used in the jungle – but the Japanese had other ideas. Even the British uniforms, said Harold, mostly light khaki for use in the Middle East, were easily visible in the green jungle environment; the Japanese uniforms were much darker; and British pith helmets were useless in the rain or for protection against Japanese shrapnel and bullets.

After reaching Singapore, Harold's Battery was evacuated on 31 January 1942, the day before his birthday. The boat they were assigned to was a very small, old Yangtse River boat with an old China hand captain; the guns were strapped on deck. Including crew, there were over 100 men aboard. They sailed south but were attacked in the Bangka Straits, off Sumatra, by a Japanese bomber, on 1 February. When they first saw the plane they thought it was American, as it was a Lockheed, and everyone began waving at it; but it was an enemy aircraft (either captured or bought pre-war) and the situation was not helped by the fact that Harold and his comrades had had a few drinks celebrating Harold's birthday, using whisky they found abandoned on the dockside in Singapore. The reality hit them when the Lockheed started firing its guns at them. Despite four runs dropping bombs and three runs firing his machine guns, the pilot missed them and left. Nobody had been hurt. However, below the waterline, severe damage had been done and the boat began to sink very slowly. There were not enough lifeboats and those that there were had been filled with bullet holes, and they sank. Using lots of wooden wreckage, and with lifejackets on, all stayed afloat save one injured man who died because he had not managed to get his lifebelt on; one large Allied vessel sailed straight past them but must have radioed their position because, after four hours, boats arrived form Sumatra and they were rescued and taken to Palembang. Harold's group consisted of only about twenty of his battery; others went on to another town.

After some days they observed a flight of Japanese planes overhead and suddenly saw paratroopers being dropped, trying to capture a nearby aerodrome; Harold and his men, together with some Ambonese soldiers (they were to the Dutch forces what the Gurkhas were to the British), engaged the paratroopers and Harold was involved in an intense battle with them. They killed about a dozen in a firefight, and quite a few others who were caught up in the trees. They did very well, in fact, to defeat this group and so were somewhat surprised when, for a reason unknown to Harold, his battery was then ordered back to Palembang and were sent south by train.

Here, in mid-February, they met up with the rest of the battery and boarded an armed Australian merchantman and were moved to Batavia in Java to be used as infantry, after a week's training, on Sten guns. They were also sent on patrols which turned out to be useful training, as the Japanese did land on Java; on one patrol, Harold recalls the Japanese using the psychological tactics of calling out 'We can see you Johnny!' – when of course they could not. The training was to stand still, as they could certainly hear you.

Eventually Harold's unit retreated to Bandung in the highlands. On 7 March they were told they would fight to the last man; on 8 March they were ordered to surrender and his CO gave every man a large bar of Cadbury's chocolate. Harold was by now ill with tonsillitis; he reckoned he had not slept under any proper shelter, nor in the same place, for more than one night since the retreat from Malaya; he felt at a low ebb and still the Japanese had not appeared, one week after the surrender. As result he was sent to hospital and was separated from his unit – who were taken to another POW camp – for the rest of his incarceration. This saved his life, as most of his unit, he learnt later, were killed on the way to Japan when it was sunk by an American submarine, and the rest were killed by the atomic bomb on Nagasaki.

POW

From the hospital, Harold was discharged to a local school that was being used as a POW camp, called Landsoproedings; it was good from the point of view of being a light and airy building with latrines and kitchens, but it was appallingly overcrowded. Also, their marching entry was re-enacted several times, as a Japanese newsreel team were filming their captivity for propaganda purposes. They were also ordered to sing and cheer.

Harold's Camp Allied CO was the courageous Australian, Dr Colonel 'Weary' Dunlop, who was greatly admired by all the men and even the Japanese for his constant care for the prisoners.

Then Harold and his comrades witnessed their first atrocity. The Japanese CO ordered the men to stand to attention, and the order was misunderstood by Colonel Dunlop, who gave the order to dismiss. The Japanese CO went berserk and began beating up the colonel incessantly and even drew his sword – the men thought that Colonel Dunlop was going to be killed there and then.

Rations at Landsop were of the usual appalling kind: mostly rice and too little of it, and virtually no protein. Being always hungry was very depressing and it is something Harold never forgot or forgave. The men all had their heads shaved by order; Harold felt this was very dehumanizing and depressed morale somewhat. He also noticed how this made everyone look more alike. However, it did reduce the lice infestations.

On 14 June the men were moved to an old Dutch Army barracks in Bandoeng. They were moved in two groups of 1,000, and it was about a five-mile hike. As the groups arrived, the Japanese had a head count each time but kept getting the totals wrong; they were running up and down with notebooks and pencils, getting into more and more of a frenzy as the numbers failed to add up. So, to take advantage, the men kept moving about, and as the Japanese got more and more angry, the POWs became more and more amused. Suddenly the order was given by the Japanese officers to fix bayonets, so the POWs got into line and made sure the figures added up. Rather than understanding that it was a joke being played on them, the Japanese assumed the POWs were just stupid – such was their lack of humour.

Bandoeng was the 'Ritz' of the Dutch camps – roomy, cool, with proper roofs, and it held about 2,000 POWs. The commandant, Suzuki, was not a maniac like many of the other Japanese commandants and lectures and activities were organized by the men. Here they were set to work for the first time but the rations were even worse than at the previous camp, the rice being full of maggots – which the men were, however, advised to eat for the protein content; the tip given was to read while eating so you did not have to look at what you ate.

Food at breakfast was a tin mug of rice pap – that is, rice porridge. The rice had reddish husks and this was particularly nutritious and saved lives in the long run – though this was not intended by the Japanese, of course. (The cooks had a privileged job, as they could always eat a bit extra, but they had to work all night to prepare food.) If they did a bad job the pap was too liquid and not filling; if made well it was thick and filling. The cookhouse would be framed in the pitch-black night, glowing orange and red around the cauldrons of food; if the cooks had had tails and horns it would have resembled a picture from hell. Lunch – or tiffin – was a mug of steamed rice with a small amount of a tough, weed-like spinach called kan-kong, whose protein content was a life-saver. Supper was the same as lunch. Those doing heavier work got more; according to Japanese ration rules, the sick got nothing. But, by agreement of all the prisoners, everything was pooled and the food was fairly shared by all. Another life-saving vegetable was katchang idjoe, a small pea that contained a lot of vitamin C. They also ate frogs and, if they could catch them, cats. Fish-heads were a luxury for the sick only. For two months of the year rice was not available, and there was only a very hard maize seed that was unpalatable; many men would eat only a little rice, in any case, as they could not bear the taste and texture; they often died first.

Much of the work was moving these very heavy rice sacks from old dumps into the camp. Gradually the men began to disobey orders as far as they could and accept a beating as a small act of resistance – for example, using two men to move a sack instead of one brought punishment from the guards.

Relations between the Dutch POWs and the British and Australians were not good; the Dutch were real colonist types, like the British in India, and not true soldiers, just militia; the British thought them arrogant and wondered openly why they had been there to defend them and their privileged way of life. On one occasion when the British refused to carry one rice sack per man, the Dutch tried to show off and picked up two per man; as the British watched them and applauded, it was a while before the Dutch realized that they had done the whole job of work, while the Tommies stood egging them on. Occasionally there were fights, which channelled aggression and were mostly about the most trivial of matters

Another job was spinning sisal for making sacks; the Japanese made money out of this by selling it locally, so if you worked well you were rewarded with an outing. Harold once won such an outing with twenty or so others, and went for a walk under guard to town. Then they were taken on to the women's area where they could accept bread and jam and other luxuries they had not seen for months, then there was a walk back, finishing with a speech from the guard about how good the Japanese were to their prisoners.

Harold also became friends with the famous explorer and later author, Laurens van der Post. He spoke Japanese fluently. He advised Harold just to focus on prison life and not think of anything else except survival as a POW; that was the way to remain alive. Another POW officer called Rintoul organized the small POW payments the Japanese gave them for work, to enable reductions from pay to be made for the sick (as the sick never received payment), thereby allowing the purchase of extra food for those who were ill, from the ever-open canteen. The Japanese favoured this scheme as it meant more business for the canteen, in which they had a financial interest, buying cheap local food and selling it at higher prices back to the prisoners. Or they simply stole the POW rations and sold them in the canteen back to the POWs. The only advantage of this was that the canteen never closed, otherwise the commandant lost money.

Conversely, the Japanese had a fear of illness; if ever a POW wished to leave his hut at night to see a friend or do some trading – strictly forbidden – he had only to say he was sick for the guard to wave him on as he cringed away.

Harold explained how the crude Japanese propaganda sheets they received enabled the men to see how the war was going in the Allies' favour, so when it said that the Americans had been wiped out on Island X, and the next time on Island Y, somebody always knew that Island Y was closer to Japan and so it meant the Allies were advancing.

All sexual urges disappeared, and pin-ups of women were replaced by pin-ups of food. Harold owned a copy of *How Green Was my Valley* by Richard Llewellyn, which contained numerous references to descriptions of meals. It was so much in demand that he hired it out, but only to be read with him present; he accumulated cigarettes in this way, to exchange for food.

When lining up for counts, the men often played tricks on the guards by deliberately moving about and so the count would never be correct when guards compared notes. This always resulted in lots of shouting, screaming and face-slapping by the officers and NCOs, at the mostly Korean guards.

On 1 November 1943, Harold was sent to Sumatra, but had first to go to Jakarta (Batavia), the exit camp for the ships, and onwards to Bicycle Camp (so called because it was a former Dutch Army Bicycle Regiment camp). The commandant here was an English-speaking sadist called Sony (later hanged for war crimes). His punishments included making men stand in the sun all day, often resulting in death; a bad sleeper, he would enter huts at night, screaming and shouting, to check the hair length of the POWs; anyone with longer than regulation hair would have it cut by him there and then, in a crude and painful manner, scalp and all. However, mosquito nets made men anonymous and the men would often swear at him in the dark, and this angered him intensely. It gave satisfaction to the men but always resulted in some kind of mass punishment next day, such as extra long inspection at *tenko*. Conversely, he never stole the men's rations as many commandants did, and the full-time Australian staff (cooks, office workers, etc.) thought he was a fair man who treated them well. Nor did he allow anyone else to hit the POWs, and the guards were as frightened of him as the prisoners were. Once, when Harold had broken some minor rule, the guard asked if he would like to be beaten by him or the commandant; Harold chose the guard. He was made to stand to attention and eight guards slowly circled him as they took off their belts; he was first kicked many times from behind at the back of his knees, then the beating began. He reckons he was beaten up to twice a week in this and other ways, mostly for reasons unknown to him.

One guard, probably a Korean, was known as 'Mad Harry'; he once offered Harold a cigarette but then punished him for smoking. Once, blind drunk, he threw a bayonet at Harold while he was weaving sisal; it just missed his back and landed, quivering, in the bobbin near his face. On other occasions the guard beat him with a baseball bat.

Sumatra

After a few weeks came the move to Palembang, Sumatra, by Hell Ship; the journey was horrific, on the decrepit tramp steamer *France Maru*. The men were locked below in appalling conditions for two nights and three days, fixed in standing or sitting positions, virtually unable to move, packed like sardines. But they still joked and sang songs and swore, and, due to the lack of food and water, few needed the toilets and so sanitary conditions were reasonable. On disembarking, Harold spied a water tap and turned it on. Out came ice-cold, fresh, sweet water. The unbelievable pleasure of drinking it is a memory that stayed with him always.

They were taken by lorry to a brand-new camp called Tanjong Priok, but, as the latrines were located between each hut, almost every man eventually contracted bacillary dysentery. Harold also contracted malaria, beri-beri and pellagra. One of the most horrific sights he saw was a POW so bloated with beri-beri that he resembled a large balloon; although not present at his death, Harold was told that his body exploded and the accumulated water gushed from him. Harold's dysentery was so acute that he had to crawl to the latrine because he was so weak – so weak that his 'mate' asked the medical orderly (who thought Harold was out of earshot) if he should take Harold's possessions now, as, if not, the sergeants or officers would grab them. There was an unwritten rule that a 'mate', in return for helping a friend when ill, inherited the possessions if the friend died. Harold got so angry that he began to sweat profusely, and this so reduced his fever that by next day he had recovered. He later 'divorced' his mate for another.

To survive, the men had different ways of arranging their meals. Hunger after evening meal was painful and miserable, and men used different strategies to cope. In the Sumatra camps, for example, using terrific willpower, Harold would only drink tea at breakfast and store his rice pap in a tin and have a cigarette – which helped kill hunger pangs. At the morning break (work was 8 until noon) he would drink tea and a small crumpet-type cake, made by the Dutch, which he kept back. Then the long midday break came (too hot for work, and the guards did not want to be out in the heat). He collected his midday meal but kept it back and ate the breakfast rice pap. Then after a sleep he went back to work until 5 p.m., then collected his evening meal and ate both lunch and supper together. Many would eat the evening meal with their backs to their friends, facing the wall of the hut. It was a private, even a sexual, moment, but there was no joy in it – it was too crucial.

Harold recalled that the roll calls were often a social occasion which was used to mock the Japanese and get some laughs. Before they were ordered to number in Japanese, they used English and so it would go 'One, two, three, jack, queen, king.' The Japanese guards were completely flummoxed.

Although there were books to read, they often disappeared because men wanted the pages for smoking papers. Even the Japanese occupation paper money, which could be used to pay the camp barber for the compulsory shaving required of all men by the Japanese, was sold for twice its face value as cigarette paper.

Other entertainments included amateur dramatics, debates and lectures; sometimes they even organized examinations in subjects such as foreign languages – many of which were spoken in the camp; even certificates were issued. Maximum use was made of any professional teachers among the POWs. Harold also said that there was a secret radio in most camps – its location very well hidden. Escape was, of course, virtually impossible, as the local population was hostile (they were anti-European), the men were too weak and there was nowhere to go; but getting under the wire to go and trade with locals was very common, especially as the Japanese guarding was very lax in most places, owing to their confidence that nobody could actually get away. Most punishments were for POWs breaking back in, rather than getting out.

The main work in this camp involved the building of an airfield nearby. Wooden stakes would be put in the ground each day by the Japanese to mark where the men had to reach by close of work; these were often moved back surreptitiously and so slowed the work down. The airfield was never finished. However, when the officers discovered in their monthly check that much less work had been completed, they would beat the guards and the guards would beat the POWs.

There were teams cutting earth, another team putting it into baskets and another taking it to the tip. Soon the Japanese put in rails to carry a skip, to speed up the movement of soil to the tip. There was a slight slope, which was as long as a mile, so the skip could be ridden to the tip area. As the men wore old straw hats against the sun, they looked a bit like cowboys. Other men acted as Indians and threw lumps of soil at them to knock them off the skip. This game was called 'The mail must get through!' This often succeeded and the skip would crash off the rails, to the intense anger of the guards, who, red-faced with rage, would come and whack the men responsible and make them push the skip back up the rails. But the amusement was worth it to provide some fun in such a miserable existence. These activities stopped, however, when people became too weak, or blows would cause injuries which reduced your chance of survival.

Another game was 'Dare'. On one occasion Harold was dared to crawl naked past the guardhouse with a bottle on his back. When the guards came out with fixed bayonets he played mad and made eerie sounds, rolling in the dust; the Japanese respect for madness led them to give him cigarettes and some food and send him gently on his way. All this, said Harold, was to maintain self respect – to show the Japanese that the POWs could get one over on them

Harold received and sent very little mail, in common with most POWs of the Japanese – his first contact taking two years. Only one Red Cross parcel was given out – and it was one between several men, probably at Christmas 1944 as far as he could remember.

On 31 May 1945, Harold was again moved by ship, this time back to Singapore. At this time he was ill, both physically and psychologically, as were most of the men, and tensions on board were very high among the POWs, not least because few knew each other in this transport; Harold himself got into one fight. But on this ship they were allowed to stay on deck. The danger now was Allied submarines, and so the ship stopped at night to avoid producing the wave phosphorescence which made ships such easy targets for submarines.

On this journey, one man went totally mad and became a religious maniac. He tied two sticks into a cross and took off all his clothes, climbing up to the bridge and screaming out 'Can't you see I am God?' – to which a cockney wit replied, 'Yes! Now for Christ's sake shut up!' Tough humour, but everyone burst out laughing and it provided an escape valve for emotion.

Changi

Here Harold experienced his most traumatic event. He went to the barber at Changi and saw himself for the first time in a full-length mirror; he was horrified. His morale and physical condition took a plunge; he found he could not even walk up stairs.

Work and entertainments petered out, and men became introverted as talk turned to the end of the war, because good news was seeping through. Harold began to lose even more weight, but then news reached them in August that the atomic bomb had been dropped. As a result, Harold began to gain weight suddenly with the feeling of relief. Happy, but very weak, they awaited full liberation; most were too weak to go into Singapore, but in this interregnum the food supply broke down. Between this time and liberation, many POWs tragically died.

As for liberation, Harold only remembers that an RAF aircraft flew over and dropped leaflets telling them to stay in camp, as troops were on their way – and they were *not* to overeat. In Karachi, on the way home, the men ate masses of food, continuously, then threw up and returned for more; Harold went from six to eleven stone in four weeks. Once at home he would eat only small amounts, never being able to finish a meal. After several months this stopped and his appetite returned to normal.

Arriving in the UK, having put on a lot of weight, the men were sent to a camp in Horsham for medicals and then sent home. Harold felt no bitterness towards the Japanese, but remembers that for years afterwards he was very restless.

T/198268 Dvr Jack Joseph Greenberg, BEM, RASC

Jack Greenberg was born in Liverpool on 4 September 1915. He joined the army in April 1940 in Edgware, London, having previously been in the OTC (Officers' Training Corps) at the Royal Grammar School in High Wycombe, attached to the Ox and Bucks Light Infantry (in 1936 he tried to join the Artists Rifles as a regular, but his eyesight was deficient). His family lived at 55 Warren Street, Euston, London.

Whilst he was training in the UK in July 1940, one of the men called Wally Hunt made various anti-Semitic remarks to him. Later they met again (see below).

On 2 May 1940 Jack was sent to Kirkby-in-Ashfield, Nottinghamshire, for basic training and was then sent to the RASC in Buxton in June; from here he was attached as an Ambulance driver to 'A' Company, 197 Field Ambulance, RAMC, at Cranwich in Norfolk, part of the 18th East Anglia Division. He was then posted to various camps all over East Anglia, and then in December to Scotland. From here he was given a week's leave in January 1941 to marry in London (where he lived and worked in the jewellery trade in the West End); then in April he was posted to the Walsall area and took part in rescuing wounded from the Birmingham Blitz. Finally, after embarkation leave and being given tropical kit, he was sent to Avonmouth, Bristol, in November 1941 and sailed on the MT *Oronsay* to Gourock, Scotland, and from there they had a US Navy escort to Halifax, Nova Scotia, arriving on 7 November 1941. From here he was transferred to the American troopship USS *Leonard Wood* and sailed to Cape Town via Trinidad, arriving on 10 December, and this was where they heard about the attack on Pearl Harbor. They were ordered to Bombay, and there they switched to the troopship *Empress of Asia*, sailing to Singapore at the end of January 1942.

With the convoy was the cruiser *Exeter* which had fought at the battle of the River Plate. Between Sumatra and Bangka Island, in the Sunda Strait approaching Singapore, on 4 February at about 11 a.m. they were attacked by Japanese aircraft. The *Exeter* was firing back but radioed that if the *Empress of India* could not keep up, they would have to leave her behind. The stokers had panicked and abandoned their posts and the soldiers had to do their work until they were persuaded to return; the ship managed to keep up.

On 5 February all the troops were ordered to be dressed with their kit and to assemble at the stern, but Jack stayed below as mess deck orderly that day. When the Japanese bombed the ship at about 11a.m. and made direct hits on the hospital and engine room, Jack was trapped below with others. He grabbed a lifejacket, a helmet and his false teeth, then banged on the watertight door, which was eventually opened, and went on deck. The anchor was dropped; the superstructure was ablaze and they were told to abandon ship. Those who could swim were told to make for the nearby lighthouse, about two miles away, and others should slide down the ropes and ladders and await pick-up. Jack slid down and then jumped the final twelve feet to the water and found a raft to hang onto with two friends – some crew tried to jump on the raft rather than hang onto its sides, so endangering them all, but Jack and his friends pushed them off. Jack's two mates, Wally Haslamph and Lou Coulsworthy, wanted to give up, but Jack said no, and encouraged them by telling them they would be picked up.

At about 2 p.m. they were rescued by the Australian corvette HMAS *Wonalong*. Jack got his two mates up the ladder but he himself was too weak and so the Aussie sailors pulled him up on the rope ladder. He was taken to the wardroom and rested and fed and given some rum, and they were taken on to Singapore.

Jack had only a towel, a pair of wet shorts, gym shoes, his specs and his false teeth – everything else was lost. They were taken in trucks to Gaylang village and billeted in bungalows, and ordered to dig trenches in the gardens. Next day Jack was detailed by the RAMC colonel to stay in post, but he refused and said he would rather help rescue any wounded men from the fighting going on nearby.

On another day he was loading ammunition and then was sent to the main dressing station to assist with the wounded; this was located in the Municipal Buildings, next to the Padang. On Sunday 15 February he was sent with an ambulance to form an advanced dressing station; he was then sent to the 2nd Cambridgeshire lines to assist, but they had no wounded and so he went to assist a Chinese family and some other wounded men and drove them back to the city. At noon he returned to the municipal buildings. Jack asked his mates where Wally Hunt (then a corporal) was; they had not seen him and just said he was 'a yellow bastard'.

At the Padang he witnessed officers forcing men at gunpoint to return to their posts as they were trying to leave. He then saw that some ambulances were going to retrieve some wounded of the 1st Cambridgeshires; he joined them and helped to rescue wounded men of the regiment, in the front line at the end of Dunearn Road, where they were lying in a house used as the advanced dressing station. The Japanese had set it on fire. Having pulled the screaming men out into the garden, his officer ordered them all to surrender, as two armed Japanese soldiers appeared.

POW

As they surrendered they were searched and then marched through an avenue of armed Japanese; at the end he saw a tall, bearded Japanese officer with a map, and one Japanese stole his watch. With 509 other men, Jack was herded onto a tennis court; there they slept huddled up together that night. In the morning, the Japanese gave them some water to drink and biscuits to eat, and some officers gave them cigarettes. That evening the Japanese asked for ten volunteers and Jack went to bring rations from a captured British food dump. They were allowed to take as much as they could for themselves too. They had a barrow and pushed the supplies back to the tennis court area. Jack kept a tin of sultanas which was used to swap for cigarettes from the Cambridgeshires.

After five days, on Friday 22 February 1942, they were marched fourteen miles to Changi gaol. Here Jack was greeted warmly by all his mates, as they thought he had been killed. In the hut he was asked by a forlorn Wally Hunt to be his mate, as he had been ostracized by his friends for his cowardice earlier in the fighting. Jack agreed but exacted half his spare food and money as the price for this. They became friends, but later Hunt was sent upcountry and Jack heard that he had died on the railway.

Jack forced himself to forget his wife as a means to try to survive. Rations were appalling and Jack was elected the hut representative in the cookhouse to see that distribution was fair; he also had to go each week in the ration truck to collect food in Singapore. Beatings and a starvation diet made life very miserable.

On 2 May, Jack was sent with a party of men to Thompson Road camp in Singapore City, to break up for scrap a number of cars and vans that the Japanese had captured and stored in various open spaces. Here they worked in parties of five, with one guard; they were marched in threes each day to the car parks and sabotaged as much as they could – putting sand in engines, for example. An old Chinese lady used to discreetly thrust loaves of French bread into their arms as they marched to the scrapyards each day, at great risk to her life. A little Chinese girl secretly brought Jack food at lunchtimes, where he had an arrangement to hide in a certain vehicle to meet her. On one occasion, when Jack had smoked a cigarette at work, he was caned severely on his back by a guard. One Japanese guard, however, was a Christian called Toby, and he showed the men great kindness.

The men were given one day off in ten – this was when they washed clothes, held concerts and often sang rude songs about the Japanese, who watched and laughed and applauded because they did not understand what was being said. This was a source of great amusement for the prisoners.

One day at the camp, some timber was stolen. The Japanese asked the British officer to return it, but nobody owned up, so the officer ordered the Japanese to beat his men to find out who was guilty – this caused intense hatred for that officer. Jack does not know if the wood was ever found and would not name the officer responsible.

From this camp Jack was able to send one card home, and he found out after the war that his wife did not receive it until September 1943 and so did not know until then that he was alive. Sometimes Jack was able to buy articles through the wire from locals (one Chinese boy was nicknamed Charlie) and sell them on in the camp, using any profits to buy extra food.

On 17 August 1942 they were moved to Serangoon camp, also in Singapore, from where they were despatched on various work parties: cutting grass, unloading ships at the docks, and working in the dock warehouses. On one occasion, to the annoyance of the Japanese guards, some German seamen on a ship gave them all some food. In late December they were taken back to Changi because the Chinese labourers complained that the POWs were taking their work and putting them out of a job. Jack recalled that at Serangoon some Red Cross parcels arrived – he was given chocolate, sugar, boots, hat, a shirt and some cigarettes. One day he and the men could hardly believe their ears when they heard bagpipes approaching – it turned out to be the Indian Rebel Army Band.

On 14 September 1942, Jack was returned to Changi and here he met and got to know the Dutch Jewish Army Chaplain, Rabbi Chaim Nussbaum, who he said gave the Jewish men much hope and inspiration. Nussbaum in turn wrote in his book that Jack was always cheerful and outgoing. Jack remembers being on burial detail on some occasions at Changi and experiencing the horror of lifting the wasted bodies of dead men at the Roberts (Changi gaol) hospital onto trolleys to be wheeled to the cemetery nearby.

On 24 April 1943 the Rabbi said a Hebrew blessing over Jack before he left on a transport to go upcountry. Many of the non-Jewish men around them, waiting in the line with their pitiful bundles of possessions, watched and took off their caps in respect for this spiritual moment, for none of them knew what lay in store for them on what they surmised was the terrible Burma Railway. Fear of the unknown gripped their starving and exhausted bodies and minds, and there was a loud 'Amen' from them all when Nussbaum finished the short prayer, which Jack is sure saved his life.

Seven hundred and fifty men were to join Force F, and next evening Jack and his comrades were sent to Singapore, then on by train to work on the notorious railway. Conditions were horrific on the journey, as described in many other testimonies. Crowded into cattle trucks, they made toilet stops at Johore and Kuala Lumpur, and Jack remembers washing in the hot water from the steam engine at this latter stop. The steel trucks were so hot that they could barely be touched, and crowding was such that the men could not lie down. Onwards they went, to Ipoh and Alor Star, being fed only twice on the whole five day journey. Finally they arrived at Prah in Thailand, where they had to get out and the Japanese made them do PT. Then they were back on the train until they got off at Bampong. Jack recognized an interpreter from Thompson Road (Fujimoto, known as 'Charlie') – he saw that the men had, with permission, brought a piano, but Fujimoto objected and harangued the British colonel, much to the humiliation of all the men who had to watch this shameful treatment of a senior British officer.

The Railway Camps

From Bampong they were sent on a forced march of 198 miles, in stages, with Force F2, in such appalling conditions that many died of starvation, illness and exhaustion. As they marched off, Jack darted into a Thai shop where he was given some bananas; later they came to a hut where the men were allowed to rest. Next day he was given a number to be worn on his hat – 7/335. Each day they would begin marching at 10 p.m. They journeyed on, via Pagoda camp, with Thais often running alongside trying to buy things from them. One Japanese guard offered Jack a woman in return for his spectacles.

The next stop was Kanchanaburi, where they were vaccinated by Japanese doctors. They slept in the open and it poured with rain; two of Jack's mates bought some rice whisky to drink, to keep themselves warm. In the jungle they marched for forty minutes and then rested for twenty minutes. For lunch they were given boiled water. Then on to Tarso camp, picking up Australian sick en route, who had been left behind from previous marches. Many were too far gone, and died. At Tarso, Jack met John Goldstein from Golders Green, whom he knew. Jack had a bad blister on his heel, a dangerous wound that could develop into a possibly fatal ulcer. Goldstein asked him if he had a watch (Jack had retrieved one from a friend who had borrowed it) and he used it to bribe a Japanese to clean and dress Jack's blister, and to give him some tea and $40 for the watch.

At the next stop, six volunteers were asked for, to clamber down to the river to get water to boil for drinking, and Jack helped. Then they arrived at Nikkei. On reaching camp Shimi Sonkrani, where cholera had broken out among the POWs, Jack and seven other men volunteered to assist the Australians as medical orderlies. Within ten days one of his fellow orderlies had caught cholera and died, but Jack never did catch it, and he believes the rabbi's blessing helped him. Jack toiled here for six months, helping to supply food to the cholera patients, and nursing and cleaning them under terrible conditions. The Australian doctors did amazing work but the bodies of the dead had to be burnt with all their effects to prevent infection. The only positive side to this was that the Japanese never came to the camp, because they feared the cholera most of all. Soon Jack caught malaria, but he struggled on; for his courage in the terrible conditions, he was later awarded the BEM.

In August he was sent to another Australian camp at Tambaya, nearer the Burma border, to help other very sick men. The death rate there was extremely high. Jack assisted the sick to excrete at night in bamboo pots, helped with amputations of men who had gangrenous ulcers, sterilized the bed pans for further use and showed men their photos of their families to encourage them to fight to live; but he watched very many of them die.

He especially remembered by name fellow Jewish soldier Sergeant Blashky, who died of beri-beri. When smallpox broke out, due to Tamils spreading the disease from a nearby camp, the number of deaths increased even more. Jack found it too difficult to describe the horrors and the intense effort it took to stay alive in these places.

Changi Again

Later he was sent south to Kanchanaburi camp. On the journey, he was in charge of caring for 400 sick men on open railway trucks. At Kanchanaburi he received his first letters (three of them) from his wife, which had gone via Japan. Then in April 1944, he was sent back to Changi to Block C, Cell 7, first floor, for the final year of the war. They were four in a cell, and each day were sent out to help build Changi aerodrome at Changi village, working in three shifts round the clock. Normally at 6 a.m. a watery rice meal ('pap') was served, then at 7a.m. came parade, then the march to the airfield, then standing in line to urinate because so much water was in the rice, and then returning to the camp after work.

After six months on airfield work, Jack's eyes were affected and the MO put him on camp garden duties and he was ordered to drink green leaf extract to help his eyes recover. On another occasion he was forced to lie on the concrete cell floor for a week when his back was sprained. On 5 November 1944, more than fifty American bombers flew overhead. It was a great boost to morale and Jack said he would never forget the sight of them shimmering in the sun. All the Japanese could do was to force the men back into their tents, as they were not allowed to look at the enemy planes.

For the final four months, Jack was digging tunnels two metres square and ten metres deep, using pick and shovel and small wicker skips to move the earth. These were probably for munitions, but some men said that they were being prepared in order to be places of execution for the prisoners. They worked in teams of five men and Jack never forgot being taught to use the pick properly by two Welsh miners. Later, on 1 April 1945, he was moved to more congenial work, sawing wood props for the tunnels at a neighbouring site, at Mandai village near Kranji camp on the Straits coast.

Liberation

At noon on 17 August 1945, the Japanese sergeant said that all men working must go to Kranji; however, that afternoon they were told by the Japanese that the war was over. The sergeant said that no celebrations were allowed, but the British colonel took over the camp and had the Japanese sergeant removed to work in the cookhouse. Jack went to see him there and said 'Churchill No. 1, Tojo No. 10, eh?', reversing the common taunt that the Japanese used on the Allied POWs throughout the war.

The Royal Navy arrived on 5 September and Jack remembers going to the harbour and sleeping one night in sheets on an aircraft carrier. It was bliss. On another night he went on to the destroyer HMS *Mings* and slept two nights there and talked to the very young sailors (from whom he heard about a new singer called Frank Sinatra). Some took him to visit captured Japanese ships, where the Japanese had to stand to attention as he boarded.

One brutal Korean guard, Toyama, was assassinated by the men in Jack's camp and he also saw the Malays and Chinese begin to fight and kill each other in the streets, because the Malays had been pro-Japanese.

FANYS (Women's Auxiliaries in the British Army – First Aid Nursing Yeomanry) came to the camp to help, but the men all hid at first, afraid to show them how ill and weak they looked. Some days later Lord Louis Mountbatten appeared and shook hands with all the officers lined up to welcome him, then called all the men to gather round him. He told them he had put General Saito in a G-string in Changi gaol, and that they would be going home soon and would be given five pounds on the boat. They would also be able to send a telegram home that day.

On 17 September Jack set sail on the SS *New Holland* to Liverpool. On the way home, they were told that each day's ration they received was equivalent to a week of nourishment, to get their weight back up; Mepacrine was also given to them to reduce the effects of malaria. As a result, Jack never had malaria again, but he did have stomach ulcers for years after. He never ate rice for twenty-five years, but enjoyed it later on. He remembers clearly that some POWs died through overfeeding after liberation.

When Jack got home on 17 October 1945, almost exactly four years to the day since he had left, he went to see Wally Hunt's mother in Poplar and they exchanged Christmas cards for many years afterwards. In 1969 he went to Singapore to try to find the Chinese who had helped him while he was in the city camps near Changi, but could not find any of them. At a reunion in Australia he was the only Pommie there, at the Brisbane gathering of ex-POWs, and he wept as he laid a wreath for the British POWs at the local war memorial. He was also a guest at QAJEX, the Queensland Association of Jewish Ex-Servicemen and Women.

In May 1980 he returned to the Far East with his wife and younger daughter and visited the cemeteries in Singapore and Kanchanaburi, and at Chungkai where Hunt was buried; he could not believe how beautiful and tranquil it was, compared to the hellhole that he had experienced there. He went on to Perth and Sydney to meet many old POW friends among the Aussies. Jack retired in 1981 and fell ill in 1983. He continued to attend FEPOW reunions and was always a loyal member of AJEX.

Jack Greenberg's BEM Citation reads:

> In April 1943 a large force of British and Australian POWs were sent deep into the jungles of Thailand to work on the Burma-Siam Railway. At one of the Australian camps, cholera broke out in a virulent form and this soldier together with seven others courageously volunteered in the difficult work of nursing the sick. Driver Greenberg did gallant services during the epidemic amongst the cholera patients, working in mud and filth and constant rain and without adequate shelter and was constantly exposed to infection from this disease which took so high a toll of the forces. He performed excellent service and for his personal bravery and steadfastness and self sacrifice he was a credit to himself, his Regiment and his country.

Jack did not know about this citation and it was given to him only many years later, after his daughter's marriage.

2364777 Sgmn Joseph (Joe) Max Hirsch, RCOS and War Artist

33. 236477 Sgmn Joseph Hirsch, RCOS and war artist.

Joe Hirsch was born on 4 March 1920 at the London Hospital in Whitechapel, one of four children, and lived in the Boundary Estate at the back of Shoreditch Church in Bethnal Green. His parents were Lewis (born in Polish-Austria) and Annie née Schoub (born in Russia) who came to England in or around 1900. Joe went to Virginia Road/Rochelle Street Primary School and then won a scholarship to Central Foundation/Cowper Street School in Finsbury. From here he won yet another (county) scholarship to study architecture at the Northern Polytechnic, Holloway.

In 1939, Joe won a travel scholarship to study the architecture in Rome and while there he saw Goebbels visiting Mussolini in a parade in the city. As he was a student when war broke out he was not called up until the winter of 1940, meanwhile serving in the Home Guard throughout the Blitz. He was sent to basic training in Gosport and then to the RCOS at Prestatyn, and he learnt Morse code and wireless operating. He was sent to the 18th Division at Stourbridge and was posted to the Far East, sailing from Liverpool in the HQ Company of the 18th Division, on RMT *Reina del Pacifico*, on 27 October 1941 – one of seven troopships with five destroyer escorts. He was made company orderly, because he could type.[1] His convoy crossed the North Atlantic to Halifax, Nova Scotia; about halfway there, a huge America naval fleet met them, while escorting a convoy bound to the UK; the US fleet took over as the UK destroyers turned about and took the convoy back to Britain – and this when America was still not in the war. Joe recalled that one man fell overboard on his ship and was lost.

In Halifax, on 9 November, Joe transferred onto the American troop ship SS *Wakefield* and they were given a huge American naval escort down the American coast to Panama, then to South America and across the South Atlantic to the Cape, their escort being refuelled at sea by an oil tanker en route. While in Cape Town for three days they had a mixture of morning route marches and afternoons free to enjoy the sights; it was here that Joe bought a sketchbook, pencils and watercolours. One night he arrived back late from leave and was arrested and given eight days labour in the galley; this proved to be most pleasant, as he could do some sketching of ship life, and ate good food with the small crew and in nicer surroundings than the mayhem of the troops' mess. They left Cape Town on 13 November, escorted by the heavy cruiser *Dorchester*; they enjoyed an excellent Christmas meal at sea and the officers staged the pantomime *Aladdin*.

Originally the convoy was due for the 8th Army in North Africa, and all equipment was painted for desert camouflage. However, while in the South Atlantic in December 1941, when Pearl Harbor was attacked, they were told that they were to be diverted to Bombay in India. Arriving on 28 December, they were sent by train to a hutted army camp on the Deccan at Ahmednagar; here their first breakfast was bread, boiled eggs and sweet tea, after a four-and-a-half-hour march to the Outram Lines depot.

During their month stay, all equipment was repainted jungle green. The huts were bug infested so the men slept outdoors in tents and mosquito nets, in their battledress, as it was very cold at night – 3,000 feet above sea level – although hot during the day. On 14 January 1942 they left Ahmednagar to return to Bombay after a four-and-a-half-hour march to the railway head and travelling all night by train. A navy lighter took them back to their ship to the same bunks as before and they sailed on 19 January with a whole convoy of troopships and escorts. As they approached Singapore, the destroyers nosed ahead to scout, and Bren guns were mounted on the ships rails for protection against air attack.

Singapore

When the division arrived, it walked straight into the fighting on Singapore Island, and the Japanese were already in control of the water supply at Johore across the strait. Joe's ship was being bombed as they entered the harbour. On 29 January, in the intense heat, his company were sent by truck to the racecourse, where they made camp. Soon Joe found himself in a ditch with his rifle and radio, facing the Japanese across a field, fully expecting to be killed. Occasionally planes came over to bomb Singapore and the men shot at them with their rifles, but on the fourth day of Joe's arrival the garrison surrendered. Joe remembers the men being ordered to pile all weapons on the road by the British officers and then they all joined the long march, with their haversacks, to Changi gaol with thousands of others. They were not herded by the Japanese and their own officers stayed with them, unlike the German experience in Europe.

At Changi, Joe's unit, under Colonel Macintosh, stayed together, and they were put in the concrete former British barracks, but there was no furniture and everyone slept on their groundsheets. The colonel ordered a hut to be built for himself with his batman. Generally the men were left to look after themselves and lived on rice (usually of low quality) and vegetables cooked by their own men. Joe remembers that the cooks seemed to find a hundred ways to cook rice and so vary the presentation of the diet at least a little.

However, despite best efforts, during the nine months that he was there illness became commonplace, especially dysentery spread by flies, and men were dying literally every day. The colonel devised a method of killing flies by building boxes made of very fine mesh and placing them over the latrines where the flies bred. They would then fly in and the bottom of the box closed, and the flies were taken away and drowned by dunking the box in water. Eventually the flies actually disappeared and the death rate among the men dropped considerably. Joe was very careful always to keep his mess tin and cutlery very clean and avoided infection in this way.

There was a cemetery where men were buried and Joe often had the job of inscribing their names on the wooden crosses, as he had calligraphic skills from his student days. This was a sad but important job. He does not recall ever doing a Star of David for a Jewish soldier. Joe also was given the task of doing design drawings for a small pavilion for the officers, though it was more of a question of keeping minds busy than anything realistic that would result from this work. One sergeant orderly, Jimmy Crooks, who was a best friend of Joe's, had the idea with Joe of organizing lectures on subjects they knew about, for officers and men together, and audiences of thirty or forty were very common. Joe prepared one on the universality of art.

Joe has often been asked how he stayed well and survived. He puts this down to doing a lot of exercise as a boy, and he was always very fit and a keen cyclist too. He is convinced that eating good Jewish home-cooked food as a child and young man made him fit, and as a man of small stature he also needed to eat less, and he believes this all helped.

Sometime in November 1942 the men were put on cattle trucks and taken by lorry, rail and on foot for three or four days up the Malaya peninsula to the railhead and then on open trucks into Thailand. Conditions were not too bad for Joe's transport and they got food and toilet stops along the way, but his memory is very hazy on this. As the road ended, their officers told them they were going to have a three-day march north to a work camp and so they were advised to throw away anything that was not absolutely necessary to their survival, as their bags were too heavy for the journey; this meant some very difficult choices. Joe was determined to take his drawing equipment – paper, pencils and eraser – but he went round to all his friends and tried to persuade them that they must make sure to take at least three books each with them so they could share – a sort of library. Clearly, for Joe this was a way of staying in touch with civilization, culture and education. He was so upset when recounting this incident that he wept and I had to stop the interview for a while; it was a very moving moment, to think that this sensitive and gentle man thought books were as important as anything else to keep mind and body and soul together in that jungle wilderness and the harshness that was to come on the Burma Railway.

Some men did take the books; some did not. Joe took a very heavy tome, *The Story of Mankind* by Hendrik Van Loon,[2] and two others that he cannot remember. Looking back, he thought Van Loon a bad choice as it was so heavy, but he did take it. They made sure to keep hats and mess tins and other essentials. Led by Japanese engineers, they began their trek, the Japanese cutting through the jungle with their pangas (bladed African machete-like tool), and waded across rivers with their kit on their heads. One night they stopped at a Buddhist temple to sleep. Many found the march too difficult and Joe witnessed sad scenes of men crying at the roadside because they could not go on. The fitter helped the weaker and nobody was left behind. Joe said that the Japanese behaved well until this point.

The Railway

On the third day they arrived at Tarso camp, on the banks of the Kwai, where bamboo huts were ready and waiting. They were issued with double-handed tree saws and axes and taken to a prepared site which they had to clear for the railway. The guards communicated with them with signs and odd English words, and were harsh and noisy, but Joe did not witness any beatings or cruelty. The work was hard: eight to ten hours a day, with a rest on Sundays only. Joe said they were just kept alive because the Japanese clearly knew that they had to have workers.

After clearing the jungle, the surface had to be levelled and embankments piled up, and they often had to cut through rock, all by hand, using long steel chisels and heavy hammers – one man holding the chisel and the other banging the hammer. Then a Japanese engineer would fill the hole with gelignite, light a long fuse and everyone would run; the resulting rubble from the explosion was cleared and the process would continue. A few Australians were with them and Joe said they were great mates.

Joe also built wooden trestle bridges. As an architect he noticed that all the Japanese engineers' drawings were carved into wooden panels and they were very cleverly made. He recalls on one occasion standing forty feet above a ravine, drilling a hole with an awl to take a bolt for a trestle, and the Japanese engineer shouting at him to straighten it out – which he did. Eventually, lorry-loads of sleepers arrived – needing two men to lift each one – and they had to be laid with gravel beneath, using great sledgehammers to push down the sleepers; then the rails arrived and had to be laid. The rails were fixed by clamps hammered in with great spikes; first Joe would tap the spike and then hit it with a huge hammer into the sleeper. The Japanese checked their work and on one occasion Joe asked a Japanese engineer how they knew the line was straight. 'Japanese eye!' he replied.

The Korean guards were quite brutal but Joe himself was not beaten and only saw some men slapped occasionally. Every morning was *tenko* (roll call) and they had to number off in Japanese. In one camp they wanted Joe to design an officers' administration hut, and then the Japanese built it. He persuaded them to have the balustrade on the veranda built lower than in Europe because they were all much smaller than Europeans. Cookhouse fires were kept alight all night, and while camp life went on Joe was drawing scenes of the daily routine (see note 3).

On one occasion, Joe and about twenty POWs were taken downriver on a boat to fetch supplies and they came to some shallow rapids; they were ordered out to lift the boat over the rapids, and meanwhile enjoyed a nice swim; they camped there that night and were allowed to cook their own food and relax with one of the more pleasant Japanese guards.

There was a lot of illness and it was here that Joe caught malaria. The Japanese supplied quinine to the British doctor and the men had their mosquito nets; Joe remembers that he had kept needle and thread to repair this, as it was essential equipment. After recovering in the medical hut, Joe was ill again about two weeks later; this happened fourteen times in a row, week after week, but each time he would go back to work in between bouts of illness.

Joe said he always doubted that he would get home; he was sure he would die there, like so many around him. In one camp that he was in, there was an outbreak of cholera when Indian infected workers bathed in the water upstream and carried the disease down to his camp; but he and most of the British POWs escaped it by being ordered by their MO to boil all their eating utensils in scalding water for ten minutes before use. Nevertheless, Joe saw over 100 men die from cholera, and he says that was when he was most frightened. Occasionally an Australian army dentist visited the camp but he was only able to pull rotten teeth, which was very painful.

Joe said that generally captivity was much worse for the married men, who missed their wives and children; Joe was not married at that time. Most of the men of Joe's unit were fortunate in being together for the whole of their captivity and so good, solid friendships developed; the books that Joe had persuaded his comrades to keep with them proved invaluable and he still keeps a list of the many he read whilst a prisoner.

Although at different camps (one he noted on a drawing was Nakon Paton), he worked on the line for two years. Only once did he ever receive a Red Cross parcel, because the Japanese stole them. As time went on, Joe's hearing was impaired, due to the doses of quinine; his officers were worried that the Japanese would think he was malingering if he did not hear orders and obey at once, and so end up getting a beating. So they spoke to the Japanese, and they tied a six-inch wooden label round his neck, and on it they wrote the Japanese characters for 'Deaf'.

At one camp, he knows that the deafness saved his life; the men were ordered to parade and the Japanese commandant said that many of them were to be sent to work in Japan. Joe, however, was refused because he was deaf. Knowing now the death rate on the Hell Ships and in the Japanese work camps, it is likely that Joe might not have lived if he had been transported. Indeed, after the war he was contacted by the niece of a friend (he remembers only that her name was Myrtle) who was taken to Japan, and he learnt that the ship his friend had been on was sunk with loss of all hands, and that was almost certainly the same ship Joe would have been on with his friend.

In January 1945, the Japanese at last allowed him to move back downriver to the hospital camp at Kanchanaburi (though his diary says Nakon Paton), where he spent the last nine months of his captivity. One amusing incident occurred when a Japanese general in a splendid white uniform came to inspect the camp. As he walked around the area, the branch of a low tree took off his hat. He had serious words with a senior officer; next day the tree was gone. Joe remembers that they carved a chess set and used to play a great deal.

Liberation

Joe had kept a diary, but this was a serious offence, often punishable by death, if the Japanese had found it. He was advised at some stage to destroy it, so he cut it into shreds with a razor and then buried it in a tin under his hut. Sadly, the hut was demolished before he could retrieve it; the tin was found and it was assumed that the strips were cigarette papers – so they went up in smoke. But on the day the war ended he started the diary again from memory, starting it on 22 August 1945. The diary was written in pencil in an old Thai children's exercise book, and he cannot remember how he got it. He had kept his war drawings, however.[3] He did these to amuse himself and also to keep a record of his experiences. On one occasion the Japanese asked to see any drawings that POWs had made and Joe offered to show his. They looked through very carefully, eventually removing some that they regarded as maybe showing them in a bad light – Joe does not really know – but they allowed him to keep most of the twenty-five innocuous ones of general scenes from the camps themselves. One he showed, made at Kanchanaburi, had a drawing of the fenced-off 'psychological' hut, where men who had sadly lost their minds were housed.

On 16 August he was playing chess with an Australian friend, Sam Welsh, when prisoner Eric Angel came running over and told him that the war was over, and he should come at once. A Japanese guard also appeared and said: 'Bang-bang over!' Another friend ran over, saying that all the Koreans had disappeared. Joe could hear a lot of shouting, but he said to his friend that they should finish the game first. They made one more move and then ran out. There on the parade ground was indeed British Lieutenant Colonel H.E. Coates, RAAMC. As he mounted the stage, he was greeted with riotous cheers of 'Good old Bert!' He held up his hands for silence. 'An Armistice has been declared in all parts ...', but he could not continue for some time as the cheers increased. When everybody had settled down again, he went on to inform the men that the Japanese colonel had said that from that day, 16 August, the IJA (Imperial Japanese Army) ceased to regard them as prisoners and that the camp was a British camp under their administration; also, that all the contents of the Jap stores had been demanded and handed over for their use and they would have their own guards and pickets. The Japanese officers had handed in their swords and left the camp, and the war was indeed over. The Union Jack and Dutch flags appeared and were waved about, and the POWs spent the day walking round shaking hands with each other and saying 'Congratulations!' and screaming themselves hoarse. A large Union Jack, previously used only for funerals, now flew over the camp.

The British were running the camp but the Japanese remained outside to protect them from any renegade Japanese soldiers who might be around. That evening there was a Christian service, and a Jewish thanksgiving service which Joe attended. The diet changed drastically as all kinds of normal food were served up from that day, and hundreds of oil lamps were lit every evening in place of the three dim lamps per hut under the Japanese regime. On 31 August, at 10 a.m., three RAF Dakotas circled the camp, then one plane zoomed down the road and three bundles fell out. There was a flicker of white and three small parachutes opened and danced gently to the ground; that was the beginning. More planes arrived and, for two hours, thirteen aircraft dropped tons of supplies, from about 100 feet, of food, clothing and medicine parcels, by parachute. Joe was covered in soap lather in the bath house and ran out naked to see what was happening. The RAF despatchers were leaning out and waving. The planes first dropped a message warning everyone to clear the parade ground for the drop in case people were hurt. Clothing bales were just thrown out, always landing with great accuracy and terrifying velocity on the main road of the camp, where they bounced higher than the huts and sometimes exploded into showers of blankets and tunics. Joe described that event as 'simply terrific!'

The first item dropped contained two letters; the first said: 'To Allied POWs with compliments of 267 Squadron sergeant and airmen messes. Hoping these few things will prove useful to you and help you to pass away the time until you are home again. It will not be long. Keep smiling. Everything is OK.' (This package contained razor blades, soap, tobacco and books.) The second letter said:

> To all Allied POWs with the compliments of 247 Squadron Sgts mess. Hope this bundle will keep you going until we can get you out. It will be only a few days now, so get your strength up and we'll meet you as we hope to take you part of the way home. Keep your chins up, Everything is OK at home. See you soon, best of luck to you, 'Lads of 267'. P.S. Look out for cigarettes, may be attached to chute or without chute.

On 1 September the first 150 men were put on a train to Bangkok. Joe was passed fit for air travel and was issued with a new uniform, underwear and a blanket, fifty cigarettes, five razor blades and soap. He needed these, as in the three-and-a-half years of captivity the Japanese had given him only one pair of shorts – what he wore was in total tatters. He was placed in Party 7, a group of twenty-five men. On 2 September at 2.30 a.m. he was woken, and at 4.10 a.m. he marched out of the camp and round a corner onto a proper passenger train at the station. It seemed unbelievable.

As the bridges had been destroyed by Allied bombers, they had to cross rivers on rope bridges and pick up trains on the other side. In Bangkok they were ferried across the river to lorries which took them to the aerodrome, cheered all the way by local Thais, and slept in a concrete aircraft hanger, formerly a hospital, with proper clean beds with mattresses.

Joe slept an uninterrupted sleep for the first time in over three years. Next afternoon, a fleet of fifty-seven Dakota RAF planes took them to Rangoon in Burma on a two-and-three-quarter-hour flight; the noise was terrific but the men were elated. In Rangoon, they landed in a huge rainstorm and were taken by lorry to a reception centre, where they were given sandwiches, and hot sweet, milky coffee, and peaches and cream, served by European girls in Red Cross uniforms, all volunteers. Then they went by lorry again, to be hospitalized in wonderful, clean beds with new pyjamas, and given thorough medical checks; they also sent messages home and saw their first up-to-date newspapers.

Next evening, following the issue of new kit and pay books, they were all taken to the cinema to watch newsreels, a Walt Disney film and a musical comedy called *Bathing Beauty*; sitting there eating peanuts in safety and security, it was like a dream come true. On 7 September Joe received a vaccination, and on the 9th several old RCOS comrades arrived; it was a joyous reunion. On the 10th he visited the main hospital and found several more old POW friends. On the 12th he was transferred with about 200 others by lorry to a transit camp, and on Saturday the 15th he was taken by truck to the docks. From here he set sail on the 18th, in very crowded conditions aboard HMS *Worcestershire.*

On entering Colombo harbour in Ceylon, at dusk on 22 September, they were greeted with dancing searchlights flashing 'V' for victory; all the shipping horns were sounded in welcome and thousands of people were on the dockside, cheering them in; bands played as they disembarked for a short stay; whole groups of pretty WRENS (Women's Royal Navy Service) in smart white tropical uniforms met them and took them into town to the shopping and restaurant area for a few hours. That evening they left, and again the searchlights danced and ships hooted; a launch came alongside with a large message of white letters on black canvas, saying 'God Speed'. What a wonderful gesture that was, remembered Joe.

The journey to Egypt was pleasant and calm and Joe did some voluntary PT to get fit, as well as eating, resting and reading. On Thursday 4 October they disembarked near Suez and were welcomed by a military band, then taken by train to a camp at Ataka and issued with winter clothes; they relaxed in a NAAFI where free refreshments were available. One man sent a cable to his sweetheart, saying he was on the way home, but she replied that she had been married for three years. He was most annoyed about the 8s 4d (about 45 pence today) cost of the cable. Then it was back on board ship and across the Mediterranean to Gibraltar, receiving diplomatic lectures en route on how to readapt to life in Britain. Eventually he arrived in Liverpool in the fog on 15 October 1945.

When asked how he felt about the Japanese now, Joe said he had put all the hatred behind him and never spoke of it. Others whom he knows took on important work helping the veterans and their dependants, but he wanted to forget it all. He did not feel that the modern Japanese bore any moral resemblance to the militaristic types he had come across and so felt he could not apportion blame to ordinary Japanese people. He remains a member of the FEPOW association and has attended some reunions, although not for many years. He does, however, always keep in touch with many of his POW friends. Sergeant Crooks became a famous diabetic consultant at Glasgow University Hospital.

When the current emperor came to the UK in the 1990s, Joe was asked by a reporter if he was going to take part in the demonstration to turn backs on the emperor. Joe said he would not turn his back on the Queen, who would be in the coach.

Joe has the 1939–45 and Pacific Stars, the War Medal and the King's Loyal Service Badge. He was awarded the CBE post-war for services to London as a town planner when working at the Greater London Council. He attended AJEX parades in Whitehall until he was unable to march. In February 1946 he married Myrtle (née Mann), a school friend of his sister, who had been a lieutenant in the ATS (Auxiliary Territorial Service); he had a long career in architecture and design, and is a proud father and grandfather.

He only once returned to the Far East, to Kranji cemetery on Singapore Island, to visit the graves of his comrades and pay tribute. It was enormously emotional for him and he wept for his lost friends.

[*Interviewed by the author.*]

Notes

1. For a detailed description of the journey, see Joe's diary at the IWM.
2. Published in New York by Boni & Liveright, 1926.
3. Copies are with the IWM but he has kept the originals, some of which are in his short autobiography, and some of which he gave to the AJEX Jewish Military Museum. The IWM also has a copy of his diary, whilst he has kept the original.

Henri Hymans: A Witness to War Atrocities

This is taken from The HopeSite HomePage of the Victoria Holocaust Remembrance and Education Society, Canada. It is an excerpt from Keeping the Memory, *edited by Rhoda Kaellis (Vancouver, BC: Vancouver Holocaust Centre Society for Education and Remembrance, 1991). Henri was interviewed in the 1980s.*

Henri Hymans: Eyewitness Account of Japanese Internment

Henri Hymans is six feet tall, a handsome strong faced man in his late forties with pepper-and-salt hair. His eyes are light green-brown and fill with pain as he recounts the details of his experience in a Japanese concentration camp in Indonesia. He answers questions directly and smiles readily. When remembrance becomes painful, the smile is alternately wistful, deprecating or even grim.

He was born in Indonesia to a Dutch-Jewish family, the youngest of three children. His father had an art shop and was arranging to have Indonesian art introduced into Europe. The Nazi invasion of Holland occurred when Henri was four or five years old, but he still remembers people talking about it. In March of 1942, the Japanese occupied Indonesia. Six months later the Hymans family were rounded up and put into a concentration camp as were other Dutch, Chinese and Jewish people. Henri was six years old.

We tried to escape into the mountains. We managed for about six months; we lived in the mountains. They were picking up everybody, especially the Dutch men. My father was deaf. He couldn't hear very much. We were afraid he wouldn't salute them properly. With the Japanese, anyone who didn't obey orders was immediately taken away and put in jail. Some Chinese friends helped us to get into the mountains. We lived near a plantation. We were growing food ourselves. There was an empty house, people who had already been imprisoned. After six months the Japanese were having military exercises around there and they found us. They came from all over. The next day they took us away. By that time they had imprisoned most Dutch people. We were some of the last to be taken.

My father was taken to another camp. We were taken to a transit camp; my mother, sister, brother and I. The men were in separate slave labour camps. Children up to 10 years old were permitted to stay with their mothers. They moved us from one camp to another, wherever they needed more labour, or for all kinds of reasons. They kept marching us from one camp to another. This happened three, four times, each one worse than the previous one. My mother and sister worked fifteen hours a day, cutting trees outside the camps, cleaning up prisons, planting things for the Japanese army. My mother mostly was employed making ropes, a kind of jute rope. They were huge camps. In the beginning there were five, six thousand. Everybody had dysentery [prisoners]. Every day about fifteen, twenty people died. The young children, starting with 6 year olds, had to work from six to ten hours a day.

We slept in long barracks. I was with my mother. I slept next to her, but I hardly saw her because she was working fifteen, sixteen hours a day. After the working day she was so exhausted. There were very few Japanese guards. I'm not sure, but generally eight or ten. Anyone who came within one-and-a-half metres of the fence, they were shot immediately. And we were so scared. So many people were tortured and killed, nobody dared to go near the fence.

In the beginning, there were huge camps with a lot of Australian women and children from Java and Singapore, and there were Chinese. Then there were many Jews, Iraqi Jews. The Nazis told the Japanese to pick us up, but most Japanese didn't know what a Jew was, really, so some Jews were put together with Christian people and they got better treatment. Others were treated very badly. The Nazis used to visit from time to time. They tried to [indoctrinate the Japanese] and sometimes they succeeded and everybody was killed. We were young boys, seven or eight years old, and we ran in gangs, fighting, and we often fight with the Iraqi Jews. I had no idea. Most of the time we were weeding – very tall grass was growing there – and also all kinds of odd jobs. We had to keep the camp clean. At times we had to catch flies. The objective first of all was to humiliate us. They were very keen on humiliating anyone. Gradually they saw no use for us anymore. I don't know why they did the things they did.

They punished us, in the end almost daily. It grew worse and worse. We were not allowed to keep anything. If people were found concealing books – once they found a children's Bible – they'd beat you to death. Anytime we offended any rule they had, not only did they beat these people to death very slowly, but we had to watch; everybody, children, women, everybody. In the sun, without water, without food. Without water you go crazy. Some of these lasted one day, some, two days. Mostly we had to bow to any Japanese who came near us, and if we didn't notice it, they got furious. They very seldom hit us children, but they hit our mothers. So we did something wrong, they found our mothers and they were beaten up. We were very scared of it. I heard my mother was beaten, but I didn't see it myself. Usually all the grownups tried to keep the children away from it when it happened.

There were hundreds in each barracks. I'm not sure exactly how many. Nice people and awful people. In our barracks everybody was speaking Dutch. I think we were separated by nationality. In the beginning we usually got a little bowl of rice every day. That was it. Then we scavenged around all kinds of insects. But as the war went on, we got less and less. And then, they started giving us what they called tapioca. It's what you make glue of. There's just so much flour and the rest is water. That was it, and that we got two or three times a day. So for thirty minutes you felt quite full, and then you were just as hungry as before. A lot of women died, but I don't think that many children did. In the men's camps it was really terrible. My brother went to one camp, he had to collect all the dead bodies. Every day there were twenty, thirty people.

Once a year we got a postcard from my father. It started, 'Long live the Emperor. I am well, thanks to the Emperor. Don't worry about me.' And everybody got the same card, so, we knew there was nothing. Sometimes people had died long before the card was received. So we didn't really know what was happening to my father. My brother, we did find out [about] because sometimes there came teams of young boys for very hard work. One of the boys was my brother's friend. We shouted to each other [information].

The war lasted three years. I was in camps for two and a half years. We were told a month after Japan had capitulated because the British didn't have enough troops to send to Indonesia and we were right in the middle of central Java. By then the Indonesians had begun to rebel against the Japanese too. The Indonesians hated the Dutch, but in the end hated the Japanese more. They were hunting Japanese everywhere. So the British told the Japanese to stay where they were and they were responsible for law and order. We noticed immediately we got plenty of food from the Red Cross, parcels that had been sent for us by Australians and Americans all along. The Japanese began to release them to us finally. We began to guess very soon, because they didn't beat us up anymore or work us like crazy. We were totally obsessed by food. But the Japanese were still in command.

We were liberated by British troops from Singapore. We were able to go out of the camps. It was fantastic. We had eaten up all the grass and leaves from the trees. We hardly knew what trees were anymore, except those we saw far away. We played with kids outside of camp.

A month later some of the Indonesians began to shoot at us, and that's because the first Dutch troops arrived. The Indonesians were very afraid of the Dutch making them into a colony again. They cut off the water supply. Japanese were protecting us. They brought in more Japanese troops to protect us.

The Sikhs didn't want to fight for the British anymore, so we didn't know which side they were on. Then more Dutch troops came and the more Dutch troops came the worse it was – the fighting became worse.

The Indonesians were shooting at us with mortars into the prison camp. One day they came down from the mountains and overran the whole camp. A lot of women and children were killed. We escaped killing just by luck. The Gurka [*sic*] soldiers who were serving the British came in and chased them away.

My father had died two months before in the camp. My older brother was with us again. In the men's camp the mortality rate was about seventy percent. The British evacuated us to another location where we stayed a few weeks, and then we were shipped out to the coast to a big city, Samara, where it was fairly safe and under British control. From there we were shipped to Sri Lanka. We were there for six months and then went to Holland. It was miraculous. My mother had nine brothers and sisters and only one was taken, and her mother and father were taken to Bergen-Belsen. My father's family, also a big family, only one brother was left. For two years we went all over the place, moving from one vacant house to another as the former occupants returned from wherever.

I think I didn't know until after the war that I was Jewish, because my family was afraid the Germans might take over Indonesia as they did Indochina. So from the age of 4, they kept the fact that we were Jewish secret.

By the time I was in high school we finally settled down in one place. I kept going back to the Far East, but I studied in Canada and Europe. I went to Japan in 1962. I react strongly to the militaristic, or fascistic, in Japan; I hate them. There are very nationalistic, very anti-Semitic [individuals] in the new Japan. The good thing I got out of Japan was my wife. There are many things I like about Japan, but many things I don't trust.

The young Japanese, this generation in Japan know nothing about what the Japanese did in the Second World War. They deny that anything like that could have been done by the Japanese. When I told some young law students about this, they never came back to school again in protest over my 'anti-Japanese propaganda'. The leadership keep drumming that Japan is top in economics but that they won't be respected until Japan is also top in military. I think they want to be a big military power.

All of this [experience] messed up my life, especially in the last ten years. I had the first what they call, concentration camp syndrome [Post Traumatic Stress Disorder] treatment, when I was just 40. In Holland there's a special treatment centre, and there are many, same thing happened. They often began to show signs of stress, difficulty in sleeping, or they began to work like crazy and they became very irritable. Then they recover. It goes well again for a short time. It's happened to me twice since I was forty. And I can't sleep. I dream all the time I'm either being beaten or I'm beating – fighting. I thought coming here to Victoria, it would be the end of it, but I can't get away from it. Sometimes it's really bad.

I have very bad memory, short term memory. It's becoming more and more difficult for me to study new things. It's caused my wife a lot of trouble. It's a shame. Some people are extremely tough. My brother was in the worse situation, but he's doing very well. My sister has some terrible things she doesn't like to talk about. She never married. She's in very bad health, physically and mentally. She worked in places where the Japanese Gestapo tortured people. She was just a teenage girl at that time. Very tough. My 15-year-old son also suffered quite a bit from my situation. He hasn't had much stability in his life. He was born in Norway. When he was 6 we were in Holland for a year and a half. Then back to Japan. He said the Japanese should pay for my pension.

He doesn't know much about what happened to me. I don't think our children should suffer or be affected anymore. With our generation, it should stop. If he would turn anti-Japanese it would make the situation for my wife extremely difficult. Someday when he gets rid of a lot of the garbage, I'll want him to know the real historical facts, but I want him to feel really Canadian first.

184947/184749 Capt. Morris 'Sonny'/ 'Jan' Janis, RASC, 55th Infantry Brigade, 18th Division

34a. 184947/184749 Capt. Morris Janis, RASC (standing, right), as a sergeant before being commissioned.

Morris was born in Mile End, London, in the heart of the Jewish ghetto, on 2 October 1913, son of Gershon Jacobs Koslowski and Alice (née Barnet). Alice was part of a family famous for their kosher smoked salmon business in Frying Pan Alley, Whitechapel. The Koslowskis later changed their name to Janis.[1] Morris won a scholarship to a famous grammar school but had to leave at the age of 14, and worked in the family smoked salmon factory. The family moved to Stamford Hill, London N16, before the war, and later to 24 St Margaret's Road, Edgware. Morris joined the army on the outbreak of war, on 18 September 1939, and was posted to No. 2 Battalion, RASC, at Bulford on Salisbury Plain. By May 1941 he was a 2nd lieutenant and posted to Crosshall, near Morley, in Yorkshire, becoming a captain by the time of his posting to the Far East. He sailed from Gourock in November 1941 to Cape Town via Halifax, Nova Scotia, and then on to Ahmedneggar in India where they spent three weeks.

On landing in Singapore his unit fought its way across the island with the Northumberland Fusiliers, fighting at Hill 85 on the north-east coast of the island for two days. After the surrender, Morris was first sent to Changi, where he experienced with others the utter dejection of the surrender. This was relieved somewhat by the re-establishment of discipline and by a 'university' for POWs (described in Alexander Oppenheim's testimony later in this book). He was then transferred to Sime Road camp as a messing officer and administrator; he worked on building the Japanese memorial at Bukit Timah. A friendly Japanese sergeant gave him a Japanese–English dictionary which Morris used to teach himself Japanese, with the help of a Lieutenant Davidge of the Malayan volunteers.[2] As a result, Morris was able to reduce misunderstandings in the giving of Japanese orders and thus reduce incidents and so punishments; also, he could explain the

POWs needs to the commandant, Colonel Tamura, who was quite well disposed towards them and warned, for example, when a Kempeitai search was about to be made of the camp.

One day the Reverend Leonard Wilson, Bishop of Singapore, arrived to inspect the camp and have dinner with the officers. Morris arrived late, somewhat tipsy from having drunk a lot of saki after negotiating an issue with the Japanese engineers, and apologized to the bishop, but His Grace replied, 'Dear fellow, I haven't tasted the stuff for three months; please come close and breathe over me.'

Thailand

On 2 October 1942, with the memorial finished, Morris and comrades were moved, via Singapore railway station, in steel trucks in appalling conditions on a five-day journey to Bampong in Thailand. The camp was flooded with monsoon rain and floating in sewage but after two days in this squalor they were moved on a thirty-mile march to Tamarkan camp.

Morris arrived at Tamarkan in November 1942, and took part for six months in the building of the wooden Kwai bridge – and the concrete piers of the later bridge about 100 metres upstream – where the SBO was Lieutenant Colonel Toosey. Morris suffered all the terrible privations and diseases described in many of the other testimonies. Morris remembered the execution of two officers and four men who had escaped but been recaptured by the Japanese; it is alleged that the Japanese officer in charge, Lieutenant Shinji Takasaki (nicknamed 'The Frog'), tried to have the sentence commuted, but failed.[3]

In May 1943 the move was made to Kinsyok; this was a poor camp with huge fatalities due to disease, especially cholera. The commandant was 'a fat, idle oaf called Lieutenant Suzuki' who cared little for the prisoners. Morris developed a septic foot and was carried by his men from place to place in order to discharge his duties.[4] Morris was able to obtain amounts of rice polishings – a potent source of vitamin B – and 112lbs of sodium sulphate and quinine for the POWs, from the camp's Japanese administration, thereby saving many lives; he remonstrated with the commandant and many semi-fit men were not sent to work as a result.

Then orders came in August 1943 to move to another camp, and Suzuki forced the already exhausted and starving men to march thirty miles; thus, on arrival, 400 of the men were too sick to start work.

At the new camp, Konkoita (on the Thai-Burma border), Morris had by then learnt Japanese well and become the Allied interpreter, and as the SBO was ill, Morris was also the acting British CO of the camp. Lieutenant Takasaki now took over from Suzuki as the new commandant, and it was he who told Morris about the 'speedo' required on the railway and how desperate the Japanese were to finish the building. Conditions were awful, illness rife, accommodation virtually non-existent and the monsoon at its worst as sick men were forced to work, with many dying as a result. There was little that Morris could do except plead for the worst cases, to keep them off work. He also managed to get fifty sick men transferred down to Kanchanaburi hospital camp, thus saving their lives. Takasaki apparently remonstrated with the Japanese engineers driving the labour force to extremes, even punishing Korean guards for their brutality to the men and obtaining extra medicines for the sick; when Morris told Takasaki the men needed beans in their rice to give them vitamin B, he obtained many sacks for him and the beri-beri disappeared within days. There was a reasonable hospital hut at Konkoita, organized by Morris, and 100 cattle were herded by twenty fit POWs down to the camp, thirty miles from Kinsyok – a most valuable food source, one being butchered every second day. But the Japanese took most of the meat, the POWs getting only the bones and some offal, for soup. Hygiene was improved by the covering of latrines with lids, which reduced the number of flies. The Korean guards were vicious but one, Okamoto, who had been born in Hawaii, was less so. It was while at this camp that Morris and his comrades witnessed many train-loads of wounded Japanese travelling south from battles with the Allies.

While Takasaki had to obey orders, Morris regarded him as an honourable man who did what he could for the POWs. Although forty of the 800 POWs at Konkoita died, mostly from cholera and other diseases, not one died from any form of brutality, according to Morris – although the Japanese were responsible for those illnesses that killed those men. But it was the executions at Tamarkan – the men were bayoneted to death – that led to Takasaki's execution after the war at Changi gaol.

After Konkoita, when the railway was finished, the camp was closed and the men were taken by truck to an assembly point at Hindato in March 1944. Takasaki and Janis stood by the track until all the men had boarded the trucks; Morris told Takasaki he had been a fair and good officer. He told Morris he had been a policeman in Tokyo and hoped Morris would visit after the war and be a friend, and that his family had two swords and he would like to present one to Morris as a token of his esteem.

There were 3,000 British and 3,000 Dutch assembled here, and Morris was the interpreter for the British. Takasaki told him that ten officers and 1,000 men were to be sent to Japan; Morris asked that he should be allowed to pick the men, and Takasaki agreed. They departed for Japan and that was the last Morris saw of the officer. Takasaki was on Morris's conscience after the war and he always wanted to find his family and tell them he had been a good man who had not deserved his death sentence.

Japan and the Diary

In May 1944, Morris was transported with about 100 other prisoners back to Singapore by train, to Havelock camp via Non Pladuk, which was in an appalling state, and then on the *Kioki Maru* ship to Japan. The ship was a filthy old coaler fitted with shelves as beds and extremely overcrowded, but the seamen were friendly and lifejackets were issued, rarely the case on the Hell Ships. Morris approached the Japanese commander, who agreed that food and water would be adequately supplied, and organized matters so that two thirds of the POWs were to sleep on deck in the fresh air, near the latrines (the usual planks-and-ropes device hanging over the ship's side), and only one third below, in a rota. Again this saved lives and huge discomfort, as squalor caused disease and that invariably meant death. A Korean guard stole the sugar ration but the Japanese quartermaster got some of it back for them later. The Japanese radio officer very unusually allowed them to hear the news that Germany was losing the war, a great boost to morale for all the men. There was a submarine attack and other ships in the convoy were sunk.

After stops at Manila (where they heard about D-Day) and Taiwan, on 26 June 1944 they arrived at Moji in Japan and were taken to their new camp. Morris was SBO at Kaugoe/Kanose, a small village fifteen miles east of Nigota, where there were 100 British and about fifty American POWs, taken at Bataan; work was in the calcium carbide plant adjoining the camp. Except for three men who died in a work accident, no others died;

the commandant, Lieutenant Aoki, had been a schoolteacher and was a kindly man who did what he could for the POWs. By maintaining discipline and caring for each other, the men survived until the end of the war. Aoki was given eight years for the deaths in the plant after the war, but Morris testified by affidavit at his trial, asking for clemency and the sentence was reduced to four years – as Aoki only supplied the labour, he did not govern the work in the factory.

Morris kept a diary while in Japan; most of it is simply his thoughts about life in general and what he planned to do after the war. But occasionally there is mention of everyday incidents which give some insight into camp life. The Japanese here had no objection to the keeping of notes, unlike in other camps, and even supplied notebooks for sale in the canteen, which men could purchase with their small earnings if they wished.

On 7 October, Morris describes the increasingly cold weather and mentions how this is good for hygiene because the flies are beginning to die off. Also, a batch of old American newspapers arrived for the men to read. On 8 October 1944 he describes how the Japanese allowed in a batch of pears for sale to the POWs, something that was greeted with great joy, and how some of the officers were very short-tempered and quarrelsome in their social relations with each other, confined together as they were – hungry, cold and bitter. On 12 October he describes a journey, with other POWs under guard, by boat to collect firewood for the camp, and the beauty of the surrounding countryside, the forests and hills, vegetable gardens and rice paddy terraces. A group of schoolchildren brought them berries after Morris engaged them in friendly chat in Japanese. They appeared undernourished and he was moved by their plight and what the future held for them. On some of these trips they were able to collect wild vegetables and fruits and spoke to local farmers; the guard even gave them sweets.

On 14 October 1944 he writes about the pleasure of being given a horse bone to gnaw on after the cooks had extracted just about every ounce of goodness in it for the soup; with a little vegetables and spice, this soup was most welcome as the weather cooled considerably. On 16 October he describes how 'Hewitt said that cat meat was delicious, but I have no desire to taste it.' Hewitt appears to have been a frustrated chef, as later Morris describes his invention of a delicious yeast made from flour, left to mature in clean earthenware jars covered in muslin. But, he says, they are all hungry, all the time, and greatly underweight. Rats compete with them for food if anything is left uncovered. On 30 October the Japanese allowed in some apples and persimmons, some of which were stored to make wine for Christmas, for the first time, and Morris notes the beautiful turning of the leaves of a Japanese autumn. He mentions the importance of the end-of-month accounts whereby each man's balance is very carefully calculated (the men received tiny amounts of pay from the Japanese to buy some little luxuries now and again).

Morris noticed how important it was to keep the less fit and depressed men off work and in the camp to chat and comfort each other; engaging them in conversation was excessively important for the morale of this group, for whom hope meant the difference between living and carrying on, or turning away and dying.

On 16 November, British Army greatcoats arrived – one per man –and were most welcome, as the weather had now become very cold indeed. On 18 November they were cooking grasshoppers to add to their rice as a protein supplement; Morris wrote: 'not a pleasant dish but I grill mine and that largely kills the taste'. Post arrived but the letters were two years old.

34b. Morris in the camp at Kanose, Japan, east of Nigota (with glasses, standing, distant left background. Japanese Commandant Lt Ayume is standing, far centre, at the end of the table, with cap and glasses).

On 26 November the camp was inspected by the colonel commanding the Tokyo area POW camps; such inspections caused a huge increase in work because everything had to be cleaned and tidied in both the camp and the factory; it even included the performance of a pantomime! By 9 December there were six inches of snow on the ground and several men down with pneumonia, but then on 10 December each man suddenly received a Red Cross parcel to himself[5] – something almost unheard of. The Japanese said that another would be given at Christmas and another in the middle of the winter cold, in February (the last time was in Thailand when they received one box between seven men). As a result, each man was asked to donate two cigarettes and the total was taken to Lieutenant Ayume,[6] the camp commandant, as a gift for his help; at first he refused but then said he would accept if they agreed he could share them with all the Japanese staff. Meanwhile, lots of trading took place as men swapped what they did not like in the parcel for what they wanted more of.

The men were able to bathe every day in the factory communal bath, an oval concrete basin with a sitting ledge and hot and cold running water, which took twenty men at a time; after over two years of river bathing, it was luxury. The British doctor told Morris that he would be mentioning in his report after the war about how Morris had been so helpful to the men. (Nothing was to come of this, however – see Morris's IWM archive.)

On 15 December, Morris went with two American POWs to collect some horse blood from a village some six kilometres away, in two large containers on a large cart. Their guard, Lieutenant Aoki, went off to visit his home and the men wandered into the village. However, at the slaughterhouse the blood was not available, and instead they were given some offal – a huge horse stomach and intestines; Morris also asked for and obtained twelve horses' feet. These were well cooked and provided delicious food for the men in the camp. One night a dog was spotted in the camp and two men went after it with murderous intent – but it got away.

Morris managed to get a great deal out of Ayume for the men, and in return he cultivated him by giving him the English lessons he so craved. But just before Christmas 1944, Ayume suddenly declared that someone had impugned his reputation about the handling of the Red Cross boxes and he decided to punish the camp by cancelling the Christmas arrangements; wood supplies for heating stopped and rice rations were reduced. He then backtracked and cancelled all the punishments after five days; Christmas was on again and everyone enjoyed the day – except that it was work as normal until 4 p.m! Each section had a Christmas tree, decorated with coloured paper and tins and card, and a Red Cross parcel was given out to each man on Christmas Eve. A church service was not well attended because everyone was busy trading!

At 4.30 p.m., the commandant made presentations to good workers at a compulsory assembly and gave a good speech with a translation prepared by Morris. The officers' meal was served at tables with sheets used as tablecloths, and leaves and card used as decorations. As well as the meal, each man had two oranges, a pear, ten biscuits and a sweet cookie; with the contents of the Red Cross boxes on the table, it looked quite festive. Ayume was photographed with each group and some of these astonishing pictures survive in Morris's file at the AJEX Jewish Military Museum and the Imperial War Museum. Ayume had coffee with them and he shared some chocolate cake that the men had made. Morris wore a tie for the first time in three years. Then there was a pantomime in two parts – *Babes in the Wood* and *Aladdin*. There was a Christmas draw and over 1,000 tickets were sold.

By now the snow was very heavy and drifts up to nine feet deep covered the camp, so it was necessary to walk between snow trenches to go from one place to another, but Morris noted how beautiful the surrounding countryside looked in the snow.

On 11 January 1945, Morris wrote that he had contracted flu. After recovering, he wrote a recommendation for Lance Bombardier Speller, commending him for his devotion to duty as a medical orderly. Throughout January snow fell very heavily but food and fuel were adequate. In the factory a large number of attractive young Japanese women were employed in labouring, and the men enjoyed playful banter with them. Work in the factory was now very easy, but in late January a Japanese worker was killed in an accident and two POWs were injured. Morris had to go to the enquiry as interpreter and give the version of what his two men saw. On 2 March, Morris wrote that some Americans had been racketeering and stealing food from the cookhouse; some were punished by imprisonment in the guardroom by the Japanese or the Allied CO, but they were, he said, incorrigible. Later some Red Cross American clothing was distributed, but who was to get what was the cause of some arguments.

On 9 March, two POWs were sadly killed and two injured in an accident at the factory when molten carbide accidentally fell from a container.[7] The Japanese provided full and proper medical care and food for the injured men, and later a cremation service with coffins and a Buddhist burial service with readings in English by a POW, named Chater, for the two fatalities. It was very emotional and the Japanese were very respectful. Thirty men were allowed off work to form the burial party and a full report was written up for the families of the men.

It was a 10-kilometre walk to the crematorium along river banks and through snow-filled woods, paddy fields and villages en route. But the weather was bright and warm and the third Red Cross parcel was given out as promised. Sadly, on 28 March, one of the injured men, Oliver Crowdell from Kilworth, Leicestershire, died of his burns, and he was cremated on 29 March.

Then the snow began to melt and Morris wrote on 9 April that work had begun on the gardens, but the downside was that flies and lice were back in force as the weather warmed up. On Friday 13 April, Morris saw the first B29 bombers pass overhead. As it was the first anniversary of the camp opening, there was a meat-filled stew for supper, followed by a huge and well-produced musical show.

A lot of work was put into ARP (Air Raid Precautions) because of the high chance of Allied bombing: oil drums were dug into the ground and filled with water; sandbags, buckets, blackout screens and light shades were prepared; and, of course, many trenches were dug.

The weight of many men was increasing with a better diet, and Morris himself was also exercising more and said how fit he felt.

On 28 April the camp commandant announced that officers would now work in the carbide factory and Morris did his stint. The work was hard and dangerous, with obsolete and faulty furnaces.

Then on 1 May the men were all called to the parade ground for an announcement by Lieutenant Azual (Azuma?). He said that in the event of air raids or food shortages, the local population might become angry and try to attack the POWs, and so, in order to protect them, the camp would receive more troops soon, as guards. Meanwhile, they were to behave very properly to local people when at work, and stick together and return quickly to camp if there were incidents or raids. Afterwards, Morris and the other officers were called to a private meeting with the commandant and they explained that they would, of course, cooperate with the Japanese, but if attacked they would defend themselves. This was understood and agreed by Azual/Azuma.

On 2 May there was a search, and all knives were confiscated and razors had to be handed in at night and were given out again next day. In early June, another half of a Red Cross parcel was given out to the prisoners and also a large quantity of new clothing, followed by a glut of fresh fish. Much of this was dried and eaten later, so as not to waste it. Morris was moved to garden work, which provided more nourishing food and helped to keep sickness at bay. A draft of ninety Dutch POWs arrived and, after some friction between them and the British and the Aussies, they soon integrated well.

The warm weather brought beautiful sunny days, and flowers covered the hills; spirits rose among the men, but now fleas and lice became a problem as temperatures rose.

The Dutch lent their books out, and classes and discussions began. Morris gave talks on citizenship. The July rains brought rapid growth in the gardens and more food for the table. A guard lent the prisoners a gramophone and European records, which circulated among the huts.

By late July, air raids were increasing and on 30 July a large Allied bomb landed near the camp and factory and shattered a lot of windows, as well as in the nearby village of Kanoye. The air raid shelters were improved as a result.

On 11 August, two POWs were arrested for stealing rice from the Japanese store, and Morris was ordered to the guardroom to witness their interrogation. One, Ramshaw, was lying unconscious with his hands tied behind him. Next day he was questioned again but denied the crime; he was knocked about a bit and then two days afterwards, two Kempeitai were brought in to question him, kneeling him down and threatening him with a naked sword blade. Morris was allowed to take food to him and notes from him to his friends. Phillips, the accomplice, was persuaded by Morris to confess and was let off; Ramshaw got away with guardhouse imprisonment. It was a lucky escape for them both.

On 15 August, suddenly all the POWs were brought back to the camp in the middle of the working day. The war was over. Food was released en masse from the stores and everyone ate heartily. By 30 August, the camp was more or less being run by the POWs while the guards protected the men from any Japanese vigilantes. (The diary ends here.)

Morris got home via Manila and Australia, sailing for part of the journey on the RMS *Empress of Australia*.

Morris's experience was such that he did not favour enmity towards the Japanese after the war and declared in a letter to *The Times* that confronting the Japanese emperor when he visited Britain was 'a stupid, theatrical gesture' – many, of course, would have disagreed with him on this point. He was a slightly controversial figure because he testified via affidavits for clemency towards some Japanese, at their war crimes trials, who he said had helped the POWs as best they could, as opposed to others who were genuinely sadistic. Some Japanese families of these men wrote to Morris after the war and thanked him for his understanding, and for the reduction of some sentences as a result of his intervention. He openly contested the evidence of some Allied ranks about which guards were bad, saying that many prisoners confused names, faces and even locations.

After the war, in 1946, he married Marta Metz, an Estonian Jewish refugee, and they had a daughter, Christine, who worked as a palaeontologist in America for many years. Morris kept in contact with many of his FEPOW friends through the Old Comrades Associations, and although he had few connections with the Jewish community for some time, in his last few years he sought out and met the local rabbi at Maidenhead synagogue, Jonathan Romain.

The AJEX and IWM archives contain many emotional letters received by Morris after the war from men whose lives he had saved, thanking him for his leadership and negotiating skills and his courage and patience in dealing with them and with the Japanese. Captain David Christison, RAMC, read at his funeral oration:

> With Morris as interpreter, organiser and disciplinarian, we realised how he had been so valuable to us. As officers it was our duty to see that as many men as possible returned home, even if in poor health, and this Morris performed admirably. Some officers thought Morris was 'Jap happy' because he could communicate with the Japanese, and I am sure this jealousy prevented Morris from receiving recognition [a decoration] by the War Office. We all owe Morris a debt of gratitude ...our camps had fewer deaths, more food ... his greatest concern was for the well being of those under his command and he used his skills to our advantage ... we shall remember him.

Morris was awarded the 1939–45 and Pacific Stars.

Notes

1. Telephone conversation with Christine Janis, Morris's only child, in Rhode Island, 28 March 2011, and communication with his niece Joy, by email.
2. Oration by Fellow POW Dr David Christison, RAMC, at Morris's funeral in Slough in November 2001.
3. Unpublished letter to *The Times* by Morris, 11 May 1998, in AJEX archives.
4. See IWM Archives on Morris Janis.
5. Morris's testimony is the only one I have read which gives the contents of a typical Red Cross parcel – probably American – and I include it here for the interest of readers: '7 packs 20 Camel cigarettes, two 4 oz bars of chocolate, 2 pkts Wrigley spearmint gum, one 8 oz tin Red Poppy salmon, one 2 oz jar Barrington Hall coffee, two 12 oz tins Peer spiced luncheon meat, 12 tablets Vitamin C, 1 lb of prunes, 8 oz Jack Frost sugar, 8 oz Borden cheese, four 4 oz packs butter, one 12 oz tin corned beef, one 6 oz jar Schinwell grape jam, 1 lb Powdered milk, 2 bars soap, one 6 oz Rose mill pate.'
6. There is some confusion in Morris's story as to whether this is just a misspelling of Lt Azuma.
7. John F. Buchan from Brentford and John Foster from Dundee.

1797718 L/Bdr Reuben 'Ben' Kandler, RA, 3rd HAA, 11th Battalion

Much of the detail in this testimony is from the splendid book by Reuben's son, Richard, The Prisoner List *(Marsworth, 2010), which was the result of several tape-recorded interviews he had with his father when Reuben was 75 years old. Reuben took some persuading to do this, but the outcome is an astonishing memoir, summarized here, with some other extraneous material, and reproduced with Richard's kind permission.*

35a. 1797718 L/Bdr Reuben Kandler, RA.

Reuben was born on 30 March 1916 in the East End, at 146 The Highway, London E1; he later lived at 43 The Drive, Edgware. He was the fifth of ten children of Jewish tailor immigrants from Galicia, Gershon and Pessel (née Kook).[1] He won a scholarship to Central Foundation Boys school in the city, where he excelled in science. He went on to study for accountancy and completed his studies in 1938, having a won a whole array of national accountancy awards. He had attended the Battle of Cable Street in 1936 to help prevent Mosley's fascists marching through the Jewish areas of the East End. With two partners he set up a business and agreed that when war broke out, two would enlist and the one refused on medical grounds would hold the fort; so it was to be, and they stayed in partnership for decades after the war.

Reuben enlisted on 25 May 1941 and was posted to Devonport to work on radar. He soon became a senior operator. He was then posted to Singapore, sailing from Liverpool on the SS *Dominion Monarch* via the Cape and Ceylon and arriving about 28 November 1941. He was billeted at Changi barracks and described how Chinese barbers would enter at night and gently lift the mosquito nets and shave the men in their sleep! It was regular PT and drill most days on the parade ground; then on 8 December the Japanese bombed Singapore. Reuben's unit of radar operators was posted to a rubber plantation just across the Causeway near Johore Bharu, with a direct link to Fort Canning HQ on the island, to warn of approaching enemy aircraft. Two nearby guns were ordered north but mock-ups were built to fool any spies who might be working on the plantation. As it was essential that the radar should not fall into enemy hands, Reuben's unit was withdrawn back to Changi as the Japanese advanced and then to the reservoirs to protect the water supply. Finally, as the enemy closed in, Reuben was moved to the Anson Road football stadium in the south near the harbour, to protect the sea lanes evacuating refugees and military personnel. It was decided that radar operators were to be evacuated as essential personnel.

An outrageous piece of army bureaucracy, however, doomed Reuben and his comrades to captivity; they would not be allowed onto a ship unless they retrieved their greatcoats – totally useless in the tropics – from Changi barracks at the other side of the island. They made the twenty-five-mile trek and simply could not locate the coats in an almost deserted barracks; on returning to the harbour, they found that the last ship had gone. Reuben expressed some amusement – rather than anger – in recollecting this tale, but when his son asked why, he replied, 'Most of those ships were sunk.'

Meanwhile, the football field, now with its two protective guns returned from the mainland, was being targeted by Japanese planes and one scored a direct hit on one of the guns, blowing all nine gunners to pieces. Reuben and his team buried the men that night; one was 2nd Lieutenant Youle, whose mother had been an acquaintance of Reuben's mother in Edgware.

Singapore continued to be relentlessly bombed and Reuben witnessed the destruction and chaos and the many dead bodies lying everywhere. Then came the order from an officer to destroy the radar and all secret documents associated with it; soon the news came of the surrender and the men received orders to march to Changi barracks. It was 15 February 1942 and Reuben was a POW.

Changi

Many locals had white sheets hanging from windows and were jeering the Allied prisoners as they marched by. The barracks, now a prison, was heaving with thousands of men who had to sleep in corridors and outdoors. After a few days the men were ordered to line the road to Singapore City while the Japanese High Command drove by in a victory parade. Later Reuben heard first-hand accounts, from the working parties sent to the city, of the beheading of hundreds of Chinese, and their heads displayed on spikes all around the city; they also heard the machine guns on the nearby beaches where hundreds more were being executed.

Saigon

On 4 April, Reuben was transported with 1,000 men and two RAMC doctors on the Hell Ship *Nisshu Maru* to Saigon. The former coal ship was filthy, and shelves – for beds – had been fitted in the holds to accommodate the men in the most crowded of conditions; it was stiflingly hot. Only one meal a day – a cup-sized ball of rice with a tiny amount of vegetable and meat, with some green tea – was given to them. Most were able to wash their hands once a day from one tap. Toilets were wooden boxes hung over the side at deck level and these, when washed down with seawater several times a day, would splash the contents everywhere in the wind. At night a metal drum toilet was put in the hold – but finding it in the dark was impossible and the smell was appalling. After six days they arrived in Saigon but almost half the men had dysentery by then.

Their camp consisted of four stable-like huts with sleeping shelves on each long side, and a narrow gangway in between. Ramshackle toilet huts were few and filthy and the overcrowding was horrendous. There were also officers' huts, an isolated hospital hut and Japanese administration huts with an open parade square, and the whole was surrounded by a barbed-wire fence. Those with dysentery went to the hospital, but most could not be fitted in and so stayed in the other huts and had to wait until the infection passed through them.

The group was to be put to work in the docks; here, Reuben and his comrades carried out sabotage whenever they could, such as spilling oil and petrol drums by loosening the stoppers, putting sand in oil, 'accidentally' dropping ammunition boxes into the sea, stealing tinned foods and medicines for the camp hospital and deliberately mishandling and damaging bulk goods being loaded for the Japanese war effort. Hoping that the local French colonists would help them, the POWs made contact with those of them working in the dock area, but sadly most were pro-Nazi Vichy French and refused to have any contact with the Allied prisoners. Some, however, were sympathetic and occasionally would throw packs of cigarettes over the fence when passing the camp.

The camp routine was 6 a.m. reveille, distribution in mess tins of a tiny amount of a rice and tea breakfast, then a *tenko* (roll call) on the square, with perhaps an announcement by the Japanese officer in charge afterwards; then the work parties would be marched off. Most of the work was loading and unloading ships and stacking in warehouses. Many men, including Reuben, were injured falling from these stacks which could be up to twenty-five feet high. Then it was back to camp for much the same meal for lunch, then back for the afternoon shift at the docks until 6 p.m. and a similar supper. Finally came evening *tenko* and free time, when most washed in the murky tap water, then went to sleep hungry. After ten days there was one day off. Quite often the men had to do extra night shifts and not finish until 2 a.m. – and then start again at 6 a.m.

On one occasion there was an attempted escape by two POWs; they were caught, beaten and exhibited, tied up, in front of the guardhouse for all to see; they were then sent to Cholon gaol in Saigon. Six weeks later, two new unmarked graves were seen in the cemetery where POWs who died of illness were buried; the Japanese commandant confirmed to the SBO, Lieutenant Colonel Hugonin, that it was the two escapers. They had been shot or beheaded. But one man, a part-black or black South African, Basil Bancroft, went out quite often to Saigon, to meet sympathetic French nationals and he even had a few girlfriends, but he was always back by dawn and brought food and medicines for the sick. On one occasion he was out for three nights and the men had a nightmare covering for him at *tenko*. When he did get back he was told either to stay out overnight only, or to stay in completely. One day he disappeared and was never seen again.[2]

Reuben's testimony, in common with many others, confirmed that the Korean guards were especially brutal and took every opportunity to beat prisoners if, for example, they did not bow low enough to them or work fast enough. For their labours, prisoners received a tiny wage but could only use it in the camp shop which was run at inflated prices by the Japanese guards, who thus profited from the slave labour scheme. The result was that the money was virtually useless, though better than nothing. But local Vietnamese were able to penetrate the camp secretly and often swapped food for items such as shirts, shorts and pens from POWs; the Japanese began to suspect what was going on and so an order was issued that all valuables were to be handed in to the guardhouse. Reuben had a treasured gold ring given to him by his father and so he bored a hole in the wooden handle of his shaving brush and hid the ring inside, packed with earth so it would not rattle, and resealed the bottom with the original zinc bottom cap. One prisoner, a former barber, used an old kitchen knife, highly sharpened on a leather strop, to give shaves to POWs as his full-time job; this way Reuben was able to have a shave every few days.

As time went on, clothes wore out totally, or were sold for food, and men began to develop all the diseases of malnutrition and the tropical climate. One local Frenchman, Emile Lienard (nicknamed Sparks because he was in charge of the Saigon electricity supply) was sympathetic to the POWs and made contact with them at the docks to pass on news about the progress of the war, but, fearing he would be found out, with men crowding around him, Colonel Hugonin forbade anyone to talk to him except Reuben and another man, Jack Bunston. Sparks also collected money from his friends to buy clothes, food and medicines which were smuggled into the camp. On one occasion he left a huge pile of shorts at a prearranged venue at the docks and the men had to wear several pairs each to smuggle them back to camp.

One very important job Sparks did was to obtain a roll of all the POWs and their addresses and to have it smuggled out of Saigon by the local Swiss consul, to London. In late September 1942, Reuben's mother at last found out that he was alive.

Although Sparks was found out in the end, the local French police commander turned a blind eye to his activities and his work continued. After the war he was awarded the King's Commendation for Brave Conduct.

When a ship loaded with captured British weapons from Singapore arrived, Colonel Hugonin, who monitored the dock area, refused to allow the men to unload it, as this was against the Geneva Convention. The men marched back to camp and the Japanese commandant arrived and threatened to have all the men shot, bringing machine guns to surround the square. The colonel relented, but when the men returned to camp that evening they saw that two tiny mud huts were in the parade ground and Hugonin and his adjutant, Captain Faraday, had been walled inside as a punishment for standing up to the Japanese. Later Hugonin acquired a stick to walk with – rather like an officer's swagger stick – and whenever weapons appeared on the dockside he would surreptitiously damage them as much as he could without being seen. Some weeks later came the infamous 'no escape' order from Japanese High Command (mentioned in previous testimonies). Like the men at Changi, the POWs were asked to sign a declaration that they would not try to escape; this was illegal and so Hugonin refused, and again the two men were imprisoned, this time in a specially built cell with no food. Eventually the POWs had to give way, as at Changi, but only after it was noted that the signing was under duress, and Hugonin and Faraday were released.

Anyone caught stealing within the camp was punished by Hugonin, but taking from the Japanese (known as 'ponging') was encouraged. A crate might be deliberately dropped in the guise of an accident, at the docks, to see what it contained and items secreted away to the camp, especially food or medicines. On one occasion Reuben found a roll of film; when he got it back to the camp, he saw that it contained the record of the British surrender at Singapore by General Percival, a few months before. He was astounded. He hid it in order to bring it home to the UK.

In the Japanese fish factory, men would cut off fish-heads, while preparing the rest of the fish for sending to the Japanese front lines, and put them in their pockets to take back to the cookhouse. This worked for some time, but one day the guards were seen to be searching POWs leaving the factory; suddenly the air was filled with flying fish-heads being thrown out of pockets. The guards were furious and beat them – this was known as 'getting a chilli', because their ears went red from the slaps they received. On other occasions, men were beaten with bamboo sticks or made to stand in the sun with hands raised, or had their pay stopped, and mass punishment was used, when a whole working party would be punished for the pilfering of much-needed food by perhaps only one man. On one occasion, when two men stole biscuits at the docks, they were tied in the kneeling position so that the movement of their feet pulled on their throats, then flogged with a sword scabbard and kicked, and had cigarette stubs crushed on their skin. After being made to stand in the sun, they were marched to the camp and given ten days tied outside with no food. Hugonin protested, but what saved them was a Japanese guard who secretly fed them at night.

Keeping Busy

Reuben knew that the Jewish New Year (in October 1942) was approaching and so he went to speak to the padre with another Jewish POW, Henry Greenwood, and the padre arranged for them to be given the day off and the time made up later. The rice paper in the men's army Bibles was used to roll cigarettes, but for many men this did not detract from their faith, and regular church services were well attended. Reuben often listened from the sidelines and was able to empathize with the sections which involved the Old Testament (Jewish Bible or Torah).

The Japanese commandant allowed a big Christmas service in 1942 and a day off; a store of tinned meat, kept back after Singapore for this purpose, was served up by the cookhouse. Colonel Hugonin made a speech, standing on an earthen mound near the square, saying that it was to be made into a new benjo (toilet) for the camp; then he added, 'I am standing on a new benjo surrounded by a lot more benjos', referring to the guards around him; there was an outburst of laughter from the men, which of course went over the heads of the Japanese.

To counter boredom and its effect on morale, Hugonin said that evening classes should be established in the camp, and Reuben set up a successful class in accountancy. One evening, guards burst into the class because permission had not been granted, and the class was stopped, but secretly Reuben continued with three officers for the rest of the time in Saigon.

Concert parties were organized – one officer made an amazing violin, others sang beautifully, and Reuben especially remembers a Sergeant Kelly singing 'Jerusalem' – and there were several drag shows. Once this backfired, however, when a drunk Japanese officer returned to camp one night demanding his own personal show, and everyone had to get up and perform and watch.

Some men literally went mad with the conditions and boredom and hunger; Reuben remembers one who went totally mute and another who did nothing but shout. Reuben himself tried to solve mathematical problems in his head while he was working, to avoid the relentless tedium of the labour. One man on the dockside went missing one day and comrades said they saw him take a dive into the river to cool off, but the men noticed fishing nets were draped in the area. The locals were alerted and the nets quickly pulled in, and there was Lofty Duggan, dead, entangled in the nets.

In fifteen months there had been about thirty deaths; a Union Jack brought from Singapore was used to drape the hastily made coffins for the funerals. Then the order came for 700 reasonably fit men to be chosen for a transport, including Reuben and Colonel Hugonin. Although Saigon had been a nightmare, fear of the unknown now gripped the men. It was also sad to say goodbye to old friends. In June 1943, they were taken in three crowded river barges to Phnom Penh in Cambodia; the boats were meant to carry 100, but they numbered 250 in Reuben's boat alone. It took forty-eight hours, with very low rations given out. They were marched to the railway station and put into crowded sealed trucks. They were desperately hungry and some Welsh miners began to sing 'Bread of Heaven'; until the day he died, Reuben could never hear that hymn without thinking of that terrible journey.

Their train crossed into Thailand, reaching Bangkok for a brief stop, and went on to Nong Pladuk, a thirty-six-hour journey, where they stopped for three days and then were sent on a forced march in darkness to Bam Pong. Here they were put on sideless carriages loaded with rails and sleepers; they dare not sleep for fear of falling off, and the journey lasted all the next day. They reached Tarsao camp for an overnight stay; the state of the prisoners in this camp horrified Reuben's party – they were walking skeletons, naked except for loincloths to cover them. In the huts were friends from Singapore, desperately ill – Reuben discovered that these men had been working on the Death Railway, and many friends they had known in Singapore were already dead.

The Railway

Next day they were again on the railway trucks and taken to barges which took them upriver to No. 2 Camp Kinsaiyok; loaded with them were rolled-up tents. That night, in the pouring rain, they slept on the crowded barges, again with no food and certainly no sleep, after several days of travelling. At daybreak on 1 July they disembarked and were ordered to clear the virgin jungle to build a camp, although they had no experience of doing this. First they set up the cookhouse with the iron vats they had brought, built of bamboo and an atap roof to protect the fires; then the Japanese guards' huts; then the sick hut; lastly they began their own huts, with the usual sleeping shelves – all around an open square – and all this had to be done without food or sleep.

By nightfall, most huts had still not been built and most of the men had to sleep in the tents, which were totally un-watertight. Lying drenched in the mud all night and with rats running over them, they had no sleep, but next morning they were sent to work. The railway was behind schedule and they were working under the tyranny of 'speedo' – the Japanese were frantic to finish the railway. The men were up at 6 a.m. for rice and tea, and *tenko*, and then a two-mile march through dense jungle, with uneven ground underfoot, to the railway, being beaten by the guards at the slightest sign of straggling. The midday meal was to be brought to them. The worst part of the journey was crossing a forty-yard-deep ravine, over a makeshift, slippery tree-trunk bridge. Often the men worked until 8 p.m; after eating, washing and roll call, it was midnight before they were in bed – and at 6 a.m. they were up again.

The work involved making cuttings through rocks and also building up embankments of soil and rock for the railway – while being wet or in the burning sun, naked, starving, ill, beaten, covered in mud, their feet bleeding, and being screamed at, insulted and bullied. Beatings were often given simply for misunderstanding an order in Japanese. Reuben was first working on embankments, moving debris in baskets alone, or on a 'stretcher' with another man, over mud and rocks, often over a distance of 100 yards. Scrambling up the embankment to dump the load, often in the pouring rain, and sliding back down, was totally demoralizing, as well as exhausting. Work often went on until late at night until the Japanese decided that a sufficient quota of work had been done, and then the men had to march back to camp, often barely able to walk. There were short breaks in the morning and afternoon, but barely any days off. Thirst was a huge problem and the half-pint of water in their bottles hardly lasted. Reuben's solution was, oddly, to drink very little; if you drank it all, he said, and knew there was none left, psychologically it was the 'kiss of death'.

Sometimes bananas and bamboo shoots could be found in the jungle and eaten; killing snakes, lizards and rats and taking them back for the cookhouse was routine and did help, but the hunger was always there.

To build cuttings, rock was removed by first drilling holes with primitive chisels, inserting dynamite and then blasting – but the Japanese engineers gave inadequate warning for blasting (this was often deliberate) and many injuries were caused to POWs. Often men at the top of the cutting cut boulders which would roll down on men below, causing further terrible injuries. Men worked in pairs, one wielding a fourteen-pound hammer, the other holding the chisel – 'hammer and tap'. The hammer work was jarring and back-breaking; men had to take turns. Reuben was very nervous with the hammer work, fearful he might hit his partner's hands, so he did it slowly, but this caused the guards to become infuriated with him. As a result he was badly beaten with bamboo across his naked back and therefore very often he was the last to finish work, as nobody was allowed back to the camp until a drill hole was completed. This meant negotiating the dreaded tree-trunk bridge, slippery with mud, in the dark, which he had to do on all fours, normally in the rain. He would return to camp, hungry, muddy, beaten and exhausted.

Latrines were open trenches, with no privacy or lids, dug at the camp's edge, which once full were filled in, but when it rained they overflowed and the resultant spillages caused rampant dysentery. Other diseases were common, and the situation was made worse by the malnutrition, so that many men became ill. But the 'speedo' madness gripping the Japanese made them send even many of the sickest men to work, and when the POW doctors protested, they were beaten. The sick were literally dragged from their beds. The worst offender was Sergeant Seiichi Okada, of the Japanese medical unit, who was known as 'Doctor Death'. Some men even died on the walk to the work site, and those who made it were beaten because they could not work, or they died on the way back. Within one month of joining the railway, 140 of the 700 from Saigon were dead.[3]

Reuben decided, one day after work, to go to the river to wash. The most direct route was through the flimsy camp fence, a route often used, as it was only a symbolic boundary. He had gone only a few yards when an angry Japanese voice called to him. Reuben turned and bowed. 'He wanted to know where I was going, so I indicated the river', he recalled. Then he gave Reuben 'a smash in the face. I'd no idea why, as I'd bowed to him. Then another whoosh and then another. Then he pointed at the guard room, and I understood. He'd decided that I wasn't allowed to go through the fence. So he beat me up there and then, and when he was finished he waved me on to the river.' Such gratuitous violence was commonplace. Reuben put it behind him.

But that evening at *tenko*, a Japanese officer told Reuben to step forward. He knew what to expect – it began by talking quietly and then it got louder and louder and then the fists would start flying, over the head and face. Ducking meant a kicking in the stomach until Reuben was on the floor, then more kicking – all this in front of the parade. Other Japanese joined in. Reuben said he had seen it so many times before. Then it was over and several comrades came to help him up. The doctors checked him; he was bruised but nothing broken. Next day was work as usual.

Kinsaiyok Main Camp

On 17 July 1943, 500 of the men, including Reuben, were marched all day through the jungle to Kinsaiyok main camp, a much larger establishment. It was in a basin and all the rainwater made it a swamp into which the latrines flooded. The huts were older and so more infected with lice and maggots. There were Dutch and Javanese prisoners here too; soon another British and also an Australian contingent arrived. Reuben described how a new adjutant was appointed – a Captain MacDonald; he had a way with the Japanese and he used to laugh in their faces. They could not quite deal with this and it appeared to reduce the excesses of the guards. But not Okada. He would often taunt the doctors for not providing more sick men for the railway work; he would make them stand in the sun, beat and kick them, scream at them. He used a special wooden sword for beating the sick men out to work. This happened daily.

Reuben also witnessed the horrific treatment of the Asian workers. Forced or duped into work by the Japanese, with no organizational structure, unlike the military, and kept in separate camps, their survival rate was lamentable. He also recalled the arrival of 'comfort women' in the camp for use by the guards as sex slaves, something that deeply saddened him.

At this camp, the main work was building five wooden bridges for the railway; this meant felling huge trees and moving them long distances to where the bridge was being erected. Starving men, of varying heights, carrying heavy tree trunks two miles to the river over undulating jungle terrain, being beaten as they went – it was insane. Elephants were rarely used, despite Japanese propaganda to the contrary.

Reuben also had to do more hammer and tap work; one day his partner was working as normal, but then suddenly stopped and he fell to the ground, ashen-faced and clutching his stomach. The guards knew at once what had happened: cholera had arrived. The man died and was cremated that evening.

Soon it was time to lay the heavy sleepers and hammer the rails into them with fixing spikes. By now virtually all the clothing and footwear of the men was worn out, just like the men themselves. Most had no hats. They had turned into the loincloth-wearing skeletons they saw at Tarso. Ulcers which developed from the tiniest of scratches became life-threatening within days and Reuben developed two on both shins, but he used some sulpha paste he had purloined from the docks in Saigon and luckily the ulcers healed. Others suffered appallingly and had to have legs amputated or else die of gangrene. But the doctors whom Reuben knew at Kinsaiyok, like the legendary Colonel 'Weary' Dunlop, saved hundreds of lives. He would never forget, however, the screams of agony of the men suffering the amputations with no anaesthetic. Still men died, and by now only rice sacks were available as coffins. Cholera patients had to be isolated because it was so infectious; the guards were terrified and stayed away, and brave POW volunteers cared for those sick men, and gave them saline injections using bamboo needles to help keep them alive.

Rice sacks were also used as sleeping bags, despite being coarse, both to keep warm during the often cold nights and to reduce the discomfort of lying on the hideous bamboo-slat shelves. Bugs, rain and cold, not to mention recurring flashbacks from the images of the day's work, meant sleep was difficult, but eventually it overwhelmed the exhausted men. Often the night was pierced by the screams of men having nightmares, or by the noise of men rushing to relieve themselves due to persistent dysentery. Then suddenly it was 6 a.m. and the gruesome day lay ahead.

By the end of the 'speedo' at Kinsaiyok, only a quarter of the original Saigon group were fit for any work – as defined by the Japanese doctors. Reuben himself, on top of everything else he had, developed bad trench foot and was told to get hold of rags, go to the cookhouse, put fat on the skin of his feet and wind the rags about his feet and stay out of the mud. Captain MacDonald saw Reuben hobbling and found him some clerical work in the camp office; here he was surprised to find that no roll of the 1,000 men from Saigon had been made over the previous eighteen months of their imprisonment, except the one given to Sparks – but that was now out of date. One day, information on these men and what happened to them would be needed, if only for historical purposes and family reasons – a kind of war diary. It might be possible to smuggle such a list to the Allies. Reuben mentioned all this to MacDonald, and next day Reuben was handed a blank exercise book. He began the amazing 'Prisoner List'.

The Prisoner List

Reuben was told to seek out the sergeant majors and ask them for lists of their men. To fit information about 1,000 men into the book meant there was no room for errors, and he wrote in small, neat block capitals.[4] Details included name, rank, number and next-of-kin addresses, including all that was known of those who had died, underlined in red, with the cause and place of death. He also went to the cemetery and drew a plan of where each grave was and put this in the back of the book, using reference numbers to place against the names of the dead in the list.

The Japanese interpreter would often drop in unannounced to check that no anti-Japanese material was being kept. The list fell very much into this category, as it described causes of deaths which would be grounds for war crimes trials, so Reuben had to be ready to hide the book quickly when it was in use, and have a safe place for it when it was not. One day the interpreter suddenly appeared in the office, drunk and with his 'seal of approval' rubber stamp in hand. Reuben and his friends gave him some ordinary papers to stamp and then quickly shoved the book under his nose and he stamped that too. This would mean there could be no objection to it if it was discovered.

Reuben sent the usual pre-prepared POW card home around October 1943 (the first one had been allowed shortly before they left Saigon), which said at the bottom 'best regards to Ross and Shona Shani'. At home in London, Reuben's teenage brother deciphered this correctly as 'Rosh Hashana Ha-shani' (meaning 'second Jewish New Year' in Hebrew) – telling his family, in other words, that at that date he was alive and well at least. Sometimes letters arrived at the camp; Reuben recalled one occasion when a man was told by his neighbour in a letter that the news of his friend's death had broken the wife's heart, and she had remarried, but his friend was sitting next to him alive and well; he had to tell him, and the man burst into tears. Reuben described this scene as 'terrible' to see. Conversely, in a letter Reuben received, news came that his brother Solomon had escaped from a POW camp in Italy. When he told his friend the news, the friend stood up and announced it to the whole hut; even though nobody knew Solomon, they all cheered.

By October the railway was finished and work was lighter, but conditions for the POWs were the same – men died, and indeed the food was even worse as it appeared that the Japanese were suffering from food shortages for their scattered armies in Asia. The rice was filthy and contained weevils and maggots, which the men ate as a source of protein.

Thai traders became less willing to come to the camp. Some men tried to steal duck eggs from the guards' compound, but terrible beatings were inflicted if they were caught. Meanwhile, Reuben continued to add information to the Prisoner List as news came from prisoners being moved up and down the line to work or hospital camps about the fate of men in different camps.

By January 1944, news began to reach the West of the awful conditions in the Japanese camps.[5] Suddenly, for a while things improved, at least in Kinsaiyok – some new loincloths and footwear were distributed, there was better hygiene and inoculations were given. The commandant asked some POWs to write essays for a propaganda radio broadcast about the 'good' treatment they were getting, and Reuben was one of those nominated by Colonel Hugonin to do this. In the event, Captain Faraday's was chosen by the Japanese and, although they failed to see it, it was a cleverly but thinly disguised attack on Japanese behaviour.

Then in March 1944 they were moved to a transit camp at Tamuang.

Tamuang

Reuben made sure he took his hidden ring, the precious film of the surrender and of course the Prisoner List. They were forced onto a train and went south. Tamuang was a large and new transit camp but conditions did not change, and Reuben had his first bout of malaria here. His friends covered his shivering body with their rice-sack blankets until fifteen were lying over him; then came stage two – the sweating. This went on for months in two-week cycles, but as he was working in the office and no longer labouring on the railway, he could cope.

One morning an American plane came over and dropped leaflets warning locals to stay away from the area as they were going to bomb the railway. This news brought some happiness because it meant that the war was going well, but it also brought dread because the men were being put to work repairing the damage, and that meant the danger of being killed by friendly fire. Indeed, Reuben witnessed one terrible attack on a train near the camp, bringing back POWs who had been working on repairs on the railway; many were killed by the American bombers who assumed that the train was full of enemy troops.

One evening in June 1944, the commandant made a rare appearance at *tenko* and told the men of the huge Allied landings in Normandy. Some knew already, because secret radios were hidden, just as they were in many other camps. Radio security was very tight, as discovery could mean death. Sometimes news came from Japanese newspapers left lying around in the Japanese offices where POWs were sent to clean. One soldier could read Japanese and this was how they heard that some of the Hell Ships taking men from their camp to work in Japan had been inadvertently sunk by American submarines. Sadly, Reuben took out the list and made the notation M (*Maru* – a word used to name all Japanese ships) against the names of those men who had been on the ships and probably drowned at sea.

In autumn 1944, a captain approached Reuben; he was known as one of the few who operated the secret radio. 'What will we do for Rosh Hashana?' he asked. Reuben was pleasantly surprised to learn that he was Jewish (he was Henry Phillips, later Sir Henry). A notice was duly posted, telling POWs of the planned Jewish service. A man whom

Reuben knew from home, John Turchin, approached him and asked if he could help. They had no prayer book and so they composed their own service from memory. John produced a wad of money to pay for food; he explained that he had been driving trucks for the Japanese, who were pilfering and selling petrol, tyres and anything else to local Thais along the railway. John got his share, and he gave most of the money to the hospitals for the sick POWs – but he had put some aside for this special occasion. It was forbidden to trade with the locals, and John, if caught, would have been executed. The cookhouse was duly paid to produce rice cakes with eggs and a substitute coffee from burnt rice. The service took place in the corner of the square and about forty personnel were present. It was very moving. Several days later, for the Yom Kippur (Day of Atonement) service, only five came, as most of the Jewish POWs had been moved on.

The men noticed more and more trains coming down the line with wounded Japanese soldiers, and few going up the line with supplies and reinforcements. Clearly the war was going badly for them. There was one particular guard who constantly taunted the men about how many islands the Japanese had captured. Leading him on, Reuben and his mates would ask him, 'What about Wales?' and he would say, 'Yes we took Wales'; 'What about the Isle of Wight?'; 'Yes that too; fierce fighting but Japanese soldiers very good; we took Isle of Wight.'

The Kempeitai were making more frequent visits to the camp, searching for the radios that they knew must be there. The men began to worry that there might be a decision to kill them all if the Allies landed. Indeed, the Japanese ordered the digging of a huge trench, ten feet deep, around three edges of the camp boundary; huge machine guns were then placed at the four corners of the camp. Then in January 1945 came another order – all POW officers, except doctors, were to be separated from the other ranks into different camps; this clearly was to rob the men of officer leadership in case of planned uprisings by the prisoners. On leaving, Colonel Hugonin arranged for the second copy of Reuben's list to be taken apart and for pages to be smuggled in the linings of several officers' rucksacks to their camp at Kanchanaburi. RSMs (Regimental Sergeant Majors) now had to act as leaders and Reuben was chosen, above some higher-rank men, to effectively become the camp adjutant. His particular job was to care for the camp cash box which contained life-saving money accumulated by various means, including illegal trading and Japanese pay to the POWs, to keep men alive by buying extra food and medicine. It was a huge responsibility.

Reuben thus gave some money to the doctors for the sick and some to each prisoner, according to how many days they worked. But the Japanese noticed that more money was being spent in the camp shop than they had given Reuben to distribute! Reuben and the RSM were constantly being called to the camp office to explain, and the answer they always gave was that men had been selling their own belongings to Japanese guards (which was permitted) to get extra money for food. Reuben also had to be careful that notes sneaked into the camp from illegal trading – which tended to be of higher denominations – were not used in the shop, as the Japanese only paid Reuben in small denominations for the men. So he made very sure that the camp shop accountants paid the Japanese in small denominations from the shop profits. Reuben always kept the cash box buried beneath his bed and, for security, it was only ever dug up at night to insert or extract cash, with POWs on the lookout. Cash smuggled illegally and daily into the camp would be temporarily concealed between the bamboo slats, so that they did not have to constantly dig up the cash box.

Impromptu searches of the huts by the guards and Kempeitai became more frequent and furious; the men would be pushed from the huts and the guards would go in with spades. Nothing was ever found, but the tension each time was extreme.

In June 1945, Reuben witnessed the raid that destroyed the Kwai bridge, visible from their camp; again there were mixed emotions, as the raid meant possible liberation, but also possible death from Allied bombs. They also heard of the German surrender. Then news came that they were moving yet again, 1,500 men in three groups. Reuben had to dump the precious film of the surrender in the latrines, as the Prisoner List took priority for secret removal to the next camp, without risk of discovery. The cash box was just thrown in the back of a lorry with other innocuous-looking boxes in the hope it would not arouse suspicion; nobody could think of any other way of hiding such a large item. They went by rail towards Bangkok, travelling by night to avoid air attack; it took five days to travel 200 miles because the Allies had blown up so many bridges, but they finally arrived at Pratchai.

Pratchai

Reuben stayed with 600 of the group; their first job was to dig again: an infamous ditch around the camp with machine guns at each corner, in which the Japanese would bury the POWs in the event of surrender to the Allies. Again an RSM was in charge, with Reuben as adjutant. The cash box was hidden, and the Prisoner List was sneaked into the Japanese administration office and kept innocuously on the shelves among other books – the reasoning being that this was the least likely place the Kempeitai would search. The POWs were to dig artillery emplacements into the nearby hills for Japanese guns. But Reuben remained in the camp office.

One day, the men noticed the Japanese burning papers in a huge bonfire near the camp office; something was afoot. Many guessed that liberation was near, and few slept that night. But Reuben had a special worry – the list was in the camp office and might be destroyed, so he got up and went to the office. Nobody was about; he saw the book, took it and returned safely to his hut. He did not know that Colonel Hugonin's copy had been destroyed months ago.

Liberation

The next day the work parties inexplicably returned at midday. They said locals had been happily thronging the route back, trying to tell them something, making hand signs that looked like something exploding. At the camp, the RSM and senior medical officer were called to the commandant's office and remained inside for some time. Reuben and the men waited expectantly in their huts. Then a bugle sounded, summoning them to the square; it was not the *tenko* call but 'Fall in', British-style, not heard by the men for three-and-a-half years. This could mean only one thing, and a huge cheer went up as they rushed to the square.

The MO stood on the raised platform where normally only the commandant stood, and as the crowd went silent, he said slowly, 'Good news – the war is over'. The outburst was tumultuous and three Union Jacks appeared from nowhere. They were tied high on bamboo poles to great cheering. Reuben said that all the men hugged and cried unashamedly; they had not believed the day would ever come.

Eventually the MO said that the men would await the relief force, but meanwhile the Japanese would guard the camp for the men's safety. On 30 August, a Captain Bradley of the Scots Guards entered the camp, smartly dressed, having been parachuted into the area. The POWs directed him to Reuben, who called the interpreter. Bradley demanded to see the commandant; the interpreter said that Captain Suzuki was asleep. Bradley shouted that he wanted to see Suzuki and all the guards, now! They arrived, and Bradley lined them up and photographed each one and noted their names.

Bradley then told the men about the atomic bombs; it later emerged that, had they not been dropped, and had Allied land invasions taken place, orders had been given from Japanese High Command to kill all POWs in the Japanese Empire.

Behind the Japanese camp office was a large hut. The RSM and Reuben ordered it to be opened, and inside were hundreds of Red Cross parcels that the Japanese had illegally failed to give the men. Men had died for want of the food and medicines here. The Japanese military code of Bushido meant that prisoners were beneath contempt because they had not fallen on their swords, but there were no Japanese doing that in Reuben's camp; on the contrary, many guards came to ask men to sign letters saying that they had been treated well; nobody did, of course, and the worst of the guards just fled, while incriminating documents were burned. One story Reuben heard was that two vicious guards from Pratchai were recognized by POWs later in Bangkok, and the prisoners cut their noses off.

Meanwhile, Reuben, who weighed just under seven stone, took a camp truck and drove with some men to the nearby town of Saburi and bought up stocks of food using IOUs on behalf of 'His Britannic Majesty King George VI'. The shopkeepers accepted them; the cash box money was no longer legal tender. Reuben gave out the Red Cross parcels and new clothes from the store. At the request of Colonel Hugonin at Kanchanaburi, a second, new copy of the Prisoner List was drawn up by a group of POWs sharing the task, and personally delivered to him.

Soon after this, aircraft appeared and dropped supplies by parachute, or supplies were just thrown from aircraft; the men had to hide in the huts, away from the square, so as not to be hit by these heavy parcels. Sadly, one hut was hit and several POWs were killed and injured; after all that suffering, they were killed by relief supplies.

Newspapers were also dropped; they were full of people and events the POWs had not heard of. Churchill was no longer prime minister and this came as a shock to the men. Suddenly a jeep drew up inside the camp and Lady Mountbatten appeared to greet them; next day came Lord Louis Mountbatten, Supreme Allied Commander, South East Asia; and the day after that, General Bill Slim, GOC of the 14th Army.

Home

Trucks arrived to take men home in shifts. Reuben had been asked to coordinate this and so was one of the last to leave the camp. Sick men always went with fitter men so that they could be cared for. Each group was asked by Reuben to write to his mother to say he was safe. At the end of September such a letter reached his mother in Edgware, saying Reuben was well. Several others followed and she could not understand why Reuben himself did not write. Then came a telegram from Ceylon, from Reuben, saying he was on his way, on board HMS *Chitral*, via Suez and Gibraltar to Southampton, where he arrived on 29 October.

35b. Reuben later in life.

Meanwhile, he had gone first to Bangkok, sleeping at Bangkok University. He saw a notice that a Yom Kippur service was being held at a house in Bangkok, led by about seventy German Jews who had been living in the city, more or less unmolested by the Japanese, who saw them as allies because they were German nationals! Then he was flown to Rangoon on 1 October for five days in the military hospital for check-ups, then embarked on the *Chitral*. Sadly, in Rangoon a brave and popular officer who always stood up for his men, Captain Kenneth Hughes, died of illness; after all his suffering since Saigon it was a bitter and terrible blow for his men and his family.

Families were not encouraged to meet the men at the ports because it impeded the speed of the processing of their arrival, but many did meet them, including Reuben's family. They were shown into a hut at an army camp in Southampton to wait. At twilight the first men arrived from the docks and Reuben's sister recognized him; he looked very gaunt but it was an indescribably happy reunion. Then the families were sent home as the men had to stay overnight for further medicals. Next day he took the train to Waterloo. On the way he saw the devastation wrought by the war, and at the station there were many people holding up placards and photos of men with the message, 'Have you seen him?' The chances that those men had survived was slim – it was a heartbreaking sight. His family took him home by car, and a sea of welcoming faces gathered outside his house to cheer him home .

Reuben had to endure constant questioning from people about his experiences but he dealt with it as best he could. After five years he was called up again for the Korean war, because of his radar skills, but these were deemed out of date. Instead, he was sent on a refresher course and then absolved from further service.

Reuben often attended the AJEX parades in Whitehall (the Jewish War Veterans), where he saw and spoke with those who remembered the Jewish services in the camps. He was for over fifty years one of the original trustees of the FEPOW and Internees Fund and later its chair, and chair of its welfare fund, for which services he received the MBE. He was also an honorary officer of the United Synagogue of Great Britain. He married Hilda (née Lowry) in 1956 at Golders Green synagogue and had two children. He died in March 2009.

Notes

1. His brother Solomon, Rifle Brigade, was POW in Libya and later made a daring 'home run' escape when in an Italian POW camp in 1943.
2. Several years after the war, Reuben was parking his car in Bloomsbury when he saw Basil; they recognized each other at once and had a long chat and met several more times at Saigon veteran reunions. From Saigon, Basil had been guided to the Chinese border, from where he managed to get to Russia and then sailed to England.
3. For crimes at this camp and later at Sandakan, Okada was sentenced in 1946 to a mere ten years for war crimes.
4. A copy of this moving and amazing book is in the AJEX Jewish Military Museum and online, and the original is with Richard Kandler.
5. See the House of Commons statement by Foreign Secretary Sir Anthony Eden.

Anka (Anna Marie) Kanturkova

The information in this testimony is taken, with kind permission, from the book Knocking on Every Door *(published in Canada by the Azrieli Foundation in 2010), which Anka Kanturkova wrote using the name Anna Voticky.*

36. Anna Kanturkova.

Anna was born on 5 July 1913 in Brandys nad Laben near Prague, daughter of Max and Hedvika (née Kohn), owners of a clothing store. They spoke German and Czech at home. In 1933 Anna married Arnold Voticky, a jeweller, and in April 1940, with ten other family members, they left for Shanghai's International Settlement by ship to escape the Nazis. This involved the acquiring of numerous visas and the giving of bribes. Others went by rail via Siberia and Kobe in Japan, due to the life-saving work of the Japanese vice-consul in Kovno, Chiune Sugihara. After Russia entered the war in June 1941, this escape route was closed.

Shanghai sheltered 20,000 Jews in the Second World War, more than the whole British Commonwealth put together. The family arrived on 10 May 1940, after many adventures en route and a month at sea. With nowhere to stay, they were put up by another Czech family who had arrived earlier, then found a hotel in the French Concession and finally, after a few weeks, a flat of their own. They lived relatively freely under Japanese control at first and Arnold opened a small jewellery factory. Then, after the attack on Pearl Harbor in December 1941, things changed; Europeans became enemy aliens, but the Jewish refugees were not immediately interned, despite German attempts to get the Japanese to act. Perhaps this was because the Japanese did not share German anti-Semitism, and also maybe because the Japanese had an inflated view of the influence of Jews financially; they especially recalled the loans made by American banker Joseph Schiff to Japan during the Russo-Japanese war of 1904–05.

Eventually, however, in April 1943, the Jews were forced into the crowded and desperately poor Hongkew district to form a ghetto. Some young men escaped to Palestine/Israel and joined the British Army, but most were caught up in the ghetto. Arnold paid a Chinese builder to construct a cement-floored house of several flats, each with one room and one bathroom, for the extended family. Everyone shared one kitchen and electricity was sparse, but they were all together.

The Japanese were not especially cruel and when Anka had to have an appendix operation she was well cared for by Japanese nurses in the Shanghai General Hospital, though most of the doctors were Jewish refugees. That winter was particularly cold and heating was almost non-existent. Her two young children attended schools just outside the ghetto; getting there was sometimes a dangerous journey, with gangs, political factions and thieves everywhere. Disease, malnutrition, rabid animals, lice and poor sanitation were rampant in the city; many refugees – especially the young and elderly – died over the two years of ghetto life.

The two Japanese in charge of the ghetto were Kanoh Ghoya and Sub-Inspector Okura, of the Imperial Japanese Navy. They controlled the movement of the Jews in and out of the ghetto using permits. Ghoya called himself 'King of the Jews' and both were prone to violence, though Ghoya less so. Keeping a radio was forbidden, as it was in the POW camps and other occupied cities, but Anka's brother Vilda hid one at great personal risk and that is how they heard about the German surrender in May 1945.

Bombing by the Americans began in July 1945 and continued for some weeks; on one occasion (17 July) they had to shelter under a mass of furniture as bombs carpeted the city. Hundreds were killed, including about forty refugees. As many families were bombed out, overcrowding became intense and many families were sharing one room divided by a single sheet to give some privacy.

Finally, on 12 August 1945, they heard the war was over. On 18 August they were liberated by the US Navy. The camps were opened and the refugees saw just how terribly the military personnel had been treated – far worse than they had experienced.

The UN (United Nations) gave out food, clothing and medicines, but still the Japanese had to be used to keep order until Allied reinforcements arrived. Some of the younger Jews saw Ghoya and tried to beat him up, but he just bowed and kept saying he was sorry. It transpired at the Shanghai war crimes trials (February–April 1946) that he had been a US spy all along.

News of the Holocaust now reached them and they were shocked at what had happened to their friends and families who never got away. Conditions in the ghetto remained very bad, with the hot summer and squalor making life unbearable. Eventually, in July 1946, they left on the American ship SS *Marine Angel* for Marseilles; they were warned by Czechs in France, on their way to Israel, that the Prague government was becoming communist, and Jews were not welcome. The trains to Prague were massively overcrowded; half of Europe seemed to be on the move as displaced persons tried to get home. They arrived in September 1946 and, after staying in hotels, rented a flat. But life was unpleasant and so in June 1948 they left for Montreal in Canada. Half the surviving 15,000 Czech Jews chose Israel.

Arnold had secretly been getting money and jewellery out of Czechoslovakia with friends travelling to the USA. The family settled and did well and took part fully in the local community there, with four generations living in Montreal to this day, as well as in other countries round the world. They were survivors of the Shanghai Japanese ghetto.[1]

Note

1. A documentary film made about the Shanghai Jews, *Port of Last Resort*, gives the best impression of what life was like for the 20,000 Jews who survived there. Many of the important sights of Jewish interest still survive (2011) and visitors can do a Jewish tour of Shanghai. See also the chapter in *Fighting Back*, by Martin Sugarman (Vallentine Mitchell, 2010), on the subject of 'The Jewish Company of the Shanghai Volunteer Corps'.

SX13978 Maj. Dr Sydney Krantz, RAAMC, AIF, 13th AGH and 2/4 CCS

37. SX13978 Maj. Dr Sydney Krantz, RAAMC.

Sydney (Syd) Krantz was born in Adelaide on 24 September 1903, graduating in medicine at Adelaide University in 1927. He later became a Fellow at the Royal College of Surgeons in London (1931) and Australia (1938). He enlisted in the AIF on 7 August 1941 and was embarked on HMAHS *Wanganella* in September 1941 to Johore Bharu in Malaya, together with 10 AGH, to later serve in the battle for Malaya and Singapore. He was described by Lex Arthurson of the 13th AGH (Australia General Hospital) as 'a brilliant surgeon who could cut off a head and sew it on'.

In December 1941 he was attached to 2/4 CCS (Casualty Clearing Station), a mobile surgical team, together with Captain Tom Brereton and six orderlies with two vehicles, based at Kluang in a rubber plantation not far from the aerodrome, in large tents. Surgeon Colleague Major Dr Alan Hobbs wrote that Sydney worked with him under huge pressure during the Japanese attacks, enduring twelve-hour shifts and saving men's lives. In one day, on 1 January 1942, the 2/4 CCS war diary recorded Majors Hobbs and Krantz carrying out thirty-five major operations between 1800 and 0600 next day. Another entry describes a flight sergeant whose shoulder had been torn open by a propeller at Kluang during the fighting in Malaya. It appeared that only amputation would save him, but Syd skilfully separated muscle from flesh and found the main artery intact. 'I think we'll give it a go!' said Syd and the arm was saved.

On capture, Syd was sent to Changi but the medical team stayed together, dealing with wounded under terrible conditions and with virtually no medical supplies. Some time in May the CCS was herded with 3,000 troops into Japanese merchant ships (one of the Hell Ships), up the Straits of Malacca, in what he said were 'indescribable conditions', to south Burma, to work on military installations. Later still they were moved to Moulmein where 'I slept on the concrete floor in the gaol' and helped build the notorious railway living mainly on 'rice, small peas and sweet potato and we could at times buy bananas, limes and duck eggs ... [every man we operated on was] suffering from starvation and malnutrition and all the deficiency diseases ... we traded our watches and fountain pens with the villagers in order to augment our meagre rations'. At one period they lived only on rice and chillies for three months. As the line neared completion, the cruelty of the guards increased; if you were tall, you were forced to kneel down so you could be more easily beaten.

In the jungle camps of the Death Railway, Sydney – like all MOs – had the terribly frustrating job of opposing the cruel Japanese guards, who demanded men for work who were clearly unfit, and standing up to the threats of these guards as he demanded that many should stay in the camp because they were too ill to work. Hobbs declared that Syd had the toughest job of any Aussie POW he knew. If he did not supply the correct number of men, he was beaten.

In the April of 1944, Syd was moved back to a base 'hospital' called Nakonpaton, fifty kilometres west of Bangkok, as the railway was complete:

> ...it contained fifty 100 meter long bamboo and atap huts each holding 200 sick, with an operating theatre with a good wood table and concrete floor. Eventually 700–800 men were admitted. With us was the amazing Dr Markowitz. Sensing defeat, the Japanese allowed some Red Cross parcels which contained life saving drugs such as novocaine, and with all the talent among the doctors and POWs there was little that could not be improvised to save lives. By the surrender, we had carried out 800 major surgical procedures.
>
> To do operations, spinal, epidural and local Novocaine was used; sutures were made from cotton and silk parachute cord, and catgut from sterilised animal intestines. Emergency operations were done by hurricane lamp sometimes with an armed Japanese or Korean guard watching over us. A stem steriliser was made and a foot operated suction machine. We had no gloves but it did not seem to matter. Our diathermy (heat treatment) was a long nail in red hot charcoal. Our only antiseptic apart from a little iodine, was 90% alcohol made by the gallon out of rice ... a tremendous achievement.
>
> Powdered milk tins were soldered together ... someone made me a small serrated circular saw which we attached to a foot operated dental drill to cut bone ... 1,500 transfusions were done ... Capt Markowitz using a bamboo brush to get the fibrin out and then draining the plasma through gauze. The classification of a donor in our camp was anyone who could stand up. Perhaps the most ingenious improvisation was the manufacture of artificial limbs with movable knee, ankle and foot joints. The limbs were modelled on the artificial leg of one soldier who had been enlisted with one leg! The MO must have forgotten to tell him to take his trousers off! We had 173 amputees in our camp – mostly the end result of gangrenous ulcers, but in some cases due to bombs dropped on our camps by our own planes. I believe there were only three cases who could not be fitted because of stump troubles ... I alone carried out 67 appendectomies ... we held clinical meetings and post mortems ... there were many unsung heroes, especially those who cared for the cholera cases, an extremely infectious disease ... morale was never at any stage shattered by the enemy ... we had disciplined team effort, based on officer control.
>
> Suddenly Lady Mountbatten arrived with two other (Red Cross) women in a jeep ... a vehicle we had never seen before ... her plane dropped supplies and then British Paratroops came and sprayed us with DDT, also never heard of.

Syd married Kathleen Lancaster late in life and had one son, Simon. Syd died on 11 August 1973.

WX7446 Pte Lazar Krasnostein (Les Krass), 2nd/4th Australian Machine Gun Corps, 8th Division

The source for this testimony is the recording of an interview with Les, made in June 1995 by his family. I thank Les for allowing me to use this recording, and thanks also to Les's son, Bruce.

38. WX7446 Pte Lazar Krasnostein/Krass, 2/4th Australian Machine Gun battalion, taken before embarkation for the Far East, with his wife and son.

Les was born in Palestine (Israel) in 1905, son of Matla and Isaac. He moved in 1923 to Perth, Australia, where he already had two brothers in the wool business. His parents stayed in Israel and died there many years later. Having had injections and drugs against malaria as a child, in Palestine, he had developed an immunity to it which later saved his life.

When he was aged 35 and married to Dolly (Beatrice, née Zeffert, from Liverpool), with a 6-year-old son, Bruce, Les joined the army in Perth in 1940 and volunteered for overseas service, as he wanted to get into the fight against the Nazis. In Perth, his rabbi, D.I. Freedman, had called together all the men in the congregation and told them they must enlist to defend Australia and fight Hitler. Les first trained in Perth and then in South Australia, near Adelaide; they were posted to Darwin. As war broke out on 7 December they were on their way to Port Moresby on the SS *Acquitaine*, but then diverted to Sydney, as Rabaul had been bombed by the Japanese. From here they were sent to Perth; unable to endure the proximity to home, he and some mates jumped ship and went home for the night to stay with their families. Then they sailed on two Dutch ships to Singapore.

Here, Les took part in the fighting for the island. He was wounded in the back by shrapnel on a beach, and his good friend, next to him, was sadly killed outright. The MO told Les that they would be back in Australia in three weeks and so he would operate on him then; the shrapnel stayed in his back until the day he died. After release from hospital he was sent to Changi gaol, four weeks after having left home.

His first working party was when he was taken by the Japanese to build the war memorial at Bukit Timah. His unit had played a major part in mowing down many Japanese as they crossed the Causeway, and his unit, among others, was deliberately chosen by the Japanese to build this shrine to the Japanese dead. At Changi he took part in Jewish Saturday services with other Allied soldiers, being given a chit to leave the Australian camp to go to the makeshift synagogue in the British section.

Thailand and Japan

In about June 1942, he was sent to work on the Death Railway in Burma and suffered terribly in the awful conditions there. However, his immunity to malaria saved his life, as he did not succumb to the many other illnesses that normally followed malaria, such as beri-beri or typhus, as the body weakened. Thus he was instrumental in assisting the sick, feeding them, smuggling in extra food and cheering them up. Dutch Indonesian POWs called him 'Santa Claus' and the Americans called him 'Mr Nice Guy'. The Australians made him honorary welfare officer. Once he worked ninety days in a row so that sick men should not be sent to work and probably die at the terrible work on the railway. A Dutch Jewish officer noticed this and ordered him to do only light duties in the camp to recover his strength, in case he fell ill from exhaustion. Despite this, he several times overheard anti-Semitic remarks among his fellow POWs in the huts; he always challenged them and dared them to fight him. They always used the excuse 'We don't mean you Kras', but he never let that resolve the matter, telling them they were ignorant and had learnt anti-Semitism from their ignorant parents. Later, most of these men became his friends, having learnt the lesson that they were all comrades together, no matter what their ethnic group.

While on the railway, the Dutch Catholic chaplain always asked Les to take the Jewish burials parties, so that the Kaddish could be said by him for the dead soldiers.

After one year he was moved back to Singapore and then sent to Japan in mid-1943, forced to work in a coal mine, at a coastal camp near Nagasaki; beatings and starvation were just as bad here as they had been in Burma. He was with Americans captured in the Philippines. Les maintained that thoughts of his wife and child were the things that kept him going. Although his wife wrote many times, he never received one letter or card from her. He worked hard here too, often going to the mine instead of the sick men, as the coal quota had to be filled every day or rations for all would be cut. Often he gave half his rations to other men to help keep them alive. After the war, at reunions, men would come up to him and hug him, saying he had saved their lives, but he rarely remembered who they were.

On the night of 14 April 1945 he was on the way to the mine and, unknown to him, a cigarette fell from the pack he carried inside his coat pocket. He hung his coat at the mine workface as usual. Later a Japanese guard, looking for a stolen lunch box, found the cigarette in his pocket and reported it; it was a severe breach of rules, as no smoking was allowed because of the danger of gas explosions. Les was called to the guardroom after work and given a severe beating with a stick (four feet long and two inches wide) on his face and head, and punched by both guards and mine officials. The bruises were very painful. The next evening before work he was taken again, this time to the beach, half-naked, and had his hands tied behind him and was pushed to his knees onto sharp stones. He was left there for twenty-four hours in the freezing cold (they had just come from the tropics and the temperature difference was huge) and, every so often, iced water was thrown over him as a cold wind blew in from the sea. He was periodically beaten during the night with bamboo sticks, buckles and leather belts and does not know how he recovered from this cruelty. Next evening, he was ordered to the office of Commandant Fukuhara, forced to kneel down and thrashed again, 100 times with a leather belt all over his naked body. In order to stop the beating, he finally had to confess that he knew the cigarette was there but had been afraid to say so. He was in the 'hospital' for a week afterwards.[1]

When the bomb was dropped on Nagasaki, Les was on the night shift at the mine (6 p.m. to 5 a.m.), but all that the prisoners saw was a huge plume of smoke across the bay. He described the Japanese as 'bastards' but as the bomb dropped they became 'kittens'. All work was stopped on the excuse that the machinery was broken, and they were allowed for the first time to go and swim in the sea; they knew something crucial had happened. When the guards tried to talk to him after the bomb fell, he spat on the ground and walked away. Another invited him to a house and then said he could sleep with his wife; he left, disgusted at such behaviour. Later, American planes dropped newspapers and that is how they found out that the war was over and that the atomic bomb was what they had seen over Nagasaki. Afterwards the POWs took a train to look at the destruction of the city, and years later Les found that he had been exposed to radiation. He had also lost half his body weight while a prisoner. The men took a train into the countryside and returned with confiscated eggs, chickens and potatoes which they cooked up in a celebratory stew. In the month during which they waited for release, most men put on a stone in weight.

On liberation, he was taken by comfortable electric train to the port at Nagasaki and then to the American aircraft carrier USS *Cape Gloucester*. In the ship's hospital Les asked if there were any Jewish personnel on board and one of the nurses came forward and said immediately that he (the nurse) was Jewish; soon afterwards, dozens of Jewish naval personnel surrounded his bed, asking what he wanted most. One of them, the war correspondent for the *Boston Post*, brought him his first request – a handkerchief. He had not owned one for over three years, but now he received several of them, monographed with the reporter's initials. The reporter also took his photograph and it appeared in the newspaper, with one of him taken before he was a POW, to show the terrible condition he was in on liberation. He went by ship to Manila and flew to Darwin by Catalina flying boat, then on to Perth in a large American bomber. There he was met by large numbers of his family and neighbours, and while his wife hugged him, his 12-year-old son looked up and pleaded, 'Please let me hug him too' – and Les burst into tears.

Later he was told that he had to stay in hospital until he had been checked by the doctors. He said he was all right and that he wanted to go home, but only if all the other men in the ward could go home too – he insisted on this, and the doctors allowed them to go.

After the war, the family moved to Melbourne. Les worked in the garment business and later for his son's retail electrical business. He was a passionate supporter of Israel and active in collecting for the Jewish National Fund. He died in August 1995, soon after making the recording of his war story.

Note

1. From Roy H. Whitecross, *Slaves of the Sons of Heaven* (London: Corgi, 1961), pp.217–18.

937872 Air Mechanic/LAC Sidney Lawrence, No.36 Torpedo Squadron, RAF

This interview was conducted by Imperial War Museum staff, with details added in 2011 by Sidney's son Michael.

39. 937872 LAC Sidney Lawrence, RAF.

Sidney was born in Hove, Brighton, on 16 June 1917, son of Harry and Suzie. He attended school in Hove and also the Middle Street Synagogue, where he was bar mitzvah at the age of 13. He studied to be an optician at night school but also joined the RAFVR in 1938 and so was called up when war broke out in 1939. He was sent for training at Cranwell, where he was attached to a unit training crews in the use and development of bombsights and related instruments, for which his knowledge of optics was useful. He was married to Irene but later they divorced, and after the war he married Pearl. After training he was posted to 106 Squadron, 5 Group Bomber Command, at Finningley in Yorkshire as a flight engineer. On one occasion while training a crew they were attacked by overzealous Spitfires and shot down, but the pilot landed safely. Then some time in 1941 he received orders to report to Southampton after three days' embarkation leave and was shipped to the Far East via Glasgow, the east coast of North America, Freetown (West Africa), the Cape, Bombay and finally Singapore, where he was based at first at Seletar.

He and his squadron of obsolete Wildebeest biplanes were then sent to Kota Bharu, where the war first began, on the coast near the Malaya–Thai border. With them were a few RAAF aircraft. He remembers clearly the first attacks on 8 December or thereabouts; he was taking a late-night shower and the tannoys announced that everyone should go to action stations. Almost immediately, Japanese bombers came overhead and began bombing and strafing. Sidney dived under his bed and to this day can still hear the bullets streaming into the hut and splintering the wood into tiny pieces. Rapidly donning shirt and shorts he dashed to the aircraft and helped load bombs and torpedoes. They watched as the aircrews took off, went out to sea and circled and dropped their ordnance, returned, reloaded, took off – until they were mostly shot down. Their heroism and self-sacrifice was astonishing, particularly among the Australian pilots. But, sadly, it was futile, as the RAF aircraft were so slow that they were easy targets for the massive flak and the fast Japanese fighters coming off their carriers in the huge fleet, which was also shelling the aerodrome with its naval guns. At the same time Sidney could clearly see Japanese infantry landing on the nearby beaches.

When only three torpedoes and a few aircrew were left in the squadron, Sidney volunteered to go on the last sortie and drop the last torpedo on the nearby Japanese fleet, even though he was ground crew and not trained to do this job. They approached the huge Japanese fleet and flew into very heavy flak. Sidney said he was absolutely terrified. The pilot was screaming in the open cockpit, 'Let it go, let it go!' The torpedo was dropped successfully and the pilot streamed away but they never knew if it hit or not. On their return to the aerodrome they discovered that the aircraft's frame was peppered with holes.

Virtually all the RAF aircraft had been shot down by then, including the few Beaufighters and underpowered Hurricanes, but Sidney testified that in any case his Wildebeest squadron had run out of torpedoes and the spare parts situation was so bad that they had been using grass and straw to fill the tyres of aircraft to get them airborne. Any surviving aircraft were ordered to go south to fight another day, leaving the ground crews to get away as best they could.[1]

As the Japanese advanced, the RAF retreated with the army down the peninsula. The aircraft leapfrogged down south, and ground crew like Sidney were left to join the ground retreat like infantry, catching up with the aircraft at aerodromes as best they could along the way.

It was chaotic and no officers were around to give orders. Sidney realized they would have to get away through the jungle, so he filled up a water bottle in preparation. Only some Indian Army soldiers were left and so he went to an Indian officer and asked if he could join them. The officer said he could not give orders to white soldiers, but Sidney told him 'Oh yes you can!' So the officer gave him a machine gun and said 'Fight with us on the beaches', and he did. The Japanese made full frontal attacks on their trenches and were mown down as the Indian soldiers with Sidney tried to stop their advance. Sidney believes he must have killed dozens of Japanese that day. One comrade in another trench nearby said afterwards that he had seen the sea and a nearby river turned red with Japanese blood. Despite their losses the Japanese typically kept coming, but were constantly – though temporarily – held back.

Eventually there was a lull in the shooting and the order came for the men to withdraw into the jungle as they ran out of ammunition. As Sidney ran he felt a blow to his side that pushed him flat on the ground and he felt a large amount of wetness on his left side. He believed he had been wounded, but felt no pain. Paralysed with fear, he placed his hand on his side and realized it was his water bottle that had been hit! It was peppered with holes, but had saved his life. Up he got and ran on. The survivors made their getaway, and using jungle lore they found water stored in bamboo, which saved their lives.

When they reached a road, some natives gave them a lift in trucks to Kuala Klie, further south. Here there was chaos as thousands of civilian refugees joined the exodus with military walking wounded and retreating troops. Worst of all were the hundreds of abandoned native children Sidney saw, something which distresses him to this day. Time and again during his time in the Far East he witnessed this horrific phenomenon; he would pick them up, dirty and bloodstained, and hug them as they hugged him, and try to calm their crying as they loaded them onto trucks to get them to some form of safety.

From Kuala Klie there were trains to take troops south again and eventually they arrived without incident in Singapore on 14 December. Sidney made his way to his old RAF base at Seletar and found it utterly abandoned and stripped of all contents. He found his old hut and bed, laid down on the springs and fell into a deep sleep. Later he caught up with his squadron in a tented area nearby. Very few of the aircraft had made it back.

Orders were for RAF personnel to be evacuated to Sumatra and so they went to the harbour and got on a Straits Settlement steamer called the *Perak*. Hundreds were trying to get away and Sidney witnessed the high-ranking civil and military personnel setting up trestle tables at the harbour entrances as all Asians were turned away and all Europeans were closely questioned as to who they were and what they did, in order to get an exit permit. Sidney was given the job of guarding the gangway to the boat, completely exposed, horrified at what he was seeing; Japanese bombs were dropping, and bodies and tables were blown to bits, often splattering him with blood. Then, as the bombers departed, tables and officials set up again and just carried on in that stoic British manner. Local people would approach him, often holding children, but they had no permit. Several times he turned away and pretended not to look as they ran aboard. It was simply too heart-rending to turn people back.

Sidney believes that quite a lot of these officials were really local European planters wearing false British uniforms in order to use their authority to get themselves and their friends evacuated, though it is true also that many, if not all, did in fact belong to the local militias in peacetime and were genuine Territorial Army-style troops.

Eventually they left harbour; dozens of ships were being sunk by the Japanese, and Sidney says he will never understand how their boat made it out into the open sea unharmed. The vessel was so crowded that a crew member had to go into the crow's nest with a megaphone, instructing people to move to one side of the deck or the other, or else the boat would turn turtle.

Finally, due to the navigating skill of the Scots captain, they arrived in Palembang, Sumatra, and began to sail up the river there. Sidney was sent to a nearby aerodrome but it was soon captured by Japanese paratroopers and he had to make his escape. He eventually got on a train, packed both inside and out (on the roof and sides), going south again. While wading across a river to escape the Japanese, he came across a Catholic priest holding his chalice and padre's equipment above his head, to prevent it getting wet. Shouting above the noise of the flooding river and strafing of the Japanese Zeros, Sidney called out, 'What the hell are you doing here, Father?', to which he called back, 'If it is good enough for a fine Jewish fellow like you, then it is good enough for me!'

At Huisthaven they arrived at the Sunda Straits and crossed in a boat to Merak in Java, and from there to Batavia/Jakarta. With other mixed service personnel, he was billeted in the King Wilhelm III school, as Batavia declared itself an 'open city' so that the Japanese would not bomb it.

Soon Sidney was being driven from aerodrome to aerodrome in Java, following the few Allied planes that were airworthy, ending up at one that was on such a high plateau that the clouds rolled over the runway. But the Japanese soon discovered them (the Javanese hated the Dutch, who by now were in command of any British military personnel left, and probably gave their position away) and sent in paratroopers to take the aerodrome. There were Gurkhas with them and a firefight developed. The Gurkha officer told Sidney and the RAF men to stay put; he said they were good men to want to fight, but only they were trained for this type of thing. Sidney was very moved by this; it was as though they were saying that they would look after them. Drawing their kukris, the Gurkhas charged the Japanese and Sidney saw something he never wants to see again. The Japanese were stopped in their tracks as the Gurkhas gleefully and courageously lopped off heads, shouting and screaming as they fell upon them. Soon it was all over. The Gurkhas returned and the airfield was abandoned.

Java

Sidney was finally captured in Java when local Dutch forces surrendered on 8 March 1942. They were ordered to a collection point at Garoet and taken to a camp at Kalijati. On arrival, they were lined up and a Japanese officer and soldier began to walk along the line of POWs. To the first man, the officer said, 'You will work for the sacred land of Japan!' The soldier swore at him in his best service foul language and said he would not, at which point he was beaten with a rifle butt and bayoneted to death. The same happened to the next man, while Sidney, who was about sixth in line, prayed he would have the courage to say no and die a decent death. Fortunately the third man said yes and so the others all followed suit. Their first job was to fill in potholes on a nearby aerodrome. Sidney went to two other camps whose names he does not remember, and all the while the food was bad and they became more and more malnourished. Beatings were numerous and frequent and invariably for no known reason, and the usual Japanese attitude of despising prisoners was prevalent.

One day two army officers decided there was a possibility of escaping, if they could get away from the camp. They did, but were soon recaptured. The first officer was brought before the commandant in front of the assembled prisoners and asked why he had escaped. The officer replied that it was due to the bad treatment they received. He was immediately decapitated. The second officer was brought from another hut; he had lived in Japan at some time and understood the mentality of the Japanese. When asked why he had escaped, he said he had craved a woman and was going to enjoy himself and then return if he had not been recaptured. The commandant roared with laughter and told him that was bad behaviour but understandable; next time he should ask and he would arrange for a woman to be brought to him.

In the last camp in Java, there was a group of about fifteen white Dutch women who were taken into a field, and the Japanese forced several of the POWs to accompany them to watch a gang rape of these unfortunate people by dozens of Japanese soldiers. It was a kind of mental torture, Sidney thinks, and as he watched he wept, and openly admits he wet himself with the anger, pity and helplessness at what he saw. To look away was punishable by severe beating. He does not know how many, if any, of the women survived the rape.

In October 1942, Sidney was moved to the port of Batavia at Tanjong Pliock and they sailed on the *Shonan Maru*[2] to Singapore, stopped for a day or so, then went to Formosa (Taiwan), then to Saigon (Indo-China) and finally Moji in Japan. This was one of the notorious Hell Ship journeys; dysentery broke out almost at once among over 2,000 men crammed into the stinking and festering holds; food and water were scarce and contaminated; and being allowed on deck was very rare, the holds being shut firmly and locked; disease from the human waste spread like wildfire. Sidney believes that only 288 survived the journey; bodies were just thrown overboard. Some men, when trying to use the wooden ledge on deck to go to the toilet, and weakened by illness, simply fell into the sea and drowned. On arrival, the ship was not allowed to dock and Japanese doctors came on board with face masks to spread disinfectant. The survivors were loaded onto a launch and they entered the town.

Japan

At Moji the men mostly did dock labouring – coaling ships, loading and unloading cement, and so on – for several months. It had the highest death rate of any Japanese camp.

When in Moji camp in Japan, Sidney suffered one of the many punishments that he and others were sometimes subjected to, which consisted of being locked in a dark hut with no food, water or latrine. If it was possible, some food was sometimes smuggled in to such prisoners to help relieve their agony. The men took turns to risk helping these prisoners and on the one occasion that Sidney tried to help someone, he was caught red-handed by a guard and beaten up; then he was shoved into the hut instead of the other man. In addition he was hung by his arms for two days. The feeling left his arms after a short while, but when he was cut down and the blood rushed back, that was excruciatingly painful.

Another method which amused the Japanese was to make two prisoners who had committed some minor infringement fight each other with clubs; refusal meant a beating. On one occasion Sidney refused and threw down his club; the other man did the same. To their amazement the Japanese shrugged and walked away. Sidney pointed out, however, that it was equally common for Japanese soldiers to be treated this way by their superior ranks at all levels and he once witnessed this himself in a camp in Japan. A soldier had been found stealing from his comrades and as his punishment he was beaten in the way that guards sometimes treated Allied prisoners.

After Moji, Sidney was moved to a camp further south which was a copper refinery, working under horrific conditions; then on to a coal mining camp and in January 1945 to a small camp which had a mixture of different nationalities. He asked them where this was and one Aussie wit said the home of Madam Butterfly – Nagasaki, camp Fukuoka 14. Here he worked at the Mitsubishi engineering yards at Saiwaimachi.

The only act of any kindness by a Japanese that Sidney could remember was in the spring of 1945 in Nagasaki, when at a camp parade they were told that the European war was over. But the camp commandant told the men that Japan would never surrender and the POWs would be killed if Japan was invaded. This would be done in different ways in different camps – some by shooting, but in his camp they were to be sealed into a tunnel in the nearby hills, which the POWs themselves were building. Sidney was horrified by the idea of being buried alive and so, speaking a bit of Japanese he had learnt, he asked a guard whom he knew well if he would shoot him first if it came to it. The guard said he would do so only if Sidney would pretend to run away, as he had to have a reason to kill. Sidney agreed and the guard promised solemnly 'on the peace of the spirits of his ancestors'.

On 9 August the atomic bomb was dropped on Nagasaki and Sidney was only one-and-a-half miles from the epicentre. He was working outside, clearing up bomb damage; in previous days they had watched American bombers dropping bombs on the city, and some prisoners had been killed that way. Sidney also found out later that the first target that had been given to the bomber was near Moji, at his old camp, but if clouded they were to continue to Nagasaki. So Sidney reckons he was destined one way or the other to witness the incident at close quarters.

The Bomb

At about 11 a.m. on that hot, fairly cloudless summer day, Sidney was working near a large pile of rubble and heard an aircraft and looked up; he says he could tell from reflections that the bomb door was opened. The next thing he saw was an astonishingly powerful flash and a huge boom, followed by a very hot wind. He recalls that he closed his right eye and covered his left eye. Being a sunny day his pupils would have retracted and so been less exposed, and this probably saved his eyesight; his only injury was a flash cataract on his right eye.

Next he remembers hearing a loud sucking noise and seeing black smoke forming a huge mushroom cloud, blotting out the sun; he was riveted to the spot in shock. A soldier nearby dropped his rifle and fled back to the camp; around him were numerous dead bodies, but he was untouched. It appears the mound of rubble may have shielded him from the blast because it stood between him and the bomb.

He walked back to the camp in a daze; there was utter silence everywhere and the camp fence had been flattened. He thought an oil dump had been hit – of course they did not know anything about Hiroshima, four days before. Trees were completely white and stripped of foliage on one side but intact on the other. Hillsides appeared exactly the same: half white on the exposed side, half green on the sheltered side. Crossing a small concrete bridge he saw shadows – one of a standing person and one of a sitting person, burnt into the cement – but no bodies remained; elsewhere there were bodies lying about looking like raw meat one side and intact on the other; and the smell of burning flesh was nauseating. He saw a little girl, badly burnt, whimpering as she sat on the ground. He picked her up and took her to a soldier who put her on a truck. Unlike bombings in London during the Blitz, or in Singapore, where he had heard noise and shouting, here there was utter silence.

Soon Sidney watched as the Japanese had to organize the burning of pyres of dead bodies to stop disease spreading, and among some of these were dead POWs as well. He helped the Japanese look for survivors and remembers saying to a soldier, 'What have we done to you?' Sidney then saw something he had never seen until then; tears rolled down the soldier's cheeks, and he replied, 'What have we done to each other?' At that moment, Sidney felt that despite all the misery and horror inflicted on him by the Japanese, he could feel sorry for what had happened to them and he was the same as them and they were the same as him. Sidney found the Japanese grateful for any help the surviving POWs were freely giving, rather than wanting to retaliate.

Fortunately, soon after, heavy rain fell and this helped to damp down the radiation dust, though at the time the victims did not know about this hazard and everyone was eating and drinking as usual. On 15 August Sidney and other POWs were allowed to listen in to the radio broadcast that the emperor made announcing the surrender. They could hardly believe their ears. Soon afterwards, the music of 'The Blue Danube' came on and an American voice announced: 'This is the voice of Chunking, China; the atomic bomb dropped on Nagasaki has ended the war; Japan has surrendered.' They, of course, had no idea what an atomic bomb was.

Sidney was aware that people from amongst the prisoners died afterwards from the effects of the fallout, but generally there was blissful ignorance of the dangers, and Sidney certainly lived into old age. By now, some of the POWs had decided to get to Tokyo, because they heard rumours that MacArthur had entered Tokyo Bay with the US Navy and they wanted to get home. Sidney declined, as he thought it would be virtually impossible to get there and preferred to wait to be liberated. He found himself virtually alone. Soon afterwards, American planes flew over and dropped food.

Liberation

Sidney travelled home via Okinawa, Manila and Vancouver to Southampton. Among the after-effects of the bomb, Sidney had outbreaks of boils which gave off green pus, and also weeping fissures between his fingers and toes and on his genitalia. No doctor has been able to explain these. He occasionally had this into old age too. At the time, the Japanese gave him an ointment to help cure this, which did work, but he has never been able to discover what it was and so the effects have stayed with him always. Sidney is very angry about the lack of psychological help available to him and other POWs in Britain after the war, and he knew that many of his comrades, who came to him for help with their eyes (he became an optician after the war), later died in mental homes. After the war he campaigned for a while with Lord Mountbatten, Percival and Toosey (see note 1) to find out why POWs like him were never screened for radiation sickness, and for a while, under duress, the government established a unit at Queen Mary's Hospital in Roehampton to do this, but it amounted to no more than a pathetic gesture.

Sidney also had terrible nightmares for years after the war but with the help of his caring wife and son, he overcame these problems and immersed himself in his work and family. He is sure that his terrible experiences made him a far better husband and father than he otherwise would have been. He died in 2007.

Notes

1. Sidney said that after the war he spoke to General Percival and Brigadier Toosey at a reunion about the way in which the men felt the officers at this stage had deserted them during the retreat, while ensuring that they themselves had got away.
2. Also known as the *Singapore Maru*; weighing 6,000 tons, it was sunk off Japan by the USS *Seal* (under Captain Turner) on 9 September 1944.

Albert Lelah, Interned in Singapore

The information here is taken from an interview carried out in 1984 and stored in the Singapore National Archives.

40. Albert Lelah, Singapore, as a young man and as an older man.

Albert was born in Baghdad in June 1913 and went to Singapore with his parents in 1920. He then returned to Baghdad in 1922, where he attended the Alliance school and learned Arabic, English, French and Hebrew, before returning to Singapore in 1938 to St Anthony's school in Victoria Street. The family lived in Bencoolen Street, in a house where the Strand Hotel now stands, and where he had relatives, to establish a small export and retail business in Middle Road. Ironically, he specialized in selling Japanese goods to Iraq. In early 1941 he married a local Jewish girl, Leah Elias, and they had a baby son; they were expecting war and so he managed to send his wife and baby to Bombay, India, in September; he stayed in Singapore. Unbeknown to Albert until after the war, their ship was bombed but did make it to India. He and his brother were interned in Changi and then Sime Road prisons.

Albert experienced the heavy bombing of the city and remembers being in an air raid shelter when the surrender was announced in February 1942; on leaving the shelter he saw the great devastation caused. The Japanese published all kinds of rules about curfew and movement, in English in *The Syonan Times*, warning everyone that they had to be obeyed.

In the early days, Albert saw many Chinese arrested, especially those wearing tattoos on their bodies;[1] they were taken away in lorries and never seen again. Meanwhile the black market started as shortages arose, especially of food. Small groups of Japanese would come to the house and demand food and possessions from them; they had guns and bayonets and nobody argued; they had no common language and just bowed to the soldiers as they were told to. After about two months, Albert was ordered, as others were, to open his store. One day a local Japanese merchant he had known arrived in uniform, with medals and sword; Albert was amazed; clearly the merchant had been a spy before the war. He asked how Albert was, calling him 'Alberto', and proceeded to take everything he wanted; he placed it in his car and drove away without payment. Later Albert and his brother moved the valuable items to their upstairs apartment as they knew this looting was not going to stop.

On pain of fines, the Japanese ordered all prices frozen; when they came to the shop, some paid in useless Japanese occupation money (locals called it 'banana money') or did not pay at all. In the third month all the Jews were ordered to register at the post office in Waterloo Street. Albert and his brother knew what had happened to the Jews in

Germany, so they declared themselves Iraqi subjects, not British. Not knowing quite what to do, the Japanese arrested them and took them to the lock-up at Beach Road police station; after a few hours a Polish Jewish teenager was brought into the cell and there they stayed for two nights. The cell was infested with bugs and they just stood by the cell doors, not daring to lie down. They watched many others being brought into other cells – mainly Chinese and Indian – and they were beaten with heavy sticks as they ran the gauntlet of Indian and local police (collaborators with the Japanese), not Japanese. Most were badly bruised and bleeding. The cells were only about nine feet by twelve feet, and fifty men were forced into one cell; they had only the three in theirs, and during the night a kindly Indian guard named Musa brought them water.

After two days, General Count Asahai and some other soldiers came to the prison, and Albert and his brother were called out to see them. The Japanese sat at a table; the two brothers bowed. They were told that Iraq was at war with Japan and so they were enemy subjects. The two brothers were shocked – they simply had no idea. They were taken straight to Changi by jeep, and were very depressed

Changi

On arrival they were registered and placed by a block leader – a British doctor – in Block C, on the first floor, and given a fork, spoon and plate, and a towel and toothbrush (in the early days, such stocks existed). Their first evening meal was mashed rice and black tea. They could not eat it and gave it away. Next morning they ate the tapioca flour bun and drank tea for breakfast.

For a while, coconuts, lemons and newspapers could be obtained, and occasional Red Cross parcels, but later everything became much more scarce. The internees organized concerts to pass the time and Albert remembers the Christmas concert as very enjoyable. He found the many professional-class prisoners and officials very nice and they adjusted to the situation among kind and well-organized people, especially the British internees, among whom morale remained very high. After January 1943, however, the Japanese confiscated the stage and lights, saying the concerts were too noisy.

Their particular friend was Harry Jackson, a company owner, who was their block quartermaster and very kind to them. They called him Uncle.

One day Count Asahi came to Changi and sent for Albert. He asked him if he would agree to become a radio broadcaster in Arabic for the Japanese, but Albert refused, saying he was not political, but a simple merchant. Asahi then asked Albert who he thought would win the war. Albert knew he had to think quickly so he said that the Japanese would, as they were all Asians together; Asahi seemed to be satisfied with this answer and Albert was allowed to go. There were no repercussions.

The Double Tenth

On 10 October 1943 (known as the 'Double Tenth'), the Japanese cracked down on all camps in Singapore, following the Australian Commando raid on the harbour, convinced there were spies and radios among the prisoners. All the men were called into the square and made to stand in the sun all day while the cells were searched, as punishment, surrounded by heavily armed guards. By the evening all were allowed back to the cells and eventually about a dozen men were arrested (see Freddy Bloom's story in this book); many were tortured and later died when sent back to the camp.

Sime Road Camp

In 1944, Albert and others were sent to Sime Road camp by lorry, in exchange for military POWs who were finding it too easy to get under the wire at Sime Road and visit the city. Changi had walls. At Sime Road they had bungalow-type huts and were able to plant gardens with crops for food; it was more open and fresh, but there was less and less food as time went on. Using secret radios they were able to obtain news about the Japanese defeats, but to cover the secret it was always said that a Chinese on a bike told them as he passed the barbed-wire fence. The camp leaders told the prisoners to continue to behave and obey at all times in case the Japanese decided to take revenge on them for the defeats they knew they were experiencing.

When the surrender was announced, the internees were allowed visits by their friends from outside who brought food and so on, and were asked by Lord Mountbatten if they wished to go and witness the signing, down near the cathedral. Albert and others went, in their shabby shorts and shirts, and he described how marvellous it was to see and how free and easy they now felt, being out of prison. After several days they were asked where they would like to be transferred, and Albert and his brother asked to go to Baghdad to see their parents; then Albert would go on to India to find his wife and baby. They sailed on the SS *Holland* to Colombo, enjoying the food and playing games, but not the swaying ship. At Colombo they were allowed ashore to enjoy the hospitality of the Red Cross and locals, and they were kitted out with new clothes from head to foot. At Port Said they were again entertained but told to report to the Iraqi consul for details to proceed to Baghdad, where they were flown to some days later, and then on to Bombay. It was a tearful meeting with his wife – his son was now over 3 years old and called him 'Daddy', even though he did not really know him.

Eventually Albert and his wife returned to Singapore in 1946 and picked up their lives anew. He reopened his business and became a wealthy man – prominent in the Jewish community and giving generously to both Singapore and Israeli charities. He died in October 1997.

Note

1. This usually symbolized membership of Chinese secret societies, which the Japanese feared as centres of resistance.

5950940 Pte Harry Leslie (aka Zilesnick), 5th Battalion, Beds and Herts Regt

Harry was interviewed and tape-recorded by the local Jewish community in Australia between 1992 and 1995, and copies are kept at the AJEX Jewish Military Museum.

41a. 5950940 Pte Harry Leslie/Zilesnick, 5th Beds and Herts, taken in July 1940.

Harry was born in Bethnal Green, London, on 29 November 1919, youngest of nine children of Russian Jewish immigrants. His father was a cabinetmaker, the son of a rabbi. Having been a tuba player in the Russian Army band, his father was often called upon to blow the shofar (traditional ram's horn trumpet, blown only on the Jewish New Year and Day of Atonement) at their local synagogue, where Harry was also in the choir. Harry left school at the age of 14, one year after his bar mitzvah, as his father died and he had to earn an income. He became a tailor like many other Jewish young men at the time. During the 1930s he was involved in the anti-fascist movement in the East End, joining the Young Communist League because, as he said, they were the only group fighting Moseley's fascists on the streets.

Harry was called up on 18 January 1940 (he was living at 50 Forest Drive, Leytonstone, London E11) and sent to Cambridge to the Beds and Herts Regiment. After six months of full military training he became the battalion tailor. Later his battalion were moved to guard Mildenhall aerodrome in Suffolk during the invasion scare. Before going overseas in October 1941, he experienced some of the bombing in the Blitz in London, where his parents lived. They were bombed out and had to live at various country addresses throughout much of the war. At this time Harry was urged by his CO to change his name in case of capture by the Nazis and so he became Leslie, not Zilesnik.

Singapore

When he left Liverpool by the troopship *Sol del Raina* on 29 October 1941, it was thought to be going to the Middle East. They sailed to Halifax, Nova Scotia, first and then transferred to the USS *America.* While they were at sea, the war with Japan broke out and they were told that they were going via Cape Town to Bombay. After one month of jungle training at Ahmednagar camp in India they re-embarked on a troopship to Singapore, where they arrived on 29 January 1942. Harry was aged 22.

Their ship was being bombed as they disembarked and there appeared to be total chaos. As they landed, other ships were evacuating civilians and soldiers. During the following days, Harry was involved in the fighting at Bukit Timah and Thompson reservoir. While he was loading stores onto his truck, it was hit by a Japanese mortar shell and he was wounded in the thigh by shrapnel on 14 February.

After a week in hospital he was taken to his unit in Changi gaol. The food was bad, and morale was low because the men had no idea what to expect. Harry was put to work making puttees into caps for the men, as few had headgear to protect them against the punishingly hot sun. Amazingly, Harry had an emotional meeting with his brother-in-law, Sam Nemko, in Changi, who had been serving with the Australian Signals throughout the fighting in Malaya (he later survived the Burma Railway). While at Changi they managed to attend some Jewish services held in the camp.

For about three months, Harry was sent on working parties in Singapore (mainly to Havelock and Valley Road camps) and the mainland of Malaya, where they worked bringing timber back to Singapore, often being able to bring back extra food for themselves too. But on these trips, he and his mates often witnessed the results of some of the appalling atrocities committed by the Japanese on local Malays and Chinese: decapitated heads were displayed on poles everywhere.

Finally, after six months, in August 1942, he was transported with about 500 others to Taiwan, to Heito camp in the south of the island, where he spent two years. He spent a further year in other Taiwan camps .

Formosa/Taiwan

Harry was transported in one of the Hell Ships, *Dainichi Maru*,[1] in appalling conditions, and battened down in the holds with hundreds of other POWs. Many died and were buried at sea.

At Heito, POWs of several nationalities passed through the camp[2] – British, Australian, American, Dutch, Chinese – and the first work he was put to was loading boulders into baskets from a riverbed and carrying them, two on the end of a pole across the shoulders, to a railway truck. From here the rocks were taken to be used as ships' ballast. Later he became the camp tailor and his life became easier. Officers worked in the vegetable gardens to provide extra food, and emptied the latrines for night soil in the gardens. Harry also worked in the nearby sugar mill, preparing heavy canvas for use in the factory.

At Christmas time, the POWs organized shows and were given some extra food by the Japanese to celebrate, when they were thought to be winning the war – that changed later on, of course. His parents did receive several cards from Harry while he was in Taiwan, which he was able to retrieve and keep when he got home.

In one incident at Heito he was falsely accused by the Japanese of breaking into the stores and stealing sugar. The whole camp was paraded by the Japanese and a search was carried out. As a tailor, and used to trading with the Japanese and other POWs for work done, Harry already had some sugar, but the Japanese concluded that he was involved in the incident and he was singled out for a severe punishment, together with his two tailor assistants, one of whom was a fellow Jewish POW, Leslie Weinfass. For a whole day he was made to stand in the sun with his hands above his head, and on the orders of Commandant Tomaki, during eight days of solitary confinement and starvation rations, he was beaten each day for an hour, and kicked and hit with rifle butts. They wanted names from him but he did not know who had carried out the theft. Fortunately a Red Cross delegation, led by a Mr Augst, happened to visit the camp. On this occasion, fake 'happy events' were set up, new uniforms given out and extra food distributed to the POWs. All this disappeared as soon as the Red Cross party had left – like the Nazi trick used at Terezin in Czechoslovakia. Only on the ninth day was Harry released

Dr O'Donell, the Australian MO, did great work in the camp, and Harry was present to assist him when he had to amputate a man's leg without anaesthetic – the man's screams were terrible. One day Heito was bombed by Allied aircraft and some POWs and Japanese were wounded and killed.

Heito was closed after more than two years, and in March 1945 Harry and the others were moved to Tyhoku, in the north of the island, which was being continually bombed by the USAAF. Harry himself was slightly wounded over the eye by shrapnel. Men died daily of dysentery and gangrene from tropical ulcers. Beatings of prisoners were common. Working on the bed rounds in the hospital he witnessed horrific sights. Between raids they worked in the fields, and one day at the end of May a special working party was formed from the fitter men, including Harry.

Taken by truck with their packs a short distance to the end of a road, they were marched across rough country along treacherous tracks, to a rope bridge, and they crossed a turbulent river. Passing a disused mine they halted in a village and rested in the school to await their next march up a mountainside on the next day. The march involved negotiating dangerous single-file tracks with a huge drop off to one side as they walked ever upwards, under vertical waterfalls, until they reached a flat-top plateau with jungle and forest. Here they were put to work on timber and bamboo felling, but all food supplies and the timber had to be taken up and down the mountain manually. Rations became fewer and fewer, sickness was barely treated, and their treatment by the guards became more and more brutal. The camp was above the clouds and they had to build their own huts and sleep on bare earth. The only positive thing was the constant supply of fresh drinking water and water to bathe in, from the waterfall.

The last three months of his imprisonment in this camp were the worst; it was known as 'the Jungle' or 'Mountain' or Yama camp.[3] In revenge for American bombing, which killed both prisoners and Japanese guards, the Japanese treated the POWs particularly harshly. Beatings were frequent, and there was less and less food. On their way to and from the work areas they foraged for and ate snails and tree bark to try and keep going. Half of the 100 men at this camp died.

Liberation

On 15 August 1945 the camp commandant suddenly announced that the war was over and everyone would go home. The guards cooked the POWs a meat meal but they were unable to digest it and were sick. They were then escorted back to the main camp at Tyhoku, where their comrades were horrified at how emaciated they had become in the short time they had been away. Harry reckoned that after another month they would all have died. The air forces began dropping food and medical supplies, but one parachute holding a canister failed to open and hit a hut, killing one of the POWs. Eventually the American and Chinese navy arrived and they were taken by truck and then train to Keyloon docks, where they boarded a destroyer. Two American sailors, one either side, had to help Harry to get up the gangplank because he was so weak. The destroyer then transferred them at sea, by hoist, to the US aircraft carrier *Santee*, where Harry was put into

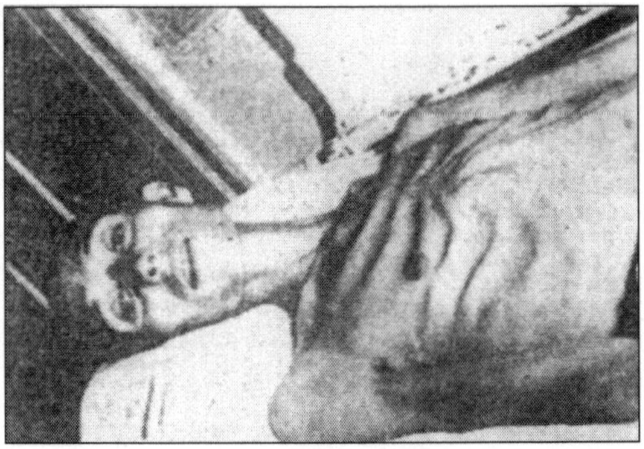

41b. Harry on liberation in 1945.

the hospital section; it was 7 September 1945. While he was in the hospital area, an American officer brought him a Jewish prayer book. All the possessions of the POWs had to be destroyed because the Americans were afraid of any infections that might be carried in the kitbags. After delousing, blood transfusions and bed, Harry was asked if there was anything he would especially like and he said orange juice, and he was given it, freshly squeezed.

On 9 September they arrived in Manila, and Harry was sent to US General Hospital 248 and given the best attention. He still had beri-beri but he managed to recover slowly and began eating normally. Others of his mates, however, could not stand the shock, and deaths still occurred in the hospital among those who had managed to reach liberation, but too late. An Australian nurse, NFX122286 Lieutenant Nancy Shepherd (née MacQueen), 2/5th Australian General Hospital, attached to 248 US Hospital, wrote to Harry's mother in England to say he was safe and well. Over fifty years later he managed to trace her and they met in Sydney at a lecture he gave, where he pointed her out to the audience as his Angel of Mercy.

By 16 September he was walking a little and on 17 September he had huge portions of bacon and eggs, even though it was actually Yom Kippur – a Jewish fast day! Gracie Fields arrived to entertain them and they all sang along.

On 9 October he left aboard the aircraft carrier HMS *Glory* to Hawaii, then on to Victoria, British Columbia, where he was entertained by a local Jewish family, the Lancasters; by 3 November his weight was up to 146lbs. He travelled across Canada by train, which made many stops to drop off soldiers on the way, arriving after five days at Halifax, Nova Scotia, which is where his journey four years earlier had really begun. From there the Canadian hospital ship *Lady Nelson* took them to Jamaica. On 2 December he arrived at Southampton, and there to meet him was his brother-in-law, Sam Nemko, who had survived the Burma Railway, and his brother, also called Sam.

Because of his malaria, he had to stay in hospital for tests for a week, but finally got home to Westcliffe in Kent for a huge family reunion. His mother had not known until 28 May 1943 that he was alive.

Harry said that he always knew he would survive; he said he dreamt of his father, who had died when he was only 13, telling him he would be alright and get home. Harry thought often of the Jewish people who had been slaves to Pharaoh, in the story of Passover celebrated each year, and he knew how that story ended – in freedom; this is why he knew he would come through. He was equally determined to adjust to normal life as quickly as possible, and he succeeded in this too.

At liberation, Harry weighed only just under seven stone (ninety-six pounds), half his normal weight, but he recovered. Twenty years later, in 1968, he went to live in Australia with his wife Julie. They had a daughter, Lauren, and a grandson, David. Harry became prominent in the clothing industry and lectured on management and manufacturing techniques in New South Wales colleges, and was involved in many communal organizations, including being secretary of the New South Wales Association of Jewish Ex-Servicemen and Women.

In February 1992, Harry returned to Singapore with veterans, on the fiftieth anniversary of the surrender. In front of a huge crowd of 1,000 people, he said Kaddish for the Jewish soldiers who had died as POWs, and wept for their loss. In 2001, now frail with age, he returned for the first and only time to Formosa, with other veterans. He hoped to help historians find the location of the remains of the notorious Yama camp, but the jungle hid it too well. However, he did meet two elderly Formosans who had been perimeter guards there, employed by the Japanese. Harry said the POWs never met the Formosan guards, and it was the Japanese who did the beatings; he found it inspiring to meet them. In later life he never bore any hatred towards the Japanese, but would never allow anyone to forget what they had done.

Notes

1. Eventually torpedoed by the USS *Gurnard* on 10 August 1943.
2. It was to this camp that General Percival, who commanded all Allied troops in Malaya/Singapore, and General Wainwright, the US commander in the Philippines, had been taken, together with other senior officers.
3. The appalling conditions and treatment in this camp have been carefully documented and appear in the book by Jack Edwards, *Banzai you Bastards* (Corporate Communications, 1991). Harry says that the detail in the book about Yama is totally true and he knew many of the men named in that chapter and experienced all the horrors described therein.

2350485 Sgmn Aubrey Lichman, RCOS

42. 2350485 Sgmn Aubrey Lichman, RCOS, taken at Kuala Lumpur in June 1941.

Aubrey (Alfred) Lichman was born on 12 June 1920 at 515 Commercial Road, Stepney, London, son of Nathan (a tailor) and Yetta (a dressmaker). He went to school at Senrab Street primary and then Mile End Central in Myrdle Street. In 1939 when war broke out, Aubrey was working with his father in their new home and shop near Oxford Street. He was called up in August 1940 and went into the RCOS in Prestatyn, where he trained as a teleprinter operator (in No.34 Teleop Section). Here Aubrey confronted his first incident of anti-Semitism when a fellow signaller named Smith referred to a Jewish soldier called Walters as a 'Jew Boy' in a conversation. Determined to challenge this, Aubrey went over to him and said that, being Jewish, he resented the term 'Jew Boy' being used. Smith apologized and said that he had spoken without thinking, and the matter was closed. Aubrey was later posted to nearby Colwyn Bay and then overseas on the SS *Duchess of York* to Singapore, where he arrived in May 1941.

Sadly, the ship had taken on contaminated water at Bombay and all the ship's company were sick. After two weeks of idleness in barracks in Singapore, there was a sudden posting to East Camp in Kuala Lumpur, Malaya, by truck, to join Malaya Command Signals. On the first night they bivouacked near a small village, and organized a night guard. During his duty, Aubrey was startled when he thought he saw the eyes of a wild animal approaching the trucks; it turned out to be two fireflies. At their next halt at Seremban, they camped with an Australian unit, where they were amazed and pleased to be invited by a major to have a drink and mix totally freely, with officers and all ranks as equals, on first-name terms – something completely prohibited in the British Army.

It was while they were in Kuala Lumpur that the Japanese attacked Pearl Harbor in December 1941, news of which Aubrey picked up on their teleprinter in clear language (unlike radio, teleprinter could not be tapped and so most messages were not coded). After this, everyone was on full alert.

Soon afterwards, Aubrey was sent, alone, on detachment 200 miles north to Sugeipatani in the rubber plantation belt in the north-west, with Indian Army Signals. Here, he was one day called to meet the OC. The white English officers in the Indian Army were very arrogant as a rule, and this particular officer, a Captain Curry, began lecturing Aubrey about the need to sacrifice one's life for King and Country. But as the Japanese invaded north Malaya, Aubrey was suddenly ordered back to Fort Canning in Singapore, British Army HQ, by train. On arrival he was told that they knew nothing about him and he was sent back across Singapore Causeway to an Australian unit at Johore, where he worked for them as a signaller. As before, the informality was a breath of fresh air and it was common to have beers while they were on duty. Sent yet again back to East Camp at Kuala Lumpur, he was kept busy, receiving constant messages about the Japanese invasion progressing down the peninsula, while during off-duty times, instead of weapon training, he was ordered to do square-bashing by officers who clearly had no idea what else they should be doing! As Aubrey said, the situation was reduced to a farce.

Yet again Aubrey was detached to return to Sugeipatani, but as his group travelled north they found the roads choked with Allied troops retreating, and the military police sent them back because the Japanese were only five miles away; meanwhile, they saw flights of Japanese bombers on their way to hit Singapore, and were receiving messages that fewer and fewer British aircraft were available to stop them, as they were being shot down and the airfields overrun. As signalmen they were, of course, sworn to secrecy and could not tell other troops the truth of what was going on.

Journeying south again, at Ipoh they stopped for a meal at an army canteen where a dramatic scene occurred. There was a lovely and kind middle-aged English lady serving them food, and suddenly a brigadier marched in and shouted, 'Mrs So-and-So, I see you are still here; you have disobeyed my direct orders to leave. You must go at once.' 'But', she replied, 'I cannot leave my boys!' 'The boys', said the brigadier, more quietly 'must now look after themselves.' Aubrey described it as a scene right out of a film; he never discovered what happened to her.

At Kuala Lumpur, all the men were paraded, and married men were ordered to continue the retreat. The others, including Aubrey, were ordered to stay to make a fighting stand at Kuala Lumpur. Godowns (warehouses) were bombed by the Japanese and one of Aubrey's mates had acquired a 1937 Ford V8 car, and had got him and some others to help him liberate some cases of whisky and condensed milk to boost rations. But orders changed again and they were soon on the way south once more, to Johore, then to Singapore itself and Fort Canning. From here they were posted to the north-west of the island, again with Australians. They received messages from the fighting units that Japanese landings were being repelled and no reinforcements were required, and at that moment Aubrey went off duty. Next morning he awoke to bedlam as huge numbers of troops and equipment were streaming back from the coast. They were also machine-gunned by a passing enemy aircraft and the terror of this, as Aubrey dived into a ditch, made him vow that if ever he had sons they would never go into the army.[1]

As the withdrawal to Singapore City began, Aubrey was ordered to stay with the rearguard signalmen but the teleprinter broke down and the Aussie officer ordered him to take his rifle and ammunition and take post in the trenches with the others, until they received orders to leave. Mortar shells began to fall nearby and a fuel dump somewhere went up in flames, but soon orders arrived to withdraw and they ended up in the botanical gardens in the city. From here he was ordered to find his unit but on the way was stopped and sent on a patrol in wooded area to look for Japanese infiltrators. This turned out to be fruitless and so Aubrey was finally directed back to his signals unit, who returned to Fort Canning. Here they heard of the surrender. Soon afterwards they were given Japanese armbands by a sergeant, and Aubrey and two other men were told to tidy up and stay behind and await orders while the rest of the garrison marched to Changi as POWs. The next day, walking back from the swimming pool in Fort Canning, which they used as a substitute for washing because the water supply had been cut off, they bumped into a

Japanese guard. This was the first time Aubrey had seen a Japanese face to face. His rifle and long bayonet made him look menacing and he began shouting at them, so Aubrey took out a cigarette case he had found in one of the offices and walked over and gave the cigarettes to the Japanese soldier, who offered them one in return. Aubrey told his mates that they should now slowly turn and walk away, and nothing further occurred.

Later that day they were sent by truck to Changi camp, where he met an old friend, Israel Shapiro, whom he knew from Redmond Road Synagogue Hebrew School (cheder)[2] in Stepney. They fell into each others arms and exchanged news.

Changi

At Changi, he and his group assembled in a hall where a concert was arranged, after which they slept on the floor. For a few days nothing happened as the men got to know the routine in the camp; he met another Jewish soldier, Len Halpern from Manchester, who asked whether they should put Jewish or British as their nationality on the forms that the Japanese gave them to complete. Aubrey told him to put British, because as allies of the Germans there was no telling what the Japanese might do. Aubrey also met David Arkush (whose testimony is included in this book) whose Friday evening synagogue services Aubrey had heard about but did not attend in the short time he was at Changi.

Food deteriorated at first and rice contaminated with sulphur from a warehouse was supplied; it tasted vile but had to be eaten. Later the food improved. Aubrey was in the former married quarters buildings of the old Changi barracks, called Temple Hill. Though not too uncomfortable, the buildings were teeming with bugs. He shared a room with three good friends who had been together with him since the arrival in Singapore, and they remained together for some time. For light they used cans of oil with a string wick. Any signals men were designated as despatch riders using bicycles to get from one section of the huge camp to another, carrying messages between the officers.

This was when Aubrey had his first brush with the enemy. The INA (Indian National Army), who were regarded as traitors, were given epaulettes and were used as armed guards by the Japanese in Changi. Orders were that British POWs were to bow to them as with the Japanese. On his first trip as a messenger Aubrey reached a gate and two British MPs told him that an INA guard was beckoning him, but he should watch his step with him. Aubrey got on his bike and went to the INA guard. He was a sergeant and Aubrey saluted him. The INA guard said, 'Why you not salute Indian soldier?', pointing to a Sikh soldier nearby. Aubrey said that he had not seen him; the two INA men had a discussion and then without warning the sergeant struck Aubrey full in the face, knocking him down. Aubrey stood up and saluted again, and again he was struck heavily in the face.[3] The Indian soldier ran towards Aubrey with his bayonet, jabbing him with it; Aubrey was not afraid, but deeply enraged. After a few moments the sergeant told him to go. When he got back to his HQ and told the story, he learned that he could have reported this to the official Japanese camp interpreter because such behaviour by guards was not permitted. Aubrey heard that two days earlier a similar incident had occurred with an Australian soldier, and the Japanese interpreter took the Australian in his car to the scene of the incident and allowed the Australian to beat the INA soldier who had beaten him.

The other major incident Aubrey remembers was the notorious order from the Japanese that all POWs were to sign a paper that they would promise not to attempt to escape. The

British refused because it was against regulations, and so all the thousands of POWs were ordered to assemble in Selarang Square at Changi barracks, an area about the size of two football fields. They stood exposed in the sun, surrounded by heavily armed Japanese; men brought their groundsheets and packs and sat around while negotiations went on. One of Aubrey's section officers, Lieutenant Corbett, was very close to his men and often left the officers' area to come to sit and chat with his them. Earlier, he had taught them in the camp how to play bridge; he preferred the company of the signallers to his fellow officers. Eventually some food and water were served by the cookhouse but conditions considerably worsened when an outbreak of diphtheria occurred and the MOs warned the senior officers that if something was not done, there would be a serious epidemic that would cost many lives. After about four days, the British agreed to sign the 'no escape' pledge, but the caveat was added that it was done 'under duress'. Tables were brought up and Japanese and British officers oversaw the signing by each man, and then the men were allowed back to their rooms.

A Jewish officer whom Aubrey met at Changi was Joseph Morris, who had come to Singapore as a tax inspector before the war and was a friend of Aubrey's brother; they had studied accountancy together. His parents owned a bakery in Christian Street, Stepney, and he gave Aubrey a small supply of biscuits he had managed to get into the camp, which he had recently received from home. Joe had been offered a commission just before the surrender, as officer friends advised that he would fare better as a POW than a civilian internee. They then made him catering officer for his POW section because he had experience in the field.

Orders came through that the men were to be moved to a so-called Red Cross camp in Thailand. A particular friend, Frank Murray,[4] tried to keep together with Aubrey but Frank developed dysentery and they were sadly separated. When they got to the railway station, the POWs were herded into metal cattle trucks, with the doors closed. The journey was hellish with only a few stops on the way to be served a little food and empty their latrine cans. At one stop the men were allowed out to get some air and Aubrey witnessed one of the rare amusing incidents in his POW experience.

Two very young British officers approached a Japanese guard (all guards had to be addressed as 'Nippon'), saluted and bowed and in patronizing pidgin English said,

'Nippon, him, me British officers' and, pointing to the cattle trucks, added, 'No good, officers travel on flat bed trucks behind, him, me, ok, you ok, him me truck.' The bemused Japanese guard stood listening to this performance, as did Aubrey and his comrades, smiling at this typical attempt to communicate with a guard by some rather inexperienced British officers. The explanation went on for some time until suddenly the guard replied in perfect English with an American accent, 'Oh I see; I guess you two want to go and sit in the truck there; that's ok with me but you will have to behave yourselves you know.' As mouths dropped open in surprise, everyone burst out laughing. It transpired that the guard had been living in San Francisco and just before war broke out had travelled to Japan to visit family and had been caught up in the war. He explained to Aubrey that Japanese Americans like him were not trusted by the Japanese authorities and were mostly consigned to guard duties.

Thailand

After a four-day journey, the train arrived at Bam Pong, which was a filthy and squalid camp that constantly flooded and was consequently full of waterborne diseases from both insects and latrine effluent. After one day they moved on to another camp (Wain Lain/Wun Lung), where they stayed in better conditions for about six to eight months.

In one incident, the Japanese asked all the POWs to parade and empty all their pockets out onto the ground. Among Aubrey's things was a photo of his very attractive sister. A Japanese guard picked it up and demanded to keep it, but Aubrey said it was his wife and he would not part with it. The guard became angry and insisted, but Aubrey stood his ground and eventually the guard gave in. Later, when he saw the cruel beatings given to POWs for much less important infringements, he realized what a terrible chance he had taken. Today the photo stands proudly in a frame in his home.

While at this camp a POW had broken into the Japanese office and stolen a typewriter. The Japanese carried out an intense search of the camp but the item could not be found. Mass punishment of the men was promised but, astonishingly, nothing was ever found or done about it by the Japanese. Aubrey said that generally it was always best to keep as far away as possible from Japanese guards, whose behaviour was totally unpredictable and not governed by any real rules of rational behaviour or military law. The Koreans were the worst and to be avoided at all costs; the Japanese ill-treated them and they in turn took this out on the POWs.

Meanwhile, the men worked on the railway for twelve-hour days. Aubrey worked in a team of six – four digging and two moving the earth on stretchers, two poles with hessian sacking to carry the earth that was being removed. Mostly at this stage they were clearing jungle between given markers and then building up the embankment to receive the sleepers and the metal rails. Most of the men's uniforms had rotted by now and many were wearing only loincloths ('Jap-happies') made from rags. After work, prisoners washed in the river and those who had kit could even shave. Aubrey's treasure was a spare pair of shorts he could put on in the evening after his hard labour, which had a very positive effect on him psychologically, making him feel less of a slave and more of a human being. Sadly, theft was rampant and both his shorts, one pair of trousers and his only clean shirt were eventually stolen by fellow prisoners. He knew who had taken one item but nothing could be proven and the matter was dropped. Blankets were particularly precious, as the temperature – which was very high during the day – fell to about 60 degrees Fahrenheit at night; they were also crucial if you were ill. Many blankets were stolen. Hats were a precious commodity too; one story Aubrey heard, although he does not know if it was apocryphal, was about when one man went into the bush to relieve himself and, while he was occupied squatting, a hand appeared from behind and stole his hat. After that all men were told to keep one hand on their hat at least, when engaged in calls of nature!

At the first Christmas, the concert party put on a show, but the Japanese did not give a day off. After work the officers organized horse racing whereby officers rode on the back of an OR (Other Ranks), for a fee, and they raced each other along a given track – carefree, if classist!

Chungkai

Aubrey was transferred eventually to Chungkai, but he had malaria, dengue fever, pellagra and dysentery, and so was taken off railway work for a time.

Aubrey later met Joe Morris again at Chungkai when they were first given cards to send home using the usual format: printed phrases which you chose to use, whilst deleting the others you did not want to use. You were permitted to write your address and signature and the date, and allowed fifteen other words in which to write a very short, innocuous message. Joe asked Aubrey to add to his own message the phrase 'please remember me to Joe Morris OK' and in return he would write on his card 'Aubrey Lichman OK'. The cards got through and Joe's wife Oona called Aubrey's parents to say the two men had met and both were alright. This was a clever ploy to reassure parents and relatives of some good news about the condition of their menfolk and how they were helping each other survive.

One day a consignment of trilby hats arrived and were given out; nobody knew where they had come from. Often the men made their own sandals from worn-out tyres and pieces of wood.

Lieutenant Joe Morris approached Aubrey and said he had an important and crucial job for him. Morris was in charge of the kitchens and he believed somebody in the cookhouse was illicitly selling supplies to local Thais and pocketing the money at the POWs' expense. Captain Strauss (another Jewish officer) and Morris were appointed to find the culprit and wanted Aubrey to go and work in the cookhouse undercover, and watch events and report back to them. Aubrey was made assistant wood-man, to help supply the cookhouse with fuel. After cutting, wood would be left to dry, and Aubrey was given an axe and he helped out. He was viewed with suspicion by the clique who ran the kitchen, and one day he was offered some extra food. He said no at first but the man persisted and Aubrey saw that this was not the way to gain their confidence, so he accepted the offer.

The officer and sergeant in charge were very nice individuals and Aubrey began to doubt that anything was amiss with the supplies. Suddenly, after about ten days, Joe told Aubrey that he and Strauss had been removed from the cookhouse post and Aubrey was on his own in the enquiry and could do as he felt best. The job was a privileged one, as he was off the railway work for a while and there was always a bit of extra food to be had. The sergeant in charge called Aubrey for a chat and said he could stay on if he wished, and so he did this, and learnt how to cook and be of general use. He is convinced that this was one of the things that saved his life.

At Chungkai a Jewish soldier called Bluestone became a wheeler-dealer in buying and selling items both within and outside the camp. He is one of several contenders for the model for the character of 'King Rat' in the movie of the same name, starring George Segal.[5] He had men 'going through the wire' at night into the Thai villages, buying and selling. However, he often acquired essential medicines for the sick, which saved many lives, but this was not widely known. While working in the kitchen, the men saw Captain Markovitch walk by and one of the men said, 'There goes Marko, great bloke'; it is not clear whether any of them knew that Markovitch was Jewish. Shortly afterwards, Bluestone walked by and the same man said, 'There he goes, Bluestone the F—g Jew-boy.' Something in Aubrey snapped. He went and spoke to the sergeant outside and explained that if there was a fight he wanted him to know that he had not started it. The sergeant said he understood. Aubrey returned to the kitchen, which had a large open space in its centre, and called out in a loud voice that he had an announcement to make. Aubrey said he was Jewish and proud of it and he had heard insulting remarks about Jews being made in the kitchen which deeply offended him. He said that if anyone wanted to discuss it further, he would be glad to meet them outside. There was silence. Suddenly there was a hand on Aubrey's shoulder; two professional boxing champions who worked in the kitchen, Taffy Hall from Wales and a cockney called Charlie Baldwin, stepped up, and Charlie said 'And what Aub here says goes for me too'; then Taffy stepped up and said 'And you can count on me too.' The matter was closed.

In another incident, one of the leading organizers of the concert parties, named Leo Britt, was with Aubrey as they were changing into clean shirts, when one of Britt's buttons came off. He remarked, 'Oh look, a Jew tailor is saving cotton!' When Aubrey objected, he replied, 'Jews always try to save money!'

It was at Chungkai that Aubrey began his friendship with David Arkush, although he attended only a few of the services there. He admired David hugely for the work he did in general but especially with the Jewish POWs.

Back and Forth to Chungkai

Having been ill again with malaria on the railway, Aubrey was sent to the hospital to recover. With limited saved pay from his work, he, like all POWs, could buy a few extra small luxuries to help himself get better, such as eggs or extra rice. He met a tough Scots soldier from Glasgow, Gibby Douglas. When Aubrey was ill, Gibby used to bring him his food and what Aubrey could not eat he gave to Gibby. It was Gibby who forced Aubrey to be self-reliant and he would not accompany him to the latrines, telling him he should go alone and get back into the normal routine to help himself recover. This psychology helped Aubrey get well. After the war, Gibby's son wrote to Aubrey and asked him to confirm the story, which Aubrey had pleasure in doing.

Feeling slightly less unwell and not wishing to be thought a malingerer, Aubrey later reported for work one day and was delegated to the charcoal party by a very obstreperous sergeant called Sergeant Major Major! But Aubrey had no shoes – they had fallen apart – and the work was thus very painful for him, walking about in the thorn-infested jungle. His feet became infected and he returned to the camp and was given treatment, and told to inform Major next day that he could not do that work again. At the next work parade the sergeant refused to accept this and so Aubrey went to see the MO and explained the problem. The MO said the sergeant was a 'Jap-happy bastard' and Aubrey was to go immediately to the hospital and stay out of the sun and away from heavy work until further notice and, furthermore, to have a blood transfusion for anaemia. Next day at the transfusion hut, Major walked in and demanded to know why Aubrey was there. The MO said, 'this is the man you are going to give blood to. Touche!' 'OK chum', said Major in good humour, 'you will get a good drop of Cornish blood.' Every time Major saw Aubrey after that, he asked him how the blood was doing.

Sadly, the New Zealand doctor who ordered the transfusion, Captain Sutherland, died later in the camp, just one week before liberation.

Across the river from Chungkai was Tamarkan camp, where Aubrey was sent for some further medical treatment. At *tenko* one evening a flight of over twenty American bombers came over to bomb the nearby Kwai bridge. The Dutch on the parade ground immediately broke for cover as they had been accidentally bombed once before by American planes, when further north, and knew what was about to happen. So the parade major shouted for all to take cover – even though there was none, on an open parade ground! Over 1,000 men scattered. Aubrey dived into one of the many hut drainage ditches and was safe.

Later Aubrey was sent further up north on the railway by train and foot with about 100 men commanded by a Major Beaver (?) to a small communications camp, under canvas, with three Japanese officers. On the way they stopped at another camp to eat and rest and set up their tents. One of the officers asked one of the cooks if they could catch a cow to cook and eat. But how could this be done? At about 2 a.m. a member of the group turned up and said they had snared a cow from the camp, owned by the Japanese, who let the cows go to the river to drink at night. One of the British party was a slaughterer in civilian life, so they stunned the cow with a hammer, cut its throat and caught the blood for cooking. Using two small knives, the butcher skinned and cut the carcass and they got the camp cook to cook the meat in return for him being given some of the meat. That night they all feasted on meat and chips to their hearts' delight, but they had to get rid of the carcass so that the Japanese would not find it. They did this by digging a large ditch near the river and burying the cow's remains. Next day the Japanese came looking, but found nothing. This was fortunate, as the prisoners did not know that the river was tidal, and the next day the ditch was partially revealed. Had it been found, the men would almost certainly have been killed. It was the riskiest liver and chips Aubrey had ever eaten.

On arrival at their destination, Aubrey was appointed as one of the assistant cooks to the Japanese. The youngest of the Japanese officers at this small camp spoke very good English and one day said to him, 'You no English.' Aubrey said he was, and explained about cockneys and Bow bells, but the officer was adamant. The officer said that in the 1936 Olympics, 'England No.10, Japan No.1.' But you never argued with the enemy. One evening the officer even brought Aubrey a cup of beer. The schizophrenic behaviour of the Japanese could never be accounted for, veering as it did between brutality and kindness, nor could the bizarre atmosphere that sometimes existed for Aubrey at this particular place, when he sat after work, after food and a beer, on a log by the river on a cool, tropical moonlit night, or as the sun set beautifully over the forest – it was an almost idyllic scene, which made men think of home a million miles away.

During the cookhouse night shift at this camp, the guards would often come around to chat. One night, in Aubrey's presence a guard came, gave his helmet and rifle to a British cook and told him to stand near the door and pretend to be on guard; meanwhile, he laid on the table and went to sleep! On another occasion, the kitchen ran short of wood and so they asked the guard for permission to go to collect some. The guard took Aubrey and two men surreptitiously to a spot near the camp entrance and showed them a pile of bamboo reserved for hut building; he said 'Bad Japanese men may come; take quickly.' So they did.

Takanun

Aubrey was then posted to a large camp at Takanun, quite a long way up the river and railway, where he spent nine months because he developed a huge tropical ulcer on his leg; this could lead to amputation if not remedied rapidly. A friend who was a medical orderly gave some initial treatment and returned after three days to find that, amazingly, the ulcer had begun to heal, although not well enough for Aubrey to return to work on the railway. Meanwhile, a sergeant approached him and said that a Jewish lad had passed away, and asked if Aubrey would help to make up the burial party; he did, but there was a sequel to this. The following morning a soldier from Manchester came to him, and said that Major Everard Woods had approached him and said that in future he was to be notified if any Jewish soldiers fell seriously ill, as he was Jewish and wanted to organize the funerals properly. Woods said that he did not want his Jewish background to be known, so this information must be confidential (see David Arkush's testimony, which mentions Woods). While at this camp, Aubrey received a rare letter from his sister-in-law, which went into great detail about how the family were all enjoying a lovely Sunday afternoon in the back garden eating, drinking and chatting. Aubrey supposed the letter was meant to give him a picture of home, but all it did, of course, was depress him immensely.

Other jobs Aubrey did included taking cans of boiled water round to the officers' huts to fill their water bottles and delivering their meals for them. Officers differed hugely in their attitude to the men. Some were very distant and wished to preserve peacetime rules of segregation. Others were just the opposite and were happy to work alongside the rank and file.

Liberation

By the late August of 1945, Aubrey was at a camp known as 'Cashew Mountain', mainly with Australians, but he is unable to identify its proper name or location. It had two airstrips that had been cut out of the jungle by the POWs for Japanese aircraft and there was a large Buddhist monastery on the hill. The men had noticed a lot of recent air activity. One evening after work, Aubrey met a fellow British Jewish POW, Ruby 'Johnny' Segal. Ruby said to Aubrey that he had heard that the war was over, but Aubrey did not believe it. Ruby told him that there was to be a special parade later to announce it.

Aubrey thought it too much to hope for and tried to sleep after his night shift, but there was too much noise as men gathered in groups talking about the news. The parade was indeed announced and the men were told that Japan had surrendered about two weeks ago; from that day, all Japanese guards would be accompanied by an Allied soldier and the blackout was ended. There had been two recent escapes, and the men involved appeared to have got away, as they were not returned to the camp. Soon some Americans arrived in a jeep and told them that about five miles away there was a camp of Free Thai guerrillas where the escapers had been sheltered. The orders given to the Thais were to attack the POW camps along the Kwai and to liberate and arm the POWs in case the Japanese decided to kill them, as many thought they would.

Soon American Dakotas arrived, bringing medical and food supplies and doctors. Air letters were made available for the men to write home. Aubrey sat down and did not know where to begin, but suddenly it all gushed out and he could not stop. His letter arrived on Yom Kippur at Aubrey's home and his brother took it to the synagogue for his father to read. Sadly, the letter was not kept.

The POWs were flown out in turn to Rangoon and given full medicals. They were warned to eat little but often, in order that their bodies could adjust to a normal diet. They stayed in a holding camp and then Aubrey set sail on the troopship SS *Corfu* and arrived at Southampton to a tremendous welcome from the crowds. Various dignitaries came aboard to talk to them, and Aubrey happened to speak to Lord Nathan, then a junior government minister. Later the men were taken to another holding camp and treated to a nice meal, but the dessert was rice pudding and it ended up being thrown around the dining area by the former POWs. The other personnel could not understand, until it was explained to them, as rice was a bit of a luxury product in wartime Britain! Next day, on Southampton station, Lord Nathan again spoke to Aubrey and he asked about the 'rumpus with the rice pudding'. Several years later, Aubrey met Lord Nathan yet again at a charity function and they recalled their meetings all those years ago.

They were then put on a train to Waterloo; Aubrey had telephoned and asked not to be met, but many of his family were there with a taxi to get him home. On arriving home there was a tremendous welcome and, typically, his mother prepared a lavish meal; but not only could Aubrey not manage it, it was rice pudding for dessert and he simply had to refuse to eat it. Suddenly everyone realized why.

Aubrey was a long-term member of FEPOW and kept in close touch with his many non-Jewish mates from the camps; he also joined AJEX, on whose marches he always paraded every year in Whitehall. In 1948 he and his wife Sylvia (they met at a Maccabi Club dance in 1946) were on holiday in San Remo and went to eat in the hotel dining room. Suddenly David Arkush walked in and there was instant recognition between Aubrey and him, and they gave each other a hug; they have been friends ever since.

After the war, Aubrey was in the tailoring and shoe business until he retired. He had a large family in the UK and Israel. He was awarded the Pacific and 1939–45 Stars, War and Victory medals. He sadly died in 2013.

Notes

1. He then glanced up at that moment during the interview and pointed to the photographs on his living room wall, noting that indeed two grandsons and one granddaughter, had gone into the Israeli Army – but that, said Aubrey, was 'different'!
2. Sadly, Shapiro died on the Burma Railway (see David Arkush's testimony in this book).
3. At this point Aubrey was unable to continue the interview as his emotions welled up and we had to stop for a few minutes as he recalled this humiliating and painful event.
4. Frank and Aubrey had been close friends since Prestatyn days and once, when Len Halpern was called a 'Jew boy' by another soldier, Frank leapt to his defence and punched the man who said it, with the words, 'My best friend is Jewish.' Frank was also a good card player and just after they were taken POW, Frank told Aubrey to give him his money and he would play cards to increase their stash and buy food. He succeeded in doing this and they purchased tinned pilchards and Marmite from black marketeers in the camp to keep for emergencies. Aubrey kept his tins for about three years.
5. Directed by Bryan Forbes, 1965.

7391556 Staff Sgt Charles Henry Lyons, 'A' Company, 198 Field Ambulance, RAMC

The information here is taken from the Jewish Chronicle, *10 November 1989, and an audio-taped interview stored in the IWM Archives.*

Called up in June 1939 as a Territorial, Charles was trained as an army nurse at Cookham and Macclesfield, and was living at 27 Sidlow House, Portland Avenue, Stamford Hill, London, N16, the son of Mrs A. Lyons. As a young Jewish lad, Charles had helped fight the Fascists at the Battle of Cable Street in Whitechapel in 1936.

He was sent to France with the BEF (British Expeditionary Force) and escaped at Dunkirk in June 1940. After this, for three months during the Blitz, Charles was attached to a full RAMC Field Hospital in Mile End, London E1, near the hospital where he had done his pre-war training as an SRN (State Registered Nurse). He was involved in many serious rescue incidents during the height of the bombing – one in particular, when St George's Hospital in Rotherhithe was hit, around July 1940. With three ambulances he crossed a docklands bridge nearby to evacuate wounded patients and staff and, on the way back, after two of the three vehicles had recrossed the bridge to return to the Mile End hospital, his was stopped by the police; a bomb had severely damaged the bridge and a vehicle could collapse the structure. But there were seriously injured on board the ambulance and so Charles asked the driver (RASC attached to RAMC) if they could risk it. The driver got out and inspected the hole in the bridge and said it was a chance, but he would do it at speed. The driver backed up, and accelerated over the huge hole, just making it to the other side, with his wheels spinning on the edge of the gap; as they drove off, the bridge collapsed.

Charles was also present at the bombing of the Coronation Flats in Stoke Newington High Street. When he arrived at the scene soon afterwards, the shelter in the huge basement had collapsed and hundreds had died, mostly from drowning when the mains collapsed; wooden hoardings were put up in the street to hide the awful scene from view. Fleets of ambulances queued up to remove the mass of bodies, which Charles said included many children and whole families, to local mortuaries. He said most bodies appeared totally unmarked and it was his worst experience of the Blitz.

In another part of Mile End, he was sent to assist when a house was bombed opposite the Charrington's Brewery on the main road. A pregnant woman and two small children were dug out alive from under a table. Charles drove them to hospital but nature intervened and he had to pull over and deliver the baby en route; though trained to do this job, it was his first delivery and he was never so nervous in his professional nursing career as at that moment. He said it was his most rewarding experience as a soldier, but he never saw the family again. Charles was anxious to point out the superb spirit of all the Londoners during the Blitz, the amazing camaraderie and 'can-do' attitude, and the atmosphere of neighbourly helpfulness.

In September 1940, Charles was posted away to Singapore, arriving in September 1941. At that time a Singapore posting meant a good life, away from the blackout and rationing in the UK. He was sent to work at the Alexandra Military Hospital. On one exercise in north Malaya, near Ipoh, they were invited for drinks at the seaside house of a local rubber planter. When Charles asked him why there were no defences visible on the beach, the planter said the beach was all quicksands, but Charles and some friends proved to the planter that this was not the case, as they could see birds sitting on the mudflats and this would not happen if they were quicksands. Charles reported this matter to his officers but nothing was done. Later they learned that the Japanese had made an important landing there during the invasion in February 1942.

Charles witnessed the arrival of about fifty Hurricane planes at Singapore naval base, but they sat in their crates because apparently nobody knew how to assemble them. He witnessed the very low-level bombing raids by Japanese aircraft from day one. He noticed that the lead bomber would dip a wing and then all the bombs were dropped at once, in a devastating salvo. He well remembers receiving the casualties from the sinking of the *Prince of Wales* and *Repulse* warships.

At the Alexandra Hospital, when fighting broke out, Charles was a staff sergeant organizing the inflow and outflow of the hundreds of wounded into the operating theatres during the battle for Singapore. Suddenly he heard firing; the Japanese had entered the hospital, shooting wildly at patients and staff. Allegedly, Sikh soldiers had been firing from the hospital, although this has never been conclusively proven. In his operating theatre on the first floor, the RAMC surgeon, his two assistants and nurses and anaesthetist were shot dead in front of Charles as he fell to the floor and played dead. The Japanese bayoneted to death the patient and two wounded colleagues on the floor; they then rushed out and another Japanese came in and ordered Charles to lie still at bayonet point. After half an hour, a Japanese officer ordered him and other survivors outside, to stand against a wall with hands up; opposite was a large group of nurses similarly lined up. The nurses were marched away and never seen again. His group were tied in bunches of ten and marched across a football field; there were three groups and he was in the centre of the third group.

(At this point, Charles broke down and the tape was interrupted.)

He saw the first group have petrol poured on them and they were set alight; the screams were appalling. He and his group then fell into an adjacent storm drain and lay there all day. Later they managed to creep away and then came across some British infantry who said the island had surrendered. Charles reported what he had witnessed to a senior officer (a colonel).

Charles was never able to discuss the horror of the hospital massacre with anyone.

Imprisoned in Changi, Charles testified that the Sikh guards were particularly treacherous and brutal to both the British and other Indian prisoners.

In late 1942 he was transported by train to Bampong in Thailand. Crammed into the terrible metal cattle trucks with no sanitation, sweltering by day and freezing at night, they suffered on the nightmare six-day journey. Then followed hard labour on the Death Railway; beatings, disease, starvation diets and death were their constant companions. Working in monsoon rains and burning heat, they toiled relentlessly. It was often bitterly cold at night and the men only had rice sacks to protect them. Charles wrote: 'Your clothes rotted off you ... you wore a loin cloth and flip-flops made from pieces of wood or bits of tyre.' Any plant or bit of edible vegetation would be used to supplement the meagre food; they called it 'jungle stew', including a python now and again if they were lucky. On one occasion, Charles remembers that he and a friend caught a twenty-foot-long python with a forked stick, and it was cooked and eaten with gusto (it tastes like chicken); one man ate it but then, having asked what it was, threw up on being told it was snake! After that it was a common food, when it could be found. The skin of this particular python was kept for years after the war by Charles's comrade.

Charles was beaten in one camp – the worst beating he ever had, among many – when they were lined up on a windy monsoon day; the atap roof of the hospital hut collapsed and he and some others left the line to go and help. One Korean guard beat him for 'damaging Japanese property' when breaking the atap roof in order to rescue the men. Such was the bizarre nature of their guards.

Among the worst illnesses affecting the POWs were the terrible tropical ulcers; these could be treated with dressings made of banana leaves and boiled old mosquito netting; maggots would be used to eat the pus if the ulcer was not too far developed, otherwise amputation had to be used, as gangrene could result, followed by a slow death. To save amputation, a man would be told to bite on a bamboo stick and held down while a sharp spoon was used to scrape away pus and dead flesh, and this was followed by a clean dressing.

The Japanese and Korean guards were ruthless: 'They were ignorant beyond belief ... if they put the speed on [the rush to get the railway built] and you weren't quick enough, you were beaten. Only visible marks were proof of illness to the Japanese, so [you had] ulcers and you were excused work, but a killer disease still meant you were sent to work and the medics who tried to protect you were beaten for their pains' – including Charles on many occasions. Frequently, men were taken on stretchers to work, merely to fill the workmen quota for the day.

There was every possible kind of disease, but only a few brilliant medical officers with few instruments. Human ingenuity, however, asserted itself: 'bamboo can be thin as a pencil or thick as a man's thigh ... we made bed pans from it, drinking mugs and even an adjustable operating table' (using bamboo pegs to raise and lower the height). Wild kapok and bamboo were used to make artificial limbs and they even held football matches between the double and single amputees. Sometimes there were reminders of the beauty of the tropical forest and Charles recalls walking along a track and passing through a cloud of millions of the most beautiful, multicoloured tropical butterflies.

But there was one small ray of light in the evil-smelling hospital huts that the Japanese avoided at all costs: the Royal Signals secret radio. This slight contact with the outside world gave the men heart.

At Chungkai, Charles remembers clearly being part of David Arkush's small minyan (quorum of ten men minimum required for Jewish prayer meetings).

Charles is clear that the threat to kill all POWs was real, and in the last few months of the war he witnessed the building of ditches and embankments with inward-facing machine guns around all the camps he was in; the guards also boasted that the caves that the POWs were building for last stands by Japanese troops would first be used to massacre all the POWs.

Then one day came a message: 'No work today; holiday'. The atomic bombs had been dropped, and soon afterwards the guards all disappeared. 'It was only the second time in my life I had been drunk', said Charles. He arrived home on 12 January 1946.

After the war Charles worked tirelessly for FEPOW, and the Ex-Servicemen's Mental Welfare Association. In March 1989 he returned to Thailand for the first time to pay tribute to those who died. He said: 'We will never forgive and never forget.'

183223 Capt. Jacob 'Marko' Markovitch/ Markowitz, MBE, RAMC (Canadian)

Jacob was born in Romania in 1901 but was brought as an infant to Toronto, son of Jeanette, of 11 Highbourne Road. Educated at Jarvis Collegiate, he graduated in medicine at Toronto University in 1926. He went to the universities of Glasgow and Minnesota to teach and study physiology and returned to Toronto as professor of experimental surgery and physiology there and at the Mayo Clinic, publishing many important medical research papers. He is reputed to be the first doctor to transplant a heart successfully of one warm-blooded animal to another (American Medical Association, 1932), worked with the discoverers of insulin, and devised a method of transfusing blood in dogs.

He gave up a prosperous medical practice and tried to join the Canadian Army, but as he received no definite reply he went to the UK in 1941 and enlisted on 10 April as a lieutenant in the RAMC, and was posted to the 6th HAA Company at Bude in Cornwall. He was then posted to Singapore in December 1941, and was taken POW in Singapore after serving upcountry in Malaya during the fighting, and continually under fire at his CCS (Casualty Clearing Station) as the British Army retreated in good order down the peninsula. Posted missing on 15 February 1942, a letter of 7 July 1942 from an RAMC officer, Lieutenant Colonel Ashdour Carr in Lucknow, eventually reached his sister, Mrs Percy Lederman, in Toronto, indicating he was still alive, and by December 1942 this was confirmed.

For the first ten days in Changi the Japanese gave the men no food and most men lived on their iron rations and what they had been able to bring with them from Singapore City. Suddenly rice began to appear – ten ounces per day per man – and this was how it remained for almost six months.

Then came the nightmare five-day journey in squalid railway trucks to Thailand in the monsoon with all its associated malaria, dysentery and bugs, and the beri-beri of malnutrition. When they arrived (at Chungkai?), twenty men were dying each day. One of the most trying things was the lack of reliable news. One day a Major Smith entered the RAMC mess and said, 'You know we have a canary but it cannot whistle without food. I have been trying to get birdseed but without success.' So Jacob waved a local five-dollar bill and said, 'Offer this to one of your thugs and I am sure some seed can be stolen.' That afternoon an enterprising Tommy acquired a battery from a parked lorry and the 'canary' began to chirp (later the radio was found, and Major Smith and his friends were tortured to make them give information). In another camp where Jacob was a POW, the radio was hidden in the bottom of an army water bottle and never found by the Japanese.

Jacob was hugely admired and loved by the POWs for saving so many of their lives from gangrene caused by the dreaded tropical ulcers. A Channel 4 film (*Miracle on the River Kwai*) features many former medical orderlies and patients who knew and worked with him. He would give local anaesthetics and in his relaxed Canadian drawl put his patients at ease and declare, 'Let's get it off!' He was known for his sense of humour and morale-boosting behaviour, in the most appalling of conditions, working with enormous surgical skill as he carried out complex life-saving amputations with primitive equipment, which was

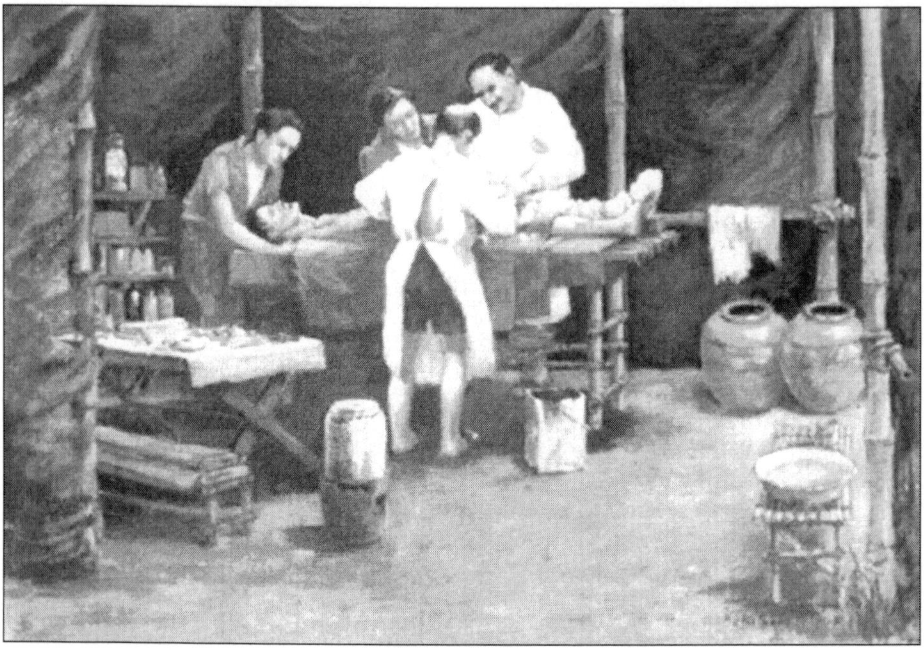

43. 183223 Capt. Jacob Markovitch, MBE, RAMC, from a painting by the famous war artist Jack Chalker, which shows him working on a thigh amputation at Chungkai camp. Facing the viewer is Col Edward 'Weary' Dunlop, RAAMC; Jacob has his back to the viewer. Though they never actually operated together at Chungkai, Chalker has depicted them like this as a tribute to their tireless efforts to save hundreds of lives of their comrades.

often locally made or consisted of domestic knives and saws. The loquacious Jacob was described as a 'Tiny Colossus' who saved thousands of lives with safe amputations.[1]

One story related by the famous war artist Jack Chalker, who knew him, was when Jacob amputated a leg and the tourniquet had flown off. Two POWs happened to be passing the hut window and were looking in aghast as Jacob was sawing off the leg and a spurt of blood covered them; Jacob's reaction was 'That'll teach 'em.' Stitches were made, using thread made of home-made catgut – and the patient survived. Chalker describes Jacob as 'that most wonderful surgeon'.

Broken bottles were used to make funnels for blood transfusions, and blood was kept from coagulating by whipping it with a bamboo whisk. Over 4,000 transfusions were carried out and not one death occurred. Blood could be taken from a fit man and used to save the life of a dying man. Equipment included a carpenter's saw, and other instruments made from mess tins; cotton thread was used for sutures, and bandages were made from towels boiled over and over again.

Jacob invented a spinal dental Novocaine anaesthetic technique which considerably reduced pain and allowed surgery to take place. He found that the Japanese doctors were brutal and gave virtually no equipment to the Allied medical teams (see Canadian press dispatch, 6 November 1945). He often gave lectures in the camp on physiology, especially to the nursing orderlies and other volunteers in the 'hospital'. In the first six months, Jacob lost two stone.

Jacob also developed the main life-saver against vitaminosis (acute vitamin deficiency) and pellagra, a very painful skin rash. He would gather any pieces of food such as rice, banana skins and fruit and use it to make a yeast, and to this he added spit from healthy prisoners to aid fermentation. He called it his 'magic brew', although prisoners were understandably reluctant to take this medicine. But it saved many lives.

When cholera broke out at Chungkai, Jacob ordered the establishment of a special isolation ward near the perimeter fence, and asked for volunteers to help him staff it. This saved hundreds of lives, by using distilled river or rain water using four discarded four-gallon petrol drums to make a still (later used to make gin!) mixed with doses of salt and giving it intravenously to the men using a needle made from a bamboo thorn. Always outspoken, Jacob once shocked the padres at lunch by saying grace as 'Jesus Christ, the same yesterday, today and forever!'

Jacob told a story during a lecture to the Empire Club of Canada in October 1946; he said there was often much to laugh about in the camps, and humour kept them going. On one occasion (it is not clear if at Chungkai or Kanchanaburi) the Japanese sold them a pig which was duly despatched and the head placed in the canteen so the men would know it was pork tonight. Soon afterwards, a British captain was stopped by a strutting Japanese sergeant major for not giving a crisp enough salute. The following performance took place:

> 'You English officer?'
> 'Yes.' (Smack across face)
> 'You proud?' (No answer)
> 'Answer, you proud?'
> 'Not proud.' (Smack)
> 'You proud?'
> 'Yes I proud.'
> 'Oh, you proud!' (Smack, smack)

Then the Japanese walked away. The English officer came over to the pig's head and said:

> 'You English officer eh?' (Smacks pig's head)
> 'You proud eh?' (Smack again)
> 'Oh you not proud , eh?' (Smack)
> 'English officer no good.' (Smack, smack)

It was, said Jacob, the most marvellous thing he ever saw.

From November 1943, Jacob kept medical records in spite of terrible threats from the Japanese if discovered. They were buried in bottles besides the bodies of British soldiers and after the war were unearthed and published in the *Journal of the Royal Army Medical Corps*;[2] they remain today a battlefront guide for military doctors. In the *Toronto Globe and Mail* of 15 January 1946, Jacob stated:

> Because I was a medical officer, I was treated a little better; the guards respected my knowledge but that did not stop them from beating me up and knocking me around ... the savage brutality inflicted on us seems unbelievable now ... the men

> of the Corps (RAMC) fashioned instruments out of anything handy ... the men deserve undying credit for this feat.
>
> Many times we were forced to sleep in open ditches because we were allowed no quarters. We dug wells by hand. When we were allowed quarters we had to lash pillar and post together with fibre from the bark of coconut trees. The roof was thatched. We had to keep a perpetual fire because they would allow us no matches. Our food consisted of rice and very little else. We found some grasses that grew there which we added to our rations because they made a good substitute for spinach. Sometimes, brewing beer was possible by mixing rice with a rotting banana to provide yeast
>
> We worked sometimes 20 hours a day treating survivors from ships who were brought to Singapore ... no news of the outside world reached us in the jungle prison camp until we were informed by the Japanese Commandant that we were no longer at war. He didn't bother to explain that Japan had surrendered. Then 25 Dakota aircraft dropped supplies by parachute and after a few days a rescue party reached the camp.

In the *Toronto Star* of 25 January 1946, Jacob stated:

> The Japanese told us [the railway] would cost 30,000 lives under the best of conditions ... that if a British prisoner broke one stone before he died he had done his job for the Emperor. We could always secure some anaesthetics by bribing local Chinese. We'd throw a stone and a note over the wire, saying what we wanted. A few days later it would be thrown back naming the terms and saying how it would be smuggled in. To fail to salute a Japanese sentry was a major crime. On one occasion our cookhouse caught fire. Everyone was running about trying to beat it out. A Jap sentry appeared and was naturally ignored in the excitement. He grabbed a British Lieutenant, who knocked him down. We never saw the Lieutenant again after he was dragged beyond the wire.

In the *Toronto Globe and Mail* of 25 June 1946 and a similar report in *Time News Magazine* of 15 June 1946, Jacob stated: 'Over one quarter of the European POWs died in the first months of captivity and an equal number fell sick ... starvation was the great risk and killer.' Even pigs, he said, could not live on what the men were given: 'When we tried to raise pigs, they often failed to subsist on our swill and their diet had to be supplemented ... we were grateful to eat ground-up peanut meal which was decomposed and gritty from admixture with earth and used as a fertiliser in Singapore'. Jacob made some medical discoveries:

> One Australian soldier whose hip was amputated was saved by a technique recorded in the Old Testament ... when Marko reached back to his knowledge of the Torah to save lives ... providing medical science with unexpected material ... he stopped breathing 20 minutes after the spinal anaesthetic (before the operation) ... we tried Schafer's method of artificial respiration which failed ... luckily there happened to be a piece of rubber tube nearby which we put in his mouth. He was given artificial respiration by the method used by Elijah when he resuscitated the widow's child.

> The patient's lips and nostrils were compressed and air was blown into his chest about 20 times per minute. After a few whiffs the heart began to beat forcibly ... after 40 minutes the patient was able to breathe and talk and use his arms. When the tube was removed from his mouth his opening remark was, 'I'm glad you guys don't eat onions; I don't like onions'! After 20 minutes the amputation was carried out, the wound healed and the patient is alive and well. Those Australians are a hardy race.

Jacob would summon up images of Florence Nightingale and the Crimean War when he later wrote of his amputation procedures in the *British Medical Journal* after the war. Orderly Cary Owtram observed Jacob at work once in Chungkai in August 1943; in this case, the patient's thigh was about half as thick as the knee joint, with two enormous ulcers on the leg; so much flesh had been eaten away that the tibia was exposed for about ten inches, and the other ulcer covered the foot, exposing the ankle bone. Several orderlies assisted, clad only in shorts and head sweatbands. Owtram described how Jacob made his incision on the thigh, giving a running commentary for the benefit and interest of all, clipping or tying the various arteries as he came to them, then sawing through the bone; he used a well-boiled tenon saw lent by the camp carpenter. Well-chipped enamel bowls of hot water with permanganate of potash contained antiseptic appliances. Swabs and bandages were bits of gauze or other material stained brown by repeated use. Dust blew in on eddies of wind, and hundreds of flies flew about, being flicked away if they landed on the patient. The operation took fifteen minutes.

Deaths from operations were negligible; not to amputate meant certain death, but patients always had to give consent first, as well as other doctors and the CO.

Jacob detested the Japanese; he said: 'they swaggered, had no restraint; instead of reprimand a man was smacked across the face; not tried but outrageously beaten like an animal; an escaper would be shot or decapitated; or a whole group punished. By this Bushido code we understood remorseless villainy, portrayed as chivalry.' When he failed to save a life, he was blamed and slapped on many occasions

For his courage and determination, Jacob was awarded the MBE, the citation for which is in the volume *Canadian Jews at War* (Montreal: Canadian Jewish Congress, 1947–48, p.15) and reads:

> As joint originator and supervisor of a successful transfusion service in POW camps in Siam (Thailand), using the most primitive and improvised apparatus, Capt. Markowitz has shown skill and ability of an outstanding degree. His training of transfusion teams, his development of simple techniques for jungle surgery, and his ingenious methods of improvisation, saved many hundreds of lives. He has shown great disregard for personal danger, and risk of brutality, in order to serve his patients.' (*London Gazette*, 7 June 1946)

An editorial in *The Maple Leaf* newspaper of 7 September 1945 describes Jacob as 'a legend among thousands of prisoners for his humane and neverending efforts to treat the sick and dying – a man who made bricks without straw'.

By the time of his liberation in January 1946, Jacob had performed over 1,000 operations in Chungkai camp alone on the Death Railway. He married Ruth and had two children, Jane and Thomas. He died at home on 29 January 1969 in Toronto. Clemson

University in the USA instituted the Jacob Markowitz Award for outstanding research in surgery.

He wrote the following short essay, titled 'I Was a Knife, Fork and Spoon Surgeon', and says it all in his own words:

'Fine surgical results you show here, doctor', observed a professional colleague of mine. He had just finished glancing through a mass of notes on surgical cases that I was compiling for a report.

His remarks were most complimentary both to myself and to the Royal Army Medical Corps, because the results he referred to were obtained under the most unusual circumstances. For eight long months in the jungle hospital of Chungkai, Thailand, I was the sole surgeon left to attend 7,000 desperately ill fellow prisoners of war – without drugs or instruments. Yet, taking stock after performing 1,200 operations under the most primitive conditions, our results were 75 per cent as good as those obtained in a modern New York hospital.

My story starts in May 1943, when Lt Col E. St Clair Barrett, RAMC, and I arrived at Chungkai under the escort of two Japanese guards. The camp was located in a jungle clearing beside the muddy Meclong River. A mile square, it was surrounded by a 12ft bamboo fence, and was constantly under Japanese surveillance. Outside the camp limits, except for a narrow road and a few small native clearings, there was nothing but thick jungle.

While being led to the Jap commandant's office, we were able to get our first close-up glimpse of the Chungkai hospital camp. It consisted of about 50 atap huts – structures without walls and with roofs of thatched palms – most of them full of sick men lying side by side on wooden platforms raised two feet off the ground.

Inside the orderly room, the Jap commandant, a 35-year-old lieutenant, who had once been the prefect of Tokyo's police force, carefully examined us.

'You', he said to Col Barrett, 'will be chief doctor. And you', indicating me, 'surgeon doctor.'

Dismissed, we found the medical headquarters where a pale young British medical officer stretched out his hand to us. 'I'm Reid', he said 'and I am glad to see you – there is a hell of a lot of work to do.'

From Reid we received a complete picture of the situation. There were 7,000 patients – British, Australian and Dutch – crowded into the encampment, sometimes many more. Our men were being forced to work at railway construction in the surrounding territory and every few days whole barges of human wrecks were floated down to Chunkai for treatment.

Dysentery was common. Hundreds were suffering with tropical ulcers – infections that ate deep wounds into the legs and were often fatal. Vitamin-deficiency diseases were everywhere. Several patients were in urgent need of surgical care. Men were dying at the rate of 17 a day and to ease the burden of the Padres and burial squads, they were being interred four to a grave.

Col Barrett rounded up the other four medical officers – all young men, fatigued and worn by disease – to plan our campaign. He asked Reid what we had in the way of supplies. 'For the 7,000 of us', Reid replied, 'we have enough purgatives and antiseptics for 50 men, enough chloroform for two people and are limited to seven bandages per month.' Then he added, looking at me dejectedly, 'In the way of instruments, all we have are a few stethoscopes.' After apportioning work to the other medical men, Col Barrett turned to me. 'I hardly know how to advise you', he said. 'All I know is that we have to start surgery in Chunkai as quickly as possible. Look around and let me know what you find.'

I found my way to the 'surgical ward', which was an ordinary atap hut, furnished with a few crude tables and benches, and there met my three assistants. Private Gordon Vaughan, a slight dark chap, had been a telegraph engineer on civvy street; Woolridge was a quiet lad with a warm face and kindly smile; and Tolson was a gangling red-haired fellow who aspired to be a doctor. None of these men had had any pre-war medical experience.

'Not much like the place I used to work in down at the Mayo Clinic', I observed looking around. Vaughan offered to go to work right away. 'I've done a bit of tinkering at mechanics', he said. 'If there's anything you need I could try to make it.'

On a piece of paper I made a rough sketch of some of the simplest tools required by a surgeon. 'See what you can do with these', I said. He was studying the paper closely when I left for my living quarters to rest up after the long trip. Where to begin?

As I lay on my cot wondering where to start the overwhelming task of repairing these thousands of suffering and broken bodies, I suddenly remembered a phrase from a forgotten book: 'Blood is all things to all tissues.' Here then, was the answer. We must begin immediately to give transfusions to save the dying and to strengthen the weak.

This brought us smack into a tough problem. Under ordinary conditions, the blood is mixed with a citrate solution just as soon as it is received from the donor. Otherwise, due to the presence of a clotting material called fibrin, within five minutes it will turn into a thick jelly-like substance. We didn't have any citrates, so how could we prepare the blood for transfusion?

From the many years spent in the laboratory as a physiologist, I recalled that we used to defibrinate blood; that is, remove the fibrin which caused the clotting by simply stirring the blood with a spoon as it was received. From a pint of blood, a clot of fibrin the size of an egg would gather around the spoon.

I discussed my plans with Col Barrett. He was enthusiastic. 'Move quickly, we are losing almost 20 men a day.' Vaughan and Tolson had fashioned a transfusion set which consisted of a few old bottles, a bit of rubber from a stethoscope and a transfusion needle which one of the doctors happened to have with him. No sooner had we set up our equipment in a corner of our hut than an emergency presented itself.

'It's one of the Dutch boys', Major Reid informed me. 'He's been unconscious for half an hour. It's anaemia from malaria. If you don't help him, he'll soon be dead.'

We decided that we were risking nothing by trying out the efficacy of our transfusion procedure. We chose a donor, after ascertaining the compatibility of his blood by a simple test, and the transfusion began. As the blood flowed from the donor, it was vigorously stirred by Vaughan. I jabbed the needle into the arm of the unconscious Dutchman and watched his face anxiously. The transfusion lasted a long ten minutes. The Dutchman stirred and then his eyes opened 'Where am I?' he whispered in Dutch 'What am I doing here?'

They were sweet words to hear! Defibrinated blood had proven itself and we had gained an important weapon in our fight to improve the health of our fellow prisoners. That was the first of 3,800 blood transfusions that were to be given without a single fatality. We gave transfusions for everything. Men dying with dysentery often recovered; men suffering from extreme vitamin deficiency became more cheerful and regained their appetite and will to live; tropical ulcers healed more quickly. Private Ball, a small Australian badly smitten by a succession of ailments, hovered between life and death for several weeks. Twenty-five transfusions made it possible for him to be home today, alive and well.

It is a source of satisfaction to know that our efforts have since drawn the attention of the medical profession to the usefulness of defibrinated blood. It is convenient for the country doctor, working in isolation far from a supply of blood in citrate solution, or, in cases of accident, when the patient can't be moved and transfusions must be given from donors obtained on the spot, it may again save many lives, just as it did in the Thailand jungle.

'Well, Mark', asked the chief one morning 'how's the surgical ward getting on?' I was able to report considerable progress. Tolson and Woolridge had canvassed the men and unearthed some real treasures. There were several pairs of haemostatic forceps which the men found handy for watch repairing and filched from heaven knows where! There were two balls of silk twine, excellent for ligature work, a carpenter's saw and a few butcher knives, sharpened and re-sharpened.

In the meantime, Gordon Vaughan was not idle. By bending back the prongs of a fork he fashioned a workable pair of retractors. He built a sturdy frame around a hacksaw blade to produce an ideal amputation instrument. His improvised anaesthesia mask plus 2ozs of chloroform completed my surgical equipment when we were called on to do our first operation.

Jock, a machine-gunner, had a severe case of osteomyelitis brought on by a tropical ulcer. Like so many others, he had been forced to work in the thorny jungle undergrowth with no protection for his legs. Cuts would develop into sores which often ate their way into the blood vessels and bone.

Before any surgery is performed, it is customary in the military forces to obtain permission from the patient. I asked the padre to come to Jock's bedside, not knowing what the machine-gunner's reaction would be to our telling him that his leg would have to come off. 'Jock', I said, 'one of you will have to be buried – either you or your leg.' 'Take it off doc', he begged, 'please take it off. It's driving me crazy.'

The other medical officers agreed and we prepared immediately for the operation. Our saw instrument, butcher knife, needle, thread, towels and bandages were placed in a gas drum full of water and boiled. The patient was placed on the table and Major Dunlop administered the anaesthetic. A gauze tent was thrown over us in a futile attempt to keep the thousands of ants and bed bugs away.

Twenty minutes later it was all over. In the days that followed Jock's operation healed nicely and his general health improved. Within a month, he was hobbling around the camp on a leg that a friend had fashioned for him out of bamboo wood and pieces of leather obtained from a Sam Browne belt.

Now that we had demonstrated that major surgery was possible, there were hundreds of operations to be performed but we were down to our last ounce of anaesthetic and the Japs refused us a further supply.

One day I found Johnny – a lad who used to give us a hand once in a while when we were over-worked – sitting in a concealed corner of our atap hut tying small pieces of paper to flat pebbles. 'Believe it or not', he said, 'these little fellows are going to fetch us some anaesthetic.' When I pressed him further, he told me that he was acting under special orders from Col Barrett. Every day at 5.00 a group of natives went down the road that passed by the camp. Some of them spoke English and when the Jap guards weren't looking, our boys established contact with them by tossing messages over the fence. 'So you see', explained Johnny, pointing to his air-mail messages, 'I've written here:- We want Novocaine. We will pay well. The rest should be easy.' And it was – easier than we expected. Hidden in the grass, Johnny negotiated successfully with the natives outside the camp right under the noses of the Jap guards. Within four days a small parcel came flying over the fence into the camp; back went our payment immediately. The rate of exchange decided upon was 3 watches for 500 doses of Novocaine. Never again did we have to worry about our anaesthetic supply.

With this bonanza we were able to start surgery in Chungkai in earnest. In the course of the next few months we performed 115 amputations – some of them doubles – with encouraging results. We erred on the side of conservatism. A regimental sergeant-major refused to part with a bad leg. 'Give it a chance Doc', he pleaded 'give it another week. I'll try and make myself eat. Maybe it will get better.' Of course we agreed – we would compromise and wait for five days. On the evening of the third day he died in his sleep.

Only a small percentage of the tropical ulcer cases required amputation. We were able to patch up huge gaping wounds by skin grafting. We would give the patient a spinal anaesthetic, scrape the ulcer with a spoon and some weeks later, when it was clean, graft skin obtained from the thigh to the wound. The wound was then covered with a home-made variety of tulle gras – a bandage that doesn't stick – improvised from shell dressing and pork fat. The whole leg was firmly wrapped in sponge rubber obtained from a mattress.

In the meantime, Gordon Vaughan's talents as an instrument-maker were developing fast. He fashioned a rectal speculum, a rib-cutter, a quadruple needle for skin grafting, a tracheotomy tube and several spinal and hypodermic needles. These same instruments so impressed the officers that liberated Chungkai two years later, that they asked permission to send them to London as exhibits for the British War Museum.

Our surgical ward became busier than ever. Numerous cases of appendicitis were turned over to us. We would keep every case for 24 hours hoping that the condition would resolve itself. It usually did. Of the half dozen appendectomies that were necessary, all patients recovered uneventfully.

Accidents were always happening. An angry elephant gored one of the men while he was laying railway ties. Another lad fell from a tree and had to have skull fragments removed from his brain. Jaw fractures, resulting from Jap beatings, were frequent.

An unforgettable incident occurred one day in July just as I finished an amputation on an Australian named Fletcher. The operation had gone smoothly despite the difficulty of operating in water up to my ankles – a condition resulting from the steady rainfall. Suddenly Vaughan tapped me. 'Look Doc', he said 'he's stopped breathing!' The spinal anaesthetic had evidently crept too far – not unusual when the general health of the patient is bad – and paralysed Fletcher's diaphragm. We didn't have a pulmotor to restore his breathing, nor could we apply the familiar technique used on drowning victims because the thorax under spinal anaesthesia is not elastic. Fletcher was slowly turning blue. 'Quick', I snapped to Woolridge, 'take apart the stethoscope and hand me the rubber tube'. I shoved one end into the Aussie's mouth and blew rhythmically into the other. After five minutes I was relieved by Tolson. For twenty minutes the tube went the rounds. Finally Fletcher recovered and was well enough to see what was going on around him. He took the tube from his mouth, smiled weakly and whispered, almost inaudibly, 'I'm sure glad you blokes don't eat onions. I hate 'em.'

We had successfully applied the oldest method of resuscitation known to man. We took our lead from Elijah, a biblical gentleman who on one occasion revived a widow's son by blowing into his mouth. This procedure saved many lives in the succeeding months.

Private Vaughan had a hand in practically every important operation performed and as a gesture of recognition, Col Barrett and I decided to award him his sergeant's hooks. 'Sorry Sir', he said when I broke the news to him, 'but I can't accept any promotion.' 'Why not?' He hesitated a moment. 'The truth of the matter is Sir, I'm listed as a conscientious objector. Whenever anyone wants to promote me I am supposed to tell them that.' I asked him about his pacifist views now. 'I was young then and I think I was mistaken', he answered. 'You can't take this sort of stuff' – indicating the suffering and brutality around him – 'lying down.' Private Vaughan became Sergeant Vaughan, army rules or no army rules.

'The doctors have gone nuts', said one orderly to another. The speaker was referring to the six doctors of Chungkai and I admit that there was justification for his opinion. The doctors were requested to come to my atap hut and perform one of the strangest chores of their army career. They were asked to file by a large tin containing a banana-smelling concoction and spit into it three times and to add the final touch of the bizarre, I sent them away with the warning 'Don't breathe a word about this to anyone.'

We were willing to be misunderstood because we were well on the way to solving our toughest problem: what to do about our inadequate diet? The effects of a practically unbroken diet of polished rice were appalling – bloated and paralysed bodies caused by beri-beri; the painful inflammation of the tongue, a symptom of pellagra; and the hideous ulcers on the mouth and scrotum resulting from a lack of riboflavin, a B-complex vitamin. One of the most important vitamins, thiamin (vitamin B-1) is normally derived from grain husk and yeast. Yeast is present everywhere. One of the familiar forms that most of us recognize is the white fuzziness that forms on a grape. Within the confines of our camp there were several jungle banana trees which yielded an unpalatable fruit. Often the fruit would become slightly rotted and exude a beery smell – a good indication that yeast was present. We made a mixture of bananas and rice but the presence of sugar is necessary to grow yeast so we spit into the mixture knowing that the diastase in the saliva would convert a part of the starchy rice into sugar. The result, after a few days' fermentation, was a vitamin-rich beer worth its weight in uranium. Our new elixir worked miracles. The swollen stomachs, sore eyes and the ulcerated mouths responded to the magic of our cure.

We discovered another source of vitamins by watching the eating habits of the cows that were pastured inside the camp and then making a thick soup from the grasses they favoured.

As the tide of the war gradually turned against our captors, our treatment as Prisoners of War steadily improved. So much so that when I was posted out of Chungkai in February 1944, medical supplies were steadily flowing in and an entire new hospital camp was being constructed.

Since coming home, I have often been asked by professional friends what general conclusion I have reached as a result of my unforgettable sojourn at Chungkai. I tell them – and I am not always taken seriously – that I don't think that basic surgery has progressed very far beyond Lister and 1868. As definite proof, I point to the hundreds of surgical operations which we carried out with results not far below the standard set by our most up-to-date hospitals. I have a picture on my desk of 18 amputees, ex-fellow prisoners, taken in a Rangoon hospital where they were being fitted with artificial limbs. Whenever the going is tough, I glance at this memento of my days as a knife, fork and spoon surgeon. It gives me a terrific lift.

Notes

1. In *Prisoners of the Japanese* by Gavin Daws (Simon & Schuster, 2006, p.249), Jacob is accused of needlessly amputating over 100 limbs (of the 2,000 men he cared for) at Chungkai, and mortality was one in three. It is implied that a Dutch colonial doctor working in other camps, Dr Hekking, used well-tried local cures for ulcers with a huge success rate, and rarely needed to amputate. However, POW doctors rarely met and were indeed isolated, and in addition had very little knowledge of tropical medicine. Jacob acted to save life, because gangrene – the cost of delay – was the real and painful killer; it was better to risk the shock of amputation, rather than certain death from not doing so.
2. *Journal of the Royal Army Medical Corps*, 86 (1946), pp.150–70, with astonishing drawings by the war artist Jack Chalker: 'Experiences with Cholera in a Jungle Camp in Thailand' and 'A Series of over 100 Amputations of the Thigh for Tropical Ulcers'.

3859180 Cpl Samuel Marks, MiD, 5th Battalion, Loyal Regiment/Recce Corps

44. 3859180 Cpl Samuel Marks, MiD, Loyal Regiment/Recce Corps.

Samuel was born on 17 November 1917 in Townley Street, Manchester, son of Abraham (born in Poland) and Sarah Marks. He was one of nine children. Sam always said he had a charmed life; when he was two days old his mother rolled over on him in bed and suffocated him, but he was brought back to life. At the age of 12 he was getting out of hand and knew no fear of anyone, and so his father said he should study to become a rabbi and settle down.

He was given a place at the world-famous Jews' College in London (now part of the University of London) and had to go down to take some entry exams. A 16-year-old cousin insisted on coming with him. Sam passed the exams and on the way back his cousin insisted on changing seats. Five minutes later they were hit by another vehicle and the cousin was killed, but Sam was able to walk away.[1]

He was called up on 18 January 1940, when he was working for his brother in a drapery shop in Cleveleys, and posted to the 5th Loyals at Fulwood barracks, Preston. From here he was transferred to the 18th Recce Corps, part of the Royal Armoured Corps. He was living at 1 Petworth St, Hightown, Manchester. They trained in various places in Britain, including Dalby Hall, Melton Mowbray; and ended in Scotland; they then set sail from Gourock in Scotland on the night of 26 October 1941, to Halifax, Nova Scotia, where they switched to an American troopship and sailed down to Trinidad, then to Cape Town, where they arrived in December. Being Jewish, at Christmas he volunteered to guard the gangway while the rest had shore leave. They sailed to Ahmednagar in India and spent three weeks in Poona, from where they sailed to Singapore to help defend the island from the Japanese invasion.

As they entered the harbour the ship was bombed and sunk by the Japanese air force and Sam ended in the water, but was rescued and taken ashore by an Australian ship. Except for his shorts, plimsolls, soft hat and shirt, everything had been lost, including their Bren gun carrier and all arms and equipment. After three or four days, they were directed to the ordnance depot to rearm, but as it was a Saturday it was a half-day; meanwhile, the Japanese were shelling and bombing, but nothing would stop the half-day closure! Eventually they were given rifles and ammunition and sent to fight at the battle of Bukit Timah. Sam was wounded in the head and shoulder; he was evacuated to hospital and revived, but some days later the surrender was announced and he was taken to Changi gaol.

At Changi, Sam suffered and witnessed many of the privations described in other testimonies in this book; he remembered being sent as a POW on working parties on the airfield and in the godowns near the docks, loading and unloading ships, and he helped build a sea wall at Tanjong Rhu, in the south-east of the island.[2] At Gillman barracks he was put to work sharpening crowbars with heavy hammers, to be used for building tunnels to store ammunition for the Japanese. Next door was a brothel and the girls would throw cigarettes to them over the wire, which they had been given in turn by their Japanese soldier clients; the brothel was well used by the Japanese guards.

When he could, he attended the synagogue services in the gaol; he said that there were always two or three guards sitting at the back, checking that nothing suspicious was going on. He joked: 'We always ensured the sick men got enough to eat, even if we had to go out under the wire at night and risk execution ... it was a lousy hotel, the food was awful, the service horrible – not recommended!'

Later Sam was transported by rail to Burma in the most appalling conditions, and then sent on a forced march, mainly at night, 200 miles to a camp at Sonkurai, near the Three Pagodas Pass in north Burma, to slave on the railway. Their huts were built on stilts to keep out both snakes and monsoon flood waters. Sam witnessed many men die of illness, and some who died just by giving up completely, physically and psychologically. He remembered finding this especially disturbing. It was common to wake up and find men dead on the bamboo bed slats in their huts. Those who died of cholera had to be cremated; he said that it was horrific to have to burn your mates. Within seven months, of the 1,680 men with him, 220 were left in the camp; the rest had died or had been sent back to so-called hospital camps.

The main work at this camp was building a three-span wooden bridge, thirty feet above and across a fast-flowing river, often working for twelve to fifteen hours per day, knee-deep in the monsoon mud to bring felled trees to the river for the construction. Beatings were constant, and guards competed with each other sadistically as to who could beat the most and the hardest; pickaxe handles were often used. If a one-star soldier hit a prisoner, a two-star soldier would also join in, but use more blows to be one-up on the one-star soldier – and so forth. There was one escape attempt by ten men but they were all captured and Sam witnessed their execution. Sometimes the funeral pyre fires had to be stoked during the rain to keep them alive; on one occasion, after returning from work, they found that the man left to do it had himself died.

Sam remembered the camp chaplain, Captain Duckworth, as a great and selfless man who did all he could to minister to the needs of the POWs and to keep hope alive and morale intact.

Sam would say little more about the war. Afterwards, he returned to Manchester (the family were living then at 1 Petworth Street). He went to Plymouth to visit his brother, who was stationed there, and met Hilda (they were married at the ancient Plymouth Synagogue), to whom he has been married for sixty-four years at the time of writing, and they are now living in Bournemouth. They had two children (Rochelle and Anthony) and many grandchildren.

Sam wrote to me:

> Hilda my wife has been my stability and the person who brought me back to the realities of normal life ... I have no photos of me in uniform at all [in fact, Rochelle

> found one] ... quite honestly I didn't keep anything, but had trained myself for over 50 years to forget and blank out my mind all of the things that happened to me. However, VJ Day brings it all back to me and I wish they did not have these activities to remind me of it all ... I have been back once to Changi, but the POW schule [synagogue] I remember, was not there any longer ... I went every year to the annual AJEX Parade in Whitehall. After your phone call I had strict instructions from my wife and daughter to write this for you, and women are very persuasive. That is all I wish to say. Some things I want to try to forget.

In the *London Gazette* of 12 September 1946 (p.4574) is Sam's award of a Mention in Despatches 'for gallant and distinguished services whilst a POW'; he also has the 1939–45 and Pacific Stars and the War Medal.

Sam wrote: 'I know I took out two Japanese machine gun nests at Bukit Timah, amongst other things. My MiD may have been for that, or robbing the Japanese canteen or maybe for marrying my wife.'

With such humour, how could we ever have lost the war?

Sadly, Sam died in April 2013.

Notes

1. In the 1980s, when Sam was driving to Plymouth on two separate occasions with his wife, they were involved in two freak road accidents (one when a lorry wheel broke loose and smashed their roof, and another when a thirty-feet-tall tree fell on their car in a storm), but they both walked away unhurt. Sam said the number of times he narrowly missed being killed as a POW was countless.
2. Sam later learned that this wall collapsed at the first high tide and heard it was the work of POWs who knew how to build faulty walls.

Pte David Saul Marshall, SVC/SSVF

45. David Saul Marshall, SVC/SSVF.

David was born on 8 March 1908 into an Orthodox Jewish family of Baghdadi Jews in Singapore, the eldest of seven children, son of Saul Nissim Mashal – a Jewish businessman of Singapore, trading in dates and property – and Flora Ezekiel, at Selegeie Road. He was bar mitzvah at 13 years old and went to school at St Joseph's and St Andrew's and later the Raffles Institute. He had been expelled from St Joseph's for taking off the Yom Kippur holiday, and from St Andrew's for punching the son of a Wesleyan minister who made anti-Semitic remarks. David almost died of TB when at school and was sent to recover in Switzerland.[1] He studied textile manufacturing in Belgium but then switched to law at London University and Middle Temple – working as a car salesman and a waiter to finance his studies – and became a barrister in 1937, returning to Singapore the same year. He experienced a lot of racism and anti-Semitism in London and this had a lifelong effect on him, giving him the determination to fight injustice. On the ship home he came across a party organized by some German passengers, and tore down the Nazi flags that were on display, throwing them overboard. He was called to the Bar in 1938, but found his Jewish background a barrier to getting a job in the law. He eventually joined the Chinese Christian firm of Aitken & Ong Siang (Sir Song Ong Siang was one of the oldest Chinese Christian families in Singapore, English-educated and bridging the white and Chinese communities; he was a friend of David's father), but only on a self-employed basis.

David soon saw the Japanese threat and especially its fascist links with German anti-Semitism, and so joined the British-formed SVC, part of the SSVF. He was insulted to find he had been arbitrarily assigned to the Eurasian Company, for he was neither European nor Asian. He bluntly told the CO that he did not want to join the snobbish British 'A' Company but would join 'B' Company, 'made up of Czechs, Poles and odds and sods'. He then discovered he was being paid half what the other Europeans were. Again he confronted the commandant, Lieutenant Colonel V.G. Stokes. He was Confined to Barracks but got his pay rise; he made it known that he was donating all his pay to the War Fund, demonstrating that he was entitled to the same pay as any man, whatever his ethnicity.

By mid-1938 they were training hard in the use of the Lewis gun, mainly at Beach Road HQ, but David found that his law practice suffered badly, owing to the long periods he was spending in the SVC instead of at work, so he changed firms in order to accommodate his military work, whereby he earned a regular monthly salary.

The Japanese threat led to about 100 of the several hundred Jews leaving Singapore, including David's brother Sonny, who went to Australia. Soon, in their family, only David and his mother were left. He categorically refused to leave and was determined to fight. When the first bombs fell on Singapore, David was on duty in the school field at Geylang Serai. Because the lights were still on all over the island, people thought the sirens were a practice. As company runner, David was busy carrying messages as the island was invaded, and on 15 February had to carry the surrender order to outlying units. He wept openly when he saw that order. David was based in the south area of the island, with the Malayan Infantry under Major General Simmons, and on 17 February they began the walk to captivity from Pierce Road to Changi gaol. Chinese civilians brought them water to drink en route along Serangoon Road, and were slapped by the Japanese guards for doing so; the guards deliberately knocked over the buckets and basins of water. It was very courageous of the Chinese to try to help them in this way.

Changi

The following morning, David's ankle was sprained and he could not walk any further; his CO came by in a car but, deciding equipment was more important than the man, took David's kit but left David at the roadside. Eventually an Australian lorry gave him a lift, after several British vehicles had completely ignored him.

The prison conditions quickly became apparent – a lack of food, water (they collected it from drainpipes), hygiene and blankets – and he began to live from day to day. Within a few months he lost four stone, living on just twelve ounces of weevil-filled rice per day. At first he would pick out the weevils but the chief medical officer in Changi ordered all men to eat the weevils because they were 95 per cent protein. Many men of course became very ill anyway, but for David it was something of a revelation to see the thin veneer of European civilization and authority unmasked at close hand, unprotected as they now were by the cloak of privilege. He also learnt later that 40 per cent of his company had died in prison.

He sincerely believed the Japanese to be totally inhuman, witnessing such terrible cruelties as he did. A Japanese guard would throw down a cigarette stub after spitting on it and grinding it with his heel, watching as a dozen desperate prisoners scrambled for it. He saw how some officers behaved appallingly selfishly, whilst a lance corporal was an angel in disguise. Sometimes brave Chinese boys would bring bread to the wire and secretly hand it to the prisoners, something which touched David deeply.

Japan

Around April 1943, David was transferred to Race Course Camp on Serangoon Road, which contained about 3,000 people, and laboured cutting grass for the Japanese cavalry horses stationed in Borneo. From here he was sent in May aboard the notorious Hell Ships to Japan via Saigon and Taiwan, with 2,000 others. His cargo ship contained 500 people crammed into closed holds; they were allowed on deck only once a day, for only twenty minutes, to bathe in buckets of seawater. The men nicknamed the ship the 'Mucky Maru'. When they docked on Honshu after three weeks at sea, they learned that they were the first ship to arrive with no deaths on board. They went by train to Hakkodate in Hokkaido in the north of Japan, to slave in an industrial complex, the first of twenty-six different camps that David would endure.

In camp, David found he was both loved and hated because he had brought three important books (*Chess for Beginners* by Alexander, *The Complete Works of Shakespeare* and the Bible) and so was besieged by bored inmates to borrow them. He did lend them, but imposed a rule of no more than one week for a loan.

Most of the men were sent down the coalmine but David had grown so weak that he was assigned, in the bitter cold, to shovelling earth for the building of a nearby airfield, and later to moving iron ore off Manchurian trains. It was always a constant battle against hunger, disease, cutting winds and temperatures often below 15 degrees Fahrenheit. The guards were deliberately cruel, and David and his comrades suffered terrible privation. He maintained his dignity, however, and took up the cause of pressurizing the MO to complain to the Japanese commandant about the horrific conditions. He remained psychologically tough and the men greatly admired him as their spokesman.

Red Cross Parcels and Liberation

The behaviour of the Japanese began to change in 1944; the war was not going well for them. Red Cross parcels had begun to arrive and small titbits were allowed in a new canteen; a new Japanese commandant had taken over. The news of Hiroshima came soon after. A POW cleaner saw a newspaper in the commandant's office and could read some Japanese. The tortuous *tenkos* and work parades disappeared, as did some of the guards. Soon US aircraft began flying low over the camps, dropping containers of food of all kinds. The Japanese lost all their arrogance and became totally docile. David decided he never wanted to meet a Japanese ever again.

Eventually they were flown to Okinawa and then for debriefing in Manila, where they were asked to detail all acts of cruelty by the Japanese during their captivity. He then requested to join his family in Perth, Australia, to recover.

Post-War Life

David returned to Singapore and set up the first democratically elected Jewish Welfare Board, which did not depend on the largesse of the Sassoons and Eliases who once controlled things in a patrimonial style, and he became its first president. He formed the Singapore War Prisoners' Association and was its honorary secretary. He became the island's most successful criminal lawyer and also became more politically involved in the island's future. As a confirmed anti-fascist, reinforced by his war experiences, he became the leader of the Labour Party and first elected chief minister of the post-war Singapore government in 1955, opposing Lee Kuan Yew. He married a lecturer, Jean Gray, in 1961 and they had three children. In 1978 he was awarded the Légion d'honneur by France. After many years in the parliament he became Singapore's ambassador to several countries and died in December 1995. His mother went to live in Israel and died there in the 1960s.

Note

1. His story is included in Eze Nathan's book, *The History of the Jews of Singapore* (Singapore, 1986), from where this information was taken.

6202787 Pte Arthur Mendelsohn, 5th Beds and Herts, 18th Division

The information here is taken from an audio-taped interview with Arthur, recorded in 1983 and held in the IWM archives.

Arthur was born in Stoke Newington in 1918 to a Jewish family of handbag makers and went to school in the area. When he left school, he worked as a clerk in a Smithfield firm called Link & Son, then the menswear shop Cecil Gee and other retail goods shops.

In December 1939 he was called up and sent to Kempston barracks to join the 5th Beds and Herts, being given the job of company clerk to the mess.

In November 1942 they were issued with tropical gear for the Middle East, with khaki-coloured vehicles; they embarked for Newfoundland on SS *Renadell Pacifica*, and then transferred to the American troopship USS *America*. What Arthur remembered most was how luxurious the American ship was, compared to the British ship. On arrival at Cape Town they heard about Pearl Harbor, and were diverted to Ahmednagar in India for two weeks; and then at Christmas they were sent to Singapore, arriving at the end of January 1942.

The first thing Arthur remembers is being issued with pyjamas. His platoon were attached to HQ whilst the rest of his battalion was dispersed across the island. He recalls being bombed by Japanese fighters while at HQ. Eventually a message arrived that all alcohol was to be destroyed; many large houses near where they were stationed had huge wine cellars, and Arthur and his comrades drank rather a lot, as it seemed a shame to them to just throw it all away. They were very drunk when the surrender occurred, and the next thing he remembers is waking in a tent near Changi gaol after being taken POW.

Changi

Near Changi there was a hutted area, which Arthur assumed had been an army camp; this was surrounded by wire and he and the men he was with were imprisoned there. Food was not bad and most of his battalion were now together. For several weeks the Japanese were absent and the POWs set up their own entertainment and educational activities.

Slowly, however, food ran out and the men were put on Japanese rations which became worse and worse, despite the gardening activities by the men to supplement the food. Trading with local Malays over the wire began, despite being dangerous and punished severely if men were caught by the Japanese.

After being moved to Changi gaol itself, Arthur vividly recalls the Selarang incident. He also remembers that the educational activities continued to flourish and there was even a Changi Parliament later, where POWs of different political beliefs debated. Many were convinced that intellectual activity would help keep them alive.

Arthur remembers clearly the Japanese asking for the officers to supply the names of all Jewish POWs to the camp commandant. The British officers, however, immediately refused to admit to any Jewish soldiers, as they were aware of the Japanese alliance with the Nazis in Germany.

The mainly Korean guards behaved in their usual brutal way. Work parties began – first to clean up Singapore after the bombing – and Arthur's treatment by the guards could be rough but was not especially brutal; some of the older Japanese were quite friendly, he found. One spoke English with an English accent and had been studying in America but caught up in the war when visiting his family in Japan; this may be the same man referred to in some of the other testimonies in the book.

Illness was now rife and men were dying of starvation-related diseases; medical supplies and Red Cross supplies were virtually unknown.

In mid-1943, Arthur was shipped out to Thailand in the last force to be moved there – F and H.

Thailand

Taken in horrifically overcrowded cattle trucks, which fortunately had the doors left open, his group of about 300 men was taken on a three-day journey to Kanchanaburi, from where they were sent north on a forced march for three weeks. It was appalling. Marching was done mostly at night in the pouring rain and thick mud, and in impenetrable jungle. Each night they had to put up shelters to sleep in, and cook some food supplied by the Japanese in cauldrons, before being able to rest. Weakness, wild animals and isolated geography all meant that escape was both impossible and pointless.

Many sick and weak men had to be carried, and some died on the way. They sometimes passed camps already established but were not allowed any communication. The Japanese issued them with one army blanket each, which was of such good quality that Arthur kept his after the war – he believes it was synthetic.

Songkrai

Eventually they arrived at their destination, Songkrai, and found huts with no roofs, which first had to be completed with atap. Then cholera broke out and vast numbers of them became very ill and over eighty died within weeks; often a man would be dead within two days of developing the disease. The corpses had to be burned but this was often bungled; drinking from streams was forbidden and all water had to be boiled until the disease died out.

Arthur developed jaundice, malaria and dysentery, and after working on the railway was moved to light cooking duties for much of the time until he was well enough to return to hard labour. Most of the work was levelling the ground to prepare for the railway sleepers in the area where he was put to work with his party of men. It was a ten- to twelve-hour hour day with very short breaks, living on rice and sometimes vegetables only, six days a week minimum. Many men were skeletal by then but it made no difference and often prisoners were carried out of the so-called sick bay, unable to walk, to be forced to work by the Japanese overseers, despite the efforts of the NCOs to persuade the guards that men would die if forced to work when ill.

Arthur had only a gold ring which he was able to trade secretly with a Thai for some food. He was fortunate to have kept his army boots, which lasted throughout his captivity and so kept his feet healthy, at least.

The most minor of misdemeanours were punished by the Japanese. Arthur witnessed one typical torture which was to tie a POW to a tree and then put honey over him so ants would bite and sting him. On one occasion, Arthur was beaten for swearing at a Japanese guard. If POWs did not work fast enough, guards would punch prisoners, or beat them with sticks; some were very badly beaten indeed, and Arthur witnessed this on many occasions. Curiously, despite all the horror around him, he never thought that he would *not* get through it all, and is convinced that attitude of mind can mean the difference between life and death. He also recalls receiving a few cards from his parents and a girlfriend – of the very few cards that he was allowed to send, some reached home only after the war.

Some news of the outside world came from secret radios but they was very carefully used in order not to raise suspicions among the Japanese.

Eventually, in late 1943, Arthur and his comrades were put on cattle trucks and sent back to Singapore – which, although bad, was far better than the primitive and isolated conditions upcountry.

Back to Changi

Arthur recalls hearing rumours about the D-Day landings in June 1944, probably emanating from one of the secret radios. He also remembers American bombers flying overhead; some of the aircrew that were shot down ended up in his camp and brought them up to date on the war.

Back at Changi, work consisted of work on the nearby airfield or in the gardens to supplement food; in addition, in the last months of the war, he was put to work on digging tunnels in Singapore City – they were never told what they were for.

On one occasion, an Italian submarine docked in Singapore and the Japanese were not sure what to do with the crew, as of course many Italians had gone over to the Allies in late 1943. They were interrogated, and those who opted to be counted as Allies were imprisoned and those who did not – Arthur clearly remembers that two officers took the latter option – were given a certain amount of freedom to wander in and out of camp and were separated from the others.

Arthur recalls that relations between officers and men were not always good; the soldiers perceived, rightly so, that officers had privileges which they were not allowed, with regard to food, work and general treatment. Many, however, behaved very bravely; it was as one would expect, a mixed bag. In addition, POWs were sometimes punished by their own officers and put into custody within the camp, for breaking rules, especially stealing or illicit trading which the Japanese forbade. Although this created bad feeling, the officers felt – and this was, in hindsight, almost certainly correct – that it was essential to maintain military discipline for morale and order, and also to keep the Japanese at bay from carrying out punishment, which would of course have been far harsher and sometimes even fatal. This would have included collective punishment by the Japanese, if the British had not themselves dealt with the actual culprits.

In addition, for those POWs caught breaking rules by the Japanese, there was the even worse prison run in the city by the Kempeitai. Arthur witnessed the state of some of these men after they were returned to Changi to recover.

The End

The first suspicions Arthur had of the end of the war was when a Japanese guard said 'Big bomb'. Work stopped and all the digging tools were collected. Then piles of Red Cross parcels arrived. Suddenly one day, a heavily armed British medical officer arrived by parachute and walked into the camp, followed soon after by Lord and Lady Mountbatten. After some days the POWs made their way home via Colombo (Ceylon).

Besides some malaria, Arthur says he suffered little physically or mentally from his experiences and held no grudge against the Japanese.

Cpl Philip Meninsky (aka Menushkin), War Artist, 5th Beds and Herts/No. 7 Commando

The information in this testimony is taken from audio tapes recorded in 1980 and a transcript from 1991 stored at the Imperial War Museum, London.

46. Cpl Philip Meninsky, 5th Beds and Herts, and war artist, taken shortly before he died.

Philip was born in Bloomsbury in November 1919, son of Bernard Meninsky, who had been born in Karotopin in 1891 and was a tiny baby when his parents fled the Pogroms of the Ukraine for Britain in 1891; he became a war artist in the First World War and served in Palestine/Israel in the Royal Fusiliers. Philip's mother left home soon after their marriage, leaving Bernard with two small boys, and went abroad with another man; she had an illegitimate child and died in childbirth when Philip was very small. Bernard had refused to take her back, and later remarried. The family never discovered the name of the baby or where Philip's mother is buried. Bernard died in 1950, while teaching at the Central School of Art in London.

Philip was brought up until the age of 16 by a retired nanny who had also been nanny to the children of Lord Samuels (a Jewish aristocratic family). He went to Hertford Grammar school, in Hoddesdon. He rarely saw his father but was greatly influenced by him and started drawing when he was 7 years old. His father, however, discouraged him from becoming an artist. Philip was reluctantly articled to a firm of accountants – Jeffries, Alfred, Henry & Marks. Marks was an astonishing man and accountant to many famous artists. He often collected paintings from them instead of fees. As a cyclist and artist, Philip rode around most of England, staying in youth hostels, sketching and painting, and he believes that the stamina he built up later saved his life in the Japanese camps.

When war broke out Philip volunteered for the RAF but failed the medical, so he joined the army. He was posted to Kempston barracks to 'B' Company of the 5th Beds and Herts; he was on coastal defences for many months in Norfolk, living under canvas and enjoying soldiering and the outdoor life which had permeated his childhood. He was a good shot and promoted soon to corporal, before volunteering for the 7th Commandos as a PT instructor, training at Felixstowe, and then he trained with them for nine months: first at Loch Fyne, where they lived on board a ship and trained on landing craft; then on to Girvan and later on to the Isle of Arran, training for cliff-climbing. That Christmas, refused permission to visit their girlfriends, he and several friends pretended to be escorting one of their group who posed as a prisoner, and they boarded the ferry to Girvan looking like an official party. They then split up to meet their different girlfriends. All went well but on the Sunday night a thick fog led to cancellation of the ferries. On returning late to camp, Philip pleaded 'using initiative' in front of Colonel Lister. But as the only NCO, Lister said he had no choice but to have him RTU (Returned to Unit) – quite a serious blow.

Within five weeks the Beds and Herts were ordered to prepare to go overseas to the Middle East, but en route were diverted to Singapore after Pearl Harbor was attacked. Meanwhile, Philip had married Alexandrina MacDonald in Scotland before he was posted abroad.

Sailing to Nova Scotia, Canada, they transferred to American troopships (Philip was on the USS *Miss America*) for South Africa, then to India where they trained for three months. Arriving in Singapore, they trained in the forests and plantations to prepare for jungle warfare, albeit with desert uniforms and equipment. On board ship, in India and in Malaya Philip worked hard as an army PT instructor. Once on night guard in the jungle, one young officer screamed at them to put out their cigarettes, but the lights he could see were fireflies, which none of them had ever seen before.

Battle

They were sent to defend Seletar aerodrome in the north of the island but it had already been bombed and evacuated and there were no defences prepared; they had to dig in. Philip patrolled the barracks, which was like a ghost town; it was obvious that all the men had just left very suddenly – food was on tables, watches and wallets on beds; even rifles were left lying around, and thousands of Singapore dollars in the officers' mess cash box.[1] While he was there, maps were issued to the platoon commanders, and they were amazed to find, when they opened them, that they were maps of the Isle of Wight! As he was known to be an artist, Philip, with two other men who had been draughtsmen, were given a map of Singapore Island, taken from a school noticeboard, and they spent a night with hurricane lamps drawing copies of this map to be given to the officers next day.

They then drained off all the airfield fuel into the sea and retreated as ordered.

At the Singapore causeway they acted for two days as rearguard, as wounded and others streamed across. Philip saw the Australians panic, warning everyone to leave; in contrast, the last of the Argyll and Sutherland Highlanders marched across in good order playing their pipes; it was very emotional and inspiring. Later, Philip and his men moved to Thomson Village.

Owing to poor leadership and the evacuation of key officers, men deserted to try to get to ships in the harbour to get away, and Philip's three companies were soon reduced to about forty men. Philip was a Bren gunner and set up a defensive line with various men from other units, which the Japanese never actually broke through because it sloped down to an open road that they had to cross with no cover. They inflicted heavy casualties on the Japanese in this battle, whose tactic was to creep quite close, then suddenly rise up and, screaming at the tops of their voices, charge full-frontal at the guns, throwing fireworks to add to the cacophony. But it failed to work.

A patrol was sent out one night and it was arranged that they should return by a certain hour from a certain direction. They failed to do this and, as a result, two were killed by friendly fire as they approached the lines. The following morning a patrol of Gurkhas went out to find some Japanese and it was reported that they had failed to return; in fact, they had got back on time but had managed to cross into the Allied lines in broad daylight without detection – which did not inspire confidence in the security of their position. As a former commando, Philip admired the Gurkhas tremendously for their aggressive spirit, but was shocked, when having a meal with them one day, when one proudly opened a tin and showed him several ears of dead Japanese that he had taken as trophies.[2] Philip did not partake in the rest of the food that day.

On the following day Philip led a patrol of twelve men to attack the Japanese line because they had captured a section of their comrades who had been on a previous patrol; they had been tied to trees facing the British lines and Philip planned to release them later. Philip and another corporal first carried out a recce and found the spot. The men crept forward and charged with fixed bayonets into the area, convinced they had not been seen, but they were cut down by a machine-gun ambush – only Philip and two others survived. He called in artillery fire and the Japanese retreated; when they found the captives they had all been brutally bayoneted to death in the testicles and stomach. Only one man had survived and was in a terrible state of shock.[3] They got him back to the British lines.

That night the Japanese attacked the line but were beaten off, and again in the morning; Philip was wounded here by a mortar shell in the ankle and elbow, but that afternoon the word came through of the surrender. Philip remembers that because they had fought well, this was shattering news – the men wanted to fight on. It was to be over two years before any of Philip's family knew he was alive, as he was posted first killed in action, then missing. His wife received a widow's pension for many months.

POW

As they handed in their weapons next day, the first Japanese appeared, on bicycles. Philip saw one grab a Chinese woman to rape her, but a Japanese officer appeared and shot him through the head in full view of everyone.

The men were marched into Singapore City and then formed up again into their own units, to march back to Changi barracks, which was to become the POW camp. It was clear that the Japanese planned to have the POWs run their own affairs with their own officers and NCOs. The Chinese population helped them with water, food and cigarettes on this march. Indeed, Philip testified that the Chinese hated the Japanese so much that they did all they could for the Allied POWs, in contrast to local Thais – though with some notable exceptions.

Philip's battalion were under canvas in large bell tents near the few former camp shops which were located at the entrance; these shops became HQ and the medical centre.[4]

A cookhouse and latrines were built and Philip took PT classes; routines were set up to keep up morale. Classes were held in languages and other subjects, and a microcosm of their previous lives emerged in the camps, and this continued later on, even on the terrible railway under the worst conditions. A small group had the courage to set up a black market by slipping out of the camp with valuables (watches, rings, pens) and trade with locals for food, money and medicines, keeping a small percentage profit for themselves (or paid others to do it for them). They were an essential part of the fabric of survival – for example, they kept the hospitals supplied with drugs for the sick, saving many lives.

Then there were the creative groups – artists, actors, academics – who put on shows, classes and concerts, and Philip helped with these. Then there were the shopkeepers who, for example, ground down burnt rice and added some sweet gulam malacca syrup to it, to sell as cups of coffee round the camp. Some were cigarette rollers and sold these too; some were tradesmen who could make surgical instruments, artificial limbs, tools or furniture from the most bizarre materials; other were just participants who looked and learnt; and a small percentage (maybe 10 per cent), according to Philip, did and contributed nothing – and he said that those were the ones who died first. He was in Changi for about two months.

On several occasions during his imprisonment, Philip woke to find that the man next to him had died. He was convinced that the key to surviving included keeping active and busy, so he put his mind to learning Dutch and spoke it quite fluently for the rest of his life; he also witnessed the power of the will, and knew many fit and intelligent men who simply gave up, turned their faces to the wall and passed away.

Philip recalled the notorious three-day Selarang incident (described in Aubrey Lichman's testimony in this book) and the crowding of thousands of men into a small barracks for refusing to sign an illegal 'no attempt to escape' promise for the Japanese. He recalled that the crowding was so great that men slept standing up, squeezed together. He remembered the frenzied digging of the latrines to try to maintain hygiene; the starvation-level food; and the outbreak of diphtheria which finally got the Allied officers to sign the promise, but only when the Japanese agreed to add 'under duress', thus in effect invalidating the agreement.

Philip was always astonished at how caring the men were. Later, when he was ill over a period of six months with malaria, leg ulcers, pleurisy, beri-neri, pellagra or dysentery, George Old, his artist friend, came every day to his hospital hut and bathed and fed him. These kind of chivalrous, life-saving deeds were very common.

Philip first met the war artists George Old and Ronald Searle in Changi when the theatre got going. Later he met another famous war artist, Jack Chalker, and it was George and Jack who remained with Philip throughout the same camps for the whole three-and-a-half years.[5] George, who resembled Van Gogh and was six feet three inches tall, was a great personality and amusing companion and brilliant draughtsman. He was very caring and always encouraged Philip to develop his skills; he was a natural teacher.

Artists' materials were usually looted or made (Philip first obtained some when clearing debris from a school while on a working party) – but not always. After Changi, Philip's group were moved to River Valley Road camp and went out on working parties into Singapore; once, clearing debris at the Alexandria hospital, Philip noticed an official Japanese war artist who was drawing British soldiers; he seemed only to draw those who had tattoos, perhaps as propaganda to show the Allies up as 'wild men'.[6] He saw Philip watching him and invited him over (speaking quite good English) and asked if Philip was also an artist. He then said that he wanted Philip to draw him, and he brought paper and charcoal. The Japanese liked Philip's drawing very much. Next day he brought Philip a supply of paper and watercolours, pens, ink and pencils.

The problem now was what to do with these. Painting could be seen as spying, so should they give the materials to the doctors for keeping records? The artists finally decided together that they would openly paint, but only innocuous subjects such as plants or butterflies – these were subjects the Japanese especially liked. One Japanese sergeant who was in charge of Philip's hut was a very cultured and charming man who always brought him flowers to paint. In return for the paintings, he would bring materials. For brushes, they used those that came in the old iodine bottles, which were apparently made of sable and very fine. In this way the Japanese left them alone to paint. Drawings and paintings of wounds, men or huts were given to the doctors to hide, and these survived the war;[7] many ended up in Australia.[8]

On one working party, Philip worked at the Kempai Tai prison; while there, they were treated very well, with food and rest periods. But he witnessed terrible treatment of the Chinese prisoners in this place and saw the publicly decapitated heads placed on poles on the bridges as a warning to others, with signs around them saying what the victims had allegedly done.

Quite often the Chinese sold back to soldiers the tinned food that they had looted from British camps after the surrender. One day when on a working party at the docks with a Japanese guard as a carer, one Chinese dealer tried to sell such tins at inflated prices. So Philip took six men aside and said that each of them should pick up one item and walk away in a different direction, on the assumption that the Chinese could not chase seven men; this they did, and when the Chinese protested, the guard pushed him away.

On another occasion, Philip organized a four-man raid on a food store they had spotted opposite the city hospital; it was an isolated building on top of a hill, a former nurses' home, surrounded by trees and hedges and a drainage ditch. Philip calculated that it took six minutes for the guard to do a circuit, and after much rehearsing in the camp, where they got the time down to four minutes, one day they took two screwdrivers, rushed the door and unscrewed the screws on the clasp lock, took one crate each out of the store, re-screwed the clasp, and hid the loot in a culvert. Each day they would sneak a few tins each back into the camp, but one of the men – Fletcher – was afterwards so impatient that he just picked up one whole crate one day and marched into the camp brazenly with it on his shoulders. Not a word was said. When they returned the following day, the rest of their stash had been stolen! As they gave away a lot of the food, it was noticed by others in the camp and so for this exploit Philip was court-martialled – as the senior rank – for looting, by the sergeant major, and brought before his CO. Philip said, in his own defence, that the rations were theirs to begin with, and sabotage was good for the war effort, but the major presiding said he had risked the lives of the other men and that was not acceptable; however, he was let off with a warning.

The Railway

In mid-1942 came news of the men being transported by rail to Siam, a journey that would take three days and three nights, crammed into cattle trucks with virtually no food and water, and no sanitation. They were allowed out once per day when the train stopped for water. The bodies of those who died on the journey had to stay in the trucks until they reached their destination.

Kanchanaburi was the first stop. At that time it was only a staging camp to the railway further north; a few very sick men from previous transports were lying around in huts which were filthy and flooded with diseased monsoon water from the river and rain. Any water available was so chlorinated that it tasted as if they were drinking from a swimming pool.

Tarsao

After a couple of days at Kanchanaburi, they were sent on a five-day-and-night march north to Tarsao. Walking in the monsoon, it was Philip's idea to keep his few possessions wrapped and dry in his haversack, which he covered in his groundsheet and had on his back, and let himself get soaked. In this way, when he arrived at night he would at least have something dry to change into before sleeping.

Trudging through the jungle mud to Tarsao camp, in the monsoon, George fell behind and slipped and fell; as Philip went back to help him, he was pulled down by George's weight and both could not get up; they burst into laughter. At Tarsao, he organized a group of men to pilfer the Japanese stores on a regular basis, as food crates were brought to the camp from the river rafts on a 100-yard chain of sick men. Philip got the men to stand just off the track to do this and every so often a crate would disappear into the bush and be secreted later into the prisoner food store. The Japanese never seemed to miss these crates.

At Tarsao they were to build their own huts and a road for lorries to bring in supplies. They began with 500 men and that later grew to 5,000. To keep out wild animals they lit a ring of fires round the camp at night, but on the other side of the river they would see hundreds of wild animals each day come down to drink – mainly buffalo and monkeys – but they did not come near the men. The fires became meeting places after dark where the men would gather and sit and talk. The Japanese had their own huts, but there was no wire because there was nowhere to escape to. The camp soon split into groups of men who were dealers, entertainers, teachers, and so on.

They began to work twelve-hour days, clearing jungle to take the railway tracks, blasting rock and building bridges across rivers and streams – from Nong Pladuk to their south and Kinsayo to their north. The work was physically very hard with virtually no food and only one half-day a week off.

The guards were mostly Korean and very cruel. On one occasion Philip watched a group of them put a rat in a wire cage and put it over a fire and then place bets on how many times it would run round the cage before dying. But even here there were exceptions, and one Korean called Yoshiamo was particularly pleasant and often allowed them long rests at work, and brought in soap and toilet paper to the men as gifts. After some time, the men began to receive some pay but it was miniscule – just enough each ten days to buy a small amount of tobacco, often rolled into cigarettes. Some of the Australians became adept at splitting ordinary paper into three thin layers, but this soon ran out, so they used the pages of a Bible, with the padre's permission – the thin pages made excellent cigarette paper, and as a result the Australians called these cigarettes 'Holy Smokes'. They rolled cigarettes for sale; this was often done by those men who were bedridden, as a way of making some money to buy extra food, instead of working. Philip usually used his tobacco ration in a pipe. As others have testified, they only saw one or two Red Cross parcels in all three-and-a-half years, and had to share a parcel between six men; the Japanese stole them, and hundreds were found in Japanese storage huts in every camp at the end of the war.

One hut was used to store the Japanese food. One night, Philip and the same men who had been on the 'raid' with him in Singapore crept under the hut and used a pointed bamboo to push through the planks to pierce a sack of much-needed sugar – which they knew was in there – and let it spill into their mugs. Suddenly some Japanese guards came in and started to play a very long game of mah-jong. The men had to lie in the dark for almost three hours until the guards were called away. When they got back to the hut, they found that the 'sugar' was salt; they were not amused.

It was while at Tarsao that Philip witnessed the brutal killing of a man by Koreans. A British POW had left the working party to relieve himself because he had terrible dysentery, but because he had not asked permission he was struck by the guard with his rifle butt and so the POW hit him back. He was immediately dragged away and badly beaten, and then made to stand outside the guardroom in the blazing sun, and in the cold at night, for three days with no food or water, forced to hold a sign describing what he had done; he was given further terrible constant beatings. Finally he was given a shovel to dig his own grave and was shot in front of the men and thrown in.

Philip also witnessed how other men, for the smallest of misdemeanours, were locked in tiny lockers in which it was impossible to stand up, and with no toilet facilities, and given minimal food and not allowed out for up to two weeks (a similar punishment was shown in the film *The Bridge on the River Kwai*).

The men in Philip's hut had a pet monkey, who was called Nelson because he had one eye, that they had found as a baby. He was the centre of much amusement and adored by the men. Often there were searches in the camp, as the Japanese were looking for hidden radios. But Philip's hut was always pre-warned, because the monkey would scream as soon as any guard came within 100 yards of their hut. Eventually the Korean guards caught on and they took the monkey and killed it. Philip never forgave them for this act of wanton cruelty.

Another event that Philip recalled concerned a handsome Anglo-Malay soldier who used to leave the camp frequently to trade with local Thais. He often brought food into the camp and he had a lovely Thai mistress. One night he was caught but, luckily for him, the commandant was away at a meeting and the Japanese doctor was in charge. The doctor's punishment was to make the soldier stand in front of the guardhouse for three days and nights, holding a large groundsheet on which was written in large English letters: 'I must not f—k Siamese girls.' This was a typical Japanese way of trying to humiliate someone, but in fact the men so admired his virility that when they passed him on their way to work they gave him a cheer.

Some men worked full-time in the cemeteries. At first the Japanese would place a food offering on the graves out of respect, in the Buddhist tradition, for the journey to the afterlife, but as soon as the ceremony was over, the food would naturally disappear. The Japanese realized this and soon the offerings were stopped.

Philip was not able to make any money to buy extra food as many did. Occasionally he would draw a birthday card for someone to give to another man, or paint someone's portrait, but that was all. However, it was soon discovered that the Japanese and local traders had a special liking for Omega and Cyma watches. So Philip was able to work at writing these brand names very neatly on more common watch faces such as Timex, for other prisoners or dealers, and so increase the value of the watch to the Japanese by ten times or more! He used very fine bamboo splinters as a brush.

After some weeks, Philip was asked by the British senior officer (Colonel Thomas) to accede to a request by the Japanese interpreter to have a British batman (servant). The CO said they needed someone inside to be able to tell them what the Japanese were up to, and to obtain copies of newspapers, advanced notice of who was going to be moved, and so on. So Philip and a former Singapore CID (Criminal Investigation Department) sergeant called Ferguson, who was half Malay and half Scots and spoke Chinese, Japanese and Thai, went off to work for the Japanese, Ferguson being batman to the camp commandant – obtaining extra food into the bargain! For three to four months they lived in the Japanese quarters, getting better food and not needing to work very hard, just cleaning and tidying. They obtained candles needed for some functioning of the secret radio. In return the interpreter would often ask Philip about certain phrases used in confiscated diaries, to gain an insight into what the men thought of their captors; once he asked Philip what the expression, 'browned off' meant; to protect the diarist, Philip explained it as getting tanned skin, and this was accepted gratefully.

On one occasion a travelling brothel of very beautiful local girls arrived for the Japanese officers; when the Japanese paid them, they secretly gave the money to the two batman prisoners. One particularly fancied Philip and tried to persuade him to go with her but he declined gracefully, being fearful of becoming infected by venereal disease.

When cholera broke out among the native workers on the railway, they ran amuck and out of control. The Japanese were just as afraid as the Allied prisoners were of this incurable disease and issued the prisoners with sharpened sticks to stop any of the native labourers entering the camp; they were ordered to fight them off. Later Philip was forced to go on working parties to bury these poor, dead native labourers; hundreds of Chinese and Indians and Tamils and Malays were taken to huge pits and thrown in and then covered in lime. Philip has never forgotten this horror, especially as many may have been still technically alive, albeit in the last coma. The Japanese flew in emergency supplies of anti-cholera drugs and all the men at Tarsao were inoculated.

It is not clear which camp he was in at the time, but Philip witnessed the joining of the railway line as the south section met the north; a Japanese general attended and a last railway spike was hammered in symbolically. The prisoners were given some fruit and cigarettes and clothes, but after they had returned to the camp, the RAF came and blew the railway section up! Next day the Japanese came and took the clothes back – but the men had already eaten the food.

Quite often Japanese soldiers would appear in the camp on their way to the front line. They were very amicable and often stopped and chatted to the POWs around the camp fires. One morning on his way to work Philip suddenly heard an English voice speak to him; he turned but could only see Japanese. It was a Japanese soldier talking – he had worked in Canada for nine years for a Japanese company, but when he went on a visit home he got caught up in the war. He said he hated the war and thought the treatment of the prisoners was appalling. He gave Philip some cigarettes and sweets before going on his way.

One day Philip scratched his old wound scar with green bamboo while working on road building, wearing only wooden clogs. The bamboo is extremely poisonous and led to a tropical ulcer the size of a saucer. Philip ended up in hospital and soon after was evacuated to Chungkai.

Chungkai

Philip spent about a year here, eight months of it in hospital; he nearly lost his leg, but it healed eventually.

At Chungkai, which was a hospital camp, the death rate was huge, mostly from dysentery; there were so many funerals that the Last Post was played only once per day following mass burials. But the camp was well organized, with a theatre, and the hospital had a well-honed routine of boiling old bandages repeatedly to reuse them as dressings. The traders soon got operating and many medicines were smuggled into the camp and saved thousands of lives. By using an ingenious steam-making machine made of old biscuit tins, a team of men were able to pass blankets and clothes through the steam and kill off disease-bearing body lice. The bed bugs were destroyed by men stripping the bamboo in the huts and passing the slats through fire to kill the eggs.

The MOs gave Philip the job of drawing wounds, so that the drawings could be used for further study after the war. At first, as a patient, he was drawing the amazing home-made instruments and artificial limbs. He was then asked to draw the wounds before treatment and after it, and then later draw and paint the healed scar – so there was a complete record of the progress of the wounds.

Later, when he was able to stand and walk, he was assigned to record secretly the operations in the operating theatre in the various camps, in the guise of a medical orderly, thus avoiding the working parties but doing a crucially important job. Men were allocated to guard the operating theatre in case guards approached and saw Philip working. The drawings were used a great deal in medical books after the war, to teach wartime surgery and medicine.

One day a native boy – nicknamed Sabu – who had been badly gored by an elephant was brought to the hospital with very serious wounds to his buttocks. The Allied surgeons saved him and Philip had the job of recording in watercolours the whole procedure and healing process.

In one theatre production that the men put on, 'Tarzan and the Apes', the men playing the apes looked like Japanese. This was totally inadvertent, but the Japanese – who liked the shows as much as the prisoners – were convinced it was meant to mock them and, infuriated, closed the theatre for three months. Lectures were always packed out.

Tamuang

Tamuang was his next camp, where he spent a further eight months. It was mainly a camp for invalids who could not work for the Japanese any longer, and Philip continued as a painter of medical procedures. There were no working parties, just those on camp maintenance. The men were allowed to grow vegetables at Tamuang and generally the guards left them totally alone, save for the odd roll call once per week.

It was here in 1944 that they began to see Allied aircraft, and guessed the war was going badly for the Japanese. A major incident occurred when they also saw cattle-truck trains arrive at the camp, packed with thousands of wounded Japanese who were squeezed into three-tier bunks, many dead and dying of their wounds, with no medical staff in sight. Such was their awful state that the prisoners brewed vast cans of tea and took them to the Japanese wounded, together with cigarettes, and the Allied doctors went onto the trains to assist in any way they could, while the Koreans stood about laughing and the Japanese medical teams in the camp refused to help. It was an astonishing exhibition of humanity and comradeship by the Allied prisoners, despite their own dire state.

However, men were being brought to Tamuang from upcountry who were in a terrible state and often dying. Philip himself was down to six stone and now had pellagra, followed by malaria and pleurisy. He was in agony and so he was given the only drug they had – a dose of opium which sent him to sleep for forty-eight hours.

One day Philip and some others were taken by lorry on a work party to fetch stores for the Japanese. On the way, the lorries stopped to allow the guards to visit a brothel; the Thai and Chinese girls came out afterwards and gave the money to the prisoners because they felt so sorry for them. One prisoner stole a rifle belonging to a guard and sold it for 200 dollars there and then to a Chinese. When the guard came out he was astonished that he could not find it and offered to pay fifty dollars, and begged Philip and the other prisoners to buy him another one before getting back to camp or he would be severely punished, not just for losing his rifle but for visiting a brothel. So they used some of the money to buy another rifle from a villager for far less, and walked away with a big profit and one very indebted Japanese guard.

On another work party, Philip and his comrades were sitting on flat cars, travelling slowly to work on the railway to carry out repairs of damage done by Allied aircraft. Suddenly a large Allied fighter bomber flew low over them and they all waved; it returned again and they waved again, but on the third pass its machine guns were blazing and they all dived off into the jungle to save their lives. It was one of Philip's closest brushes with death while a POW.

Nahkon Pathom

This was regarded as a Red Cross camp, and the prisoners rarely saw the guards except when they came to the prisoners' concerts. There were no working parties. It was quite different from other camps – near the town and surrounded by a high wall with machine-gun posts. It was well organized by the POWs. But the Japanese, expecting to be invaded, separated off all the officers except doctors to another camp, so that the men would be leaderless in the event of fighting breaking out. It was here that it was decided to carefully disguise the medical drawings and other papers kept by the officers and smuggle them out with the officers who were leaving.

Between camps, the drawings were hidden in hollowed-out bamboo bed-legs and they were put in a false section of Philip's rucksack, especially designed and made for him by a friend, wrapped in oilskins or army gas capes. Philip explained: 'If found, it was a death sentence to sketch the operations they conducted ... in all 140 horrific drawings ... were wrapped in oilskins and buried secretly in old ammunition tins', said Philip. 'Amazingly we were able to rescue them after the war ... they keep the memory of those fearful years alive and finally played a part in shaming the government into paying compensation ... though what we wanted was payment from the Japanese ... and a genuine apology.'

Liberation

Philip's foot eventually healed. He was still at Nahkon Pathom when he and his comrades heard about the end of the war: one morning at about 10.30 a.m. – a moment Philip said he would never forget – men came along the rows of huts telling everyone the war was over. All 3,000 to 4,000 men gravitated without orders to the centre of the camp area where there was a small stage they had built for performances, and the three Allied flags – British, Australian and Dutch – were being hauled up. The small camp band was playing 'There'll Always Be an England' and 'Land of Hope and Glory'. There was much weeping, even by the tough Aussies. Philip described it as quite unbelievable. Then the three national anthems were played and the senior officer, Colonel 'Weary' Dunlop, officially announced the end of the war. Orders were given not to leave the camp yet. They returned to their huts, and then went to the Japanese stores and took all the food to the cookhouses, and a huge meal was made for all the men.

Philip was at this stage asked to enter the Japanese HQ to make lists of names of survivors for the Red Cross. Suddenly half-a-dozen heavily armed Allied paratroopers, with bush hats, ammunition bandoliers and Sten guns, entered the camp, just like a scene in a war film. The POWs discovered later that these men had been hiding in the nearby hills overlooking the camp for three months and their mission had been that if the Allies had to fight their way to the camps, they were to attack and arm the prisoners to fight the Japanese guards in an insurrection, to prevent a massacre of the POWs.

The senior para officer was a major and he told them about the dropping of the atomic bombs on Japan. He had a huge supply of cigarettes, and next day the RAF dropped food, blankets and medicines by parachute.

Soon the Japanese left, after an official handing-over of their swords. The Koreans had been put in a compound to protect them from the Japanese – in fact, the Japanese just gave them some money and told them to go back to Korea; no revenge was taken, as there had been no ill-treatment at this camp. The Japanese were ordered to take Philip and his comrades to Bangkok by train, walking only where the bridges had been blown; at Bangkok they were put on barges and taken to godowns on the riverside to be kept and fed for a few nights before being flown to Rangoon. There was an incident here involving Australian soldiers, and the Japanese guards were killed quietly in revenge.

In Rangoon, when Philip was in hospital, he and George Old were asked by a Colonel Ransome to continue their drawing work, for use in the coming war crimes trials, and they did this while recovering, for about three months. Work in colour was better able to convey the horror than black-and-white photographs. Philip completed about forty such works, but he never discovered what happened to these drawings.

A nurse taught him, in thirty minutes, how to drive and he was issued with a jeep to go into Mingladon to buy art materials. In their spare time they went to visit the temples, ate out in Chinese restaurants and befriended Chinese locals. His relationship with the nurse eased him back into the normal world and helped him to readjust, something for which he was always indebted to her.

Then his unit, which had been working on identifying war graves and cemeteries, as well as the war crimes work, were flown to Calcutta to recover further, as guests of the viceroy.

Home

Returning to Britain, Philip met for the first time his baby daughter Elaine Olive, who was born just after he was posted abroad. He and his wife had twin sons later, but divorced in 1969. Philip then married Elaine Wells, and became official artist to the Royal Ballet in 1990. One son said he was scarred always by the war but 'kept his suffering in a separate compartment of his mind', always being 'marvellous company'. In 1995 he was reunited with a fellow war artist, George Old, at an Imperial War Museum reunion of war artists; they had not met for fifty years.

Philip died in New Malden in 2007. Seventy of his sketches of the POW camps in Burma/Thailand, done under unspeakable mental and physical suffering, are kept at the

Imperial War Museum and were used as evidence in several war crimes trials of the Japanese. He left a large family, including seven great-grandchildren.

Notes

1. This money was left, much to Philip's later regret, as it would have been extremely useful in the prison camps for buying food and medicines.
2. This was quite common practice in Asian cultures in wartime in those days.
3. He recovered, and Philip saw him later in the prison camp at Tamuang.
4. A painting of this area by Philip is at the Imperial War Museum.
5. By a bizarre coincidence, Philip's father turned down an application from Ronald Searle in the late 1940s to become a student at the Central School of Art where Bernard Meninsky was teaching. Ronald said he never held it against Philip!
6. This seems somewhat odd, as there is a large ethnic minority in northern Japan whose men and women wear tattoos as part of their culture.
7. George Old had ignored this advice and thus his paintings were eventually stolen by the Japanese; he was heartbroken about this incident
8. Philip later discovered that several of his and Jack Chalker's drawings were used in Colonel 'Weary' Dunlop's published diary, years after the war.

NX10477 Sgmn Neville Hyman Milston, AIF, Efficiency Award, POW no. 2471

Neville was born on 31 July 1920 in Parramatta Road, Leichhardt, Sydney, Australia; his family lived above their shop. He attended Crystal Street school, then Coogee Public School. He left school at the age of 14 and worked in his parents' gift shop business, joining the militia in 1938 at the time of Munich, and the AIF in early1941. He trained as a radio operator at Tamworth, NSW; he was sent to Malaya with 8th Division Signals in September 1941, on SS *Sibajak*, accompanied by the cruiser HMAS *Sydney*. They travelled via Fremantle to the Sunda Straits and were joined by three other troop carriers – one named *Aquatania* – until arrival in Singapore. Neville was on duty as a signaller on the bridge for much of the journey, using lamps only for safety, and no wireless.[1]

On arrival in Malaya they did some jungle infantry training and were posted to Malacca with their radio truck and then to the east coast at Mersing. During this time he saw many Japanese aircraft flying over to bomb Singapore, after 8 December. By 12 February they had withdrawn to Singapore Island and were converted to infantry, supporting the 4th Anti-Tank Unit and guarding twenty-five-pounder guns. When one bomb landed close to a nearby police station, they went in to see if they could help but all the officers were dead from bomb blast with not one visible sign of an injury on them. After this they were constantly shelled and had many near misses, but had dug slit trenches to shelter in, as per army orders. Then suddenly, about 8 p.m. on 15 February, there was silence. The senior Corporal asked Neville and another signalman to go to investigate down the road and they came upon some British officers with a white flag, ordering them to put their rifles away; the army had surrendered. When Neville went back and told his comrades they could not believe it.

Neville was captured in February 1942 and marched to Changi gaol, where he stayed until 14 May. Conditions there were crude and rough – not nearly as bad, however, as what awaited them. Then he was shipped with 3,000 Australian POWs, known as 'A' Force, on two rusty Japanese freighters. His was called the *Toyahasi Maru* and 2,000 men were crowded aboard in the most appalling conditions. Machine guns were placed to prevent any mutiny, and toilets were flimsy bamboo boxes hung over the side. After some weeks at sea, always in danger of being sunk by Allied submarines, they arrived at Tavoy Port, south Burma. Here they were forced to help move a large steamroller in the docks but they 'accidentally on purpose' pushed it into the harbour. Then about 800 of them were marched twenty-four miles to the town of Tavoy to work on repairing the airfield, which still had the wrecks of RAF planes on it.

It was here that eight Australians tried an escape but were caught just five miles away. They were brought back and, despite the protests of their officers, were shot by firing squad in front of Neville and his comrades.

In September, Neville and his group were taken by coastal boat to Moulmein and then by truck to Thambuzyat, the start of the railway at its northern end, to work on the Death Railway. Neville was to spend over three years on the notorious Burma Railway at its northern end, on starvation rations, working twelve-hour days in the excruciating heat and rain. Most of the work was building the embankments, cutting through rock to

47. NX10477 Sgmn Neville Milston, AIF, taken in September 1941 with (left to right) Bill Le Beau, Jim Whitley, Ken Brown, Geoff Sheppard, and Neville and Thomas Carradine; inset: Neville Milston, taken just before his death in 2007.

excavate cuttings and then laying sleepers and fixing the rails. They also had to build many wooden railway bridges over deep gorges, often working at night with the real risk of falling through the sleepers into the gorge and river torrents below; this remained a nightmare Neville never forgot.

From July to December 1943, Neville was at Khonkhan (55 kilometre) camp in Burma, just fifty-five kilometres from Thambuzyat – a so-called 'hospital' camp. Neville said that the death rate here was particularly high. Virtually all the POWs were ill; 400 out of 600 POWs died while Neville was there, despite the amazing efforts of the Allied doctors.

Then Neville was sent to Tamarkan, where the work was mostly maintenance of the now completed railway, and cutting wood for the trains because coal was not available. Allied planes often bombed them and their clothes wore out. However, the war appeared not to be going well for the Japanese. The Japanese ordered the digging of pits around the camp and it was known that the they planned to massacre all POWs if an Allied invasion came. It was Neville's worn and grubby Jewish Singers's prayer book,[2] that had been given to him at his bar mitzvah in the Great Synagogue, Sydney, in 1933, that was used in the Tamarkan camp bar mitzvah service in September 1944, described in the introduction to this book. Neville was too ill to attend and was often tempted to sell the prayer book as cigarette paper, to buy extra food, but he never did. His faith and the book both survived, and he brought the prayer book safely back to Australia. At Tamarkan he met and prayed with another Jewish POW, Harry Courts of Golders Green, London. Harry used to quip that Neville's Hebrew was good, 'but for that bloody Australian accent!'

Neville remembers that ANZAC Day (April 1944) at Tamarkan was especially moving. The Japanese had given permission for a small ceremony, and thirty to forty Aussies took part; they were First World War veterans and were serving in the 2nd AIF as truck drivers; in their slouch hats they marched a few metres and the Last Post and Reveille were sounded.

Neville recalled how on one occasion he thought of drowning himself in a pool where the POWs used to come and wash after work, to put an end to his suffering, but he felt it would so grieve his mother that he could not bring himself to do it. An occasional Red Cross parcel was given by the Japanese, but although meant for one man, they were shared – one parcel between eight men. Often they would boil rice to make a coffee-like wine concoction; it had to be sweetened with sugar, and this was stolen from Japanese stores at the risk of execution if caught. He witnessed many amputations of the limbs of POWs who had developed the terrible tropical ulcers, often from a simple bamboo cut; he remembers that in one four-week period there were 160 of these operations, with little or no anaesthetic, and men were forced to get drunk on the rice wine to numb the pain. On the few occasions when mail arrived, the letters were often 'for blokes who had already died'.

Liberation

On 15 August 1945, Neville was in a small work camp in south Burma, cutting metal railway lines with a hacksaw. Some excited POW was shouting that the war was over but they had heard rumours before; an atomic bomb was mentioned. It turned out to be true. They were taken on the railway south to Tamuang camp, where Allied planes dropped huge containers of food and clothes to them; they were warned to stand clear, but one excited Aussie ran out into the open and was killed.

Lady Mountbatten arrived at the camp as the Japanese guards began to disappear. Neville remembers her saying, 'Can I be frank? You all smell!'

On 8 September, Neville and about twelve other Jewish POWs, with the help of the Methodist RAN chaplain, Keith Matheson, held a short Rosh Hashana service in the open air, their first in freedom since 1942, using Neville's prayer book. While in Singapore, Neville received a backlog of eighteen letters from home, written between June 1942 and March 1944, of the sixty his parents had written. Others took one or two years to reach him. Meanwhile, both his grandparents had died.

He finally got home via Bangkok and Singapore on the SS *Moreton Bay*, arriving in Sydney on 2 November 1945. He needed extensive rehabilitation after liberation from the camps.

At a Jewish war veterans' party in 1949 he met and later married Madge Grouse, a Jewish ex-servicewoman who had served in New Guinea in the 2/1 Australian General Hospital, Port Moresby, and they had three children – Michael, Richard and Leah. They lived in Castlecrag (Sydney) where he was active in local charity, political (Labour Party) and Jewish religious affairs all his life. In 1981 he took a BA degree, 'to get the education I missed out on' when young. He died in July 2007.

Notes

1. From letters sent home by Neville to his parents.
2. The standard Hebrew and English prayer book used then, and largely now, in synagogues round the English-speaking world.

9137762/913662 Sgt Maurice Hillel Minchom (aka Mincovitch), MiD, TD, RA

With thanks to Maurice's son and daughter, Antony and Anthea, for the details in this account.

48. 9137762/913662 Sgt Maurice Minchom (aka Mincovitch), MiD, drawn in Thailand while on the death railway, 1944, artist unknown.

Maurice, son of Emmanuel and Rachel Leah (née Merskey), was born in Sunderland at 11 Claremont Terrace on 7 October 1917, one of fourteen children, and the family later moved to 31 Elmwood Street, when his father was a private in the 4th Durham Light Infantry in the First World War. Maurice attended the famous Bede Collegiate Boys School and joined the Territorial Army in 1938 in Sunderland, and in 1940 was transferred to the RA (125 AT Regt). His father ran a furniture business in Frederick Street. Maurice married Leonora Sophia Gillow on 15 October 1941 at Ryhope Road Synagogue and they lived at 17 Belle Vue Park. Maurice was sent overseas in January 1942, travelling from Scotland on the HMT *Andes* to the USA, and then from the USA on the USS *Wakefield* to South Africa, arriving only ten days prior to the surrender of Singapore.

Maurice's debrief sheet shows he was at Changi and then deported on the 'Hell Trains' to Kanchanaburi on 15 March 1942; on 20 March 1943 he was sent to Wampo for two weeks, then on to Kinsayo, where he was kept until July 1943. After this he was at Chungkai until May 1944, then Tamuan until May 1945; he was finally liberated at Nahom Nayok in August 1945.

Whilst a POW in Changi he met Captain David Arkush. He remembered the amazing shows the POWs put on and how professional they were, particularly one Welsh doctor who had 'the voice of an angel, drilling a motley crew of singers'; sadly, he died in the camps and Maurice does not mention his name. He recalled the Anglo-Australian Test Matches organized by the former *Daily Telegraph* reporter E.W. Swanton. In one camp – possibly Kanchanaburi – he recalls a tall RAMC colonel (possibly 'Weary' Dunlop), whose skill had saved thousands of lives, taking part in a 'Race Meeting' with jockeys, owners, trainers, bookies with real odds – using real hard-earned money – and men acting as horses. The railway-building was over, and the colonel chose the smallest man as his jockey who was the favourite, winning the race easily by half a field; the winner was listed as a 4-year-old stallion called 'Bed bugs, out of slats by Midnight' – referring to the terrible and insatiable bugs that infested their bamboo slat sleeping shelves.

Worst of all were the daily lines of handcarts carrying the dead, wending their way ceaselessly to the cemetery, carried by comrades, many of whom soon followed in their wake. Maurice recalled the grim determination to stay alive by the many sick; the burials of the dead, head-to-toe in rice sacks; officers bravely risking severe beating or even death to prevent their sick men being forced to work, by defying the brutal guards; he recalled the brave 'black marketeers' who went under or over the wire in places to bargain valuables such as pens and lighters with locals and bring back valuable food and drugs. They did well but they also shared. The 'cigarette barons', however, got themselves into non-working dysentery wards and 'employed' patients to roll their tobacco into cigarettes for sale for them, ruling the 'wards' much as cigarette barons do in some prisons today.

Maurice always said that the camps brought out both the worst and the best in the men around him. He was awarded an MiD for courage, gazetted in 1946.

After treatment in Queen Mary's Hospital, Roehampton, he became an active member of synagogues in Sunderland and later in London, where he went to live. He had mental health issues as a result of what had happened to him, but his wife was a bottomless well of love and support; sadly, she died very young in 1965 and Maurice's world collapsed. But his brother, a GP, was able to assist him in his recovery.

Maurice died just before his ninety-third birthday in September 2010. Besides his MiD and TD (Territorial Decoration), his medals included the 1939–45 Star, Pacific Star and War Medal.

106368 Maj. Dr Coleman/Colman Nairnsey, RAMC

49. 106368 Maj. Dr Coleman Nairnsey, RAMC.

Coleman was born on 10 September 1897 in South Africa, son of Jewish Russian immigrants who fled persecution in Lithuania and had first gone to Britain and then on to Johannesburg. They owned a farm at Rustenburg where Coleman was born, and where he attended Jeppey High School. After dabbling in law, he opted for medicine and went on to Trinity College, Dublin, to do his medical degree in the 1920s. He returned to South Africa to work but, being a restless man, decided to take a job at St Mary's Hospital, London, in 1932. Here he met his first wife, Marjorie Ann Isaacs/Sandelovich/Sandell, sister of a good friend, and they married in the same year. He then took over a practice at 1 Cholmley Gardens, north-west London, from where he volunteered in 1939 when war broke out, in his forty-third year.

At first he served as MO in the Royal Devon Yeomanry and he mixed with many of the titled people who populated that pukkah regiment, stationed on Drake's Island; he was often with them on shooting and other posh social events during the Phoney War. He was posted abroad, leaving Liverpool by ship and sailing via Cape Town. Here he met up with his uncle, who gave him a cigarette lighter as a gift. This lighter played an important role in an incident later on. From Cape Town he went via Bombay and Colombo to Singapore. When the city was falling he was offered a rifle, but he said he was a doctor, there to save life, not take it. Thus he was taken POW.

In the first camp, probably Changi, he was taken ill with malaria and so missed being sent upcountry to the Burma Railway. It may have been here that Coleman witnessed the incident when two Australians escaped and had been recaptured. The whole camp was told about this and ordered to parade. Two sacks were brought out onto the parade ground and the bodies of the two men, tortured, beaten and dead, were thrown out for all to see as a reminder of what happened to escapees.

Soon afterwards, in May 1943, Coleman was transported to Japan via Formosa and Okinawa, on one of the Hell Ships, to work in a coal mine run by Commandant Suribachu for the Mitsubishi company, for three years. It was located next to a steel mill near to Nagasaki – it was called Camp 6 Orio. In his camp were about 1,200 POWs of all Allied nationalities. He was in charge of the camp, together with a Dutch doctor, and they had constantly to decide who was fit enough or not to work, keeping as many men as possible out of the horrific mining conditions, while enduring and combating the ruthless Japanese demands to send even the sickest POWs down to work.

As a senior officer he had a room separate from the other men . Nobody was allowed to wear shoes in the barracks, by order of the Japanese. One rest day (they were allowed one day in ten) he was dozing on his straw mat, when a Japanese NCO, Takahashu, came to his room. Coleman pretended to be asleep. He heard him rifling through his possessions in his cupboard, and then heard him leave.

Immediately afterwards, without thinking, Coleman went to the commandant and told him a Japanese had been to his room and stolen his lighter. At that moment Takahashu entered the commandant's office, and the two Japanese had a conversation; then the commandant said to Coleman that he should return to his room and see whether maybe he had made a mistake about the lighter. Coleman returned to his room and suddenly Takahashu came in; saying nothing, he took the lighter out of his pocket and gave it to Coleman and left. Coleman was in a quandary; if he went back and accused him of stealing the lighter, and then returning it, his life might be in danger, as the Japanese would give him hell for the rest of his stay as a POW. He decided that he should just go back and say he had made a mistake and that he had found the lighter under his bed. When Coleman went back he told the commandant this story. The commandant told him to stand to attention and then he drew his sword, shouting how dare he accuse a Japanese of stealing! He then beat Coleman all over the body very severely with the side of his sword for some considerable time. Coleman managed to get back to his room and collapsed on his bed. After a few minutes, Takahashu burst into his room. He said nothing, but gave Coleman two pears: his way of thanking him for not pursuing the matter of the lighter, which would have brought loss of face and disgrace on him. Coleman was ill for three weeks, and Takahashu looked after him. Such were the vagaries of being a POW of the Japanese.

One day the camp commandant called Coleman to his office and said he had acquired two ducks and asked for advice on how to fatten them up for the Japanese to eat. Coleman said he should give them lots of vitamin B because he knew that Red Cross parcels contained jars of Marmite (rich in vitamin B), and the Japanese had the parcels but probably did not know what Marmite was. He said that Marmite was excellent and asked if the commandant had any. The Japanese officer left the office and returned with jars of the stuff, and gave the ducks to him to feed. Needless to say, the Marmite was given to the sick prisoners and the ducks were fattened on grass and rice, and everyone was happy!

The mine regime consisted of three shifts: morning, afternoon and night. A box of rice was given to each POW to take down the mine to eat during their short break from the gruelling work. At Coleman's camp, men who died or were killed were cremated, and both Coleman and the other doctor would accompany the bodies, to witness the cremation about two miles away from the camp. There was one Japanese sergeant in command and he was particularly brutal. The men tried to think of a way to kill him, but could not mange to do it. After liberation they heard that he had been killed in Okinawa, but it is not clear if it was in the fighting or by execution for war crimes

One typical piece of Japanese sadism was when they let two dogs loose in the camp and told Coleman that this was the evening meal for the 1,200 men, and they had to do the rest.

Coleman ran a sick parade in the morning and evening; over the whole three years that he was incarcerated, he tried to see that as few men as possible were sent down to work in the terrible conditions. However, this meant that those who did not go to work, already on starvation rations anyway, had them reduced by half, as they did not go to the mine.

In the summer of 1945, they knew something had happened when one of the shifts was told not to go down the mine, and others returned early. On 9 August, Coleman claims he saw a huge flash from the direction of Nagasaki some miles away. He later realized that this was the atomic bomb. They became aware of a lot of Allied air activity, of the Japanese scurrying around and listening to their radios, and eventually leaflets were dropped telling all POWs to stay in the camps for their own safety – help was on its way.

Later that morning, while conducting sick parade, Suribachu called for him, as senior officer, to go to his office. There he found some other Allied officers, including a Major Humber, with Corporal Takahashu sitting to one side shaking with nervousness. Suribachu said that the war was over and they were at peace, and he hoped they had treated the POWs well and that they would be treated well in return. Coleman said, 'Now you will get the hell knocked out of you!' Suribachu threatened to cut his head off. Coleman replied, 'No, you are my prisoner now!' Suribachu did not reply.

After the bomb fell the attitude of the Japanese changed completely. Some POWs went on looting sprees in nearby towns and took swords as mementos; some took kimonos and even postage stamps. This was their small revenge. Suribachu had handed the keys of the offices to Coleman and when the stores were opened they found Red Cross parcels which they gave to the POWs. Coleman found British uniforms in the stores and he put his on with his major's crown and demanded that the Japanese guards should come out and salute him. If those parcels had been given to the men earlier, with the food and medicines they contained, many lives would have been saved. Coleman weighed just sixty pounds.

On the night of the surrender, the mine owner invited Coleman and two other officers to dinner, and he wore his uniform. He was offered wine and beef, and on that tiny glass of wine he became drunk, because he was so unused to alcohol. On the way back to the camp, he ordered the Japanese guard of six men to turn out to salute and present arms, which they did with great ceremony; clearly they were all afraid of what might become of them, because of the ill treatment they had meted out over the years to the Allied POWs.

After a few days the Americans arrived and were horrified at the emaciated state of the men. Coleman warned all the POWs to eat little and eat slowly, as their stomachs could not cope. Some ignored him and died from overeating.

Later Coleman was taken to Okinawa, where they stopped to recover for a few days and then on to Manila, Honolulu (Hawaii) and San Francisco, all the while being fed with the most amazing fare. From San Francisco they were taken by hospital train to Vancouver. Coleman's wife and two children still had no idea that he was alive. After spending the war in Canada, where his wife had been brought up in Winnipeg, they had returned to the UK in May 1945. They had only ever had one postcard from him in three years.

After some leave, Coleman continued to New York, and after some time there he went on the liner *Queen Mary*, taking five days to reach Southampton, arriving on 18 November 1945. The cable he had sent from the USA, saying he was on his way home, never arrived. On the evening of 19 November he went to the house in Cholmley Gardens, where he had had his surgery before the war. The family thought it was a former patient knocking at the door. Elizabeth or Ian went to the door and came back saying there was a soldier outside asking for their mother; they were amazed when they saw them kiss! They had a joyful reunion. Ian remembers that although Coleman was unkempt and weary, after about an hour he took out a map and showed them all the travels he had been on since liberation. He was remarkably clear-headed but very thin. He was unwell but stayed at home and was finally demobbed in November 1946, with a permanent war pension.

Coleman went back to being a GP, but sadly his wife died in 1963, so he went to live in Saskatchewan, Canada, for twelve years, returning to Willesden, London, in 1975, and later retiring to Florida. He had seven grandchildren. After retirement he qualified as a homeopathic doctor in London and practised in Harley Street.

He had volunteered to serve Britain – he would never have been called up, because of his age, and being South African he could have claimed exemption anyway, but it never entered his mind to do so. Rightly or wrongly, Coleman never forgave the Japanese for what they did and would never greet one socially if he came across them. He lived to be 95 years old and died in 1993.

MZ19224 Lt Col Dr Sydney 'Syd' George Nardell, RIAMC and 11th Division, MiD

50. MZ19224 Lt Col Sydney Nardell, RIAMC.

Sydney was born on 11 April 1913 and at the very early age of 15 entered the Medical School at King's College, London; here he did his BSc in Physiology and won a prize to continue at St George's Hospital in Hyde Park, where he became a renowned surgeon. On qualifying he spent a short spell as a ship's doctor on cruise ships. He enlisted on 1 May 1939 and was in the Territorial Army, living at 2 Fox Lane, Palmers Green, London N3.

From Sydney's book, *Jungle Medicine and Surgery* (Book Guild, 1999), we learn that by 1941 he was overseas, with his wife Eileen and baby son David, and was serving in the military hospital at Shillong in Assam (Gurkha training school). Leaving his family, he was posted to Lucknow (17th Combined General Hospital) and then to Tanjong Malim, fifty miles north of Kuala Lumpur in Malaya. From here he went to No. 5 CCS (2nd in command) at Bedong on the Thai border. He did not see his wife and baby again for five years, as they were shipped back to Britain at the height of the submarine war.

Bedong was a tiny village and the CCS was just a few huts hidden in a rubber plantation; most of the doctors and staff were Indians of the Indian Army. The colonel was a typical 'Blimp' figure: a First World War veteran and a disciplinarian; racist towards the Indian staff; a red-faced, hard-drinking confirmed bachelor. But Sydney had a good working relationship with him. Attached to them was a Sikh Ambulance unit for transport.

On the morning of 7 December 1941, Sydney was shaving on the veranda of his hut when he looked up and saw a flight of twenty-seven bombers overhead, flying south. He said to his Indian officer neighbour – who was cleaning his teeth with bamboo – how good it was to have some decent aircraft for once defending Malaya. Suddenly there were massive explosions as bombs landed all over the area. Within minutes, casualties came pouring in. For ten days and nights Sydney did not undress and worked constantly with only snatches of sleep. Eventually two new doctors arrived to help and one was Captain Dr Jacob Markowitz from Canada.

Meanwhile the Japanese were advancing rapidly and the CCS was ordered to start packing and be ready to move with very short notice. At last the order came to meet a train to go south, but the officers knew it would never come and so they used the ambulances and moved out all the wounded, staff and equipment, on their own initiative, to the Civil Hospital at Ipoh. After that it was constant retreat down the peninsula from location to location, towards Singapore – Tanjong Malim, Kluang and Rengam, to name three – and they were frequently strafed and bombed by the Japanese air force. At Tanjong, Sydney watched as an RAMC officer, Colonel Hurd-Wood, stood in the open, filming the Japanese dive-bombers with his cine camera, whilst everyone else dived for the slit trenches.[1] He had been at the battle of Jutland in 1916 and said this was nothing compared to the screams of the shells in that great naval battle. Captain Markowitz commented: 'I hope we are not going to have to learn Japanese' – but of course they did.

On 4 February, Sydney's CCS was ordered to cross into Singapore and billeted at a school. Sydney went to visit a surgeon colleague at the Alexandra Hospital and as it was so crowded with wounded and had such good and modern facilities, he asked to be transferred there to help. He was soon busily at work, causing some amusement by wearing his tin helmet while operating.

The Massacre at Alexandra Hospital

On the evening of 12 February, the Alexandra hospital was in no-man's-land as the Allies had retreated beyond a perimeter into Singapore City. The CO invited all the staff to a farewell drink and then ordered all stocks of alcohol to be destroyed. On Friday 13 February the hospital was being hit by mortars from both sides and a great deal of damage was being done. Water was scarce and electricity cut off. Large red crosses and white blankets clearly marked the building as a hospital. But in the afternoon, on the pretext that they were being fired at from the hospital by Indian troops, the Japanese entered the building and began throwing hand grenades and firing indiscriminately at both patients and staff. Sydney was in a darkened room with several other staff, feigning death, lying on their faces; he looked up and saw a Japanese bayonet pointing through the door and then disappear. Meanwhile, ten medical officers had been murdered in the corridor outside. In the operating theatre, Captain Smiley and all his team were bayoneted and shot in the middle of an operation, including the patient, though Smiley was only wounded and fell beneath the bodies, the only survivor. Over 350 doctors, staff and wounded, were then marched out of the building and massacred in the grounds of the hospital. The Japanese then put a guard on the perimeter of the hospital and left the survivors to fend for themselves over the weekend. On Sunday evening the surrender was announced and firing ceased. Sydney was in a ward with dozens of other wounded lying on beds and floors; corridors were crammed full, blood was everywhere and the building was badly damaged.

Over the next two weeks, Sydney and the staff tried to get the hospital back in working order as best they could. One day a Japanese soldier stopped Sydney in a corridor and prodded him with his bayonet, looking at Sydney's watch, a gold one given to him by his wife as a wedding present. Although the Japanese was already wearing three watches, Sydney did not argue and handed it over. Soon afterwards, they were moved by truck to Changi gaol. Most of Sydney's personal possessions had disappeared and he was left with just what he stood up in and some spare clothes.

Changi

A hospital was set up in Changi and with the help of some RE men, accommodation for the medical staff was added, including later a workshop for making artificial limbs; for his work in this, Sydney was later awarded an MiD. Over the next few months the hospital became quite efficient. One man attempted an escape by swimming the straits of Johore; recaptured and struck by a Japanese officer's sabre-blade across the back, he managed to get back to the camp where Sydney, even after several days, was able to successfully treat him.

For the next eighteen months, Sydney learnt many new skills from fellow surgeons while caring for the sick and wounded and those injured while on working parties. One night a sick Dutch soldier needed treatment for abdominal pain but the Japanese would allow no lights, as they were carrying out blackout practice. Waiting at the patient's bedside until first light, Sydney tried to operate next morning but it was too late and the burst appendix killed the man; Sydney was distraught because had he been able to work the night before, the man would have lived.

Meanwhile, POWs came and went – mostly to work on the Death Railway. A canteen was established, entertainment was organized, some small amounts of pay for work reached the men, and all kinds of education courses started up. A few captured Chindits also arrived at the camp and were able to tell the men of their commando-style raids on the Japanese behind their lines.

Small groups of prisoners were allowed to go swimming on the nearby beach, as long as the officer in charge carried a special Japanese flag and saluted the sentry on the way in and out of the camp. Work also started on the establishment of a proper cemetery, and work parties were sent to break stones for the graves.

As equipment wore out or was used up, Sydney and his medical teams had bare hands and feet in the operating theatre, wearing only their 'Jap-happy' loincloths. Sometimes the occasional Japanese soldier would come for treatment at the outpatient section.

In June 1943, the Japanese ordered that all officers who had served in Assam had to go for interrogation at Changi civilian gaol, where the Kempeitai had their offices and frequently used torture. It was suspected that this was to gain information about the area, as the Japanese were about to invade India via Assam. So they decided to send an officer who knew nothing of military value – Sydney!

At the prison, Sydney was shown into an office and questioned politely by a senior uniformed officer, whilst standing in the background in civilian clothing, nonchalantly swinging a golf club, was a Kempeitai member. He was asked about the buildings and countryside in Assam but was able to sidestep and give little away. He was soon allowed to go, much to his relief.

Thailand and the Coolie Camps

Soon after this incident, in September 1943, Sydney was moved by train – in old coal trucks, in the usual filthy and crowded conditions – north to Kanchanaburi in Thailand, in a medical group of about 100 staff, divided into small teams, to care for coolie labourers on the Death Railway. His experience was unusual in that he was not at first in POW camps, where conditions were better, and he witnessed at first-hand the brutal treatment by the Japanese of the coolie labourers – Indians, Malays, Chinese and others.[2] Sydney was at Kanchamaburi until October 1943, then Konkoida until February 1944. He was then sent to Tadeim until August 1944 and Kriangkri until December 1944, finally being returned to Kanchanaburi until March 1945. After this he was at Chungkai until June 1945. The coolie camps where he worked were in the most appalling condition, with hundreds left to die in so-called hospital huts, with virtually no medicines supplied by the Japanese. One needle sufficed for hundreds of injections and when it finally broke, Sydney was told by the Japanese orderly that 'Doctor no good!' When the Japanese felt like it, they would simply announce that no treatment was to be given, and on that day all the medical team would be ordered out to work on the railway. The only positive thing about this was that it allowed them to barter for extra food from locals outside the camp. Needless to say – and as is well documented – the death rate in the coolie camps was much higher than in the soldiers' camps; tens of thousands of them died.

Beatings for coolies were brutal; one practice Sydney witnessed was coolies lined up naked and forced to hold stones or logs above their heads in the heat of the day whilst passing Japanese hit them with rifle butts; such conditions often caused madness and suicide among them, in the latrines, the forest or the river. After morning *tenko*, the coolies were made to do gymnastics. In the same way that the Japanese seemed to be unable to understand that filthy latrines, flies, starvation diet and disease caused deaths, they could not see why sick people could not do gymnastics. All the sick were forced into this horrific 'game' and made to take part, whilst a Japanese medical officer nicknamed 'Pappy' walked up and down the lines, swinging his sword to mimic decapitations.

Sometime in October 1943, Sydney's medical group of about twenty personnel were moved by barge for several days up to Nike, near the Three Pagodas Pass on the Thai-Burmese border, where there was a coolie camp alongside a POW camp. One day a Japanese propaganda film crew arrived to show how well the native labourers were being treated. A Japanese medical team in white coats and gloves were filmed treating a coolie with a huge leg ulcer and gently bandaging his leg. As the film crew left, so did the coats and the medicines. Sydney also witnessed the use of elephants in moving cut tree trunks for railway building, and one day an actual Japanese doctor appeared and asked Sydney to accompany him to advise on an injury to an elephant handler. Sydney saw that it was a huge pus-filled abscess on the buttock and painlessly evacuated the pus, solving the problem; but the more sadistic Japanese doctor insisted on needlessly scraping the cavity clean, causing the man horrendous pain.

Another doctor nicknamed 'Four Eyes' seemed to be in charge in this camp and was involved in an unpleasant incident in December 1943. All of Sydney's team, including himself, had malaria and dysentery and yet were constantly ordered out to work by the Japanese. When Sydney, wrapped in a greatcoat and on his bed, explained that all the men were ill, the doctor flew into a fury, slapping him several times and knocking his gold-rimmed glasses to the floor. Sydney knelt with difficulty to pick them up. He told the doctor that he was going now to complain to the commandant, but Four Eyes only laughed. The commandant was no more sympathetic and shouted at Sydney that all English soldiers were lazy and would get no food, then punched him in the face, cutting Sydney's lip with his ring. Sydney had totally miscalculated, but stood his ground, turned and walked out, to the congratulations of his men for making a courageous protest.

On two occasions that Sydney remembered, the Japanese threw dynamite into a river and thousands of fish came up, and the POWs were allowed to gather them and gorge on fish for a few days.

In January 1944, with his POW team now numbering ten, Sydney noticed a change in the Japanese guard, with a new medical officer who seemed to be most kind and friendly. An unusual event occurred. A young Tamil labourer was brought to the camp hospital with a severe intestinal problem. The Japanese doctor asked Sydney to assist him in operating on the man and produced from nowhere a decent operating table, mosquito netting, washbasins, bandages, instruments and so on, the likes of which Sydney had never seen in the camps before. The Japanese team crowded around watching and a British officer was brought in as anaesthetist. As they were about to begin, the Japanese doctor declared that Sydney would operate instead, and he would assist him. Then the Japanese doctor declared that another British doctor would assist and he, the Japanese, would just observe. Clearly the Japanese doctor did not want to lose face in front of his men by botching the operation, so he was allowing the British doctors to do the job. The operation, however, was a great success and the Japanese applauded; the Japanese doctor took no further interest – it had a just been a show. However, Sydney and his team were aware of the high chance of post-operative infection and obtained permission to visit the Tamil in his hut regularly – usually not allowed by the guards in case a conspiracy to rebel was hatched. Sydney fed him cautiously each day on small amounts from his precious personal supplies of tinned food. He was determined that the man should survive, as a matter of British surgical prestige – and he did, making a full recovery.

Burma

In September 1944, six of Sydney's team were taken north on open rail trucks with him to another camp, over Three Pagodas Pass and into Burma itself, to 84km Post – that is, eighty-four kilometres from the start of the railway at Thanbuzyat. Here they were marched under guard into the jungle along a path and slept out in the open that night; a light rain fell and Sydney and his men were at a very low ebb, ill with malaria, exhausted and starving. Sydney said he was prepared just to give up and die there and then; he felt he just could not go on. Then, as the sun broke through at dawn, he felt a surge of life and they were marched the final two kilometres to the new camp. It was like all the others – a clearing, near a stream and not far from the railway. They were led to the camp administration office and an interpreter said that they would all die there, as there was cholera, but Sydney knew it had died out along the line so he simply shrugged his shoulders.

Medical facilities in the coolie camp were of the usual appalling standard. The Japanese remained heartless – they had quinine to cure malaria, as they controlled most of the world supply in the Dutch East Indies; they had adequate supplies of vitamin-B-rich rice polishings which would cure beri-beri; they had dressings and potassium permanganate to cure tropical ulcers – but medicine and food was deliberately withheld by the Japanese, and POWs and coolie labourers died in the thousands as a result. Sydney had the job, among others, of visiting the huts where the labourers lay dying and recording the number tag of each dead man, then proceeding with the daily list of five or six to the senior Japanese NCOs office to report the losses. He was astounded at their reaction as he bowed to offer the report; often they would laugh to hear that men were dying of such multiple illnesses. On some days he would receive a slap in the face, but on others he would be given a few beans to augment the soup of his team.

One day an order came that they were to salute all guards whenever they went to the latrines; it was bizarre, as they had already been saluting guards for two years. One Aussie soldier told Sydney he would not do it, even though Sydney pleaded with him, reminding him that it was an existing condition anyway and warning him to take care of the crazy unpredictability of the Japanese. As a result, the soldier was seized and badly beaten, but survived.

In October 1944, while bathing in the stream with his men after work, a huge four-engine bomber flew over. Bombs were dropped on the railway and the area was machine-gunned whilst the Japanese fired back with useless rifles: it was the Allies. Many other raids followed. In November, Sydney went to deliver his usual daily report to the NCO and saw the guards sitting around a large Japanese newspaper laid on the table and a map of France with arrows and front lines; it was clear that the Allies were in France and advancing east to Germany. Sydney was riveted, but the NCO, seeing Sydney staring at the newspaper, rapidly closed it up. Sydney left, trembling with excitement, to tell his comrades.

A few days later a friendly Japanese guard whom Sydney had got to know, and who always respectfully called him 'Doctor', asked him to walk with him along a path, for a chat. Some time before, this soldier had sold Sydney a pair of good shoes when he saw the state of his old, worn boots. They arrived at the 84km Post railway hut and the soldier went inside to deliver some message whilst Sydney stood around outside near a large group of about 100 open railway trucks. It was a hot humid day, completely silent save for the noises of the jungle. Curious, he took a look inside the trucks; to his horror the trucks were full of hundreds of dead and dying Japanese soldiers, none of them making a sound. They were clearly battle casualties, in full combat uniform, and there had been a massive Japanese defeat somewhere. Some days later, all of Sydney's five comrades were moved on and he was left alone as the sole POW in the camp.

Alone with the Japanese

Sydney may have been the only Second World War POW of the Japanese to be alone with a force of 200 engineers and a coolie camp to tend for. He was given a living space, separated by a thin wooden partition, in their long, spotlessly clean hut, into which no coolie was ever allowed to step. As an officer with his badge of rank on his breast pocket (one pip only was allowed) and his Red Cross armband showing he was a doctor, he received the utmost respect and the same rations as the Japanese. They ate the same food but in much larger quantities than the prisoners. He did not tend to their medical needs except in one respect – he was ordered to give them injections of vitamin B or camphor in oil, before they visited a local brothel, even though it is hugely unlikely that it helped their performance! For their part, the Japanese always gave him a small gift of food after such injections.

The camp was on a sandy beach along the river; each evening a large oil drum raised on stone stilts on the sand would be filled with water and heated by a bamboo fire to a warm temperature. The Japanese would undress and soap themselves and then rinse with cold water. Then, in order of seniority, all 200 would immerse themselves in turn in the warm tank for a few minutes, and then get out, dry off and dress. Sydney would always be invited, but not wishing to get in water used by 200 others, he would politely decline and say that he had already bathed in the river. In all, it was the most pleasant three weeks of his three-and-a-half years of imprisonment. In January 1945 he was posted back south to Kanchanaburi, which he had last seen in September 1943.

Kanchanaburi and Tamuan

Sydney had accumulated some money and decided to buy a watch in case the money was stolen, so he would at least have something to exchange for food. He also had some anti-dysentery drugs for use for the sick men but they were stolen from his rucksack, as they could be sold at great profit to local Thais across the wire. It was a despicable act, as it deprived sick men of a life-saving drug.

News filtered through of VE Day in Europe, and air raids from Allied planes were now common; some POWs were killed as a result of these, in particular in one raid on Bampong, further down the railway.

In the summer, Sydney was moved again, this time from Chungkai to Tamuan, a proper POW camp with a moat and guard towers and a fairly well-equipped hospital, compared to the places he had been in for the past two years. He worked for the benefit of the prisoners, and the food was better; some pay meant the ability to buy extra food from local people. Then on 17 August came news that the war had ended; nobody could believe it. The Japanese had been seen dismantling their anti-aircraft guns and surrounding their compound with barbed wire in case of a POW revolt. POWs still continued to arrive in the camp from the railway, in a dreadful condition, and continued to die – it was so sad, when the end was now in sight. But still nobody knew exactly what had happened so far. Large signs were laid out on the ground and rooftops, saying 'POW', and soon air drops of food, medicine and clothing arrived. Many got drunk on purchased Thai liquor. Soon a young British SOE officer from Force 136, armed only with a revolver, arrived to announce liberation and tell them about the atomic bombs. Then slowly the evacuation began, with the sick going first. Telegrams were sent – Sydney's wife had received only one card, in December 1942.

After six weeks of good food and rest, the men were taken by train to Bangkok and then by Dakotas to Rangoon, where they had extensive medicals. Here Sydney met again his old CO of No. 5 CCS, looking fit and well as he had not had alcohol or meat for three years; Sydney's last view of him was seeing him play leapfrog! Sydney arrived in Liverpool on 20 November 1945, and thence to Worthing where his wife was. After a time with malaria at Mill Bank hospital, he was finally sent home cured.

Post-War Life

Sydney took a job as a consultant at New Delhi Military Hospital in 1946. Here he met again Colonel Hurd-Wood, mentioned earlier in this testimony, who had filmed the Japanese dive-bombers with his cine-camera. He was taken on a ward round by a young officer and at the end was told there was just one more ward to see. Sydney was taken to a 200-bed ward – full of Japanese wounded prisoners. He was flabbergasted. They were awaiting transport back to Japan. He had been their prisoner, and now they were his. He left that ward to junior doctors and never returned there again.

Sydney returned to the UK and was consultant surgeon at Whipps Cross Hospital in Woodford for thirty-two years. Sydney died in May 1998, survived by his wife Eileen and his sons David and George. Sadly, he died before his book was published.

Notes

1. Sydney met him again in Delhi after the war, but forgot to ask him if his film had survived.
2. Native labourers were often enticed to work for the Japanese by empty promises of good food and wages and ended up as slaves on the railway.

Ezekiel 'Eze' Nathan, Civilian Interned by the Japanese

The information in this testimony is taken from The History of the Jews of Singapore, *by E. Nathan (Singapore, 1986).*

Eze Nathan belonged to the small but long-established Jewish community of Singapore that mainly originated in Iraq and India. During the 1930s, Eze's family in Singapore had already become active in organizing boycotts of German goods and fundraising for Jewish refugees who had escaped to Singapore as anti-Semitism in Europe increased. But sadly, when war broke out in 1939, those who were German were interned by the British and even deported to Australia, even though they were Jews who had fled Europe; the same thing, of course, happened in Britain. Some left for Shanghai but some, due to the usual bureaucratic foul-ups, were simply left free to live on the island.

Eze's family ran an electrical business in the busy Orchard Road area and lived at Amber Mansions near Fort Canning; he was a JP (Justice of the Peace) and also a member of the Singapore Jewish Welfare Board.

By December 1941, Singapore itself was threatened and Eze volunteered to fight, but was turned down. The reason was that Jews in the Straits Settlements were considered by the British as racially Asiatic and thus exempt from call-up. The British later relented on this and many Singapore Jews were accepted.[1] On 8 December, as Japanese bombs fell on Singapore, Eze witnessed the deaths of a Jewish family of four in Joo Chiat Road, blown to bits in their home, and no remains were found for burial in the Jewish cemetery. In Waterloo Street, Hannah Moses, Maurice Ezekiel, Joseph Elias and Albert Deborah were all killed in the bombing. In this case, proper burials were possible. Meanwhile, blackout was imposed, sand collected for sandbagging and shop windows boarded up.

Eze's firm maintained all the military electrical equipment and so was kept very busy, day and night, during the siege of Singapore. His sister Dorothy and brother Ivor were refused evacuation by the Australians because they were not white enough for the White Australia immigration policy of the time; eventually they got tickets on a P&O ship, and left. His cousin Meyer Nathan was flying recce for the RAF; he survived a crash landing at Johore and made it back to the island. He came to say goodbye as he was being taken out by the RAF. Jacob Ephraim of the Cleansing Department of City Hall witnessed the burning of five million dollars in notes so that the Japanese would not get them.

But even during the chaos and tragedy, life went on; at the Maghain Aboth synagogue, Helen Elias was married to Raymond Tax, and Major Dr Philip Bloom (RAMC) married Freddie Retz, an American journalist.[2] Meanwhile, everyone could watch the aerial dogfights taking place above, and whilst bodies from the bombings lay in the streets, hotels advertised for visitors and tourists. Sales soared as soldiers spent their money, and vets worked overtime as people queued to have their pets put down. Children and babies, who had been tragically orphaned by the bombings, arrived at the hospitals.

Obituaries

Ezekiel Nathan

JEWISH CHRONICLE MAY 21 1993

Ezekiel Nathan, who has died aged 86, devoted himself to serving the small community of Newport, South Wales. He died three days after learning of an attack on the Jewish cemetery he had tended for 20 years.

Born in Singapore to a family of Iraqi origin, he ran a successful electrical business in the island's busy Orchard Road, and was active both in the general and Jewish community, serving as a JP and president of the Singapore Jewish Welfare Board.

During the Second World War, as a prisoner-of-war under Japanese occupation, he was a pillar of strength to the Singapore community, keeping Judaism alive even in the brutal internment camps. His youth work in the camps was recognised with the Scout Medal of Merit.

He met his wife, Nina French, a member of a well-known Newport family, in 1946, when he was released and sent to England, where the women members of his family had earlier fled.

He returned to Singapore with his bride in 1946, working there until the riots and emergency situation of 1957 sent Nina back to Britain with their two sons.

Eze Nathan came to see his family in Wales for three months annually until he finally sold his business and joined his family in Newport in 1966.

He immediately entered the life of the community, meticulously attending services, giving hospitality and attending to funerals and the cemetery. As the main support of the chevrah kadisha, he was a familiar figure to every municipal official.

He was synagogue treasurer for most of his time in Newport and president for 16 years. When he retired in 1985, he was made life vice-president.

In January this year, he attended the Chief Rabbi's Awards for Excellence ceremony as a nominee for the lay leadership award. His selfless devotion to the community was acknowledged by all.

His deep interest in the past was shown by his publication in 1987 of the history of Singapore Jewry from 1830-1945 and his careful recording of the Newport community's activities and plans — including all cemetery plots and reservations.

In an introduction to the book, David Marshall, Singapore ambassador to France, said: "He was recognised by all in our community as the high priest of the community's welfare."

He is survived by his wife, two sons and grandchildren.

51. Ezekiel Nathan, Singapore.

On Saturday 31 January, the military breached the 1,100-metre-wide Causeway to slow down the Japanese invasion and water gushed through the 20-metre-wide gap. The governor addressed the people and warned of grim days ahead and advised them to be ready to meet the enemy face to face. Bizarrely, while the colonial secretary residency was being shelled, servants continued to polish the glasses for lunch, and cinemas played to packed audiences. From 7 February, having used balloonists as observers over Johore, Fort Canning (just behind Amber Mansions), was shelled by the Japanese. Refugees carrying bundles and pushing carts filled the streets; communiqués were posted on walls.

On 9 February, Eze was at RAF HQ in Sime Road when he was ordered to dismantle all electrical and telegraphic equipment for shipment to Sumatra, and destroy the rest. The Japanese were so close that much of this was done under direct fire. Troops of the Allied rearguard, tired but cheerful, appeared in the courtyard of Amber Mansions to rest, and the tenants gave them food and drink. They returned to battle next day and said that if they did not come back, it would all be over. In contrast to this, Eze witnessed one British officer rebuking two exhausted Indian soldiers for not saluting him; bystanders jeered at him for his stupid behaviour in such times.

On the night of 13 February, Eze saw brilliant red and blue flashes to the north, a flamboyant signal that the Japanese Chrysanthemum Division had crossed the repaired Causeway. As the water mains were destroyed, the hospital staff used soda water to wash their hands. Eze and others tried to dig for water but hit rock after two feet. The river seemed to be on fire as burning oil flowed into it and sets boats alight. Eze heard people singing 'There'll Always Be an England'.

On 15 February it was all over. On its way to Government House, the governor's car stopped outside Eze's shop, with a puncture. While it was being repaired, Eze approached him and asked if the Jews would be singled out by the Japanese. The governor simply said that they should obey all the enemy demands.

Eze first saw the Japanese victors from the balcony of his flat; they were ragged and worn out, scarcely recognizable as soldiers. Curfew was imposed from 6 p.m. to 6 a.m. and the order came that everyone had to bow on passing any Japanese soldier.

Occupation

Within days, Eze witnessed the brutality of the Japanese. Decapitated heads of Chinese and any others who were labelled dissidents were seen around the city, stuck on poles; random massacres of civilians were carried out and the bodies were left in the street. One day, Eze was ordered to go the Alexandra Military Hospital with the Auxiliaries. What they found there defied description – the notorious massacre of staff and patients by Japanese troops. They were ordered to place the bodies on trucks to be taken for burial, and ordered to do the same thing at Malacca Street and Raffles Place. Later they were ordered to evacuate the General Hospital of patients in order that Japanese wounded could be brought there. Local patients were sent to the mental hospital some twenty miles away, whilst the mental patients were sent to a nearby island facility. No medical supplies were to be removed, but they managed to smuggle some out. One Jewish woman, Thelma Guston, was among the sick, but she survived the ordeal. Meanwhile, the wounded Allied soldiers were housed in the Cricket Club, Victoria Memorial Hall, the Singapore Club and the Fullerton building, and then all were transferred to Changi gaol, including Philip and Freddy Bloom. Chinese and Eurasian members of the SVC were ordered to assemble at the YMCA. Some were taken away and never seen again. Eze's shop was requisitioned and all his stock stolen.

The Japanese allowed only one English newspaper to be published, *The Singapore Times*, which they renamed the *Shonan Shimbun*. It had to be read because it contained daily orders to the population: all radios had to be handed in; cars were confiscated and all public transport ceased; Japanese were to be referred to only as Nippon. One bright event was the arrival of three members of the tiny Jewish community from Penang, including Isaac Manasseh, on the Malaya–Singapore Railway, which the Japanese wanted to be kept running for military purposes.

The Interim

On Tuesday 17 February an order was issued to all citizens of countries at war with Japan to assemble on the Padang (cricket pitch and sports ground) in the centre of Singapore City, to prepare for internment. Citizens were told to bring clothes for ten days, but could rest assured that international law concerning internees would be respected by the Nipponese. Most of the 1,279 who gathered were expatriates, and this included a number of Jews. Eze remembers Bill and Shooker Abbott, M. Lelah and Maz Woeltz. They were marched to camps in Joo Chiat, and crowds gathered to jeer them, though some came forward with food and drink to help the elderly or sick. After several weeks of chaos and appalling conditions, they were marched to Changi gaol in the north of the island. The women especially showed great spirit and marched into the camp singing 'There'll Always Be an England'; tears flowed.

Eze, his brother Ivor, his friend Sassoon Sadka and others remained free, but on the evening of 1 March, when Zeke Elias and Ivor went to visit Eze, a car stopped them. A Colonel Hayashi demanded their passes, but they explained that none had been issued. Furious, Hayashi hit them with his riding whip and ordered them to the Kempeitai (Gestapo-type police) HQ at the former YMCA to obtain passes immediately. They were sent on a wild goose chase from building to building, without success. Returning to Eze's flat, they were visited by an officer who ordered them back to the Kempeitai offices, where another officer examined a file and, looking up at Eze, told him that if he valued his freedom he should never come back again. Eze never saw the others again. When he asked the Japanese authorities to investigate their disappearance, they refused to interfere in Kempeitai affairs.

Incongruous things happened: Eze was at the docks and witnessed some Japanese using rifle butts against some Allied POWs; some German soldiers were nearby and angrily intervened on behalf of the British.

Days later, Eze was working on the electric fans at the Kempeitai HQ and had a good look round the building to see if his compatriots were there. The chief of police invited him to have a glass of brandy with biscuits; this was always an offer that could never be refused. When asked if he had a car, Eze said it had been requisitioned. The officer told Eze to follow him and drove him to the car dump on the esplanade, where he found the car, a Morris. But Eze wanted no favours from the police and so he said that the car was so blocked in the centre of the dump it would be impossible to drive away. The officer said 'Buggaro!' (bloody fool) and Eze walked away.

Random arrests occurred. Two young lads, Elias Albert and Elias Ezekiel, were stopped by a Japanese lorry in Short Street, taken and never seen again. Abram Pinhas, Aaron Isaac and Charlie Ezekiel were picked up in New Bridge Road and roughly questioned in an interrogation centre in Anson Road; a sword was brandished over their heads and a revolver placed by a Japanese on the table in front of them. From elsewhere in the building they heard screams and shots being fired. Abram was eventually given a cigarette and taken away by a Japanese officer; then a shot was heard. He was never seen again. The other two were released. Meanwhile, David Marshall, now a POW, got a message out via an Indian National Army (INA) guard, asking about the welfare of his mother, who was taking refuge in the synagogue area with other Jewish refugees.

Discrimination

On 16 March an order was issued for all Jews to muster outside Orchard Street police station next day. While waiting, they elected Moses Cohen as their spokesman. Each person was called forward into the station by the Japanese, one at a time, and had to complete a form. They were issued with a white armband with a one-inch red stripe with Japanese characters identifying their name, a number and their racial origin. Then they were told to return to their jobs. The armbands had a curious effect on some people – the Chinese, for example, regarded them as signs of collaboration and were very suspicious of the Jews as a result. On the other hand, one day Eze was stopped in the street by a Japanese official. Looking at his armband, the official asked what it was. Eze said 'I am a Jew.' To his astonishment, the official grabbed Eze's hand and shook it energetically, saying, 'Werry good; in world three great 'J's – one Japan, two Julius Caesar and three Jew!'

Some Jews, like other people, were reduced to poverty and began hawking goods. One friend of Eze sold a mock Rolex watch to a senior Japanese officer, who later angrily discovered it was a fake. He found the hawker, who pleaded that his family were hungry. The officer not only drove him home, but stopped on the way and bought food for him to take home with him.

The synagogue remained a centre for the exchange of news about evacuated relatives and others; money was collected to help those in need. Moses Cohen was ordered to prepare a paper on the Jews for the Japanese commander Yamashita, by the minister of religious affairs, who called a meeting at the Adelphi Hotel of the heads of all religious communities. He was impressed and attended the Shavuot (Feast of Weeks) service, later assuring the Jewish community of his full cooperation in all religious matters. Soon afterwards, permission was given for Moses Cohen to go and fetch an etrog for the festival of Succoth (Tabernacles), from the garden of the former house of Sir Manasseh Meyer, Belle Vue, but the tree had died and the Japanese had taken over the house.

Then came the first Jewish funeral, of Abdullah Shhoker, aged 93; he had been turned out of his house (1 Oxley Road) and was living at the synagogue when he passed away.

Meanwhile, Eze discovered he could pick up British India radio and the family listened in secretly each evening using earphones, so obtaining real news of the progress of the war, rather than the nonsense of Japanese propaganda. Nearby was a Japanese interrogation centre, at 3 Oxley Road; they passed it every day and the compound was open to public view. On one occasion Eze saw a former neighbour of his (a Singaporean) supervising the torture of a Chinese; the victim was strapped upside down to a ladder and his head immersed periodically in water. At night they could hear the screams of prisoners.

On the corner of Eber Road was the quite different house of the Japanese 'comfort women'. Each morning they were brought out to pray to the emperor, facing east.

One evening two Japanese arrived at the flat while they were playing cards on the verandah. They spoke excellent English and asked to be taught bridge and chess, and also to borrow English books, especially Shakespeare. Eze and friends were suspicious. The Japanese visited very frequently and began asking questions about the Jewish community and their supposed wealth, how they earned their living. Moses said that few were rich and most were barely getting by. The Japanese replied: 'How do you account for so many Jews spending time and money in "The Great World" [an entertainment centre]? Come with us tomorrow and we will show you.' Next day the community was warned to stay at home, but some ignored this advice, and some were seen spending money unwisely. The Japanese took Moses to witness this. Moses agreed with the Japanese that it was unwise of people to behave in this way, and although nothing more came of the incident, it increased the vigilance of the community leaders towards the Japanese.

One evening a Japanese naval officer came to the apartment and ordered Eze to come with him and remove a radio transmitter from a damaged ship in the harbour and to repair it in a Japanese workshop. The parts needed were simply not available, so Eze came to work each day and pretended to repair it. One evening the officer arrived at Eze's home when the Kempeitai men were visiting; the naval officer left quickly and the next day he paid Eze to leave. Clearly he was involved in selling salvaged equipment illegally to locals and was not happy to find out that Eze knew Kempeitai officials.

On another occasion, Freddy Isaacs, the optician, was arrested on suspicion of being a spy and was kept and interrogated by the Kempeitai for three weeks at their centre at 3 Oxley Road. Only the courage and persistence of his wife visiting the police got him released.

Crime increased for a while, as the civilian prisons were not able to control discipline in the first few weeks of the occupation. Some escapes occurred and, sadly, one afternoon Cecil Reuben was murdered in his shop in De Souza Street when a gang of Chinese broke in to rob him. In another incident, a Malay police officer friend of Eze's called to question him about an allegation that he had been hiding a Dutch woman in the flat. Eze was angry and showed him the empty rooms, and the police officer soon left.

For a year, things for the Jewish community remained almost normal, as long as they remained inconspicuous, denied knowledge of the outside world and always bowed to the Japanese.

Internment

In the early morning of 5 April 1943, Jews in Singapore were awakened in their homes by a terrible clamour. Their doors were broken down by Japanese soldiers who held them at bayonet point, thrusting leaflets into their hands, informing them that they were now classed as enemy subjects and were to be interned. Stuffing a few personal belongings into bags and their pockets, they were taken in trucks and stopped periodically to pick up other Jews standing outside their homes, under arrest. Packages and bags broke open in the street as people were roughly bundled on to the trucks. At first only men were taken. Eze was taken with about 100 others to Changi, leaving about 450 others awaiting an uncertain fate in their homes.

At Changi they were dumped inside the yard but were met by representatives of the civilian internees and shown to their quarters, a large former rice shed (godown) made of brick and metal, which they called 'Aldgate' (see Sion Elias's testimony in this book). The women's section was separate. It was extremely crowded, but those who had been there for a year had done an amazing job in organizing life so that it became tolerable and bearable, using all the varied skills people had to reduce the squalor. The Japanese rarely entered the prison and left the internees to organize themselves – their welfare, discipline and health, and the purchase of food and other requirements, financed by a loan that internees paid to the camp and then drew from the fund to buy what they needed from outside for as long as the fund lasted. They set up committees to deal with crimes, medical affairs, work rotas, hygiene, supplies, and so on – all elected by the population for three-month periods. Each prison block had a representative (Moses Cohen for Eze's block) and each internee was given an aluminium strip with their name on in Japanese, and a number. Malay was the main language used in dealing with the Japanese. Eze found himself in charge of hygiene and control of insects. The Red Cross were refused access, despite Japanese promises to abide by the international law for internees.

At first, overall control was in the hands of comparatively humane and able Japanese officers – Asahi (Enemy Aliens) and Naito (Camp Supervisor); but the more sinister Lieutenant Touchitana (of the Military Police) soon arrived with more aggressive Japanese and Sikh traitor guards. Very soon Asahi and Naito were replaced by the more brutal Susuki and Tominaga.

The Sikh sentry who guarded their quarters was very irritating, demanding maximum subservience, especially in bowing, and hurling abuse at them whenever he wished to. To teach him a lesson, twenty inmates lined up to pass him at regular intervals, each one bowing low; he eventually gave up and asked them not to keep bowing. He later disappeared and the guard was never replaced.

Another brutal guard sported jackboots and was nicknamed 'Puss in Boots'; every night he would enter the barracks, drunk, and beat men or women for not bowing to him quickly enough, and many were injured in this way.

Aldgate shed led to a yard, which in turn led to the kitchen and the library and laundry; this was also the bathing area, with water coming from two large concrete containers. Each block floor also had one tap, and waste water was used to flush the latrines, which had about thirty squats based over the main drain. Inmates called this area the Wailing Wall – an insult that the Jewish prisoners learnt to live with. Urine was used to fertilize the vegetable patches.

The hospital was relatively well equipped and the elderly were cared for in a block called Cripplegate. Beyond a twenty-five-feet-high wall was a football pitch, the use of which cheered people up, though few had the strength to play, owing to the starvation diet. Food was collected by members from each block in large cauldrons called *tongs*, and usually consisted of a porridge called *congee* and black tea for breakfast; rice, thin soup with perhaps a tiny piece of fish for lunch; *congee* and tea for supper. Sometimes there were the odd vegetables thrown in, and if possible the meals were varied slightly if there was a birthday to celebrate, to break the monotony of the consistently boring diet. The Jewish internees managed to observe the main dietary laws – for example, they never ate pork. If anyone was given extra provisions from outside, these were supposed to be handed over to the committees for general use, but this was more honoured in the breach than in the observance; there was one vicar who kept a large stock for himself.

There was a camp tuck shop supplied by a lorry owned by an Indian merchant, which was nicknamed 'the Blue Angel' when it arrived periodically, bringing great joy with its cigarettes, tinned foods, biscuits, salt, soap, and so on. Sometimes a few vegetables arrived from the gardens; a piece of spinach or sweet potato could make all the difference to a boring meal. Once, while collecting palm leaves for brooms, Sol Elias, Ellis Ezekiel and Harry Sassoon spotted a stray chicken and secreted it back into the camp, but Eze noted that their mates had to watch while they cooked and ate it. Jacob Gazal had been a butcher and made *patcha* – a Sephardic version of haggis; using tinned ingredients he conjured up one such meal on one occasion, which brought great delight to the block.

Soon afterwards, news arrived from Shooker Abott's family in Britain that Eze's womenfolk had arrived safely in the UK.

One sad event was the death of Abraham Ezekiel Abdullah, a saintly old man who had owned a crockery shop in Beach Road. Arrested early on and separated from his wife, he had become ill. The doctor said he was dying and Moses Cohen went to appeal to the Japanese to allow his wife to visit him; she arrived just in time, and the Japanese also agreed that he should be buried according to Jewish rites in the Singapore Jewish cemetery. Some seriously ill patients were taken to the Japanese Miyako hospital, Joe David and E.J. Nathan being two who later returned, cured, to the camp.

The camp fund was running low; each person received five dollars per month, and an anonymous donor – probably the brothers Adolph and David Storch, who had Chilean passports and had escaped interment – had kept it going. Representations were made to the Japanese, who told the civilians to ask the Jews. The Jewish community loaned $100,000, interest free, to be repaid by the British after the war. When the good news was announced to the camp, some shouted about 'bloody Jewish money lenders' and Jews in the prison were besieged with requests for huge loans and offered IOUs. It was Eze's first experience of anti-Semitism and greatly upset him. All Jews were now regarded as stereotypically rich and as controlling the black market in the camp. This was untrue; the black market had been developed in Changi long before the Jews were even interned. Attempts by the real villains to involve Jews were always rejected by the Jewish community.

Eze adapted to the work routine: fatigues, wood collection for the boilers, and so on. He worked in the furniture party making items for the elderly. He and David Tweg also became barbers and were often in demand. As weeks went by, new inmates arrived, non-Jewish friends and Jews alike – including Annie Silverman, who had escaped from Singapore but had been recaptured. Sunday was the day off and on Saturday they held Shabbat service for Jews, led by Moses Cohen and Sion Daniel. After three weeks, Passover arrived and Moses was sent to 'Pharoah' to ask if they could have a lorry to collect the unleavened bread (matzoh) from the synagogue. At first he was refused but at the last minute permission was given. A guard accompanied them to the matzoh makers in the synagogue yard where bakers were at work; they asked for extra to be baked so all the internees could have some. Only three hours were left before the lorry would return to take them back, but the task was completed. Meanwhile, a crowd had gathered at the schule to ask about their loved ones in Changi, and all looked underfed and ragged. Moses reassured them all was well, and the lorry took them back to the camp.

That night, 112 people celebrated the Seder (Passover meal) at Aldgate in a space created in the centre of the barrack. Some Jews joined them from other blocks – including Lazarus Rayman, a former chair of Singapore Municipality, and Mr H. Weisberg, colonial treasurer of the Straits Settlements. Some eggs and fish were purchased and it was a memorable evening, as it was to be the only time they were able to do this during the three years that followed – later food became very scarce. Months later came Rosh Hashana (New Year) and Sion had brought his precious Shofar (ritual trumpet or ram's horn) to use in the service, as well as at Yom Kippur a few days later.

On some evenings, Moses and Mr Rayman would study Gemorrah (commentary on the Bible) and inmates would gather and listen to the discussions. One amazing visitor was Rev. Robert Greer, a young Presbyterian minister who was a Hebrew scholar and who offered new insights into translations. Other friends included the crew of the *Empress of Asia* who had been the unexpected guests at the wedding of Freddy and Philip Bloom. Simon Hayasad kept people amused with his impersonations, Leon Simon was the joker, as well as doing valuable work as Camp Committee secretary, and Silas Thomas quoted Shakespeare for any occasion. Some people played sports – especially cricket – and there were camp entertainments: Sunday concerts by the five musicians of the Raffles Hotel, three of whom were Jews; bridge and poker competitions (though gambling was strictly forbidden by the Japanese authorities); plays (in one of which Eze played a dancing girl quite convincingly); lectures and classes on every conceivable subject (many academics were among the internees); and of course the library. Eze, who had been active in the Scout movement, also got together with the pre-war commissioner, Frank Sands, and set up a camp Scout movement to teach scout crafts (map reading, first aid, and so on) and generally involved the Scouts in the welfare of the camp.

However, real life could be brutal, and when Joe Isaac applied to collect some rolls of canvas from his shop, he was interrogated and badly beaten for referring to the material as 'his' property; it was Japanese property now!

The Double Tenth

Very early on 10 October 1943, the internees were ordered to muster in the main yard; there was rumour of repatriation as a Swedish ship, the *Gripsholm*, was seen in the harbour.

But suddenly all the gates were locked and armed guards appeared. Forty names were called out, and Eze heard the name R. Nathan and so he stepped forward. Moses pulled him back and whispered, 'You are not R. Nathan.' This probably saved Eze's life. The forty were taken and the rest were left standing in the sun with no food or water for about eight hours. Then block by block they were allowed back, and guards followed and searched every inch of the barracks, scattering belongings everywhere; diaries and photos were confiscated as well as prayer books, although these books were later returned.

Apparently the Japanese had believed that a resistance movement had developed among the internees and an uprising was planned, and that hidden radios were receiving and giving messages. They feared the internees more, because they reasoned that they knew the country and languages better than the soldier POWs did. An ambulance driver, John Long, was kept for six months by the Kempeitai in a cell; he was released in an emaciated state and allowed to recover, but then shot, apparently the only internee to be executed. Usually the Japanese preferred men to die of exhaustion, so they could not be accused of murder. Freddie Bloom was another victim, but she survived and wrote a book about her experiences in the form of a diary that she kept (see her testimony and story above). Of course all the victims were innocent.

Following this incident, privileges were removed and only the fatigues continued, with punishments added – no concerts, no school for the children; rations were cut, religious services limited, camp funds cut, armed guards increased; mail was stopped until Christmas; hospital visits were stopped.

In January 1944, Eze saw wood and atap huts being erected on the football pitch by Chinese labourers. Soon he and his friends at Aldgate were moved into them, and exposed to the oncoming monsoon weather. On 1 May they were told that they were to be moved to an old RAF base at Sime Road, and again rumours spread about repatriation. But of course nothing came of this.

Moses Cohen was called for interview by the Japanese commandant. He was asked to become an informer but gracefully refused. He was dismissed, with, surprisingly, no repercussions.

Sime Road Camp

As lorries came to pick them up, they saw British POWs being brought in to take their place. The two groups cheered each other on. The Jewish group looked forward to trees and grass at Sime Road, away from the grey walls of Changi, but on arrival they were sent to a dilapidated hut with excreta on the floors and a caved-in roof. That night the Jewish internees chose to sleep outdoors on the concrete path. Next day the camp representative offered them a different hut (No. 106 – by amazing coincidence it was where Eze had worked installing telecommunication equipment for the RAF before the invasion) with a cookhouse attached. So they became a self-catering group, collecting rations each day from a central point for cooking in their own ovens.

Four hundred yards away was an overgrown garden near huts where the women were quartered. For all the prisoners there were basic toilets, showers and running water, plus a primitive laundry and hospital.

Following the Double Tenth, discipline became more relaxed, there were rest periods after chores were done, evening study groups sprung up, and a Sunday half-hour visiting time was allowed for families to be reunited, if only briefly; on these occasions, somehow little gifts were found and wrapped up and exchanged; but it was still lights out at 10 p.m. Small groups gathered for prayers occasionally. Then came the birth of a baby boy, and a *brit milah* (circumcision), performed by Dr Landor, brought great joy. For *yartzeit* (prayers for remembering the dead) a minyan (quorum) of ten men were allowed to go to the Jewish cemetery, but towards the end of the war, burials had to be made in Sime Road (they were reinterred after the war). Sadly, Eze Manasseh was the first to die of dysentery.

The Japanese then instituted a 'Grow what you eat' policy, so grass was dug up and crops such as tapioca were planted; however, whatever vegetables the internees grew, the Japanese deducted from the rations, so the net result was no extra food at all. Nutritious papayas were grown but few matured and they were eaten as vegetables. When Eze tried to grow tomatoes, they were stolen one night, just as they had ripened. A new hazard then developed – the tough grass (lalang) which grew around the base harboured the typhus-bearing parasite carried by rats. There were some fatal cases but no epidemic, as the internees managed to spray the grass with disinfectant, while covered in protective clothing. But malnutrition brought other diseases – dysentery, ulcers, beri-beri and pellagra – and there was always malaria. All medical supplies became more and more rare, despite the smuggling in of some items (Eze himself smuggled in some drugs and bandages when he was sent to acquire some spare parts for a Japanese radio set in town).

Other dangers included snakes (one fell from the roof onto the mosquito net of the sleeping Eliah Sharbanee, and was later killed with a stick) and there were many scorpions and centipedes. Bandages soaked in rubber from some nearby rubber trees made good plaster, rubber seeds were given for debility and swallowed, palm oil was used for ointments, rusty nails shaken in water gave iron tonic, old toothbrushes were made into spectacle frames, wood ash was used for toothpaste, crushed onion was applied to the feet, and garlic was swallowed to soothe dysentery. One internee was a herbalist who extracted all kinds of cures from plants, which saved many lives.

Eze's carpentry fatigue worked very hard making bunks and tables from any kind of scrap they could find, reroofing huts, and also surreptitiously – as they moved about the large camp – passing on and receiving messages from among the prisoner community. On one occasion the roof-menders were repairing a Japanese storeroom and saw sacks of soy, salt, onions, and so on – and risked their lives by sliding down and secretly bringing large amounts back to the huts.

In June 1944, Eze was given the task of repairing, under the guard of Kobyashi, a radio receiver for the Japanese in the camp administration building near the gate, called the Green House; it was a minor fault but he made great play about it and the Japanese left him alone, so he could secretly pick up 'All India Radio'. This is when he heard about the Normandy Landings and the fall of Paris. He told the Japanese he needed spare parts so he could get back to the radio another day, and so he did. In addition, some medicines were smuggled in among the spare parts by Eze's friend on the outside. This time, towns in France being overrun by the Allies were named and this was duly and secretly reported back to the imprisoned acting colonial secretary, Mr Blythe. But on the way back to the hut, Kobyashi had pestered Eze to buy Eze's Rolex watch from him. Eze said he would ask first at the hut, as he believed it might open opportunities for those with valuables to be able to get extra food to survive, by using Kobyashi as an intermediary. It was agreed, and Eze was the main dealer on Kobyashi's insistence. For some months all went well – the Japanese took the valuables and the official lorry brought in extra food and medicines to a pre-arranged spot where trusty inmates unloaded and gave out the food and other supplies. However, on later inspecting the women's cooking quarters, Lieutenant Suzuki noticed that large numbers of eggs that had been brought in recently (by Kobyashi). Suspecting who was behind this, Suzuki had Kobyashi beaten and the trading stopped for a while. But as the health of internees deteriorated and as Kobyashi needed more money, trade soon resumed and some items were wrapped up in local newspapers, so news came in too.

In early November the camp woke one day to the drone of aircraft; Allied bombers were hitting the docks and oil depots in Singapore. There was real excitement, but also danger from friendly fire, so the internees were ordered to dig trenches all over the camp as a precaution; indeed, on one occasion a shell fragment came straight though the roof of their hut, No. 106. One day Eze was on a trench-digging fatigue when a Japanese soldier decided they were working too slowly. So he took off his shirt and dug like a madman, showing them how tough he was, while they watched and held his rifle for him, and heaped praise on his splendid efforts.

In December 1944 they were ordered to work on building six more huts, and new internees arrived – men and women – from Malaya, and all the remaining 450 Jews in the city were finally gathered in. The positive side was that at least families were together again. One or two newcomers would not listen to advice and tried to break camp to get back into the city; on being caught, the whole Jewish group was put on half rations for two weeks.

By March 1945 the Allied air raids were very frequent, and even though weak with hunger, illness and fatigue, the internees walked with a jaunty step and gloated at the obvious discomfort of their guards. One dark moment came when on 12 April a drunken Kobyashi blurted out that Roosevelt had died.

Bombing continued, but one day in August, Lieutenant Susuki ordered a sergeant (nicknamed 'Gila' or 'Crackpot') to take Eze to his apartment in Adam Park to use his short-wave radio and obtain news of what was really going on. Here he heard the BBC announce the declaration of war by Russia on Japan. Ordered back to the camp, Gila made Eze some breakfast of eggs and rice; the Japanese were clearly wanting to ingratiate themselves with the prisoners because they knew the game was up.

Eze returned to the hut and gave the good news.

Liberation

On 13 August, no fatigue parties were sent out; on 16 August a passing POW group shouted out through the wire that a peace had been signed. Red Cross parcels began to arrive. The Japanese were fearful, but still there was no official confirmation. Meals suddenly improved, and on Sunday 19 August such huge portions of rice and fish were issued that nobody could finish their plate. The last of the Double Tenth prisoners were released in a most pitiable state. Guards and internees were edgy and sleepless. They heard about the atomic bombs but there was no explanation. Via the camp senior representative, C.E. Collinge, the Japanese ordered everyone to avoid incidents and said that order was to be maintained until further notice. The Japanese claimed to regret the food shortages and deaths. More fresh food arrived.

On 1 September, Eze and some others, unable to wait any longer, broke out of the camp and got a bus to Albert Street. As they got off, a crowd of violent Chinese approached and beat the local Malays with hammers and sticks, in revenge attacks; Eze and friends escaped to the Fire Station on Hill Street. They managed to get back to the camp and, tamed by the horrors they had seen in the city, stayed put. Friends began to visit and one brought a radio at Eze's request; it was 2 September and the day of the formal Japanese surrender. They heard the ceremony on the BBC and many wept openly.

The Allies dropped leaflets in Japanese telling the guards to treat all prisoners well, and in English telling prisoners to remain calm and obey orders. On 3 September, the Union Jack was hoisted over the camp and the prisoners assembled and sang 'Land of Hope and Glory'. On 5 September the Allies landed and fresh bread arrived, cooked by the sailors on HMS *Sussex*. Then the camp gates were opened and the island reverted to Singapore time (all Japanese conquered territory had to use Tokyo time during the war), and one day the band of the Royal Marines marched into camp to play 'Beating the Retreat'. Eze, meanwhile, knew that things would never be the same after the experiences he had been through; it would be impossible to blot them out.

On 9 September the local government gave them the status of Repatriated Allied POWs and Internees (RAPWI) but little changed except the food was better; Lady Mountbatten visited to see their conditions and ENSA (Entertainments National Service Association) arrived to perform some shows. Then one day Eze's cousin Meyer Nathan arrived with invading British troops (he had escaped Singapore and served in the RAF) from India, with the best news – of surviving family in India, UK and Australia. The nightmare was over.

When 12 September was declared Surrender and Victory Day, many of the internees decided they would go to the ceremony. Nobody had arranged to take them so they struggled on foot, a strange procession of scarecrows dressed in an assortment of tattered clothes and threadbare shoes, or bare-footed, but they felt like princes as they passed dejected Japanese soldiers sweeping the streets. They joined the crowds in front of the municipal buildings and watched the Japanese High Command as they were marched by, flanked by Commonwealth troops. Then they returned to inspect their pre-war homes. Like many, Eze's flat and shop were intact but totally empty. Their amahs (servants) had remained loyal and had hidden a few of their possessions; the synagogue was also intact. Next day Lord Mountbatten arrived and they gathered at Sime Road to meet him and cheer; Eze kept the worthless ten Occupation Dollar Japanese note that Lord Louis autographed for him.

Many Jewish internees had nowhere to go; some homes had been bombed, and others occupied by refugees who would have to be paid to leave, but nobody had any assets or money. The British authorities caused great confusion and few knew where to turn. Those who asked for repatriation to the UK or India where they had family – including Eze – were refused. Rich European internees, with whom they had worked and helped in the camp, became aloof and made racist remarks about Eurasians and Jews. At the Dutch Club in Cairnhill Road, Eze was having a drink with some ex-internees and someone shouted a remark about Jews and the black market. Leon Simon and Eze went for him, but were restrained. In another incident, Eze went to the Cathay building to try to claim a lot of radio equipment stolen from his shop by the Japanese, used in the radio station and now taken over by the BBC. A Captain Dumerique refused and said he had to prove ownership, even though his letterheads and nameplates were all over the equipment. A year later, Dumerique came into Eze's newly opened shop and noticed some special new-style microphones that Eze had brought back from a business trip to the UK, and wanted them for the BBC; he said pompously that he would send an order for them, and Eze had great pleasure in telling him they were not for sale.

On 1 October 1945, Eze was repatriated to England on the SS *Antenor* after some help from a colonial officer friend. Here he met and married his wife, Nina French, who was from the Jewish community in Newport, South Wales. They lived between Newport and Singapore until finally settling in Newport in the 1960s. Eze died in 1993.

Notes

1. Ivor Ellis in the Navy, Terence Sassoon in the RAF (later Killed in Action), Reg and Meyer Nathan and Ned Clumeck in the Malayan RAF and later RAAF, Moshe Gareh in the 2nd North Lancs, Leon Simon in the Auxiliary Police, and Meyer Grand in the Medical Auxiliaries. There were dozens more.
2. See the testimony of Freddy Bloom in this book.

Rabbi Chaim 'Nuss' Nussbaum, Dutch Forces Chaplain

For personal reflections by Rabbi Nussbaum on his incarceration, see his book Chaplain on the River Kwai. *It is written largely in the form of a diary and I have paraphrased many excerpts from it below. He also did an audio-taped interview, available at the IWM, and information from it is included here.*

Chaim was born in 1909 near Auschwitz, Poland, but moved with his family to Scheveningen, near The Hague in Holland, in 1914; he married his wife Rachel in Haarlem in 1936. They lived for a short while in Telshe, Lithuania, where Chaim studied for the rabbinate, as well as studying physics and heading the Hebrew Studies Department of Yavne Teachers' College. But the Soviet invasion of 1940 made life there inimical for many Jews and so they moved on to Kobe, Japan, awaiting transit to Canada, or Curacao in the Dutch West Indies. While they were in Japan, the Dutch consul advised them that teachers were in demand in Dutch Java and also told them that the Jewish community was without spiritual leadership. Chaim took a post at Jakarta's Roman Catholic Teachers' College, intrigued by the mystery of Java, and also became rabbi to the communities in Jakarta (Batavia) and Bandung.

Here he joined the Home Guard and was stationed at Batavia Airport when the Japanese invaded; he witnessed several aircraft take off to fight but all were shot down. On his return to Bandung his CO offered him the chance of a place on an aeroplane with his family, to Australia, but he insisted that he should stay as chaplain to his men. So without having the chance to be a proper battlefield chaplain to his men, he became a POW on 8 March, despite being a rabbi, as the Dutch capitulated.

52a. The Field Grave of Lt Elze, with Star of David, at Changi.

Internment at Bandung

All men of military age were immediately interned by the Japanese. Chaim's wife was pregnant with their third child and was staying in Bandung with the Rappaport family, old Jewish friends from Holland. In his camp, a former Dutch anti-aircraft barracks, the men found an underground telephone exchange and were able to use this to speak to their families on the outside. Just after Rachel had the baby in early March, and Chaim was able to speak to her, the telephone was discovered by the Japanese and cut off.

Rachel managed to get kosher food sent into Chaim's camp by having it smuggled through a hole made by the POWs in the back fence. Notes were also smuggled in this way, and replies sent back. When visits were allowed, only hands could touch across a narrow space. Some east European Jews were still free because Russia and Germany were still allies and regarded as not at war with Japan. On 28 March, Chaim was able to bribe a Dutch marine to let him do the rice cart duty, to go to fetch rice from the town; he told Rachel he would be there and she met him with the children as the guard turned a blind eye. He began giving lectures in the camp to keep the men busy; they were well attended. As Passover approached he began to think about how he would be able to arrange a Seder (Passover meal and prayers) for the Jewish POWs. Chaim crept out of the camp via the back fence and contacted a local baker, Jan de Baaker; matzohs (unleavened bread) were prepared and the Seder took place on 22 April. Twenty Jewish personnel took part with over 200 non-Jews, and a candelabra and white cloths were smuggled in as the service was held behind large drawn curtains in a hut. The sermon on the theme of 'Freedom' (the Exodus of the Jews from Egypt to Israel) was readily understood by one and all.

On 14 May, Chaim wrote in his diary that the honeymoon was over; the order came that all POWs were to shave their heads. One evening, three young men, known only as Robert, 'Houdini' and Adrian, broke out of camp to have a liaison with some local girls; this was quite common and usually punished by intense beatings if anyone was caught. But on this occasion, the three men, who were caught coming back, were tied up against the fence for execution. They bravely refused blindfolds. The POWs were made to watch as three volunteer Japanese bayoneted them to death as they writhed in agony, screaming in pain. It took many minutes. This was the first atrocity.

On 17 June rumours began that the POWs would be moved. Chaim and his comrades were on a work party at Bandung freight station and he got a message to Rachel, who came with the children. Again the guards allowed the families to meet and they sat on the grass talking; his son Mottie and daughter Brenda had brought cakes for him; the baby Shifra was in her mother's arms. But as Japanese officers appeared, the guards pushed the families apart and Chaim was hit in the hip with a rifle butt. Mottie protested and was shocked, demanding that the soldier should stop. They waved goodbye. They would not meet again for over three years.

Tjilatjap

On 19 June 1942 Chaim and the other POWs were suddenly woken at midnight and ordered to pack 'just essentials' and were taken on a twelve-hour train journey to the small port of Tjilatjap in south Java. They were taken to an old Dutch army camp nearby. Among them was Caun, the Dutch Jewish soldier interpreter who was with them at Bandung. The Japanese ordered them all to sign a document to say, among other things, that they would not try to escape, and those officers who refused were handcuffed. The Japanese were so enraged that they made all the prisoners stand for two hours on the parade ground, screaming insults at them, kicking haphazardly at random men. They were then ordered to continuously scream out an apology in Japanese while jumping into the air, old and young alike, and when the Japanese commandant was satisfied that the humiliation had gone on long enough, the POWs were suddenly dismissed, exhausted, and the Japanese had 'saved face'.

It was decided that obtaining a secret radio was a high priority in order to get news from the outside world. Near the warehouses where the men were working was an abandoned Dutch aircraft and it contained an old radio; it was decided that while others covered their movements, some POW mechanics would approach the wreck and obtain bits and smuggle them into the camp. This was done, but later one of the men – an older Dutch sergeant – was caught smuggling tinned food in, found in the warehouse. All the camp was paraded and he was severely beaten in front of them. However, he stood up each time he was felled. The guards were even more enraged at this defiance and courage, and beat him each time, until he was finally unable to get up. He was ordered to be taken to the hospital. He survived.

Alva

On 2 July the men were all moved a short distance to a second camp, whose sadistic commandant they named Alva, after the cruel Spanish admiral sent to attack Holland in the sixteenth century. The camp was, in fact, an old Coast Artillery base across the bay. On the way a Japanese guard tried to force Chaim to sell his watch to him, but Chaim refused. The Japanese pointed his rifle at Chaim's face and a fellow POW quietly advised Chaim, as they marched along, to say that the watch was a gift from a loved one, and that would stop the pestering. Chaim did this and the guard stopped nagging – but, instead, the guard drew his finger across his throat, pointing at the new camp as though to say, here they would all die.

Guards at this camp were particularly nasty; on 10 July Chaim wrote that he had a bayonet pushed into his navel for not saluting well enough, and he saw two Dutch soldiers made to stand bareheaded in the sweltering sun, saluting up and down at nobody for an hour, as a punishment for not doing it properly, while a smiling guard nearby lazed in the shade of a verandah, watching, and screaming if they stopped. When it finally ended, Chaim and others rushed to them to give them water and take them to their bunks.

On another occasion, a Japanese colonel wearing spurs on his boots walked along a line at *tenko* deliberately kicking his spurs into the prisoners' shins.

On 10 August Chaim wrote in his diary that while he was on a working party loading paint tins onto an old freighter, a guard saw Chaim and a colleague taking an unofficial rest. They were both viciously assaulted by the guard, who brought his hand down in a crushing, chopping blow onto their heads several times. They were both concussed and the other man, Pete Trom, had to do light duties in camp for two days as a result.

On 8 September Chaim wrote in his diary that a rumour had spread that a large group of men were going to seriously provoke the Japanese by gathering to celebrate Queen Wilhelmina's sixty-second birthday, without Japanese permission, and – even worse – sing the Dutch national anthem. It was an act of rebellion against the hellish conditions of disease, hunger and slave work in the camp. The men gathered in a quiet, tree-shaded corner of the camp and Chaim led the address to the men, but then they were warned that guards were approaching. Thinking fast, the men all stood and then bowed to the Japanese. The Japanese saluted and waved them to sit down, and Chaim continued his story of the Queen Wilhelmina's life and her courage against the Nazis, both in Holland and in exile in London. Thinking this was a religious service, the Japanese rose, saluted and went on their way; the POWs breathed a sigh of relief. They quietly sang the anthem.

Soon a new and more amiable commandant arrived, nicknamed 'Opa' by the prisoners (Dutch for grandfather); things slightly improved and he encouraged the planting of a field of peanuts in a nearby field to help feed the POWs.

On 26 November Chaim wrote in his diary that a Jewish POW, Louis de Vries, had carved a Magen David, which hung over his bed; one day the Japanese officer (second in command) saw it on an inspection and shouted that it was 'a sign of Jodaidjo, Jews, who wanted the war against Japan...Jew Roosevelt...Jew Churchill'. As a result of this outburst, the Dutch camp authorities banned the use of Jewish religious symbols, for fear, they said, of retribution from the Japanese. Chaim ignored this and organized a Chanukkah service with a wooden menorah carved by de Vries.

On 4 December Chaim wrote in his diary that posting friends at the entrance to the barrack room to warn of approaching guards, Chaim stood the menorah prominently on a table made from his bed board: 'the light made from coconut oil ... donated by the chief cook, a devout Protestant. The flame was small and the smoke blue ... our own present miracle', he said.[1] Ten Jewish boys came and all sang the Chanukkah songs; slowly the room filled. Other friends came and stood and stared into the tiny flame ... asking questions about the festival and the Jews. One commented abruptly that the Jews had a lot at stake in the war. Nussbaum said, 'Don't you?' After a long pause he answered, 'We all do', buried his face in his hands, and left.

In most of the camps that Chaim was in, the cooks tried very hard to serve him kosher food as far as was possible; he was always moved and touched by this.

Then orders came to move again; they were perhaps going to Thailand, said 'Opa', to work on the Burma Railway. First they went by train on 27 December, via Bandung, to a transit camp in Jakarta, known as 10th Battalion Camp; it was filthy, but food was plentiful and there were no working parties. Then on 16 January 1943 they were taken to Tandjongpriok port by truck and herded onto a Japanese freighter for Singapore, where they arrived three days later.

Changi

From Singapore harbour, trucks took the men to Changi, where they were greeted by British officers. This camp was a pleasant change, as the British ran Changi barracks and the Japanese guards were outside and not constantly harassing the men. They were shown to large ward-like buildings with beds and given tea and rice, and slept well that night. A few days later Chaim was reading a noticeboard on which was announced the Saturday Jewish services; a British POW, Alf Levy, saw him and asked if he was interested, and told him that Sergeant Victor Blashki, an Australian, organized them. These men were thrilled to meet Chaim, a real rabbi who could lead the services, and Chaim told them about Private Yits Lisser, a Dutch soldier who had been a cantor at the synagogue in Jakarta before the war. Chaim began by organizing the visiting of the sick in the multinational hospital in the camp. Now even more regular Sabbath and Festival services took place. Chaim also took full part in the lecture programme.

But food was very short and the Japanese asked for volunteers to go to build the Burma Railway, promising good, new conditions in the camps upcountry. Lisser and others decided to go. During Purim, in March 1943, Chaim spoke in the service about the angels that protect the Jews in the diaspora, promising they would overcome their enemies. Meanwhile, Chaim became disillusioned with Changi, saying that he had little do as a minister, that many of his friends had gone north, that conditions were appalling, and that maybe he should volunteer too, to work in Burma and Thailand. On 22 May 1943, Chaim left with about 300 others to travel north for five days on trains to Bampong, on the Kwai river, Thailand.

The Kwai

There were frequent stops and the food was not bad. On arrival at Bampong, however, some Dutch POWs, who had rather 'cushy' jobs as lorry drivers for the Japanese, told Chaim and his friends that on the march north they would be so exhausted that they would begin to throw away their belongings, and local Thais would even try to steal them from them, as they would see that the POWs were too weak to resist. Therefore, the Dutch POWs were offering to take their possessions now, and sell them for a commission. Chaim was shocked at the uncaring attitude of these black-market exploiters, and walked away.

From Bampong they were marched further north, on the night of 31 May. Most had no shoes and were hungry and ill. Chaim and four friends managed to get some shoes at Bampong but these were even more painful to walk in than using their bare feet. On 2 June they reached Kanchanaburi and rested, and (amazingly) were injected against cholera – the Japanese clearly wanted men alive to finish the railway. Part of the journey continued on railway flat cars, open to the intensely cold rain and the burning sun. Then there were hours more of walking, until on 5 June they reached Tarsao camp.

Tarsao/Tonchin

Here Chaim fell exhausted into a tent and was given a hot drink and fell into a deep sleep. On the march, he had been dutifully calculating the days of the Omer (the fifty days between Passover and the Feast of Weeks or Shavuot); he reckoned it was the forty-fifth day. For him, this was an anchor to civilization. It was hard, amidst the daily deaths and beatings and illness and hunger, to keep his mind on such things, but he did it anyway, to help to give himself a sense of some control over time and life.

Building work continued: cutting jungle, laying ballast, laying track. Tarsao was ironically called a 'hospital camp', but death was everywhere, officers and men all struggling to stay alive. Prisoners were sent down from the north half dead, revived with some food and aspirin, and sent back up the line to die.

From Tarsao, they were marched to nearby Tonchin. Cholera broke out and in one day, 13 June, Chaim was told 130 British soldiers in the camp opposite had died of this terrible illness. The following weeks were harrowing. Gaps in the line occurred, as Japanese engineers could not get the lines to meet; POW sabotage was partly to blame. The daily march to and from the camp to the railway was itself exhausting and deadly in the awful monsoon-induced mud. Chaim vividly describes the walking British skeletons, and how one officer fell face down into the black mud, and another soldier, not seeing him, trod over him, until a second man came and lifted the fallen man and two others carried him between them on the route.

As spoil was taken from holes to make the embankment, ladders had to be used to extract it, whilst another ladder had to be negotiated, with the heavy basket-load, to reach up to the embankment – and all the while the Japanese were screaming and beating the men with bamboo canes across their heads, which meant that the spoil was dropped in most cases and the whole process had to be repeated. Chaim had severe dysentery while this was going on which added to his weakness.

The descriptions of the dying Tamils, wracked with cholera, defy belief. In one heart-rending scene, a young Tamil man placed an old man by a tree, gently smoothing a space for him, propping up his head with an old shirt, leaving him food and water – he then stood, and both looked sadly at each other – probably a father and son. The young man was called by his friends to move along with them; the older man was left to die quietly in the shade. Chaim wondered at the brutality of the Japanese and he watched as the bodies of both soldiers and coolies were taken and stacked, ready to be cremated.

The tents they lived in leaked, of course, and men were up most of the night, unable to sleep in the cold and wet. Chaim had severe dysentery and was unable to work, but tried with other sick men to help those returning from work to eat; they, however, wished only to sleep. Because of the terrible 'speedo', many were working all night and beatings were frequent – the Japanese were in a frenzy.

Then suddenly it was over; the work slowed as the railway was declared finished; the cholera ended; the locals returned after the cholera with their boats of produce, and men were able to buy extra food as health improved.

Tamarkan

In mid-July 1943 they were moved by marching and by train, this time fifty miles to Tamarkan camp. For the first time on the railway, they were caged in by wire, which came as an ugly shock to the POWs.

When Chaim arrived at Tamarkan from Tonchin camp, he and his group were stopped and sprayed with disinfectant, but he noticed that the men were quite healthy:

> There are no hungry, thin eyes, no empty begging hands. Our Commandant [Commander] is a born leader [Col Toosey] ...True it is a closed camp but there is civilisation here ... a canteen ... almost everyone walks as if proud ... large well-ordered huts. Because of our lice we must leave all of our clothes in a huge pot of boiling water at the other end of the camp ... [in] our hut we washed every bamboo stick with boiling hot water [an unbelievable luxury in a Japanese POW camp] ... but I hear that the lice will return in about four to six weeks.

Toosey was a hero to Chaim – always adamant that sick men would not be sent to work and standing up to the Japanese very courageously at risk to his health and life.

On 19 July, Chaim met Sol Sternfeld, a fellow Jewish POW from Tjilatjap, and, sadly, was told by him that Private Lisser had died en route to Tamarkan. He had weakened from severe dysentery and simply faded away. They had not all seen each other since Changi; Sol told Chaim that he said Kaddish (the Jewish prayer for the dead), helped bury Private Yits in his mosquito net and placed in the grave some pages from his siddur (Hebrew prayer book), his Haggadah (book used on the Passover) and a piece of paper with the

Hebrew word *Bittachon* – faith. Sol was at least happy that Chaim was still alive. Chaim was deeply upset and just sat for hours in a symbolic shiva (Hebrew for seven, a state of mourning meant to last for seven days). That afternoon they held a special service (*minchah*) for Yits.

On 27 July the camp theatre group performed the First World War play, *Journey's End*; it was very emotional, and appreciated by the men, who easily identified with the story and for whom there was little of cultural value; even the Japanese officers came and enjoyed the show, although they could not understand the English.

On 16 August, Chaim wrote that there had been a big increase in the black market among POWs. Some men who had access to the drug store had been selling life-saving quinine to local Thais at a huge profit, endangering the lives of the sick. The officers appeared to do nothing. Even the British camp military police were involved in a huge theft of tools, which they sold to locals when the camp was left almost unguarded when the Japanese all went off to Bangkok for some Japanese national holiday. The British imagined that the Japanese would help to cover this up, because they profited from the deal and also would not want it to be known to their superiors that they had left the camp unguarded. But when the theft was discovered, the guilty British party were taken to the guardhouse while the remainder of the POWs waited and listened. The Kempeitai were investigating and the screams of the suspects could be clearly heard. The men were divided; some said it was a duty to steal from the Japs, and in any case a lot of the profits went to the canteen for everyone's benefit; others said that it would only result in collective punishment for everyone, and the culprits would keep the profits. In the end, the Japanese imprisoned the men, and the British were insulted by being made to share the camp police function with the Dutch POWs, as it implied that they could not be trusted.

On 20 September 1943, they were ordered to move again. The Kempeitai searched all belongings and confiscated Chaim's lecture notes on talks he had given, and his diary which he tried to hide between two pieces of card.[2] He was devastated by this act.

Kanchanaburi

They marched to Kanchanaburi, arriving on 26 September. Chaim met up again with British POWs Alf Levy and Reg Warshawsky. Alf was very ill with severe dysentery and remarked jokingly that with Chaim's arrival he was now at least assured of a Jewish funeral. Reg sold a shirt to buy some drugs for Alf, but the doctor said that Alf needed vitamins such as those contained in eggs. Reg and Chaim obtained some for him but he could barely eat part of one. Chaim said that it was known that other sick men stole the food of those unable to move, and even medical orderlies sold drugs for cigarettes, so Reg stayed alongside Alf's bed to watch over him. On Saturday 28 September, Alf told Chaim he wanted just one final Shabbos (Sabbath day) with his mum's home cooking, and he would be happy. On the following Thursday it was the first day of the Jewish New Year and Chaim visited Alf to tell him about the service the Jewish POWs had held. Alf rallied a little in the following days but, sadly, died on the morning of Sunday 6 October.

Alf had wanted a Jewish burial and, despite some objection from another Jewish POW, Samuelson, who said that it might look odd that when so many died without the proper rites, the Jewish POWs should hold a large gathering, Chaim went ahead with his plans and ten men (a minyan, or quorum) took part. Using Alf's British Army Hebrew prayer book, Chaim intoned the ancient Kaddish prayer and the party moved off, with many watching the proceedings; six Christian burials were also taking place, led by the camp chaplain. Four of Alf's friends carried him, with others walking behind as Chaim led them to the cemetery; the Japanese guards watched, surprised at the unusual decorum, jumping up and saluting as the funerals passed by. The Christian service took place first – the chaplain had buried 100 men in recent weeks and it was difficult for him to show the emotion that would normally be displayed. Chaim's service was thus longer; there were tears as a friend laid Alf's head gently in the shallow grave. Shadows fell as the sun set; psalms were recited.

The next day, another death, this time from malaria: Michael (Lewis) Singer. Another Kaddish, another funeral. He left a wife and three children. Later Chaim helped Eddie Rappaport remove boils full of pus from his back, using a bamboo scraper. He went to the kitchen to obtain a mug of boiling water to clean his patient, but a new sergeant was there and demanded to know who had given him permission to enter the cookhouse. Chaim explained, but the sergeant hit the mug of water out of his hand and pushed him onto a nearby cooking fire. Another man helped him up quickly and the enraged sergeant fled in panic. Later an apology was given.

The death rate remained high; many men arrived already too far gone to be saved by the hospital staff. Often a man would die in front of Chaim while he was doing his hospital rounds, talking to the sick; one died in the middle of a quiz event organized by an officer. The officer stopped the quiz, but then thought he had best continue to keep up morale and finish the game; was it right? Was it a heroic defiance of death or disrespect? Who knew the answer to such things?

Sime Road Camp, Singapore

In December, Chaim and his comrades were moved south by rail to Sime Road camp on Singapore Island – smaller than Changi and far more pleasant, resembling a fenced-in golf course. Chaim celebrated Chanukkah with the Jewish POWs, using soya beans to make the latkes.[3] In January, classes began on all kinds of subjects, Chaim running a series on Judaism which attracted many non-Jewish participants. On 10 April 1944, they were moved to Changi

Changi

In May, Liverpool Jewish POW Jack Greenberg was due to leave on a work party upcountry. He asked Chaim to bless him. Chaim recited the ancient Jewish priestly blessing: 'May the Lord bless you and keep you, and make his face to shine upon you and be gracious unto you' – and as they heard this, the many non-Jewish prisoners within earshot removed their hats and bowed their heads in prayer with Jack, accepting Chaim's words too, as men going into the unknown. As he finished, a loud 'Amen' followed from them all; what a moving sight that was.

In their absence, gardens had sprung up at Changi, where POWs had been able to grow some fresh foods (some of the plantation owners and workers in the gaol gave advice and help); there was a library again and two tents served as a synagogue, but conditions and food were still bad and the death rate was still high. Chaim gave lectures and lessons and started a Jewish magazine, *Ha Beemah* (The Prayer Platform), to which non-Jewish artists such as Ronald Searle and Otto Cramer contributed. Although a store of paper was discovered in an old office, it was still a rare commodity, and as there were so few non-smokers ready to give up their cigarette paper to exchange for paper for the magazine, consequently, not many magazines could be printed.

Chaim also began to rewrite his diary here. Scottish soldiers built a small hut for use as a synagogue and it was named 'Ohel Yaacov' (Jacob's Tent), probably the only synagogue ever built in a Second World War POW camp.[4] After first using the old YMCA hut as a synagogue, in August 1944 Scots soldiers of the Royal Engineers constructed a small, purpose-built hut with the help of the Assistant Chaplain General (ACG) Lieutenant Colonel Lewis Bryan; it became a haven of peace in a vastly overcrowded gaol and even many non-Jews attended. A service of dedication was held on 26 September and many senior non-Jewish officers were present. Chaim said that Joe Bernstein, the 'soft-souled young man from Glasgow', sang beautifully; Chaim himself gave the sermon to a riveted audience.

Dave Birnbaum was a Canadian serving in the British Army who had his leg amputated after being wounded in the fighting for Singapore. Chaim started to teach him Hebrew because he wanted to go to Israel after the war and find a useful job on a kibbutz.

Soon news reached the prison of the D-Day landings. But the Japanese were not fazed, and boasted to the prisoners that the Germans and Italians were cowards and could not fight, but Japan would fight on for a long time and were 'Number 1 soldiers!'

Jimmy Bines asked Chaim if he could convert to Judaism, and Chaim explained he was already Jewish if his mother was. Bines was overjoyed. Yom Kippur was on 1 October and, to Chaim's amazement, twenty-five Jewish soldiers decided to fast, despite their hunger.

52b. Seder in Singapore 1946, Wilhelmina Camp, with Nussbaum seated centre in white shirt, with his wife Rachel to his right and daughter Shifra to her right; to Nussbaum's left are his two other children, Mottie and Brenda.

In November 1944, Chaim's good friend Lieutenant Elze became ill with blackwater fever, an incurable disease. Blood transfusions worked for while but his condition worsened. Chaim visited him, sometimes several times a day. On 10 November he died. The whole Jewish community and over 100 Dutch officers turned out for his funeral. Flowers were placed on his coffin as it was taken on a trailer along a narrow road to the cemetery. Clearly Elze had touched many people. Chaim recited Psalm 90 and the Kaddish, and the Senior Dutch officer made a moving speech about the important engineering work Elze had done before the war. It was probably the largest funeral seen in the prison.

On 21 December, Eddie Rappaport received a letter dated June 1944, telling him that his father had died in May at Bandung. Chaim organized a symbolic one-day shiva (Jewish period of mourning), and Eddie took off his shoes and sat on the floor; there was much weeping, and the Kaddish was said at the next service.

In 1944, Chaim asked himself, as American and British bombers flew overhead and sometimes bombed near Changi, 'Who is our friend ... must we fear both sides?'

In January 1945, East End Jewish POW Cyril Wernick developed TB and was quite ill; in camp conditions this disease could be fatal. He had a good bed and mattress in the low, dark TB hut, and received regular visits from the doctors, and extra food. A handsome and literary man, he wrote angry poetry, his wife's photo at his bedside together with his writing materials.

By April, men were still dying of the many diseases; working parties of only healthy men were sent to maintain and build the airport nearby. As bad war news reached the Japanese, the punishments became less harsh.

The End

Around 8 August 1945, rumours spread that a huge bomb had been dropped on Japan, killings hundreds of thousands of people. The news came via the secret radio, and the sources had to be careful that the Japanese did not realize that the POWs knew something. By 9 August the story was confirmed, but there was still no sign of Japanese surrender; there was a real worry that the Japanese might kill all POWs in retaliation. Chaim wrote:

> Suddenly, the mood changed and the Japanese guards became much kinder. On the 10th the secret radio was openly used and many came to smile excitedly and view it, saying they knew there had to be one hidden all these years. They gazed at the man with the earphones as he scribbles down something on a scrap of paper; the guards keep away. A deep voice reads out, 'a second atom bomb has been dropped ... Emperor agrees to unconditional surrender'. Uproar drowns out the rest of his words; men hug and cry.

The guards kept away. The men remained confined and the Japanese armed, but nothing happened; the guards now protected the POWs from Malay nationalists outside the camp.

Rachel, meanwhile, had been interned with the children at Adek camp near Batavia since 1942 and it was there that Chaim went to liberate them as soon as he was able. Here the Japanese were being used by the British to keep control over Indonesian nationalists – how ironic! But Rachel encouraged Chaim to visit the other Indonesian camps and assist where he could in his official capacity as Rabbi and chaplain.

52c. Left to right: Pte Lou/Loet Velmans, Rabbi Chaim Nussbaum and Eddie Rappaport, Dutch Jewish POWs at liberation.

From there they went to Singapore, to live for a while at Wilhelmina Camp, and here Chaim reopened the synagogues and held a beautiful Seder (Passover meal) in the spring of 1946, but heard about the true horrors of the Holocaust in Europe. He fell ill again with recurrent dysentery. He was officially made a rabbi of the RAPWI (Rehabilitation of POW and Internees) and assisted the wives and children of POWs, but also had the sad duty of reporting to the widows the deaths of their men. Finally, in May 1946, a Dutch ship (*Oranje*) arrived, and Rachel and the children went home to Scheveningen, Chaim following two months later on the SS *New Amsterdam*, and on this journey he wrote up his official POW report. They had three more children in Holland and a seventh was born in Canada in 1953, to where they had emigrated, to Toronto. Chaim became a teacher at a Jewish college, and a congregational rabbi. He was a frequent visitor to Israel.

Chaim prayed that the survivors would all be better people for their terrible experiences.

Epilogue

At the end of his book, Chaim wrote some priceless notes on the links he maintained after the war, and the fates of some of the men with whom he suffered. I give some extracts and information from these notes below, beginning with the four non-Jewish artists.

The famous artist Ronald Searle wrote:

> I certainly look forward to reading your diary. I remember our encounters with pleasure ... there is a bond of friendship through mutual experience. After the war I made a point of going to Poland in 1948 and seeing Auschwitz ... for myself. In

> 1961 I spent a month in Jerusalem attending the Eichmann trial. I wanted to understand ... why it happened. It has helped to make me a better artist.

Otto Cramer became an artist and teacher in Holland and died in 1980. He had illustrated the Passover issue of *Ha Beemah* in Changi and helped decorate the synagogue.

George Sprod did most of the illustrations for *Ha Beemah* and in the 1970s was a writer and cartoonist in Fleet Street. He returned to Australia and his work is on display at the Canberra War Museum.

Kees van Willigen returned to Holland and his war artist work is on display in The Hague. He wrote to Chaim:

> I remember with pleasure and gratitude that it was you who introduced me to the Jewish community in Changi Camp, Singapore, and enabled me to spend many hours in the little schule there reading and drawing ... and I did drawings for 'Ha Beemah'. And then I also remember how we went on long walks in Singapore ... right after the Japanese capitulation, talking about ... what it really meant to us to be free again.

Joe Bernstein organized the choir at Changi and throughout the camp, men – Jews and non-Jews alike – went around whistling Hebrew melodies.

Lou Velmans was an editor of *Ha Beemah* and on his return to Holland wrote about his POW experience.

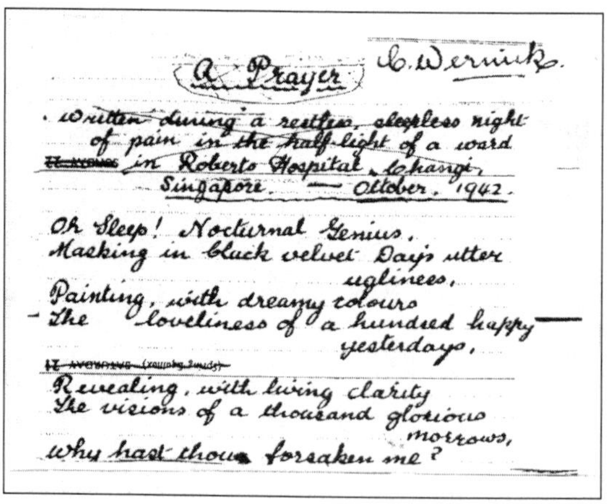
A Prayer C. Wernick.

Written during a restless, sleepless night of pain in the half-light of a ward in Roberts Hospital, Changi, Singapore. — October, 1942.

Oh Sleep! Nocturnal Genius,
Masking in black velvet Day's utter ugliness,
Painting, with dreamy colours
- The loveliness of a hundred happy yesterdays,
Revealing, with living clarity
The visions of a thousand glorious morrows,
Why hast thou forsaken me?

52d. Cyril Wernick's original poem (see Frontispiece), as given by him to Rabbi Nussbaum.

Eddie Rappaport moved to Israel and lived on a kibbutz and then moved to the USA.

Dave Birnbaum/Goldstein went to live on a kibbutz in Israel after the war, married and worked at translating Hebrew stories into English. In 1956 he returned to Leeds in England, his wife became ill and died and he became depressed. Chaim was unable to contact him after that.

William Herman died in a work camp in 1945.

Cyril Wernick, whose poem appears at the beginning of this book, survived the whole war in Japanese POW camps, but sadly died of his illnesses on his way home to Britain and is buried in Madras, India.

J.A. Van der Bent was present at Chaim's Seder in April 1942 and wrote to him:

> I still remember you ... with your prayer shawl and green army cap ... one evening you were teaching Psalm 130; your way of treating the text fascinated me so that I could have spent hours speaking and listening ... as a believing Christian I am plagued with guilt feelings about the Jewish people ... whenever my wife and I read verses regarding Jerusalem and Israel, I cannot help but think of you. Whenever you read the Priest's Blessing (Numbers 6), please think of me.

J. Glaser remained in military service after the war, retiring as a lieutenant colonel in 1965. He prepared a report for the Dutch government about the Jews of Indonesia. He remained involved in Jewish community services in Holland, and all ex-POW matters.

Notes

1. Chanukah celebrates the miracle of the Candelabra (Menorah) in the Temple, burning for eight days instead of one, at the time of the revolt of the Maccabees in the first century BCE in Jerusalem, Israel.
2. He rewrote this from memory a year later in Changi.
3. Pancakes fried in oil, to commemorate the Holy Oil used in rekindling the ancient Temple Menorah after its desecration by the Syrians in the first-century Maccabean Jewish revolt.
4. It was painted in colour by Australian non-Jewish POW artist George Sprod and a copy can be seen in both the AJEX Jewish Military Museum in London, and the Changi Gaol Museum in Singapore.

130501 Gnr (Sir) Alexander 'Oppie' Oppenheim, SSVF

The information here was taken mainly from Alexander's recorded interview at the Singapore National Archives, with kind permission.

53. 130501 Gnr Alexander Oppenheim, SSVF, taken in his later years.

Alexander was born in Salford, Manchester, on 4 February 1903, son of well-to-do orthodox Jewish immigrants from Kovno in Russia (his mother's initials were H.J., of 29 Rectory Road, Crumpsall, Manchester) and was one of six children. His first language was Yiddish. He attended Derby Street Primary school and Manchester Grammar School. He attended Hebrew school every evening and at weekends. From a very early age he was fascinated by numbers. He won a mathematics scholarship to Balliol, Oxford; he became chess captain, got a First in 1924, and became a tutor at Exeter College. He later went to the University of Chicago in the USA and got his PhD in 1930. He married an American, Beatrice Templer Nesbit (they had one daughter, Judith, and were later divorced), then went to teach at Edinburgh University and eventually became professor of mathematics at Raffles College, Singapore.

He joined the SSVF in September 1938 (he served in the same unit as David Marshall) and became a gunner; his wife and daughter later got away to the USA but Alexander was made a POW of the Japanese when Singapore surrendered.

He remembered well the first air raids when the city was still totally lit up; the harbour was badly hit, where he was stationed at the Keppel Harbour Yacht Club on coastal protection and anti-aircraft duty with eighteen-pounder guns, often running messages to nearby Fort Canning HQ.

Changi

On capture, his men were sent on a forced march to Changi barracks, carrying kit bag and boots, watching as the victorious Japanese marched in to take over the city. They carried as much food as they could bring with them. It was very arduous and took one-and-a-half days; and many dropped out but were forced onwards. Along the way, local villagers gave them water and coconut milk to help them.

At Changi gaol, Alexander suffered the well-known privations of all POWs, especially falling prey to dysentery on many occasions. Food was short at first but later it improved as the camp became more organized, though rations were always at starvation level. There were problems convincing the medical staff that it was all right to boil rice in seawater, a practice used since ancient times in the East; this saved on using the scarce supplies of fresh water. Morale was very low following the surrender; there was a lot of blame put on the military leadership and rumours of liberation around the corner – all false, of course. At first Alexander was sharing with five other men in what was a bathroom; overcrowding was appalling and, before long, lice and fly infestations began. Work parties were organized and his first task was to bury the dead of recent shootings by the Japanese on the beach nearby – mostly Chinese, allegedly caught looting. Some were still alive and the British smuggled them back to camp and tended their wounds and got them back to Singapore later. The rest were put into horrendous mass graves.

Alexander witnessed the Selarang incident, and said that the Japanese were not aware that photographs were taken and drawings made of what happened.

Those on working parties could smuggle food back in from Singapore for a price, but often IOUs were used in lieu of scarce cash and redeemed, in many cases, after the war. Some men even broke out of camp under the wire and obtained food from locals, and smuggled it back in. Sometimes even the Japanese guards took bribes to look the other way or sold items themselves to prisoners. Cigarettes were a special form of currency. As he had stopped smoking, Alexander at first gave his away, not realizing their value, and wondered why a certain sergeant major always befriended him with a visit when rations of them were given to the men! In Thailand he saw men give up food and die, in order to obtain a cigarette in exchange. They also made alcohol from the fruit of the cashew nut plant, using human spit to help the fermentation, and the nuts of the rubber tree could be burnt for fuel; the seeds of the durian fruit were good to eat when roasted or fried.

But he was of indomitable spirit, and helped to set up the Changi 'POW University' and was made dean by fellow academic prisoners. After convincing very senior Allied POW officers of the value of such a project, the men persuaded a particularly sensitive Japanese officer, Lieutenant Okazaki, to allow the collection of about 500 books from the Raffles College library, and this gave a sense of purpose to many of those in despair. The Japanese agreed because they thought the project would help maintain discipline. Alexander was allowed to go with several other former lecturers in a lorry under guard and choose the books from Raffles College, which had been ransacked. As the books were so valuable a resource, they were never lent out and were only for use in the library building itself. Pieces of chalk were dug up and blackboards were made from doors; open-air lectures were given under trees or in rooms, using home-made benches. Bits of apparatus were made from waste collected from a nearby army dump. Toilet paper was used as notepaper, and some men had kept notebooks when captured. Pencils and other items could be scrounged or even stolen – often at great risk to their lives – by POWs who were working as batmen (army servants) to Japanese officers. A timetable for lectures was organized, covering a very wide range of disciplines. Subjects were on demand – that is, what the POWs wanted to hear and learn about; Alexander remembers giving one lecture fifty times. Lectures began early and went on until as late in the day as possible, but both staff and students were very keen that it should be so. The lecture time was deliberately kept to forty minutes so that as many as possible could be fitted into the working day.

As the dean, Alexander was also the main administrator and had to deal with all sorts of personnel and logistical issues. The Japanese did not interfere, but frequent searches were made in case forbidden subjects were dealt with – military matters, for example, or map skills which could mean that escape was planned. Lectures were given with zest, between army and other chores in the camp, so normal work was not really affected. Sundials were made so that they could tell the time, because watches were very scarce; cigarette packets were used as library cards. Alexander was in the rather unique position of being a lowly gunner who often lectured to very senior officers far less educated than he was.

Students who applied were interviewed by a committee, as not all were suitable or bright enough; over 5,000 applied and over 1,000 enrolled. Rank was ignored and chances were given to those who had no experience of higher education but were highly motivated. The Army Education Corps also ran courses for those who were interested and tried to take over the university courses – but the staff stood firm and remained independent throughout. But underlying it all was the problem of ill health and poor food; this affected all staff and students alike.

Alexander also said that when Red Cross and other supplies arrived there was corruption among officers and ORs as to how these were distributed, and a lot of selfishness and bad behaviour occurred, but he would not allow this to be recorded on the tapes.

There was a secret radio; some bits were hidden actually in the Japanese officers' quarters by the British batmen selected to work for them – this, of course, being the last place the Japanese would search for radios. If a search was carried out by the Japanese and they seemed close to discovering something important, various distractions would be organized, such as a fight or brawling, until they were put off the scent. Some false news was given out deliberately so that the Japanese did not suspect that accurate war news was somehow getting to the POWs, or a loose-tongued POW might talk to a guard, thus setting off alarm bells to the Japanese intelligence officers.

One officer known to Alexander even kept his dinner jacket, and another had a piano shipped to Changi and it was taken to Thailand later – a great source of morale for the men who organized the shows in such terrible places. Many had musical instruments, and chess became very popular, with hundreds wanting to learn how to play and many others carving beautiful sets from all kinds of materials. Prisoners discovered they had hidden talents in leather-working, woodwork and so on. Alexander knew one young soldier, who had been an accountant, who discovered a way of making very valuable sandals from old tyres and bits of linen, that in effect never wore out.

Thailand

The university was disrupted but not stopped when Alexander left and was sent to work, like thousands before him, on the notorious 'Death Railway' in October 1942. The journey, in horrendous conditions, took five days and nights. The filth in the crowded cattle trucks meant many became ill – mentally as well as physically. Food was appalling and the stops infrequent. After detraining, they were sent by lorry and on forced marches, in monsoon rains, forced to keep up at bayonet point by heartless guards. Alexander remembered that at one stop they rested in a temple, and a Thai villager offered him water in a brass bowl.

At the first camp, Tarsao, the conditions were indescribably filthy. One Thai tried to sell them food under the wire, but he was taken by the guards and tied to a tree and severely beaten. Alexander thinks he died soon afterwards. They later went to Kanyu, where work on the railway began. Water had to be brought up by hand from the river far below the camp, in treacherous conditions. A march of several kilometres had to be made before reaching the work point on the railway. Food was appalling – mostly watery, sour rice and jungle stew. Men who were ill were forced out to work and many died on the march to work, including close friends of Alexander. The routine was a very weak cup of tea at dawn, before they were given a small ration of rice and jam and marched to work; then hard labour for ten hours on the railway, with a tiny midday meal and break, and the march back, arriving exhausted but still with camp duties and personal cleaning to do. There was some leisure time but the men were so tired, dispirited and ill that lectures and games were out of the question, except perhaps for some chess.

Very small amounts of pay were given by the Japanese, but card tokens were made and used by the POWs in lieu of cash. There was the usual black-market trading with Thais when men got out at night to do deals. Some men even broke into the Japanese stores and supplied items to the hospital hut.

After some time, Alexander passed the test to become a medical orderly; he was asked to look after a dying man and clean him up. He did not flinch, and the sergeant in charge gave him the job. At one stage he was in charge of a hut with eighty very sick men suffering from every conceivable tropical illness. At one stage all four other orderlies were sick and he was on his own. Slowly his health deteriorated in the appalling conditions, and he caught dysentery; he stopped counting after forty when men died around him in the sick hut. He also witnessed conditions in the coolie camps and was reminded that, despite how badly off he was, their situation was even worse. Only the thought of his family kept him alive. Men would watch to see who would die next and then there was a rush for his rations, so hungry was everyone.

Eventually, in August 1943, he was sent back to various base and hospital camps at Chungkai (where he stayed until May 1944), Tamuan (until June 1944), Bampong and Kanchanaburi, and then back to Tarsao, which was a haven compared to where he had been. To survive the journey he was given an injection of morphine. He was down to about five stone in weight and it was thought he might die. One Japanese officer stopped to inspect the men and apologized to Alexander for the conditions, explaining that they were forced to cram so many men into one train because they needed other trains to send supplies to their soldiers in Burma.

At Tarso he saw evidence of how the Japanese organized machine-gun posts in such a way that they may have been planning to kill all POWs should the order be given as the Allied advanced. Here they heard about VE Day and he also met and saw the remarkable work of Captain Markowitz, the Canadian Jewish RAMC surgeon. Food was better and duties were light.

They were allowed to take a lot of books with them and lectures did continue in some form in these camps. Alexander always said that one could never blame all the Japanese for the atrocious treatment given to them by a few. Yet while in the various camps he was asked by the senior British officers to give lectures on the Jewish faith to POWs, to combat Japanese anti-Semitic propaganda.

Liberation

Alexander remembers that this occurred at Nakon Pladok, and was marked by the sudden distribution from Japanese stores of Red Cross parcels, and the appearance of a British colonel to tell them that the war was over. Documents were gathered and some reprisals took place – but Alexander would not elaborate. Then they were put on a train to Bangkok; from here he was flown to Rangoon and shipped by a Dutch vessel to Liverpool. He went home to Manchester and met his five sisters and his mother, who of course hardly recognized him. He had only received one card from them in 1944, despite the many others they sent, to tell him that his father had died that year. None of his cards ever got home. He had lots of treatment and debriefings in local hospitals for some months, but eventually he was reunited with his wife and daughter in the USA in January 1946, but his wife was mentally ill. They divorced, and his daughter went to a school in Brighton.[1]

In 1947 Alexander returned to Raffles College, married a Chinese girl and had two more children; he became vice chancellor of the University of Malaya and worked in many universities round the world, retiring to Henley-on-Thames in 1977. He married Margaret Ng in 1982 and died in 1997.

Note

1. His daughter Judith married a Nigerian lawyer who was taken prisoner when he was in Biafra in the civil war in 1966. He said that the stories he heard from his father-in-law (Sir Alexander) about his experiences as a POW of the Japanese were what kept him alive in Biafra and determined to get home; and indeed he did get home.

1754077 L/Bdr Edwin/Edward Joseph Ornstein, RA

The information here is taken from an affidavit written by Edwin on 9 September 1945, held in the IWM Archives.

54. 1754077 L/Bmdr Edwin/Edward Ornstein, taken in his later years.

Edwin was born on 26 May 1908. When war broke out he was living at 6 Ranelagh Court, Glendale Avenue, Edgware, London. He was captured in Singapore and spent most of his time as a POW on the island at Changi, but he spent eight months on the Death Railway; on 14 May 1943 Edwin was shipped north with 10,000 others and by the end of the year half of these men were dead.

Changi

Owing to the smashing of the sewage system during the fighting, dysentery broke out almost at once in Changi, and Edwin fell victim in March 1942. Buckets were toilets, there was no running water or disinfectant and the dysentery wards were overflowing with thousands of sick; the stench was appalling, and sheets and blankets filthy with faeces. No electricity meant that the MOs worked by hurricane lamps at night.

For three days Edwin had nothing but salt water to consume, followed by two days of tea, then some rice. He was then released, but suffered attacks for months afterwards. He was lucky – others just wasted away because the lack of food often resulted in beri-beri, and then eating brought on the dysentery – a vicious downward spiral. But generally Changi barracks prison was relatively clean and dry, spacious and had sea views – compared to what came soon after in Thailand/Burma. Edwin was made a clerk in the 'Bureau of Records' during this time and so had an essential role in compiling a list of all the men captured, so invaluable after the war for the information it gave on their fates.

In June 1942 a friend of Edwin's escaped with two Gordon Highlanders; this friend spoke Malay and knew the country. They crossed the Strait to the mainland by boat and proceeded north, intending to meet up with known guerrilla bands fighting the Japanese. Some Malays helped them on their way, but one betrayed them. The friend and one Gordon escaped, although the Scot was wounded, and hid as the Japanese rushed them, but the other was captured and beheaded in front of their eyes. After four days they gave themselves up; the wounded man died at Changi, and Edwin's friend got off free.

Leaving Changi

On recovery, Edwin was shipped out with a group of prisoners in May 1943. With almost thirty men and their kit packed into each steel cattle carriage, they travelled for five days through Malaya, bruised and baked and crammed in. There were occasional stops for toilet and food and at one place the engine water-feeder was turned on and they all scrambled naked to wash – much to the amusement of local male and female bystanders. After two days at Bampong transit camp, which was a mass of filthy pit latrines, mud floor huts and hysterical locals trying to buy their army kit, they marched for two nights to Kanchanaburi/Kanburi, with the promise that their kit would be sent on; it was never seen again. Many men were already sick and weak but morale at this time was quite high and attempts were made at singing while they marched, to keep spirits up, as the fit helped the ill.

Tarso

They spent five days in the open in Kanburi, again sleeping in the mud and pouring monsoon rains; Edwin was lucky to find a place in the sergeant's tent – forty men in a tent meant for twelve; when he woke after a sound sleep, he was in six inches of water. Then they proceeded by open train-trucks up the railway to Tarso camp, where Edwin was sent in a detail of seventy men to build a new encampment in the jungle. This involved mass clearing of thick jungle, levelling ground and putting up leaky tents in which they built bamboo shelves for beds – excessively uncomfortable and harbouring lice – and where they lived for several weeks.

The Railway

On the second day, work began – twelve-hour days, seven days a week, carrying heavy sacks of gravel from the riverbed to cement mixers for bridge building. The distance was four miles, through four deep ravines, each with steep, jagged, slippery sides. Not only was this dangerous but it also poured with rain most of the time; each journey was a martyrdom. Edwin fell three times, badly cutting his legs; his finger was broken when he failed to move his hand in time when a log was lowered; his spirit sank to the depths.

At midday came a two-hour break, with a miserable portion of rice and greens, then work until sunset. On arriving back at the camp, there was another miserable meal of rice and dried fish. And so it went on, tiny portions at breakfast, exhausting work, starvation rations, continuously soaked clothes and boots, and the only release was sleep. When the sun did shine, they were dehydrated within minutes. Even worse was the humiliation and utter boredom of the work; they did not even have the energy to speak to each other of home; their minds became dulled.

One morning Edwin was picked to help with the concrete mixer – an easier job, but he was constantly running about at the beck and call of the Japanese guards, who screamed at him if in his weak state he made a mistake or could not lift a sack. They shouted at him to hurry ('Speedo!) and derided his emaciated body while they showed him their muscles, continuously shouting at him in their bad English: 'No F—g good!' After two days he wangled his way back to the gravel-carrying; better to be exhausted than have a broken spirit.

Within a few days his cut legs and broken finger began to ulcerate, so he was moved to 'light work', putting river gravel into baskets for chain-ganging to the ravines, knee deep in water in his ragged clothes. Cholera began in neighbouring camps but did not reach them, but Edwin, like others, had dysentery and malaria. Edwin is one of the few witnesses to say that his particular Korean guard was not too vicious.

Tenchang and Tamarkan

After six weeks the bridges were completed and the men went to Tenchang in July 1943; from here many moved to other camps, but Edwin was so ill that he was allowed to stay, so that he might recover. Despite treatment, his ulcers grew but he still had to do one appalling job: help carry dying Tamil cholera patients to the edge of mass graves, wait until they died and then throw them in.

Whenever the Japanese insisted on taking sick men to work, which was invariably the case, the doctors would protest, and would be beaten for doing so. The swarms of flies made their conditions even worse. Soon, parties of sick were taken south, but some were housed with cholera-ridden Tamils and died of the illness.

Edwin's cattle truck took him to Tamarkan; arriving at 3 a.m., they were harried by ill-tempered guards for *tenko*, and generally slapped around. There was a so-called hospital of atap and bamboo huts crowded with hundreds of men. The death rate was huge, as all kinds of infections spread like wildfire with virtually no medicines available. Once Edwin was eating his small food ration when the man next to him gave a death rattle and died; Edwin covered the soldier's face and resumed his meal, so common was this awful occurrence.

Edwin grew worse with malaria, hunger and then beri-beri; his friends despaired for him, but he insisted on playing a rubber of bridge – and won.

At Tamarkan, lice were the worst pest, together with rats, which often woke patients as they tried to eat the rice poultices from their ulcerated wounds. Edwin's malaria and ulcers slowly got better, thanks to an Aussie orderly whose only training was treating sick cattle and horses in the Queensland outback! But morale remained good and one night a heavy rainstorm blew down the ulcer ward and the men lay there in the rain, with the ruins around them, singing 'Roll out the Barrel'.

Kanburi Again

In mid-October 1943, Edwin was moved with others back to Kanchanaburi 'hospital', but it was much worse as rain made it a quagmire. Often their only drink was half a pint of tea in twenty-four hours– agony for malaria patients. They only had a wash if a mate brought them water from the river, up to two miles away.

Deaths were up to eight a day. The Australian surgeon, Dr Fegan, excised up to twenty-five major ulcers a day; his skill and patience and kindness saved many a life and many men from despair. He would gently ask a man about an amputation, suggesting that it would be best and probably save his life; the men always replied that they trusted him and that it was his decision, and in this way they had a say in what was to happen. Suddenly, in mid-December, they were taken to Singapore.

Changi

In April 1944, Edwin was moved to Changi old (convict) gaol (not the barracks), where civilians had been kept previously. Though conditions were much better than on the railway, food was always in short supply and 90 per cent of it was rice (which the ingenuity of the cooks turned into various hashes or pasties, to make it more palatable) with tiny scraps of fish, or greens occasionally, usually home-grown in the POW gardens; as the Allied blockade took hold, however, there was less and less food. Even at the beginning, anyone who was sick got either nothing or very little from the Japanese, and so all rations were shared. The results of all this vitamin deficiency were beri-beri and pellagra, which opened the body to other diseases because resistance was so low. Life-saving items thus included Marmite from Red Cross parcels, rice husks which the Japanese threw out, and weavil-ridden ICI fertilizer called Vita-Meal. In addition, the grey walls, rafters and pipes were totally ugly and depressing; 6,000 men crowded into a gaol meant to hold 600. Three men shared, with one on the concrete bed and one each side of it. Others slept in the dining rooms, on and under concrete tables, in workrooms and on the wire mesh which ran down the centre of each floor level.

There were minimal Red Cross parcels (four in three-and-a-half years) and Edwin unfortunately missed out on two in 1942 because he was ill with dysentery. After one year he shared one with seven men, and finally, after one more year, they all had one parcel each, which seemed like a goldmine, but the officers wisely decided to spread these over ten weeks, and so twice a week they had a Red Cross night, with one parcel between twenty men. These were evenings to remember and although they did little for the men's health, they had an amazing psychological effect on personal well-being. The powdered milk was invariably given to the TB patients, because as the saying went amongst them all, 'there is always someone worse off'.

After four weeks Edwin was transferred to the atap huts outside the gaol, where it was bright and tree lined. Here he was company clerk and adviser to Sergeant 'Buck' Taylor. The main work was building the aerodrome at Changi village, begun in August 1943, and by April 1945 it was in use by the Japanese air force.

Rackets

Many NCOs and officers were involved in cartels involving stealing – all as a result of scarcity. One RASC major openly admitted to Edwin that he was making a lot of money selling medicines which should have gone to the hospital for the common good; one MO and a padre were in a food and drug racket together but Edwin did not name them publicly, at least in his statement. Another MO pointed to a sergeant major and said he was living proof of how good Marmite was! A cook could easily be picked out because he was always fit and fat. And officers were all much fitter than other ranks, because they had a better diet and did less work. One MO examining liberated POWs said that some cooks were fitter than men he had examined as recruits in 1939!

When it was decided to give a 1944 Christmas present to the civilians in the neighbouring Changi gaol, the POWs were asked what they would like to receive, if they had the chance; all ORs said food and the officers and NCOs said sewing kits, books, and so on.

Some ORs also did well but they took greater risks because they traded illegally when on working parties with locals, in Singapore; if caught by the Japanese, it could mean death. They also often shared their earnings with the hospital patients and bought them food, unlike some of the officer cliques.

Stealing from the Japanese and selling it on, or selling one's own items, was allowed, but persuading men to part with food for cigarettes, for example, was a menace, as it would eventually make them ill and use up precious drugs and medical attention. Others stole from their own men and then sold it on; this was heavily punished by imprisonment within the compound, and a rice-and-water diet. But most POW's engaged in some racket or other to survive; some gambled and made or lost money; others issued IOUs at high interest and recouped it after the war and made money that way – in many ways it saved lives, as the signatory was able to buy food and survive.

Edwin had one flutter by obtaining old notepaper to sell as cigarette paper – a valuable commodity. But an officer saw him and threatened to hand him over to the camp police; it never transpired, and Edwin ditched the paper and never got involved again. A starvation diet in the prison might have killed him.

Selling clothes for food, ripping them up for bandages and dropping items on long marches to lighten loads meant that most men were in rags by the time of the liberation. At one point, all clothes above a basic minimum were declared 'Surplus Camp Property' and ordered to be handed in for distribution to those who had lost theirs. Edwin got a sleeveless shirt out of this system, but noted that officers were exempt from this scheme and could keep as much as they wanted of their clothing. A man who lost any minimum kit or failed to get it repaired was punished.

Towards the end, most men were like skeletons but many actually craved rice as the main food and life-saver, as other food simply disappeared towards the end of the war, and many even continued to eat rice after the liberation.

But the hospital at Changi was much cleaner than when Edwin was first there in 1942. There were real beds, and bandages and drugs, though the lice remained rampant. Ingenuity was amazing; toothbrushes were made of coconut husks, soap from wood-ash and palm oil, and gramophone needles from barbed wire.

Outside Contact: June 1944–May 1945

With no radios, newspapers or contact with locals allowed on pain of death, and only twenty-five words in the rare letters received from or sent to home, it was bleak indeed. To feel in touch, Edwin often used a latrine nearest the fence so as to watch normal life in the Chinese village, or looked at an old 1941 magazine showing people in shops, a baby, a white woman, a queue for the cinema – anything to get away from the same faces, the grey walls; even a 1941 stock market list was light reading. And being cut off from home was heart-wrenching; were their families worrying about them? Were they safe from the V1 and V2 bombings? What had happened since the previous letters (which had been several months old when received) were written?

A radio was in operation in Changi almost continuously – known as 'the canary' (as opposed to rumours, known as 'boreholes', because the latrines were from where such chatty rumours often spread through the camp). Edwin was in touch with the operators because one radio was hidden in the office where he worked and so he got headline news, but often, as it was passed along, it became garbled; so 'surrender of Germans in Warsaw' became 'surrender of Germany' by the time it reached many men. As result, when the big news of Japan's surrender came, it was received quite unemotionally, as many feared it was another borehole or garbled canary. Under a mosquito net, Edwin often heard through the single earphone a strain of a waltz or a Spanish news bulletin; if an alarm was called, the radio was whipped away and hidden in a tool shed until the danger of a passing guard had gone.

The worst enemies were boredom and depression, although these were not very common. One man committed suicide by throwing himself into a borehole – an appalling way to die. Edwin noted that great efforts were made to remedy this debility; he himself kept a diary and produced educational quizzes and question evenings for his men, as well as giving lectures in basic German, advertising, 'Jewish characters in English literature' and writing; he organized discussions and was briefly 'Chancellor of the Exchequer' in the 'Changi Parliament'. He also organized a newsletter which contained excerpts from letters the men received from home with interesting titbits.

The concerts were famous, with sets by Ronald Searle, the artist and later famous inventor of 'St Trinians' (who died in January 2011). As the war worsened for the Japanese, however, they ordered that the concerts must be stopped, out of vindictiveness. There was a circulating library and this was a 'life-saver' for many – but many books were spoiled by page extractions being made for cigarette and toilet paper.

Most thrilling was the first air raid by the Allies, on 5 November 1944, when 100 B29s raided Singapore, against which the Japanese anti-craft fire was ineffective because the aircraft were so high.

Work on the airfield continued and was hard, but not compared to what Edwin had done in Thailand. At the end of the war, it was very useful for receiving parachute drops of medical teams, and was renamed 'Liberation Drome' by the men.

Other work that Edwin did was cutting logs, but he never discovered what they were for. Other men were sent to dig tunnels for what they thought would be guerrilla warfare by the Japanese – but it came to nothing.

Liberation

On 11 August 1945, the news came of the Japanese surrender. Initially there was disbelief, but a man with whom Edwin had quarrelled and whom he hated came into the hut and grasped his hand and thrust a cigar in his mouth, saying, 'We're going home.' And that was that.

After the war, Edwin wrote a novel based on his experiences, *Shadow of Hanaki* (Angus & Robertson, 1981), and he became a top advertising and marketing executive.

917084 Bdr Jacob 'Edward John' Pearlman, RA

55. 917084 Gnr Jacob Pearlman, RA, taken in his later years in Israel.

Born on 17 June 1917 at 6 Belvedere Road, Sunderland, Jacob became an upholsterer but enlisted in May 1939 before the outbreak of war. He was posted to the 125th Anti-Tank Unit, serving around Britain until the end of 1941. He was then posted to Singapore, travelling by ship via Canada, Cape Town and Bombay. From there they went by truck to the east coast of India and then by the SS *Empress of Asia*, a former coal cargo ship, to Singapore.

Singapore

Outside Singapore the ship was dive-bombed by the Japanese and many were killed as the order to abandon ship was given at about 1 p.m., in shark-infested waters. Few had lifebelts, there were no lifeboats and the water was freezing. After four hours they were picked up by small boats from Singapore, but within ten days they were POWs.

Jack remembers that the men were aware of the Japanese reputation for cruelty and so fear was very real. They were marched to Changi gaol. It would be a year before they were allowed to write a card home to tell loved ones they were alive. While working in the docks at Singapore, Jack took part in minor sabotage, puncturing Japanese fuel drums awaiting shipment to Japan.

Jack remembers well the day when the Japanese demanded that every soldier sign the notorious order to agree never to try to escape, contrary to all international laws of war. When the men refused they were placed in overcrowded barracks and left to rot until disease meant the agreement had to be signed. This is well told in David Arkush's testimony.

Food was two small bowls of rice per day; Jack lost half his body-weight by the end of the war. Disease and hunger killed thousands of POWs. After a short time at River Valley camp, he was shipped to the Kwai in October 1942.

The Kwai

Jack was sent with hundreds of others to Kanchanaburi to slave on the Burma Railway; transport by train in steel cattle wagons was horrific and has been described many times. On arrival they had to sleep in tiny damp huts on bamboo mats or bamboo shelving. The noise of frogs and insects at night meant sleep was rare. Shoes wore out within weeks and they had to go barefoot for years; if you had shoes, it was likely that they would be stolen by another prisoner; clothes disintegrated until only loincloths remained.

Jack said that the guards only ever addressed them in barked orders and were bullying and cruel. Even worse is that since the war they have denied what they did and never paid compensation. He will never forgive them for what they did.

At first, the men he was with were employed in woodcutting for building huts and for fuel, but by October the slave work began on the railway. For the breakneck speed they demanded, the Japanese devised a ten-day week with one day off. The Japanese, said Jack, behaved just like the Nazis – men died like flies from hunger and disease and the enervating climate, and back-breaking work cutting trees and rock. As if this were not enough, Jack remembers being strafed and bombed by Allied aircraft trying to stop the building of the bridges; often he was blown into the water by blast, and then percussion from bombs in the water would sweep them onshore or along the river. Many died through friendly fire. Later Jack was at Chungkai until May 1944 and then sent to Linsin camp.

Jack said that survival depended above all on attitude of mind and good friends who helped you through the worst of times, by sticking together, keeping a sense of humour and caring for each other when you were sick; you supported each other and kept up morale. You talked of your families and the determination to get home. They always sang in the evening and always whistled on the march to work on the bridges.

For Jack, although close friendships formed, they did not, for him, last long after the war; memories were too painful and life had to go on.

Liberation

Sometime in August 1945, while at Linsin, after the Japanese surrendered, the camp Jack was at was evacuated by train, but with no guards. Clearly something was up. They went to Rangoon and were met by British forces there, and put in a recuperation camp. Jack had a brother in the army in Rangoon, and when he heard about liberated POWs his brother went to look for him. Jack was in a cinema and his brother just came in and shouted his name. They fell weeping into each other's arms. Jack then journeyed home by ship.

After recovering, he met and married Sonya and they had four children. In 1980 they went to live in Israel in Kiryat Ata, two of their children already having emigrated there.

99729 Capt. Adj. Henry Ellis Isidore Phillips, MBE, 5th Battalion Beds and Herts, 18th Division

My thanks to Henry's son and daughter, who provided two papers that Henry wrote after the war for his regimental magazine, and his war crimes affidavit, which is reproduced below.

56. 99729 Capt. Adj. Henry Isidore Phillips, MBE, 5th Beds and Herts, at his marriage to Vivien in 1940

Henry was born on 30 August 1914, son of Harry and Rachel of 233 Fordwych Road, Cricklewood, London, grandson of Rabbi E.P. Phillips of Glasgow, and he attended Haberdashers' Aske's Boys School in London. He did his BA degree at University College, London, in 1936 and his MA in 1939. He lived at 9 Lincoln Road, Oxford, and married Vivien Hyamson in Oxford in 1940 (officiated by Dayan Gollop, Senior Jewish Chaplain to the Forces), but had already volunteered for the army in October 1939. He was commissioned into the 5th Battalion Beds and Herts regiment. He was posted overseas on 29 October 1941 and eight troop ships landed at Halifax, Nova Scotia, where they transferred to US ships (Henry was on the USS *West Point*) to be taken via Trinidad to Cape Town. The circuitous route was in order to avoid the U-boats. They arrived on 9 December, two days after Pearl Harbor. Due to go to the Middle East, they were diverted, and continued to Bombay, where they stayed for three weeks, and then to Singapore, arriving on 29 January 1942.

They docked in the midst of an air raid and the Causeway had already been blown up. As an intelligence officer (IO) of his regiment, Henry was painfully aware that they had absolutely no knowledge of the Japanese deployments, fighting skills, uniforms or weapons or the Malayan terrain. As adjutant to the colonel, Henry well remembers the night of 14 February when the colonel returned from HQ with a bottle of whisky to fortify them all, ordering Henry to command the withdrawal to Singapore City and Fort Canning HQ; within hours they were ordered to surrender.

POW

After the surrender, Henry was in camps in Singapore but in November 1942 was moved upcountry to the Burma Railway camp at Tarsao, where he was the POW camp IO, and also IO for the whole Burma Railway. As adjutant, he was responsible for keeping the records, but it had to be done secretly, usually at night, as the Japanese wanted no notes kept of the POWs' experiences. His diary remains a permanent record of the horrendous conditions the men suffered and the death roll that resulted. He was also involved with the important morale-boosting gathering of news from the outside – a capital offence if discovered – on the secret radio. He listened at night to the BBC on a secret radio; in order that the Japanese should not become suspicious, only a select few knew of this radio's existence and what was broadcast. In the first few months it was relatively easy to smuggle radio parts into camps and hide them in ingenious places. This network of radios and written/spoken messages existed along the whole line of the railway camps and Henry was one of the key members of the group who relayed news to POWs in the camps where he was incarcerated.

Knowing that some prisoners spoke Malay and Thai, Henry used other methods of news-gathering, and grouped around himself men who would get hold of local language newspapers, when on working parties outside the camp – from local people, for example – or brought them in to camps when delivering food; also those who would translate war-related material; then at meetings they would assess what was important enough to pass on to the intelligence movement in the camp, for transmission to the POWs. This activity was regarded as essential in order to keep hope alive and to keep up morale in the horrific conditions in which they lived, especially as few letters from home were ever received. Only a spoken report would circulate, and it was recited only by a select few; anything in writing was memorized and soon destroyed. At each camp they were sent to, Henry performed this important and dangerous role. Radio batteries were smuggled in by friendly locals in hollowed bamboo, rice sacks and vegetable crates and some were 'liberated' from Japanese vehicles. In addition, his work ensured that the Allies knew what was going on inside the dreadful Japanese camps, and where the camps were located. In one camp a POW corporal who worked in the Japanese officers' mess would 'borrow' Japanese newspapers and they would be translated by Japanese speakers among the prisoners, and secretly returned to the mess. Some POWs went 'under the wire' at night and obtained food and medicines from local Thais and Chinese, as well as newspapers. These were paid for with a fund from officers' wages.

Another dangerous task was the job – taken in turn by various officers, including Henry – of overseeing the building of bridges on the Kwai, specifically at Kanchanaburi. This was because, throughout 1945, Allied planes would bomb without warning and POWs were often caught in these attacks and killed or wounded.

The Japanese knew that information had to be getting into and out of the camps and a sudden search was ordered on the same day along the whole Railway Line camp complex, by the Kempeitai. In Henry's camp a disgruntled Korean guard warned them, and nothing was found.

In early 1945, officers were separated from ORs. Henry was sent for on 21 February by the Japanese Kempeitai (Gestapo-type police) at Kanchanaburi, who used torture to extract information from POWs about how they gathered news. They were getting nervous about the approaching Allies and how the war was being lost by Japan. Somebody had leaked Henry's name – probably an officer who had been bringing news-sheets from a camp where a radio had been found. He was taken to a small prison and thrown in a cell. During interrogation, one Japanese officer smashed Henry on the head with an axe handle, seriously wounding him and knocking him unconscious. He needed several stitches on the wound afterwards. He gave nothing away under this brutal treatment, but cleverly confessed to receiving news-sheets, so satisfying the Japanese that it was true. They stopped their enquiries and other secret sources were never found. Henry was later released.

Later in February he was arrested again by the Kempeitai and sent on 7 March to Bangkok prison, where various POWs accused of so-called 'crimes' were kept. Many were beaten badly. Several weeks later, on Good Friday in April 1945, Henry was taken to be court-martialled.

Trial

Henry would not be intimidated. He was allowed to speak on his own behalf and told the Japanese what he thought of their horrendous methods and morals, their cruelty and their disregard for the Geneva Convention. He was given two years in gaol; the other POWs told him that they did not know how he got away with not being executed. It was clear that he would not serve two years as the prisoners knew from secret sources that the war was coming to an end; they even saw the Swastika flag at the nearby German Embassy lowered after VE day.

Not content with sentencing him, the Kempeitai accused him of holding secret prisoner meetings in the prison, and sent him to Singapore in July, in a cattle truck, sitting on sacks of rice; the unidentified train was frequently attacked by Allied planes. In Singapore he was sent to Outram Road gaol, known as 'the Belsen of the East', an old colonial prison where he shared a cell with two Dutchmen. He was put to work in the gardens, but was still gathering clandestine news from among the POWs there. It was there he heard about the dropping of the atomic bombs.

Overnight, the Japanese became friendly and the starvation diet improved, with the distribution of condensed milk; they were then sent back to Changi gaol in the north of the island. Here he celebrated Rosh Hashana (Jewish New Year) before being sent back home in October 1945. He was awarded the MBE for gallant and distinguished service as a POW of the Japanese.

Post-War Life

Henry moved to Nyasaland (Malawi) as ADC (Assistant District Commissioner) for the Karongo area and later became a senior Treasury official in the Federation of Rhodesia and Nyasaland in 1953, and was appointed CMG (Companion of the Order of St Michael and St George) in 1960. He became a close and trusted advisor to Hastings Banda, the first prime minister of Malawi. He left Malawi in 1964 and was knighted for his service to the Commonwealth. He married Philippa Cohen in 1966, his first marriage to Vivien Hyamson having been dissolved. He died in December 2004, leaving children David, Angela and Robert.

The affidavit that Henry gave to the Japanese War Crimes Tribunals in 1946 is reproduced here, courtesy of his family.

AFFIDAVIT

I, HENRY ELLIS ISIDORE PHILLIPS, Esquire, Member of the Military Division of the Most Excellent Order of the British Empire, Assistant District Commissioner of the District of KARONGA in the NYASALAND PROTEOTORATE, and sometime Captain and Adjutant of the Fifth Battalion The Bedfordshire and Hertfordshire Regiment, make oath, and say as follows.

I was taken prisoner by the Japanese at the fall of Singapore on 15th February 1942. After passing through various prison camps in Malaya and Thailand, I was posted to Number 7 Camp, Thailand on 23rd January 1945. During my time in Thailand I was concerned in activities designed to bring into the camps by secret means news of the outside world. As a consequence of a leakage of information I was arrested by the Japanese Military Police on the 21st February 1945, and was taken to the Headquarters of the KEMPEITAI at KANBURI.

On my arrival I was interrogated by a Sergeant-Major through an interpreter whose name I afterwards discovered to be TAKAHASHI. I have been shown four plates of photographs of Japanese, and I now recognise the man portrayed in photograph number 19 on plate number 9 as the aforesaid Sergeant-Major, and the man portrayed in photograph number 20 on plate number 10 as the aforesaid TAKAHASHI.

Within a few moments of my arrival my head was laid open by an axe-helve thrown by the aforesaid Sergeant-Major. I was subsequently incarcerated in a wooden cell measuring about 7 feet long by 3 feet wide, in which were three other men, two of them natives. Neither blankets nor mosquito-nets were provided, and the cell was alive with bugs. There were three slender meals of rice per day. In common with other European prisoners in adjoining cells I remained thus for five days, during which time I was allowed no washing facilities whatsoever and was released from the cell only for interrogation or in order to go to the latrine, an act which had to be performed in handcuffs.

While I was in the cell I watched the beating of the following Europeans, to wit, Captain McDERMOTT, Lieutenant PEARSON, Flying-Officer KINGWELL, and Private FIELD. They were beaten with bamboos, thrown several times to the gravel, and kicked mercilessly while they were on the ground. Captain McDERMOTT sustained a severe injury to his groin which made it difficult for him to walk thereafter. The aforesaid Sergeant-Major, assisted by the aforesaid TAKAHASHI perpetrated these outrages. Later I discovered that Lieutenant-Colonel LILLY, who was in the next cell to mine, had been made to stand at the door of his cell with his hands tied through the bars and he was kept thus without food or drink for 36 hours. I subsequently saw lacerations on his back indicative of a recent whipping, and he had also sustained an injury to his foot.

During the course of this five days of incarceration the Major commanding the Kempeitai, and another senior officer whom I recognize portrayed in photograph number 21 of plate 10, used often to walk through the courtyard.

The case against me and my associates was completed without our being given an opportunity to make a statement in our defence, in spite of representations on this score made personally by me to the aforesaid TAKAHASHI. We were then transported to the Detention Cells at BANGKOK arriving there on the 8th March. I, together with nine other Europeans, was locked in a cell about twelve feet square, in the corner of which was a native-type latrine with a totally inadequate box-container emptied from the outside. There were other similar cells adjoining mine, some containing other European prisoners, and some Japanese prisoners. The maximum number of the latter allowed in a single cell was 5. These cells were in the charge of two Warrant Officers who were on duty on alternate days. I recognise these to be the men portrayed in the photographs numbered 176 and 177 on plate 20. The prison guards were the persons whom I now recognize as portrayed in photographs numbered 178, 179, 180, and 181 on plate 20, and in photographs numbered 182, 183, 184, and 185 on plate 21. They did tours of duty on alternate days. I do not know the names of any of these men, and so, for convenience, I shall refer to them hereafter in this my affidavit by their photograph numbers.

Of medical attention there was next to none, and, in the three months of my stay in these cells, a doctor came on only three occasions. The most usual method of treating any sort of ailment was by cutting down rations. One of my associates, Private Christian DILLING of the Netherlands East Indies Army, was taken seriously ill with malaria, and two out of the three doctors' visits were made because he collapsed. His treatment was to be put into a cell by himself where nobody could help him. I saw very little food being taken to him and nothing which appeared to be medicine. Early in May he was sent down to Singapore. I could see that he was in a very emaciated condition, was unable to stand without support, and was obviously unfit to travel. I saw him later at OUTRAM ROAD GAOL, Singapore, and he died on August 12th.

A Chung-king agent was imprisoned in a neighbouring cell. His prison number was l05. I personally saw Warrant-Officer number 176 take a chain covered with cloth into his cell, and I subsequently heard sounds indicative of a severe flogging. This and other tortures, of which I can submit only aural evidence, were meted out practically every day when this Warrant-Officer was on duty. I personally saw the agent on more than one occasion through the bars of his cell, and can bear testimony to the fact that the upper part of his body was covered with lacerations and bruises. On 5th July he was removed from his cell before dawn, and was taken away to be shot.

Of the guards the cruellest was no. 178. I was beaten by him about the head and face, and kicked by him on numerous occasions, and I also saw him inflict similar punishment on many of my associates. Guard no. 179 and Guard no. 184 were also accustomed to lay about them with their hands and I was hit several times by them. Such treatment usually arose out of linguistic misunderstandings. I am not prepared to accuse any of the other guards individually; but one and all at various times refused to listen to our complaints and in particular would make no effort to supply adequate medical treatment. Lieutenant MITCHELL, who was next to me in my cell, had an ulcer on his leg for many weeks, and had to treat himself clandestinely with rice poultices saved from his own rations. Never once did an interpreter come to the gaol after we had had our preliminary interrogation there; and several requests for pencil and paper with which to send complaints to an interpreter were ignored by the two Warrant-Officers to whom the requests were addressed.

Air-raid precautions were inadequate. A mudded air-raid shelter measuring internally about 6 feet long, 3 feet wide, and 5 feet high, had been constructed; and into this on two occasions all 26 European prisoners were crammed. When a warning was given, the practice was to open but not to remove the padlocks on the cell doors. Whether or not we were taken out to the doubtful protection of the shelter depended upon the Warrant-Officer on duty. No.176 made it a rule not to do so; no. 177 occasionally took us out. The Japanese prisoners in the gaol were invariably taken to a shelter.

A brief tour of inspection was carried out from time to time by a Major, who was also the prosecutor at my trial by Court-Martial. He never asked to speak to any of us and communication with him was impossible because of the lack of an interpreter.

I left BANGKOK for SINGAPORE on 5th July in the company of 10 other Europeans. The train journey lasted twelve days. Our arms were secured by ropes behind our backs for the whole of this time and we were also handcuffed in pairs every night and for the whole of the last 36 hours of the journey. We were accommodated on the top of rice sacks, up against the roof and in the furthest corner of steel cattle trucks, in conditions of almost indescribable discomfort and squalor. During the whole journey we were given facilities for a proper wash on only one occasion. From PRAI to SINGAPORE we were made to travel sitting bolt upright by day and by night.

Our rations averaged between a half and three-quarters of a pint of rice per day and an infinitesimal amount of stew. The guard-commander in charge of us was a Sergeant of the Bangkok KEMPEITAI whose photograph I am unable to recognize among those exhibited to me.

Among my associates at KANBURI and BANGKOK were, Captain W. McDERMOTT of the Indian Army, Lieutenant C.W. PEARSON of the Intelligence Corps Malaya, Lieutenant W.S. MITCHELL of 18th Divisional Royal Army Service Corps, and Private C. FIELD of the Straits Settlements Volunteer Force. Members of the Netherlands East Indies Army were Major P. GERSON, Captain F. MUSCH, Lieutenant P.J. KOETS and Sergeant M. TRAP, and all these four also accompanied me on the train journey to Singapore.

When the photographs referred to in this my affidavit were shown to me, the names and personal particulars of all the persons portrayed therein were permanently obscured.

SWORN by the said H.E.I. Phillips Esquire,
at KARONGA this 13th day of February 1946
BEFORE ME
A COMMISSIONER FOR OATHS

2361728 L/Cpl Abraham (Alfred) Posener, RCOS

57. 2361728 L/Cpl Abraham Posener/Posner, RCOS (standing, far right).

Alfred was born on 15 October 1915. His father died when he was 18 months old and his mother Sarah (née Bernstein) remarried and became Mrs Ike (Isaac) Klein. The family ran the Still and Star public house at 1 Little Somerset Street, Aldgate, London, which is still there, at the time of writing, next to the Aldgate bus station. Alfred attended the Westminster Jews Free School in Hanway Place, London W1, and grew up in the Tottenham Court Road area.

When Alfred joined up he was working as an electrician and was attached to the 66th Heavy Anti-Aircraft (RA) in February 1941. His family were then at both 41 New Church Road, Hove, and 86 Second Avenue, E12.

He was sent by troopship to the Far East via South Africa and arrived in Singapore on 13 January 1942. At some point during the Japanese invasion he was evacuated to Java and was in hospital with a cataract in one eye. The hospital was not evacuated as fast as other troops and he became a prisoner. Twice during this period he was officially lost, presumed dead. On 15 February he had been able to send a telegram home to his mother from Batavia in Java, and it was she who told the War Office that he was alive. He remained at Tanjong Priok camp until October 1942 and then was shipped to Fukuoka (Moji) Japan and was there by February 1943.

Alfred's experiences at this camp were similar to those described in many other POW testimonies.

Soon after he arrived in Japan they called Alfred in to fix a light, as he was an electrician. He did not understand them, so they hit him and broke his nose. Later they realized their 'mistake' and as compensation he received half a cigarette. He was working on a lathe and was shown how to reduce the size of steel rods to make railway axles. When the guards were not looking he would deliberately over-cut them to make them useless. He spent some time in punishment for that.

He was working in the docks. Food was scarce for everybody. If prisoners could grab some rice from a sack, they would give it to the guards, who would cook it and return half. Occasionally they would find a rattlesnake, which would have been a real treat to eat. One of his favourite stories concerned a Japanese guard who spoke English; on discovering this, they held a conversation within his hearing to describe the terrible things the Germans (their Allies) were doing to the Jews, but describing the Jews as 'Orientals'. This was to plant in the Japanese minds the fact that the Germans actually hated the Japanese too. There may have been a lot of truth in this German dislike anyway, as the Japanese were far from the German racist ideals of 'Aryans'.

In the docks they were loading and unloading ships. If they wanted a break, they could grab a handful of dirt and stones and throw it in the crane's gears. Most cranes would then fail, and this would take a few hours to repair. Unfortunately, the Japanese had purchased some British-built ships before the war. The cranes were working twenty-four hours per day.

When the Americans began to bomb Japan, the prisoners were put on fire service. They were taken to the towns where the fires were raging and given manual water pumps, with two prisoners either side pushing down in turns. They soon found that if all four pushed down together, the pump would break.

The Japanese were almost relieved when their town was bombed as they knew the bombers would not return.

Alfred saw the second atomic bomb go off in Nagasaki. Ten days later he went through the flattened ruins to an America ship, and was on his way home.

When the Japanese surrendered, the prisoners lived in the camps with the gates and doors open so they could come and go freely. The camps were marked 'POW' so they could be seen from the air and not bombed. The biggest shock the Japanese had was not the bomb, but the fact that within forty-eight hours of the surrender, the bombers were dropping containers of food to the POW camps. For a cigarette or two, they could get an ex-guard to clean the camp. One Japanese guard drew a pencil portrait of Alfred. One of the worse things that happened was prisoners gorging themselves – their bodies could not take the food.

Alfred was six feet tall and down to seven stone or less. From Japan he went by ship to one of the Pacific Islands, where the Americans gave him a medical, and then on to San Francisco. The troops were then taken on a six-week trip across America/Canada to 'fatten up' before coming back on the *Queen Mary* and arriving in the UK on 5 November 1945. As the troops disembarked, the ladies from the Red Cross gave them each a pack of ten cigarettes.

Out of the 200 soldiers with whom Alfred enlisted, only four survived. In February 1946 he married Fay Polishack and the couple moved to 35 Sigdon Road, Hackney, London E8. After the war, Alfred sold some of his Japanese postcards he had sent home to a stamp dealer, to raise money to buy himself a camera, his own having been stolen by a Japanese guard. He visited Israel many times, a country he passionately supported, and was chair of his local AJEX branch in Hackney/Stamford Hill.

K34771 Sgmn Jack (Jacob) Rose, RCCOS, CEF

58a. K34771 Sgmn Jack Rose, RCCOS, taken before he went overseas.

Jack was the son of poor, working-class Jewish parents, Benjamin Rose and Jean (née Robinstone), born in Vancouver on 4 January 1920. Jean and Benjamin met in Scotland when Ben was serving in the Canadian RCOS during the First World War. Jack's family lived at 2356 14th Avenue West. He went first to Tecumseh School and Lord Tennyson School, and then to Kitsilano High School, one of only three Jewish students out of 2,000. He excelled in gymnastics and swimming and aged 15 he won the British Columbia high-diving championships. He secretly became part of a trapeze act in a circus, performing without a net and without his parents' knowledge. He joined the Jewish school fraternity, 'Aleph, Zadick, Aleph' or AZA, as Jewish children could not get into the standard fraternities, owing to racism.

He graduated with honours from school in 1936 and wanted to be a chartered accountant but anti-Semitism blocked his way and he got a job instead as a messenger boy with his father's employers, the Canadian Pacific Railway (CPR), where his father was a telegraph operator. He bought a bike for one dollar from Woodward's and earned three cents for each telegram delivered; he then studied to be a telegraph operator himself.

In April 1940, Jack enlisted, as his father had before him in the First World War, and became a telegraph operator in the RCCOS. His unit, the 10 Fortress Signal Company, did basic training at Victoria and were then posted to coast watch at Point Atkinson lighthouse and York Island, guarding the approaches to Vancouver. Later they were posted to the old Vancouver Hotel in the centre of the city, manning radios and training new recruits. Eventually they were issued with tropical kit and sent overseas on 19 October 1941 on the Australian vessel *Awatea*, escorted by the destroyer RAN *The Prince Robert*, via Honolulu and Manila. They arrived in Hong Kong on 19 November, very poorly trained as soldiers, and marched to Sham Shui Po barracks. He enjoyed the exotic sights of the Orient, with many nights out on the town, and he and some friends even employed a local servant for a few dollars per week, to take care of all their personal needs. Within four weeks they were POWs of the Japanese.

Hong Kong

On the day the war began, Jack and his team of five men were camped in pup tents at Skeats Ground near Castle Peak Road, on an exercise. They saw planes overhead and thought they were Americans until the bombs dropped. Chaos ensued and they were ordered to retreat to Kowloon and cross by ferry to Hong Kong Island. Here he saw his first dead bodies. Dodging enemy shelling, for many days they lived in abandoned houses and foraged for food in a bombed NAAFI, as there appeared to be no organized rallying points. At one stage in the battle they were standing on the balcony of an abandoned

58b. Jack as POW in 1943.

house and Japanese shells were incoming. They hurriedly took cover at the foot of a nearby slope when one shell made a direct hit on the house where they had been standing. When they returned to the house, they discovered the twisted bodies and equipment of several RAF men who had taken shelter overnight in the downstairs room, unbeknown to the Canadians. Their dead bodies were covered in dust, their rifles bent totally out of shape. Eventually Jack and his comrades reached Victoria barracks and it was here that they heard of the surrender. An English sergeant had called out that 'they were packing up' and Jack and his men thought it meant packing to leave; it meant, of course, in English slang, that they were surrendering.

In emotional shock, they were sent back to the mainland and then marched as prisoners to Sham Shui Po barracks, which had been totally looted by locals during the fighting. Jack had kept with him a whole carton of cigarettes he had taken from the bombed NAAFI, and he just gave them away, realizing later that he could have sold these for extra food in the camp. Conditions in the camp were appalling, and the poor diet and tropical disease began to take their toll on the health of the men. Locals came to the fence and trading started – food for watches, pens and anything else the prisoners had to sell.

After about two months, Jack was moved to North Point camp. Among his fellow prisoners was Jewish comrade William Allister, and much of William's story parallels Jack's (see William Allister's testimony in this book). They were put to work building an airfield. Anyone who approached the camp was executed by the Japanese, and Jack and his comrades saw many bodies hung from telephone poles as a warning. On one occasion he saw a woman and child in a small sampan, which inadvertently passed by the camp fence near the sea wall; a Japanese guard simply shot and killed them both with his rifle. Although things were better organized than at Sham Shui Po, still beatings, hard labour, disease and starvation became their daily lot and Jack soon found himself ill with dysentery. For this he was given only chalk mixed with water. From the start Jack decided to keep his mind active and focused and set out to learn Japanese – in fact the Japanese helped him by supplying him with books. However, there was a downside to this because he often acted as interpreter for the prisoners and found himself on the receiving end of beatings by guards when they were angered that a certain instruction had not been carried out to their satisfaction. On one occasion, he said to a Japanese interpreter, 'Winston Churchill number one, Roosevelt number two and Tojo batman', and he got a beating for that.

After a confused start, the officers soon re-established some discipline, with the proper sharing of food, organized activities and scrounging parties; burial parties were also organized to bring in the Allied dead. At Kai Tak airport they were levelling the ground to improve the landing strip. They not only had the very heavy work, but had to walk to the ferry, cross to the island and walk to the airport – then back again at night. Using picks and shovels they moved the waste dirt in carts, pushed them to the sea edge and dumped the contents. The poor diet was one small cup of rice with some weak soup, three times per day, and the sick had to work till they dropped.

While in the camp, many men died around him; Jack would find men dead in their bunks in the morning, from diseases such as diphtheria, or from starvation. He had a bad bout of dry beri-beri, but survived. In the photograph taken at the camp, his beri-beri had so swollen his face that he resembled a Japanese soldier! For a while he was sent to Bowen Road hospital to recover but very little medication was available.

With a few exceptions, the officers behaved badly, showing little interest in the welfare of the other men; they lived in a separate compound and took more food and had very little contact with the other ranks.

Jack was able to send a few letters home and these were heavily censored by the Japanese (his family still have several); but in one he fooled the Japanese by writing that when he got home he was looking forward to a nice holiday at Oakalla. This was a notorious prison in Canada so he was getting the message out that they were being badly treated. The family only received his first letter after one year and so did not know he was alive until then.

Japan

Around January of 1943, Jack was deported to Tsurumi near Yokohama (Camp 3D) in Japan to work in a shipyard as a welder in the pipe-bending shop for Nippon Kokan Kaishan. The men had first been given medicals by the Japanese to ensure they were fit for work, as well as various vaccinations against cholera and typhoid; the Japanese were probably afraid that the men would bring tropical diseases to the Homeland. Jack saw the transfer, with William Allister, as a positive move because the climate in Hong Kong caused a high death rate from disease and this was less prevalent in the cooler climate of Japan.

The journey by Hell Ship was horrendous, with hundreds crowded in the dark hold for days on end with appalling food and sanitation facilities; on the audio tape, Jack found it very hard to describe the filth and horror. On arrival in Japan, they were taken by train to the shipyard camp.

Forced to illegally work on gunboats, the 500 men earned a tiny ten cents a day, enough to buy only one packet of cigarettes a month. Again, Jack had a very low opinion of the officers, who not only did not work, but also made sure that they got a lion's share of the limited food. However, the conditions in the camp were much cleaner and more spacious than in Hong Kong, and the food was better, although there was still a two-mile hike to and from the shipyard each day. They also had access to a large, communal pit hot bath, though these were taken in rota by groups of men, as the water became progressively dirtier and colder as it was used. Working in the pipe-bending forge meant that they could at least keep warm in the very cold winters, and obtain hot water to take back to the camp for tea, from the forge ovens.

The Japanese workers often gave them extra food and cigarette butts; as a result, Jack never really bore the Japanese people ill-will after the war; the guards, however, were quite brutal. The prisoners were, as usual, on a starvation diet, and on the very few occasions when Red Cross parcels were given to them, these were shared – one between several men. Because Jack spoke and understood Japanese, he was able to do black-market deals with the local workers and this brought some extra food to him and his friends – but such deals were heavily punished if discovered by the guards.

Jack's knowledge of the Japanese language also meant that he was given a job in the important tools distribution shed. This was because serious misunderstandings and conflicts arose between the Japanese managers and the POWs about what they should be doing at their work stations each day, and Jack could ease tensions by translating clearly what the men had to do. As he often had some time on his hands, he used this to fashion a brass ring from an old pipe and shone it until it looked like gold; he then sold it to a Japanese guard who was being shipped out and used the inflated fee of five packs of cigarettes to buy extra food for himself and his friends.

On 3 October 1943 he and others were asked to make a broadcast to Canada on Tokyo radio. He was able to get a message to his family and also give the names of POWs who were alive, and these messages were picked up in North America by amateur radio hams and the news was sent to their families. These families wrote to Jack's parents to thank him for what he had done, and Jack's daughter Valerie has the letters to this day. By late 1944, some letters began to reach him from home but they were heavily censored and did not say much.

Sendai

In May 1945, Jack and about 250 other Canadians were shipped by rail to work in a coal mine in the frozen northern mountains at Sendai, near Onahama; Jack described this as a 'hellhole'. There were already some Dutch and Javanese POWs working there and they looked ill and starved. While at Sendai, Jack had a small box containing a ring which his mother had given to him when he joined up, which she had kept since she was 13 years old.[1] When they were ordered to strip to have their clothes replaced, and items such knives and boots confiscated, Jack forgot that the ring was in his trouser pocket. Desperate, he made the dangerous decision to approach the Japanese commandant and ask if he could go to the pile and take back his mother's ring; to Jack's amazement, he agreed.[2]

Wearing only straw shoes, loincloths and headlamps – with leaking acid batteries strapped to their backs, which caused awful acid burns in their skin – they worked in intense heat, 5,000 feet underground. There was no lift, and coal carts powered only by donkeys carried them for the agonizing journey down to the coal face on zigzag sloping inclines. For ten hours a day they mined with obsolete equipment in extremely hazardous conditions, on the usual starvation diet. But at least there were no tropical diseases. Jack said: 'they were always slapping us around for no reason ... [they would] tell us the mine was dangerous and then grin at us and give us a few kicks and punches and tell us to get to work'. A daily quota had to be met and work did not stop until it was filled. At the day's end they could at least wash in a deep pool of hot water down in the mine before returning to the surface. In the evenings Jack tried to think of his family and home in order to remove himself psychologically from his nightmare life, but after a while he could not achieve even this – the cruel regime was so pervasive, robbing him of his memory.

A brutal guard – a sergeant known as 'the Frog' – constantly harangued them at work; when Jack tried to explain that certain jobs could not be done because they did not have the equipment, 'the Frog' beat him.

Liberation

Jack and his comrades remained oblivious throughout as to how the war was going. When it was over, the first that they knew of this was when American B29s dropped leaflets over their camp. The Japanese guards suddenly disappeared. For leaving the camp to explore a nearby village, without permission, Jack was then locked up by his own officers for three days! He could not believe it. While he was locked up, American planes dropped supplies in oil drums, but they crashed through roofs and split open. Jack and his mates were terrified that one would hit their gaol hut and they screamed until they were let out; in some camps men were killed by falling supplies from the Allies.

Jack was asked to accompany sick prisoners with Japanese stretcher bearers to hospital in Yokohama, as he could speak Japanese, and Japanese civilians were ordered to get out of the carriages in order to get the stretcher cases on board. On arrival at Yokohama, Jack was greeted by American soldiers and knew he was free at last. He was in shock. Their first job was to be deloused in specially prepared tents.

On liberation, Jack was taken with about 800 others by the ship USS *Ozark* to San Francisco via Guam. They ate and ate on the journey; he went from 120 to 170 pounds in six weeks, but it was painful because they ate and threw up a great deal at the start, as their stomachs could not cope with twenty-four-hour feeding – which is what they were doing.[3]

From San Francisco, Jack went by train to Seattle and then boat to Victoria, where the men were given a slap-up meal, with silverware and white tablecloths – it seemed like a dream. Finally he went to Vancouver to be reunited with his family

Home

Jack was traumatized when he got home; he could not bear to have too many people around him and had lost so much confidence that he would break out into a sweat. Until the end of his life, his feet ached from the after-effects of beri-beri, especially during the winters. It took him a long time to adjust to the real world around him that had changed so much. For thirty years he had nightmares that people were chasing him and he was cornered in a cul-de-sac, and he would wake in a cold sweat. He would speak Japanese in his sleep and dream he was in prison again. He deeply resented the attitude of the Canadian government, tardy as it was in paying him such a poor pension for the suffering he underwent, whilst others with the same illnesses received more compensation.

In October 1946 he married a Jewish girl, Sarah (Shirley) Davidson and went to work again with the CPR. Later his father-in-law gave him a job selling clothes door-to-door but he hated it and finally found his way into the insurance business, where he stayed until he retired. He and Shirley had three children.

Jack did not get too involved with the Hong Kong Veterans' Association because they only talked of the past at reunions and Jack wanted to live only for the future. He became active in the campaign for redress for Japanese Canadians who had been interned during the Second World War. He saw a strong link between their treatment and that of Jews suffering from anti-Semitism.

Both Jack and Shirley remained members of the Beth Israel synagogue in Vancouver until they passed away. Jack's medals were the 1939–45 Star, War Medal, Defence Medal and Canadian Volunteer Service Medal with Hong Kong clasp. He died in 2006.

Notes

1. He used to wear this ring, but it had been damaged in the fighting in Hong Kong.
2. Jack's daughter still has the box in which that ring was kept.
3. More detail on the terrible experiences of the Canadian POWs with Jack and William can be found in Burke Penny's book, *Beyond The Call*, published by the Hong Kong Veterans' Commemorative Association, 2009.

27005/92365 Maj. Harry Edward Schulman, TD, 5th Royal Norfolks, Malaya

The information here is taken from an IWM audio tape, recorded at South Creake, Norfolk, in 1998, and some further personal information from his daughter Sarah.

59. 27005/92365 Maj. Harry Shulman, Royal Norfolks.

Harry was born in London at 99 Westbourne Terrace, near Hyde Park, on 3 November, 1910, fourth son of a Jewish couple, Heinrich/Henry Schulman and Lucy Augusta Heine.[1] Henry, born in Prague and educated mainly in Zurich, was a professor of mechanical engineering who established the engineering faculty of Sao Paulo University in Brazil, and also owned coffee plantations. Lucy, daughter of a general merchant in Liverpool, married Henry in Manchester in 1898, where they had met.[2] Henry said that his father had visions of being 'a Rothschild'. Later the family moved to Tidmarsh in Berkshire. Harry went to the Dragon Preparatory School in Oxford and after a brief time at Haileybury School in Hertfordshire, including the OTC, he succeeded in getting a place at Cambridge university, aged only 16. But his father said he could go when he was 18, and meanwhile he was sent to spend two years in Brazil, learning the coffee growing business and cattle-ranching as a cowboy. On returning to England, he went to Trinity Hall, Cambridge, in 1929, to study agricultural science and forestry.

India

When Harry refused to return to Brazil after Cambridge, his father cut off his allowance and so he had to find work in London. Through a friend, he got a job with Williamson's & Sons, a firm with interests in India. He sailed for Assam in 1933 to help manage and work on a tea estate. At the European Club he led the life of a member of the Raj, playing polo and socializing in the bar with other Europeans. He also spent time fishing, and sometimes hunting using an elephant with bearers. On one occasion he had to lead a party to shoot a rogue elephant that had killed many people in an area where he had responsibility for some of the tea gardens and its workers. He only wounded it and it took two further separate shootings to finish it off over several days. The local Naga then ate it, and Harry got the 100 rupee reward from the district commissioner but gave it to the native who had guided him to where the elephant had been lying up.

Meanwhile, Harry joined the Territorial Force as a Trooper of the Assam Valley Light Horse at Dibrugar, which had 300 men, each equipped with horse, 303 rifle and sword.

Training consisted of occasional exercises and rehearsing boarding trains to deploy to possible flash points, cross-country marches and mock charges.

England and France

Having always wished to farm in England, Harry left India in 1938 and became an arable farmer at Abbey Farm, North Creake, Norfolk. He joined the TA at Swaffham, as 2nd lieutenant of 'A' Company, the 5th Norfolks, based at Dereham HQ; he was called up in August 1939 as company intelligence officer and was moved to coastal protection work in huts at Wayburn during the invasion scare and a very severe winter.

Skiers and Commandos

Because he had lived in Switzerland and skied there with his father on many long holidays, Harry, now a full lieutenant, was ordered by the War Office to join the 300 men of the 5th Battalion Scots Guards as a ski instructor to be attached to French Alpine troops, and to be prepared to be posted to assist the Finns against the Russians; but he would have to resign his commission – though remaining on officers' pay – as all the British men being formed into this group were officers, and he was to be demoted to guardsman. Among these volunteers was a large number of very tough French Canadian soldiers who were skiers too.

First he was sent to Borden near Aldershot for training, then later, via Dover and Calais, to Chamonix near Mont Blanc to the French Chasseurs Alpins troop centre; here they learnt about battle survival in the snow – specialist tents, use of sleighs, and so on. They had French rations, including horse meat. After three weeks the guardsmen could cross-country ski and they were sent – via a destroyer to Dover and then a train to Scotland – to Gourock. They boarded the captured Polish ship, the MV *Battory*, but the mission was cancelled and the skiing unit was disbanded.

Harry was delegated to be posted to Norway after training near Thetford, but after they had returned to Gourock the Norwegian mission was also cancelled because Norway had fallen to the Nazis. He was picked for the Commandos and agreed to join fifty men of No. 1 Commando in Southampton, based at the Dell football ground, training also on small boats in the Solent. On one exercise, they had to capture the Isle of Wight, but nobody had informed the local Home Guard, who put up a fierce resistance thinking they were real Germans. Nobody was hurt.

Posted to Cornwall, Harry was sent with the Commandos on a raid near the village of Ardlow (?) south of Boulogne, travelling on RAF speedboats. Their job was to extract stragglers after the Dunkirk evacuation. They penetrated the village but woke up the German army, and so had to withdraw. On the journey home, they encountered German E-boats and dive-bombing Stukas in a running battle, but were rescued by the RAF.

After further training, aimed at recapturing Guernsey airport in late 1941 – helped in the planning stage by Guernsey refugees – they set out on two destroyers, followed by small speedboats. About five miles from Guernsey, in the dark, they transferred to the small boats to make the landing. A diversion Commando party was to distract the Germans, having come from a different direction, but they failed to land as the boats hit rocks and several men were killed. As the Germans were now alerted, Harry's party came under heavy fire and his boat was sunk. He was picked up as dawn came, on a choppy sea, to discover that the landing was abandoned. The destroyers had left because of a submarine scare, so the small boats made their own way back to Ramsgate, which was a long distance.

Posted back to the Norfolks, Harry was promoted to captain and second in command of 'A' Company before being posted abroad (with the 18th Division) on the SS *Duchess of Atholl* on 9 November 1941, sailing from Liverpool to Halifax in a huge convoy of battleships. They were transferred to the US Navy ship *Mount Vernon* to go to the Persian Gulf. After four weeks they arrived, via Jamaica, in Cape Town on 9 December, to find that Pearl Harbor had been attacked.

Harry was ordered by the general commanding to be in charge of all the 100 British divisional police and 100 South African police, to keep law and order among the 10,000 troops for the first twenty-four hours in port; he set up HQ in the Queen Victoria hotel, with nine ambulances. The Cape Town population was very hospitable to the troops, but Harry knew there was also a large American presence and he asked the local American military police (MP) commander for his cooperation. It worked well and there was little trouble. But, exhausted, Harry forgot to collect in all the MP armbands he had given out; when he got home after the war in 1946, he was billed for thirty-six missing armbands by the War Office!

After leaving Cape Town, the convoy with Harry's 53rd Brigade was diverted to Mombassa for Christmas 1941, and then was sent at high speed, unescorted, to Singapore, leaving on 28 December.

Singapore

When they arrived in the pouring rain on 13 January 1942, the island was being heavily bombed by the Japanese; the men had no jungle training and were lightly armed, with no transport, and were put into a transit camp at Woodlands and given some very brief tropical warfare training and pep talks. After two days they were sent upcountry by truck to defend a line between Mersing (on the west) and Batabahat port on the east (some 100 miles north of Singapore). 'A' Company was at Ayer Hitam and then sent to relieve an Australian company at Mersing, and there they dug in, in slit trenches, in the jungle. Still without any contact with the Japanese, they were sent back to Ayer Hitam. But 'D' Company was in action and lost its commander and two other officers, so Harry was promoted to major and told to take over 'D' Company at Batabahat, who then took over positions where the Cambridgeshires had taken some casualties in a big battle and left a lot of dead Japanese. After they had spent two days waiting for a Japanese attack, the enemy had infiltrated the lines and so Harry and 'D' Company had to withdraw south towards Singapore, acting as the rearguard for retreating Allies. Harry was ordered to blow the bridge at Senggarang to stop the Japanese, but before he got there his CO had blown it up with many Allied armoured cars and troop carriers stuck on the wrong side; the CO had not known what had happened to Harry's company. Totally unexpectedly, an order was given by an Indian Army brigadier, named Shalon (?): 'Every man for himself.' The Malay Scouts volunteers could not believe the tragic order to destroy their vehicles. They wanted to attack the Japanese and make a fight of it, as did Harry and his men.

Harry decided to take two men and make for the sea and then try to get to Singapore along the coast. But on the way he picked up hundreds of stragglers in the jungle, and he was the senior officer. Trapped by mangroves, the Japanese were also amongst them, but they got to a village and used their radio to call Singapore. Two shallow draft boats were sent to pick them up – they now numbered many hundreds. But even though the boats could not get closer than about half a mile, they managed to get everyone aboard and made it to Singapore.

In Singapore, Harry was sent to guard the naval base, as the enemy rushed the Causeway to the north. They were constantly shelled, and looting by troops and locals was widespread. After about a week they were ordered back into the town, with Harry's company as rearguard again. After they had set up positions, and while Harry was inspecting his machine-gun posts with his RSM and orderlies, a Japanese plane dropped a string of bombs and killed several of his men and blew him and others into a ditch. Harry was badly shaken up, but carried on. Then orders came to Harry to move and hold an area on the perimeter of the town at Braddell Road, in a Chinese cemetery. Suddenly, the order to surrender came. They were amazed. An order also came that five men from each unit were to form escape parties, as it was traditional that in such situations some men from each unit would try to get away in order to form a core to re-establish the unit, once free. He remembers that his CO very quickly chose the first five men around him and ordered them away to the collecting point at the YMCA. Harry said that the Norfolks group did escape to Sumatra and got away to the UK via Australia.[3] Meanwhile, the rest of the men met at a rally point and then were marched to Changi barracks gaol.

Changi

So far, no Japanese had actually intervened with the men. Despite desperate overcrowding, units soon got organized, but food began to run out. Sikh guards who had turned traitor appeared and were especially brutal to the men.

Soon after their arrival, most of the officers were taken for individual interrogation by the Japanese at the main Changi gaol. For about an hour, under armed guard, they were asked – by fluent English-speakers – very general questions about what they thought of Japan, the Japanese, why they had joined the army and what were things like in Britain. In one answer, Harry said that he had been led to believe all the Japanese were short and wore glasses with thick lenses; they were not pleased. Eventually the officers were dismissed.

For twenty-one days after the rice ration was distributed, as the army rations ran out, Harry was severely constipated as the change of diet took hold. Many found that added salt or sugar would enable the body to cope with rice better and enable it to be swallowed more easily.

At Changi, Harry's agricultural knowledge was of great use and he was heavily involved in managing the gardens, planting fast-growing vegetables to provide vitamins and variety in the diet; the most common was kankong, a kind of water spinach. As much land as possible was used on the edge of the camp. At one point Harry was trapped in a minefield when attempting to widen the planted area. Work in the gardens was a good ten to twelve hours per day.

Salt was obtained by Harry's team by burning bamboo as fuel to evaporate seawater; it produced a dirty brown residue and was used in cooking and flavouring.

In addition, Harry was head of the agricultural department of the famous 'Changi University', where he lectured about his farming experience in Britain to hundreds of prisoners. Some books were smuggled in by working parties who were sent into the town to work.

Harry also remembered the horror of the Selarang incident, which he said was started because some men had been caught trying to escape. Some Allied officers were taken to the execution place and forced to watch the men being shot.

One of Harry's first acts of resistance was to assist in the gathering of secret radio parts into the camps wherever he found himself, beginning at Changi. Whenever they moved, he took part in the clandestine transport of bits, an offence which carried capital punishment if discovered by the Japanese. He also used his forestry skills to advise the POW doctors on which plants were good for gathering for medical purposes or for vitamins for the starving men, as well as advising which greens to grow in the gardens, to give maximum nourishment. His care and devotion to the men was widely acknowledged and talked about among the prisoners who were with him.

The Railway

On 6 November 1942, 300 officers were put on board one of the notorious, crowded steel cattle-truck trains, to be sent north to work on the Burma Railway. Before departure the men were all searched and all paper documents were confiscated, as the Japanese did not want any notes written down. Thirty men were crowded into each truck and, with many suffering from dysentery, the conditions were shocking; Harry said it was the worst experience of his life.

At Singapore railway station they were harassed by traders trying to sell them items but the Japanese guards kept the locals well away from the prisoners. On the train, Harry found himself squashed between two 'Richards' – one with a double first from Oxford, and the other secretary to a colonial governor; they tried to stay together, whenever possible, for the next three years. At Kuala Lumpur they were briefly allowed out and given a meal of thin soup and then travelled northwards via Ipoh. The journey to Bampong in Thailand took four terrible days and there was no space to lie down.

At Bampong, the prisoners were assembled and a table was brought out and put in front of them with two chairs. After some time a small, spectacle-wearing Japanese officer appeared wearing a huge purple jacket and matching gloves, baggy trousers and big boots, with a huge sword dangling from his waist; he had two orderlies with him. He climbed onto the table, with his interpreter, so that he would appear taller than the prisoners, and began telling them – at great length – how they were going to help to form the Great Japanese Co-Prosperity Sphere by building a railway. After each harangue, the interpreter would give a very short version of what the officer had taken minutes to say, in a few seconds of translation, concluding with, 'His Excellency says you are a shower of bastards! And you will be shot if you try to escape!'

After one-and-a-half hours of this, the interpreter announced that in case they had not understood all that had been said, they would be given it in writing. At that moment several more Japanese soldiers appeared with piles of paper. A huge murmur went up from the POWs and Harry said you could hear the rifles of the surrounding Japanese being cocked – they thought there was going to be a mutiny. But it was a murmur of pleasure and amusement, because the lieutenant colonel commanding the POWs had dashed forward and smartly saluted the Japanese, saying, 'Thank you, Excellency, for talking to us. It will be my personal duty to see the essence of your talk reaches the very bottoms of my men.' It would make wonderful lavatory paper.

From here the men were sent on a forced march in the pouring rain to a filthy base camp at Kanchanaburi (also known as Kanburi in other books about the war) – about twenty miles, taking several days; on one occasion they slept in a field of tobacco, which Harry described as 'a smoker's dream'. The nearby camp was Chungkai but it was full and so they marched on.

On arrival at their next camp, Wun Lun, about six kilometres from Chungkai, the Japanese commandant gathered the POW officers on the parade area and told them: 'You will work.' They responded that it was against the Geneva Convention for officers to work. He insisted that they would work, and a succession of more and more senior Japanese officers kept arriving to repeat the work order, and each time the POWs said no, but they did compromise by saying that they would do camp chores. Finally, one Japanese ordered a volley of shots to be fired over the POWs heads. The POWs replied: 'We will work!' They were sent to build the Whampo Bridge, a viaduct that clung to a steep cliff along the river.

The men worked a twelve-hour day at this task. Harry was sent to cut timber, but as he was known to be a forester, the Japanese asked him to choose the wood for the bridge; he ensured that it was always the kind most prone to wood-boring insects, which helped to ensure that the bridge had a short life. Later a Japanese engineer found out what Harry had done and 'there was all hell to pay' – but Harry did not elaborate on what happened to him.

Cutting timber for the viaduct was extremely difficult, as the forest was teak and they had only hand saws. So Harry contrived to have the men cut the soft wood which had the wood-boring insects in it, thus both damaging the bridge and easing the work. They got away with this for some time. Occasionally elephants would help drag the wood to the bridge site using tree-creeper rope, but mostly it was all done by the men. The men then used an adze to hand-smooth the timbers. And hand augers to drill the holes to receive the pins.

While building this bridge, Harry had a brush with death when using a wood auger to bore holes in the timbers to receive the metal pins. It was often necessary to bore the timbers from both sides and the prisoners always ensured the holes never aligned. The Japanese engineers always banged the pins in, and on this occasion the hammer bounced up into the guard's face and knocked him down the riverbank. Harry judiciously 'disappeared' into the mass of working men, to escape the wrath of the Japanese engineer.

It took 7,000 men to complete the viaduct in seventeen days, but it was bombed fourteen days later by the Allies, and collapsed.

One man left a working party and was caught by the Japanese with a Thai woman; he was beaten badly and it was expected that he would be killed, but Harry and some other officers went to the commandant to plead for the man's life. They argued that he was only doing what a Japanese soldier would, and he was doing it in his own break-time, not affecting the building of the railway; he was freed, and survived. Such was the unpredictability of the Japanese.

The POWs observed that the Japanese put bombs in the rivers to kill fish for food. Later the POWs discovered that many dead fish fell to the bottom of the river, rather than floating to the top, so at night they would wade out and pick these up and have night feasts while the Japanese were sleeping. On one occasion, Harry stole a tin of Del Monte tomatoes from the Japanese store; they sieved the seeds out of the pulp and planted them, and this provided tomatoes for the rest of the war as they took the seeds from camp to camp.

In one camp, three radios had been smuggled in, in the bottoms of hospital water bottles. With these they could pick up the real war news from Radio Delhi. One radio was found, and five POWs appear to have been executed, according to Harry; they certainly disappeared and were not seen again. It was always arranged that the Japanese commandant had a British officer as his batman so that valuable radio parts could be hidden in the commandant's belongings whenever they moved camps, ensuring the bits would never be searched for or found! Batteries and aerials were smuggled in by Thais working on the river boats, usually in hollowed-out bamboo or pumpkins. Care had to be taken because some camps had stool pigeons (POWs who had turned traitor and, for extra food, gave information to the enemy)

Harry's worst job was taking his turn with the men who were terminally ill in the sick hut; he had to sit with them until they died.

In another incident, in 1943, a Chinese offered Harry help to pass money to the POWs, who then used it to buy essential life-saving foods and medicine. Hidden in his pack, Harry had his Westminster Bank chequebook, issued near his home at Fakenham, and he arranged the transfer of funds with the Chinese helper via his personal cheque. The Chinese gave Harry ten ticals (Thai currency) for a pound and so the deal was done. Harry asked the Chinese why he took the cheque; he said that if the Japanese won, the money he had would be useless anyway; if the British won, the cheque would be honoured – and so it was. They both took a great risk as Harry's name was of course printed on the cheque and if the Chinese had been caught, both would have faced the death penalty. The cheque eventually made it all the way to the UK, and was honoured; it was kept by the bank as an historic Second World War document that helped save many lives. After the war some POWs went back to collect all the many chits and IOUs the Thais had accepted, and honoured them.

In other camps further up the track, Harry was working on cuttings where explosives were used to blast rock which then had to be moved. By now he had malaria and beri-beri and was very ill. As the monsoon arrived, the working conditions in the mud and ever-poorer rations led to many deaths. Harry testified that besides the many good and courageous officers, there were many who were thoroughly bad and did not care for their men. He would not, however, say any more on the subject.

When he was at Camp 211, cholera broke out. Harry testified that this was caused by men eating infected molasses (gulam malacca), which was brought into the camp in open tubs, contaminated by flies. The death rate was shocking thereafter and an isolation camp had to be built.

It was not until 1944 that Harry's family even knew he was alive (the three cards that he remembered being allowed to send home never arrived until after the war). His older brother, Charles, worked at the Department of Economic Warfare in Whitehall, and had heard Harry's name mentioned on Vatican Radio. This was because the wife of the Vichy French ambassador in Bangkok (Mme Millet, an Anglophile) had heard about Harry through one of her Thai couriers, who had obtained a list of Allied POWs from a camp on the railway. This list was then passed to the Vichy French in Saigon, who passed such details to Vichy, who sent them to the Vatican. One of these couriers was the Thai loyalist known as Boon Pong (see David Arkush's testimony, note 6), and he helped Mme Millet smuggle food, money and medicines into the camps along the whole Burma Railway throughout the war.

On several occasions, Harry was sent to drive some cattle up and down the line to camps to be used for food; they were heavily diseased, but all the way others were trying to steal the animals, especially locals. On some occasions Harry would assess if it was possible to escape, by wandering into the jungle, observing the plants and animals, and seeing how long it was before he was missed. He found that one whole day was the maximum that he could remain unmissed from the camp, but no more than that, and with his illness, his European features and high rewards offered to locals for handing in POWs, he judged it was impossible to escape. At the same time he took advantage of enjoying the incredible flowers and bringing them into the camp to place near his hut; on one occasion the men woke to see the valley full of millions of butterflies – a sight that Harry said was one of the most beautiful he had ever seen

Harry noticed that as the men under his command were mostly Norfolk farm men, they seemed to have an inbuilt store of vitamins from a good childhood diet of vegetables and meat; he is convinced that this accounted for their high survival rate, although he noted that mental attitude was a powerful survival factor too.

In late 1943 a battalion of Japanese artillery came through their camp, the guns pulled by ponies, and two of the soldiers came and engaged Harry and some other men in conversation. Though the POWs were guarded by Koreans, no Korean would challenge a Japanese soldier. The Japanese spoke quite good English and said they had been engineering students at Tokyo University and conscripted into the army. Harry asked them why they were going to fight, given that the propaganda newspaper the POWs got from the Japanese told them every week that the UK and USA navies had been destroyed, Europe had been conquered by the Germans, and London had been flattened. One Japanese lad replied: 'You lucky, hear this every week – we hear it every day!'

Harry also witnessed the trains bringing wounded Japanese down the line; he said the Japanese medical and other soldiers treated them very badly, as though they were simply expendable and not worth saving. The Allied doctors had to step in and treat the Japanese wounded themselves. On occasions he saw Japanese officers kick their own men in the face for minor infringements, in front of Allied POWs.

At Tamarkan, Harry witnessed the destruction of a big bridge, while he was standing underneath it, by six very low-flying American Liberators, bristling with machine guns and bombs; he does not remember why he was there. The first attack left the bridge intact. An hour later a second attack was made by one high-level bomber and it blew the centre out of the bridge.

Near the end of the war, he was at Kanchanaburi and was told with colleagues to dig a trench, twelve feet wide and nine feet deep, right around the camp, with machine guns at each corner. They understood that this would be their grave, should the Allies invade.

Liberation

By August 1945 their wireless batteries had run out, so the first they knew of the atomic bombs was from pamphlets dropped by the RAF. Most of the Japanese then disappeared.

After liberation, Harry and four fellow officers had heard that a group of Norfolks had been left leaderless in a notorious camp on the Kra Isthmus, on the border of Thailand and Malaya, some 200 miles away. Despite their own ill health, they asked their CO for permission to go and bring them in; they ordered the Japanese commandant to give them transport and two armed guards, with a letter guaranteeing them safe passage because there were thousands of heavily armed Japanese on the route, waiting to surrender. They made their way with great difficulty to the site and brought the men home, knowing full well that this would delay their own much-needed repatriation.

At the Kra Isthmus camp, Harry found that the men there had just finished building an airstrip. As they reached the men, an SOE officer emerged from the jungle with a radio and several Thai aides; he was wondering how to get out of the area, back to his base, so Harry asked him to ask Rangoon on his radio to send aircraft to bring the POWs out, and he could go too. Within a few days the American air force arrived, but the first Dakota initially refused to land because a whole heavily armed Japanese regiment was sitting alongside the airstrip – but in surrender mode, unknown to the pilot. Eventually Harry persuaded the pilot to land and they were rescued. He said the Americans were amazing, and landed and took off in the most appalling monsoon conditions.

First they sent the Dutch native troops out, but were told to stop as the planes were going to British Singapore, not Indonesia. The Aussies – who had to be restrained from killing the Japanese, who had surrendered, with their bare hands – also had their evacuation stopped as on arrival too many were getting into fist fights in Rangoon and damaging the base. Finally the British were evacuated. At first it was about thirty men per aircraft, but it was as many as thirty-six by the end – and all were safely evacuated.

Harry put himself on the last plane. On arrival at Rangoon, he saw a copy of the *Daily Mirror* in the base camp, and on the front page was a photograph of oil drilling in North Creake, next door to his farm in Norfolk! He imagined he might become a millionaire.

Harry's first thought was not food but a good bath. After being put in hospital for check-ups, he and a friend decided they could not abide hospital life, so they walked out and found some Norfolk soldiers recently arrived in Rangoon from the UK; the CO allowed them to live in their mess and they were looked after like royalty.

The CO then said he could get them home by air, land or sea. They chose to travel by boat and arrived at Southampton to a civic reception, but the meal consisted of rice and caused a riot among the thousands of ex-POWs; the mayor was most put out, as rice was a luxury item in the UK!

Harry returned to Norfolk, but still in the army. This is because he refused to sign himself as A1 fit, as he still had beri-beri, malaria and dysentery. Two years later a lorry arrived with a sergeant and escort to take him to hospital as a sick man to be checked over; they would not take no for an answer. But Harry refused to go and said he would then sign himself as fit. This was agreed and so the army allowed him to be demobbed.

After the War

Harry married Peggy and they had two daughters, Prudence (Finch) and Sarah (de Chair). He was known widely as a progressive and innovative farmer, growing mainly foods for the London market. On one occasion, while doing some experimental work for ICI, he was asked to receive an agricultural delegation from Japan. He agreed but said he would not entertain them afterwards.

He also worked on many committees for the well-being of local people, for which he received the MBE in 1990.He died aged 100 in November 2010. He had had very little, if any, connection with the Jewish community, and in fact was a churchwarden in his local village for many years.

Notes

1. Lucy was daughter of Augustus and Louise (née Samson) and was born in Rusholme, Manchester, in 1874.
2. Daughter Sarah de Chair's telephone conversation with me, October 2011, and ensuing correspondence.
3. These evaders reported Harry as killed and that is all his family knew until 1944 (this is mentioned later in the testimony).

119924 Capt. Dr Harry Silman, RAMC

The information in this testimony is taken, with the kind permission of Harry's family, from their various interviews with him over a period of time, and excerpts from his diary kept while a POW at great risk to his life, which was transcribed by his daughter Jackie. The dates have been put into bold type, but the diary entries have not been altered in any other way. My comments are in square brackets.

60a. 119924 Capt. Dr Harry Silman, RAMC, sitting, far right, Changi.

Harry Silman was born on 21 December 1910 to an orthodox Jewish Russian immigrant family of tailors, son of Reuben and Fanny (née Landman) of Grange Avenue, Leeds. He attended Leeds Grammar School and Leeds University, where he studied medicine. He was a GP in Leeds for forty years, but the events that framed his attitude to life were his experiences in the Second World War. He volunteered the day war broke out and his first posting was to Newcastle, to the 9th Battalion of the Royal Northumberland Fusiliers in December 1939. Soon afterwards he was sent to Europe, and was present at the battles around, and later at the evacuation of, Dunkirk, in May and June 1940.

His company were defending a ridge that was becoming surrounded by German tanks and so they were ordered to withdraw; meanwhile, ambulances were arriving at his medical HQ, a large commandeered house, with wounded and dying, but the Casualty Clearing Stations had all moved back towards the beaches – it was total chaos, so Harry commandeered a truck to transport the sick and injured to the waiting boats, taking over a cafe to temporarily locate his patients. He delivered three babies to local women as well as attending the wounded. He decided to drive the ten miles to La Panne over the sand and get his wounded onto the boats. This task of driving on the beach was not easy, as many soldiers had temporarily buried themselves in the sand with only their heads showing, in an attempt to avoid the shelling of enemy bombs while awaiting rescue. Harry made several journeys, picking up wounded stragglers on the way, until he ran out of fuel. He finally walked the last stretch alone and got away in the destroyer HMS *Malcolm*, accompanying the wounded whom he had rescued.

Some sixteen months later, on 14 October 1941, his division was sent from Greenock via Canada, the USA and Cape Town to Singapore, arriving just ten days before the fall of Singapore on 15 February 1942. Following the surrender, Harry and the senior padre, Cordingley, were allowed to go over to the Japanese lines and bring back the dead and

60b. Harry, standing, right, Changi.

wounded; this was Harry's first confrontation with the Japanese. His impression of them was 'that they resembled apes with beards and fierce expressions and very long blades, bayonets, watching us closely; they appeared cruel'. With 100,000 POWs he was marched seventeen miles to the notorious Changi gaol, where he spent the next three-and-a-half years, except for eight months upcountry.

Some four months after he was captured, he was allowed to send one card home but it got mislaid under the house doormat and was not discovered for some weeks.

There is no doubt that Harry's sense of humour helped to combat the low morale caused by the dreadful conditions and deprivations in the prison – disease, humidity, poor water and sanitation. Stray animals would be caught and cooked; soap was made from wood ash and palm oil; rainwater for washing was harvested from roofs; vegetables were grown in primitive gardens from seed; and shoes were made from old tyres, or clogs from wood, as boots rotted. At great personal risk (because what he did was punishable by death), Harry hid the components of an illegal radio in hollowed-out chair-legs and the sink pipes of his medical room.

The Selarang Incident

After four POWs tried to escape, the Japanese insisted on the notorious signing of an agreement that no further escapes would occur. Up to 20,000 men were crammed into Selarang barracks, including many sick and wounded, because the British and Allied officers refused to allow anyone to sign this illegal document. In retaliation, the Japanese cut the water supply, food queues lengthened and there was barely room for men to lie down to sleep. Sanitation broke down as men tried to dig through concrete to build sufficient latrines. Harry had to use the ambulance van as his surgery and only acute cases could be seen as the health situation collapsed; rations were cut to a third. Then Allied senior officers were forced to witness the brutal shooting of the four escapers by an inefficient Sikh firing squad on the beach under a blazing sun. It took twenty shots and twenty-seven minutes before the four died. The British had to give way after three days and the 'parole' was signed under duress, thus making it invalid in Allied eyes.

Going North

The only break in Harry's time in Changi was when he was sent with a party of 7,000 to work on the Burma–Thai railway with a group designated as 'F' Force. The Japanese pretended they were going to better conditions in pleasant work camps, with a cooler climate, better food and no marching. There was almost a party atmosphere. But they were being sent to work on the notorious 'Railway of Death'. Harry said that even some of the sick Japanese in the hospital were happy to go, so deceitful was the propaganda. In his unit he was made the MO in charge. Before leaving, he hid his priceless film of the surrender in the roof of a mosque, but when he came to retrieve it many months later it had been stolen and was never seen again.

Many died on the horrific five-day journey in cattle trucks. Harry took the temperature in the steel carriages and it was over 105 degrees Fahrenheit. There was virtually no food, water, sanitary facilities or ventilation; the men were packed in like sardines. Harry said the men realized later that, despite these conditions, the hardships of the train journey were nothing when compared to what was waiting for them in the camps. The men he was with were put on a 200-mile forced march over three weeks, many with bare feet as boots wore out in the jungle conditions. The Japanese laughed at them for all the equipment they had brought; it all had to be abandoned, including even precious medical supplies, as only food and essentials could be carried. Many men, already sick from Changi or the journey, simply collapsed and died at the wayside.

Harry described the march into Thailand from Bam Pong as a journey that defied description, 'a tale of hardship, suffering, hunger, thirst and disease'. His conduct on this march was typical of his overriding concern to help others. There were rest stops at various points on the way, and while the weary men tried to find shade from the sun or shelter from tropical rain, Harry tended to the sick, who were suffering from dehydration, malaria, dysentery, tropical ulcers, horrible blisters, and fractured arms and legs from falls off the narrow paths. By the time he had managed to treat most of those who needed help, it was time to go on again and he had had no rest himself: 'On the march, we struggled through tropical storms, sometimes waist high in water; one shoe was torn off so I discarded the other one and completed the journey barefooted. Every step was agony; I never realized before what it meant to go through a night of mental and physical torture.'

At Tanbaya, the British senior officer was very strict with black marketeers and thieves. He beat them himself with his fists and sticks. Harry said it was probably against King's Regulations (KRs) but 'it was the only way to deal with these chaps'.

A sapper officer, James Bradley, was on this march with Harry and pays tribute to him in his book, *Towards the Setting Sun* (J. Fuller, 1982). Harry himself was inspired by the tireless work of Padre Duckworth, whom he specifically mentions in his taped interview.

Much marching was done at night because it was cooler; as the sick had to be carried, even more equipment had to be discarded. Many fell off cliffs or off bridges along the course of the river and they were even attacked by marauding Thais. Hundreds more died, and Harry became ill with dysentery and fatigue. A Japanese MO passed Harry as unfit to march, but as he was essential to the group he was put in a truck five days later and driven some distance, together with other sick. Harry described the journey:

> Travelling by truck sounds good but going on a heavy five tonner through a Thai jungle, up and down very steep slopes, across dangerous swaying bamboo bridges which broke under the wheels, clinging onto the baggage on which we were so precariously balanced, bumping incessantly up and down till all parts of our bodies were sore, seeing the back wheels skid on the slippery edge of a precipice as we looked down from the truck straight onto the water 50–60 feet below, going across the sides of a quarry on a rough road made of shifting stone, pushing the truck several miles when it broke down, climbing on and off to push it out of muddy ruts, keeping back the shifting machinery in the truck, red hot from the sun, blistering our backs and hands, seeing two trucks overturn going over narrow bridges and dropping into the valley below – that's what going by truck meant!

Harry continued on foot, passing through many railway camps; but at one stop they came across a camp where there were cases of cholera, something he had never met before, caused by faeces left on the ground. When the Allied officers asked the Japanese to lend them spades to clear it, they were told to use their hands. Harry tried to isolate the sick by moving them into the nearby jungle, waiting three days without shelter or medical supplies before a truck came to take them onwards.

After five weeks they arrived at their railway camp; the condition of the men 'beggars description, no boots, tattered clothes, sunken unshaven cheeks, [but] the relief of being able to lie down and sleep at night surpasses all else'.

The camp was often a sea of mud during the five-month monsoon and so sanitary conditions were appalling, the men sleeping on the usual two-feet-wide bamboo slatted platforms, but after Harry treated a Japanese officer he was given his own hut as an MI [Medical Inspection] room. Harry often worked a twenty-four-hour day, and was sad that in spite of this he could give only limited help, having no proper medical supplies.

The Japanese guards demanded a minimum number of men to work each day. Harry had the terrible task of deciding which of the sick men were least sick, as the guards would use sticks and weapons to drive out everyone they thought could put in even the smallest amount of work. Harry's diary records with great sadness the daily death toll, which was immaterial to the Japanese because they had huge numbers of prisoners to use for their labour force.

Once settled in their camp, many men suddenly became ill with severe stomach pains, which Harry thought were caused by gastric problems connected with the poor hygiene and dysentery. But they were symptoms of cholera, spread from the native (mainly Indonesian) slave labourers working nearby, and soon there was a major epidemic among the Allied POWs – over 200 men affected, with others having malaria, beri-beri and other illnesses. The men needed fluids to stay alive and to prevent dehydration, and so needles made of bamboo shoots were used to provide them with fluids – boiled cooled water – in makeshift intravenous drips. Survivors testified how Harry saved the lives of hundreds of men by using this method. Sadly, many also died, which Harry estimated at a one-in-ten death rate, despite the fact that he was able to isolate the cholera cases in a separate area. Harry described his ward rounds in the cholera ward as horrific, and there was little he could do to help many of the men as they died, shrunken and in agony, lying in their own excreta.

Each day a funeral pyre was made to burn those who had died from cholera; it was like Dante's Inferno – sometimes twenty or thirty a night, sometimes friends. One became hardened, and the line between life and death was thin indeed.

Meanwhile, others died of severe malnutrition as the Japanese stole the Red Cross parcels which would have saved hundreds of lives. Instead, the POWs were given made-up occupation money, which was nearly useless, to buy food from local people, who often would not accept it, even when it was possible to find people with whom to trade. Most simply parted with their precious possessions, and Harry sold his watch, wallet and cigarette case – even a piece of gold filling from his tooth. Inevitably the MOs themselves became ill, and on one occasion Harry was carried on his ward rounds on a stretcher after he had fainted. His weight was down to seven stone. He was caring at one stage for 300–400 cholera patients alone. The smell of the dead and dying was appalling. Diphtheria, smallpox and typhus were other illnesses for which Harry had no answer; tropical ulcers (often caused by infected scratches from bamboo) also resulted in a high death rate, caused by the necessary amputations from which many did not recover.

In one case, a very tall and fit warrant officer from a Scots battalion came to see him and said he could not go on. He had been a tower of strength, helping everyone and especially helping Harry, as a doctor, with anything that needed carrying or lifting. Harry told him there was nothing wrong with him, but he took to his bed, refused food and became morose. Harry would visit him and show him photographs, which were in the officer's wallet, of his wife and children, and say that he must surely want to go back to them, but he would not rally. He just slowly faded away and finally died.

Sometimes the Japanese would come to Harry for treatment because they had very few of their own doctors in the POW camps. In return, Harry would obtain gifts from them – one item he asked for was Marmite, as it was full of vitamins for the sick POWs.

Harry said that the Japanese remained relentless, no matter how ill the POWs were. They demanded the same number of slave workers each day. Harry described them as semi-human monsters – they could not care less for the lives of the prisoners. When he remonstrated with them that men were too ill, they would threaten him with a bayonet. Many were brought back from work dead. The men were often beaten with wire whips and bamboo sticks, or made to stand in the sun holding rocks above their head, and even if a man was ill, the Japanese simply said 'No work, no food'; such men had to be kept alive by using food from the already meagre rations of active men.

On one occasion, Harry went to buy six dozen eggs with hospital funds for his patients. A Japanese guard saw him on his way back and forced him at bayonet point to take a lift on a wagon, but then deliberately drove at speed past the camp. Harry had to jump off, trying to protect his rucksack with its precious contents. That night the sick men enjoyed a huge omelette of broken eggs, shell and fluff from inside Harry's pack.

Return to Changi

After eight months the Japanese brought Harry's unit – or what remained of it – back to Changi, in December 1943, on the railway they had built. Harry spent the remainder of the war there. It happened to be his birthday as the truck they were in arrived in Singapore and the cathedral clock struck midnight (on 20 December). Harry said to his fellow officers

60c. Harry, standing, sixth from left, Changi.

in the truck that it was his birthday, so they all began singing 'Happy Birthday' and the Japanese soldier stopped the truck and asked what was going on, thinking that all the British were quite mad.

Harry often used his army Jewish prayer book in services at the Changi synagogue (he knew the Dutch Army Chaplain Rabbi Nussbaum, who was a fellow prisoner), and also in other places where he was sent, to pray with fellow Jewish POWs. His faith was important to him all his life. Like David Arkush, he agreed that the Japanese were not anti-Semitic in any way and nobody suffered for being Jewish while a POW under Nippon.

He remembered well the Seder (Passover) of April 1945. Nussbaum had made matzoh (unleavened bread) from sago flour, wine from fermented rice, *charoset* (symbolizing the mortar) from grated coconut, and *maror* (the bitter herbs) from mint leaves; spinach was substituted for boiled eggs, and coffee was made from burnt rice. The age-old traditional song 'Next Year in Jerusalem' was sung with special gusto.

As the Allies closed in, there was a danger of being bombed by their own planes, so in the camps where Harry worked they always tried to put big red crosses on the top of the huts to warn the pilots. By the end of the war, half of 'F' Force were dead.

Harry kept the very basic medical kit he carried with him throughout the camps when on the Burma Railway; he mused that he might need it for the next war! His own strength of character helped him survive this dreadful period. He was liberated in August 1945 and returned to Leeds, where he met his beloved wife, Dorothea, and was happily married for fifty-nine years. Only after he re-visited Singapore in the 1980s did he begin to talk to his family about his experiences. He was the most dearly loved father of two daughters (who later lived in Leeds and London) and a son (in Manchester), a caring grandfather and great-grandfather. The intelligence, humanity and dry sense of humour that he had shown throughout the war years stayed with him for the remainder of his life. Harry Silman died on 9 May 2005, aged 94, much missed by all who knew him.

The Diary

This diary was written at Changi POW Camp, Singapore.

Thurs Feb 5th 1942

60d. Sketch of Harry's billet at Changi.

Arrived at Singapore after an exciting morning with the Jap bombers. Two bombs hit the upper deck but did very little damage. Six killed & about 14 injured were our total casualties. I was in the aft RAP and had a busy time for the duration of the attack. The Empress of Asia was bombed and caught fire, but all the crew and troops were taken ashore with very few casualties. The pilot of our ship was Capt Newing of the Glenbeg [merchant ship on which HS served in 1936 as ship's doctor]. Had short talk with him after the excitement had died down. Came into dock at night without tugs or shore help, & bumped the boat a few times but finally managed to get in.

Aussies were waiting with trucks to take us to Hill 85 where we were to camp out. Had a bad cold and sleeping on damp grass without covering didn't help any. Our baggage party unloaded through the night.

Friday 6th February

Developed a really bad cold at a very awkward time but managed to carry on with my work. Shared a tent with Padre, Archie and a new man called Hindmarsh. He was attached to us from the Malayan police force as liaison officer and his services were invaluable. He hails from Northumberland and was at home with our chaps. He also knows several of the passengers I met on my sea trip who live in Kuala Lumpur.

He brought a little open two-seater with him and was always willing to run us into Singapore or wherever we wanted to go. As he knew his way about very well, he was useful for shopping expeditions etc. He brought a coconut husk at night in the tent to keep mosquitos off – the smoke kept us all out until it caught fire and we had to die it out quickly.

Sent a cable home giving my postal & cable address. Malaya very optimistic.

Saturday 7th February

Set up MI Room in tent near Main road and got unpacked as the packages were brought up by bit by bit from the docks. Only one bale of blankets was missing.

The officers mess was a rather Heath Robinson affair consisting of bamboo poles, palm leaves and string. It made good cover and was well camouflaged. Several times during the day we had to jump quickly into slit trenches as the Jap bombers were overhead.

Our seats at table consisted of ammunition boxes and the table made of planks resting on similar boxes. I never inquired if they were full but when bombers were overhead, I took no chances. Several fellows received letters from home mostly Air Mail. I have had nothing except a cable in Deolali and it looks as though it will be months or years(!) before I get anything.

Sunday 8th February

The companies took up various positions round the camp and one company was detailed to join its brigade. My cold had improved considerably and I was able to wander about and get my medical arrangements working smoothly. Liaised with a nearby field ambulance for the evacuation of possible wounded. Newspapers brought up from Singapore kept us in touch with local and foreign news. We heard that the Felix Roussel [a ship] had got away safely packed tight with civilian evacuees, mostly bound for Java or Australia. It is very hard for families to leave their houses, possessions and businesses, and go out into the unknown with just the money they can lay their hands on. Their furniture and chattels will be non-existent when they return.

Monday 9th February

News received that the Japanese had effected a landing on the west coast and were advancing. This part of the coast had been defended by 'West Force' composed mainly of Australians who had fought on the mainland. They did not appear to offer much resistance to the landing forces. Consequently the Japs were able to land large forces without much hindrance and advanced eastward.

The advance was preceded by a terrific barrage which we heard in the early hours of the morning and sounded like a continuous rumble.

Went to General Hospital with Hindmarsh in his car. Saw Prof of Pathology and had a talk about malaria and the routine quinine which had been advised. Started it off but owing to shortage of quinine had to stop. Major Bloom, c.c. [in charge?] Hospital, was at the Rotunda [part of Leeds Infirmary building] with me in '37. Called at a milk bar and had an ice-cream soda but interrupted by an air-raid.

Tuesday 10th February

Small arms fire was heard about 4 am so we jumped into our clothes and stood to in the trenches until dawn. It seemed incredible that they could have penetrated so far so quickly but it may have been due to over-eager British forces. A lot of the firing on our part was unnecessary and due to 'nerves' – setting off the rest of the area into a flap. The Jap exploding bullets helped to create a state of confusion. In pm went with Major Leech & Hindmarsh to reccy new area, but spent most of the time lying flat and watching shells pass overhead both ways.

CO & Major Leech left to command Tom Force (3 battalions) so Major Flower took command. That afternoon we moved to our new position, HQ & reinforcements and the latter were detailed to defend Div HQ. We formed a defensive perimeter round our camp area near the village of Paya Laibur.

Wednesday 11th February

Situation became incredibly serious as the news of the Jap advance came through. The RAP in one corner of the perimeter was dug down to a depth of 4 feet so that we could get stretchers under cover. Positions were manned constantly and patrols were sent out.

I had a small blue tent for myself which was resituated three times until a suitable site was found – very comfortable little affair. The mess was below ground level and the batman helped by the Padre spent more of the day digging it deeper so we could sit and eat in spite of shelling overhead. Our table was a solid soil centre. Jap bombers were frequently overhead, but fortunately did not unload over our neighbourhood. British AA gunners put up a heavy barrage and I began to wonder how long the supply of shells would last.

The water situation began to be a little difficult as it all had to be collected by hand.

Thursday 12th February

After the CO had spent a morning in conferences, he came back at lunchtime and told us to drop everything – take what we could carry and a packet of biscuits and prepare to move off. There was no cause for all this sudden flap except that Div were moving during the afternoon sometime. However we obeyed orders and I left all my cherished kit behind in my tent and took my webbing and my 'knitting bag' filled with necessaries and put them in Hindmarsh's car.

After a wait of 2 hours we moved off, leading the convoy of 9 trucks filled with food and ammo. Part of the convoy later took the wrong turning and we waited for them near Government House which was just then receiving attention from the Jap guns. We could feel the shells as they burst on the far side of the road. Finally we moved on to new area and joined CO at Malcolm Road next to Brigade HQ, he having left Tom Force. Parked the trucks in the fading light and then looked for a place to sleep. Found a tent and turned in with my medical orderlies lying alongside.

Friday 13th February

We had a morning of scares and shelling but managed to get a good breakfast served in the house which served as the mess. Had a wash & shave in the bathroom – a luxury – and helped myself to some clothing and underwear – justifiable looting in wartime. We were then ordered to form a defensive perimeter round 18 & 11 Div HQ at 838132 [a map reference] at junction of Thompson & Whitley Road. Travelled with Sandy in CO's car and led convoy which we parked behind police station. Our mess was a nice little bungalow on top of the hill and Div occupied 3 adjacent houses just below.

Beamish gave me the ground behind HQ for my RAP, so I pitched a tent there and moved my personnel in. The Company was split up into sections which manned the perimeter constantly and the mess was in the centre. The 'fridge' and gas cooker were both working, and we lived like lords for a day, and had lemonade, whisky and soda etc. There was a full larder so we made good use of it. We spoke to Company commanders at night and suggested that a small party of volunteers penetrate the Jap lines.

Saturday 14th February

Very restless night on the lawn. CO and Major Leech with six ORs left during the night on the special party but believe that most of them didn't get away as the building in which they were waiting was shelled. We heard that the ship originally booked to take them was filled with Australian deserters who refused to get off. As the Japs have control of Sumatra it is doubtful if they could get far, even if lucky enough to get away from Singapore.

Very busy day for me as the casualties came pouring in and the ambulance I had picked up was a godsend. The Padre stayed in the RAP and did good work cheering up the wounded. We buried one man behind our tent. Intense mortar fire made us crouch down most of the day and our tent was ripped by shrapnel many times. Meals in the mess were interrupted by bombing and a lovely evening meal was spoiled as we had to take cover for over an hour – being dark when we returned. Decided to get a good night's rest so slept in bed with sheets!

Sunday 15th February

Very momentous day. In morning had a good deal of attention from planes, guns and mortars and started to dig a deeper trench in the southern pit of the RAP. Numerous casualties including about 8 shell shock cases – the latter were packed off into the deep shelter. The ambulance worked splendidly getting the wounded away and the driver SB [Stretcher Bearer] did very good work dodging shells and shell holes. One SB (Fundy) was seriously injured with W Company.

At 3.30 was told of the 'ceasefire' at 4.00pm. General astonishment at this unexpected ending – but shortage of water, mounting number of casualties (military and civilian), shortage of shells for big guns were contributing factors. Unfortunately 3 W Company men were seriously injured an hour after the ceasefire and several bombs were dropped near us – sequel to AA fire by uninformed gunners. Had a good meal, wash and change at the mess and sat on comfortable chairs for a change. Feeling of relief mixed with feeling of anger that we had to surrender such a large force with arms and equipment.

Monday 16th February

Had a good night's rest – went behind the Jap lines with the Padre and a burial party. Very difficult job as the bodies were swollen black and sizzling in the heat. The men in the Bren carriers were mangled and burnt making identification impossible. We found one sergeant in a dug-out badly wounded, and by using a door managed to get him on to an ambulance. Saw a group of bodies covered over. We used slit trenches as far as possible as there was a lot of work to do. Took wounded Sergeant to Sophia Road Hospital which was crowded out with cases. One Brigade including X Company were rounded up by the Japs and were herded in a field.

Death of Watson, Besley and Willis – the last two while leading a charge against a Jap MG [machine gun]. Very severe loss of gallant officers. Afternoon spent in clearing up the mess in the RAP and dispersing the shell shocks back to their Companies. I burnt all the souvenirs of the ships and also my diaries so conscientiously written up each day, covering the whole duration of the voyage.

Tuesday 17th February

After a morning of conferences, orders and cancellations – we were finally ordered to leave at 3.40pm and march to Robert's barracks at Changi. Padre and I went to and fro on the ambulance taking up the officers' kit and mess stores. The 2nd journey we scrounged mattresses, library books and cases of food. We gave Sandy a lift and very fortunately so, as the advance party officer was still sitting waiting for instructions. We were given very cramped quarters with a promise of re-allotment.

Stanbury put petrol in the water can and boiled it setting fire to the ambulance which was put out by excited helpers, who squirted each other with extinguishers. Had bully beef biscuits and tea on the verandah, laid our mattresses out for a sleep. Someone pinched my blanket and cushions – there was a good deal of petty pilfering and one had to keep a constant guard on one's property. There was very little sleep as the companies were coming in throughout the night and officers groping about in the dark trying to sort themselves out.

Wednesday 18th February

Began to get organized and the NF [Northumberland Fusiliers] officers moved into the ground floor of our building to the RASC. RAOC and RE officers slept on the 1st floor. The rear portion of the ground floor is being used as a mess. All the cooks are lending a hand but the meals are mostly bully and biscuits. The pints of tea are a life-saving measure and keep us going (& sweating). The weather is very hot – no rain and we perspire profusely.

The companies are spreading out a little and occupying better billets. Muster parades are held to find out our total strength. Mark Ward and part of his platoon is still missing. 3 officers killed – four wounded, OR casualties not yet determined as men are returning from hospital each day.

Our RAP held on the grass in front of the billet is very popular with all units, and my medical orderlies are working from morning till night tending the sick and wounded.

Thursday 19th February

By begging, scrounging and borrowing, have managed to collect a reasonable amount of personal kit which I have stowed away in my locker and have borrowed a blanket to cover my sofa-cushion, which on a table top serves as my bed. Found a camp chair and a small wooden table so am quite comfortable for the time being.

The flies are a pest especially during meal times and with the open latrines there is a danger of rapid spread of any intestinal disease e.g. dysentery. There is very little malaria in this part of the island – which is quite a healthy spot.

We go sea-bathing from Changi beach each day – very pleasant except for getting over the barbed wire to get into the sea. The sight of motor boats lying about sets men thinking. The amenities as POW are very good and except for curtailed liberty and no news, it is just like a peace-time holiday.

Friday 20th February

Moved my RAP from the open space into a small concrete hut which is much cooler – still large numbers, but men of other units are being referred to their own MO. The sanitation and water are big problems as water is rationed to 2 pints (1 water bottleful) a day from the water cart which brings water each day from the Div water points. Horrocks test shows that the water is not pure, so that we chlorinate the water using the set that the Z Company water duty man managed to save. Half a gallon a day for all purposes is the ration until the REs get the water going again.

The 198 FA [Field Ambulance] has set up a hospital and has commandeered all medical stores and equipment from unit MOs, as they left Singapore with no kit. I managed to keep a few dressings and instruments and am keeping my RAP going. The RASC MO was a GP in KL [Kuala Lumpur] and only joined the army a short time before the end! He has lost all his practice house and furniture.

Saturday 21st February

A meeting of all MOs was convened by the ADMS [Assistant Director Medical Services], who arranged four MI rooms in our Div area. I am with RE and RASC MOs and we are using the NAAFI as the MI room for our troops.

198 FA has supplied us with a small amount of stores to start off with. The Reccy Doc [Reconnaissance Regiment Doctor] was killed and one other MO was wounded during operations. The ADMS congratulated the medics on their valuable services and on the way they had functioned so well during the combat. The total Div casualties are about 3–4,000 but this is only a very approximate figure as lots of men are still unaccounted for.

Arranged times for medical attendance and arranged for the MOs to work two mornings and then one afternoon in succession.

Sunday 22nd February

Our first sick parade in the new MI room seemed to be an enormous success as a large crowd turned up – so have arranged for afternoon treatments with two teams of orderlies, and a medical officer on call. The Field Ambulance is getting swamped out with new cases mostly enteritis. We are making provisional arrangements to start a sick bay later.

It is very pleasant to sit out in the cool of the evening on the verandah or on the grass in front of the billet. We discuss the cause of our failures, individual deeds of bravery and otherwise; our future, rumours which are numerous and often far-fetched. Some most remarkable rumours spread round and are believed by many of the troops e.g. the Russians are fighting on German soil, there has been an invasion of England, we have landed in France, Rommel has been beaten etc. It is impossible to trace the source of these 'authentic' rumours.

Monday 23rd February

There have been lots of promotions to fill vacancies and Flowers who is now OC [Officer Commanding] 9th RNF is a Lt Col and Burn is a major (OC W Company). Thornhill, who had the worst of an argument with a Bren carrier, has now returned to his Coy. It was at first thought that he had a fractured skull and leg but actually all he had was a scalp wound and concussion. Two new captains have also been made.

There is a Div conference each morning after which the CO has a Company Commanders Conference which I have to attend – breaking off from my sick parade. New orders come out each day about water, rations, equipment etc. The food is causing many grumbles as the midday meal consists of tea and three biscuits and there is nothing till 6.30 when we partake of a bully beef stew and more tea and biscuits. Most of the fellows are feeling the pinch – short rations which may get even smaller later on when the Jap Administration staff arrive.

Tuesday 24th February

My days are settling down to a definite routine of work and most of my time is occupied. The other officers and men find that time hangs heavily on their hands – this accounts for the numerous complaints about food and other things. Idleness is a curse and is one of the worst features of the POW camp. It gives men too much time to sit and think and brood. I am only too glad to get into a chair or bed to rest after a long session in the MI Room.

In the evening we have a regular four for bridge; Sandy and I take on Mick Burn and John Thornhill. We hope to be several million up by the time we are released. Unfortunately the other two seem to get all the good cards and we have to rely on good play to keep our end up.

Occasional showers are welcomed as we take off our clothing and taking a piece of soap, step out into the rain and give ourselves a good clean up.

Wednesday 25th February

A few days ago, our Adj [Adjutant] and numerous other senior and staff officers went to Jap HQ for questioning. Sandy said that the bloke who questioned him was amiable and asked very reasonable questions. Sandy asked him why Japan went to war and he replied that the Dutch EI [East Indies] refused to supply Japan with sufficient oil for their needs. Now neither side will get the oil as millions of £s of oil and installations have been burnt or blown up by the Dutch. GI was kept overnight after questioning and groundless fears for his safety were felt.

The GOC is feeling the position very keenly and is looking very worried. He is greatly to be pitied as he never had an opportunity of using his Div as such. Brigades, Battalions and Companies were taken away and the whole Division was scattered into odd bits which were thrown into the battle piecemeal.

The Japs were surprised to get such a quick victory in an 'impregnable' fortress.

Thursday 26th February

The Army Commander of the Japs made an inspection of the troops and they were all lined up very smartly three deep along 2 miles of road while a long procession of new cars went slowly by. Col Flower brought us all to attention as he saluted the many cars. We had a wait of 1½ hours before the procession and an hour's wait afterwards so we were thoroughly browned off by the whole business.

The men were undoubtedly smart. The [Japanese] C in C [Commander-in-Chief] later sent his congratulations on the performance. The retinue consisted of a few cars containing camera men who took films for propaganda purposes – very humiliating. We were all glad when the gloating party was over.

Friday 27th February

Attended my first army religious service [Jewish] – strange that it should be as a prisoner of war. I was the only officer, 1 Sergeant, 1 Corporal and about 16 ORs. No-one had a prayer book so we had a short stand up service – three men each reciting a bit of a prayer that they could remember. The service was held at 8pm in a rubber plantation behind some distant billets. We couldn't sit down owing to the ants. I acted as OC and tried to get the thing organized and arranged for certain men to write out the prayers they could remember and next Friday night we could have a better show. We discussed news and rumours for a short time and then dispersed.

[No entries for 28 February or 1 March]

Monday 2nd March

The naval C in C wanting to emulate his army counterpart ordered a parade – similar to the previous one. There was a flap on about our MI room and we (doctors) spent the early part of the afternoon discussing moving into the new MI room where we would have charge of approx. 9,000 troops. I was afraid that sick parades would cause terrific congestion – hundreds of new cases of diarrhoea are occurring each day. Septic sores are very common and are not healing and of course there are the war wounds.

Went to the General Hospital (Roberts) and saw Titus who has a nasty wound of leg and was even threatened with amputation. I saw the procession from the balcony of the hospital – as there were so many carts with naval officers in them it was difficult to distinguish the Jap C in C from other officers.

Tuesday 3rd March

Enteric diseases are very prevalent and 20–30 dysentery cases are sent into the FA from our MI room each day. It is difficult to find a cause for it but a strict water discipline is now being maintained. We have chlorinated our water all the time and relatively our numbers are much smaller than other units. The FA and hospital are finding it a big problem to receive the large numbers of sick.

Each evening between 9.30 and 10, the medical section prepares 'medicine' which the MO samples with unfailing regularity. It is a life saving measure and removes that 'empty' feeling one gets later in the evening.

Wednesday 4th March

Today all officers were ordered to take off their pips and badges of rank and to affix one star to the flap of the left breast pocket. The CO had a general muster of the men and explained to them this new move and also that discipline and obedience remained as before.

Went bathing on Changi beach again after a lapse of a week with Sandy, Wilkie and the Padre. The Padre was unfortunate enough to choose a patch of oil to lie in and we had to scrape his back with sandy water to get the oil off. It was impossible to get Wilkie out of the water as he is as happy and excited as a dog in the sea.

Thursday 5th March

Liberties are now being curtailed – no sea bathing until further notice was the first order. The second one was that all the troops on the far side of the main road would have to move over to our side and live in the woods – after the way our fellows have cleaned up the billets, built latrines and cook houses, it must be heart-breaking to have to leave it all behind.

In the afternoon a large naval force came into the naval base consisting of four capital ships, one aircraft carrier and numbers of destroyers. The sound of exploding mines was heard occasionally during the day. It seems that the Jap army and navy are moving into Singapore and that they are going to use it as a base.

The food question is rather important. We are getting almost ¾lb of rice a day in various forms but mostly boiled – sometimes curried and sometimes with stew. Some of the meals consist of sugarless tea and tasteless rice – causing discontent. The cooks are being changed in the hope of getting tastier meals. There is too much eyeing the other fellows rations and the complaints about other units getting more or better rations are numerous and groundless. There is a mad scramble after the meals have been issued for the small quantity left over in the dixies. It amazes me how the men can consume such large quantities of rice by itself. They can't possibly be hungry after a meal. I'm full of rice but the chaps come back for more.

Friday 6th March

Attended the second [Jewish] service which was held in a small sports pavilion way out in the woods. It started in a typical fashion 30 minutes late – one man had a siddur [Jewish prayer book] and acted as a chazan in a rather strained way so that we had a fairly good service. Two men with good voices acted as the nucleus of a choir and led the singing. After such absence from services it will take a little while before the words become familiar again. A large typical Eastender who escaped from Greece and Crete and Dunkirk with a hospital unit was the main talker during the service.

During the day was accosted by a familiar-looking face wearing a smile and a Red Cross – Metcalfe of Lynn, Dentist. He has been out here since August 1941 with a dental unit and until war entered this island had a good time.

Saturday 7th March

The news came through definitely today that the move must be completed by Monday. We will be enclosed in a barbed wire enclosure and the exit gates will be guarded by Japs. Fortunately the MI room and the officers remain in status quo unless the latter have to move out for the sick. Sent in my first sick return for very many months. Examined 174 patients this morning – sick parades are reaching very high proportions due to diarrhoea, tinea, boils and septic sores. Was working continuously from 9.30 till 5 o'clock – am working far harder than I ever worked in England.

Sunday 8th March

Had to move the MI room from the NAAFI to the small room beside the officers mess – very useful for the duty officer who can lie in bed awaiting the call. The third compartment has been taken over by a dentist which will be very useful for us.

Six members of the 198 FA came along with authority and threw our orderlies out. This caused much discontent and argument and finally the mistake was discovered and the FA blokes moved on. There is too much of this state of affairs – chaos and confusion. The RNF orderlies slept in the ambulance behind the MI room.

Monday 9th March

Companies busily engaged in wiring us in our enclosure. Others were occupied in collecting coconuts outside the area – large opportunity. The minor industries that go on are legion. Simpson has his yeast factory, Sandy his vegetable gardens. I make fly covers, fly traps (that are not very successful).

Midnight last night the Japs decided that they wanted a representation from each Bde and the RNF to reconstruct the battle. Archie was sent off in the morning after spending most of the night getting map references rather audibly. He took with him a large shopping list and loads of dough. I hope that he is able to fulfil his shopping mission.

After bridge sat playing spelling bee and 'my auntie went to town' [a spelling game and a memory game] on the verandah till bedtime.

Tuesday 10th March

Like a bombshell last night came the news that sick parades were to start at 8.00 hrs. Awoke early and without a shave toddled along to the MI room for 8 am. The REs turned up 1 hour late through being uninformed. Stopped at 9 for breakfast and then carried on again from 9.30 till 12 when the gunners turned up – hundreds of them – sent them to the gunners MI room and had them sent to and fro until we won – CO198 FA authority.

In pm carried on with my fly-paper experiments – very enjoyable as it involves cutting up coconuts. Rather shocked to find a convalescent dysentery patient cleaning dixies in the cookhouse – one good way of spreading disease.

Wednesday 11th March

The rumour that Germany had given in was very strong and new light was appearing in the eyes of the fellows as they thought about getting home soon. As most of the good rumours are based on wishful thinking, they will be disappointed. I understand that the same spate of rumours were current in POW camps during the last war.

Spent the afternoon making a portable shower bath – unfortunately it leaked continuously so will have to think out a better idea. Most of my furniture is very sticky following my experiments on fly catching fluids. Am using various combinations of castor oil, coconut oil, cachou nut resin and the scummy stuff from the Jack-fruit tree. No success so far.

Concert in evening on lawn in front of officers' quarters – well attended and quite amusing.

Thursday 12th March

Half the meat supply to the camp was condemned as being bad – this was due to the refrigerator in Singapore going off. The meat should have been used up quickly – it was a heartbreaking sight to see all the carcasses being buried and men eating plain rice. The diet is bound to have an effect on the general health and teeth in the long run.

Saw Captain Phillips of the Beds and Herts re Friday night services – not very helpful – I had a scout round & tried to find somewhere suitable but in the cramped perimeter there was nothing that we could use. Our Padre promised to find something – but in any case tomorrow night's service will be off.

Y Company is going to Singapore tomorrow for reconstruction work so supplied Henry with money and a shopping list in the hope that he may be able to get out and purchase necessities.

Friday 13th March

As part of the blitz against dysentery, all mess tins etc have to be dipped in boiling water and sterilised before and after meals. Looking after the cookhouse is a full-time occupation now, what with fly measures, sterilization, rice cooking etc.

Rumours still fly fast and furious – Italy is supposed to have given in and the Germans are supposedly being overwhelmed. However the only true and bitter fact that emerges is that Java has fallen and that Australia is immediately threatened. The position is very serious & must be causing Wavell a bit of a headache.

We have had heavy rainstorms for the past four days and have found the ARP coat very useful for my visiting practice.

Saturday 14th March

As Roulston was taken to hospital with dysentery, had to work all day today. He was in a bad way last night. The C in C came round today – not very impressive personality – it is said that he was a sick man during the last few days of the fighting – may be homesick.

The stocks at the hospital were reduced by 75% yesterday evening and the supply of purgative medicines is running low. How we shall treat the dysenteries when the medicines run out is quite a baffling problem.

Limed rice was an issue yesterday in place of plain rice. It is said that the lime acts as a preservative – it gives an eggy taste to the rice and is not very palatable. It is medically harmless and can be blown out by a foot pump.

Sunday 15th March

The natives are running a very lucrative business (if paper money is still any good). All the cigarette and food stores that they stole from the barracks are now being sold back to the soldiers ($1 for 10 cigarettes). This has to be done surreptitiously as the Japs would shoot both the vendor and emptor if they saw the transactions.

During my tour of the companies I noticed several small ovens being constructed – this was for small individual parties to make pasties illegally from the flour and foodstuffs that they have bought.

I sympathize with their efforts to provide themselves with more food but the whole idea is officially frowned upon. I get very hungry myself especially about 9pm until bedtime. The move to provide tea at night doesn't seem to have borne fruit. Spent the afternoon lying down on my bed reading Sapper's Bulldog Drummond once more – have read the stories numerous times but they are always rereadable.

Monday 16th March

Starting sick parades early has its advantage as I get a really good cup of tea with plenty of milk and sugar in it – not like the watery unsweetened stuff that they put on the table for meal times. I am almost used to it by now and never thought that I should ever be able to drink 3 or 4 cups of unsweetened tea at a time. The meals today have not been good consisting of 99% boiled rice. The cooks should try some other method of cooking it.

The senior Padre called in on me in the MI room and offered his little church near Div HQ for our Friday night services. Very decent gesture. I have been put in charge as the other officer is reluctant to associate himself with the move. We start next Friday night. I am writing this in the MI room and it is pouring with rain outside – there is a large queue of sodden soldiers lining up patiently waiting for dressings.

Tuesday 17th March

Yesterday was a bad day for meals – consisting of 99% rice but as it was Sandy's birthday, Stanbury, at Wilkie's instigation, made a cake out of pound rice and cocoa with a greeting superscription and also a realistic representation of an important part of the male anatomy. It was a private party that started very surreptitiously at 10pm and consisted of Sandy, Wilkie, Padre, Richard Simpson and myself. I went out in the pouring rain to borrow a tin opener and stepped in a puddle and had to change and unfortunately just that night a crowd of chaps stood outside Sandy's door talking.

We opened some sardines and tongue which we mixed with cold cooked rice left over from dinner and swilled it down with Padre's rum. After 'lights out' at 10.15 we had to talk in whispers but the party was voted a great success. I went to bed without that hungry feeling that is so common after a diet of rice.

Rained most of the day. Archie came back yesterday and brought a few items of news and 20 cigarettes each. He wasn't able to buy the other things on the shopping list so bought a little food illegally from the natives.

Wednesday 18th March

Command is getting a bit of a nuisance – twice in four days they have demanded an immediate complicated return and as luck would have it – we burnt all our sick reports up till last Friday. They want all the cases of diarrhoea etc from Feb 15th daily – impossible to do – no records kept!

Last night, through the good offices of a Company Commander, managed to get 80 cigs (cheap variety) for $10 – the rest to follow today in foodstuffs. The purchaser has to be careful as no natives are allowed inside the wire. Prices are going up rapidly as the demand increases and the natives are making a fortune. $5 for jam, $7 for syrup, $3 for bully beef etc. However as my money is no good to me here I am going to spend all I have. Rupees are even being changed at an exorbitant rate.

Spent the afternoon making myself a pair of slippers from an old tyre – very hard work cutting through the thick rubber. Also cut up (& ate) a coconut doing further experiments on fly paper.

Thursday 19th March

The rain has been pelting down almost without any interval for the past four days and the ground is soft and messy. The weather makes any outdoor parades or work very difficult. I went up to the hospital in the afternoon to see the sick officers. They are improving – except my fellow Doc who appears to have caught a bad dose of dysentery. We have now arranged to get up early alternate mornings for sick parades as there is a falling off in the numbers attending, possibly due to the large number that went to Singapore.

Some frozen meat has arrived and gives added proof to the rumour that two food ships have arrived at Singapore. The meals are getting impossible and there is a change of PMC [President of Mess Committee] very shortly as the present one is getting slack and resents criticism. In evening watched battalion play football and beat the Cambs [Cambridgeshire Regiment]. Sat on the verandah at night and chewed a coconut, surreptitiously acquired.

Friday 20th March

Weather has turned very warm again. Had a good night's sleep last night – unbroken for the first time. We are now getting sweet ground rice porridge for breakfast – excellent dish. Tea is served without milk or sugar – soon it will be without tea. The milk and sugar rations one supposes go into making the other foods. Meals today have been excellent – meat 3 times and 3 courses for dinner: soup, roast beef and 2 veg and a banana caramel cream sweet. The diet has its ups and downs.

The DADMS [Deputy Assistant Director Medical Services] came round for another urgent return about dysentery – this appears to be bothering command quite a lot.

Attended a service in the evening held in the new 'hall' near Div HQ. There was a good attendance – Phillips turned up and everything passed off very satisfactorily.

Saturday 21st March

The 'official rumours' were very encouraging this morning – thousands of bombers over Berlin, landings in Sicily, Pantelleria and Norway. Russians took Danzig, Brest-Litovsk. I only hope that some of them are true. A victory in the west would shorten our period of incarceration.

Vit B, in rice is dissolved out by washing and is the only source of the vitamin – important topic which will have to be taken up by Divn. There is a possibility of beriberi and other vitamin deficiency diseases if we live as we do on an almost pure rice diet. The medical chiefs have submitted to the Japs a minimum scale of rations for a balanced diet – received 'favourably'. Japs shot 3 gunners who were caught in plain clothes trying to escape. 120 others wandering beyond the perimeter were kept without food or shelter for 2 days as a lesson.

Spent the early part of the pm examining Companies, and the latter part making myself a side cap from a borrowed pair of battle dress trousers – very hard work and will take me 3–4 days. Played bridge with new fourth as the old one has dengue – several cases about the camp.

Sunday 22nd March

After sick parade went with Emery to Signals HQ to see the new fly trap which is said to be so successful – it was quite ingenious. Capt Arkush, Dental Corps, came along to see me about Friday night services. He says that there are a large number of Jewish doctors in the depots whom he will try and rope in. Pesach [Passover] is a week on Monday so will hold a special service without the concomitant titbits.

In pm specialist (surgeon) from Command accompanied by ADMS inspected all our wounded. Only one was required to attend OP for X-ray ?piece of shrapnel in wound.

'Official' rumours continued to flow in – Musso has fled – landings in Italy, large naval battle with loss of Nelson and Rodney. Excited crowd of officers gathered in latrine late at night discussing and exchanging news. Thoughts of being home for next Xmas now uppermost in their minds.

Monday 23rd March

Saccharine tins were scrounged for the officers mess and the morning porridge has twenty times too much – nobody ate the porridge – waste of food and sweetening material. Ground rice porridge is a popular item on the menu as the rest of breakfast consists of stodgy flavoured rice.

Am putting in a lot of work on my cap – never realized I would have to do so much sewing for such a small article.

The news about the raids or landings at Kiel and Amsterdam caused a flutter of excitement. It will be very interesting after the war to see how much of the rumours is true.

Tuesday 24th March

I acquired 90 more cigs making a total of 230 for 25 dollars, very expensive – still it is our only means of getting them. The lads take risks getting through the wire and one was even shot at. The RASC are getting tins of food from Singapore when they go for the rations – this is causing some jealousy. I am getting used to the short rations and have lost that continual sense of slight hunger but am definitely below par in general fitness. Bending down, giddiness and climbing stairs is an exertion.

Had terrible bother getting the sick up to hospital as the Japs suddenly forbade anyone going out of the Div area without a Jap pass so had to wait till late afternoon for the medical ambulance to arrive. We share it with the Gunners.

Wednesday 25th March

General Heath, Commander III Indian Corps came round the lines and just popped into my MI room. He followed this by a lecture to the officers and sergeants which lasted over an hour and the heat and flies were very troublesome. The lecture was a summary of events from the beginning of the Jap invasion of the mainland until the siege of Singapore. The 'war was lost in the air'; our few bombers and fighters were no match in number or performance. They had prepared all details very efficiently beforehand. A Chinese smuggler had guided the first Jap landing party. Two ammo ships could not be unloaded at Penang owing to lack of labour. There was only water for 12 hours when we capitulated. If only a line could have been held on the mainland until air reinforcements arrived, the result may have been different. We expected that soon the tide would turn and we would be in action again. The GOC Division and his staff were all present at this lecture.

Thursday 26th March

I have to write a diary of military medical records from Feb 1st to 17th. How they expect me to remember so far back is not quite understood but it is very fortunate that I have kept a diary. Another high ranking Jap army bloke inspected our Division today and they were lined up in the heat of the day for 1 ½ hours. I was 'on duty' and so escaped this parade again.

Very busy in pm. COs conference followed by large sick parade after which I lectured to the colour sergeants and fly NCOs (who are now in each cookhouse) on dysentery, flies, cleanliness etc. I was almost hoarse at the end of a long talk. There was a terrific storm during this and we had to close windows and doors to keep the rain out. Played bridge on verandah as usual but extended the hours of play till 10pm by using my hurricane lamp. Drawback is the large number of little insects and animals that collect around the light.

Friday 27th March

Japs are getting very particular about the wire boundary and have issued orders that they will shoot anyone going outside. Two more men were shot yesterday. The Japs are proposing to allow merchants up from Singapore to sell toilet articles and other small essentials. The difficulty will be finding the money to pay.

There was a meat scare in the pm as it was thought that bad meat had been refrozen in the fridges when the electricity was repaired. However very little of the Bn [Batallion] meat was condemned.

The officers mess bought 20 chickens locally and are keeping them in a hastily erected hen run at the back. They look rather small so that it will be months before we get an egg for breakfast. There was bread for breakfast this morning from the Div bakery using sour dough instead of yeast. It tasted very nice with a little pat of butter.

Attended [Jewish] services again at night. 40 persons present – excellent attendance. Wore my home made forage cap for the first time and no caustic comments.

Saturday 28th March

It's strange to think that we get up at dawn which is at 8am – due to Jap time. It gets dark again about 8.45 pm each day. I finish the first part of sick parades just in time for breakfast. As my batman has cut his hand I did my own washing today – just smalls. Fortunately it was a sunny day and the clothes soon dried on the verandah.

After the pm session filled in lots of army forms and also began my war diary for the ADMS. I am afraid it won't be very accurate. Sat out sunbathing for a short time before dinner – sun still very strong and one has to be careful of sunburn. Many men have blistered backs which turn septic very easily and are a devil to heal.

The Padre developed the usual complaint and I had to dose him heavily with chlorodyne to keep him fit for his services tomorrow.

Sunday 29th March

After sick parade went along to hospital with the sick. Had to wait at the Gunners MI Room for the armband and generally wasted about 1 hour. The whole business is ridiculous as a sick man may be stopped going to hospital if he hasn't the necessary armband in the party. Visited our two sick officers and the RF [Royal Fusiliers] Doc. A little No 1 news and all the landings etc are fiction – but the Russians appear to be doing well in the central sector.

Spent the pm on the verandah completing my medical war diary. It is very difficult to write up a diary accurately so far back.

Chickens bought for the mess are dying off rapidly due to croup. It is possible that the Malayans, finding the disease spreading among a brood, sell the seemingly healthy one off cheaply and after the incubation period of the disease the birds die. With the new messing officer in full swing the food is getting much better and complaints are less heard.

Monday 30th March

This being our day to see the Div concert it was bound to rain, so taking our macs and folding chairs we wended our way to the leaky barn where the show is put on nightly. In spite of the prop and lighting difficulties they produced a very good show of its kind. Many topical sketches went down well and the show was aptly named 'Rain and Shine'.

After the show, the CO was informed that the Bn had to leave for Singapore the following morning. The usual flap developed as nothing had been arranged and two pasties per man had to be cooked as haversack rations. The cook houses were working till after midnight before all were completed.

The shortage of aperient pills is causing grave difficulties and I have to send the constipation cases for OPs as I have no way of treating them! There are only two classes of men, those with diarrhoea and those with constipation!

Tuesday 31st March

Awake very early as the departing officers started packing at 6.30. They left at 10.00 for a 14 mile walk to Farrar Park. The MO of the Reccy Bn has gone with them. I wouldn't have minded going except for the long walk. I hope Sandy is able to make some local purchases. The Japs are often very helpful. The RASC drivers last night had their purchases taken off them at the gate by the Jap guards and were knocked about a bit.

My batman roasted some rice grains which I have ground into a powder and which when mixed with boiling water tastes like good class coffee. My thermos is being filled each evening and placed in the padre's room where I hope to have my nightcap. The mess is very quiet today with most of the officers away and the dining tables appear empty.

Wednesday 1st April

A very busy day for me. More forms to be filled in and several letters to be sent off. Working in MI room till 5.00. Armband business is causing hold ups and Indian Guards now on duty are very particular. We have to salute these guards who are said to have gone over to the Japs. Heavy thunderstorm during the pm – all had a vigorous shower.

1st night of Pesach – as I walked to the service I thought of the Seder table at home and felt sure that they would at the same time be thinking about me and under what circumstances I was living. Fortunately I hear that they have now been informed that I am safe and a POW. Had a short service and concluded with Hallel [Jewish prayer] read by Capt Phillips. Capt Arkush of ADC gave them a short talk afterwards about Jewish life in Singapore. He hopes to give them similar talks after each service. It is a good idea and helps them to remember their faith and upbringing. Acquired a pillow case from a fellow officer – definitely a great day!

Thursday 2nd April

In the absence of Col Flower, Major Burn is taking over as CO and conferences and the atmosphere is so much easier. The rumours about Norway and Kiel still persist and now we hear that there was a landing at St Nazaire. Difficulties in the cook house caused the meals to fall off today – mostly badly cooked rice and rice cakes. We took in the ration store twice the amount of potatoes instead of M & V [meat & veg] – not good policy.

Completed one slipper in pm and then continued with my German lessons which I am learning from a book 'German self-taught'. I am afraid that I have forgotten nearly all my Russian that I learnt on the ship. Heavy rains during the afternoon caused a plague of white ants to come flying on to our verandah – so in the evening we remained indoors and played rummy. Someone produced a coconut which we shared and then taking a biscuit which I had saved 'against the day' along to the Padre's room I was pleasantly surprised to find that my batman had produced some tea with milk and sugar.

Friday 3rd April

Last night I used my pillow case for the first time and slept much better – a hankie keeps on ruffling up. I still have to get up once or twice every night due to the amount of water taken in with the rice. I compiled a list of my losses today – comes to well over £100 – very little possibility of getting anything back. Fresh meat ration very abundant and a tasty pie was produced. The cooks are still experimenting with biscuits consisting of various combinations of rice, flour, sugar and coconut. Some are successful and others are not even put on the table. The morning porridge (ground rice) served with diluted sweetened milk is a success. The ingenuity shown in cookhouses is remarkable – one rice washer consists of a large round can, perforated and revolves in a shallow trough and is worked by a handle.

Walked to evening service in pouring rain – the old chazan [cantor] is back again and gave a very homely rendering tonight – reminding me of old Samuels [probably a reference to his own cantor in Leeds].

Rumours of an optimistic Churchill speech.

Saturday 4th April

Letter from ADMS re case of diphtheria in hospital – had to investigate and examine contacts. Most of them are in Singapore. The anti-tank Bn have gone to S [Singapore] for 6 months. News about mail rather depressing – not very hopeful of sending or receiving letters so I will have received no letters all the time I have been away.

Survivors of the Exeter are reported to be in Singapore and they give hopeful news of the European situation. The Bn that went to Singapore today for 6 months is said to have embarked on transports ?destination. There is also much activity down at the docks – rubber boats, bicycles etc are being loaded on ships. What all this portends is a mystery.

At the nightly meeting in the Padre's room he offered me a lovely crust of bread with jam on and I succumbed to the temptation and ate it – the conscience pricks are very faint these days. It is these odd bites and frequent bits of coconut that keep us going between meals.

Table 1: Notes. List of my Personal Possessions and their former or present owners.

Wristlet watch	Loaned by Archie
Blanket	Loaned by Eva
Pillow case	Given by Addy
Shoes	
Topee	
Towel	
ARP raincoat	
Hankies	
Blue shorts	Borrowed from Mr Reeves of Singapore (without his knowledge or permission)
Coloured shirt	Given by Thornhill
Short pants	Taken from some house in Singapore
2 pairs stockings	COs kit left behind
2 pairs KD shorts	
2 K shirts	Taken from Singapore laundry
1 stick	Converted golf club
1 Forage cap	Made by me from battle dress trousers of Webb
1 pair slippers	Made from old tyres
1 haversack	Given by Macreath
1 spring bed	Supplied by hospital
1 long sofa cushion (used as mattress)	Mr Reeves
Mosquito net	'Found'
1 folding camp chair	'Found'
1 table	'Found'

Sunday 5th April

Easter Sunday – brilliantly hot – up early for my first session of sick parades which are getting smaller – occasional flare-up of diarrhoea here and there amongst the Companies. All the mess staff in the off. [officers] cookhouse have been changed – including Stanbury so that we must now wait and see whether the food gets better or worse. My batman who was also assistant cook will also leave and will therefore be able to resume his former duties. Last year this time we were travelling down from Scotland to Sandbach.

I managed to acquire the armband in the afternoon and took several 'sick' officers up to the hospital to see Titus and Webb. They are both improving but still complaining about the food. The place looks cleaner but still has that foul clinging pus-odour. Had to salute the Indian sentry going in and coming out. Terrific storm broke late at night and the sky was lit up by vivid flashes. Had to go out and see an RAOC cook who was rolling in agony – got a cup of 'chah' [tea] as compensation.

Monday 6th April

The food is much better and is very tasty – especially the biscuits have improved. The latrines and ablutions at the back now have running water so that we are almost civilized again.

1,000 men are reported to have left the island by sea for Saigon where the Japs have established a large prison-of-war camp.

The Div University will be starting in about a week's time and the favourite subject appears to be Economics. Other faculties are Law, Maths, Theology, French, German etc. The entrance standard is Matric and there are lectures each afternoon. Although not on duty I saw three cases in the evening, one appendicitis which I sent to hospital late at night. It is very difficult to distinguish between the abdominal pain of dysentery and a true appendicitis and one must be careful not to miss the latter.

Tuesday 7th April

News that Col. Thomas and the fusiliers with him were safe and that Col. Dillon is back was received from Div. Were all glad to hear that the CO was free. There were a lot of news rumours today – several bad ones e.g. about Malta and Burma.

A nasty accident to a Div Sig officer caused me to seize a 'bicycle' and pedal furiously off at 5mph. After weeks of cajoling the RAOC officers to get some paraffin for my hurricane lamp I got the precious fluid but found some swine had pinched the lamp.

The native vendors and merchants have agreed to a long term policy for the canteen and will be paid by the British Government after the war. For the sweet course at dinner we had 2 'chipattis' cooked by the natives locally and for which we paid or rather owed the PMC 10c. In India the men were forbidden to buy cooked things from natives and yet here we eat them. They are small cakes fried with plenty of oil tasting of garlic.

Called out after 9pm by the RAOC for the 3rd night running.

Wednesday 8th April

Heard that Col Dillon was on a small fishing vessel 800 miles from Colombo when he was intercepted by Jap tankers, taken on board and treated very kindly [this was a possible escape attempt]. The large supply of fresh meat went bad today and we will have two meatless days. It was sheer stupidity keeping that meat in lukewarm water overnight. We have to be very careful of every morsel of protein and vitamin. Spent an hour in the officers cookhouse with the messing officer discussing washing of rice, wheat and serving turnip as sweet. The order came out today that rice was not to be washed but cooked once in its own water till it was all evaporated. The difficulties are time, fuel and cooking space. Recovered my lamp from someone who had 'borrowed' it and we were able to play cards till 10pm.

Thursday 9th April

A soldier from Southern Command was found outside the wire two nights ago and was shot by the Japs. They have punished us all by stopping our next meat, cigarette and soap ration. This is doubly hard in view of the last meat ration being wasted. We will have to live almost on rice alone for the next week.

The 3 companies arrived back from Singapore in the late pm. The first arrival, Bowen, brought back a sorry tale. One man from the Royals died of typhus – CO in bed all week with dysentery and 4 other officers all off sick. When the CO came back – thin and exophthalmic, he was in a hell of a state due to an argument with an Indian sentry who threatened to beat him with a stick.

Flossie put him off by saying he was a sick man but spoilt things by telling him that soon thousands of British planes will be over the island. This propaganda set the Japs on his trail and he is a worried man. I sent him straight to bed as he has a temperature. The others look bronzed and healthy as some working parties had a good time and could buy food.

The sanitation was very bad and the flies were everywhere leading to dysentery. Sandy remained behind as Camp adjutant as he was on good terms with the Japs. Thornhill brought me back a terrific amount of food with the 50 rupees I gave him. At night various small parties were held all over the mess. We had bully beef and bread – tasted very nice!

Friday 10th April

RNFs had an enormous sick parade this morning, mostly blisters and diarrhoea. CO still in bed. Tidied up my cupboards and kept all the food in a locked bag as petty pilfering is very common. Broke my thermometer but managed to get the last one from the hospital. Very busy most of the day dashing from officer to officer, 5 of whom are unwell or off colour.

Service at night – new chazan with excellent voice. Walked back with Morris 2nd Lt RASC who only arrived in Singapore July 41 from London as Income Tax Assessor under the Government. When crisis came he joined the army. He left England for a 'rest' and was torpedoed off W Coast of Africa and finished the remainder of the journey by plane – very interesting trip. He is giving lectures to officers on Bookkeeping.

At night shared my bully with Carters bread. 3 of us had a good tuck-in. Then the padre felt hungry so opened and shared a tin of herrings with him. An RASC officer brought in some yam (turnip) and some mouldy Indian biscuits so there was quite a party.

Saturday 11th April

For the first time slept through the whole night without waking so it seems that insufficiency of food may be the cause. We have collected tins of food from most of the table – I seem to be the chief donor and the padre dishes out the supplementary rations if the meals require them. Today we put some anchovy sauce on the plain rice: it was a really tasty dish.

ADMS came into the MI room this morning and chatted away for ¾ hour about nothing in particular. It seems that the Japs stopped our meat because no-one owns up to missing the two fellows that were shot a few nights ago. They also heard through Geneva Red Cross that Jap civilian internees in India were not well housed. Punishing a whole PoW camp for the sins of one is contrary to the spirit of the Hague convention.

The Padre has a fresh loaf every Sat night to cut up for his Sunday services. The crusts are not used so were consumed with a liberal supply of jam followed by a tin of pineapple.

Sunday 12th April

Most of the officers including the CO were better today and were taken off their diet – sick officers reluctantly returned to ordinary rations.

After lunch supplemented by my bully beef I lowered my mosquito net and read a whole Wodehouse novel. Sunday afternoon finds most of the officers either dozing on their beds or reading on the verandah.

In the evening there was a hurried collection of $2 per officer as a monthly mess subscription to buy extras e.g. coconut oil, flour, cocoa, vegetables. They raised about $200 and this will help the messing officers to supplement the small rations allowed us. One difficulty was that some chaps were completely broke and a loan fund had to be established by the very few fellows with any money. Several other messes have the same scheme so that we are much behind.

Monday 13th April

CO was much better but still was not fit to attend the daily conference. In the matter of medical returns the DADH (Deputy Assistant Director of Hygiene – in the camp) surpassed all others by wanting a tremendous weekly return involving the nature and quantity of rice issued to each man, how much vegetables etc the gardens produce, the occurrence of obscure types of deficiency diseases. It would have been more helpful if he had sent out the early symptoms and signs of beri-beri, scurvy, intestinal anaemia etc.

After a tour of Z & W Coy lines with Major Harper RE, he agreed to build two soak pits for them and do a Serbian Barrel Disinfection for me [method of disinfection against typhus invented by Serbian and British doctors in the First World War]. There is no way of disinfecting the clothes of scabies, lice etc and it will come in useful for tinea which is increasing daily.

At night had quite a party with Tom, Jack and Peter round my little table. Some cold rice saved from dinner with cold baked beans and followed by tinned pineapple. It is amazing how tasty these snack meals are!

Gave the 1st of a series (?) of bridge lessons to some would-be players.

Tuesday 14th April

The breakfast meal was the best I had since being a POW. The porridge was supplemented by condensed milk & sugar provided by 2 officers, & then the boiled rice & watery sauce had added to it a tinned herring & tomato sauce. I got up feeling comfortable & happy.

A W Company soldier died yesterday from dysentery & his funeral is being held today. I only hope that this will bring home to the rest of the fellows the menace of flies & uncovered food.

It is difficult to find a quiet place to rest in the afternoon as the mess has been taken over by the Varsity for lectures every afternoon, and I can almost attend by lying on my bed two yards away. The lectures are now in full swing & the whole place has a studious atmosphere with the fellows walking about with books & pads under their arms. Even the General is attending the theology lectures.

The evening party was marred by the fact that we had purloined someone else's rice, and they were rather annoyed. Had jam on rice followed by pineapple.

Wednesday 15th April

Started a fly-drive using posters as propaganda, & giving lectures to the troops. 25 deaths in the past 4 days is a high mortality rate, and all measures possible must be adopted. We still meet chaos in some depts, e.g. the swill, – in spite of the order that swill should be burnt in the incinerator, they sent the stuff back to put on the garden! Petty difficulties arise due to non-co-operation between different units & sections, making my work very difficult. The REs won't lend us tools to put flytraps on our latrines for some minor reason.

Heard that whiskey & cigarettes had an increased tax by the latest budget. It doesn't affect us very much now.

The newly formed 18th Div Officers Club had its first meeting & will put forward claims by officers for compensation, loss of kit, staff pay, etc. There is a membership fee of £1/1/0 and an annual subscription of 10/-.

Kept my dinner biscuit(?) till later & ate it with Marmite. Had violent heartburn so drank chlorodyne most of the night.

Thursday 16th April

The canteen from Singapore has started but is not very useful as the margarine is too expensive. Soups are sold by the dozen & the only other article is brown boot polish! The mess has however bought a few dozen tins of soup besides the flour etc purchased in Singapore. The meals have fallen off a bit lately & require brightening up. By dropping a few hints, I was able to get my lunch biscuit liberally smeared with some melon & lemon jam – a very tasty combination.

Each afternoon I change into my blue shorts & home-made slippers and potter about doing odd jobs, reading or learning German till about 6 o'clock when I have a shower at the back in time for dinner. The latrines are now working, but being Indian style it is rather uncomfortable till one gets used to it. Still it is a relief not to have hundreds of bluebottles biting one's bottom, the damp seats, the smell & crowded accommodation of the deep trench type.

Friday 17th April

Our early morning sleep is disturbed twice a week by RASC officers leaving to collect rations, but last night there appeared to be a mug-throwing competition. Actually I understand that some rat-catchers were at work, but still don't know why they have to kick tins & buckets at short intervals.

Arrangements were made today for the forthcoming St George's Day. I suggested that the concert be put off till then. This was decided on & fixed up, together with several other amusements.

Major Agnew sent for me about the evening services, as the hall is occupied each evening until 8.30. He is trying to find other accommodation for us, & was very helpful. The congregation were rather excited as a 'news rumour' was spread round that the Germans had capitulated. This was backed up by 'bets' & stories of its authenticity. As I had been hearing the same thing for weeks, I took it with a grain of salt.

Morris told me as we walked back how he nearly escaped from the island the day after capitulation, but turned back owing to mal-de-mer on a Chinese junk!

Had a big evening party – 6 of us – bully beef & rice.

Saturday 18th April

Very heavy rain woke us up early & blew in through the wide open doors. After sick parade, filled up the numerous forms that go into the ADMS each Saturday, and also the monthly return to the Japs.

Fly propaganda is going on nicely, & all men have been lectured, & numerous gruesome placards adorn cookhouses & billets. A padre died yesterday from dysentery which is claiming its daily toll of 5-10. Very annoyed with W & Z companies for having lukewarm water for sterilising, after all my efforts.

Went to first meeting of Changi Medic Society. About 150 Medics there, British & Australian from the whole area. The first lecture was by the President, Julian Taylor, Command Specialist in Surgery. Lectured on Trotter [British surgeon] – quite good, but he read it as though he had been doing it for years & was getting bored.

Went to see Titus afterwards – took him some cigs & matches.

A small outbreak of encephalitis is disturbing the medics as the mortality is very high. No armband to come out of hospital area, so after waiting a short time, just marched through, saluting the Indians smartly, & there was no bother. The guards themselves realize what an unnecessary farce the whole business is.

Sunday 19th April

A new order from ADMS – all cases of diarrhoea – febrile or not – have to be admitted to hospital. This is to enable the bacillary types to be diagnosed early & lower the mortality.

Padre has opened a tin of melon & lemon jam sent up by Sandy with a pile of other tinned foods. We are smearing our biscuits etc with it. They hold promise of many parties to come.

A second man in the Bn died today – Pyrexia NYD [not yet diagnosed]. The hospital got permission from the CO to hold a PM [post mortem] before the burial tomorrow. The Japs are said to have about 5,000 dysentery cases in Singapore & are worried by the high mortality. It is possible that the terminal coma may be a deficiency disease caused by the starvation & diarrhoea for prolonged periods. The cause is being carefully sought.

18th Div officers played RASC officers at hockey this evening – a colonel led each side. The play was more wild than clever, and the spectators enjoyed it.

The evening beano – bully & rice – is getting to be quite a habit & I follow it up by having a little snack in the Padre's room later.

Monday 20th April

The nights are very hot & one wakes up with the pyjamas wringing wet. The morning cold shower is very useful & removes that sticky feeling. Roulston has started work again, so I will be able to sleep later 2 mornings out of 3. Have arranged to lecture to all the Companies on Health, Diarrhoea, Skin disease, Vitamins etc. There must be quite a lot of these important subjects which baffle the average soldier.

The canteen is a failure & one can only buy soap or Epsom salt. I am buying the former, and the mess is clubbing together to buy the latter in bulk for the officers. The soap is very cheap. I paid $1 & got an armful of several varieties. It will probably last for several months.

Letter from Major Agnew allowing us to use the Div Sigs Sergeants mess for our Friday night service. Rather a pity that we have fixed place for the service, but the hall we have previously used is occupied by education classes until late at night.

Tuesday 21st April

I started giving lectures to platoons today – two in the morning. Rather nervous at first as it is a long time since I addressed a gathering, but after the first five minutes, I regained my old confidence. 'Stage fright' is a terrible affair as one's mind becomes a complete blank & one is unable to utter a word.

A lot of 'news' at the conf. mostly good – heavy bombing of German industrial districts & also Japan. Russia claims air & tank fronts with Germany – good sign. Churchill said question of supplies to POW in the Far East is 'under consideration'.

Amid much excitement Sandy paid a flying visit with loads of food which the padre took charge of. The 'wolves' gathered round, but the Padre held them off. He has given me a jolly good share of $14 worth including a large tin of sweet biscuits & some strong Chinese cigarettes. He is returning to us on the 27th. The Bn is going down to S again next month.

Had a party at night in Padre's room – herrings, bread, biscuits & jam, & then followed it up with one in mine – bully & rice & marmalade. Felt rather sick.

Wednesday 22nd April

My early morning again. Feel quite fresh after a cold shower. The bowels being more regular & the extra food all help. Went round with sanitary serg. from Div who states that the DADH has got symptoms of beri-beri. Also hinted about exciting news.

More canteen stores have arrived. I just managed to scrape together sufficient money to pay for some shaving cream, a tooth brush, boot polish etc. All my S money is now gone & Sandy has 70 rupees of mine which I hope he will change.

Ordered to X & HQ & was asked some awkward questions – based on inefficiency of cookhouse fatigue men who cannot be constantly watched.

Bought a lock & key for 10 cigarettes (Chinese) – the food will be safer under lock & key. Started the bridge competition again. Major Charmand & partner – using Jack as mine until Sandy returns. Playing best of five rubbers. As usual he turned up 10 mins late. We stopped before dinner leading 2 rubbers to one.

New order re flag came out – only one flag for whole sick of the Div. Fortunately this is kept at the gate, & sick can get through hourly.

Thursday 23rd April

St George's Day dawned bright & clear (& very hot). We all went home & made roses in our hats – made out of red flags & white bandages. Everyone appeared to have a sparkle in their eyes. This morning at 10.45 I left the sick parade & formed a background to the General who addressed the men lined up in a square in front. He made a very sensible speech & said that a decoration will be awarded to Col. Thomas for saving the Felix Roussel. He praised the men for their part in that action on the island & hoped that we would soon show our true worth. Our coming to the island was a political & not a military fault. He then went to a special church service, followed by taking a salute as the troops marched past. It all went off very well.

In the afternoon there was a hockey match v. NCOs in which I could not play as I went to a very important lecture & demonstration at the hospital on beri-beri & vitamin deficiency. The complexity of this problem is enormous & we are just trying to unravel some of it as there may be a deficiency of any combination of the vitamins.

Special dinner with tablecloths. Sharp was guest – toast to the King & absent friends drunk in rum (kept from our original stores). Div concert 7-8.30 – very good – then a little celebration in Wilkie's room with Padre, Wilkie, Richard & myself.

Friday 24th April

Slept till nearly 9 am – must be the rum I drank last night. Wet day. Gave 2 lectures to Z Company in the morning – the fellows are very interested as it affects their health, and as they are frightened of some 'terrible tropical' disease I didn't tell them what the symptoms of beri-beri are as I didn't want to put ideas into their heads when they found out what a good mixed diet the patients are getting.

Poured heavily all day without stopping. After pm sick parade went to HQ to give my lecture to 2nd half, but found that 1st half had been collected again. Very annoyed, & orderly serg. was put on a charge.

Went to service with Morris – new place. Dursiga (?) Sergeants Mess is quite suitable & off the beaten track. We took names & addresses of the congregation for keeping a record. Morris cashed me a cheque for a guinea ($9). He has been pestered by his colleagues & has had to refuse many, causing some ill-feeling.

Party of six in the evening – mixture of sardines, salmon & rice.

Saturday 25th April

Very hot today. My early morning for work. Sent my batman to hospital with diarrhoea – increase in no. of cases today, including one officer. Gave evidence against the orderly serg. who was let off with a reprimand. The NCOs are getting very lax and will have to be severely handled, for the laxity in discipline is spreading to the men.

A large number of gunners have left today for Singapore & will be there indefinitely. It is possibly part of the Jap plan to split us up into small groups.

More lectures in the afternoon – completing my circuit. Japs have delivered nearly 2 million vit B tablets to hospital – probably to be used only for established cases. Rumours that Allied forces had taken Rangoon with 20,000 prisoners.

Meals very bad during the past two days due to the fact that no meat was supplied with the rations & the rice cut down to 12 oz/day. The last tin of margarine we bought comes in useful for adding to plain rice or biscuits, but has a sickly, oily taste & one has to get used to it.

Sunday 26th April

First daily working party went to Singapore (100 men/day). Very small sick parade, finished by 10.30. DDMS Command sent a letter round describing the early symptoms of beri-beri, and asked us not to treat cases in the MI room!! The danger appears to be sudden heart failure without any definite warning symptoms & occurs in 2 classes of cases (i) those who have lived in Malaya for several years (ii) after a long bout of dysentery or other severe illness.

Did my own washing today in the absence of my batman. After lunch, finished the first round of the bridge & ended up victors 4 to 1. We had all the cards and didn't make the best use of them.

An officers' hockey match before dinner was curtailed by rain, and there was an excellent dinner with a slice of bread on which we heaped marg & jam.

Monday 27th April

As it was my morning off, I did my washing – v. arduous toil. I longed for Rinso to do the work, but in its absence I just scrubbed away. After my row with the messing officers, the MOs are doing a week each looking after the officers mess cookhouse.

Sat out on grass after sick parade, sunbathing and doing my German lesson – progressing very slowly. Sandy returned from Singapore with large quantities of foods etc. He changed 50 rupees for 20 dollars for me so was able to make more purchases, including gym shoes, shaving cream, soap, jam, oranges, bread etc. He did very well having bought whisky, radio malt [a brand of malt extract], boiled sweets, bananas & many other valuable necessities. The Japs treated him very well & allowed him to enter shops himself – fancy a German POW shopping in Regent Street. He even hired a rickshaw one day! On St George's Day, the Japs threw a party for him, and they finished up by singing Auld Lang Syne in Japanese & God Save the King in English. In some respects the Japs are very good to us, and don't regard us as POW.

At night we all gathered in Sandy's room & drank some fresh coffee while he recounted his tales. The 'other crowd' had saved some fish and rice for me which I hurriedly bolted after lights out.

Tuesday 28th April

It seems that Sandy, Padre & Wilkie had insomnia last night for they had a party lasting till 5.30 am. The former stayed in bed for breakfast & brought round a cup of coffee to the MI room at 11 am & showed me his 'Singapore feet'. This is getting common now, and I am careful about showers & have my own footboard – marked Sergeants Mess. In view of the rapid spread, it is possible that the fungus responsible is air borne & contaminates bottles etc.

My batman came out of hospital – v. mild case of diarrhoea – & started duties again.

After dinner played second round of bridge tournament with Sandy as partner. One rubber each when we chucked up for the night. Pineapple party round my bed & then one in Sandy's room till 11 pm. Fortunately he brought 2 gallons paraffin back with him so my lamp will be very useful.

Wednesday 29th April

The Japanese Emperor's birthday today, so there was a good issue of tinned pineapple for each man. Today was a holiday – no work, no fatigues & the daily party to Singapore was cancelled. Sat in Sandy's room most of the morning drinking coffee & deciding what he should buy next time. Letter from Henry saying he had sent two haversacks of cigs etc to Titus – so Sandy & I went up to Hospital in pm & saw Titus & Brooks. Two respirator haversacks with cigs, tobacco & matches – no list of prices & no names – it will be difficult to decide how to divide it all out. We are sending down for some more haversacks for him.

Coming out of hosp gate, Indian sentry turned us back because our party was eight strong instead of a minimum of ten – v. annoying to have to wait in the rain.

Fertiliser used to make a sweet for dinner – full of vitamins (so it is said). Small party in Sandy's room at night. Argued until early hours of morning why water climbs up a capillary tube. Tried to remember my physics – problem remained unsolved.

Thursday 30th April

Inspection by GOC of billets etc. Spent morning going over area & finding faults. Becky (Major General Beckwith Smith, CO 18th Div.) questioned the men re clothing, books, illness etc. He caught out a cook about rice & vit B – on which subject he is very keen. Very hot day – perspired profusely & was glad when inspection was over.

At conference after lunch came news that in 3 days time we are to go to Singapore for a month – exact destination unknown. I haven't received any instruction yet, but started to prepare myself.

Started sick parade very late & was kept busy till late in afternoon. After dinner played another rubber in bridge competition – very slow – took 2 hours. Followed this with terrific party full of vitamins – tomato juice, pilchards, salmon, with bread, marg & marmalade-sweetened orange juice & a tin of pineapple each. Felt very full & contented.

Was allotted 200 cigs from Henry's parcel. Don't know the value but may sell a few. An officer brought up some rubber from the docks, so my batman resoled my slippers using a rubber solution borrowed from ADOS [Assistant Director of Supplies].

Friday 1st May

At morning conf. was told that Doc Lennon of the anti-tanks was going as RMO (Regimental MO) to the party & also a Major RAMC with a full hospital party. I was excluded – mixed feelings – sorry to part with the officers, but glad that I hadn't to move to doubtful billets in a malarial area. Actually, great feeling of relief. Bought up tinned goods of officers who were going to Singapore.

Went to hosp. with Lennon & fatigue party & helped him collect large quantity of medic. stores and anti-mosq ointment. Saw Titus for a few mins, and sold 50 cigs for $3.50 to hosp patients. Also went round wards interviewing men wounded in action, & prepared a list of men who are permanently unfit with a view to repatriation. Unwanted RAMC personnel (not 18th Div) may also be repatriated. Exhumed the 'body' [this could refer to a secret radio] today with John – terrible mess – everything very damp & will require a helluva lot of cleaning up. Wilkie is fixing up the future destination – excellent spot.

After dinner went with fairly top Div HQ & played bridge against Richard & Phillips (DADMS – Deputy Assistant Director Medical Services) – a good game. They have tea & biscuits at 9 o'clock. Returned at 10.30 & had a pukka farewell party for Sandy in Padre's room. I am certainly getting sufficient vitamins these days.

Saturday 2nd May

V. hot – awoke at 6 am & heard the hens calling (or the ducks) still two hours before dawn. Advance party astir early – loading medic stores & baggage on trucks. Two cases of beri-beri in MI room today – 160 in hospital with 10 deaths.

At conference learned that party will prob. move by MT on Monday & go to a food billet by the harbour – may get paid. Men are now quite keen!

Went round camp with sanitary inspector – a reliable source of rumours. News about Burma appears to be very bad. Spent two hours in afternoon cleaning up 'the body' [secret radio] & had to give up owing to shortage of tools. It will probably require expert attention & will cost several pounds to repair. F. (?) ruined & found a grave in a deep trench latrine.

Completed my series of lectures today so that everyone has now been warned of the dangers.

During Sandy's absence I have borrowed his long cane chair, & with the aid of my 'mattress' have converted it into a luxurious piece of furniture.

Mosquitoes are now biting at night & there have been a few cases of malaria at Changi.

Sunday 3rd May

Padre had a row with the gong striker, who struck eight o'clock at 7.45. As they possess no watch, they have to ask the time from passers-by, and the whole position is farcical as the time is only correct to 10 minutes each hour!

Party will not leave until Wed. Fresh meat, however, has been delivered to S. so that they will probably get ½lb/man in the advance party. The news that men in working parties are getting paid is a great incentive, & even the lame and the halt are asking if they are fit to go.

Sat out on verandah late in pm & watched Div HQ play the RASC etc. The average age was about 55 years, and all we could hear was the grunts and groans of the old colonels as they tried to follow up the ball. GOC came & sat beside me & roared with laughter.

Small pineapple party in the evening with the big crowd round my bed. Felt a bit off colour, & did not have my usual voracious appetite.

Monday 4th May

Disturbed night, but after a cold shower felt refreshed. Sick parades are not very big, and are usually finished fairly quickly. The DADH came round our cookhouse, and played hell because of the flies. He said that it was the worst he had seen! We have put in a tremendous amount of work trying to clean the place up, but the Sappers won't part with any fly-proofing material or cement. However, this time they will have to do it or else.

Started inspecting the Singapore party this afternoon & worked through 2 companies. One of Z Company's men who has the necessary tools & experience cleaned and oiled 'the body' – it took him nearly 2 hours. It looks a lot better now, but still rather bedraggled. The padre was informed that only 2,000 men will be left in this Div after all the parties have gone down, and the Japs appear to have plenty of work for our fellows to do. They don't mind doing it is as long as they get well fed and/or paid.

Tuesday 5th May

At conference this morning – good news of Singapore rations – nearly 25 oz rice per man. It is an enormous quantity, and should satisfy them. The meat ration is also increased for them.

Beri-beri cases mounting up steadily – nearly 200 now. One camp in Singapore has been inoculated against cholera for some unknown reason.

The rumour that the Japs are evacuating Sumatra due to Yellow Fever must be baseless as YF is unknown out here. That is rather strange when one remembers that Dengue & YF have the same carrier – Aedes Egypti.

The other officers were busy darning, sewing & packing in readiness for the morrow. We had a farewell beano (I supplied most of the food!) & the chaps turned in early. One of those unbearably hot humid nights & my pyjamas were soon soaked in perspiration.

Wednesday 6th May 1942

Main party left for Singapore today & were astir from 6.30 onwards, so didn't get much sleep after that hour. Had an early cup of tea with them & waved them good-bye – not knowing how long they would be away. The trucks were late in arriving, but the party got away safely. Wilkie is left in charge here & 6 other officers.

We cleaned up the billet & divided it into three: (i) library (ii) sleeping (iii) ante room. There are several armchairs & tables available & numerous old USA newspapers (bumpf) so that it resembles old times again. I have acquired a large mirror & a table, so am well set up.

Sat in ante room in pm & read a medical tome on tropical diseases – find difficulty in concentrating for long periods – other people have the same difficulty & are more irritable than usual. In evening we all sat round in the mess & Wilkie produced some tea looted from S. Had a spot of food later in Padre's room.

Thursday 7th May

The morning sick parades are very slack & we finish quite early. The pm sick parades are enormous due to Tineas, so in pm went up to Hosp to see skin specialist & piled questions on him. Not very helpful, but managed to get some salic. acid in spirit – only useful treatment.

Saw Metcalfe looking fat & hot – he is now the Colonel's batman. Some canteen stores arrived today & from my meagre funds I purchased some margarine & tomato puree for the table. The men have an enormous vegetable ration issue – bananas, cucumbers, papaya, beans etc, so are well away for vit B. We in the mess don't seem to get so much, but possibly will get our share later.

In the evening sat round the table in the mess sipping tea (surreptitiously acquired) & discussed previous incidents during the journey until well after lights out. Since the other officers departed, we have become a small homely circle.

Friday 8th May

Wakened early by noisy RASC officers so lay in bed reading Penguins Hansard vol 3. Churchill's speeches in Parliament during the difficult period May – Aug 1940 make very good reading.

Had a blitz on the rear party who are tending to get slack & leave the area messy. Started to make a pair of wooden shoes (as per Div orders) – most of the work was done by Laybourne & Simcocks. My shoes are worn very thin & I am wearing the shoes bought from Sandy.

The meals are improving each day due to the large amounts of vegetables we are now getting.

A large party of RASC is leaving for S in the morning & a party of 8,000, mostly Aussies, but 300 RASC & 150 REs are leaving the island during the week & going abroad. This will deplete our camp & mess considerably.

Attended Service [Friday night] – only 11 present – may have to stop soon if more men leave the area.

Saturday May 9th

About 600 RASC left for S early this morning. The small rear parties of units like ourselves are going to be merged into one large battalion for admin purposes, & also to do fatigues & work up here. Sick men from S have also to be replaced immediately. Our MI room is closing down & we are going over to the RA MI room – there are only 400 gunners & 2 MOs left.

The ADDH had a meeting at Div HQ of all the MOs in this Div, & talked about anti-dip. & malarial measures – all well-known. Dysentery figures are being halved each week – v good. Malaria is increasing slightly.

News of big naval victory off coast of Australia [Battle of Coral Sea] cheered us up, helped by 200 parcels of foodstuffs from Henry In S.

Afternoon & nightly teas are being laid on – one advantage of a small mess – and we are no longer assailed by pangs of hunger. Henry's parcels consisted mostly of soap, toothpaste, etc, so will be well off in that respect for many months.

Sunday May 10th

Woke up this morning & found that both forearms were badly bug-bitten, probably from a wicker chair I sat in last night – I must try to remember that chair in future. News about the Bn is very good – comfortable billets, electric lights, showers, no flies, good food etc. I hope I get an opportunity to join them later.

Watched a 'big' cricket match – Aussies v. an English XI. Barnett was their stumper and Capt Chipperfield also played – nothing helped the ball to turn. They scored over 200 for 4 declared and we were all out for 60. It was amazing to see in a POW Camp spectators sitting around in long cane chairs, some with sunshades, and watching 'test' ... [there is a gap here as half a page is torn out]

Monday May 11th

Started in new MI room with gunners – very good quarters. We work 2 mornings 8.30 – 9.30 and then one afternoon. The work is quite light.

Weighed myself at RASC – 10st 10½lbs – satisfactory as I have lost only about 6 lbs. Went along to RAOC where I rummaged amongst the salvaged clothing and collected shirts, pants, towel & long trousers. They will need disinfecting and boiling as they were dirty and stained. Still they come in very handy.

Before lunch 3 staff cars with big Jap naval blokes came to the mess. Seemingly they had lost their way – someone suggested that they were scramming following the news of our …

[Again there is a gap, as half a page is torn out – probably to avoid the guards realizing that the POWs had an illegal radio with news of the war.]

Tuesday May 12th

Early sick parade with Roulston in new quarters. Finished by 9.15 – rest of day free. Julian Taylor examined our permanent unfits with a view to repatriating if the Japs agree. He said that he was a little uncertain what cases to include as the Japs had given him no lead.

June Awards announced. CO gets bar to DSO, several other Cols also get a DSO. Considering that most units did not put up a good show, it is rather surprising.

Henry sent a parcel to Richard containing bacon & steaks etc so that a few good parties are in the offing.

Owing to the rapid diminishing of numbers of men up here, unit educational & vocational training has stopped and the Div varsity may also have to stop. It will be a terrible shame as a vast amount of work has been put into it to get it going, and books have been collected from S for each faculty. However we are POW and have no say in the matter.

Wednesday May 13th

Still very hot – long spell without rain – flies are increasing again. 36 bottles of Jap vit B tablets have been sent to the MI room for our use – directions are in Japanese, dose about 6 a day. Henry sent 4 jars of Marmite up to RS (?) – I bought one for 75c and will use it on the table.

Went up to hospital & saw Titus & collected a food parcel Henry had sent for us – unfortunately ¾ of it had been looted. Felt very hungry so Padre & I opened a tin of Kraft cheese and consumed half before dinner. News that the body of Mark & his batman had been found in Burkiluna [this could be Bukit Timah battle] and been buried. He always had a premonition that he would die in action and he was particularly depressed in Deolali.

Electric light was installed in the mess and used at night – lovely scene in the mess as light helps the long evenings. ADOS came and complained about the beacon effect so we put shades on.

Thursday May 14th

A large party left for S this morning exc. Div troops & 8,000 Aussies inc. 700 sick left S to go abroad to an unknown destination.

Walked round the lines with the 'CO' Wilkie – fairly neat & tidy. Conf. notes – new Jap commander & troops have arrived in S – rather officious & may search rooms etc. All OK here, Wilkie having completed his task satisfactorily. Rushed through sick parade hoping to get a cup of tea in before the lecture, but some of Padre's friends arrived &

spoiled things. Talk 4–6 o'clock by Div. ord. [Divisional Orders] to owner drivers – small groups each with a serg. in charge. Very interesting but tiring – will take a chair next time.

Yesterday I saw the Dentist who, after scraping the tartar off for half an hour actually exposed the teeth – now look like Ivory Castles. Pyorrhoea was setting in. I must remember to go regularly.

Friday 15th May

DADH came round inspecting our workhouse again. Stinking report – giving out rockets right & left. We are having a blitz today, but without the help of the sappers, we can't do anything with it. The whole place wants tearing down and a new place built.

I am making a potent alcoholic drink, ostensibly for its vit B content, but raisins, sugar & pineapple juice, if allowed to stand for a few days, make a drink containing more than vit B. These yeast drinks form a very interesting experiment, and consequently more research is required.

Went to evening service – 14 chaps there but no Siddur, & both our chazanim had gone to S. so we sat & discussed rumours. It will be rather a blow if our services have to be discontinued owing to absence of books. There were about 6 initially.

A letter from Div to officers asking them to make out lists of our losses was very hopeful as there is a small possibility of getting compensation.

Saturday 16th May

Had a verbal tussle with the acting CO and managed to get a little action out of him. Went round inspecting other workhouses and comparing them with our own. Ours compares very badly, and we have now got the sappers' promise to do something very soon. It is a strange & paradoxical fact that although we have the worst cookhouse from the hygiene point of view, we have the lowest dysentery return. There must be lots of other factors involved.

The second lecture to owner drivers was more interesting & comfortable as I took my own chair along. I now know how to decoke a car but I doubt whether I would practise on my own. I wondered how my Hillman & Standard were getting on in England. They will be very useful after the war as cars will be very scarce.

Some trucks came up from K Harbour [Keppel – main harbour of Singapore Island], and our MT chaps were driving them – very happy and satisfied with conditions down there – good food & fags. Compared to other labour camps, they are very lucky.

Sunday 17th May

The number of flies has increased owing to the absence of rain for the past 10 days. Very annoying during meal times as one has to eat with one hand & keep the flies off with the other.

Large sick parade in spite of it being Sunday.

Spent a lazy day in my armchair laughing over AA Milne – never realized before what a clever humourist he is. Also reading Haldane's Science in Wartime. I don't like the way he mixes science & Socialism. Everything that's new, good or progressive he attributes to

Lenin, Marx & USSR, and anything vaguely bad or wrong he blames Tories, Chamberlain or Capitalists. It is remarkable for a man to use politics as examples in a scientific book. It lowers the value of an otherwise excellent handbook.

Padre's birthday at night. 'Silman's Special' was the hard drink, & a mighty fine brew it was – I must try it when I get home. We had a four course dinner ending up with bread, marg. & cheese. Continued after lights out, & ADOS came down to give us a polite hint.

Monday 18th May

The number of cases of diarrhoea is going up again due to the raw veg e.g. cucumber now being eaten. All the deep trench latrines have been closed so that theoretically there should be no dysentery.

Without previous warning the RA blokes moved out of our joint MI room, and left us in sole charge. They very kindly left a lot of fittings & drugs which will be very useful.

Replenished my brew with raisins & sugar borrowed from the Padre & the brew quickly started fermenting again.

Message received to send one officer & 11 men to S. for driving purposes ?destination. It seems they will have a good job driving for the Japs – they may even go up to the mainland. Simcocks has been chosen, and that leaves us light. Wore my 'new' long trousers this evening. Rather strange after all these months – drainpipe effect not exactly Kilgour French [Savile Row tailors].

Tuesday 19th May

First bathing party for many months. Wilkie took about 120 men RNF & RASC to the new place past Southern Command – a long walk about 2 miles – not very pleasant water, rather dirty & barbed wire on concealed bottom step made it a dangerous business. I cut my knees & legs in a few places. Still it was a good change & I got a little exercise out of it. It was an awkward time between 12 & 2, & lunch had to be delayed. Rested in pm after my exertions.

Simcocks & his crowd started off 1½ hrs late – feeling is that they will return each night for a few nights till they finally move down to S. for good. They may drive staff cars. There is also a feeling that many more will soon go to S. or abroad & that the whole Div. will leave this area. However sufficient unto to the day etc.

Weather very much cooler in evening – rainy & windy. Didn't sleep much as all the doors were banging & rattling. Got out of bed twice to try & fix them – in vain. In evening gave out another dollop of Silman's Special which is getting to be quite a successful venture.

Wednesday 20th May

Wed. is a 'dry' day by order of the DADH. All water containers have to be emptied and dried – anti-mosquito measures. Went round the area with Wilkie – official tour of inspection. Also inspected RAOC lines myself. Very hot work – looking in and under billets & examining old tins to find accumulations of stagnant petrol.

News from S. that 3 of our officers were down with 'fever' – otherwise they seem to be having an excellent time.

In pm went up to the hospital with Padre & visited cases. In officers' ward came upon a concert party & stayed to watch. Sent some rupees down to Henry via Titus hoping that he will be able to send some more Marmite. Our evening meal syndicate managed to put together about $4 with which we bought several tins of food – this will last us about another week. The evening meal is getting quite an institution & is a pleasant affair in our little 'mess'. The electric light of course makes all the difference.

Thursday 21st May

Am reading a very interesting tome – 'Science of Life' by Wells & Huxley. Medical matters made easily readable & understandable to the layman. While having a little snack in the Padre's room, the General came in for a short talk with him. I hurriedly crammed a slice of bread & cheese in my mouth which made conversation rather difficult.

Lecture on electrical apparatus in a car from 4–6 pm – not very good – lecturer talked vaguely about specific gravity in a battery – obviously didn't know what he was talking about.

Div. concert after dinner – quite good – they improve with each new series, and the rough edges are being knocked off – the whole thing proceeds smoothly. No electric light tonight as one of the wires appears to have snapped. I am not surprised considering the very amateurish way this overhead wiring was put up – reminiscent of Heath Robinson. Argued till late at night on the merits of the British Army & how they actually did well in Malaya under very difficult circumstances.

Friday 22nd May

The Australian Medic. Services have produced several bottles of Flexuer vaccine which will be used prophylactically on the RASC. It is thought that the Aussies did not have much success with it. We have had 1 or 2 cases of diarrhoea each day – am trying to find the source but it is very difficult.

I had my shoes mended at the cost of 10 cigs – using strips of inner tubing for soles – not waterproof but serviceable. Found and mended a one-wheeled stretcher which we hope to use on Sunday to bring Tate down here for lunch. He has been in hospital all the time and it will do him good to see how we live and what we eat!

The General came to dinner tonight and a posh do was arranged for him at the top table & all the unit reps were invited. They had a 4-course dinner (we only had 2 courses).

Shavuoth [Pentecost]. Went to service at night in pouring rain – stockings got soaked – just managed a minyan. Brodie is back again at the hospital & acted as Chazan. He has a good voice but unfortunately he has almost forgotten how to read [Hebrew]! I lent him a book to rehearse during the week.

Saturday 23rd May

Two years ago today we were engaged in a bloody battle at Sternbeck. We often sit & discuss the days two years ago as we retreated through Flanders to Dunkirk.

Letter from Flossie saying that six officers are down with dengue. It is becoming prevalent & we have to send in a dengue return each week. 3lbs of peanuts per unit were on sale & we rushed off to collect them. They were dirty so had to be washed & sorted & will try & get them roasted.

Car lecture in pm on gear boxes – main business finished about 5 pm & for the remainder of the time, the lecturer reminisced.

Another party of 500 RASC have been warned to stand by to go overseas next Tuesday.

Richard Sharp gave a gramophone recital in the dining-room at night – well attended & well received. Afterwards he came into the 'mess' & over a cup of 'chah' we had our own selection. Quick snack when he had gone – cold soup, pasty & rice.

Sunday 24th May

Whit Sunday – or as is probable in England now – wet Sunday. It did rain a little here in the afternoon so we felt quite at home.

4 SBs & Webb went into the hospital with a wheeled stretcher & brought Tate here for lunch & tea. He was very excited & pleased to be up & dressed. It is the first time he has sat at table for about 4 months, so one can imagine what his feelings were. We put on an extra course (soup) out of our fast diminishing reserves, & for tea a spot of food was again produced. With the aid of a gramophone & some newspapers, quite a festive air was given to the afternoon's proceedings. The roasted peanuts had their usual fascination – one kept on eating them until they were removed.

The upstairs blokes are now having proper feeds at night – with table servants etc. It would be funny if after all they did not leave as they had hogged all their food.

Monday 25th May

Whit Monday always reminds me of Wetherby Races [Yorkshire]– I don't expect that there'll be any racing this year. I got up early this morning by mistake thinking it was my turn for duty. I was very annoyed to find two MOs present already in the M.I. room. There is not much medical work to be done & in view of the imminent departure of the party overseas, there will soon be only sufficient work for one man. The cases of tinea are still the most numerous. Beri-beri is being kept in check & sporadic cases of dysentery are still occurring.

A remarkable sight in the pm – a saloon car driven by a Jap soldier drew up in front – two British officers jumped out – carried 14 bottles of beer from the back of the car upstairs, and then one of the officers climbed back into the car. The Jap and this officer waved goodbye & they set off again. In view of the purely carbohydrate diet during the day, at night we borrowed a stove and fried some bacon & then had bread & cheese afterwards.

Tuesday 26th May

At our table we spend a good deal of time arguing the ethics & morals of using other people's tinned food left behind. The padre & I have to pay 50 rupees – a debt left behind –& think that as the food cannot be sent down, it could be used by us. It is most unlikely that the Bn will come back.

Went to the hospital in the afternoon & saw a miraculous cure. A W Company man whom I sent in raving & delirious had lapsed into a coma for 3 days & was thought to be dying. Today he was up & dressed, smiling & talking rationally. The MO thought originally it was a case of malaria (cerebral), & only towards the end gave him injections of vit B, and when he could swallow, large doses by mouth. If it was a pure case of vit B deficiency, it was a startling syndrome. Usually the encephalitis (Wernicke's) causes a lethargic state & this man was violent & had to be held down. This probably put them off the diagnosis. There will be a great amount of useful knowledge obtained from this mass experiment on this island of vit deficiency.

Wednesday 27th May

The companies under canvas have closed in to our area and are all living in huts or billets. We are now about 160 strong. Went round with Wilkie on tour of inspection – dry day. Big flap on – someone had stolen some Jap cig. rations outside the wire. There was a general roll call at noon & at first we had 4 men missing. They turned up 10 mins late.

In pm examined the feet & between the toes of the men looking for Singapore foot (not a pleasant pastime esp. after lunch – rather malodorous). Found about 25 new cases – rather a large proportion in our small community – largely due to shower baths & men walking about in bare feet. This is now strictly forbidden.

Played a few rubbers of happy-go-lucky bridge in evening – several opening bids of 3 no-trumps. Started on our 'looted' rations at night – bully beef stew & pasty. The evening meal is definitely the best meal of the day. The padre brought a gunner major in – v. nice bloke – fought up country using 4.5 Canons as mortars from 300 yards – v. uncomfortable. One battery was surrounded & wiped out.

Thursday 28th May

Rained incessantly all day. Orders came through for the overseas party which is leaving on Saturday. We are providing 1 officer & 40 men. The Shaw play 'Arms & the Man' which Southern area were producing has been cancelled today owing to rain – open air show.

Motor lecture in pm – dopey lecturer – he seemed to fall asleep in the middle of sentences & was constantly dropping things out of his hand – must be pathological.

Special menu in evening for overseas party only – duck & green peas. They had a dinner worthy of the Grosvenor. It is a strange POW camp that allows us to turn out such a magnificent repast. We heard that the C-in-C has been kept in a cell for 4 days without food – no comforts or furniture – & was questioned & requestioned in the hope that his resistance would break down. The Japs go to both extremes, and one can never be certain whether one will be treated very kindly or biffed over the head with a rifle.

Friday 29th May

The canteen now has large & varied supplies of canned foods & foodstuffs inc. Mackintosh's toffees & Bovril. We bought a jar of the latter as the Marmite is running short, and some soups & sardines. Only one tin of margarine was available for the unit so we let the men have it & we shared an old 'combined' tin.

Brody, who is our new cantor & is working at the hospital, sent down a note saying he is having the service in the Cinema, & 'could we come through' the wire. As this was impossible at a moment's notice, I had to cancel our service here. Fortuitous circumstances as we had an excellent musical concert in the evening in the dining hall provided by the Div. artists. They mixed high-brow & popular items, & had a stage with lighting specially erected for them.

The Japs sent a large quantity of fish in the evening – all sorts & kinds – catfish, dogfish, eels, shark etc. Most of it was going bad & the rest was gutted by the light of the moon & covered over till the morning – hoping!

Saturday 30th May

A serious outbreak of diarrhoea among the RASC occurred – was investigated & nothing found. There must be numerous cases of 'faecal' diarrhoea due to dietetic upset & not due to enteric bacilli. It was a strange coincidence that it occurred when the party was supposed to go abroad. It has now been postponed till the 2nd week in June.

A high-ranking Jap M.O. is coming round & we are to display our wounded & emaciated fairly obviously. After a lapse of some weeks, the Changi Medical Soc had another meeting – Symposium on Australian Medic. Practice, but I went to Car lecture instead – picked up some useful hints on brakes & braking systems.

After our usual nocturnal feed we sat on the verandah in the moonlight & discussed war. Padre is firstly of the impression that war is not inevitable & will soon cease to exist as men become more civilized. I argued that history always repeats itself – there will always be wars & that in some countries civilisation is going backward. The nature of man cannot be altered & the bad side occasionally comes to the top with will to power, leading to war.

Sunday 31st May

Shaw's play 'Arms & the Man' is being produced by Southern area – large party from 18th Div inc. myself marched over to see it. Open air show in grassy hollow – scenery made out of twigs & leaves – v. ingeniously constructed portions fixed into slots in the ground. Actors had remarkable variety of makeshift uniforms & dresses (our military 'order' was the 18th Div Sigs). Although the play was written 40 years ago, many passages were strangely topical & could have been applied to the Jap-British fight. Five minutes before the end it started to pour down, so that the play ended prematurely & in some disorder. The audience scattered & ran back to their respective areas. I have never seen rainstorms to equal it except on the films with Jon Hall [1940s actor] etc. The rain tore down trees & wires, & the main roads were ankle deep in water. In spite of my raincoat, my underclothes were soaking wet. I had a bath, rub-down, tea with a good dollop of rum in it, & felt a lot better after my battle with the elements. Frequent lightning & thunder provided the necessary background for this electrical disturbance.

Monday 1st June

The weather as befitting the day was glorious. The malarial drains etc looked a sorry sight after the previous day's havoc. Managed to acquire from the Padre a new pair of long KD trousers (Australian). It seems that both in quality & quantity the Aussies' equipment & clothing was far superior to ours. They sent bales of new & old clothing to ADOS where they are now being sorted out & distributed to our men, some of whom have only one shirt & shorts. Many of them have no boots, blanket or groundsheet. It is rather hard on those going abroad as they will have no opportunity of obtaining them.

Tried to get the trousers shortened on the sewing machine, but in the absence of the special needles required, all mending is being held up.

Some cad pinched my only cellonese [a synthetic fabric] pair of pants off the line. Fortunately Brooks produced a pair from his kit (which he had saved). Henry sent up two parcels, one for Richard containing a lot of bully beef etc & one for us containing soap & toothpaste.

Tuesday 2nd June

When we opened Henry's box from S. which had been packed ready for dispatch for the past 3 months, it was found that the packets of fruit pastilles had melted & mixed with the soap, sugar & toothpaste to form a conglomerate sticky mass. However much of the contents were rescued, but what we are to do with the enormous quantity of soap is not yet decided. I am trying to exchange some of it for bread which an officer is trying to sell.

There was a good meat ration today 7oz/man & a live cow for the end of the week & vegetables in galore. The difficulty is finding containers for cooking the veg – mostly spinach-type i.e. vegetable tops. We had new potatoes at night, roast beef & marrow. Pineapple & peanut flan to follow. A good meal.

Rumours are still scarce but heard that 1,000 bombers had flown over Cologne the previous weekend. We are eagerly looking forward to some sign of activity up this way, but all is quiet on the Eastern front.

Wednesday 3rd June

All Z & W Coy tents & stores have now been moved from the old site & when Wilkie & I did our tour of inspection this morning, there were only a few drains left to mark the site of the tents. The unit gardens which are left have just started to produce potatoes & cucumbers, so will still have to be tended.

The WD [medical device for ear examinations] auriscope battery had corroded & was immovable, so had to get the ordnance workshops on to it. It had to be drilled out as it was so firmly fixed. I have put the box away for the duration, as it is impossible to get batteries to fit.

The electric light plant was taken from a large house outside the wire, & brought in by a party of RAOC chaps who fitted it up themselves. They have done an excellent job of work & have a little powerhouse going constantly, either charging batteries or giving us light which is on from 8.15–10.45. The Japs now know about it and have cast covetous eyes on it, so that if they think they may require it, they may whip it all away.

Thursday June 4th

As a climax to my suspicions re my batman's activities, he was seen going to our officers' cupboard during mealtime while we were all away. The cupboard was later found to be unlocked & to contain foodstuffs & toilet articles. When I taxed him with it, he denied having taken anything & made some excuse for being at the cupboard. I have missed little things like a cigarette from my locker & have often caught him looking through my things when no-one was about. It is a shame that after having a batman for 2½ years, one cannot trust him. During the last war, the batmen must have been made of different stuff for they were loyal to their officers & stuck by them through shot & shell. This war their only object appears to be to steal as much as possible without being found out. Being a POW seems to have killed the finer & decenter feelings & actions of the men. I was very disappointed with L. & have taken action accordingly.

Friday June 5th

In a flurry & whirl of excitement & hurried questions & answers, Sandy paid a flying visit. The Jap guards were kept amused & given a cup of tea while Sandy told us about their troubles with Flossie etc. He brought a letter from Tom to me giving me the latest props. In typical Sandy fashion he had brought all the things we asked him to bring i.e. Marmite, vit B tablets, Horlicks, herrings, & there were also tins of jam, bully, a cake etc. We had a jolly good food binge at night with our new hoard.

L. asked to resign his post as batman. I accepted eagerly & he left under a cloud – not a pleasant ending to a long association. Appointed McFall in his place – shy, but steady & trustworthy.

Just managed a minyan [quorum of ten men for prayers] at night. Arkush was cantor, & we discussed the possibility of having future services in the hospital area as it was central, and we could always count on the walking wounded & the medical orderlies. But, in view of a possible move, matters were left as they are.

Saturday June 6th

My new batman started his duties – had to show him round & explain everything. Helluva flap when a letter arrived at 12.30 asking for a return by 13.00 of personnel fit to go to S. Had a muster parade & only pushed four out to remain behind. It certainly looks as though a move is imminent, although there is nothing known officially.

Eggs at 10c each were obtained from the Local Purchase Canteen & at night we had a fried egg on a bulky beef rissole – really hot. It is about 4 months since I last tasted an egg – one of the dishes that I really miss.

The car lecture in the afternoon was very poor – back axles & differential; too much names & theory, & too little practical advice. On the whole I have picked up ... [There is a gap here, as part of the page has been cut out.]

Sunday June 7th

Today is Corps Sunday for the RASC, so the Padre preached a special sermon. We wanted him to deliver a bitter denunciation of the Japanese, but it was decided that the time was not appropriate.

The officers played the NCO at football – the former seemed to spend most of their time on the ground. Some of them had sticking plaster on various parts of their anatomies in the evening.

The CO RASC delivered a special address to his men & officers.

Went up to hospital to see Tate who had his heel explored the previous day. A few bits of leather were salvaged – the piece of shrapnel was not found. It is now nearly four months he has been in bed – all due to a bit of metal in his heel.

Monday June 8th

The epidemic of lock-breaking & cupboard forcing is getting very bad. The 3rd in a week was Ken's. His property was in a fearful mess & jumble. He suffers from a bad habit of squirrelling all sorts of odd junk that he has collected over a period of years. It took us three hours to sort the wheat from the chaff – hang his numerous suits up to air & lock away his personal papers. It is remarkable how a man can be so careless with his possessions. A large quantity of tinned food was locked away in case he ever needs it.

In view of the sad fate that befell a Naval yacht carrying Adm. Spooner & about 40 Navy & Army personnel – they landed on a little island – after a Jap destroyer had taken their petrol. They were smitten by blackwater fever, and only 20 dying skeletons were rescued after 3 months of suffering & privation. The people who got away were very lucky as this story indicates.

Tuesday June 9th

Dengue is on the increase – due to the increased number of mosquitoes. The DADH sent a long signal message asking us to smash & hack away all vegetation & undergrowth etc. In view of the large areas which are now uninhabited, it is a very difficult task. We still have one or two cases of diarrhoea each day in spite of all the precautions we take. When & if we move, we may have to start all over again from the beginning.

Wrote to Tom & went up to the hospital to post the letter. Visited some of the sick – one man who is mental (?Vit B deficiency) believes that he has electricity in him & keeps on touching the ground to 'earth' himself.

I am improving the shining hour by delving deep into the library – mainly English literature, geography & history. I am rather pleased with myself that I can now sit down & enjoy a book that I formerly would have called 'heavy'.

Wednesday June 10th

We have a large bunch of bananas hanging up, waiting for them to ripen. Bananas & pineapples are very cheap in this island, & only a nominal sum is paid for them for carriage etc. The mess gets a delivery of about 30 pineapples once a week. Unfortunately after they have been peeled & the cooks have had their rake-off, there isn't very much left to go on the 'pineapple flan'. We also bought a further quantity of eggs, but owing to domestic difficulties, had to postpone the cooking thereof.

Final lecture in motor car course by Peter Everett (now a Major) & he issued a useful little booklet – Hints on buying a second-hand car. Fortunately I have, or believe I have, a fairly good car waiting for me at home when this war ceases.

Rumours about Wake Island & Guam – not very strong but very optimistic.

Thursday June 11th

During the night, had to get out of bed & mend a lavatory cistern that was constantly flushing. Took me about ¾ hour before I could stop the water flowing.

Heard that our party overseas will not be going with the main party, but will accompany us to S. if & when we go.

Laroque paid a flying visit & collected a lot of officers' kits. We didn't mention the food which has ceased to prick our consciences. He parked a Jap officer in the mess, gave him some tea & generally flurried about. Wilkie was very annoyed with him for failing to report officially to the orderly room. He left a note asking for his kit etc to be packed. What they'll say when they find no food enclosed I can't imagine. I am glad we are getting rid of all the surplus kit as it would have been difficult to return it safely to their owners indirectly.

Eggs at night caused a lot of controversy. Finally it was decided to bake them in a warm oven – owing to absence of fat or heating apparatus.

Friday 12th June

Eggs for breakfast – provided by the hens at the back. We have waited nearly 2 months for our turn, & it will probably be another two before our turn again. The money for paying out came from Div. to Camp today but will not be handed out till tomorrow. This is rather a pity as it is Canteen day today & the men could have spent their $3.

We, in Changi, are very fortunate to get paid as the Japs say that as we do not work, they won't pay us. Personal intervention of the General & the Japs agreed to pay us every 10 days, retrospective from May 1st – $7.50 officers, $4.50 NCOs & $3 for the men. This is delightful news for the men but unfortunately it may start the Black Market all over again, i.e. the fellows will pay high prices for cigarettes. If this does happen, all the money will be paid into a PRI [?] account, & food bought for the men.

Special treat for evening gorge – bananas condensed with sugar.

Saturday 13th June

For the past few days I have awakened early & have been unable to sleep until 7.45 when I feel very tired – like Norman [HS's brother] who always slept soundly only when his morning cup of tea came in! It may be due to lack of exercise during the day as most of the day I sit in the billet reading except when I do sick parades or have a short stroll in the evening.

The men were paid out today – amid much rejoicing, & after much juggling of figures.

I went to a concert on Southern Area at night – had to gulp my dinner down. Webb, who was taking the NFO [Northumberland Fusilier Officers?], dashed off without them, following the wrong party, & later came back for them. The show was a musical comedy – a first rate production, even compared to a West End show. It was slick & went with a professional swing. The best individual was a female (impersonator) who talked, sang & danced like a girl. The whole show was deliciously dirty & reminded me of a Ben Travers show with Tom Walls etc. The best joke was the definition of an Indian army officer.

Sunday 14th June

In the afternoon we paid our usual visit to hospital to see Tate each taking a biscuit or a pasty or a pineapple. Now that the hospital patients are being paid, they will be able to purchase the tinned goods they are craving. The food at the hospital has for a long time been a source of discontent & Becky himself has investigated conditions and appointed regimental officers to each block to hear complaints. 5% of the money received by Command is being used for hospital purposes.

We heard today the good news that we are not leaving Changi but that the overseas party – now named the mainland party as they are proceeding up country – will be leaving on Thursday. Webb & about 40 of our fellows will be in the party. I have come to regard Changi like a home & will be very sorry to leave, partly from the comfort point of view & partly the health, as this corner of Malay is TG [thank God] almost free from malaria.

Monday 15th June

Busy day for units – getting men pronounced fit & making lists of kit deficiencies. The Japs are going to make these up to full scale before the men leave the Island. This is a decent gesture. It also looks as though the remainder will in time get kit up to a similar scale.

[Illegible] Y Company, admitted to Roberts Hospital from Singapore, died today – perforated gastric ulcer & general peritonitis. In such a welter of dysentery cases, it is easy to overlook an acute abdomen unless one is very careful.

Gave 30 cigarettes to Webb for his Motor Manual – excellent little handbook on the working of a car. As the cigarettes are damp & stale, and burn my throat, I was quite glad to get rid of them. The rest will go as presents to the batman, hairdresser etc. Chinese & Jap cigarettes are obtainable 20 for 25c – rather strong but smokable. Buying canteen & local purchase stuff for the unit is getting to be a very complex job & requires a lot of book-keeping.

Tuesday 16th June

Several Malayan officers have volunteered to go with the Mainland party IOs [Intelligence Officers]. They are probably keen to see how their homes & rubber plantations etc have fared. There is another party going off consisting of 500 men, not necessarily fit ?duties, & may consist of some of our fellows.

There is trouble in River Valley camp – men are getting out of hand & discipline is going to pot. They are getting rid of some of their bad eggs by sending them back to Changi 'on medical grounds'. I think that the best way to punish them is to withhold their pay – that would soon stop the nonsense.

We have had no word from Y Coy for a long time, & wonder whether their letters etc have been intercepted. As there is no reliable postal service, news from S is very scarce.

Wednesday 17th June

The meals lately have been very poor due to the absence of meat – today a cow was killed & we got fresh meat. The leaves which form our veg are almost tasteless & it is like eating grass. The absence of salt is felt, especially in the veg. The MI room has now acquired a petrol stove & the orderly makes tea constantly – & is very welcome during the early morning sick parade – before breakfast. The canteen list which came out today was a bit of a fiasco as all the good foodstuffs were ordered in advance (unofficially) last Sat. when the men were paid. Consequently our chaps will get nothing till next weekend. It seems rather unfair. The milk is being rationed – 1 tin to 5 men, 7 cigs per man. These amounts will only whet the appetite – no more.

Weighed myself again – 11st 1lb – so have gained nearly 5lbs in as many weeks. I have developed quite a pot – due to rice & this will account for the increase in weight.

Thursday 18th June

The first half of the mainland party left today after much speechmaking & farewells. CO of RASC spoke to his men on the green, & then Becky addressed the whole lot on the gun park, saying it was not good-bye but adieu, as they were only the advance party. One assumes that they may be going to build a camp for us to move into later.

We are now seven. Eva fell down 30ft into the hold of a ship in S. & only fractured an ankle – he was taken to Roberts Hospital, after which he will join us – in a few weeks time.

The canteen stores arrived today & there was much confusion as preference was given to the party leaving. This caused many men to go without their ration of milk & pineapple. These can never be repeated.

The Japs have supplied postcards for us to write home – they were given to the mainland party before they left. We will get ours later. They are blank cards with just a little Jap writing on the address side & are of good quality.

Friday June 19th

Since the party left, sick parades have been very small, & the morning session is usually over shortly after nine. Work will be slack until some more people come back from S. Two hospital companies consisting of about 4 of our officers & 40–50 men are leaving for the mainland, one this Sat. & one next week – they will set up a 200-bedded hospital, presumably for these parties going away. Capt Arkush is going with the 2nd party, & we will miss him & his siddur at the evening services.

We held our last evening service in the Sergeants Mess & from tomorrow the services will be held every Sat. afternoon in the cinema in the hospital grounds.

I received a letter from the ADMS asking me to volunteer for the overseas party of 500 leaving shortly. As none of our crowd are going, I refused the invitation. There is no point in sacrificing home comforts to go out with a strange crowd of fellows.

Saturday June 20th

Wrote the postcard home in block letters, being very careful not to include double meanings, Braille etc. Simple childish sentences were the order of the day – I wonder what they will think at home when they see my effort. Still the main thing is to get some sort of message home telling them that I am fit. Letters from Hong Kong prisoners, I hear, were dropped over New Guinea during a bombing raid. I hope they drop ours over India!

Went to a lecture by a Capt. in the Marines who served on the Prince of Wales – he described the Bismarck affair & one tended to sympathize with the latter as they were hemmed in by the whole British Navy. Their gunnery was excellent in putting out the Hood & Prince of Wales (who had to retire) in the initial stages. He said nothing about the end of the POW except that future construct of battleships should differ ie battle carriers as the war is mainly air versus sea & the ships are vulnerable to repeated torpedo attacks.

Sunday June 21st

Yesterday afternoon we had our first service in the cinema – 3 officers & about 30 ORs including 6 Aussies. One Aussie is called Silman & went from Lodz (Poland) to Australia a short time ago. They lived in Manchester for 2 years during the last war. We probably have common progenitors. By a coincidence they have an officer, also called Silman, who is a gentile. There was a Leeds chap there who is still a hospital patient having lost a leg (?called Goldstein).

I had a long chat with a Jewish Gurkha officer who was also in the last war – now aged 49 – his family live in Poona. He was very bitter about the British troops as the Gurkhas fought & fell where they stood while they saw British troops streaming back. There are only 3 officers & 20 men left out of a full regiment. He thinks that under better leadership we could have kept the island. I doubt this as the Jap numbered superiority was bound to tell in the end. The Gurkhas are being badly treated now – great pity after their grand show & loyalty.

Monday June 22nd

Heard bad news about Libya – it seems that Tobruk has fallen. A substantial victory now would help the morale of the allies who have so far heard only of evacuations & retreats. In spite of all the bad news, we signed bankers orders to pay our annual subs to the 18th Div. Officers Soc. after the war. This complete faith in victory is one of the reasons why we will win this war in the long run – the inability to conceive of defeat.

The monthly report for the Japs was sent up as usual. Figures for illness are much better, & this was put down to increased fruit & vegetables in the diet. Acclimatisation is also an important factor & I feel that the fellows will continue to improve their health & put weight on, except a few unfortunates who have had several recurrences of dysentery.

Hall of Z Coy came out of hospital – burnt in action during Feb, he looks like a living skeleton with a thin layer of scarred skin over his face & hands.

Tuesday June 23rd

Rained constantly the whole day – usually it rains for only an hour or so, and the rest of the day is fine. There are not many men now under canvas – huts & buildings are being occupied. The second half of the mainland party left this morning, and the Division is now virtually divided into two Battalions (i) Div troops & (ii) Brigades (mixed). We are about 1,000, & the number slightly increases each day as sick men & bad lads drift up from Singapore.

Quite a scare at night when the Japs came round & ordered a complete black-out from 9 o'clock. This means that we will have no lights at night until proper black-out screens are put up. The pity is that they were all torn down when we first came here, little thinking that we should ever have to use them. It may be a good sign – we don't know yet!

Wednesday June 24th

There is a suggestion on foot for us to move out of our billet & for it to be converted into a large ante-room for all officers & properly blacked out at night. A difficulty will be ventilation when the nights are hot & stuffy. Most people will probably want to sit on the verandah.

Many of the postcards have been returned as the writing is illegible & the senders have broken rules. In future it will be limited to 24 words clearly written.

3 out of 4 medical orderlies are down with dengue which is very prevalent, but fortunately is of a mild type.

Thursday June 25th

Today was definitely a bad day.

1. The news was bad esp. Libya & Russia
2. No biscuit for lunch
3. Archie sent some insulting & threatening letters to Willie, Richard & myself via Sgt Leach. His letter to me was very curt & demanded his watch back, which he had loaned to me (as he has a good one) or $5 immediately. The watch cost him 5/- in Cape Town. His letter to Willie was the last word & accused him & the combine of breaking open his cupboard & eating his food – he demanded the full value at once otherwise he was going to 'report it'. We did him a good turn by salvaging his rifled locker & had already paid him one instalment on the food we had eaten. The attitude adopted caused the Padre to write a stinking letter to Archie & he is also going to write to Sandy clearing up the matter. The Singapore crowd owe us about $36 in debts etc & they have several camp beds of the men here, so the grumble is on our side. This incident reveals Veitch's character in all its mercenary, malicious cupidity. It is a pity we can't write to Veitch & tell him exactly what we think of him, but the Padre suggests that we do not descend to his level in a vituperative exchange.
4. No lights after 8.30 pm so we sat on the balcony at night, supping tea & discussing the letter to be written to S. Willie missed his rice, & it was altogether a black day for him.

Friday June 26th

Several letters have gone up from S. especially from Flossie & Burns so we have prepared a statement of account & forwarded it on to S. We are paying V in kind & getting him to draw $5 from the debt.

The canteen stuff arrived today & we got a good haul – sardines, margarine, 8lbs beans, etc. There was an issue of 25 cigs & 3 cheroots a man, but as most of the men were fully spent up, the officers managed to get a full whack of smokes. The cheroots are good & cheap – 20 for 25c – almost as cheap as cigarettes.

Symposium on Dysentery at Hospital opened by Col. Bennett & Aussie Col. V. interesting discussion on types of dysentery & morbidity. Aussies had 16 deaths out of 2,000 cases. 3 cases of meliodosis. Vit. deficiency diseases: encephalitis, sprue, pellagra, etc were also discussed. The chronic dysenteries with pathological colours were very difficult to cure. More virulent types of dysentery now occurring especially in Singapore.

Saturday June 27th

News of outbreak of cholera in Singapore was received today. I took immediate steps to see that all water, vegetables & fruit are boiled. I hope it doesn't spread up to Changi otherwise we are all in a helluva mess. No word of prophylactic inoculation – I doubt whether there is sufficient supply on the island.

Tried to go up to hospital in pm for Service, but as big Jap wigs were floating around on a visit, we were turned back from the main road. First service I have missed since they started. It is strange how men have turned to religion for mental comfort while a POW. The Padre has services morning & night, & crowded services on Sunday. It gives a certain feeling of relief & satisfaction to lift up one's voice in hymns etc. It is a pity we haven't sufficient prayer books & a good chazan so that we could do that.

I am reading a book called Light of Peretz by Noah Elstern which reminds me vividly of Friday nights & Yomtovim [Jewish festivals] & all the strange idioms like 'lie in the ground – not'.

Sunday June 28th

Morris RASC came over from Singapore area today to visit me – he looks a bit washed-out as he is just recovering from dengue.

Parcelled up all the S. officers' kit very neatly & sent the letters off via the hospital when we went up in the afternoon to see the sick. Tate has now a walking iron & will be out soon. A sergeant has just returned ill from Port Swettenham where they have been erecting sawmills. He says that there are many Jap wounded in the hospital there, & also many British P of W in a camp near them. There must have been large numbers of men cut off & taken prisoner later – though from accounts we had heard, the Japs were reluctant to take prisoners during the actual fighting.

As the electric light is only on from 8.25 till 8.30, we have our evening 'fress' [Yiddish for meal] on the verandah by the light of the full moon. It is far cooler than closing up windows & doors in a small room.

We were paid a further $2.25.

Monday June 29th

Letters about cholera are beginning to pour in from HQ, DADH & ADMS, & all instructions re sanitation have to be carefully observed.

We had a little conference at the back of the MI room and decided that the ducks must die – also appointed Emery to see ADOS with a view to disposing of them. We feel sure that the plague of flies & little pests of flies are caused by the fowls & fowl habitation at the back of the cookhouse.

Some of the best laughs are raised by innocent remarks e.g. Jack, when the talking turned to old schools, said 'our gym mistress ...'. His mixed schooling must account for his present condition. We are a cheerful little crowd & have pleasant sessions in the evening. Tea in a container is laid on from the cookhouse morning & night so that things proceed very smoothly. Tempers are more even, & there is a better feeling than there was during the first few months.

Tuesday June 30th

The cholera scare we found out was caused by one case on board ship coming into Singapore. This man was quarantined & all emergency measures including inoculation were suspended.

The fall of Mersa Matruh came as a shock to us – each defeat & each evacuation increases the duration of our stay here. It is a remarkable thing when one comes to recount British victories on land in this war – there is only Ethiopia against the poor natives. How we always muddle through to victory is amazing.

Changi public school has started. It gives education up to army certificate standard, & is required by all sergeants in peace time. There is a large response in spite of the very few numbers here. The varsity term is now finishing & will recommence in a fortnight's time.

Wednesday July 1st

Rather envious of numerous drawings of the Padre's room etc which have been made for him. I have started drawing in fine perspective style my portion of the billet – it will always be engraved on my mind, but the drawing will help. Went to a Southern Area concert at night by same team as last time. Not such a high standard, but 'Gloria D'Earie' is more feminine than a girl. The Apache dance was a perfect one & 'she' was always graceful & light. The humour was still as low as ever, especially the birth control scene & the baby photographs. The last item was a full orchestra on the stage including a xylophone – as a POW camp, this is a huge success.

Thursday July 2nd

Figures published recently showed that there have been 1,760 cases of beri-beri in hospital. This is a very high figure but includes all doubtful cases. There are now 1,900 cases of all types in hospital; dysentery is decreasing except that a virulent form is still coming in from Singapore.

The first cholera inoculation was given today to about 200 men, cooks and men who handle local purchase goods. The MOs were also inoculated. Although the reaction is very slight normally, I was off-colour during the evening and developed a temperature of 99° during the night. There should be no general reaction, but I being hyper-sensitive shoot up a temp very quickly.

We had electric light again at night as the black-out restrictions were raised, but we were not a merry party at night. For the first time we had no meal at night, only tea. J & W were rather depressed as the canteen account is $4 down. As one has no extra money to put in, it is going to be very difficult getting things straight.

Friday July 3rd

Crawled out of bed rather weary this morning after a restless night – arm not very sore but temp up. Canteen stores arrived today incl. powdered milk, bulls-eyes, marmite etc. There is literally nothing that cannot be bought. The amazing part is that the Japs are printing unnumbered notes & giving them to us as an amenity grant, & we are buying the stores which were left behind by the British or were looted from stores. The whole position is ludicrous. It is difficult to ascertain who is making a profit out of these transactions.

Letter from Flossie to Wilkie giving all the dope. It seems that a neutral ship is taking the allied diplomatic staff from Japan to Lorenco Marques via Singapore. This ship may take our mail – I hope it does. The Japs have the habit of arresting the orderly officer who is going out in the evening to inspect the British guard that is mounted to prevent us escaping! Sounds like Gilbert & Sullivan.

Medical notes

1. A fusilier vomited up a roundworm 8" long – he says it gave him a shock! Worms are an oft unsuspected cause of abd. pain, nausea, vomiting & possibly loss of weight. Worms are quite common in this country.

2. Ulcers tend to be indolent & for some unexplained reason heal up to a certain size & then refuse to heal any more in spite of all treatments. Lately we have tried cleaning with lusol, touching up the edges & base of the ulcer with copper sulphate & then Ung. Hyd. Ammon. on lint for two days. The treatment has been very successful.

3. Prophylactic inoculation of dysentery vaccine (Flexiner) has not proved to be of any value. 300 RASC men were inoculated, yet 10 got dysentery or diarrhoea within the month.

4. Otitis externa with a temperature still occurs and is very troublesome. The annoying part was that the hospital only treated them as Ops. Pain & loss of sleep – marked features.

5. There are no cases of backache or fibrositis now as occurred when we first came here. This must be due to acclimatisation.

Saturday July 4th [assumed]

[First 5 lines of this entry and following entry have been cut out of diary.]

... little chat with him before the lecture – most the campaign & the evacuation is shrouded in secrecy. The Glorious was sunk by Scharnhorst, & with no other allied ship in sight, the poor survivors died of exhaustion in the sea. The rest of the fleet was 12 hours behind. The deck was crowded with Hurricanes, and no spotter planes up. The men in

Norway suffered incredible hardship & one section went two days cut off without food or warm clothing. The German planes did as they liked, and even attacked solitary DRs [Destroyers?]. The sight of (the only) two Gloster Gladiators in the sky caused all the men to cheer & gave them new heart. There were no AA guns & the Huns often machine-gunned positions for hour after hour. The effect on the nerves can be imagined. He was evacuated on the Lancastria – which met her end three days later. The work of the Navy was praised – especially in the speed of the evacuation.

Sunday July 5th [assumed, for same reason as previous entry]

...propaganda. He thinks that by Nov the issue should be settled in our favour. We shall see.

The combined efforts of the MOs caused the deaths of ten of the ducks for dinner tonight – a good portion for each officer with roast potatoes, veg & stuffing, followed by a doughnut & cream!

Three Jap guards came into the mess during the morning & I had to entertain them with tea & biscuits. Their knowledge of English was confined to Thank you, and we were thanking each other very often. They were amiable, and except for throwing the tea dregs on the floor, behaved very well. They offered me a cigarette & when I refused they started mumbling to each other – I wondered whether I had insulted them.

Monday July 6th

My first drawing for twenty years or so at last concluded – general appraisal. It has exceeded my most sanguine expectations. I am thinking of doing a complete series – given time & materials.

The Japs have chopped off the hands of several Chinese & Malayans & decorating them with ribbons have hung them up in public places as a warning to looters & rioters. This puts the clock back several hundred years.

The attitude towards the Aussies has changed, and now the Japs are very amiable – hoping to break Australia from Britain.

Tuesday July 7th

There is a series of lectures on general subjects being given at Changi Cinema. Yesterday I attended a lecture on 'Earthquakes'. Most hopeful part is that Japan has a big quake about once every 20 years – her time is now up. Some buildings in Tokyo are made quake-proof by being built on rollers or pontoons. This district is fairly free from major shocks. They occur mostly where mountains meet plains e.g. Himalayas. Buildings and articles fall towards the shock. A basement 30ft deep is the safest place.

I purchased two sheets, one pillowslip & a small towel from Brooks for £1. The note was useless to me, but the sheets are invaluable in this climate. Sleeping on a rough, sticky blanket is no fun. Either the nights are cooler or I am getting acclimatised, for I no longer wake up 'vi in a Torch'. [Yiddish expression]

Wednesday July 8th

Aerial activity today – over 60 bombers flew east this morning – perfect formation. AA fire in the distance must have been target practising as we saw two 4-engined seaplanes towing targets. All this is very exciting for the men who try to see in all this an omen.

OAE 53 Brigade was captured in the Banka straits in Feb (?18) & was brought back to Singapore with a few naval ratings. He was put on a ship anchored off Singapore & told to mend the engines under pain of death. He just got them mended in time and was then made skipper. In spite of protests he was confined to the ship & made to do fatigues etc. It would be very interesting to find out exactly how many of those who got away from S actually managed to escape. Those going west through Sumatra to Ceylon had more chance than those going south through Banka Strait to Java.

The aerial demonstration I understand is for the diplomats on board Conte Verdi which is now in S harbour & which is proceeding to sea for exchange with the allies.

Read a copy of the Syonan Times. Forecast an early fall of Alexandria & complete break-up of the British Empire. The most amazing part is the advert of the show in Singapore: 'all singing, all dancing' Military 10c.

Thursday July 9th

Lecture by Percival (C-in-C) on BEF in France. He was on the staff of II Corps. Not very good delivery & not much personality. The material was good – lucid, clear and to the point. He ascribes our failure to 3 things:

1. Heart of French people not in the war
2. No reserves available
3. Political intrigue e.g. Laval.

The French had only 400 fighters & no modern bombers. They had spent millions on the Maginot line, and had no money left to spend on improving the army. Therefore some of her first line transport was still horsed at the beginning of the war.

Reasons why we went into Belgium:

1. to shorten the front
2. to keep Lille out of the fighting area – important industrial town for war materials.

The Germans knew this plan because of the fake attempt in Jan. They pushed 6 armoured Divs through the Gap & cut off the N armies. The French in the south, instead of attacking, merely lined the banks of the Somme. Percival put in a good word for the TA Divs who fought although only half equipped.

Friday July 10th

Canteen supplies came through very well. Out of my meagre pittance I paid $1.05 for a tube of Brushless Shaving Cream. I could have bought a small Marmite for the same money. A good buy is a large bottle of Soya sauce at 44c – it contains a high proportion of salt of which we are now short. The Japs threatened to stop our supply of salt & Soya sauce, but it seems to be arriving just the same. Jap tinned salmon is the price of well-known brands & in the interests of economy we have to buy it.

I obtained the key of the room of Padre Ross's (who has left) – it is only a small unfurnished room, but I use it to swot for two hours during the morning – French & German. The only drawback are the '4 engined bombers' [probably a type of beetle] but they keep out of sight during the day unless disturbed. Our padre has a helluva time each night in his room as he has to kill several before he feels safe. They are about 3" long with armour-plating, big wings & long antennae. They can fly, jump, climb and are a devil to kill. They come from the drains and hide in dark places like cupboards. Their appearance is quite frightening, but actually they are quite harmless.

Saturday July 11th

Had good attendance at [Shabbat] service, about 30 including 5 Aussies. Unfortunately there are very few officers among the numerous Jewish officers who care to come. Morris and I are the only regular attenders. One sees round here a trend towards religion as part of a reaction in a POW. This tendency has not made itself felt amongst the Jewish officers. There was scheduled a lecture on the Western Desert, and as I was leaving the service, a brass hat got hold of me & poured out a few indignant phrases – that he was going to jail – I don't know in what capacity. I hoped he would have a good time.

We did the 2nd cholera injection (ICE) Japanese stuff – clear water coloured fluid. As I have had no reaction at all, I suspect it may actually be water.

General's birthday today – many gifts & congratulatory messages sent to him. He was very touched & issued a special message of thanks. First night of Dover Road [play by A. A. Milne] – huge success.

Sunday July 12th

Further to yesterday's remarks about Becky – the general opinion of him has gone up by leaps and bounds. He has certainly done everything he could to make our stay here as pleasant as possible & he has actively identified himself with all the educational & university faculties. Given a fair opportunity in war, I am sure he would come out very well.

Walked over to S. [South?] area in the morning to see Morris – borrowed a German dictionary & a German novel from him. S. area has good billets on the side of a wooded slope, very green & pleasant. Very few ORs, so that most of the work is done by officers e.g. flag carrying, wood-chopping, cooking & bathing. [Rest of this day's entry has been cut out of diary.]

Monday July 13th

Addy sent up for his German book so I settled down to copying the important parts out. Books in modern languages are very scarce here, so that the University is starting over again with Modern languages this term. I have sent in my name for German only as the pm sick parade may interfere with French. I am reading a simple French book, Contes et Legendes – a book I read over 20 years ago. On paper 20 years seems a long time, yet when I read the book I am vividly reminded of my schooldays. [Rest of this day's entry has been cut out of diary.]

Tuesday 14th July

I purchased some Gula Malacca today – they consist of solid brown sugar with a nutty flavour. They are very pleasant to eat but expensive at 5c each. The eggs we are getting now are of the large duck variety, but unfortunately a lot of them are bad. Their natives probably collect them for a few weeks before they sell them. When hard-boiled they have a peculiar taste & smell, and are better fried.

We have a small petrol stove in the billet, and at night we fry minced up sardines & rice or fried bread (when we get any). It makes a tasty dish with Soya sauce added.

Wednesday July 15th

Today was the 'big day' as the C-in-C, 3 generals & lesser fry were assembled at St George's Church awaiting the Bishop of Singapore who was supposed to attend a special confirmation service. Unfortunately the Bish. failed to appear, & after hanging about for 2 hours, they all went in to tea. A marvellous tea had been laid on – all kinds of sandwiches (real bread) & even tea with milk & sugar. This more than compensated for the former fiasco. I don't know whether they can afford a 2nd tea when the Bish. actually comes.

In evening went to play Dover Road by AA Milne, produced & played by 18th Div officers. Under the POW circumstances, it was excellently done – scenery, suits & dresses were first-rate. Wilkie was the best individual performer. Dillon appeared to be wondering what came next – this spoilt his acting.

The Japs wanted full particulars of all MOs – reason unknown.

Thursday July 16th

Paid again today $2.25. This just about covers my debts. The hospital patients especially the officers pay large sums towards extras for the hospital & only get 75 cents for themselves. There are bound to be grumbles especially by the chaps who do not get extras, but someone has to pay for the thousands of eggs, tinned food etc which the hospital buys daily. It is a very difficult problem.

Much excitement today when it became known that two large parties are going to Japan next week consisting provisionally of old sappers and senior officers down to full colonels. This news which was supposed to be secret got round so quickly that at the S. area concert at night, the programme seller called 'buy a programme to read on the boat'. It was the 2nd time I had seen the show and I enjoyed it just as much. It was a big night as the C-in-C & many other red hats from Command were there. The show was as funny as ever & even Percival's face broke into a smile now and then. I suppose his mind was on the forthcoming trip.

Friday July 17th

The main topic of conversation was naturally about Japan – although our unit is not itself involved. The 8 MOs in this Division had to put Yes or No against their names – we all put No except one bloke. So Roulston, MO of the Sappers was chosen also to go with his men. He didn't appear to be frightfully keen, although Japan has many attractions to offer.

The trip is causing havoc with all the various side-shows ie deputies have to be found for main parts in the play 'Dover Road'. University lecturers are going away. All Command Div & Bde staffs are going. In a way it will be a change, but whether for better or worse is a moot point. The people who are going are indulging in much wishful thinking.

Saturday July 18th

Interviewed the OC Modern Languages & was informed that in view of this big move & the possibility of losing many lecturers, the next session's start would be postponed for a few weeks. This was a bit of a blow as I had been slaving away at my German for 4 or 5 hours each day in my little room. I have nearly finished copying out the German Grammar Book so that it can be sent down to S.

Service in pm – 28 attended. Brodie is going to Japan so we will have to rely on our Paylisher [Polish accent] cantor in future.

Visited Sgt Armstrong in hospital – he has had a bad dose of beri-beri. I am not satisfied with his condition.

Lecture on the Western Desert by a Col. who accompanied Major Bayley in 1930 when they went exploring the desert in search of an oasis. The talk was more geological than geographical & those who had come expecting a resumé of present events in Egypt were disappointed. An interesting fact is that less than 2,000 years ago the area was habitable due to good rainfall. Evidence of previous habitation is abundant. For some unknown reason, the water supply is getting less & less, & travellers have to depend on the oases.

Sunday July 19th

For some unknown reason I wake up about 5 am & am unable to fall asleep again. As I don't sleep during the day & do a few exercises every now & then, I am at a loss to explain it. On the whole the weather has been cooler & the nights are not sticky. There have been severe wind & rainstorms in the early hours of the morning & I have had to get up to stop the doors banging. If it is a cloudy morning there is barely enough light at 8.15 to see to shave.

A large supply of bananas came in today & we buy them at 25c a lb – this makes the little ones cost about 1c each. As they were ripe I ate rather a lot today & developed bellyache at night – however I had assimilated the vitamins. We are having large duck eggs once or twice a day. They have a peculiar fishy smell which makes one think they are bad. When one overcomes the initial nausea, they taste quite nice. Tinned beef dripping (Australian) helps us to fry them at night.

Monday July 20th

All the overseas party have had to undergo a rigorous examination. Thousands lined up on the gun park & there were several teams of doctors & assistants, Jap & British mixed – all working as fast as they could.

Men were examined for VD & then a rectal swab was inserted & a smear taken on an agar plate. The lobe of the ear was then pricked & a blood film taken. The organisation necessary to prepare, number & examine all the specimens is enormous. Besides this, all the luggage will be put through a disinfectant to 60°. Personal kit carried in a pack will be sprayed with carbolic & the person himself will be dipped in disinfectant. They are certainly sparing no effort to make sure that we take no disease into Japan. What amazed me was to see a small Jap MO standing on tiptoe to prick the ears of the tall British Tommies.

Tuesday July 21st

The medical examinations continued all today – they had to wait 3½ hours in the burning sun in the pm for the Japs to turn up. There was a parade in the morning for the new Jap C in C who arrived recently. There was a wait of 1½ hours for him.

I acquired a nucleus of a buggy camp bed which I attacked vigorously – disinfected & obtained new canvas for. I do not intend to use it until when we have to move. It is always best to be prepared. I am afraid that my mattress would be too bulky to take. The deportees are collecting warm clothes & battle dress. Even officers SD [Service Dress i.e. formal dress] will come in useful.

Wednesday July 22nd

The party that was supposed to leave for Japan has been put off for a few days. This may be due to the fact that all the cultures & specimens have not been examined. There was a rumour that the boat taking the Aussies some time ago had been sunk & there were no survivors. The Chinese appear to be doing very well in the coastal sector near Foochow. This will be a menace to Jap shipping if they get any airports. It is thought that if Britain had given full aid to the Chinese a few years ago, the latter could have wiped up the J[apanese]. They must be getting a lot of supplies from USA now.

Simcocks came up today for the first time. He has a party of men in Singapore driving trucks. They live underneath a house occupied by J[apanese], & until 2 weeks ago were getting splendid food ie 1lb meat/day. A change of command brought down their rations. He has a photo of British prisoners & Jap guards posed as a group.

Thursday July 23rd

Didn't feel so well when I got out of bed so took my temp & found it was 100°. Assumed I had Dengue & climbed back into bed. My being in bed caused much mirth & laughter as I had been very 'firm' with the others when they had Dengue. I had an intermittent stream of visitors during the day, making a quiet sleep & a sweat impossible. There was a crowd of 8 round my bedside [at] teatime, & I shared my lunch biscuits with the first few. The rest I offered APC.

Brooks lent me a dressing-gown & a suit of pyjamas so that I could sit up in comfort. The symptoms of dengue have not been very bad so far – merely a general malaise & hot burning eyeballs. The pain in the limbs & back are missing.

[Interval of 1 week in diary at this point.]

Friday July 31st

I am writing this propped up in bed in U3 ward in Changi Hospital. Much water has flowed (in every sense) since I wrote the notes on the opposite page. I had a pukka attack of Dengue with a saddleback temp & slow pulse (58), complete anorexia – nausea, smarting eyes, and on Monday when the final phase was starting with a rise of temp & a terminal rash, I was looking forward to getting up soon.

On Tuesday (Black Tuesday) my temp was falling & I was feeling a little like some food so my colleagues (?friends) who had a half tin of sardines – opened the previous day – gave me a fish. This tasted a bit dry but not bad. I also chewed a few pieces of coconut during the morning. About lunchtime I felt a bit sick – refused any food & soon brought up the morning food. This developed very quickly into a serious state of vomiting & diarrhoea – large copious amounts of fluid coming away simultaneously from both ends. This was combined with extremes of exhaustion & collapse. I had to be helped to & from the latrine. Covered with a cold sweat & had an aching pain on the epigastrium. Ineson was similarly affected & took to his bed, but none of the other three (Padre, J Brooks & Williams) were affected. It was a typical acute gastro-enteritis – undoubtedly caused by the fish which somehow spared three participants.

As we were getting rapidly dehydrated in spite of drinking hot tea between the bouts, the doctor called for the ambulance & sent us to hospital. There was much consternation amongst the officers and spectators who saw us being taken away as we looked almost moribund. After a wait of ¾ hour in the reception office, we were carried up 3 flights of stairs by four non-co-operative stretcher bearers & my life was in danger of being suddenly ended. Fortunately the attack very soon settled down & though a slight diarrhoea and a belly ache remained for a few days, the whole show has cleared up.

Very little treatment was required except a stiff dose of Mist. Alba. I was on a fluid diet for three days & am now on ordinary diet & still have a poor appetite. Brooks was very good during the acute stage & looked after me like a mother. He lent me his dressing-gown, pillow, pyjamas, extra blanket etc. I am afraid I must have made a helluva mess at the back, but at the time I didn't care as I was concentrating on [not] dying.

During the dengue do, the other lads were very attentive & scrambled eggs for me & once the Padre cut it up small & kneeling down by my bed, he fed me like a child. Wilkie would come in at night & boom out in his parade voice – How are you? and walk away without waiting for an answer. McFall worked hard & came in several times a day to make my bed & hang the sheets up on the line to dry. Emery & Roulston came in once or twice a day to feel my spleen & look for a rash. I had a more severe attack of dengue than the general run.

Saturday August 1st

I have been seen by the MO this morning & he is discharging me tomorrow. I do not like being in hospital – the hard bed takes some getting used to & I haven't slept very much. There is a lot of noise & chatter at night – no lights except a hurricane lamp which an orderly flashes up & down the room as he walks. Jack & I would sit out on the balcony from 9–10 looking at the stars & the lights glittering in the distance & think of our home town. Being in hospital has a tendency to make one more home-sick & to turn one's thoughts to those at home.

The orderlies are mostly Anglo-Indians & are very efficient, willing & keen. The food is not as good as in the [camp] but is certainly not as bad as has been painted. It is served up nicely and in huge quantities. An orderly comes round later shouting 'more rice'.

Ineson went out today – presumably the dengue attack set me back a bit more. The ward is half empty. There are very few other medical cases except an odd malaria.

Monday August 3rd

Bank Holiday Monday. Back again with the unit. I was discharged from hospital yesterday & given 5 days 'excused duty'. Before I left I was asked if I had any complaints about the food or service. When I said I had been perfectly satisfied they were dumbfounded. It seems that it is the usual thing to complain about something or other. McFall called for my things so that I walked out of hospital quite free & unencumbered. I felt a little weak at the knees, but otherwise quite fit.

I weighed myself today – 10st 7 lbs – I have lost about 8 lbs during my illness. My face is thinner but I hope it will soon fill out again.

Sgt Leach had changed my 60 R[upees] for $32.40. $6 of this went to Willie to pay up an old debt. I still have a substantial sum to buy things I need e.g. towel, talcum powder etc. During my absence, the billet was rearranged & all the bed & furniture closed up to the top of the room. My cupboards had also been changed. As my bed was 'on top' of the centre table, I moved it back 2 yards.

There are many new faces in the mess – RAOC & RASC officers who have returned from Singapore. Sandy wrote from S that they have been very slack – although the only time he went on a ship, he had a food binge with the Japs.

I am starting the German Univ. class today – a week late. I have been doing bits of hurried swotting up in an endeavour to catch up. It will take my mind off moping – post dengue depression – and wondering what the folks are doing today at home – Bank Holiday. They won't be able to use the car very much as is our usual wont.

Tuesday Aug 4th

I do not intend to make this a day by day diary as heretofore, but just jot down notes when anything occurs. I have had 2 German lessons so far & am trying to catch up with the others. There are about 10 in the class & they can all speak German to some extent. Today he read out a large slice of Edgar Wallace in German – we had to take it down as dictation & later translate it – not very easy.

My flat stomach amazes me – it must feel the same after a woman has a child to regain her normal outline. My little pot has completely disappeared. No doubt it will return after a few rice meals.

I gave my camp bed framework & some canvas to the sappers to make a pukka bed for me. The S folk took a trip to Paya Lebar to see what they could salvage. They found several empty cases & trunks (none mine) & two ties of mine! – all that is left out of a complete kit!

Saturday Aug 8th

The week of my AUC [?] 7 days is now ending – having spent most of the week in a deck chair reading 'Kleiner Mann – was nun?' [novel by Hans Fallada]. I have to look up every other word in the dictionary but am improving.

Being the wealthiest fellow in the mess (pro tem) I purchased several mixed tins of foodstuffs from the canteen e.g. beans, spaghetti, tomatoes, sausages etc. $10 have already been spent so that it looks as though I will run through the money fairly quickly. A tin of polimalt – a local poor substitute for ovaltine we put in our porridge, but today have bought some powdered milk & this should be better.

Discussion on beri-beri yesterday at the hospital. Important points were:

1. Cases of oedema probably not due to vit. B def – there are other causes eg protein or salt deficiency, climate, exercise – for lots of cases get better just by rest in bed.
2. Skin diseases e.g. scrotal are being treated with Marmite. This cures the condition which is also cured in our MI room with Ung. Zinci alone.
3. Metatarsalgia due to deficiency of ?B has occurred due to sagging of ligaments of the foot.
4. Many cases develop beri-beri on a good mixed diet & other people on a poor diet don't.
5. It seems to attack men who have been resident in this country for many years.
6. The increase in proteins e.g. eggs & meat has cut down the cases of B-B considerably.
7. The cardiac & neuritic are true cases & these alone should get the Marmite etc.
8. Jap vit B tablets have in some cases proved useless.
9. There is a deficiency of all the vitamins & this is bound to complicate the picture.

Monday Aug 10th

CQMS Ormston was sent up from S yesterday afternoon & was seen dressed & carrying his kit. In the evening we got a note saying that he had died from acute faucial diphtheria. The fatal nature of diphtheria is apt to be forgotten & if the MOs in S sit on cases of ?tonsillitis, they are endangering the lives of the men. There is a shortage of antitoxin & bacteriological facilities

Two ships, Conte Verde & [Asama Maru] have returned from Lorenco Marques with the exchanged Jap diplomats. We are all hoping that they have letters on board for us. The ships may take the Jap party B away or may return to LM & pick up 10 Jap diplomats who recently left England.

Simcocks was up again & told us about the English guards on the patrol who are armed with sticks to prevent the Japs getting at it. The Jap guards and British guards often stop & hob-nob together. A strange POW camp!

Started work today – not very busy. Am feeling fit again & am putting on weight.

Diary

Changi PoW Camp
H Silman Capt RADC
9th Bn RNF
Commencing 13th August 1942

Friday August 14th

I bought this little book in Cape Town and as my diary is now used up – will make use of this. I wonder whether the '42 diary is on the High Seas having been dispatched as requested. My new batman is turning out to be very hardworking and industrious – for the past two days he has been sewing a canvas top for my renovated camp-bed. It should be like new when completed.

A lecture was given at the Hospital the other day by an officer who was at large for 2 months after the capitulation. He lived with Chinese guerrilla bands on the mainland and was well treated. There are numerous groups of Chinese and British soldier now living in the jungle – into which the Japs do not penetrate. The Chinese supply them with food etc but they cannot do very much from a military point of view. They all get malaria sooner or later, as mosquitoes are very prevalent in the swampy regions. Am spending my money very quickly – my cupboard is filled with raisins, peanuts, dates, Gula Malacca, milk powder, bananas, jam, margarine, eggs etc. The meals are improving too, so that we are now getting a good mixed diet.

Sunday August 16th

It is now six months since the capitulation – half a year of captivity which on looking back has passed very quickly and not too uncomfortably. Originally we thought that conditions and food would rapidly deteriorate but on the contrary the food and rations have improved and we can buy almost anything we like from the canteen.

The Padre introduced me to L/Cpl Durrer RASC – a Swiss boy who came over to England and was naturalised. He speaks German and French fluently but the former with a Swiss accent e.g. he says 'w' for 'v' and a guttural 'ch' in ich. He has forgotten quite a few words but is very useful to me as an instructor. I have arranged to spend an hour each day in conversation, one day French and one day German. In my opinion conversation is the main thing and I intend to concentrate on that. I may even give up the University if it affects the hours mutually suitable to ourselves. I had my first 'Stunde' lesson yesterday and I pride myself that in the few weeks I have studied the language I have mastered sufficient knowledge to keep a conversation going.

This morning and at long last, the first Japanese party left. It included from 18 Div colonels and upwards. All the troops lined the roads and gave the departing brass hats three rousing cheers. I went along also to bid my adieu to the chief who was obviously affected – tears were in his eyes as he leant out of his car to respond to the cheers of his men. It was a touching scene – a general saying goodbye to his Division – whom he may never see again and whom he had such a raw deal with. He issued a farewell message copied on the opposite page.

The brigadiers were crowded on open trucks sitting on top of their luggage – an unusual sight – but they took it in good part and smiled down at the troops.

Special Order of the Day
by Major General Beckwith Smith, Commander 18th Division

On my departure for Japan I wish to take what may be my last chance to thank all ranks for their cheerful serious and loyal support on many shores and seas during the two years in which I have had the honour to command the Division.

I regret that I have been unable to lead you to the success in battle to which your cause and your sacrifice is entitled and although I leave you with a heavy heart I carry with me many precious memories and a sense of real comradeship such as could only have been inspired by the trials and disappointments which we have shared in the last few months. Difficult days may still be ahead but I know that the spirit which today animates all ranks of the Division will prevail and will form the corner stone on which one day a just and lasting peace will be founded. God grant that that day many not be long delayed and that we may soon meet again.

Meanwhile GOOD LUCK, HEADS UP – KEEP SMILING
Signed M Beckwith Smith
Major General
16th August 1942

Friday August 20th

Julian Taylor today re-examined the 'permanently unfit men' with a view to repatriation. We have seven men with the unit and a few in hospital. Frankly I can't see the Japs sending these men back home as they may convey important information. Still there is a small hope.

We received the Red Cross supplies viz. 11 vit C sweets, 9 vit A caramels, 1 lb of jam. Maize, dried soup and 6oz of jam remained with the mess. They will improve the menu and we are very grateful to the donors. We hear that a further boatload is on the way.

We were very excited when we heard about Dieppe as we thought that it was the real thing i.e. second front. It was quite an anticlimax when we heard that it was just a large scale Commando raid. That in its way is a big thing and is a pointer which way the wind is blowing.

Yesterday all the P of W were assembled to line the route and then listen to an address in Japanese by the new C-in-C (General Fuku). RMOs and Padres were exempt for which I was very thankful as the sun burnt down from a tropical sky. The men were out for about four hours and had to march down to II Div to hear the speech which was translated by the interpreter. Unfortunately the majority of the 14,000 assembled there were unable to hear so started talking amongst themselves. This must have caused a bit of bad feeling. The men have no sense whatsoever. The Jap General made a very encouraging speech saying that it was his wish that we should return home healthy and fit and therefore we should cooperate with the Japs and obey their orders.

The sick rate is very high. During this week 104 men out of 165 present in camp have been at least once to the MI room. Tinea is the main cause. Diphtheria and malaria cases occur about one a week. The Padre is down with dengue – mild attack.

Sunday August 22nd

Yesterday afternoon I went to the Service again after a lapse of a few weeks. There were about 30 present. We discussed the arrangements for the forthcoming Yomtovim. I went later to a talk by the Australian wicket-keeper Barnett on his Test Match experiences. He spoke for quite a while about that exciting Test match at Leeds, when they won on the 3rd day and I knelt on the gravel all day watching them. He had a fund of good stories to relate and it was a very enjoyable talk.

Looking back on the past six months I see how I have got gradually acclimatised e.g.:

1. I rarely get up during the [night] now
2. I never perspire at night – at first I used to wake up wringing with sweat.
3. I use a hankie a day now and at first a hankie would last me a week.
4. I rarely drink any water now during the day.
5. I feel more active and lively as distinct from that heavy lethargic feeling we first had.
6. I've lost that constant state of hunger that I originally had, and made some people into animals.
7. The bowels are regular and easy. Initially there was often an interval of 10–14 days.

Thursday August 27th

I have just learned that there was a Jewish funeral last week – 1st 5th Forresters. The BCF [Batallion Chaplain to Forces] had about an hour's notice and dashed about trying to get someone to take the funeral. Fortunately a Jewish fellow in the same battalion rose to the occasion and conducted the funeral. At the Padre's conference today the ACG [Assistant Chaplain General] tried to put it on a regular footing and is getting a nominal roll from all units of the Japanese personnel. And I am going to get a list of names of men who are prepared to conduct a burial service when necessary. The difficulty will be getting someone in each area.

The BCF with whom I had a chat this afternoon says that he is going to get our fellows off for Yom Kippur. I have tried to book the cinema on Monday 21st for the Service. There will not be many who fast this year and am myself not encouraging it leaving it to each man's inclination. The notice about the Saturday Service is also to be repeated. I seem to be the spokesman for the Jews here – as the majority are 'non-practising'. The ACG says that the percentage of Jews in the army is 0.5% – much higher than I thought.

The sick are still flowing up from Singapore and scrotal dermatitis is spreading apace. The bad cases from Singapore are in a helluva state and it amazes me that they can walk about at all. We now have moved the S. platoon and the isolation cases over to the former Z & W site where they are now entirely separated. Cases of diphtheria are still occurring due to contacts arriving from S. and also by carelessness here ie contacts who should be isolated, sleeping and mixing with the other men. The position is at present chaotic but hope to get straightened out shortly.

The news about the Generals is rather vague but the general opinion is that they were transferred from the small 3,000 ton boat to a larger boat – after they had protested about the accommodation – and that the food was better and that officers and men are sleeping together in the holds in tiers – the men in the lower ones and the officers in the top ones. The returning diplomats were fêted by the Japs in Singapore and taken for a tour. It is said that another ship has left Liverpool for S. and may contain mail. We live in hopes.

Sunday August 30th

Yesterday at the Service I met Dr Braham of Leeds (youngest) who came over here in Nov/41 rather unwillingly. He is now in S. area and detached to a coastal unit. I also heard from a signaller that Dennis Cohen who was killed in a motorcycle crash in Penang had a Jewish funeral and that the local civilian Jewish population turned out. A nominal roll of all Jewish personnel has been called for – this is purely for record purposes and not as some men yesterday thought possible ie [for] axis purposes. I have collected about six names of men in different areas who are willing to undertake Jewish burial services should they ever occur G.F. [God forbid].

I spent about two hours visiting the 45 patients we have in hospital. The diphtheria wards are chock-full. They only give ADS Co the bad cases, the mild cases get NAB! [needle aspiration biopsy].

I hear that the Japs are flying a plane from Tokyo with more serum. The mortality is about 10% – there have been 16 deaths – very high for an adult community. If it's not one disease it's something else. Dysentery figures are going up again and also the accursed cryptogenic metatarsalgias which are characterized by being more painful when the patient is off his feet and especially at night in bed and leads to insomnia. It therefore can't be orthopaedic by origin. It is very baffling – large numbers of cases of oedema of feet are still occurring – ?Vit B deficiency as they occur mostly in S [Singapore] men who are on a good diet and especially in men who have been out East a long time. I think there must be something more e.g. a cardio-vascular or circulatory disturbance – cause unknown heavy work in a humid climate.

Scrotal dermatitis caused 2 deaths last week – some cases are very bad and the men roll about in agony on their beds ?vit B6 deficiency. Marmite seems to help some cases. Yet strangely enough these cases occur mostly in S men who are getting 6oz of meat a day and can buy all extras. There is a vast field of research here.

I had my shoes mended by the RAOC and they made a first class job of them. They rebuilt the sole with leather and put an outer sole of rubber on. They should now last me ages.

Sunday September 6th

We have taken part in an incident unparalleled in Military History. The events in order are:

August 30th The Japs gave us the opportunity of signing a certificate stating that under no circumstances would we on our honour attempt escape. Naturally we all refused to sign as it is against British Military Law to give parole voluntarily. Also the question of honour being involved we could not sign in case a good opportunity ever occurred. On the night of 30/31 Col. Holmes gave this verdict to the Japs.

The following day the Japs changed their former request to an order, which the C-in-C refused to obey pointing out as above that we are not permitted to sign. On the night of Sept 1st/2nd, the C-in-C was sent for by the Japs and told that by 6pm the following day we all had to move to Selarang Barracks which was to be a prison for the whole P of W Camp and 'measures of severity' would be applied to us till we signed.

Consequently the following day there was an early morning conference and a flap lasting till midday when the order came to march off about 3 pm. Not knowing whether we would come back or not, all stores as far as possible were taken, reserve rations were given out, library books and tools were scattered among the men, medical stores were loaded on the ambulance which also towed two full water tank trucks – loaded full, with kit, sick men, tables etc.

The motley procession that set off was a remarkable sight. Ovens, stoves, containers were all taken including firewood and rations on small trucks loaded to a height of 10–12 feet. Men carried their belongings slung on bamboo poles and the move was carried out in good spirit, the men being very cheerful and light-hearted. The C-in-C complimented the men on this in an Order of the Day. We didn't know what to do with our livestock so the ducks were carried upside down by officers and the chickens were left (we had a good duck stew the following night).

There was a march of about 2 miles to our new home in the former Australian area at Selarang. This consists of barrack blocks on 3 sides of a square with a concrete courtyard in the centre. Normally it holds two battalions – now it was to hold 15,000 troops (equivalent to ¾ million per square mile).

18th Division had 2 blocks with the threat that if the Hospital also moved, one block would be taken from us. It was estimated that three to five hundred men would die if the hospital were moved – just by the move alone. The Japs postponed this drastic step. As it was, each man had 6 x 2ft of space – just enough for a man to lie down without moving.

Thousands slept outside on the concrete with no protection from the rain. We were fortunate as we were on a verandah on the 1st floor facing outwards and had about 1ft of space between camp beds. Batmen slept at our feet in the narrow space left. The officers on the ground floor verandah were badly off as the malarial drains that ran past their heads were used for excretion purposes both liquid and solid. Consequently the first night the stink was terrific.

There were two water-points for the whole area water limited to one gallon per man for all purposes – queuing up for 2–3 hours was common. No washing or shaving was the order.

One small MI room was available for all the sick. I was appointed MO for Div Troops 10.30–12 each morning. But as the position was chaotic and impossible, I took all my stores to the ambulance and rigged up a little MI room there. Only the urgent and acute cases were seen. No ordinary dressings or treatments were done. The Aussies fortunately had set up a hospital tent in the middle and were doing dentistry and operations. There was an acute appendix done the first night by lamplight as the Japs said 'no communication with hospital'.

Flies were increasing in number and swill etc was accumulating. Latrines were dug all through the first night; the men worked like Trojans, digging long rows of trenches 16ft deep while the sappers made latrine tops from doors of billets, men were sleeping on the roofs and all over the courtyard except for space for cookhouses and latrines. The road running round the billet was out of bounds and the Jap sentries beat up anyone treading on the road. A machine gun was mounted on one corner and surrounded with barbed wire.

The worst feature of the day was the shooting of four men who tried to escape. Two Aussies, one RAOC and one Gordon Highlander (two of them were in hospital with malaria and were dragged out of bed) were taken to the seashore and kept waiting for 30 minutes in the boiling sun. A firing party consisted of 4 Sikhs. The Division Commanders and the ACG were summoned to watch. The ACG was allowed a few minutes with each man first. He said that their bearing was wonderful, stood up straight, refused bandages to the eyes and although wounded in the arms and legs by the first few shots remained standing till mortally wounded. Not a groan escaped their lips. It took about 20 minutes and 27 shots were fired – a beastly and brutal affair. The Sikhs then had to bury the dead. This demonstration was supposed to cower us.

A medical section was sent from Roberts Hospital – given an hour's notice to pack etc and arrived without food or cooking facilities – even the fatigue men who pushed the trucks were kept in. The important question was a medical one, whether we could keep on living under such circumstances. All the latrine space was bound to be used up fairly soon and men were already making a mess over most of the area. Without proper washing facilities diseases spread by direct contact as well as by flies were a grave danger. There were diphtheria contacts scattered among the men and there was no antitoxin for treatment.

The first day there were about 40 cases of dysentery and 22 were allowed by the Japs to go to Roberts Hospital. There were constant conferences between our chiefs and the Japs – we were trying to get the certificate altered to show that it was an order to sign. The Japs stated that they had orders from Tokyo to get the original certificate signed at all costs and every means would be used. Rations would be cut down to a third, water by a half and the accommodation by a third until we gave in.

It became obvious that we couldn't hold out very long under those conditions and reluctantly the C-in-C agreed to sign under duress. The Sappers were united in not signing until actually ordered to do so. They may have been misled by the wonderful spirit of the men. It was an amazing sight. The courtyard looked like Hyde Park with its flickering lights, men sleeping all over the place, concerts going on in every corner, pianos playing for sing-songs, men feverishly digging latrines by lamp light. One impromptu concert had a hilarious 'slap-team' comedian.

The officers queued up with the men for meals from the cookhouse and balancing a hot plate and a mug in one hand, tried to negotiate the steps against a surging downward tide. The Padre held a short evening service on Thursday night Sept 3rd – anniversary of the outbreak of war.

On Friday 4th we cut down our rations in case we had to stay at Selarang for some time. Fortunately the canteen stores which had been brought along lasted out – we got tinned herring, soup, vegemite and cigarettes. This helped to supplement our meagre meals. I was the only MO on duty and had a short session morning and afternoon – altogether in the 3 days there I sent 13 men to hospital.

On Friday we all signed two copies of the Jap form. The C-in-C issued a special Order of the Day (No 3) in which he outlined the events leading up to it and the reasons why he ordered us to sign. It was obvious to any sane man that we couldn't live long under such circumstances. The medical authorities estimated that after a month there would be up to 50% dead from dysentery and the remainder would probably be too weak to carry on the essential fatigues, e.g. fetching water, cooking, cleaning, making latrines etc. The question of the dead was difficult as they would have to be interred in the courtyard as that was the only area available. If there were many dead and dying the place would be a shambles.

To spare the lives of his men, Col Holmes signed and on Saturday morning we were told to pack. A heavy downpour of rain made things a bit difficult. However by 2pm the main party left and the trucks followed and returned again for a further load. I do not envy the Aussies having to live in that billet now as it must stink like fury. A few days hard work will be necessary to clean it up. It was estimated that 400 tons of soil and cement were dug up by the men during the two days of hard work – very good show.

We had a good view of Changi Gaol where the civilians are interned and from our bed we could see the lights go on and off every night. We wondered what the poor civilians were doing and whether they were getting similar treatment or not.

The fusiliers were the first back and what struck me forcibly was the quietness and stillness everywhere, where formerly was noise and chatter. There was very little missing from our billets and what was missing must have been taken by our rear party. We soon settled down to normal, brushed out the dust, unpacked and had a shower and a clean change of clothing and everything was again ante bello. The organ and prayer desk were safely brought back by the Padre and his willing band of helpers. The only ones to be put out by our return were the birds who appeared to have inhabited our billet during our absence.

(Roberts Hospital, I hear, had a large influx of 'sick' officers the day we went away and these were all discharged the day we came back.) The hospital patients all signed the same evening as us and rations were then delivered to them. The hospital was very quiet as fatigue men and staff had been cut down to a minimum. On Sunday we had a medical conference at Division and we agreed that after a few weeks there, it would be impossible to carry on due to the dead, dying and those incapable of carrying on by reason of debility, shortage of food and water. Medical suppliers would soon run out and there would be virtually no treatment for the sick. However we are now back safe and sound and it is a closed chapter.

Wednesday Sept 9th

On Monday afternoon I had the largest sick parade ever, 186 men attending for examination and treatment. I took them in bundles of 10 and was a limp rag when I had finished. Scrotal dermatitis is affecting 1 in 3 and there is an acute shortage of medical supplies. Ung. Zinci which was so very useful is now finished and they are all having PP. Although it is supposed to be a B6 avitaminosis we can't give Marmite to everyone. Also the 'metatarsalgia' and foot aches are now said to be [due to] a vitamin deficiency. They are causing us a lot of trouble.

A medical report from Keppel Harbour shows that they are badly off as skin diseases of all types are common and stores are negligible except for what they buy locally. They are using cod liver oil on ulcers and getting good results and this suggests a vitamin A deficiency. I knew it must be something internal as the ulcers resist all forms of external treatment. Conjunctivitis which is also prevalent may have as its cause a vitamin A deficiency. Malaria cases when discharged are now divided into two groups A and B. A are in good condition and would live through another attack, B are in bad condition and may get a deficiency disease and sink after another attack. This latter group will have atebrin twice weekly to ward off attacks. There have been a number of sore throats and pharyngitis and two cases seemingly minor in nature when admitted to hospital have been D.I. Diphtheria after a few days. The lowered resistance of the men has made them less immune than normal individuals.

Friday September 11th

A medical screed has finally recognized the existence of an eye condition caused by vitamin A deficiency. If not tested in time it causes permanent blindness. It starts off with pain and blurring of vision; opacities in the cornea extend to Bowman's capsule and subst. propria and may cause permanent impaired vision.

A book I have just read called 'Vegetable Gardening in Malaya' shows that Kane Kong contains large amounts of vitamin A as well as iron. Also peanuts contain 60% fat, 20–30% protein, vitamin B and vitamin C, so that the two together form an excellent combination of foodstuffs. A PT class of remedial exercises has been started and as I watched the instructors being instructed I thought of the Orizaba [a ship?] days. The mild PT exercises for the convalescent patients looked to me as though they would put a healthy man on his back. Capt … has been admitted to hospital with ?beri-beri. He appears to have spasm of his muscles – very queer.

Stomatitis due to B2 deficiency is often very severe – starts at the corners of the mouth and affects the lips and gums making swallowing painful. Marmite is suggested by the hospital doctors but they fail to say where we are to get it from. Similarly with metatarsalgia – we are awaiting instructions for treatment. Big flap today – meeting of all MOs and Padres at Division as the Japs want our particulars for the 6th time. It is suggested that it may be associated with the repatriation scheme.

Wednesday September 15th

The first day Rosh Hashanah last Saturday we had a record attendance, nearly 50 being present and 5 off. Owing to the limitations imposed by the prayer book it is difficult to give a true service but we managed to put bits and pieces together and all in all it worked out very well. We are trying to get more English into the Service as the Aussies are not the very religious type. The chazzan reads his Ivrit [Hebrew] with a plus Pailish [East European] accent reminding me of the old Chevrah Shass [Orthodox synagogue in Leeds]. I think that an anglicised service would go down better with all concerned.

There has been a hurriedly summoned meeting of non-combatants ie Padres and Doctors, as the Japs wanted full particulars so that they may carry out article 21 of the Geneva Convention. Naturally we NCs were very bucked and hoped for the best. There were 3 possibilities:

A. The permanent unfits were going away with all doctors and padres

B. A selected few were going away

C. Conditions would be ameliorated ie better food, living conditions and unhindered passage between areas.

We will see whether anything comes of this latest flap. Large numbers of men came up from Thompson Road Camp suffering from nutritional oedema (as we now call it). The hospital would not take them in and didn't issue any vitamin B for treatment, so have written to OC hospital for advice.

Rice polishings are on the canteen list 10c a lb. Command who can get almost unlimited supplies of this are messing about as usual trying to make a decision. So we may expect an answer sometime before Xmas. On Saturday night I went up to Div (C mess) to play bridge, my first game since the Battalion went away. I really enjoyed it in spite of the row made by the other young members of the mess as they indulged in their usual gay, frothy chatter.

I miss the nightly games of bridge we used to have when the Battalion was together.

Sandy came up on Sunday to find out about signing the parole forms. He is looking quite fit and says the men are keeping well. I hear that Burn has refused to sign and that he is now waiting a Japanese escort to take him to prison. It is a different position for Flossie, who will probably have to order him to sign. Most of the other S. camps have signed by now.

Simcocks, who was up yesterday, said that the Japs were very apologetic. They are getting on very well with the Japs and mix with them for occasional meals. Some camps are good in that respect but a few camps like Thompson Road are really bad.

There was news of the mainland party – Webb's crowd who are said to be 50 miles from Bangkok and are building a railway into Burma and constructing huts for 15,000 men. The latter has given rise to much speculation and it is generally feared that it will be meant for us. But the Japs have also taken prisoners in Burma and elsewhere so we are not the only possible tenants.

Tinned food especially herrings and meat is now getting very scarce and the S. folk say that it is almost unobtainable. Our fresh meat issue now consists of mutton and this is taken to be a bad omen as the mutton fridge was said to be the last one. How we shall manage when all the British stocks are exhausted I don't know. Rumours about officers pay are flying around and we may get it at the end of this month.

Saturday September 19th

Yesterday about 1,000 men and officers, mostly RAF but some Army and Navy arrived here from Java. The RAF were formerly at Seletar in Singapore and were evacuated a few days before capitulation. 15 officers are sharing our mess and we have closed in. They arrived with closely cropped hair, dirty and thin. They had been cooped up in the hold of a ship for four days.

They had a bitter tale to tell, especially the padre who is sitting at our table and took part in all the adventures. They had a helluva time in S. as their planes were outnumbered and outdistanced by superior Jap planes and large numbers of pilots went out to their doom against impossible odds. 52 Hurricanes landed on the island in Jan and 11 were flown down to Java. The rest had been shot down or damaged. The Brewsters were ridiculously slow compared to the Japs. It seems that a large number of planters etc joined the Volunteers and were given positions of importance e.g. in the intelligence section. They had no knowledge of modern aerial warfare and contributed to the eclipse of the RAF. Another factor was the total lack of PAD [P. Air Defence?] etc on the aerodromes.

When it was impossible for them to carry on they went down to Java by sea, being bombed all the way down and in Java they found no resting place as they were sent from one place to another by the Dutch who had no army. A small air force and a small but gallant Navy that was annihilated when it sallied forth to do battle with the Jap Fleet. A glorious end, but what a waste of what might be a nucleus of a bigger and better Navy. There were a few American and British soldiers in Java but no serious clash took place. The few Hurricanes that were there swept up and down the beaches as the Japs landed and killed hundreds but could not stop a mass landing.

They went down to the south coast hoping to find a boat to take them off but the Japs were sinking all craft trying to approach or leave Java. It was rumoured that the Warwick Castle might take them to Australia and the Padre says that life belts bearing the name Warwick Castle came floating in with the tide. Very ominous if it is true. It is certain that thousands were drowned trying to escape. No-one will know till after the war the full extent of the tragedy. They went back into the hills and settled down in a hill station after much journeying to and fro, and waited for the Japs.

They have not been treated as well as us and have been constantly under Jap supervision. Beatings up were frequent and four officers were shot after being beaten up each day for a week. They were caught trying to escape.

The Dutch prisoners were treated very badly, they say, as the Japs appear to hate them. Large demolition parties destroyed the oil wells etc before the Japs came and there is only the canned oil available for them.

They heard definite 'news' that Singapore had been retaken by the British and that we were attacking Borneo. It seems that these rumours were Jap inspired but with what purpose it is difficult to say. So we must beware of believing similar rumours. They were able to buy stores etc and thus keep their strength up.

The Japs then selected all officers below the rank of Sq Leader and all non-technicians to go on this party. They are proceeding up North ?destination after a short time here. It is strange that Colonels and upwards of our lot went to Japan so it seems as though the ones left behind will go to Japan and that we will accompany the present party to their future destinations. The officers think that it is like Paradise here, freedom of movement, good accommodation, excellent and varied food, light, water and with all mod cons. In fact we don't realize how lucky we are.

Yesterday I went to a clinical meeting at the Hospital. There was a new type of case on show which is believed to be due to vit B deficiency i.e. progressive demyelination of the nerves in the brain especially affecting the motor nerves and causing a spastic paralysis. There are many other symptoms e.g. fits, change in mentality, eye signs, difficulty in speech varying on the area of grey matter involved. One bad case was improved by Marmite but our knowledge is limited. Home-made media for culturing KLB were on show, and a case of tropical sore of the noise, toxic hepatitis ?cause.

Officer i/c scrotal dermatitis insisted that it was a pure vit deficiency condition and that a fungus infection may be superadded but was not the primer factor. He advised rice polishings to be added to the diet and a bland ointment to the scrotum. Talks are still going on with Command and we hope that large amounts of peanuts and polishings will soon become available.

On Thursday night I went to the opening night of 'I killed the Count'. The scenery and decor could not have been better even in a London show. The Sappers made the scenery and converted old furniture into modern stuff. The acting was of a very high standard and of course Wilkie was the star performer. As the Chief Inspector he has a very long part to learn and was word perfect. They have been unfortunately stricken with illness and each day some member of the cast gets ill and they have to rehearse it all over again with the new addition. This must be very tiring and very annoying to the old members.

The Japs have now asked for full particulars of all Prisoners of War including height, weight and chest measurements, and medical remarks, so it looks as though I am to be kept fairly busy during the next few days.

Dysentery figures are going up again – this may be a sequel to Selarang so may be the reduced resistance to infection.

Tues Sept 22nd

Yesterday was the most remarkable Yom Kippur I have ever spent. We had two services, a long morning one and a shorter afternoon one. There were over 70 present in the morning including a few RAF newly arrived. Capt. Phillips had prepared an anglicised service. We read a portion about Isaac in English (probably Rosh Hashanah parsha [portion of Torah reading]). Our paylisher cantor put a little more wailing into his voice in a vain attempt at Chazanus. There were about 10 officers present – many of whom I had never seen before. One of the men asked us if we could finish by 12.30 as they had to back for 1 o'clock for dinner! A few observed the day strictly, but those that didn't can be forgiven for it was terrifically hot & sweat was pouring out of us as we stood listening. In the afternoon, Phillips read out a portion of Jonah (without the necessity of paying for the so-called privilege).

We had a little Neilah service but didn't have the usual shofar [ram's horn] blowing. I wonder whether in England there will still be the same stream of cars after the Service & whether Mam gets her annual headache. It is at times like this that one thinks of one's family & vice versa. We are not having a succah [small hut for festival of Sukkot] this year, but will have plenty of palm leaves for a lulav [symbolizing a fertile harvest].

The RAF are now settling down – there have been quite a lot of minor illnesses & I sent an officer to hospital today with dysentery.

Cases of painful feet are going up to hospital in hundreds for the polishing prescriptions, and corneal troubles are getting red palm oil (vit A). The out-patients department must be very busy these days.

Friday Sept 23rd

There has been a lot of talk about officers pay lately & it seems as though we are to get $30 a month, $10 of which is to go to the hospital ?a loan. $60 the Japs keep for board & lodgings & the remainder (varying with the rank) is deposited in a Jap bank in Singapore for our future use. This seems to depend on whether Britain will pay Japan accordingly. Meanwhile we are ordering goods heavily and getting into debt. Men will still continue to get their amenity grant – unless they are in hospital & they will get nothing. So the officers have to support the hospital.

I have bought a lot of food lately including 60 pieces of peanut toffee at 3 cents a piece. In view of the vitamin shortage, I am eating a lot of peanuts.

The position in regard to vitamins is very serious. About a third of the Battalion is suffering from one form or another of deficiency, & some have been turned back from hospital saying that the unit will supply rice polishings. We only have 6lbs of RP [rice polishings] for RNF & RAOC, and are doing our best. I now hear that the so-called RP consists of floor sweepings & is not true polishings, containing no B2 and is therefore useless for treatment. If this is the case we are foxed as there are no other remedies to use. The outlook is very grave unless the Japs will provide a good source of B2 for us. One RAOC man developed complete paralysis of his legs in 24 hours – motor & sensory. How little do people in England realize the importance of vitamins in the diet.

I called a meeting of the two messing officers & discussed future policy. A memo came out recently stating minimum amounts of peanuts, RP, palm oil & whitebait to be included in the daily diet.

Wednesday Sept 30th

Today the Japs had an inspection of our kit, and we all stood by our beds. One Jap came quickly through our billet, scarcely pausing and passed on. So all our preparations and precautions were in vain. Last night, the padre and I had a helluva fright as the Japs came in at night, and lying behind me on the table were the things ready to be disposed of. Fortunately the lights went out just as the Japs came in [but] I thought it was all up with me. I managed to push everything under the mattress & later we had stealthy fun & games. Actually they were the cycle patrol and their torch went out. They came in to get it mended.

Burn was released yesterday after spending 2 weeks in the Sikh camp for refusing to sign [the Escape clause]. Harris (Col OC Div) went over & talked to him so finally he signed. He had a bad time to start with, as he was put in a barbed wire cage with no protection against the elements, and had to work 12 hours a day chopping down trees etc. He complained to the Jap CO that officers could not be compelled to work, so he was put in solitary confinement. He looked poorly when he was released & had a temp of 100° – probably dengue. He is better today & wants to take over from Wilkie who I imagine is extremely reluctant to hand over his command which he enjoys in all senses of the word. If Adrian comes out soon, he will be very junior here.

[whole page cut out here]

... departure has been put off till the 7th. Yesterday after the Service (special Succoth Service) we collected names & addresses & gave them to the 2 chaps who are due to leave with the repatriation party. I gave my home address to the Leeds chap – Goldstein who lives in Harehills Lane – & hope that when and if he gets home, he will visit 22 G[range] A[venue]. I also gave him the Anglo-African address in case he is left in S Africa.

Sandy was up yesterday & said that they have already been paid, so that we may expect our pay shortly. Jap rations are short this period, so it seems as though the Red Cross stuff should be arriving to supplement them.

My vitamin clinic on Thurs pm was very well attended. I worked flat out from 2.15 till 5.45 seeing cases of painful feet, eyes, tongue, mouth, scrota etc. They are all [needing] rice polishings, and we must hope for the best.

Thursday October 8th

Red Cross parcels and supplies have been the main topic of conversation during the past week. There were about 90 personal Red Cross parcels from South Africa. One of the RAF officers in the mess got a parcel which contained clothing and common medical and toilet necessities. I hoped that Betts [a relative of Harry] might have sent me a parcel too, so I was a bit disappointed. It will be grand to get mail or parcels from home. We are well supplied with clothing so that cigs and food would be most welcome. The general Red Cross supplies were delivered to the camp this morning and we all get a fair share, but only a half of what Keppel Harbour people got. The ration for us is approx. 3¼ tons of Bully, 1 ton of milk, 65 cigs, M+V [meat and vegetables], biscuits, sugar, cocoa etc

Most of it will go into the mess to improve our meals as we have not been paid for nearly two months. The meals are suffering somewhat and as we are all heavily in debt, we can't buy lunch from the canteen till we get paid. Command are still arguing and haggling instead of paying us.

For the past two weeks I have had arguments about the cross of the Jewish boy who died. The REs who make the cross will not remove the cross bar although they have agreed to paint a magen david on it. They say it is a Div order which cannot be altered. The SCF [Senior Chaplain to Forces] however says that it only applies to C of E etc and was not meant to apply to Jews. I'm leaving them to settle the quarrel amongst themselves. Burn has taken over from Wilkie. I don't know which is the lesser of two evils. They both have their faults but the former is so unreliable and makes sweeping changes without thinking. He has no sense of proportion and makes an issue of minor matters. He seems to have got over his prison experience although he is a little thin. Still a lot of trouble with the Sikhs – ADOS and DADOS [Deputy ADOS] had to do press-ups on the ground for failing to salute the right person. I hear that they have been forbidden to slap faces but are making things difficult in other ways. It shows that when some people are given any power or authority at all they immediately abuse it.

I've caught my first cold since being on the island. Unusual for me to go so long in England without a cold. Don't know the cause – possibly due to wearing a towel instead of pyjamas at night.

Saturday October 10th

There was a medical meeting this afternoon and many subjects were discussed. Diphtheria is on the increase especially in southern area so that all intercommunication has been stopped ie no concerts. They decided to keep the Windmill Theatre going. There is going to be a sack of rice polishings for each MI room which should ease the position. The number of vitamin deficiency cases is increasing, 180 from RNO [Royal Northumberland?] and RAOC last Thursday. The painful feet are not responding very well and some are developing spastic paralysis ie pyramidal tract lesions, which is a grave symptom. Sgt Armstrong went in with these symptoms – rather ill.

RAF left yesterday. Our mess looks rather empty now. Brown of Reading, very nice middle-aged solicitor from Reading who had been sitting at our table, exchanged addresses with me. He was profuse in his thanks and gratitude for what was in fact very little. They were all sorry to go as they were appreciating this rural POW atmosphere. Their destination is unknown ?Japan ?Bangkok. They got their share out of Red Cross parcels just before they left so were very lucky.

We had a large mess meeting yesterday to discuss what we should do with our parcels. We agreed unanimously to put all the stuff in the mess, excepting 65 cigarettes given out individually. We now get cocoa and biscuits at night, biscuits for tea, milk rice puddings and sugar on our porridge etc so that life is more bearable. When we get our pay we have agreed to subscribe $5 each month so that breakfast can also be made a good meal. The Japs took over the distribution of these supplies from the Red Cross Commissioner who it is rumoured has resigned. They gave them out very unevenly. Some camps did very well and others very badly. They wouldn't allow any of the food to go to the Indian prisoners – a bit hard but nothing can be done about it.

Americans have arrived from Java and also some chaps from Kuala Lumpur who are in a bad state physically and hygienically. The Japs were in the hospital yesterday and told Col Craven (OC) that the Japs would be prepared to send all the medical officers home if the number of such was small but while it was large we had to stay here. Extremely doubtful if anything would come of it. Only extreme optimists base any hopes on it.

Last night we had a quiz against the fusiliers – Padre, Willie, Owen, Jack and self. We were beaten very badly but as the paper was prepared by Stanbury, I suspect that he was not quite neutral. We will have a return match with a neutral officer this time.

Tuesday October 13th

I am just recovering from a slight pyrexia following my cold. I had to get up this morning and parade on the gun park with all 18th Div officers so that the Japs could count us for purposes of pay. We are still waiting for it and are hopeful of getting it. It was very tiring standing over an hour in the strong sunshine waiting for the Japs to come. One of the things that after the war I won't forgive Changi for will be these periods of waiting. Japanese unpunctuality, one of their national vices, is done on purpose to POWs. A second large issue of Red Cross parcels has been given out and every ration store is bulging with food. We should be well fed for months to come. The Red Cross have done a good job of work here.

My batman, who is doing very well, spent most of the weekend washing and ironing hankies. As I only have 8 hankies and used one every 5 minutes I was using them faster than he could wash them.

A few of the officers and lots of the men have had a close crop. It may be more hygienic but it certainly makes one look a terrible sight for a few weeks.

I managed to get the burial cross question settled. After sweeping away a lot of red tape etc, RE agreed to cut off the projecting parts of the transverse arm, thus satisfying all parties.

We have been ordered to set up a camp reception station near the MI room to accommodate 12 patients mostly simple diarrhoeas – thus reducing the number admitted to hospital. Tate is coming out tomorrow so it looks as though the hospital is keen on reducing itself considerably and thus enabling the Japs 'to send them home'.

Sunday October 18th

Following my cold in the head I started to grow a 'tache and with thoughts of Errol Flynn as I cultivated it. As day succeeded day, it looked more like a rat-bitten toothbrush so this morning it came off mid much wailing and moaning.

We finally got our pay $20 – $3 advance – $5 to mess = $12 – $2 to batman = $10 – all went to Willie and am now about $3 in credit – which will have to last me till next month's pay. I don't seem to be much better off even after this eagerly awaited pay.

Talking to a young officer yesterday at the Service – small attendance – he came out to India last year and went through an Octu [Officer Cadet Training Unit] ASC [Army Service Corps] and is now a captain RIASC [Royal Indian Army Service Corps] and is getting about £2 a day. This pay is all very well in India but out here we are all living under the same circumstances and it does seem a little unfair.

Tate came out of hospital on Wednesday after being in for 8 months. He looked fitter than we did and hops about quite gaily. He is looking forward to being repatriated. DADH [Deput Assistant Director Hygiene] made an inspection of RNF and RAOC cookhouses etc and sent in a stinking report about flies and dirt – a grossly exaggerated report. There is still a lot of dysentery about so we are tackling the problem very seriously.

Painful feet and various other neuritic conditions are causing me bother especially amongst the RAOC. So this morning I am giving them a talk about taking rice polishing 12 dessertspoonfuls a day – that will shake them. Simpson has written a thesis on 'Tea being the cause of vitamin deficiency complaints' and gives a number of instances where peripheral neuritis has been due to excessive tea drinking without milk to 'neutralise' it. Phillips has sent the paper to the medical specialist. Last night another quiz against the men. We lost again. I did badly.

Friday October 23rd

Rumours of a move received official confirmation today. Large parties are moving off up country starting from the 29th. 4,500 are going from this Division. My fate is as yet uncertain.

Hundreds of Javanese are arriving in the camp on their way up north. They are all shades of colour from white Dutch to black natives. They have no colour question, everyone being equal. This solves a lot of difficulties and one which the British are finding very hard to solve with their coloured and Eurasian bar.

I bought a linen coloured sarong today for $2.90, as my pyjamas are falling to bits and are all patched and torn.

Sandy, who came up today with Tom and Veitch said that he bought a silk one for $1.60. That is one advantage of being in Singapore. They also do well from the ships they are unloading and from local shopping.

They are all looking well and there is no confirmation of the rumour that they are moving. I had a ride in their truck up to hospital to a meeting of the Changi medical society. It was a pleasant change to be saluted instead of saluting. The Jap on the truck was very patient and allowed them to stay a short time in the mess. I have extended the 4 spoonfuls per meal of rice polishings to the RNF as about 12 of them are not improving. The RAOC are causing me a headache as about 80 of their painful feet are no better after a week of large RP doses. I hope next week shows a major improvement.

Two interesting papers at the meeting today – gastroscopy and prostate resection. Jack and I are going to Southern area twice a week to do Russian. It brings back memories of the Orizaba.

Tuesday October 27th

175 officers and ORs are going up the mainland from this Battalion next Tuesday. There are seven parties of 650 men leaving every day from Thursday till Tuesday. They all go in trucks to Singapore then a 3 day train ride in trucks, 31 to a truck and at the far end there is a 2 day march and 10% of the men can be taken on trucks ie the sick and lame. I am sending most of the painful feet cases and other cases of vitamin deficiency in the hope that the change of air and food may cure them for frankly there is no cure here – the polishings in the majority of cases has no effect. There is no precedent in medical history. Braham, whom I saw at the Service last Saturday, has suffered for 6 weeks and no treatment has helped at all. Going into the hospital I was pulled up by a young Sikh for no reason except that he didn't like the way I saluted. He made me sit down on the grass and watch while he demonstrated. I then had to go back and repeat his performance. We have to put up with a lot from these chaps and I hope they get their desserts at the end of the war.

Most of the Jewish chaps are going including Morris, but we will carry on the services as long as we can. We probably will follow the others fairly soon! I was asked by Phillips if I would like to go and when I said I didn't mind either way he chose Emery. But Roulston at the last moment decided he wanted to go so Emery and I are staying. I, being senior, will be responsible for Div TR [?] area. Our MI room is full of Dutch officers and men who come for treatment all day long. A large number of them have gone to hospital with dysentery. I am rather worried about them spreading the infection. Two Dutch MOs are attending to their men and send very funny chits in to Etheridge about lotions etc they require. 300 RAF men left for Japan this morning.

Sunday November 1st

Large parties are going away daily and the area is becoming gradually denuded except that the Java people are still here in large numbers. 2 or 3 come in the mess each evening and the Dutch MO who speaks English well also plays a good game of bridge. We have had 2 games this week – very enjoyable. I had a long talk with him about painful feet and he told me that Quinine in small doses gives relief. I tried it out on Braham and when I saw him after the Service yesterday he said he had the best night's sleep for months. It seems as though it works. Whether it is just a relief or also a cure remains to be seen.

We had a bare minyan yesterday and it seems as though we will have to discontinue the services as there will only be about 5 or 6 available next week. However we have had a good run of almost unbroken services for almost 6 months. Braham and I seem to be the only 2 officers left. He is now out of work as his men have gone up the mainland. Stories that reach us suggest that conditions are not bad – 100/hut and officers are separated from the men. A good story I heard yesterday was about a Jap cycle patrol that captured some prisoners during the fighting on the mainland. The prisoners were made to get on the step behind the bikes while the Japs pedalled. But as it was rather hot the Japs stopped and made the British pedal while the Japs rode behind.

I have been very busy this week – working morning and afternoon. I took over the gunners MI room and treatment of them and now have quite a few RA officers to visit. My morning round of RNF, RAOC and RA lines occupies a lot of time. Sorting out the troops to travel and those to stay is a full-time job especially when at the last moment some men fall ill. They would not let vit B def eye cases travel, so we had to change four RAOC men in a hurry just before they left. I am glad that most of the ordnance have gone – they were a bit of a burden.

Wednesday November 3rd

The fusiliers have now gone leaving 6 officers and about 60 men. Jack I. is going on Friday and that will leave us with only 1 fit regimental officer ie Wilkie. Adrian and Titus are both unfit. It is rumoured that Friday's party which includes 20 MOs and 450 staff from the hospital is going to Saigon to set up a base there. No-one knows definitely. The hospital is having mass discharging of patients and cleared 3 wards in one day. It looks as though the policy is to send all the British up North to IndoChina or Siam and bring all the Dutch in here. Every day sees thousands more Dutch in Changi. We are moving upstairs tomorrow and the Dutch officers are moving in downstairs. We are feeding upstairs – this will be a bit of a nuisance.

There have been several concerts lately, one last Sunday night at Div Sigs mainly for the overland party. I got a seat in the front row among the big noise. The compere, with a serious face, announced during the show 'Is there a doctor in the house?' I knew the old joke but as I was pointed out he asked me the introit 'How do you like the show Doc?' There was a tremendous burst of laughter and cheering, and now everywhere I go I am greeted with that phrase. The compere was from the Houston and he told some of the dirtiest jokes I have ever heard in my life. The men enjoyed them but the colonels were very annoyed.

Last night a revue at the Windmill 'Gentlemen only'. I went to the first night mostly officers – very well done with West End style. First class band with a Sid Seymour [1930s dance band] touch to amuse us. A skit on a Russian play was the highlight. I'm treating the producer Bradshaw who is a professional actor for a septic arm and phlebitis.

Sat Nov 7th

The parties have all gone & although there are many rumours floating about us all going shortly, we are still here. The few remaining RAF officers living in the big billet downstairs moved up with Titus & myself, & we are living on the front balcony where we are very comfortable. I took up a large wooden cupboard & a goodly assortment of furniture so that I am well set. The only drawback is T with his gramophone a few yards from my ear.

It is very pleasant on this balcony – no flies or mosquitoes & I sleep very well at night. The food has gone off a bit lately, but we are expecting some more Red Cross supplies today.

The night before Jack left, we had a large feed of cold meat pasties & jam tarts in the black-out downstairs. The Dutch officers are occupying our quarters but are not very sanitary-minded. The Padre had a bad undercarriage & is hors de combat now.

Fri Nov 13th

We are now all on the balcony except Padre who is still suffering badly from his undercarriage. Div HQ have moved into RA lines & the area has closed in. There is now one

MI room, 4 of us working. I do RNF, RAOC, RE & Div Sgs before breakfast & 55 Bde afterwards. There have been about 40 of them, but a lot will go to the Skin Clinic now starting. Painful feet is still very common & Nicomide injections give temporary relief only.

Braham has gone up country – I heard the following day that he had been court-martialled for selling Marmite (that had been prescribed for him) to his mess. He got a severe reprimand & reduced 50 places. This came as a great surprise to me as he had never mentioned a word about it.

I went last Saturday to the 'Service' but no-one was there so I cancelled it sine die [till further notice].

We are very comfortable up here & have entertaining evenings with other officers playing quiz & word games. There have been several black-outs lately & this has relieved the monotony. I have played bridge several times against 2 very good Dutch players – their Smith 4NT [a bridge card game term] convention baffles me & also them at times.

The news from North Africa, ie the landings of the Yanks in Morocco, is very cheering. I read a full account of it in the Syonan Times & they reported it fairly, but only gave a few lines to our terrific attack in Libya. If the North Africa situation is completely cleared up this winter & the Russians hold the Germans, we should be in a good position next spring to launch our second front offensive. It seems as though USA is not bothering very seriously with the Pacific just now & will turn a bit later when the more important job is done.

Thurs Nov 19th

NaCl is a very important problem just now. It is being prescribed by the eye specialist for the numerous cases of corneal & optic neuritis complaints that are now occurring. Unfortunately there is no salt to be had so that the treatment is impossible. The rationale is as yet obscure & it may be that everything is being tried of which we are short, & NaCl is one of the important ones. It can be got from the sea in small quantities, but it is not sufficient for general use. We are feeling the lack of it at table as the rice etc is tasteless without salt. The meals have been rather poor lately due to the diminishing stocks of Red Cross food & the absence of pay. I hear that pay has come in today & that our starvation may be relieved. I have smoked very little lately but this has been due to lack of money & fags.

There has been a bit of trouble among the Dutch, from cases of kniving. There is also an abnormally high percentage death rate from dysentery in hospital as some of the Dutch have lost the will to live & just lie down & die.

We heard that the Generals etc are in Formosa & are receiving gifts from Japan.

Owing to the general closing-in of the area, the Windmill Theatre was closed down. Last night I went to an open air show by the Div Concert Party in RASC lines. Quite amusing, but had to sit on sand as some cad had pinched my chair.

News about N Africa continues to be very good & the news about the USA naval victory over the Japs near the Solomons is very cheering to us. We are very optimistic & are basing our hopes on a victory about next spring.

I have just finished The Sun is my Undoing by Marguerite Steen, one of the really long books that one can enjoy like Gone with the Wind. We have been fortunate with literature so far. I hope it continues.

Sunday Nov 22nd

We heard yesterday of the death of Becky – heart failure following diphtheria ? in Formosa. This is a terrible blow to us in the 18th Div as he was very much respected & loved, and we all hoped that when we got back to England, he would act as our spokesman & explain the raw deal that our Div got in this battle. There is to be a unified Memorial Service on the hockey pitch in front of the mess this afternoon. All the Div, including myself, will be present. Our Padre is conducting the service. He was a great friend of Becky's & will miss him more than the rest of us.

I read the Syonan Times today & it gives a front page column to Becky's death & explains how everything was done for him by the Japs, and the homage that was paid to him as he lay in state by Japs & Allied Commanders alike.

All officers have now to parade for roll call at 8.15 & 7.30 each day, & 3 times a week in the evenings, the Japs come round & check our figures. Unfortunately they keep us waiting from 7.30 till about 8.15 when it is almost dark, & figures can only be seen as vague shadows. Jap ORs run about counting aloud, getting in each other's way, running off to report, salute & return running to do some more counting. There is a lot of saluting, counting & recounting in the half-light & the moon is often visible rising over the tops of the palm trees before the Japs have finished & we are dismissed wearily to our billets.

I have had several chats with Dr Berkoman (Dutch MO) & have picked up quite a few useful tips on tropical diseases & their treatment. He lived among natives for 5 years & knows many strange tribal customs.

Sat Nov 28th

Last Sunday I sent McFall to Hospital with dysentery, a recurrence of an earlier attack or possibly a reinfection. As I was left without a batman & also a waiter, we had to choose someone who hasn't had dysentery. Weir of W Company who was formerly assistant pro at Hexham Golf Club (+2) was the only selection. He has fitted the bill admirably. He is clean, neat, efficient & well-spoken. We had to collect a lot of kit for him, as most of his stuff had been stolen while he was in Keppel Harbour. I asked him about golf lessons. He says he is only too willing to teach me, so one of these days I will start reducing my girth & my handicap.

Last night the Dutch thought-reader made a return visit. I bet the Padre 20 cigs beforehand that it was all a fake & was done by using an accomplice in the room. I decided on a test & gave sealed instructions to Doc Emery who was acting as the medium, so that only Emery & myself knew what he had to do. After a lot of fumbling, he performed the task written down. This completely shattered my disbelief & scepticism. With my scientific education, I have hitherto refused to believe anything supernatural, anything that could not be explained by ordinary laws of physics, chemistry, etc. The existence of God came into the argument with the Padre, who argued that just because he couldn't prove scientifically the presence of a God, it doesn't mean to say that there isn't one. The experiment last night shows that there are still lots of phenomena that cannot be explained in our present state of knowledge & there may be something in spiritualism etc. I have lost my scepticism & am now prepared to believe anything, even Santa Claus & Bogeymen.

The bosun of the Dalhousie, a cargo ship of 7,000 tons, arrived here as a prisoner-of-war a few days ago. He was interviewed & a report has been circulated. His ship was sunk in the S Atlantic by a German raider – they were transferred to a tanker acting as a prison ship 'worse than Thrallmark'. There were 2 women & 2 children there, the sole survivors of the Gloucester Castle which was on its way to the Middle East with war material & wives of officers. The tanker refuelled a German sub, & collected lots of prisoners & brought them to Singapore. The technical experts were sent to Japan, & the rest interned here.

Some news about April last was interesting. The loss of Singapore was a big blow to England, & he says that the opinion was that the Army wouldn't fight. We don't believe that this opinion was general in England. The sinking of our two battleships (*Repulse* & *Prince of Wales*) & the escape of the German ones (*Scharnhorst* & *Gneisenau*), were shocks to England too. USA promised to get Singapore back for the Allies.

Sunday Dec 6th

Last night we had a duplicate bridge tournament against the Dutch officers who live below us. They beat us by 1,700 points due mostly to one table that gave away 3,000 points on 4 hands. We held our own except for one serious lapse on my part that cost me the game. The Dutch provided sweetened strong coffee – of which I had 2 cups, which accounted for me not sleeping much, or maybe I was going over my mistake.

This morning at sick parade, a man from Div HQ told me where I live, & complained about the large postal packets I used to get. He was our postman. What a small world!

It is raining like hell today & the Dutch have their Santa Claus Sunday celebrations today. The sports have been cancelled but the procession of Santa on a decorated wagon to Church took place in a downpour. Poor old Santa's beard was very bedraggled & the clothing of the gaily dressed attendants was soaked. The festival has something to do with St Nicholas & is celebrated by all the Dutch.

We are getting used to having the Dutch in our midst – all notices are printed in the two languages & the concert now showing has alternate nights for Dutch & British. The main item is Cortini, the illusionist, & the band fills in most of the remainder.

I am occupying my afternoons by golf lessons & learning Nippon-Go from a self-teaching handbook. The former is progressing better than the latter. I can now hit a reasonable straight drive. How long I can do this is a matter of speculation. Japanese is a difficult language & I am concentrating on learning phrases.

The Medic. Soc. has been fairly active lately – there are two lectures a week on malaria given by the MO of no. 6 Malaria lab (very interesting & containing a lot of new material) & there was a well-attended clinical meeting. Cases of painful feet have been found to be associated with a high BP. Amyl Nitrite eases the pain temporarily. Sod. Nitrite med. over a period doesn't reduce the BP much & causes nitrite poisoning so that treatment was discarded. It is believed that the pain is central in origin ?subthalamic. I mentioned the quinine treatment to Cruikshank & he is going to try it out experimentally. If we can discover the cause, at least we should be able to do something to relieve the pain & sleeplessness.

I am now starting brewing for Xmas. I mixed a tin of pineapple with some Gula Malacca & some bakers yeast. It is bubbling away merrily – the difficulty is feeding it with sugar as the latter is very scarce. Still we must have some alcohol to stimulate us.

Percival has had his article on 'why we were defeated' published in the Syonan Times. He says plainly that our troops were unseasoned, couldn't fight for long periods, lack of air support & absence of coastal defences on the N shore of Singapore. He also mentions the speed of the Jap advance which probably was too fast for our slowly moving army minds. British C-in-Cs never believe in being hurried & always insist on having the weekend off. 500 years ago they would have been a terrific success.

Thurs Dec 10th

There has been incessant rain for the past few days, at times reaching hurricane heights. Local residents say that as far as they can remember, it's never been like this before. The balcony on which we live, so far immune from rain, was turned into a swimming-pool yesterday, so we had to fish all our belongings and bedding out of the balcony in to the main block where amid sarcastic remarks & cheers, we proceeded to dry out & bed down. To complicate matters further, the past 2 nights have been a complete black-out, making our nocturnal manoeuvres more difficult.

I have taken over a double locker space & have moved all my belongings in, and am now fairly comfortable. The difficulty is getting clothing & towels dry, for this atmosphere is so humid that things are perpetually damp.

The malaria lectures had to be postponed for the noise of the rain drowned the speaker's voice.

We received an advance of $10 yesterday – am eating loads of peanuts.

Wed Dec 16th

This morning I had to get up at 7 am (pitch black) to attend the Jap medical inspection of the infantry. They swabbed the throats of 200 men & cultured them on agar for diphtheria. No-one quite knows what the idea is as diphtheria is now going down. They may be looking for carriers & intend going through the whole PoW camp. The MOs were not needed so were turned away after a little while. My early breakfast was unnecessary. Still I hope we won't be disturbed again.

We were paid a further $10 yesterday, $2 of which goes to the mess to pay for the extra Xmas fare. I have been trying to get a cake made for my birthday but as white flour is essential & is so difficult to obtain, I may not be able to get one made. Still I am trying hard to put on something of a 'do' next Monday.

Tomorrow morning, all the troops have to line up for an inspection & tour of the ranks by the new Jap General. Fortunately we MOs have been excused. Have had a helluva row with SH Jackson who is 'MO i/c' MI room. He is throwing his weight about & wrote me a letter unprecedented in Medical History. I showed it to the ADMS who thought it must be due to the spirit of pettiness now prevalent. He thought that J was frightened of me muscling in on his brief authority. I told J exactly what I thought about him, and am now going to carry on as though he doesn't exist. The power of authority in irresponsible hands (e.g. also Archie) is a bad thing.

Recently in the evening I have been playing card games & paper games with the Sappers. They are a good crowd of officers & we pass the evenings very pleasantly. I borrowed a Mah Jong set from the Dutch last night, & the Sappers are going to copy it & make a set.

There are several letters for distribution – am afraid that so far there is none for me. We heard that a PC [postcard] from Mrs B to Becky states that she had heard from Col T, so that it looks as though part of the escape party got away safely.

Weighed 10st 12lbs.

Tues December 22nd

Well I'm now 32 years old – having spent a very pleasant birthday yesterday. I got gifts of cigarettes & cheroots from Titus, Simmie & Bill, and also a 'book of poems by Damon Runyon' from the Sappers. The joke is taking a long time to die down. We held a supper party in the Padre's room for the Fusiliers ie Hedley, Wilkie, Titus, Simcocks, Padre and myself. Titus did the catering in his usual wholesale style. We had M & V [tinned meat & vegetables in gravy] pasties, chipped potatoes, pineapple tarts, coffee with Gula & a TIN OF PLAYERS that I had faithfully kept for over a year in my case, without being tempted. The Players were the most enjoyable part of the meal. After supper, Doc Emery, Enali Jones & Bradley (Sappers) & Ireland joined us.

I had a large cake made by the officers mess cookhouse (rather expensively). This unfortunately collapsed in the morning & had to be supported with string & bumph – tastefully cut to make it appear decorative. I also produced my famous home-made brew ie Panther's Breath which, although there was a small portion per person, was very tasty & alcoholic – not however sufficient to make us tipsy. The later guests joined us in cake & 'wine' & my health was proposed with the latter. The air became very thick in the narrow confines of the room & at 11.30, most of the guests went up on the roof. I found a late guest – PMC Homes who got out of bed to go down & finish the cake. All in all it was a good night's work & my first birthday in a POW camp. I wonder how many more there will be. None I hope!

This morning the Japs did some more throat swabs. We MOs turned up & were turned away again – not being needed.

All the Singapore people (10,000) are returning to Changi. Simcocks with his 10 men returned yesterday & is full of praise for the Japs for their treatment to his men. He must have been lucky to find a good crowd of Japs. One poor Sapper here was knocked down 3 times by a Jap sentry for failing to salute at night someone approaching on a bicycle from behind!

It is incredible we will have to start spreading out again – this will affect the infantry Brigades who may have to go back to the swamps & jungles. Simmie found a Tidy [a device for storing test tubes] in S & casually produced it for me when he came. The Dutch are planning big Xmas celebrations e.g. a mixed late supper for us all.

Dec 25th 1942

As my supply of diaries has now run out, I am compelled to use this partly-used Bill of Lading book. Today being Xmas Day, it is a day of mutual compliments and bright atmosphere. There is a tremendous menu arranged which will satisfy every gourmand and gourmet. We are entertaining the Dutch to coffee at 12 o'clock to reciprocate their former party.

Last night on the green in front of the mess there was a grand sacred concert – lights by RAOC. It looked like a crowded fairground from our position on the balcony. It ended up by the fire ceremony, which is a symbolic affair – the fire represents the spirit of life and freedom which keeps burning, and the Dutch 'broadcast' a message to their wives and families in front of the fire, and then each one took a burning rod and lit a candle in his own tent. The association of flame with life is a very common one in many religions. It was quite a dramatic affair, as all the lights were doused, and only the flickering fire played on the announcer's face as he spoke in a loud voice to the families at home. Our padre translated it into English for the benefit of the non-Dutch.

The midnight communion in the church next door prevented sleep till nearly 1 am, but as there was no roll call this morning, we could stay in bed late. Several more officers came back from S [Singapore] yesterday, and we are now filling up the mess. Some fiends have been stealing our fowls, and last night while we were all at the concert, they stole four, including our last cock. All the others had just been killed for our Xmas dinner.

Xmas makes me think of home, and there are many sad faces today as they read the various messages issued by various commanders. They all have one thing in common – hoping that by next Xmas we will be by our own fireside – tropical humour. Well, this morning was the first time in my life that I have had a cold shower on Xmas day.

Dec 26th

The meals yesterday were excellent – no rice on the table. At night we had chicken soup, fish creole, roast chicken, and 3 veg inc new potatoes, Xmas pudding, savoury, wine (home-made). At midday we entertained the Dutch officers to coffee and cakes, and tried to fix up a small Dutch conversation class for a few of us so that we can talk to them in Dutch, instead of relying on speaking English as the universal language. It is an amazing fact that nearly all the Dutch officers can speak our language passably well.

Their party at night was attended by about 25 British officers, including Harris & Dillon. We paid 75c each, and the Dutch provided soup, sandwiches and a sweet. Mutual congratulation speeches were made by the respective OCs, and we stood in silence for various reasons on numerous occasions e.g. Dutch royal family, Dutch wives, allied casualties, etc. Harris made a very optimistic speech in which he said that by June '43 we would be victorious, and that next Xmas we would be out of here. Time will tell. I exchanged my Xmas card on my plate for a Dutch one, and collected a few illegible signatures. There was an outdoor concert on the green in front of the mess after dinner at night.

It was interrupted in dramatic fashion by cries of Help! Thief! We dashed through to the back where the cries were coming from, and discovered that two Dutch men had been caught in flagranto delicto, while trying to steal our poultry. The British officers who caught them laid about them so lustily that the two unfortunate men had to be taken to hospital, one with fractured ribs and spitting blood, the other with a skull wound and fractured hand. They were noted bad-hats, but still it cast a damper on the festive proceedings as they were taken bleeding into the room fixed up for the party. I hope it will stop the fowl pinching racket. We have lost the major portion of our collection by their small nightly raids. We even had to keep the Xmas cockerels in the mess for a few days prior to Xmas.

Diphtheria is still taking its toll. An RASC major died in hospital today – post diphtheria paralysis. He had been ill a long time with KLB ulcers. As we are going to be crowded out again soon with this new influx from S, there is a danger of it spreading.

Dec 31st

I am now in the throes of a heavy common cold – my 3rd since being on the island. I had thought that a tropical climate would reduce the number of colds, but I have been worse here as the cold is often accompanied by a temperature. There have been a large number of colds in our crowded mess during the last few weeks, and I knew it was only a matter of time before I got mine. One disadvantage of a big mess is the lack of privacy and quietness. With streaming eyes and nose, I have to answer countless enquiries about my health. One only has to look at me to see how bloody I feel.

To celebrate the New Year the Japs have given us 2/7 tin of pineapple and 1/10 bottle of wine each. What the latter will be we don't know yet, but probably will be some local brew.

One hears many stories concerning the Japs from officers who have returned from Singapore and I will try and put down a few of them. Australian soldiers in charge of a steam roller for mending roads used to call at Jap HQ every day to collect 4 gallons of petrol. This went on for 4 months before the Japs realized that a steam roller doesn't need petrol. Meanwhile the petrol was sold to the Chinese at a good profit. In one camp the Jap guards were taken away and not replaced. When the Tommies found out this state of affairs, they just wandered out into town, seeing the sights and shopping, and Japs guarding other camps used to watch them go by.

Jan 1st 1943

I am feeling tons better this afternoon, after a very peculiar morning. I woke up full of cold and feeling lousy. In spite of the fact that the Japs declared today to be a holiday, there was a sick parade – so I toddled along and had a large mixed cocktail of medicines to help me throw off my cold. For the rest of the morning I had the lightheaded, happy feeling that one gets after several double whiskies.

Whether this sensation was due to the potent medicine, or was a delayed hangover from last night is a moot point. Five of us (exc. Wilkie who is agin anything) had a New Year's Eve party in Padre's room. We finished off the meal by a toast to the New Year, drunk with a mixture of the Jap 'brandy', Padre's communion wine and pineapple juice. The Jap stuff was almost raw spirit, and we got about ½ cupful each. It made the troops very hilarious, and at midnight they surrounded the Padre's room, and we went out and shook hands with all and sundry. There were the usual songs e.g. Auld Lang Syne, followed by Blaydon Races, We're Here Because We're Here, etc. Led by a bloke banging on a tin box, the men marched up and down outside till nearly 2 am singing and carousing. The Dutch downstairs also entered into the spirit of the thing, and altogether we passed from the old into the new year in traditional style.

The Japs complained about the noise, and when an officer who knows only about six words of Japanese was brought along to talk to them, he was stuck after a few minutes so the Japs struck him, thinking that he was being obstinate.

The general opinion seems to be that this year will see the end of the war in our favour – let us hope and pray that this will be so.

Jan 4th

On Sunday last I went to a meeting of the Lancastrian Society which also welcomes Yorkshiremen. An Australian Sapper WO who was in Libya and Syria gave a talk on his experiences. He mentioned the RNF and gave them a word of praise. The sandstorms were terrific affairs and blotted out all vision completely, and they had to wear respirators to breathe. The fighting in Syria was more than is generally thought. The French Foreign Legion are good soldiers and excellent snipers, and it was often difficult to dislodge them.

Typical examples of mistakes made were:

1) landing 500 parachute troops in the middle of a strong enemy camp, in mistake for the allies – only 40 escaped alive.

2) Allied planes when they finally arrived bombed the British.

3) the Mediterranean fleet which was supporting the coastal action lost a destroyer by bombing, so asked for air support. None was forthcoming, so he withdrew to Egypt. The French fleet then stepped in and shelled the Allied position. The number of allied planes in the air at one time could almost be counted on one hand.

During the 40/41 advance into Libya there was very little, if any, air support. It amazes me. I often wonder what happened to all the thousands of British & US planes turned out. Now the troops there are getting marvellous air support, but it has taken three years to get, whereas the axis has had it constantly since the beginning of the war.

The Dutch Governor General and the Dutch C-in-C arrived in S today. 15,000 more Dutch troops are expected in bunches of 5,000. The Sikh guards have been taken away, and there are many rumours of trouble between the Indians and the Japs. It is also suggested that the Sikhs and coloured Dutch may clash, so they have removed the Indians. Various stories [are circulating] about Japs arresting Australian and Command senior officers ?giving information to Chinese ?cigarette or food racket. There is very little true information on the subject, but all Victory V cigarettes were called in yesterday.

One of the Sapper officers has a small portable typewriter and he is teaching me touch-typing. It takes a lot of learning, like the piano, so doubt if I will become very proficient.

Jan 9th

A very exciting auction sale of the effects of Major Kennan RASC, who died of diphtheria, was held this afternoon. A few tins of food brought nearly £6 – I bid £5. I had quite a collection of goods at the end, including sheets, pillow-cases, stocking, pyjamas, dressing-gown, blanket, pyjama-case and cushion – for which I gave a cheque for £12–£13 payable after the war to Malaya Command who will, one assumes, pay the money to the widow. The auctioneer, an RASC officer, did not emphasize the charity aspect sufficiently – I feel sure he would have been offered larger sums.

To show the difference in values out here, a new greatcoat only brought in 10/-, while a raincoat was sold for 10 guineas. The difficulty of having a lot of kit is getting it all into a small space when we move. The Dutch in this area are leaving very soon, and a few days ago I was called on to help the IJA medical examination: a medical orderly (Jap) inserts a glass rod into the rectum and makes a smear on an agar plate. Looking very bored and blasé, all I did was to fuss round, and when he said anything, to nod or shake my head according to the expression on his face. It is extremely doubtful whether the cultures are examined, but they are carrying out HQ orders.

Several hundred Aussies have arrived here from Timor via Java. They were attacked by Allied plane & sub between Timor & Java. An officer who reported sick this am said that there are about 700 Aussies still fighting in Timor doing damage to the Japs, and are in an impregnable position. They are reinforced and supplied by air and sea – they are putting up a good show.

120 Aussies reported sick this morning, nearly all suffering from vitamin B deficiency – eye, mouth, lips, scrotum & feet. As we have run out of rice polishings and there is very little Marmite, the position is very difficult. The food situation is very precarious as the vit and protein intake is very low – the Red Cross supplies being nearly exhausted. All the symptoms of last autumn are coming back again, and I hear that Command is forming a Central Messing Committee and is purchasing additional food with money to be deducted from our pay. The problem is a serious one and must be tackled efficiently. The fusiliers were doing very well, and at one time a few weeks ago down to 2 vit B deficiency cases – they are now going up rapidly.

Yesterday I went to a lecture on diphtheria at the Hospital. There have been 960 cases of diph to date, with 46 deaths, mostly due to intercurrent disease e.g. dysentery or malaria. About 40% of the cases were skin or scrotal diph. All the serum ran out fairly early, but it made no difference to the occurrence of delayed paralysis which was present in over 30% of cases and possibly more due to the lack of follow-up. The Aussies only had 70 cases and this is said to be due to the higher age group. One lecturer said that peripheral circulatory failure was a cause of collapse. In England, antitoxin is given immediately to all cases – here it is amazing how they clear up with rest alone.

4 MOs from the hospital were very lucky to have a broadcast message from home, picked up by the Japs and passed on to them. They in turn sent a message back to England. It will probably be used for propaganda, but still a POW doesn't care as long as his folk can be told of his welfare. The Japs were very annoyed recently by the action of USA and Canada in their treatment of Jap internees. I read a Jap paper in English in which they devote several pages to US atrocities, and say that they must stop immediately. There is also a paragraph about the British inhuman treatment of German POW, and Britain is warned that the Japs will treat us badly unless this is altered. I hope the quarrel is settled satisfactorily without a lot of us being shot as reprisals.

Jan 12th

We have been paid once more our $20 monthly allowance. Among the canteen goods this time were some mangosteens – very sweet and juicy, but not much for the money. I haven't smoked since my last cold and am trying to give up completely, thus killing two birds with one stone, ie save money for food and also keep clear of the foul local tobacco. They are getting some good Virginian cigs from French Indo-China, but they are rather expensive.

Two Australian officers are eating at our table – one an RC padre who looks like a weather-beaten old trooper. I have sent nearly 30 Aussies to hospital suffering mostly from deficiency, amblyopia or malaria. About 90% of the Aussies have some form of vit B deficiency – probably a sequel of the poor food they had while on Timor. I get nearly 100 each morning on the sick parade, and on the whole they are a good crowd of fellows.

For the past few evenings after roll call I have been out – walks with Adrian & Wilkie to the top of the hill from which one can see the sea for miles all around Changi. A Jap cruiser was seen on the way to the Naval Base ?at anchor or ?lying aground. On the way back in the evening, the bats are a frightful nuisance as hundreds of them swarm, zoom and 'dive bomb' in and out of the pedestrians. One is afraid of getting one in the face as they come past so close. I had to walk with my head down and my stick waving in front of my face on one occasion.

Col D'anbue RA has been giving an excellent series of lectures on the causes of the war, and has opened our eyes to what the public didn't know, and what Baldwin couldn't tell when he said 'his lips were sealed'. When we had the bother with Italy re Abyssinia, they had 120,000 troops in Libya and we had 17,000 and TWO anti-aircraft guns. Our planes were also outnumbered 100 to 1 in speed and performance. We just couldn't think of going to war with anybody in those days.

Jan 18th

This afternoon Major Hunt, Australian medical specialist, gave a lecture to 18th Div MOs, OCs and Messing officers on Vitamin Deficiency Diseases. The main points were that these diseases were forecasted in March and Command was notified that evidence of vit B def. would appear in April on the restricted diet we had to live on. This came to pass and at varying intervals of time, beri-beri, scrotal dermatitis, angular stomatis, glossitis, painful feet and def. amblyopia succeeded one another.

None of these diseases are new, on the contrary he quotes 1935 reports on Chinese and Indian prisoners in Singapore who had all these symptoms when in jail on a diet of polished rice. They were given unpolished rice and were cured. The serious part is that some manifestations e.g. amblyopia, spastic paralysis, cardiac beri-beri may lead on to death or permanent disablement, ie some nerve damages are not reversible. The eye changes may spread back as far as the occipital lobe. There has been no evidence so far of any other vit. deficiency. Vit. E def. results will only be observed in the fullness of time after we are released and as the lecturer said, it would be very interesting to see what the reproductive capacity of the POWs will be. The only answer to our problem is rice polishings, once a day for everyone, and twice a day extra for all with vit B def. men. The B2 complex is thermo-stable and is therefore not destroyed by cooking.

Jan 21st

Following the last lecture, I have been stimulating the cookhouse to use RP [rice polishings] and to record the amount given each day so that I can inspect the record from time to time, to make sure that the stuff is being taken. I typed out letters myself and also the monthly General Health State in which I stressed the increasing number of vit. B def. cases, and also the lack of RP. It seems that the purchasing parties have great difficulty in getting supplies from the Japs. Soya beans are supplied in lieu, but I don't think they are as effective.

The Aussies left this morning and I must record that we were sorry to see them go as men and officers were smart and agreeable. One man, (Forster) who is a collector of coins, spent two evenings with me showing me his collection and also helping me to buy some from his mates. I have enriched my humble collection but the main drawback is the lack of money. I could buy any number, but the limited amount of money I receive I want to spend on foodstuffs which are valuable to me now. However I spent 80c on a Mexican silver dollar and 25c on 4 Palestinian coins. My best buy was exchanging my US silver dollar for 3 very old East India coins, 1 marked VOC 1790 was made in the early days of the Dutch landing on the islands. There is also one of the British East India Company (rather worn). This Australian lad has a large representative collection of coins from most of the Middle and Far East, gathered mainly from POW, British, American, Dutch and Australian, whom he met in Timor and Java. I shall try to build up a good collection by begging rather than buying.

Last night I went to a boisterous, lively pantomime, Jack & the Beanstalk, presented by the Serangoon Camp. It was very good entertainment and surprisingly clean. I acquired another shirt and 2 vests from Ordnance and am now quite well off in clothing.

Jan 24th

Last night I went to a lecture given by Major Wild who is interpreter & liaison officer and who lived in Japan for several years before the war. He gave a racy account of Japanese history including the story of Adams, an English shipwright and pilot, who sailed with a Dutch vessel to Japan, built ships for them and was deified by the Japs. [Adams was the main protagonist in James Clavell's novel, *Shogun*.] About 300 years ago when they had a large navy and merchant fleet, they invaded Korea only, and let slip the opportunity of taking the numerous islands in the Pacific ie Singapore, Java, Philippines etc. They are always kicking themselves for this lapse.

Then followed a period of several centuries during which Japan was a closed country, entirely shut off from the outside world. Europe progressed – went through the Renaissance period while Japan remained enclosed with her medieval customs. In 1870, with the restoration of the monarchy, they formed a new constitution – started trade with other nations and began to progress rapidly. Unfortunately an active member of the Army and Navy had final say over the conduct of foreign affairs, and this has led to the present war. They are grossly overpopulated, with a population of 80 million and increasing by 1 million every year. They believe that the soldiers are divinely inspired and immune from danger.

The majority of the non-military were pro-British before the war, and they distrust Germany for making an alliance with Russia after signing the Anti-comintern pact with Japan, and then attacking Russia without notifying Japan. He [Major Wild] believes that when the Japs crack they will crack quickly once the myth of invincibility of their armed forces has been exposed.

Yesterday morning we all had to remain indoors from 9.30 till 1 pm behind closed doors and windows – no movement allowed. It is believed that a member of the Jap. Royal Household was making a tour of Singapore. The Japs themselves have to look at the ground whenever a Prince passes. I spent the morning playing bridge with my 'permanent' partner Poulter who is a good individualist who doesn't seem to stick to any one well-known convention. We have however won in the large majority of games.

A Medic Soc lecture on Amoebic Dysentery the other day was very interesting. The only way to diagnose is by sigmoidoscopy and microscopy, for often 6 or 7 stools are negative. A man may feel perfectly well and still have ulcers in the sigmoid. There may be a very acute onset – like bad dysentery, but the stools of amoebic dys. are usually v. foul smelling and contain faeces. Patients should not be starved more than 24 hours because the poor food with restricted protein which is our normal diet cannot build up again the tissues which are burnt up for energy. Water should be almost forced down in large quantities. Stovarsol is the best form of treatment – the last Red Cross left plenty, so that we are well off. Emetine & Yatren are also useful.

I still haven't solved the problem of what happened to Roberts, the former Red Cross bloke, who is said to have gone to prison for 8 months for ?dealing with the Chinese or ?passing on information through the Chinese to Australia. We get rumours of fresh Red Cross ships in every day, but so far nothing has turned up. A few letters are delivered each week to the Camp – as they are censored very slowly. Nothing for me yet – but I live in hope. Some fellows have had as many as 5 or 6 letters while the vast majority have had none.

Jan 29th

The Japs have ordered us to give 2 doses of TAB and 2 doses of anti-dys. vaccine to every man. I started giving my TAB yesterday and found that it had a similar effect to the English variety in spite of HQ's advice that there was no reaction. The Infantry are rather difficult as the only time they can spare is after evening roll call.

The number of cases of glossitis, palatal erythema and stomatitis is increasing daily. I managed to wangle a bag of rice polishings out of 54 Bde messing officer – it is just like gold these days and we can carry on treatment for a few more days. Eggs have come back into the canteen again after a long lapse of time and I have bought and eaten 2 or 3 each day. I have been very short of money this month and had to borrow dollars from Tate, as I spent $1 on a toothbrush and about $2 on foreign coins.

Feb 4th [written in pencil]

I am such a chump. I tried to fill my fountain pen with thick black Indian Ink; am writing this in pencil while my pen is being rebored. The reason for the special ink is that I was copying out a sketch map of the POW Camp made for the Padre. It is an excellent map drawn to scale in a humorous vein. I borrowed all the necessary paper and tools from the Sappers and have made a fair copy with a few alterations for myself. I will have to take my fountain pen to small pieces and flush it out well and rebore it before it will be fit again for use.

I broke my clinical thermometer the other day and the same day we got a letter from Div saying that these would not be replaced owing to an acute shortage in Hosp.

Last Sat. night, the first Changi Spring Meeting was held on the Padang in front of our mess. The RE made a Tote which was illuminated, likewise the course. The horses were men with a wooden horse fore and aft, and the whole affair passed off very successfully. There was the usual excitement and feverish betting – in fact one quite forgot that this is a POW camp. I, being broke, borrowed a few cents and backed all the losers. It was a good thing that I didn't have much money to play with.

During the week I went to the Dress Rehearsal of 'Badgers Green' – a play about village life that was very successful in London. Bradstow produced it and Wilkinson was as good as ever. The whole cast played well, and in the congenial surrounds of the hosp. cinema, recently renovated, repainted and lit up, the show was a grand success.

I am now in the throes of the second TAB inoculations and am disturbed by the 'bottom-sticking' parade of the whole 18th Div which is taking place this weekend on the Gun Park. The idea, one assumes, is to find dysentery carriers, but as the knowledge is useless, the whole affair seems to me to be a waste of time. I have to go up on Sat pm and may get a Jap orderly or a British MO to do it acc. to which queue I get into.

The Service started last Sat. but I didn't go as I didn't find out till Sunday. If I get away from the Gun Park in time, I may go to the next one. I had .25 u [units?] of TAB last week; it didn't upset me very much.

Feb 9th [written in ink again]

My pen is now OK again after being 'rebored'.

It was my fate last Sat. to be a sticker instead of a 'stickee', ie for 2 hours on the Gun Park I was inserting glass rods into large bottoms pushed into my face, obscuring all other views. All went well till a Jap came along and told me to use both ends of each rod. I did until he disappeared, and then went back to the cleaner way. The whole proceedings were very unpleasant, and the sight of the colonels and officers with their trousers down was rather humiliating. Everyone took it in good part and joked about it for several days afterwards. I was the butt of numerous jokes. I don't know when my turn is to be done, but I understand that all 18th Div ones are now completed, TG.

All 2nd TABs are finished and in a few days time we start on the first Dysentery. Last night, owing to a hitch with the ordnance needle sharpening department, there were no sharp needles, and poor ADOS had the bluntest needle ever used. I haven't heard the last of it yet.

I have developed an ulcer of the axilla which appears to be spreading – I think I am suffering from over-treatment. I have been doing a lot of sun-bathing recently on the roof. I don't think that this has made it worse. Scrotal dermatitis is on the increase again and 25% of the RNF are bed down– this is a very high percentage, but supplies of RP [rice polishings] are few and far between.

I have taken over the job of officers' canteen officer and get all the stuff from Tate to distribute out to the few officers of our unit. We all put $10 each into the pool which I hold and I pay out for each order. There are plenty of peanuts, bananas, etc in the canteen so that there is no shortage of food for those who have plenty of money. We pay $7 messing each month – this doesn't leave us with much.

I went with a party to a revue in 11th Div. the other night. Quite fair, mostly pinched from Noel Coward. An elderly producer insisted on appearing on the stage from time to time and was quite out of place and he didn't even trouble to learn his part.

Feb 14th

It is now a year since we capitulated that fateful Sunday when the 'impregnable fortress', 'the bastion', 'bristling with guns' etc gave in to the Japs. A year seems so long when one looks ahead, but this year has passed quickly and judged by the prevailing POW standards, very comfortably. The Padre held a memorial service today for those fallen in battle – there was quite a long list of fusiliers.

The past few days have been completely beggared up by a desire of our captors to film us en masse. Consequently we were awakened at 6 am – pitch black – pouring with rain – early breakfast – stand out on parade at 8 am and were just about to start off when it was cancelled. The following day the same procedure occurred but we got out to the Padang about 1 mile away where 15,000 troops were drawn up in files. After waiting in the wet for 2 hours, the filming was postponed owing to bad light. The following morning the parade was cancelled before we started off, but took place in the afternoon – when I was fortunate enough to be excused. The other fellows were out for 5 hours, most of which was spent ankle-deep in pools of mud and the Japs are believed to have taken a quick film of their backs – so much for the week's excitement and activity.

A fresh party from Java has arrived, including 200 British Gunners and their force. Two MOs in their party came up last night to our mess, and we exchanged stories. One of them was in an original party to Bizerta when Weygand was arrested by the Germans and the raid fell through, otherwise we would have been in N. Africa about a year earlier.

Last week I went to four shows – all very good, and the intervening evenings I played bridge, at which game I am becoming quite a devotee – having originated and put into operation the 1–4 club convention, which is so complicated that even my partner can't follow it. However it has worked very well on numerous occasions and netted up several hundreds above the line.

Feb 20th

The Japs celebrated the anniversary of the fall of S. [Singapore] with a special edition of the *Sinbun* [newspaper], speeches, holidays etc. They definitely mark the fall of S. as the opening of the Greater East Asia movement. Many scathing comments about the British surrender occur frequently with the mention of the white flag which has been sent to Tokyo to be preserved, a reproduction of the special oil-painting of the Anglo-Jap meeting pending capitulation shows Percival clearly and the Union Jack with the white flag almost on top of it. We received no mark or token of the day, and we passed into our second year of captivity without fuss.

Last evening a batch of sailors came off a German Raider which sank 3 ships in the Atlantic and Italian Oceans – 1 Greek, 1 USA and 1 British. The Greek and Slav POWs could not be landed here as Japan is not at war with those countries. The survivors have landed without any kit, having had to jump into the sea when the raider attacked. It carried four 8" guns, torpedo tubes, two seaplanes and two motor torpedo boats which were used to finish off the attacked ship.

The wireless room and bridge were usually smashed first, then the boats and rafts and finally the cabins. Those that were left living jumped into the water and were picked up after their ship had been sunk. They were well treated and given quite good food. There was enough on board to last 3 years. The medical facilities were good – operating theatre, Xray plant, etc. The American boat – a new boat of 100,000 tons – had only recently been launched.

I have been to several medic lectures lately as there is a 6 weeks course in Radiology and also Mental Diseases. Yesterday a meeting of the Medic Soc. discussed 'Painful Feet'.

Although we know that rice polishings and rest cures the condition, there are so many accessory and conflicting factors that the condition is still far from being understood. The vaso-motor nervous and psychological elements are being sorted out. It is believed that absence of the riboflavin portion of the vit B complex causes SD and ang. Stomatitis.

The eye conditions are not responding well to Marmite, good food and a darkened room, and are a source of worry to me.

I have been trying to print some of the negatives we took down-stairs, but owing to the poor developer they have not come out well. They will have to wait until we're free.

Feb 25th/43

Yesterday we sent off our second postcard after a lapse of about 8 months. We were allowed 24 words. I wrote

Feb/43. Still fit, cheerful. Normal medical duties. Received cable but no mail.

Hope all well. Don't worry. Looking forward reunion. Fondest love all. Harry.

I hope the cards get home soon. It will be interesting after the war to find out exactly when they did arrive.

Medically speaking – things are getting rather bad amongst the N.F. re dysentery. There were 5 new cases last week and yesterday Roger Mosgrave was sent to hosp with amoebic dysentery with ext. ulceration of bowel and which has probably been present for several months. I have sent him to see the Medic Spec. twice and he has been an in-patient for two days – discharged 'diarrhoea'. I have been treating him for 'mucous colitis'. I have been suspecting amoebic dysentery for some time but my hands were tied.

CSM Brooks also went in yesterday (bac. dysentery) so that we now have no WO. The cook's stools are being examined in the hope of finding a carrier – faint hope!

Scabies is increasing rapidly and with no sulphur or benzoyl benzoate we have to use a local preparation of derris root which leaves an irritation following treatment making it difficult to determine when the cure is completed.

Inoculations are now finished. I had TAB but not Dyph as I am not satisfied that the latter is any good and I cannot afford either the time or health to be ill now. Looking after 55 Bde and Div [??] is a full-time job. Some amusing incidents occurred at sick parade recently. I said to one group of fellows who turned up for inoculation 'Are you 2nd dysenteries?' They said 'No, 1st Cambridgeshires'! One American seaman (from the raider) was put down as 'Dennis' on the sick report. When he came in he said his name wasn't Dennis but Atlopozcyki (or something similar) – he had toothache and he wanted to see a 'dennis'.

The padre went to hospital two days ago with abscess over the jaw – secondary infection of tinea ?actinomycosis. It is always one's colleagues, relatives or friends who get obscure complaints as all doctors in private practice know to their cost. I don't think there is any danger with the Padre but he has been very run down recently and needed a rest. He is getting good food and fresh air on the balcony in hospital and looks far better already.

A Dutch OR who used to teach German in schools has started teaching me German in the afternoon for one hour. He doesn't know much English therefore there is a mutual gleaning of knowledge.

The entertainments still continue in the various theatres. Last night at the Hippodrome I saw 'Roll out the barrel', a review full of life and pep – an excellent effort. On Saturday night I have got a seat for the Palladium for the new command show 'The Admirable

Dyeton' which is said to be a true skit on Malaya Command. It seems to be a bitter farce. This British characteristic of laughing against oneself can go too far.

March 5th

There is a very strong rumour that there are several mail bags at the Gaol and we are all eagerly awaiting our first letters. I'm afraid it will take several weeks for them to be sorted and censored.

A few days ago I took another glass-rod parade and yesterday found myself at the other end of the rod. It is not painful but rather undignified among a crowd of thousands. The rash on my bottom has broken out again and I have to carry a cushion round with me as the wooden slats are too hard.

My German lessons are progressing satisfactorily. My conversation is becoming less halting and 'ja-ing'. At the hospital this afternoon there was a demonstration of skin cases, including a group of 'vesicular' dermatitis associated with vitamin B deficiency signs, loss of weight and pigmentation. No name or cause has so far been found. Sefton gave a talk on Dermatology but owing to the room and his way of talking it was nearly all inaudible.

There is a great controversy raging whether we pay an extra $5 a month to the central messing fund for the benefit of all ranks, as it is essential to provide more and better food. I am in full agreement, unfortunately some of the officers for selfish reasons do not agree.

The stools of all the cooks are being examined and we found one of our cooks was a carrier of amoebic dysentery. The PMC has been changed after a long and successful 8 months reign. Major Harper has taken it over and it is easy for the first few weeks. Then it gets difficult to think of new menus. There has been a great shortage of fats and oils lately – palm oil is almost unobtainable and fried food is getting scarce.

Dysentery among the NF is now settling down – only about 1 case a week.

March 10th

An outsize in flaps is now on us as we (18th Div) have to provide 1,800 out of 5,000 men to go up country next week. The Japs originally wanted men who are fit to work hard and who have never had dysentery or malaria. As it was impossible to provide this number of fit men, Alt A, B & C are all going – RNF are providing 100 men (out of 120 – 30 of whom are sick). Hedley is the one officer appointed to go with the party. There have been numerous alterations of the figures owing to units complaining about their sick and necessity of leaving some fit men behind. Things are now I believe finally settled and our fellows leave on the 18th with the infantry. I spend most of my days now doing skin inspections – the best time is after the evening roll call when all the men are assembled.

Last night I went to see 'Two Masks' at the Coconut Grove – plays serious and merry. It included 'The Monkey's Paw' – although I have seen this before it held me spellbound in the dramatic last scene.

I watched the American seamen playing baseball on the Padang and hearing the US twang and expressions seemed like a scene in a film.

The 'rumour' about the mail was true – there are several thousand letters now being sorted and censored – 18th Div has priority as they are going up country. They are censoring about 300 a day so it won't be long now!

March 16th

I have been infernally busy with this darned up country party. We have just completed giving everyone TAB and dysentery [injections] when we had to start on the up country party and give them two doses of cholera, two doses of plague, vaccination, bum inspection for dysentery and numerous visits to the fun park for malaria testing. The fellows should be kept healthy now – I know my fingers are blistered from holding the syringe. The RNF leave next Sunday – ceteris paribus [all things being equal] and Japs willing.

The letters have all been sorted out now and are being delivered as they are censored. Some officers have had 2 or 3 each. So far I haven't had one but am told that there is one for me so am living in hopes.

Last night I attended a demonstration of Hypnotism by a Dutchman downstairs in the Dutch officers' Mess. A young medical orderly was hypnotised after only 30 seconds and was kept under for 30 minutes. He was as stiff as a log and was suspended between two chairs, and a heavy man jumped on his middle without bending him. I examined the boy while he was under the influence and was much impressed. He was in a state of tonic contraction, (except his arms which were flaccid), his pupils were fixed and dilated – insensitive to pain – P and R normal. He came round quite easily and except for feeling very tired there were no after effects. He described a sensation as of electric shocks going through him and he was helpless and unable to move. He then remembered nothing till he woke up.

This demonstration of Hypnotism and the previous one of Thought-Reading both by Dutchmen have made me a believer in the occult. I am now willing to believe anything!

March 18th/43

This is a red letter day or should I say postcard day, as today I received my first mail from England since I left home 18 months ago. It was in the form of a postcard carefully written by Dad, presumably conforming to some regulations issued. But the annoying part is that several officers are getting 6–12 letters – closely written pages – all items of news and interest and some have photos included! I can appreciate Dad's efforts to make sure this PC got through but he might at the same time have sent off a fuller and more detailed letter. Still I was very pleased to hear that all are OK at home.

A very distressing scene took place this morning just outside the mess. Hedley, trying to be helpful in his capacity as amateur interpreter, put his foot in it with a Jap Corporal who proceeded to give Adrian a hell of a bashing with his fist, feet and rifle. This went on for several minutes and a large crowd of onlookers gathered round. It is very humiliating mentally besides the physical damage and Adrian behaved splendidly and bore it all with very good grace. It has certainly dampened his enthusiasm to learn Nippon Go.

March 22nd

Yesterday morning I spent several perspiring hours on the Gunpark watching the two parties of fusiliers leave – all tightly packed in trucks with one wee Nip soldier in each truck. Adrian left in the first party, L Col Carpenter in command, with whom I am afraid Hedley will have several differences of opinion. It was rather sad – as all these farewells are – but we are now left with only about 30 fusiliers – all the rest are scattered up-country.

Wier, my former batman, went with. I was not sorry to see him go as he was completely useless as a batman. I had to give him three demonstrations before he learned how to iron properly. Invariably when he washed my things they came back dirtier than when they went away – either with black rust marks or smudges of tar – sometimes even large tears in the clothing. He was good at his golf and I probably learned a good deal from him there but otherwise he was impossible.

I have taken my former batman, McFall, back again – he has had two bad spells of dysentery – including a visit to the diet centre. He is looking quite fit now and I hope he keeps fit – he is very willing and a clean worker.

I have given up the infantry and have taken over camp, Provost and Sigs in addition to RAOC and RNF. I started off this morning in good style by a minor epidemic of diphtheria among Div HQ sergeants – 2 cases.

Extracts from letters are being published and there is an amazing amount of information to be gleaned from them. The instructions issued to next of kin in England to send 1 sheet and no enclosures and to send only 1 letter a week [seem to be ignored]. One picture paper published a photo of a POW camp and one relative recognized a man and wrote to him accordingly.

There is still some doubt as to the time when we ceased to be 'missing' and became officially listed as 'P of W'. The government announced one month after the capitulation of the units involved and we were all notified as being missing. The large numbers of officers and men who got away safely to India with 'official' and 'unofficial' parties even after the capitulation is surprising, considering the fact that the island was surrounded by Jap forces. It was fortunate that N Sumatra was free from enemy troops so as to enable our chaps to cross over and embark from the other side. Col Thomas & Beamish are in India, and from what we hear, the escape chaps are having a 'really good time'. There is still no news of Major Leach and other members of our party.

March 27th/43

Am writing this in the MI room during evening roll-call. I am permanently 'on duty' so am excused parade – a small recompense for my arduous toil during the day.

This afternoon I went to the Service in the Hosp area – the first time for many weeks. What a difference since my last visit. There is a Dutchman who is a proper Chazan – wears a large black & white tallis [prayer shawl], a yamulki [skull cap] on the back of his head and reads out of an enormous Siddur, just like in the Chevra Chass. There was a large congregation present including an Australian Padre visitor and a CSM (Australian) Blashki [who] keeps order like a shammas [synagogue beadle] and gives visitors a Siddur. He read the Prayer for the Royal Family in English and then the Chazzan read one for the Dutch Royal family in Hebrew. He then gave a short spirited sermon in broken English & Hebrew which consisted mostly of blessings and good wishes for the future. On the whole it was a very creditable performance. There was a large sprinkling of Dutch and Australian amongst the congregation which numbered about 50.

I have been unable to attend regularly owing to inoculations etc, but hope to go more frequently in the future. The number may be reduced as there are two parties going overseas next week.

During the week I saw 'Dancing Tears' at the Palladium – an excellent show on Cochran lines. The staging, scenery, costumes, lighting and band were first rate, and one is amazed to find such a high standard in a POW camp. They are fortunate in having [Ronald] Searle (a sapper) who is expert at designing scenery etc. An Australian called 'Judy' Garland impersonated a girl very successfully and dances very neatly. The number of female impersonators grows each week – there is the normal talk regarding these men and I am afraid that there is an element of truth in some of the rumours [i.e. homosexuality]

A snack bar was opened during the week – run by British and Dutch, and cooked for by the latter who are expert at making all kinds of tasty delicacies. We now have sweet coffee with milk (coconut) in the mid-morning (3c), all kinds of sandwiches, fishcakes, sweets etc. The half-caste [mixed race] Dutch are born cooks and make full use of all the locally grown produce. They have decorated a hut with palm and coconut leaves, made pictures on the walls, put in tables & chairs and now there is quite a social gathering each day. We buy sandwiches and cakes for our tea, and as long as the money lasts, we should not be hungry.

The Japs have agreed to give officers another $10 a month, but it goes straight into the messing fund and we don't see the money.

I am getting on very well with my German classes and understand him perfectly when he rattles away in German – but I am still finding difficulty in expressing myself. My memory, like everyone else's, is very bad, and I often can't remember the easiest words.

Eye cases are getting more frequent and we still haven't discovered a cure for them. Rice polishings are certainly not the answer, nor for stomatitis which often need Marmite before they are cured. It is the riboflavine element that is missing.

April 2nd

Have just recovered from an axillary abscess – a sequel to the tropical pemphigus which I suffered until a few weeks ago. There was no need to treat it as it burst spontaneously. McFall, my former batman, who had only been back at work a week, got another attack of dysentery and had to go to hospital again. Meanwhile I am sharing a batman.

The mail has died away but I understand that there may be some more yet to come, so have not given up hope of getting a full letter.

Various odd parties are still going up-country, and one party is said to have gone to Borneo. A daring attempt to escape by two Dutchmen in a sampan was nipped in the bud as the Japs spotted them and took them prisoner. They had 2,000 guilders with them.

I had to rescue a British soldier from the clutches of a crowd of angry Dutch who had beaten him about and left him lying in a faint on the floor. The fellow had taken articles from the Dutch for sale, but never brought back any money and later denied having received the articles. He was finally recognized and chased. He is now down for a well-deserved court-martial.

I had a chat at sick parade this morning with a CMP who was wrecked on an island with the Air & Navy chiefs for 3 months. The first case of malaria occurred after 2 weeks – they had no quinine and 25 out of 52 died. Death must have come as a welcome relief to their suffering and agonies.

April 10th 1943

We are nearing the end of our stay in Changi. A party of 7,000, half British and half Australian, is going up country 'to a cooler climate' in a week or so's time. It is believed that North Malaya is a probable destination. When divisional HQ decided to go, everyone thought that this was a good thing, and much intrigue is going on to be included in the party. The padre at first wanted to stay. However, when he found out that the division was going, he had to alter his tactics, and by pulling strings etc, he is now included.

I was detailed by Col Houston today, so I have heard – I've had no official intimation – that I am going up country. There is a schoolgirlish atmosphere as the officers discuss their trip and future abode, as though they were going away on holiday. The Japs seem to be very lenient with us, and all luggage and baggage is allowed to be taken with us. It has even been suggested that we take our lighting system. The camp is said to be a new one, probably just bare boards, so that everything we take will be useful.

The Javanese Dutch are going away next weekend, and the people remaining in Changi after we go away are to concentrate in the Selarang area with the Australians. One assumes that everyone will be sent up country in due course as the Japs confess that they cannot feed us properly here.

My batman, McFall, is still in hospital. I went to visit him today, and it is doubtful whether he will be fit enough to come with me.

I am starting a souvenir scrap book and am tapping all local talent. I had a good looseleaf binder made by the bookbinding department and I am asking all the artistically inclined officers to contribute. Rogers, a lithographic artist in the RAOC, has done two drawings for me.

April 13th

Col Houston, the ADMS who is in charge of the up country medical party, wanted to include me in his party, but Command cut his numbers down so that they could send some surplus medical officers from the hospital. Houston wouldn't stand for this and fought for my inclusion tooth and nail. Finally he had to give in before superior forces. He broke the news to me with tears in his eyes and thanked me for the 'noble work and unselfish labour' etc. during the past twelve months. I replied suitably and then tried to say goodbye, but stood on one foot by the doorway of his bedroom for two hours until he paused for breath, and then I slipped out.

Yesterday we had the usual bum-prodding and malaria testing on the carpark and Houston returned to the attack, this time enlisting the aid of Col Harris. Today I heard, after much yessing and no-ing, that my name had finally been included and so I go – I think.

April 18th

Some cards have come from home, and they don't give any news about what's happening in Leeds. They just say what they hope and feel that I'm OK. It seems to me to be a wasted effort to send a missive thousands of miles with just two or three words on it. However they may be following to the letter instructions issued at the time. But seeing the other chaps get long letters makes me feel very envious.

I have been very busy preparing for the up country party, and Windmill, my medical orderly, has been working like a Trojan. He cleaned out a wooden medical box which he painted and stencilled with my name, and I have now converted it to my own use. He also made me a small, white, first aid box with all the necessary contents included. My haversack or pack he converted into a rucksack that clips on in front, and is adjustable. Two ammunition pouches that are stitched to the sides are very useful for holding toilet articles, etc.

The RAOC made me a canvas bag like a lilo in blue. I have stuffed it with flock from my large and luxurious mattress which, unfortunately, is not portable, and I have made a neat looking mattress which folds up into a very small bundle. My pillow, which was a car seat, I can't take with me, so I have stuffed a pillowcase with some of the flock left over.

On the eve of the departure of the first party, a bombshell was dropped. The Japs have ordered the Aussies to proceed first in the first five trains, and divisional HQ to travel last. Col. Harris went along to Japanese headquarters today to protest, but to no avail. What the cause of this is, no-one knows. It has certainly piqued the divisional staff who were hoping to get there first, and fix all the good billets, hospital hygiene, etc. It is a strange reversal of the normal order of things, as divisional HQ are supposed to be administering the party.

Yesterday we had a talk by Col. Rose, the eye consultant, who told us how to treat corneal ulcers, which is an acute eye condition which might appear on the way up country. The question of medical supplies for parties going up country is very acute, as there is a grave shortage in hospital. Emetine is almost non-existent. Fortunately there are plenty of quinine tablets, so that we are OK in the malaria line. We are now busy giving the usual injections for plague and cholera, but not vaccinating this time, as none of them took when the last party went up country.

April 21st

The dates of leaving here have been changed time and time again. I believe, although I have not been informed definitely, that I am leaving on Easter Sunday, probably with the fusiliers. The medical officers tend to be neglected, and I have to go about seeking information.

There was a cinema show in the mess last night. One of the MOs showed films he took in Malaya, Australia and New Zealand. They were very interesting.

I took an 8mm cine camera with me when I left England and I took films on the journey over to the Far East. I also took some very interesting films during the battle and the immediate aftermath. I was very careful with the camera and the films. I sewed the camera and the films up in a canvas bag, and I decided to hide it somewhere until the end of the war. The padre took over a little mosque, next door to our billet. I climbed into the roof of the mosque into a circular dome. I pushed the camera and films into a very dark recess and I hoped I would reclaim them after the war. Before I went up country, I went to check everything was there. Unfortunately, I found that someone had stolen it all. How someone knew it was there, or how it was spotted in the darkness, I do not know. I mourn the loss of the film more than the camera. The former are irreplaceable.

I have been very busy these last few days, making a valise, changing and mending everything, arranging my medical stores, in the intervals of speculating where our destination is.

(Now there is a long gap in my diary. It restarts when I have time to spare to write of the horrors of the journey up country.)

[There is now a complete change in the tone of the diary as the horrors begin on 'The Railway'.]

Sunday, May 23rd
HELL IN THE JUNGLE

I am now seated in my own small portion of a leaky atap hut, during a storm in an unknown camp near the Burma border in northern Thailand. The journey from Changi to here almost defies description. It is a tale of hardship, suffering, hunger, thirst and disease, in conditions which could never have previously been imposed in a civilised society and which have taken their toll on the health of the men. Truth is stranger than fiction, and we have come up against conditions which would seem unbelievable in a novel.

We left Singapore and travelled for five days in a metal truck, 27 men to a truck. No room to stretch one's legs, and one meal a day. There were numerous cases of heatstroke and dysentery. It was impossible to take a man's temperature as the air in the trucks was about 103°. The floors of the trucks were filthy and oily. There was no way we could rest. Rice and stew covered with flies formed the main meal of the day.

We reached Bampong, our first stop, which I think is about 70 miles from Bangkok, and we were told to leave all our heavy baggage at the station, and take what we could carry on our back, as we were to march to Thailand. Little did we realize then that over 200 miles would have to be covered, marching by night and snatching a little sleep by day when it wasn't raining. The officials at Bampong laughed at our pianos, lighting sets, etc that we had been told to bring. We marched to the first rest camp where we were searched, numbered, pushed about, and herded altogether in large huts. Numbers of men fainted on the short march after the exhausting rail journey.

The first MI room that we came to had about 200 patients, which each succeeding medical officer took over for one day before he started off himself again marching. I will try and confine this first part to generalisations, and later will go into more detail.

Each night march covered about 15 to 20 miles. This may not sound much, but when only two thirds of the men are fit when leaving Changi, exhausted by a long, stifling train journey, having to carry all their kit on their backs, struggling through jungle paths and swamps, and often through tropical storms on very poor rations, with diarrhoea rife, struggling on night after night, with blistered feet, very few medical supplies – all these throw a different light on the picture. At the rest camps, there were no tents. One just had to find shade under a tree. If it rained, one waited, shivering, for it to stop, and then build a fire to dry out one's clothing.

Some rest camps provided good food, and one could buy from the natives things like coffee, eggs, bananas, etc. These kept the men going. They raised money by selling their kit to the natives who paid good prices for KD. I sold a few surplus items but stuck to my blankets, sheets and mosquito net which I later had to throw away, with my little hand bag which had been my constant companion since the beginning of the war. This annoyed me very much later because I could have sold the stuff I threw away, probably getting about $30–40 for it. Instead of which, I had no blankets and no money, so was not able to purchase much food.

I was medical officer in charge of my train party and marched at the tail end of the column which often straggled out for several hundred yards. Consequently, at a halt, I treated the stragglers and by the time I caught up with the main group, the halt was almost over. I had really the world's worst job of assisting the sick, lame and the weary, supporting them, carrying stretchers, and helping with their kit, in spite of my own kit which I had to carry. The fifth night march took its toll of my health. I had been losing weight, had diarrhoea, and tenosynovitis of my tendo-achilles. My mental and physical condition can be imagined as I struggled on through a tropical storm, through water, swirling and rushing waist-high on two occasions. One of my shoes was torn off, so I discarded the other one and completed the journey barefooted. Every step was agony. I had never realized before what it meant to go through such a night of mental and physical torture.

The rest camp reached the following morning, Tarso, was a big Jap headquarters and contained a lot of British, including several of the former up country parties. Major Harris and Major Wild were waiting at the camp to receive us. I must have looked pretty bad as Harris and GI each came fussing round, bringing me tea, etc.

(I must mention that there are gaps here and at various other points in my diary. I had to cut parts out later on when I realized that I had included bits of up to date news given to me by those brave POWs who operated an illegal radio, using the battery which I carried for my portable lighting set. They knew that if they were caught they would be shot.)

Hutch and Dillon treated me very well when I lived with them in their tent for the following five days. I was left behind to act as the medical officer in charge of the MI room in Tarso, partly because of my poor physical condition and partly because there was need for one there.

I saw the Jap MO who passed me unfit to march. He was very strict and examined all the men who were left behind who were marked unfit to travel. He sent on cases of diarrhoea, dysentery, septic and blistered feet, etc. It made my blood boil to see the weary men with large, swollen and septic feet, infected blisters and ulcers, having to hobble away at night. I can feel for them, as I know what they went through.

Jack Brooks came round to our camp on the first day I was there. He was a friend indeed. He helped me with my sick parade. He took me round to his billet where I had a hot bath, a change of clothing and a good meal. My dirty clothing was left behind to be washed by his batman. I also saw William and Flossie, both looking quite fit.

Hutch very kindly arranged for me to continue the journey by truck so I could arrive at the final destination fairly fresh, and receive the incoming parties. Little did I realize that each stage of the journey was worse than its predecessor.

Travelling by truck sounds good, but going on a heavy five tonner through a Thai jungle, up and down very steep slopes, across dangerous swaying bamboo bridges which broke under the wheels, clinging on to the baggage on which we were so precariously balanced, bumping incessantly up and down till all parts of our bodies were sore, seeing the back wheels skid on the slippery edge of a precipice as we looked down from the truck straight onto the water fifty or sixty feet below and going across the sides of a quarry on a rough road made out of shifting stone, pushing the truck several miles when it broke down, climbing off and on to push it out of muddy ruts, keeping back the shifting machinery, red-hot from the sun, blistering our backs and hands, seeing two trucks overturn going over narrow bridges and dropping into the valley below – that's what driving by truck meant, and I went forward in the truck to take the injured men from the first overturned truck to the base hospital which was fifteen miles away when our truck broke down with a broken differential.

We were stranded for twenty-four hours by the roadside with a query fractured skull and numerous cases of minor injuries plus one query intestinal injury. Finally we were marched two miles back to the last camp. The Australians were there with a small MI room. Major Stevens and Major Hunt, Australian MOs, were there and I lived with them in their tent for two days. I was very comfortable there.

Then my troubles started again. Some Burmese coolies, who were doing roofing, lived amongst the Australian troops, and they were dying like flies in and around the cookhouse and were fouling the ground. It was only after a couple of days had elapsed that it was discovered that cholera was the cause of death of these coolies. By that time, it was too late, as cholera had clearly spread to the troops, some of whom had only had one injection before leaving Changi. Inoculation was feverishly given, and all the troops were moved out to another camp. I moved a few hundred yards down the road with the sick, and was parked in the jungle with numerous sand-flies and mosquitoes.

Parties came and went and it was only after three days, alone in the jungle, with no covering or medical supplies, that a truck was provided to take me and my little party forward. We reached base headquarters where we found the other British train parties settled in. A hospital had been established, but the huts had no roofs. A very small stream provided water for all the camp, and conditions generally were not good. The only good feature was the Jap commanding officer who worked hard on our behalf. We stayed in the camp for two days, and I was just about settling in when I was ordered to move on with train 7 to their final destination, 15 kilometres further along the road.

Once again I set off on a night march, leaving at 1 am, and getting to the new camp the following morning. We passed the Australian camp after 10km where they were having trouble with numerous cases of cholera and were short of medical officers. We were certainly glad to reach our final destination after so many so-called 'final' ones. And so ended the long and weary trek through 200 miles of wild jungle to our new home where we may stay till the end of the war. The condition of the men as they arrived beggars description – no boots, tattered clothes, sunken cheeks, unshaven faces. It was often impossible to recognize old friends. However the relief of being able to lie down and sleep at night surpasses all else.

I will now try, three weeks after our arrival here, to fill in a few details as they occur to me. They are bound to be disjointed, but will bring back certain interesting events to the memory. One of the most amusing was the sight of men completing their march by hailing cycle carriages, and finishing the journey in style. Some even stopped to buy fruit from vendors by the roadside while the guards waited.

This particular camp sold fried eggs, meat, etc, and the men tucked in with gusto. Selling kit – one's own, and some that I suspect was stolen – went with a swing all along the line of communications. The natives paid good money, especially for shirts and shorts. I sold one or two things, and managed to scrape up five dollars.

On one or two occasions, Major Barber who was OC our party managed to stop a truck for the lame and sick. There was a disgusting scramble for lifts by fit men. At times like this, I was very disappointed by the behaviour of some of the men who were very selfish, so that towards the end of the journey, in spite of the difficulties, I ceased to have much sympathy for them.

I used to have difficulty sleeping on the hard ground or strange bed. The few minutes rest we had at each halt would see me throw myself down where I stood, on the muddy road, and fall fast asleep till awakened all too soon, stiff and weary, to continue the agony. I made the mistake originally of trying to take too much kit, and suffered from the heavy pack on my back. It felt much lighter when the blanket and mosquito net were discarded.

The MI room in the rest camps were hastily improvised shelters, and often had no covering at all. The MOs left behind to look after the sick had a difficult job with no supplies and very few facilities for treatment. In most of the camps I did my own sick parade, often very large, while the rest of the men and officers were sleeping.

The job of a medical officer was very arduous, and I often went for days without a proper rest. The annoying feature of the march was the number of men with bad feet, etc, 33% unfit that we brought with us never having anticipated such a march. I had problems with these men again and again. Most of them were scattered along the various rest camps on the way, and may catch us up later.

Looking back on our journey, the five days in the train when we left Changi which then were thought to be the height of discomfort, were at times almost unbelievable. The heat from the metal roof was burning to the fingers. The absence of a cooling draught of air made us look like dirty rags after a few days.

The one meal a day left us terribly hungry and thirsty. Fortunately the Nip guards allowed us to descend at most of the stations and buy food. The hot sweet coffee and hard boiled eggs that I bought at dawn at Alastare will always be in my memory as one of the most welcome meals I have ever had. We stuffed ourselves with papaya, bananas and eggs at each halt, so we managed to keep going. We were able to wash when we stopped by a river. At one place we had to do PT!

At Haadya where we were held up for nearly a day, I tried to get a bad dysentery case sent on, but they put another coach on for the sick man and the contents. There was the usual Nip fuss and rushing about and disinfection notices, isolation, etc etc. Haadya was full of Siamese on holiday, in holiday mood and clothes, and when we reached the station, it looked as though we were back again in civilisation.

Major Wild had a very difficult job and succeeded very well. Evidence of partial destruction of bridges, etc by the British during the retreat was still visible, but repair work had been efficiently carried out. The causeway which was supposed to have been blown up was as good as new. The Chinese were quite friendly, but were obviously afraid of the Japs.

The final journey by truck was the most exciting, and for my green medical case to come through unscathed after being hurled twice into the ravine, was a miracle. I had a two volt battery lamp, a dinner plate and a mirror, all unbroken, and good testimony to the packer. Elephants are very useful, pushing heavy tree trunks out of impossible positions, and can do the work of fifty men.

The most annoying part of the journey was the way four or five Nips would all shout out different orders together, whenever the truck got into difficulties. They are very excitable and flare up very quickly. If any part, e.g. a petrol pump, is being repaired, one of them watches the British driver doing the job, and then takes the bit himself and repeats it, often very badly.

It pained me to see the thin bodies of men heaving stones from the side of the quarry and they are expected to keep going on poor rations. How they can do this hard work is beyond my comprehension.

For two nights we shared tents with Nips. They are very neat and tidy but take most of the room, and we were closely crowded. On the whole, the guards who accompanied us treated us fairly well. It was the Nips at the halts who beat the men and made officers do general fatigues and cookhouse duties. Beatings were all too common. I am looking forward to the day when it will all be repaid with interest.

Thursday, May 27th

I am writing this before the first part of the story is completed while the matter is fresh in my mind. We arrived here six days ago, very weary and tired, with large numbers of sick. Two large unroofed huts were given to the men and a smaller atap hut was allotted to the officers. At the far end, the Japanese soldiers are sleeping. It is a camp sited on the far side of a slope overlooking a clean, fast-flowing stream, and has the makings of a good camp. Unfortunately for the first 48 hours it poured incessantly. It was the start of the monsoon season. Parties kept arriving and I was kept busy from dawn till dusk.

Gillies and myself at first, then with Lowe, and finally Hanbury and Turner (MOs) arrived later to help. Sanitary conditions are very bad, and there was no accommodation for the MI room or the hospital. The ration Jap became a patient of mine, and he let me use the ration hut as an MI room. This was a godsend. He has been very good otherwise and gave me a large quantity of sugar etc when I asked him for it, as Wilkie is a sick man and has a bad attack of dengue. Other Nips have been very good and offered us food. One caused a stir the other day by slapping Gillies in public. This cursed cholera is the cause of the trouble. It has followed us all the way since the camp where the Burmese were dying from cholera in our cookhouse. It has spread to the Aussies who had only had one inoculation. They were moved out to another camp, but the curse has followed us.

So far we have had about 30 cases and twelve deaths. The number of cases is increasing daily, and we are holding conferences, etc. A hygiene officer has been appointed, and everything possible is being done. At first the isolation place was an unroofed small hut in which patients lay soaking and froze to death. Now we have a roofed isolation place and the two far ends of the big hut are used for hospital accommodation. There are over two hundred patients with diarrhoea, dysentery and various fevers. Most of the MOs are sick themselves, with either swollen feet or fever, and it is a hell of a job to keep going.

I was sent back to Force HQ three days ago as a special messenger to carry information and to collect cholera vaccine, medical stores and comforts. The Manchester regiment, which has been a constant source of trouble – lack of discipline, etc – only had one inoculation, and some not at all in Changi. Fortunately my mission was highly successful. Every point I raised was satisfied. Col Banya, the Nip CO at HQ, is an elderly, kindly man who is doing what he can to help us. He was furious when he heard about the face-slapping and promised redress. He is going to provide a rest day and shorten working hours.

The Aussie working parties are working till late in the evening and coming back well soaked. They have to eat in the darkness and try and dry their clothes before they get to bed. Major Hunt with the two MOs has got his hands full in the Aussie camp. They had 40 cases of cholera up to two days ago, and this has been increasing daily. I tried to get decentralisation to hospital as the impassable road conditions preclude any daily ambulance service. Banjo was against it and said that he would keep the road open for the ambulances. I brought back a good selection of medical stores, comforts, vaccine, cooking pots, etc and everyone was highly satisfied.

We walked most of the way there except for two miles when we tried to get a lift on a passing truck, but as we spent ages pulling the truck out of swamps, pools, ditches, etc, we decided it was quicker to walk. Harris is moving his HQ 4 miles further back where the hospital will be sited. The present HQ will be abandoned as it is unroofed. They are very worried about the six to eight hundred sick who are scattered in MI rooms along the line of communication with very little covering.

While I was there a truck came in with ten men on, one dead and nine men seriously ill. Fortunately it was not cholera. They have had no fresh cases since. We have a gruesome job here, burning the dead cholera patients. Today, six had to be buried. It is a terrible, tragic end to die of disease in a jungle camp like this.

Major Stevens, an Australian officer who is second in command at the hospital in the absence of Col. Houston, is temporarily i/c hospital. He is very close-fisted with his supplies and comforts. He seems to think that Aussie lives are more precious than British ones. He got 11 jars of marmite by searching a party of British who were engaged on unloading. Two weeks afterwards, 400 bottles of marmite were looted.

$2,500 of canteen stores, carefully packed and crated, were smashed open and looted one day in Bampong by British troops. At times the men have behaved more like animals than human beings, and deserve very little sympathy. They don't think twice about stealing another fellow's entire kit, flogging it for fags or tobacco. Some British troops actually pinch the kit of Aussies who are cooking their meals. We have been issued with mats, blankets and mosquito nets of large size to hold 8 people.

The food is not bad. We have a collection of rather emaciated yaks, two of which are killed off each day, and we have a tasty stew at night. Otherwise the meals are rice and a thin stew of onions or beans. A large number of men, especially officers, have gross oedema of the legs, due mostly to deficient diet.

May 28th

The number of deaths from cholera is increasing, now 23, and is spreading to other units. The isolation centre is working more efficiently now. We are handicapped by having one intravenous infusion apparatus. Sanitary conditions are very primitive, and the task of looking after the patients is unenviable.

Gillies has been sent back to look after a small Australian camp near here. I was nearly sent off instead, but ADHS very kindly interfered on my behalf. I have a full-time job looking after the hospital of 350 patients and also having large sick parades. Things are now getting organized medically, and although there is a hell of a lot of work to do, it's being done systematically. A hygiene officer has been appointed to do full-time work, looking after latrines, disinfection, etc. I brought back several ideas from the Aussie lines which are being put into practice. The Japs have issued us with large mosquito nets and blankets, and together with a stretcher I brought up, I hope to make myself quite comfortable during the rest periods that the work allows.

I sold my wristlet watch yesterday for 20 Burma rupees to a Jap, and I also sold some of the other things I brought up with me, e.g. cigarette case and wallet. The natives are reluctant to take money and prefer clothing and goods instead. Yesterday we bought 16 bananas for $1.

May 30th

The position is going from bad to worse. There are over 100 cholera patients and there have been nearly 40 deaths. There are 18 bodies waiting to be buried today. Labour shortage for essential tasks in hospital is acute. We have no men to clean up the place or carry the food and fluids to the patients. Some cholera patients have waited 20 hours without a drink to be admitted. I have found that looking after the MI room and the hospital is really far too much for one MO, and so I am concentrating on the hospital.

I appointed a messing officer for the hospital to keep up the supply of fluids to the patients. It is tragic to see all the dehydrated patients lying for hours without a drink. The one bucket that the hospital possessed has been stolen.

HQ had a conference with the Nips this morning and tried to get a moratorium on the working parties building the railway until the illness peak has passed. I am afraid that HQ will get very little sympathy from the Japs as they are blinded by the God of Road and Railway. They will find out their mistake very soon when the number of fit men is decreasing daily to almost none at all. The Japs in their usual efficient manner are doing bumstabbing but offer no medical help or assistance. I have never been so weary in all my life. I just drag one foot after the other, leaning on a stick.

June 2nd

The death roll continues to mount. Yesterday the number was 62 dead and about 110 cases in hospital. The isolation place is grossly overcrowded, and the Nips say that they will build an extension later on. I have nearly 400 patients in my hospital, mostly with diarrhoea, and 4 cases of diphtheria for which I have no treatment. Yesterday an officer died. He was getting on in years, and I am not sure what the cause of death was. It is often very difficult to make a definite diagnosis. Nearly all the cases of malaria are given quinine because of the dangers of malignant tertian malaria.

The rain here has been continuous for the past twenty-four hours, and the ground outside is like a bog. The Japs gave all the men a two day holiday from roadwork, so that essential indoor tasks could be completed, e.g. latrines, swill pits, pathways, railway clearing areas etc. There is a lot of work to be done, and the problem has ceased to be a medical one, but is now a general one. The whole place is a filthy mess, full of vomit and faeces and cholera germs, and I fail to see, how under these conditions, we can ever get rid of disease. Meals are very poor, and I am still in a poor state.

June 3rd

I have been over to the cholera centre. It looks like a scene from a film, completely unreal. There is a long, dark, atap hut, with over a hundred thin skeleton-like beings, writhing on the long platform, vomiting and passing motions where they lie. Groans and cries are the only noises to break the silence. Two or three orderlies with masks over their mouths were giving intravenous injections of saline, using Heath Robinson contraptions. About nine corpses lay outside covered with blankets and groundsheets, and a little distance away, the smoke of the pyre where the corpses are burning could be seen.

The death roll is now about 87, and in my own isolation hut, there are still lots of cases requiring treatment, but unable to be evacuated due to shortage of accommodation. An astonishing scene is the looting of the dead men's goods by the orderlies and working parties who sell the patients' possessions to the natives and then keep the money. This is trafficking in death.

Col Dillon came through yesterday and brought a lot more milk. It is given out to the orderlies to supply the patients with their nourishment. I hope it gets as far as the patients.

We have had an impossible Jap living here with us. He is a medical NCO, pretends to be a doctor, and has come to take charge of the cholera outbreak. It seems to have got the better of him, for he has left today, 'sick'. He was up all last night, and kept going through our billet during the night, kicking up a fuss. He dismissed the cattle guard so the cattle came into our hut to be near the fire. He dismissed the firemaker so the fires went out, then he played hell with everyone.

There is a very large number of officers who are sick, mostly with dengue and malaria, diarrhoea and oedema of the legs. One officer has definite cholera. I hope it does not spread any further. My day is fully occupied, and I get very little time for rest. The worst curse of all is the disturbed nights, like we had when we first arrived in Changi. I have to get up three or four times a night, go out in the pouring rain, through inches of mud, sliding about in the darkness to see the patients.

My appetite has come back, and at times I eat a tremendous amount of rice. The meals are sometimes quite good, and last night we got an excellent meat stew, but often we get just a thin soup with rice.

We heard yesterday that John Excell died. It was a terrible shock. He was the life and soul of any mess. He had started designing my house which he called Bedside Manor! He had hoped to finish it later. Still he has provided me with the germ of an idea for the house if I ever get one built.

June 4th

Deaths have now topped the century – a hundred and six. There are a hundred and twenty cases in isolation, and about twenty awaiting admission. There were four deaths in my hospital amongst the waiting list. There was one death from diphtheria today. One feels so helpless. With these cases there are no antitoxins, just rest, and the men are choking to a slow and painful death.

June 6th

Sweeping changes have taken place overnight. The Japs have bumstuck everyone, and a hundred and seventy-four men are off with 'possible' results, and have been sent off to the far side of the camp into isolation. The weather was atrocious and it poured down all afternoon so the move was postponed. The Japs got annoyed and ordered it to take place forthwith. So it took place in pitch darkness until the early hours of the morning in a terrific downpour. The helpless patients were floundering about in the mud. Some of those who fell down, stayed down, until they were found in the mud by other people.

A further order that all diarrhoea patients had to be segregated was carried out at first light this morning, and the hospital was split up into two parts – a fever department in the lower hut, and a diarrhoea department in the upper one, but all diarrhoea men have to be segregated. Special latrines, buckets of antiseptic, bags of lime etc have been provided, and a Jap officer from Moulmein is coming today to inspect the camp.

Theoretically, the move is very sound, one I would recommend. All washing and bathing in the river has been stopped, and our source of supply of water for washing etc, cut off. A large number of men have escaped in organized parties, complete with food, medicines, etc, and they are heading westward to the coast forty miles away, where they hope to get a boat to India. I hope they succeed, and they can carry back the terrible story of our hardships. I hear that two ships leaving Singapore and travelling north were attacked by the British. One was captured so it looks as though they have a favourable chance.

Yesterday some Manchester men got half a yak and a cow by doubtful means. We got a mouth-watering meal that night. A runner arrived yesterday from base HQ with medical supplies, and I hear that we can send tomorrow for some more, for it seems as though the Nips have had a consignment. I have organized all the fever cases and given all the malaria cases doses of quinine for a whole week. There must be a lot of malignant tertian malaria in these parts.

June 9th

I think the corner has been turned. The number of cases for admission is falling rapidly, and the number of deaths is falling more slowly, I feel sure that the boiling of all water and the prevention of washing or bathing, a measure which I advocated a long time ago, has helped considerably. The isolation of the diarrhoea cases is also a great help. Unfortunately the poor men in isolation are living in such deplorable conditions, they have had five deaths already. One Nip has got cholera, another is in isolation. We heard the good news that four Nips in a camp further up had died from cholera.

The two camps higher up here are also badly hit. In one camp with Dr Emery they've got malaria and dysentery, and he is ill, but he is carrying on from his bed. At least we here have had relief in case we feel sick. I've had two cases of anuria leading to uraemia for which nothing can be done except, unfortunately, to watch them die.

The deaths from cholera are now about one hundred and thirty-five. Because of the escapees there are now morning and evening roll calls, and these take place in the rain, and this makes it very uncomfortable for the poor chaps.

I hear the escapees have a good chance of success, except for the climb through the Pagoda Pass, some five thousand feet up, which will tax their strength. The food here is still quite fair and the officers' messing officer has been appointed.

June 11th

Wilkie and I have been doing a little business with the Nips. I gave a wallet, and Wilkie a fountain pen for half a bucket of brown sugar. We are very lucky to get this as it is very difficult to obtain and comes in useful for the morning burnt rice porridge which is now our daily breakfast. I also sold a cigarette case for a hundred and fifty cheroots. It is a sort of Burma cigarette using a leaf instead of paper. My cheap wristlet watch was sold a few days ago for 20 dollars, so we are now quite comfortably off.

Yesterday we had a cholera inoculation, a fortifying dose. No reaction today. After two days, no fresh cases of cholera. It flared up a few days ago with ten fresh cases. The number of deaths is now about 60. One man in ten has died of the disease here. It is a terrible tragedy, and one that could have been prevented. A long report has been made out and it is being sent to Changi.

The officer who went to collect medical stores from base HQ reported that cholera is now spreading down the line and has almost reached Tarso. It's harder to think what may be happening at some of the MI rooms at the rest camps. It seems that the camps up here are now getting on top of the cholera, but malaria is still prevalent. I am taking two tablets of quinine every evening as I must keep on my feet. I have a full time job of work to do in the hospital looking after the malaria and diarrhoea cases. Fortunately I am feeling a little fitter each day, although I tend to drag myself round a bit.

June 15th

I am still travelling. I seem to be very unfortunate in that respect. Two days ago, owing to the bad state of the roads and the poor ration provision – food rations were cut down to one meal a day. Ration parties are being sent out to bring food from a camp eight miles away. On my journey to the camp to pick up supplies, the truck stalled, and with the brakes not working, we ran backwards downhill at a fast pace and ended up in a stream at a very acute angle. We had a very lucky escape as a tree trunk prevented the truck from overturning. We couldn't get it out, so five of us spent the night by the roadside while the others walked on to the camp. They sent food out at one o'clock in the morning. Marvellous food – fried rice and whitebait! The first fried food I've had for ages.

The camp where we went to seemed to be a paradise from the food point of view. Later an elephant and fifty men appeared, and after a great deal of tugging, we got the truck out, put our kit on it, and finished the journey on foot. This camp is not so good – boggy, low-lying and dirty, but the rations are very good. For tiffin today we had an enormous portion of meat, stew and beans, about four times anything served previously.

The diet is very deficient in some part of the B complex. Scrotal dermatitis is on the increase again, and also beriberi and pellagra to a smaller extent. It is very similar to our first summer here after capitulation, only there are no rice polishings now.

Lowe and I expect to go back again very shortly as soon as the Japs realize that Hanbury has been left to look after the other camp by himself. I have a lot of packs to look after when I move, and I hate all this moving around, but who can tell with these devils. They may abandon the idea of a railway and concentrate on keeping the road open for food etc to get through. They never expected such difficulties when they made their grand plan.

I don't know how long the good food will continue here but I hope to take advantage of it while I can. I have only 40 patients – quite a change after the 500 I looked after at the last camp. Cpl Ullman reached the former camp and is acting as interpreter; he is very good and self-taught. Ullman should be a great help to the camp as he has a better approach to the Japs than Dorfman who bluffs and laughs his way through.

June 22nd

Since writing the last entry, I have been on the move again and have now returned to the river camp having made a tiresome journey over a very boggy road. The Nip guard had malaria and got a soldier to carry his baggage, and I got another one to carry mine as I have so much darned stuff to carry, ie all the medical supplies and equipment, as well as my own kit.

I was very happy at the last camp. The food was excellent, the work moderate and quite comfortable quarters. All that changed during the last few days and I was glad to go. I waited two days for a yak party to come, but none came. This camp sends up a human yak party to the other camp every so often to bring back food, etc as the road has been impassable. It is now a lot better and trucks have passed through today. It is a terrific task, laying great logs down along the track and bridging the very boggy parts as the men are not fit and there are comparatively few men available to do the job.

We called in at the Aussie camp on the way back and had a good lunch. They are also on a good diet. The food here has been cut down by a half, and so we queue up eagerly for the meagre rations, and we are very soon hungry again. I brought back some raw rice with me and have been browning some with each meal to make a good drink of coffee with my fast diminishing supply of sugar.

The Aussies have two cases of typhus and there were three cases of small pox here when I arrived back. Banjo is rather worried about the latter, and he is sending up some stuff for vaccination. Cholera is settling down now at all the camps here, and there have been very few admissions. Bumstabbing is done almost every other day, and today there was a malarial slide too.

Everyone is still feeling very weak, and the sick take ages to get going again. I have my ups and downs also. Actually I should have left the other camp sooner, but I fainted while doing a sick parade, and then had to be helped to the latrine. I thought I had got cholera, but fortunately it passed off and except for a little squitters for a day or so, I haven't felt too bad. I still feel a bit weary, but thank God I am on my feet.

There were instructions out today about what to do in case of an air raid. It seems that the Allies are getting closer, and the Nips seem worried.

A large snake, five feet long was killed in the officers' quarters yesterday, and was cleaned and boiled. Those brave enough to eat it said it tasted like chicken!

June 27th

Things are far brighter now. There have been no new cholera cases for over a week, and diarrhoea and malaria cases are fewer. Also there are no new cases of smallpox.

This camp is now a food supply base and is being stocked with rice, fish, beans, etc. There are now two diets – one full one (600g of rice per day) for workers and men on essential duties including MOs and medical staff, and half rations for sick men. We get rice, fish, cucumber and meat stew, often with beans. Breakfast is still the same ie burnt rice porridge. Wilkie is still very swollen and I get his rations with extra liver whenever I can. I have bought some cured fishes for a dollar each – very tasty.

Col Harris was here recently and is optimistic regarding the future here. I can see no cause for rejoicing and will be only too glad to get out of the place. Our thoughts and conversation turn frequently to England and good food as we lie awake at night on our mats and gaze into the fire lighting the darkness of the hut. I still get intermittent attacks of diarrhoea but am eating anything I can get hold of.

July 1st

Things are moving again. The Nips are setting up a hospital camp for 2,000 patients, forty kilometres from Moulmein. Five hundred men, two MOs and forty RAMC personnel are going from this camp. One of the MOs is myself.

Wilkie still has gross oedema but is also going, so we will still be together. It is said to be a two day journey, but with the state of the roads it will take much longer.

The men working on the roads are being very badly treated by the Nips. They are mostly sick men, and have to work a full day, carrying heavy logs. We are being paid, but can't buy anything so the money is useless,

July 6th

Great difficulty today in raising sufficient fit men for the daily working party, so the Japs went round the hospital picking out men they thought were fit and increased the number by 50. The MOs had the difficult task of sorting out the convalescent cases into fit and unfit.

Yesterday, a bombshell! When we got up we found that Col Wilkinson, Robbie, Jack Feathers, Bill Anker, Jim Bradley and three other officers had slipped off during the night, and made a daring bid to escape. The Japs were very annoyed and created a hell of a flap. We were on a roll call parade for two hours and all closely questioned. Col Banyo [Banjo?] imposed a penalty of five days starvation on all the officers, and said that next time, Col Hingston would be shot. Fortunately the ban on food was lifted for MOs and officers on working parties. Food was secretly slipped in to the others during the night.

Our kit was searched, and all our money was confiscated. I felt bloody awful, but I found some tins of soup which I had hidden under the floor boards, and Wilkie and I drank this. I thought how marvellous it must be to reach safety and freedom, but I don't think that the risk is worth it.

[The following comment was added later by Harry.] *I didn't put this in my diary, but I had given Jim Bradley some M&B 693 tablets previously from my small supply, and I had guessed that the escape was being planned.*]

July 12th

Because of the escape, the present relatively friendly Nips are being replaced by a group of 30 unfriendly Nips and an officer. They patrol day and night. We have now been warned that if anyone else tries to escape, we will ALL be shot. Roll calls are frequent and last hours. Yesterday we were on parade from early morning till 1.30 am as the roll call didn't tally.

Col Banyo saw the emaciated chronic diarrhoea cases in hospital, and one died as he was touched to answer for roll call, I think that even he was moved by it.

Six died during the past twenty-four hours. We are having difficulty getting the two hundred men for the working party each morning – many are very ill, and often collapse and die on the way back.

July 15th

The rainy season is here and it often rains for days at a time. Deaths from dysentery are increasing and more than nine died in one day recently. Major Barber was sent into hospital with diphtheria – he has been very unfortunate as he has been ill for a long time.

July 23rd

Reveille is at 6 am. The men have to parade at 7.30 am, and in the half light, the MO's have the heart-breaking task of sorting out the semi-fit from the very ill to send out on the working parties, working 13 to 14 hours a day, from early morning till the evening.

We heard today that the escapees have been captured near Moulmein and that Col Harris has sent a letter pleading for their lives to be spared.

The death roll here is now 280, nearly a fifth of the total.

Sgt Armstrong, my medical sergeant from the beginning, died a few days ago from dysentery. He had put up a hard fight. I was very sad, and I was with him when he died peacefully.

September 1st

I am writing this in retrospect about the past month as I have just received my diary that Wilkie brought up as I was given only ten minutes notice to leave the last camp.

We left number 2 camp on Friday, 30th July in pouring rain and marched to number 5 camp where we stopped overnight. We left there at 3.30 am and marched in pitch blackness for ten kilometres. Many men fell into rivers, ditches, etc, and some men lost their kit. I was unable to see clearly in the dark, and had to hold on to the man in front like a blind man. We trucked to Kandoo, and then by train to Ronshi. We marched from there as a bridge had been swept away by the storm, and we arrived at Tanbaya where the hospital is stationed.

Burmese coolies were still living in some of the huts, and they came and sold us eggs, cheroots, coffee, etc. There were a few yaks and we killed about two a day for a short time and had very thick meat stews. In fact the place promised to be a land flowing with milk and honey. Unfortunately it didn't last long. The vendor ran out of eggs and we went back to the ordinary diet of rice and gippo (a thin soup). We spent the first week sawing wood, digging drains, etc, and I felt very fit. The good food and the outdoor life put inches on my chest and stomach! After a week, sixty patients arrived, and then there was an inexplicable pause of three weeks before the next batch arrived.

I was given charge of the first ward, and had quite an easy life, except for the fact that three consultants, so-called, used to see my cases, also making my position rather awkward at times. Hunt, an Australian MO, tends to treat the medical officers as students, and not as fellow medics. It is annoying to be constantly pulled up and questioned about silly little points, but one has to put up with it. He is officially senior medical officer, but he appears to have taken over all the administrative side of the hospital too. Hunt insists on every man eating all his food, even if he has to be forcibly fed. It is hard for a man with no appetite to push down a monotonous diet of rice and gippo, but the majority of the patients are living skeletons, due, according to Hunt, to our previous policy of starving them. It is a fact that the Aussies have had far fewer deaths than we have, but the food consumed is not the only factor.

The railway is working well, and steam trains are rolling to and fro carrying rails and sleepers to push the completion of the railway on. I hear that the date for it to be finished is September 16th.

There are numerous cases of malaria amongst the staff, and we are hard pushed to find sufficient men to do the essential tasks. The Jap guards who were here first were a nasty bunch who led us a hell of a life. They cut down the rations, no salt and very little of anything else, constantly interfering with us, and demanding men for jobs like stoking fires for their bath-house etc.

We heard that cholera had broken out amongst the Bungs [possible slang for Burmese locals] at Nikki and that may have been the reason for holding up the parties. However after three weeks, the patients started to roll in again, usually in the middle of the night, and fires and flares had to be lit. But even then it was no easy task, taking the helpless patients off the trucks and carrying them into the camp hospital.

The condition of the patients was terrible after a nightmare journey for three days in open trucks and trains, often in pouring rain. Several died on the journey, and some were brought here dead. This morning, out of a party of 114 that left no. 2 camp, 14 died on the way here and were buried en route, and 6 were brought in dead. It seems to me to be a criminal act to send these dying men on a long journey for no purpose at all, and sending no fit men or medical orderlies with them. It is absolutely amazing.

About ten days ago we got a new Jap administration consisting mostly of kindly-looking elderly Japs who are in stark contrast to their predecessors. They left us strictly alone, increased our food, and generally allowed us everything we asked for. The new officer, although he looks stern, is very fair, and listens attentively to all Hutch's requests, and does not give a propaganda speech or ask about whyfor etc.

Aircraft occasionally appear overhead about midday, and once we were told by the Japs that they were Allied planes. I doubt it.

Reports of parties that left Changi after us, reach here from time to time. Col. Houston is now at no. 2 camp, and has taken over medical charge there. One party of thirty medical officers have had a rough time, and some of them have had to look after thousands of Bungs who have had cholera and dysentery, a lousy job. An officers' working party of three hundred men are having a bad time too.

No. 2 camp, since I left, has become a charnel house, so I hear, and is having 15–20 deaths a day. The death roll is over the 600 mark out of the original 1,600, most of the deaths now being due to chronic dysentery and malnutrition. The men who are arriving here daily are barely alive and every bone showing, Even the Japs are amazed when they see these living skeletons.

No. 5 camp and Major Gillis are in Major Hunt's black books, as in fact all British patients and medical orderlies seem to be at present.

The food has gone off a lot lately, and our meals consist of very watery yak gippo and rice, three times a day. The Japs don't seem to have increased the ration from when we were a hundred here to when we are now twelve hundred.

Wilkie arrived yesterday with my kit and Pringle who is looking very fit. I now feel happier having all my goods and chattels around me. The nuisance is carting all this stuff around when I move. Two officers from no. 2 died of dysentery during the past two days.

September 4th

Today a party of 200 arrived here, with Hingston, Auld and numerous officers, and they looked very travel-stained and hungry. They were told that they were going to open a new convalescent home! but no arrangements were made, either on the journey or at their destination, which seems to be our hospital. Hundreds more are arriving soon, leaving only 700 fit men at no. 2. What the Jap idea is in sending all these men here, no-one knows. We certainly can't fit them all here. Most of the men are reasonably fit and will be useful here.

They saw four of the escaped officers being taken back under escort, Bradley and Anker looking well after their incarceration of one month. The Japs say that the others died in the jungle and were buried, and that these two will be shot.

September 7th

Several more large parties of fit men and officers have arrived here in search of a mythical convalescent home. Most of the officers in our group of no.2 have come here. A difficulty is finding jobs for them like assistant wardmaster. It's not pleasant sleeping in a ward, especially a dysentery ward.

The food has been very bad recently, there having been only two rice meals in the last two days. The thin, chronic dysentery patients will never improve on this diet. We are having about eight deaths a day. Our total number here has now reached 1,600, and accommodation is very crowded. The Japs have built a high wooden tower in which a sentry stands and looks over the whole camp. We will need good food very soon if we are to save hundreds of lives here. In my dysentery ward of 200 there are also 50 cases of associated beri-beri. The sudden deaths of these are due to cardiac beriberi.

September 10th

The staff officers have just made a good move from the general staff hut to the officers' hut who were moved out against their wishes. It is peaceful and quiet, and far from the rest of the hospital. The food is now brought round to the hut, and I have a third share in an Aussie batman who collects my meals, which are now vastly improved. He also washes up the dishes. The Japs have promised us rice polishings and unpolished rice, but the only question is when will we get it.

September 12th

Hunt is meting out rude justice to offenders – corporal punishment for black market offences, and running the gauntlet for thieving. Some of the wardmasters are rather squeamish about inflicting the punishment, so Hunt just sails into then with his fist and stick. It is the only way to deal with these chaps, although it is probably against King's Regulations.

I had a third cholera injection two days ago and developed a slight fever ?malaria. As malaria is rife in this ward, I have started taking quinine on the off-chance. The first supply of canteen stores arrived today. A good supply but totally insufficient to supply the needs of 1,700 men and officers. I hear that our pay has arrived in the camp, a very suitable occasion for it to arrive.

September 18th

This is my eighth day of fever, certainly malaria. Ten quinine tablets a day. Whether it is the illness or the treatment, I don't know, but I feel bloody awful – no appetite, nausea, dizziness, cold sweats, deafness and tinnitus. It is very annoying to lose ground like this, just when the rations are improving, and I was beginning to fill out again, like in Changi. However two more days of treatment should see the worst over.

The canteen sold me half a hand of bananas the other day, one of the few things I could look in the face. I would give anything for a few Sunkist oranges now. My batman is very good, washes my things, makes me tea twice a day, and sees to my needs.

Stores and food supplies are now beginning to flow in, and we recently had a thick meat stew with potatoes and vegetables. Mosquito nets, blankets, and various diverse cooking utensils have arrived too. Also two pairs of white shorts. This latter will solve the clothing problem for several weeks. There was an issue of 61 Jap Red Cross cigarettes per person yesterday to all the camp. This must have cheered up the chaps considerably.

Houston paid a flying visit to the camp two days ago, and he looks very ill with malaria. He brought cooking pots, medical stores, etc. but we are far from him because of the distance involved. The Japs appear to be getting kinder each day. Is this a sign of the times?

Our death toll is now on the two hundred mark. Extensive, non-healing ulcers are now being treated by amputation. It seems such a tragedy to lose a leg following a scratch a few weeks previously. It is the only treatment as they otherwise die of toxaemia and inanition.

September 22nd

This is my first day up after recovering from my fever, ?malaria, ?tropical typhus. It was certainly a strange form of malaria, if malaria it was. I was treated with quinine, of course, in the absence of any other form of treatment. I am very pleased to have finished the quinine as I now can look a bowl of rice in the face again. There are a large number of unexplained fevers, similar to mine, occurring almost every day.

The three escapee officers who were recaptured went through here yesterday, back to Rangoon where they are to serve a long prison term. Things will be bloody awful for them, but at least they will still have their lives, which none of us expected would be spared.

I saw my former batman, Windmill, RE, today. He is in the ulcer ward, and looks a wreck of his former vigorous self. He pleaded with me with tears in his eyes for his leg to be taken off, but as he is such a bad operative risk, I understand it will not be done. He says the pain is terrific, twenty-four hours a day, and at night he is afraid of losing his reason. What a terrible thing a tropical ulcer is. Amputation is the only treatment and relief, and the post-operative mortality is very high, owing to the general poor condition of the patients.

Thirteen out of the original twenty-five Northumberland fusiliers are known to be dead, and there are a few unaccounted for. We have lost over two hundred men already, with no slackening in the daily death toll.

The food has improved greatly, but in most cases it seems to be too late.

Scrotal dermatitis is very prevalent especially amongst the officers. Grass extract is being given to provide the essential amounts of riboflavine, but it doesn't seem to hasten the cure, and the cause of it is rather obscure. It appears to be seasonal for there is a severe outbreak each year about this time of the year. I have it in a mild form but manage to keep it in cheek with talcum powder, I am sprouting my own towgay to try and keep up my vitamin B intake.

October 1st

I am again in bed after being back at work for a few days. This time it is a 'cold' on the chest which is rapidly settling down. Dillon turned up two days ago and stayed for twenty-four hours. He brought two months pay – 90 rupees per officer. Our monthly pay has gone up as the Japs have reduced the cost of food and lodgings from 60 to 25 a month! Our pay is now 50 rupees a month. The value of the rupee is less than the Thai tickel, and some natives are giving five rupees for one tickel, but in our case everything is par.

The railway is almost completed, and the rumour of a move back to Singapore is now a certainty. Yesterday we were asked to supply the Japs with the number of men who could not be moved at all, and those who could be moved only on a stretcher. This is very significant. It is possible that the Burma offensive has already started.

We intercepted a large yak cart convoy with five thousand rupees of canteen stores yesterday. These came in very handy, coinciding with our pay. I also got some eggs from Anoquin (a little village nearby) to help build me up.

My nights lately have been very much disturbed by ants in their millions who climb up the posts at the back of the bed, and scatter in search of food. Ants in the hair are enough to make anyone go mad. I have killed thousands, but it makes no appreciable difference to the numbers. They go particularly for goola which is a sugary substance.

The news of the death of Foster Haigh was very distressing. That almost completes the demise of all the concert party that left Changi with F Force.

I have acquired a pocket chess set – home made, and have played many a game with my bed fellow, Dr Lowe.

The death toll is rising by about ten a day and is now over 350. The Japs have supplied a few bottles of vitamin B tablets which may help the bad beri-beri cases. Cardiac beri-beri must be the cause of apparently healthy chaps suddenly pegging out. The food is improving daily, but unfortunately it is too late to help the sick.

October 8th

Today has been rather eventful. A large load of canteen supplies arrived this morning, including a few thousand eggs and lots of goola.

Later Col Harris arrived with a month's pay and the news that we are moving back to Kamburi at the end of this month, and later to proceed from there to Malaya. Rumours that Sumatra is our final destination are now current.

The four escapee officers have been brought back again to no. 2 camp. We heard that they had been twenty-one days without food before they were captured. The CRE [CO of REs] was the first to die.

Our death toll has now topped the four hundred mark, and our daily death toll is still very high.

Last night we listened to a talk on crime in Malaya. Very interesting. He mentioned Somerset Maugham's letter based on an actual happening shortly before the last war.

As I mentioned before, there is a strange type of fever, prevalent now, that lasts ten to twelve days, and does not respond to quinine. It is being treated as malaria in the absence of any other treatment. I wonder whether this is dengue.

Meat is very scarce, and we are having a lot of stewed prawns, perfectly nauseating. Food lately takes a lot of pushing down.

The Japs agreed to let us mark out a large Red Cross on the ground, after mentioning that at Tanbazar, the Red Cross was bombed by the Allies. Actually, what happened was that two reconnaissance flights were made before the Red Cross was laid down, and the bombing squadron had its orders to bomb after the Red Cross was down. Unfortunately, orders were orders, and 60 Australians were killed. The Japs of course took the worst view and used it as propaganda against the Allies. Now that the rainy season is over, the bombing out this way will be intensified.

October 14th

The dry season has its disadvantages. The creek dried up and the cookhouse had to be moved to the nearest river, half a mile away. We also have to walk that distance for a bath. However it is worth it as one perspires very freely these close sultry afternoons. As the drought continues, most of the rivers round here have dried up and well water has to be used.

In the evening talks are given by officers, either in our ward or the officers' ward to which we go. A very interesting series has been given on the British Empire. The Aussies are good at describing the virtues of their own country and their talks are an excellent advertisement,

Black market activities came to a head yesterday when the Japs caught a soldier quarter of a mile down the road. They also caught a Bung with British clothing. The usual threats followed ie all supplies will be stopped etc. It would be rather a pity as the canteen supplies are coming in now in great quantities. Five thousand eggs were delivered the other day and we all got three each.

I think the death roll is now about 430, but the beri-beri shows no signs of a cure. Odd cases of mumps are breaking out and extensive pustular skin rashes are getting to be the curse in addition to infected scabies. It seems that very few patients sleep at night.

October 27th

The death roll is now over 520, and 300 of the 750 who left no. 2 camp have died. The food is still improving but is having no appreciable effect on the beri-beri cases. Rice polishings have been issued by the Japs, and we are giving it to the worst cases of beri-beri. There is a little left for the MOs. The river for bathing is keeping fairly high and is deep enough to swim in, very popular these hot sweaty afternoons. A further large batch of men have had amputations performed, including Windmill. There is only one alive, unfortunately, of the previous 17 men.

There are talks by officers and other ranks almost every other evening on many interesting topics, and it is really the only way to fill in the evenings in the absence of lighting. An Italian submarine commander who was taken prisoner when Italy capitulated gave a lecture last night on modern submarine warfare. It was an amazing thing to hear the cheers at the end for a man who recently was helping to sink our merchant fleet. In his opinion, the aerial escort of our ships had entirely prevented their subs from attacking convoys. After a bad period for us and the USA, January to March '42, called by the Axis the 'Golden Age for Submarines', when on average of over one million tons were sunk each month, it is amazing we were able to carry on in spite of such losses. Only bitter experience taught the Americans the value of planes, and now the ships have a constant aerial escort.

My knowledge of Italian is progressing and I am now able to converse with my Italian patients who come to me with the most obscure complaints.

It is now two years since we left England, and the general opinion is one more year. The skin conditions are still a curse. Infected scabies, pustular dermatitis, sweat rash – all combine to make life unbearable for the poor patients who are already dying from other viral diseases.

November 9th

The death roll is now over the 600 mark and still rising. Col Houston arrived yesterday with news of the forthcoming move to Singapore and we are leaving behind those too ill to travel. Windmill died a few days ago following amputation of his leg. What a tragic end for a healthy young lad. The last series of amputations are doing better and most of the patients look like surviving. Two supplies of various medicines have been received from the Japs for which we had to pay. What a racket! All the supplies for the cholera patients unfortunately arrived too late to help.

November 19th

Action at last! The first party leaves tomorrow for Kanburi, and after a two day halt carries on to Singapore. I am on the third train. It is amazing that a heartbreaking march of three weeks on the way up should be accomplished in two days on the way back.

The death roll is now 640, and a small medical and administrative staff are being left behind with those too sick to move. In my dysentery ward we had a bad day recently, eight patients dying within twenty-four hours. The men are as thin as ever. Malaria and beri-beri on top of dysentery ultimately finishes them off. The rice polishings don't seem to have helped very much.

November 28th

I left Tanbaya about one week ago, and after a very uncomfortable journey by train which lasted four and a half days, we arrived at Kanburi. We were crowded 33 to a truck and I sat on my pack as there was no room for my feet on the floor. The sores on my buttocks became inflamed and made sitting an absolute torture.

We were greeted at the station by a large bevy of female vendors of eggs, bananas, coffee etc, and we indulged in an orgy of spending. Up until then we had very little opportunity of spending our Thai money. We marched to the fit camp where we met the officers and ORs who had been left behind. Yesterday I was transferred to this hospital a few kilometres away where H Force had collected and which is to be the home of F Force sick until they are well enough to be sent to Singapore. We left behind in

Burma about 250 men who are unfit to travel. Probably time alone will solve the problem of their ultimate fate.

Conditions here are cramped – latrines and water supplies are bad. Mail has arrived at Kanburi and I am eagerly awaiting any post for me.

December 12th

I received three letters – all from David and nothing from home! There are about 1,500 men here and the death rate is still high (8–10 a day).

My valise arrived intact recently, complete with bed, mattress, sheet etc, and I am now much more comfortable. The temperature has dropped and everyone dresses up to go to bed – it is very cold at night. I am eating about six eggs a day and have caught a cold which is rather prevalent in this camp. There is a lot of literature about – quite a change after an absence of any reading material up country.

December 22nd 1943

All the Kanburi patients fit to travel have now moved down to Singapore. I arrived back yesterday and to my surprise was taken back to Changi (11th Division area). We never imagined we would get back here! Our old area is being converted into an aerodrome, and the hospital is in the Selarang area. I was sorry to leave Kanburi where the conditions were good and the food excellent. I envied the officers and chaps who stayed there all the time.

The trip down here was bad as usual with the following additions: an initial search revealed my Jap blanket which was confiscated. I had a hell of a job getting my heavy valise from place to place. My kit bag was sat on heavily and frequently so that my store of brown sugar spilt all over my clothing. A new lamp I had bought was trodden on and crushed. We bought tons of fruit of all types on the way down and we were almost constantly eating bananas, mangoes, mangosteens, pineapples, pawpaws and pommeloes.

Three men died on the train during the journey and many men were very sick with malaria and dysentery. We were brought from Singapore station by truck at 12.45 am on December 21st, my 33rd birthday. As the truck, with us balanced on our baggage, sped through the silent dark streets of Singapore, they sang 'Happy Birthday'. I was happy to be back safely. I could not celebrate my birthday in any way this year so the day passed uneventfully.

I have a good corner site and have made myself very comfortable. We live in wooden huts with electric lights and showers. It is good to be back in civilization again and to keep in touch with things. The canteen is poor here and the prices are high. A large portion of our pay goes to the CMF [Central Mess Fund] and the meals are very good.

I have acquired a pass to visit the hospital area and hope to see Tate and Bradley who have been admitted from prison with severe beri-beri. The officers here were shocked when they heard of our experiences up country, and said that after the war, this would not be forgotten.

I have no job as yet and am taking things very easy. I even had breakfast in bed this morning.

We are in quarantine for a week because of a case of smallpox in another party on the way down, and we are now writing postcards home this being the third. No loads of mail and Red Cross parcels as we expected. However we are making good arrangements for Xmas.

January 5th 1944

Christmas and New Year have passed again. We subscribed to the pot for Christmas and got good meals that day. We were given some chickens for New Year's Eve and had a late [New Year's Eve] dinner. The troops had an open-air concert and almost achieved traditional NYE gaiety and fun until midnight.

A malaria centre for ORs has been set up owing to the large number of BT relapses occurring daily. A large proportion of officers are now in hospital, mostly malaria. I stopped my suppressive atebrin two weeks ago and I'm waiting to see whether I develop malaria or not. Many of the other medical officers have gone down with it. I am in medical charge of the British Battalion since F Force split into Aussies and British, and still have large sick parades. A medical classification of this whole crowd had to be done recently – a very complex affair.

I visit the hospital quite often and had a long chat with the escapees. They are very lucky to be alive after their most amazing adventures. They had to hack their way through thick jungle and climb high ranges of mountains daily. This slowed their speed and they ran short of food, and for the last three weeks of their journey they starved. Four officers died, one by one and were left – a tragic end somewhere in the jungle. They built a raft when they came to a river but the current was strong and carried them over rocks and falls and finally it smashed in midstream. They were flung into the water and had to swim for their lives. They later saw some native hunters up a canal who looked after them for a few days but finally gave them up to the Japs. Bradley is very ill. Ankers is not too bad.

I went to a British pantomime with Tate and am going to the Aussie one tonight. It is good to get back to civilisation again with good shows etc.

February 11th 1944

I received eleven letters a few weeks ago – about a year old but none with the last batch of mail which dates up to July. The first official news of me was January '42 when I was reported as being a POW.

The food and canteen situation is bad and threatens to become worse and prices are rising every week. The small amount of money left after paying the PMC [President of the Mess Committee] is only sufficient to buy bananas and sugar. I have given up smoking – firstly, the price is exorbitant and secondly, I would rather buy food with the money. The daily diet consists mostly of beans and rice, and occasionally a little meat. F Force is recovering its health slowly but malaria relapses are the curse. Almost everyone is going down with it and the period of treatment is too short, consequently they are relapsing very soon afterwards. I am still MO to the British Battalion and the main part of my job appears to be clinical and attending conferences re categorisation of the men as to the state of their health. We have to find men to work in the garden daily.

I have been going down to the sea, bathing frequently during the past few weeks and have enjoyed getting away for ½ day. It was recommended by the Japs for skin problems and debilitated men but it's healthy chaps who go down and bring back wood for the fires and coconuts for the Mess.

'Hay Fever' by Coward was produced here excellently. A John Wood who is a professional takes the main part and puts over a polished performance. The whole show would do credit to a West End stage.

We had to write an essay of 150 words for the IJA (Imperial Japanese Army) on an incident during the campaign. Some were returned as 'unsuitable'! There have been IJA broadcasts in England recently and my name was mentioned as being in the audience two weeks ago. I wonder whether it will reach the folks at home.

March 3rd 1944

Malaria relapse rate is still very high in spite of all treatment. Men are returning to work every 2 to 3 weeks, mostly BT (benign) and very rarely MT (malignant). Other diseases e.g. beri-beri and dysentery are much less pronounced. The Nips held a sudden inspection of all our unfit men – to what purpose? The examination was very cursory and almost agreed with our classification. The men are, on the whole, improving in physique and are regaining some of their former sparkle.

American Red Cross parcels arrived a few days ago, 1 parcel to 7 men. Not much but sufficient to make the meals tasty. The Yanks certainly know how to put up a good parcel. The tins contained variety – a mixture of health foods and vitamins. The meals otherwise have been very monotonous, mostly beans and rice. Canteen prices are rising and our monthly pay doesn't buy very much.

A number of officers moved across the road to Selarang leaving us room to expand. We are also starting a 'long-term' garden policy. I have joined a small Italian class for beginners – it will help to keep my mind occupied as there is a shortage of good literature.

April 3rd 1944

No lessening of malarial incidence. Fresh cases of MT are recurring. Have just completed two weeks 'holiday' – Hanbury having done my locum. It gave me the opportunity of going bathing and also doing my share of gardening. We are cultivating the ground all around our billet.

A new general (Bacto) has taken over this area and has issued loads of new orders – mostly restrictions. There is no CMF, rations look like being short and officers' pay is a bit down. The Nips have moved in amongst us and have turned out several houses. There will be now overcrowding. There is a guard house just behind our billet and patrols are frequent through our lines. Red Cross parcels have all been consumed and there is no sign of any further supplies. Concerts etc still go on. An outbreak of chicken cholera has killed off hundreds of chickens.

May 3rd 1944

Having just completed a 'good' move to an officer's house near Selarang, we are now preparing for a major move to the jail and its environs. This house has all mod cons and for the first time for a long while, flush latrines etc. We are very comfortable and have dining and sleeping accommodation in the large downstairs room.

The garden which the Dutch left behind provides us with 10lbs of greens a day. We also get coconuts to increase the bulk. Food appears to be getting better as the rice ration has gone up a little and we may be using up reserves. Pay is 30 dollars this month but I am still broke as I am heavily mortgaged with the debt incurred buying towgay. This has been very useful increasing the tiffin and providing the necessary vitamins.

The remainder of the up-country parties have now arrived back looking very fit. The death roll for F and H Forces must be now nearly 5,000. Out of the 1,600 who went to Number 2 camp there are only 185 alive. This figure should make someone in Tokyo sit up.

Padre C has to spend 2 days in a pit handcuffed to A/S (Acting Sgt) Wolfe for carrying letters from one camp to another containing alleged news. A large number of officers were diagnosed as cardiac beri-beri and travelled down BD [BR? Burma Railway] but there has been a certain amount of scepticism regarding these cases. Hanbury finally died in K [Kanburi Camp] from dysentery ?amoebic.

The future arrangements re medicals are rather vague but it appears as though non-combatants are going into a special compound. I am now looking after B group and having one morning sick parades and two each evening for the working parties.

The aerodrome is progressing very slowly and I doubt whether the Japs will have it ready in time. We have so far seen no allied planes here but the K folk say that it was an everyday occurrence up there.

Padre C brought back with him a strange bible containing an illustrated account of ancient coins. I see that my collection is mainly centred round this 100 BC to 50 AD era, their value is at present unknown.

Malaria is now settling down – quinine is still scarce. Bacillary dysentery is having a bit of a run but is well in hand. Stomatitis ?cause is not so prevalent and is still a mystery. All salvage and anything useful in our new billeted area by the Jail is being carted daily by trailer parties and huge dumps of sinks, containers etc are being gathered. The civilians are being moved to Sime Road after 2 years in the jail – poor chaps. They include 400 women and children.

May 13th

I never imagined when seeing pictures of the Nip House that I would one day come to live in a jail. Two days ago I was ordered to move with a party of 1,000 British to Changi Gaol. I was very sorry to leave my last Mess as food and accommodation were very good. We marched here and settled down inside the jail, the men going to small cells and the officers going to large hospital cells holding 14–20. Initially there is a depressing atmosphere about the place which takes some time to get used to. Two high walls surrounding the jail guarded by Nips. All windows are enclosed with bars and wires and one gets a feeling of being hemmed in. I feel it particularly after the open air comfort of the last billet.

My first morning sick parade was an absolute shambles. 700 sick fighting their way in simultaneously.

We are now divided into two companies and I take one company out in the open air when the weather is fine. There was a concert in the open last night, very well attended. Absence of bulbs is causing a blackout of all passage corridors and mortuary cells so that moving about at night is a dangerous business.

The food here is all cooked on steam boilers and consists of hashes and stews. The officers are trying to get their own cookhouse going so that we can get fried pasties etc. The civilians lived on stews and hashes for over two years!! I hear that following a search in jail last October and the discovery of 2 WRS (wireless radio sets), conditions here were very strict and many were taken away to another jail in Singapore and some have since died from their treatment. The bishop of Singapore was also taken to jail and flogged for passing money to the civilians.

There is not much to do here when not attending sick parades so I lie on my bed and read or mend my sheets which are rapidly falling to pieces. Huts are going up very rapidly round the jail and the remainder of the men and officers are expected here by the end of the month.

F and A Forces totalled 10,300 of whom about 5,200 died from disease up-country. Of the 7,000 F Force, about 4,000 are dead. When I first arrived in this Gaol, I had a short chat with some of our Italian allies – officers of 2 Italian submarines who were in Singapore when Italy capitulated. They were given the choice of pro- or anti-Mussolini, and about 1/3 chose the latter, in spite of possible consequences. They have been imprisoned ever since, and are being moved from here to Changi.

I heard from the Italian officers that the Navy is very royalist. The death of Prince Aosta was a blow as he was the potential future ruler in Italy – a good man.

June 5th

I have now moved to one of the warders' houses outside the jail. Only ORs (other ranks) are now inside, and all the officers (over 1,000) and the remainder of the ORs are living in huts and coolie quarters round the jail. I am living with 15 other Aussie and British MOs in comparative luxury, although very cramped quarters. The officers in the huts are lying cheek by jowl and some are double berthing in order to give themselves more floor space.

There is a spate of remarkable orders as all beds have to be dressed by the right, all clothes lines have to run north & south, no easy chairs outside huts etc. How far our own staff are to blame, I don't know.

The food is quite good and the rice ration is now 7oz a day. Letters are trickling in very slowly. Any news about Leeds etc I have to get from other chaps' letters as the letters I get are masterpieces of lack of information.

July 6th

I have made one move since the last entry and we now occupy three houses in 'Harley Street', one of which is for sick officers, one for working MOs and one for officers' stores by chiefs [?]. We have now all settled in very well. Gardens are springing up all around. The officers have built themselves a theatre. There is a library for officers etc.

The only drawback is the poor food that our mess is getting due to several factors. We have had a few mess meetings. Things are being slowly altered. The billet supply is really quite good, 17½oz rice per day as well as vegetables and fish in good quantity.

Dysentery has broken out again and our house has suffered rather heavily. I had it for a few hours but I dosed myself with M&B and carried on.

I recently got a windfall in the shape of extra kit. A maximum limit was fixed for officers' kit, and all surplus had to be handed in. Fortunately I got at Hecht late PO [Procurement Officer], M/C [Medical Corps] and got several very valuable & useful articles of clothing. I am now almost fully equipped again.

I have now paid off all my debts for towgay etc and may now start smoking again very soon.

82 bags of mail have come in and have started to be sorted. I started Spanish thinking that it would help my Italian but as it only confused me I had to give it up after a few weeks. I am progressing well with Italian. We have an exam next week. Last night here was a very good debate on 'If we want peace, we must prepare for war'. The motion was carried.

July 24th

There is a great wave of optimism on now and I have 2 bets that this war will be over by Xmas. Some of my colleagues even think that I'm being pessimistic! Wild rumours are rife and we believe almost any story that sounds good. There have been major upheavals in the camp. Following the unauthorised visit of an officer to Singapore where he was picked up by the Nips, Eb Holmes, Black Jack and all the chiefs have been sacked. Lt Col Newey replaces them. He is an SSVF officer and gets on well with Takahashi. I hope the change is for the better. It is rather a shame that the senior officers should be sacked for no fault of their own and then to have a junior officer put over them. He is said to be a martinet so I hope that he stops some of this petty thieving that is constantly occurring.

A party of 200 is going to Singapore tomorrow ?destination and ?work. The Thailand parties are coming down country and there are large numbers in Singapore but what the future is, is rather obscure.

The letter from Dad breaking the news gently about Norman's engagement came as a shock. I keep on forgetting that he is a grown-up man and not a child.

I'm progressing favourably with Italian and went to hear a lecture in German by one of the Italian officers on submarine warfare. It is a strange state of affairs when we hear one of our former enemies talk about his part in the undersea warfare against our shipping.

Re health: malaria is still recurring as often as before – no treatment available here seems to be able to stop attacks of PT recurring each month. Dysentery had a flare-up a short time ago and has now settled down considerably. Scrotal dermatitis has recurred similar to our first summer 1942 – green leaf extract is being used to supply the extra Vit B. The food is sufficient in bulk but is poor in quality. There has been no meat for several weeks now.

August 30th

Last Saturday night I was a spectator of a rather unique ceremony – the dedication of the recently built shul in the gaol called Ohel Jacob. Everything was built and prepared by the boys themselves. A representation of the tablets above the ark was painted realistically and a decorated (suitably) curtain hung in front of an imaginary ark. A Dutch Chazan called Chaim Nussbaum conducted the service and gave a long discourse on the ego and the psyche. His metaphysical wanderings went above the heads of some of the audience which included various gentile guests of CO, 2 l/t, & the chief padres British and Australian.

The synagogue is used for lectures etc during the week and it is rather amusing to see the location of a maths lecture given on the board as Ohel Jacob. We should have good services during the forthcoming Yomim Noraim [High Holydays].

Medical: The diet is very deficient in some part of the B-complex for scrotal dermatitis is on the increase and also beri-beri and pellagra to a smaller extent. It is very similar to our first summer here after capitulation only there are no rice polishings now.

September 26th

I have just recovered after a bad spell of ill health. I have been off-duty 3 times during the past month. First time with a cold, second time with laryngitis and fever and the last time I was admitted to hospital with acute bacterial dysentery, the first and I hope the last attack. I treated myself with M and B and felt so lousy that I asked to go into dock. I ate nothing for 3 days and then pap 5 times a day. I was discharged yesterday and am looking a lot leaner than usual – under the 10 stone mark. However my appetite is good and I'm trying to put on my lost weight. I quite enjoyed my stay in hospital. I lived on the balcony, no mosquito net and it was very cool in the evenings. I am now having a week's alt C before I resume work in the MI room.

I was unable to go to shool [synagogue] for Rosh Hashanah but hope to go tonight for Kol Nidrei [evening of Yom Kippur]. There is a batch of recent mail in – up to May – but so far have not had any from home. Miriam and David seem to be the most frequent correspondents. I'm afraid my first card did not get home. I heard that a plane carrying mail from POWs crashed in flames and only a few of the cards were saved.

Parties of officers and men have left River Valley for ?Japan and only Lever is left in Singapore. We heard that a camp up-country had been bombed by the Allies and 96 men killed. No confirmation yet. We have a permanent brown-out here and switch to black-out when the siren goes.

28th September

The Kol Nidrei service almost transported me back to the old days of the Chevra Shass. In the small Gaol shool but by candles the Chazan sang a typical Kol Nidrei that Rev Samuels might give. I, being the Senior present, stood by the Chazan but without a Sefer Torah [Torah scroll]. Nussbaum gave a short sermon in English and read through the whole service. For the hour or so one forgot that we were PsOW. As I was recovering from dysentery I did not fast this year but hope that next year we will be amongst our own folk once again.

I am still getting cards that say nothing. The message is now getting to be stereotyped – rather annoying. This morning I went bathing for the first time for several months. It was very pleasant except for jellyfish which have a nasty sting.

December 22nd

Well! Another birthday spent on foreign soil has come and gone. I gave a little party for 10 people in the billet – Wilkie, Titus, Simmie, Padre, Dixie, Bill English, Hecht, Burgess, Delamos and me. Owing to the scarcity of foodstuffs and the expense I had a pie made of tapioca root, tomatoes (own grown) and 2 tins of veg soup and numerous extras. Ted – our Messing officer – made it himself and made a jolly good job of it. It was very tasty and we all felt happily replete at the end. I provided coffee and gula and cheroots. It cost me about 30 dollars and as I was heavily in debt, the loan of 40 dollars by the Padre came at a welcome time. I must remember to pay him back after the war or sooner, £5 (he wanted only £4). We all had a very good evening under the circumstances. I got a few presents including a pair of pants from Wilkie and a box of powder from Bill. I keep hoping that each birthday here will be the last one but I have now ceased to speculate about the future but if we are still here next year this time I will be rather annoyed.

I have not entered anything up in this diary for several months now for several reasons, the main one being procrastination. Actually very little startling has happened. We had our first air raid on November 4th, about 30–40 planes and have had an occasional reccy plane over since. We are hoping for big things in the air but aerial activity appears to have died down recently. We have had a spell of cold, rainy weather for the past week that would do credit to a bad English winter – out here, it is almost incredible. We go about in shirts and pullovers when we are not lying in bed curled-up under a blanket.

K and L forces are supposed to have left Thailand on 6th December and we cleared a hut in anticipation but they have failed to materialise. No-one knows what has happened to them.

Christmas is almost on us once more. The officers had planned a grand fairground in Coconut Grove but the Nips put their foot down and said that Xmas Day was a day of prayer and solemn thought especially during war time and it was cancelled. The batmen are however putting a show on in the evening and a tremendous menu has been planned. We should be getting enough to eat even though the variety may leave something to be desired. I have been very lucky with my little tomato patch and have been eating 6–8 tomatoes a day for the past 2 months in addition to 2lbs for the pie. The big medical officers' garden is also coming on apace and we have regular work hours. This morning in the half-light before dawn while it was raining and cold, I had to get up and collect buckets of urine and take them along to the garden. This afternoon I am on a trailer party to collect lalang for compost.

Since my Italian tutor went into the Gaol (for their own protection!) I have ceased to do any study. I tried to learn Spanish from a self-teacher but found I couldn't concentrate very long so gave it up. I am reading quite a lot and belong to 2 libraries. I play bridge early every evening with Titus as my partner and we are over 30,000 up this month. We have won over $3. He plays quite well but is rather impetuous at times and I have to undercall consistently to balance his optimistic calls.

An occasional game of small poker for cent chips has been a source of small change.

The meals generally are quite good, there being sufficient bulk – about 1lb rice/day and plenty of greens from the garden. The canteen prices are high and sell very little, the main things are gula melacca, tapioca root, cheroots and toilet requisites.

Over 300 toys were made by the fellows here and sent along to Sime Road for the internee children.

Col Newey (rep officer) holds regular and frequent court to try offenders whom he sentences to periods of imprisonment (!) up to 3 months. He is justifiably down on drug selling but also on trading and black market. In spite of this, most of the officers have sold something e.g. watch or clothing. Otherwise I fail to see how some officers can smoke cheroots a whole month. There are some ORs with thousands of dollars – profits of illegal trading which is still going on in spite of all the restrictions and punishments.

Medically things are quiet. I am still working in the MI room behind the Gaol and am not very busy. The outbreak of Changi balls has now died down completely and except for the odd spot of beri-beri and malaria there is not much disease about except general debility. Most of us have lost a little weight. I weigh 9st 12 which is about 5 lbs below my normal weight.

January 1st 1945

Two years have passed since I started this diary and the worst pessimist here won't predict another full year. There is a strong feeling that the war will finish this spring or summer. Xmas passed as usual with the traditional stuffing of oneself. There was an enormous meal on Xmas night after a big tiffin and tea and consequently most of it was left. Some including the Xmas cake was eaten the following day. Titus collected a basinful of doovers [hors d'oeuvres] here and distributed them to the fusiliers in hospital. I think that in a POW camp where rations are poor, the extra food should be spread over a few days and ease the shock of going back to ordinary meals. Actually today (New Year's Day) the meals have been increased considerably and are very good. The mess was able to buy a good stock of palm oil recently and I bought a lb with which I made a sambal [chilli sauce] using blachan [shrimp paste] and garlic. It is a fairly potent mixture and only needs a small quantity to flavour a meal.

Last night (New Year's Eve) passed off uneventfully. Six MOs including self played an MI room staff and were narrowly beaten after an exciting game. We provided coffee and managed to bolster up a festive spirit. I went to bed before midnight and was asleep when the old year ended. There was a scare of an early morning search but this turned out to be a false rumour. The latest flap is about repatriation and lists are being hurriedly revised, new cases put on and old ones taken off. We hear that a Red Cross ship has left Japan on its way here. Whether it picks up our repatriates is very doubtful but I hope it contains some of the Red Cross stores for us that appear to be piling up in foreign parts.

We hear that about 1,000 Indians have arrived back in Singapore from 4,000 that went to Timor originally. Parties also appear to be going up to Saigon to River Valley.

March 9th 1945

We have had so many ups and downs during the past few days that I am almost suffering from manic depressive psychosis. The main up is the arrival of a White Cross ship with thousands of individual parcels on board – our first Red Cross stuff since the small USA food parcels over 12 months ago. The boxes are unloaded and are now in Singapore waiting distribution. It is also rumoured that a second ship is on the way. To offset this however, we heard that a large proportion of the parcels contain clothing, toilet goods etc. Our rice ration has been cut from 450g a day to 275g a day and the vegetables and greens also cut.

The Playhouse Theatre was first ordered to be pulled down and then altered to close down because certain items in the present show appear to have offended the Nips (it is only an excuse again). This is a great pity for the good theatre is our main source of entertainment.

We had to sign a non-escape form (2nd one) today – why two no-one quite knows. Nips are mounted on the gaol tower with machine guns etc so it looks as though they are getting a bit worried. It appears as though we are in for a lean time before the end. There has been almost a daily reccy by a B29 and 2 weeks ago in 2 raids the docks, station and naval base were done over. In Singapore they only dropped incendiaries so that injuries to civilians were reduced to a minimum. It was an exciting time seeing wave after wave of heavy bombers come sweeping over the island. We hope that they know where we are and don't give us the works by mistake.

I had a slight tiff with the powers that be when I was caught cooking over a 'private fire' by 2 CMPs and I told them to mind their own business. The whole position is farcical as although private cooking is officially banned one is doing it. The officers with electric heaters are very well off as they do their cooking in style. I do mine over a smoking wood fire making fried tap root or drovers. I have heard no more about the matter so assume that it has been dropped.

I attended my first funeral as a POW when I went to Jack Baker's funeral yesterday. He was a boy I got to know up-country and he did me many good turns. I helped him now and then with money. His father who was an internee in Sime Road owns the Cathay cinema and restaurant – one of the finest in the Far East and Baker had invited me to a big dinner there after the war. He was very ill up-country, bacterial diarrhoea etc and developed TB after arriving back here in Changi. Nussbaum was very good and sat at his bedside for several hours every day.

When he died all the Volunteer officers and many of the men of his unit in addition to all his Jewish friends turned up for the funeral. Takahashi and Newey stood outside the Gaol and saluted the cortege – the Nip guards presented arms and the long procession moved slowly out of the Gaol to begin its two mile walk to the cemetery. It must have been strange for the gentiles listening to a Jewish funeral with its traditional wailing melodies and Nussbaum gave a short funeral oration. Baker's brother was the chief mourner and said Kaddish [prayer for the dead] by the graveside after which the buglers blew the Last Post (British and Nippon). His father was informed the day after his son died. What a family tragedy – imprisonment and death.

I gave a lecture about a month ago in the shool to the Jewish community, mostly on medical topics. I have been asked to give another talk in a fortnight's time and will probably talk about 'humour in medicine'.

For the past 2 to 3 months I have done no serious work but am now taking up French and German again, doing a little each day. One of my medical orderlies is a French boy. I am having a weekly conversation class with him. I have dropped Italian for the time being and now and then glance at medical books which are months out of date. If the Red Cross supplies contain any medical books I may do a spot of catching up. The gardens are coming on apace and large areas are shooting up – tap root – which will be ready for consumption soon. We are fortunate in having such a large source of greens and root to supplement our meagre diet. I put in about an hour's work every afternoon and once a week I get up before dawn (7.30) and put a mixture of water and urine on the garden.

30th March

On the move again? 3,000 fit men are leaving here to form 3 camps of 1,000 each, 1 in Johore and 2 in other parts of the island. The idea I believe is dispersion of fit men for Nip security reasons. There is the usual crop of rumours, orders and last minute cancellations and speculation is rife as to who are going as MOs. Changi is splitting up and providing the personnel for 3 base camp hospitals. I was nearly included in the 1st party and am preparing all my gear in case. I am also put on the next party at the last moment. This leaves ?tomorrow ?destination.

There is the usual bustle and excitement early morning as each party leaves, reminiscent of F-force ill fame. Our nights have been disturbed on the past 2 occasions by alerts and bomb droppings. Fires were started last night and it has been suggested that carrier-borne planes are now operating off the coast of Sumatra. I hope so. We are awaiting them impatiently.

2 nights ago I attended a Seder (Passover) service in the Gaol shool (Synagogue) . It was a remarkably successful affair. Nussbaum is a genius for overcoming difficulties. We had matzo made out of sago flour, wine made from rice water and fermented towgay. The 'cosiose' (glasses of wine) consisted of coffee owing to the shortage of wine. Charoseth [mixture to represent mortar for building] was grated coconut and sugar, maror [bitter herbs to represent bitter years of slavery] was mint leaves, and for hard-boiled eggs [to represent the renewal of life] we had spinach leaves. Somehow Pesach doesn't seem the same without hard-boiled eggs. However it all made me feel quite homesick and over coffee and smokes at the end we sang the usual songs and broke up after a very merry evening.

As a token, 1/20 of a Red Cross parcel per man is being given out for consumption on Easter Sunday. Not much but it is a start. We have clubbed together and bought a small electric heater which is very useful. We make stews (very tasty), fried tap chips, coffee, etc to supplement the very meagre rations. Everyone is feeling rather weak and easily tired as we are living on about 1,600–1,800 calories per day. There is an average loss of weight of about 6lbs over the past 3 months. However a few weeks good feeding will soon put that back. We sent off a card about 2 weeks ago. I have taken bets on getting home before the cards – I don't mind losing these bets!

April 25th 1945

Having scrounged some ink, I have dug out my old pen once again and pressed it into service. Since my last entry we have had Red Cross supplies twice a week – about 5–6oz a time. This consists of approximately 2oz meat, jam, butter, sardines, salmon, sugar and coffee in small amounts. It is just enough to provide one good meal, supplemented with tapioca root, grown in our own garden. A few cigarettes and 'comforts' have also been distributed. I got a toothbrush in the 'draw' and I swapped some paper for a new blade. I was given one stick of chewing gum by a recipient who loathes the stuff!

The working parties have now all left and we remain about 8,000 strong. I was nearly included in the parties on a few occasions but I believe that F force MOs were given preference to stay behind. Hecht went on the last party although he was just getting over an attack of malaria. Burgess and Delemos have also gone and that reduced my little circle considerably. My bed is now a metal one and I have acquired many goods and chattels left behind by the departing officers. We hear that they are getting full scale rations and are not badly off re working conditions.

We are living on about 7oz rice and 3oz maize and small amounts of fish, palm oil etc and are still below the BM [body mass] amount. Primary malaria is on the increase especially MT and it is rather risky to sit outside at night chatting. On two occasions recently we had midnight roll calls extending over a couple of hours due to suspected escapes. No-one had escaped and on the second occasion the missing man was found dead in a bore hole [usually a sign of a revenge killing by other POWs].

Last Monday was St George's Day so the five of us, Wilkie, Padre, Titus, Simmie and I decided to celebrate. We had a pie made and as Titus was responsible for the size there was more than enough for 10 men. It was a delicious pie and only cost about $4 each. I however was not used to eating so much and felt sick and had to sit out in the cool of the evening for a short while to recover before I could totter homeward. We rounded off the meal with coffee, santan [a local herb] and sugar and a big cheroot apiece. All in all a memorable occasion, gastronomically and mentally.

I am still doing MI rooms and have now taken over part of the British (mostly 18th div and fusiliers) who are living in the Gaol. I had a short spell with the Aussies and found them very pleasant to deal with and easy to handle. It is extremely difficult these days to find sufficient fit men to go out on the aerodrome and a complicated system of medical categories is in use showing the varying grades of fitness. Unfortunately the office cannot be funded by this system and have to keep up their quota irrespective of the health of the men. Also men staying in camp get less food and pay, so sick men volunteer to go out to work. The job of an MO is no sinecure!

May 8th

After all my precautions for avoiding the malarial bug it has finally claimed me as a victim. I started a week ago with headaches, pains and fever and had negative slides till Friday. I thought from the symptoms that it was dengue again but the slide showed just a few parasites actually only in the first 150 fields. The intensity of symptoms bears no relation to the number of parasites in the bloodstream. I was admitted to hospital feeling pretty weak and was put on to quinine as I already had headache and nausea, the increase in fever which the quinine caused made things a lot more uncomfortable so I asked to be allowed to go on to atebrin and produced my own tablets. I am now feeling very much better, fever down and eating like a horse. Hospital food is lighter than food in the lines but is served up hot and tasty. Conditions are very good in hospital and patients are well looked after.

I have had 3 cards from home Oct 44! Still newsless. They seem to be receiving my cards. Further medical comforts are coming in and I received a tube of shaving cream (brushless) which I am now using. It is difficult to be economical with it though! I cashed a cheque the other day £25 for $100. This will enable me to buy the few things that the canteen is still selling e.g. tap root, cheroots and soap.

June 9th

I have now recovered completely from my attack of malaria. I had 7 days atebrin and 5 days platinoquine. I hope that will prevent any relapses but as this area is a hotbed of malaria (BT and MT) there is always the possibility of a further infection. As a change from MI rooms which I have done since capitulation I am now MO/gardens. I go out about 5 times a week. The work is very light treating cuts etc and the time is spent reading, drinking tea and smoking. On some gardens an excellent midday meal is given. I try as far as possible to get in on it, but there seems to be a fair amount of competition from senior officers. I like the change and will keep on with this role for another few weeks.

We are still getting 1/ 20th Red Cross parcel twice a week. It doesn't seem much but it makes all the difference to the meals. Unfortunately there is only enough left for 2 or 3 more issues. Two large parties have come in from Pelambang. It is not known how long they are staying. Some of the men are terribly thin – just skin stretched on bones – their general condition seems to be worse than F force.

One of the Dutch MOs with the party (who sleeps opposite me) was found with his throat cut early in the morning a few days ago. He died shortly afterwards in the operating theatre. It appears that he was under a cloud and had been in some shady transactions. However nothing is definitely known. He left a large and varied medical and surgical kit behind which will be very useful to the hospital.

Simmie was taken ill suddenly a few nights ago. I was called hurriedly to see him and diagnosed appendicitis and had him taken to hospital where he was operated on the following morning and a nasty appendix was found. He is now doing splendidly. Titus and I are making tapioca root and vegetable pies 2 or 3 times a week to supplement the meagre rations. The price of canteen foods is shooting up almost daily. 1lb of tap root now costs $3.60 which is about 15/- at the present ratio of cashing cheques. Sugar $33, rice £6/1lb. The only items keeping fairly low are cheroots 25c each and palm oil $2/pt. I am taking about ¼lb palm oil a day – it is a food source of vitamin A and also provides the calories and fats necessary for a balanced diet.

I got a shirt and a pair of shorts from the American Red Cross, both good quality and well fitting. Also a pair of thin khaki socks which don't look as though they will stand up to heavy wear.

I have started taking lessons in Ivrit (Modern Hebrew) from Nussbaum. I meet him in the shool where we have a private session. He is a master of many tongues and can converse in about 5 languages including Latin. He gave me a Siddur (Jewish prayer book) and a Bible (American Red Cross) suitably inscribed, which I read from time to time. The officers' library was recently enriched by several hundred new books received with the Red Cross stores but no modern medical books. There was a tragic death recently of a young MO who was with a Singapore working party and sent back here to hospital with beri-beri. He developed pneumonia and died. John Diver was up country with H force and did a good job of work there.

July 5th

At last Newey's reign has come to an end and Lt Col Dillon has been appointed in his place. Both moves were received with general acclamation. As Takahashi is leaving – Newey is taking this opportunity of laying down his seals of office. His has not been an easy task but he has not called in other senior officers for advice etc and has often acted in a hot-headed fashion. He is said to have a bad temper – very easily lost, not a good thing for a man dispensing justice daily. Many of his verdicts and awards have been criticized as being too harsh. What will happen after the war is a matter for conjecture although many victims swear revenge, reprisals etc. I think it will all die down.

Large parties have left the Gaol recently and there are rumours of lots more. We are now just over 5,000 in camp.

I have had a varied medical life recently. I was taken off the gardens and put in care of TB ward with Prothero who in fact contracted the disease. The work was very interesting and for the first time I was really struck by obvious physical signs in chests. I worked there for 3 weeks and was then transferred to the malaria ward where I am now pro tem. It is rumoured that the hospital may soon move into the Gaol.

The cost of living goes up daily and it is only by cashing frequent cheques that I am able to supplement the rations. I estimate my cheques will come to £150–200 before I am through here. However I think it is less than my mess bills would have been under other circumstances. I am now beginning to show something for my purchases for I have put on 6lbs in weight during the past few weeks and now weigh more than Tate or Simcocks who were both 12st before captivity. I have a late evening snack most nights with Dix and Walker (2 dentists living in the same hut). The Red Cross rations have now come to an end and there appears to be little hope of any more coming in. I received 4 postcards with the last batch of mail – all dated August to December last year and all completely newsless.

There have been no air raids recently or even the daily reccy plane which has ceased to visit us now. We hope it is the calm before the storm.

All shows etc were cancelled last week following the escape of an Aussie. So far he has not been recaptured. It is so foolish to attempt to escape at this stage especially when reprisals fall on the whole camp afterwards.

I am having weekly lessons with Nussbaum in the Gaol shool. He is trying to teach me to speak Modern Ivrit. I will have a smattering of many languages when I leave camp but master of none. However it all helps to pass the time and to occupy the mind.

August 25th 1945

The event we have been looking forward to so eagerly for the past 3½ years has now taken place and with a suddenness that surprised us all. Early morning of August 11th found most of the camp up, in little groups discussing the rumour that the Emperor had agreed to accept the terms of the Potsdam conference. When this was confirmed later in the day we began to make plans for celebrations and an early evacuation.

Most of the tapioca root in the gardens was pulled up and consumed over 3 days and our little conclave Dixie, Doug Walker, John McNeath and I had 3 successive evening parties when we opened our reserve tins of food (fish and meat), condensed milk and emergency chocolate. We also pulled up all our greens, root etc and had continuous orgies. But day succeeded day and we still had Japanese roll call parades and had to salute the Nips and generally carried on as previously. This was a bit of an anti-climax and we had a bit of a shock when the local commander said that he intended to fight on and defend the island. However an order from the emperor to cease fire ultimately reached the island after 10 days of ups and downs of spirits.

By a gradual process the local Nips became convinced that the war was over and then restrictions etc were slowly relaxed. But during the past 3 days since Dillon has nominally taken over the camp, food and clothing has poured in. We each have been supplied with a complete outfit of clothing – cheap Nip stuff – the rations have been pushed up to 3,000 calories and fish and meat. Red Cross parcels are piling into the Gaol. 1/20 parcel per man per day is being issued and cigarettes etc at frequent intervals. Nussbaum put on a party in the shool and a thanksgiving service is being held shortly.

We now have official printed news bulletins after years of whispered extracts and are awaiting Lord Louis' [Mountbatten] arrival eagerly. The outlying working camps have concentrated round the Gaol area and we are packed like sardines. The hospital has over 2,000 patients and there is a severe outbreak of gastroenteritis to complicate matters. We are pushing vitamin B etc into patients but there are several cases of sudden death due to heart failure which are baffling the experts.

We follow all references to POWs in the news very carefully and there is much speculation as to our future as each successive speaker seems to have different ideas on the subject. It seems however that we will be taken back home as soon as possible by air or by sea as transport becomes available.

During this exciting waiting period we rush to the doorways every time we hear a plane in the sky hoping to see for the first time for over 3 years the tricolour that means liberation and freedom for us.

[The following passage is a later addition by Harry.]

> F Force comprised 3,600 Australians and 3,400 British who went up country to work on the Burma-Thai Railway. Our furthest camp was in Burma at Tanbaya which is north of the Three Pagoda Pass at the Burma-Thai border. 3,900 men returned, a death rate of 45%. F Force suffered such large casualties mainly because of the cholera epidemic combined with the harsh conditions and a diet which had a severe deficiency of vitamin B.
>
> Looking back, it seems inconceivable that we believed the Japs when they told us the move would be to convalescent and rest homes – to the extent that a third of the Force were unfit men looking forward to a chance to recover their health! We happily transported lighting sets, pianos and other musical instruments in the expectation of continuing our concert parties in our new camp. These were soon dumped along the way during the dreadful march north.

Epilogue – First Letter Home

Sept 6/45

My dear Mam, Dad and all the family
How can I begin to express in this, my first letter home after 3½ years durance fairly vile, what has transpired during that time. So much has shaken the Silman frame that when we are gathered round the fireside in the old homestead and pass round the brandy, I will recount to you such tales that will make you rock on your heels, and you will begin to wonder whether or not you had given birth to a second Dumas or Munchausen.

I expect that so much will be written by so many by the time I get home, that it will only need my corroboration and a few personal details to set the seal. However let me start with a few facts about me. I am exuding rude health and rice from every pore, and during the past four weeks have put on over 20lbs in weight. I spend most of my waking hours taking out the tucks in my pants. I am now back to my old weight and have regained my manly figure and vigour. One naturally loses weight on a restricted Asiatic diet such as we have been pushing back for the past four years, but now with the superabundance of good food that is pouring in on us, it will soon be necessary to dig out and dust down the old Roussel [?] before adjusting same.

A POW life teaches one to appreciate the simple things of life e.g. crêpe suzette, duck à l'orange etc. I don't think I will turn up my nose at boiled chicken on Friday night in the future.

We have now had the ball and chain hacked off, and ships and planes are daily unloading supplies and troops of all descriptions. We are slowly getting used to a non-POW life, and still feel embarrassed when a Nip salutes us. We are hoping to leave here in a few days and proceed direct to the UK, but things are very confused at the present and we do not know exactly what is to happen to us.

As you probably know, in spite of the strict censorship prevailing at the time, I was nabbed by the Nips in Feb 42, and have been resting here all the time except for a short trip up country to Thailand and Burma lasting 9 months, during which time we constructed a model railway.

This interim period of waiting before we are bunged home is being made very pleasant by all sorts of organisations doing their best to provide us with food, clothing, toilet requisites etc, and we look like being re-equipped from head to foot before we are through with these blokes.

I hope that all at home are in the pink, and although I have been away from no. 22 for nearly 4 years, expect to find it unchanged except for the books not being tidied and replaced in the bookcase, and Mam's teeth possibly in a different glass due to the other one being worn out. I imagine with the concentration of forces at no. 22, conditions may be a little crowded, but I wouldn't notice that now – interval for a spot of bitter laughter. I anticipate getting a few months leave (with pay) after my return to England, before I recommence the daily toil. You might cut out the odd advert from the Wanted column – if anyone wants a doctor who has been sober and out of mischief for the past few years. I have done all sorts of jobs so am not very particular now.

Writing this letter has its difficulties as I am shooing away mosquitoes with one hand, stopping would-be purloiners of my ink bottle with the other and trying to concentrate in spite of 2 loudspeakers blaring out the news and events of the day. One of the things we appreciate most is listening to the BBC again, and getting a regular up-to-date news service. The Nips neglected to lay this on for us.

I expect that when I have sealed this letter I will remember things I wanted to tell you. However they will have to wait till we meet.

Give my regards and love to all the family. I will write again as soon as poss.

Love from

Harry.

Letter after Dunkirk

Lichfield (4/6/40)

Dear Betts & Louis
I am writing this letter from a 'rest camp' in the heart of the Midlands where we have come to recuperate. The weather is glorious, and we eat, sleep and lie about on the lawn all day, enjoying doing nothing.

We have had some exciting times during the past two weeks, but none more exciting than when we embarked at Dunkirk. The sky overhead was black with German bombers who wheeled and dived continuously, bombing and machine-gunning.

We walked 200 yards along a stone jetty to a destroyer which was waiting to take us back to England, under constant bomb and shell fire. It is a miracle that so many have been able to get back. I drove a large truck backwards and forwards along the beach from La Panne to Dunkirk to help our chaps to move along more quickly, and I could hear the shells and bombs bursting all around. It was a nightmare ride that I shall not forget in a hurry.

Another occasion when we went into action against the Huns, we were actually surrounded by German tanks and got away by a miracle.

Our lads have had a tough time of it during the past few weeks and deserve a rest for a little while. The Germans have turned out to be a much tougher proposition than we originally thought. The speed-up in our tank and plane production will however level things up.

We will not be here much longer, and in fact we are expecting to be moved any minute. We may get leave now or after we have all been sorted out. Nothing has as yet been decided. I am feeling very fit now and look as tho I have just returned from a long holiday by the sea. I hope you are both keeping well and that the family are thriving.

I will tell you my adventures in detail when I see you.

Yrs affect
Harry

Author's Note

A letter reference from Montague Burton [the clothes manufacturer] comments on HS's 'bravery, self-possession and calmness in the firing line'. He drove a truck back and forth along the beach, trying to collect all the sick and wounded men to take them to the boats.

5775523 L/Cpl Abraham 'Abe' Silver, RCOS

The information in this testimony is taken from the IWM Archives for Abe, who wrote to his family (Sam, Len, Ann, Fay, Jack and Esther, at 110 William Road, Regent's Park, London NW1) at liberation in September 1945.

61. 5775523 L/Cpl Abraham Silver RCOS.

Abe lived at 163 Gloucester Avenue, London. He was captured at Singapore and was first incarcerated at Changi, until April 1942. Here there were at least some concert parties, and organized food distribution. He was then taken to the camp at Bukit Timah where he was kept until November 1942.

From November 1942 to September 1943, however, he was transported to work on the Burma Railway, in solitary camps on the river; these were Tarsao until April 1944, Tamuang until November 1944, Tardan until December 1944, Tamarkan until June 1945 and Nakon Pladuk until liberation. He was constantly on a starvation diet, and endured an almost complete lack of shelter from monsoon rains, with murderous guards, almost no clothing or footwear, and huge disease problems including massive tropical ulcers, and he witnessed massive deaths among the Malays, Tamils and Chinese prisoners as well.

Low-level bombing raids by Allied planes also killed POWs. Conditions in the camps were so bad that men just ran off into the jungle, and when brought back were shot, bayoneted or decapitated. The Korean guards were especially brutal. On forced marches between camps, or carrying supplies, many POWs died of exhaustion.

In March 1944, Abe was at Tamuang and was fit enough to play basketball against rival national teams from Holland, Java, the USA, Australia and New Zealand. There were 10,000 POWs there. Perhaps it was at Tamuang that the following incident, described by Abe, occurred (the item is not dated and the location is not given):

> Six thousand came to attention as the command, 'Keyoski' was given by the British camp commandant. Not an officer or man among the British, Australian, Dutch and American prisoners of war standing on that field dared show any signs of movement, as 'Yuki' – 1st class Korean private and virtual governor general of No. 4 Group camp – was coming on parade to take roll call. Snatching the parade state [command?] from the Colonel, without acknowledging the salute given him under IJA instructions, Yuki turned towards the battalions formed up for his inspection. Alternately biting his lower lip and grinning in his supercilious manner, he stood staring at our closely packed ranks, his eyes blood shot with

> saki. There was going to be trouble for someone today and it was not long in coming ...

In October, Abe was moved upcountry to work on a large bridge alongside the railway; this project was abandoned after fifty Japanese and twenty-two POWs were lost to Allied bombing raids. While POWs were camped just fifty yards from the bridge, Allied planes would come in low and bomb and strafe with Flying Fortresses, whilst the Japanese anti-aircraft batteries fired back from the rear of their camp. As a result, the POWs had to run for the jungle to take cover, and if the bridge was damaged they had to work all night to repair it.

In the last three months of captivity, Abe was attached to a Hard Labour Unit of the IJA, and acted as a pack mule carrying arms and ammunition, rice and stores over mountains and through swamps and paddy fields, probably walking over 2,000 kilometres in all. The Japanese guards were 'the most uncivilized bastards imaginable, forcing us in our bare feet over rocks and thorny scrub, beating us unmercifully, if we stopped for an unofficial rest'.

One document he typed almost certainly refers to this episode:

> Drugged by the heavy jungle scents and the foetid air in the hut, I stumbled into the teeming rain to join the endless queue for a messy gluelike slop that we called breakfast. It was good that it was still dark – maggoty rice does not look appetising. I washed my mess tin out with half of the mug of the warm muddy water which was served in place of tea, then drifted off to join the hopefuls in the sick line. I was glad of the gaping hole in my heel. This should guarantee me at least one day from work. But by the time it came to be my turn to face the MO, the wound had filled with mud, which dimmed the redness of the wound, and with strict orders for a 100% work turn out that day, I joined the miserable, drenched, starving, malarial band of cripples who would face another day of torture in the hills and paddy fields of Nacom Nyok.
>
> From one day to the next we never knew what our task would be, and it was best not to think of today. For two hours we slipped like drunks along a muddy, rock strewn road which led into the hills where battalions of Jap troops were hidden from the ever searching eyes of the allied air force. Our Jap guards gave us no rest and laggards on the march were viciously dealt with. We made it in two hours and taking what little shelter there was, under the dripping trees, we awaited the task for the day.
>
> A pile of ammunition boxes, picks and shovels and sacks of rice stood in the open. Each man pretended not to notice these but each man had made those careful calculations known only to POWs as to what his share of the burden was going to be. I chose six picks and six shovels and lived to regret it. Calamity came to me almost as soon as we had moved off, as a sharp bamboo stump hidden by the mud drove its agonising way into my injured bare foot. I am not ashamed to record that I screamed with the pain ... after an hour of tortuous progress through the jungle we came to a river which we had to cross. The Japs had thrown a rope across and we were told to hang on with one hand and support our loads with the other. In parts the river was over six feet deep and in angry flood and I would challenge

> any normal man today to cross that river by these means, let alone starving cripples burdened as we were. Cunning, developed by three years of POW training, came to my assistance. I volunteered to help the small chaps across and the Nip sergeant grunted his assent and I made three trips across adding two picks and shovels of my own load to the joint load each time we crossed. There was hardly one successful crossing and many a sack of rice slipped into the water. Strange that the Japs did not seem to mind that. We knew why later when these same sacks were issued to us as rations.

Abe says he received twenty letters from the firm and his family, though he knew that many more were sent in the almost four years of his captivity. No Red Cross parcels ever reached him and there was virtually no medical care from the Japanese. He mentions appendix operations carried out with no ether and only a sterilized razor blade.

On liberation, Abe volunteered to work on the staff of the American and British paratroopers who dropped into the camp to protect them, and to assist in the evacuation of other POWs home, via Bangkok and Rangoon.

Two poems are in his archive, reproduced here for the sake of what they say and in remembrance:

Rattle, Stutter, crackle, splutter, Lewis guns and Brens all round.
ack-ack, bofors, mortars, barrage help to swell this hellish sound.
Overhead the Nippon war planes fill the sky with angry roar.
'Lie down flat you silly bleeders!' This is what the world calls war.

Now upon their bellies creeping, through the rubber and the palm,
Parched and hungry and bereft of sleeping, knowing not a moment's calm.
Wading through the swamp and marsh and clothing stiff with mud and gore,
On they go these helpless victims, sacrificed to Gods of War.

All around them men are lying, fathers, brothers, husbands, sons,
Some are dead and some are dying, victims all of bombs and guns.
Gasping, groaning, crying, moaning, 'Is this Nature in the raw?'
No it's simply bloody murder, history books just call it war.

Blackened, bloated, stinking corpses, lie unburied all around.
Ants and flies and loathsome maggots use them as a breeding ground.
They have died to save the Empire – don't for God's sake ask, 'Who for?'
They were simply slain and butchered in a democratic war.

While at home in London's clublands, see the ticker tapes flash in,
'Sharp Decline United Rubber, down five points Malayan Tin.
'Dammit, I shall lose a packet, lucky I've got plenty more.'
'Waiter bring me a double brandy.' Yes my Masters, this is war.

Who was to Blame?

Singapore we hold no more
Our struggle was in vain
We had the guns
We had the men
But not a single plane

The battle raged from left to right
Bombed by day
Shelled by night
But still there was no plane in sight
We were in a bloody plight

We were told to hold our ground
Dead our comrades lay around
Though we had no sleep that night
Onward went our men to fight

Our water supply was getting low,
Singapore was all aglow
Hour by hour we held our ground
Still waiting for that droning sound

Though the odds were very great
And we knew not our fate
Singapore we knew must fall
Who was to blame for it all?

Abe's debrief sheet reveals that at Tardan and Tamarkan he made many forays into the jungle to scout paths for possible escape; discovered once by the Japanese, he was severely beaten, together with Sergeant J. Staddon. While on bridge-building work, he and his comrades constantly dropped tools 'accidentally' into the river. They stole timber for the fires in the cookhouse, and stole and sold large bolts to the Thais, thus slowing down the completion of the bridges. He especially mentions for bravery a Jewish sergeant, Maurice Mincovitch/Minchom (whose testimony is included in this book), who ordered the men in his work party at Nakon Nyak camp in July 1945 not to carry Japanese baggage containing rifles and ammunition, despite the many threats made by the guards to beat him.

5564 Pte Charles Simon, 1st/3rd Battalion, Machine Gunner, SSVF

More information is available from an interview with Charles held at the Singapore National Archives.

62. 5564 Pte Charles Simon, SSVF, taken in his later years.

Charles was born on 14 September 1914 at 2 Eber Road, Singapore, one of four children in an orthodox Jewish family, and educated at Makassar, Celebes, Batavia and Singapore (the Raffles Institute) as his father ran businesses in those locations and the family moved around the Dutch East Indies. At Raffles he joined the Cadet Force. The family, like most of the Jews of Singapore, originated in Iraq in the nineteenth century. Charles had many senior jobs as a salesman all over the Far East with some big international companies, but later became manager of his own electrical supplies import-export company, Bolter & Simon, and a member of the Singapore Jewish Welfare Board.

Charles enlisted in July 1939 and when war broke out he was based in Penang (Malaya) with his battalion, which he joined on 1 May 1941, after being mobilized on 1 December 1940. He manned a machine gun in a concrete pillbox. By 16 December the Japanese had already reached south of their positions by using boats and constantly leapfrogging south-east down the Malay peninsula, forcing the British to consistently retreat. Ordered south, his battalion reached Singapore by boat, being machine-gunned a lot of the way by Japanese aircraft, and Charles was assigned to the commander-in-chief's HQ at Fort Canning, with General Percival. From here he was sent to man a pillbox near the Roxy cinema in Katong, and then to Orchard Street and Cairnhill Road to support the retreat of the Allies towards the city.

After three weeks he was attached to the 1st SSVF, a Scottish machine-gun company, which on 1 February became a rifle company, and was then sent to Goodwood Park Hotel to dig defensive trenches as the Japanese advanced across the Island of Singapore. Ordered to spike their guns, they took their rifles and formed a defensive line along Stevens Road and Newton Circus as the Japanese advanced from Bukit Timah.

Charles was taken POW. He volunteered to defuse the mines near the Roxy cinema and was then ordered to the drill hall in Beach Road, and sent to Changi. Here, in Block B, he met up with fellow Jewish soldiers Alex Oppenheim, David Marshall, and Max Bolter who later died at Kinsayok in September 1943. He did not meet David again until August 1945, after liberation, in Manila, in a PX (an American army store).

On 26 June 1942, Charles was sent to Bampong on the Kwai. He suffered five terrible days in sealed cattle trucks with no ventilation or hygiene facilities on the journey there, at the end of which two of the thirty-four in his carriage had died. Ablutions were performed in the carriage and when they did stop, to clean up, many were simply too weak even to get out. On the Death Railway he witnessed the appalling death rate from disease, beatings and hunger. Cutting bamboo to clear the railway and build camps meant that the men were frequently and deeply scratched by the broken branches and shoots; unstoppable infections led to leg and arm ulcers which, if untreated, caused death through gangrene. Charles worked on the blasting of rocks to push the railway through.

In October 1942 he was sent to Chungkai. He suffered several bouts of malaria, dysentery and beri-beri, so he was moved to Tamarkan 'hospital' camp, from working on the railway to 'light duties' in the camp itself, including the awful and sad burial parties. It was here that he met another Jewish soldier colleague, Maurice Gareh, who was shrivelled with illness, covered in ulcers and on his way to Chungkai. Lorry-loads of sick came south to the camp, as did the many bodies for burial in the hard soil; graves could usually be no deeper than three feet because of this. After two weeks spent putting up telephone wires, Charles was sent to Nompladuc camp, where treatment by heavily built Korean guards was sadistic in the extreme. Again he was blasting through rock for the railway cuttings. His eyes suffered from intense sunburn. Here too, in October 1943, he witnessed the effects of the dreaded cholera which killed within hours of being contracted; the Japanese did absolutely nothing and would not even enter the camp, and many POWs died. By this time Charles's weight had dropped from sixty-three kilograms to forty-one kilograms.

Transported to Japan

In March 1944, Charles was sent to Non Pladuk camp and in May he was sent back to Singapore, in a five-day nightmare journey in a locked cattle truck, where again thirty-four men were packed into a small space. Stops were few and the conditions – no sanitary facilities and foul food – were horrific. Arriving at River Valley camp, he soon embarked with 750 other POWs on the small cargo ship, the *Asake Maru*. They were battened down in the holds below, but after a near-riot against the guards, the men were allowed on deck. On a night stop at Manila they observed many Japanese dead floating in the harbour and assumed there had been an attack by US forces.

They did not know the destination but en route a ferocious storm washed away the galley and they had no food, water or sleep for three days as the vessel was tossed about like a cork. The ship finally grounded on the coast of southern Taiwan and began to break up in the waves. Discipline broke down and the POWs broke into the Japanese stores and took food, Red Cross parcels and medicines.

After two days the crew and POWs were rescued by a Japanese destroyer (the *Oyer*) and taken to Keelung in north Taiwan, and put onto another ship, the *Hakusan Maru*. Finally, after seventy-three days of hell since leaving Singapore, they reached Moji in Japan; over twenty men were dead and all were debilitated after such an appalling journey. From here they were taken to Amagasaki camp near Osaka, where they spent the winter working in a factory making marine engine parts. Charles managed to sabotage a large planing machine and put it totally out of action. Frozen pipes often meant no water for days, and their uniforms were merely cotton so they froze constantly in the awful weather. After arguing with a guard about his workload, Charles was given severe beatings around the face, and as a marked man he was then beaten by any guard who passed him for many days after. Charles said that for the guards it was a kind of sadistic sport. Rations were meagre but in December they were given Red Cross parcels, although only one between nine men. The Japanese simply stole the parcels for themselves.

During the winter of 1944–45 they were moved to work in Nagasaki. Again they froze, with only cotton clothing in the freezing weather, but they did have access to some Japanese newspapers and a secret radio and got to know how the war was progressing. When they left that camp the radio had to be destroyed because searches were made as they were shipped out.

By April 1945 the Allied bombing intensified, so the men were moved to 'safer' camps on the west coast, and 100 British, including Charles, were sent to the small port of Toyoma, their camp being in a rice paddy, where they worked as dockers with American and Australian POWs. They loaded soya beans, coal and salt, and the salt was especially unpleasant: loaded on their naked backs in baskets, it chafed their skin and the pain gave most men sleepless nights. They worked twelve-hour days and often went on all night in order that cargo ships could be rapidly 'turned round'. The poor diet of barley and rotten vegetables and one rice ball per day meant that many were ill and died. Only the hope of freedom kept living skeletons alive, but news was very sparse and as nobody knew how the war was progressing the men became very demoralized. Treatment from the Japanese got worse and beatings were very commonplace.

Liberation

On 10 August 1945 a voice came over the loudspeakers in the docks and some of the men who had learnt some Japanese understood that it was the emperor, who spoke for about ten minutes, announcing the surrender. There was bedlam as work stopped. The men were herded into trucks to return to the camp, the guards shouting and screaming as usual. Back at the camp, work stopped and cigarettes and new clothing were given out by the guards to the POWs. The American senior officer, a major, announced the surrender formally to the POWs and said that armed guards would remain outside the camp only, to protect them against possible local hostility, but the men would run the camp themselves. The next day a lone American plane circled three times and then dropped food canisters filled with excellent US K-rations (American army one-day combat rations, named after its inventor Dr A. Keys). Charles kept a piece of the parachute for years afterwards.

Charles still weighed just forty-one kilograms. A few days later the POWs were taken by train to Nagoya, then to the beach by truck where LCTs (landing craft tanks) took them to a hospital ship offshore. They were given full medical checks, deloused and given new clothes and their first proper western meal in three-and-a-half years. This included ice cream, which made Charles vomit straight afterwards!

Taken to Tokyo, they were flown by B24 bomber to Okinawa, where they rested for two weeks, and on to Manila, where the Jewish POWs were able to celebrate Rosh Hashana and Yom Kippur at the Rizal stadium with 5,000 American Jewish troops and POWs. Here Charles met up again with Jewish POW David Marshall.

Hearing that his mother had been evacuated to Britain, Charles sailed to Britain on the aircraft carrier HMS *Implacable*, with 2,500 liberated POWs; this was in December 1945.

There was a riot when the first meal was rice! They sailed to Vancouver via Pearl Harbor, then travelled by train across Canada to Diburd, Newfoundland, and by ship to Southampton on the liner *Isle de France* – recuperating all the way. Charles was in the UK for six months, meeting up with family and friends. He arrived in Singapore in June 1946 and discovered that his brother Edward had died in a Japanese camp. His mother went to live in Israel, where she died in the 1980s.

After the war, Charles was president of the Singapore Ex-Services Club.

1449697 Gnr/Bdr Moss Simon, RA, 239 Glamorgan Battery, 77th Welsh HAA Regt

63. 1449697 Gnr/Bdr Moss Simon on his return to Cardiff in 1945. Left to right: unknown; wife Franny; Moss; mother Tamara.

Moss was born at 83 Duffryn Street, Ferndale, Rhondda, in June 1916, the son of David Simon and Tamara (née Sussman), Lithuanian Jewish immigrants.[1] He was the youngest of six children and trained and worked as a jewellery dealer before the war, when the family moved to Cardiff in 1930. During the war his mother was living at 48 Dartmouth Road, London NW2. Moss became a Territorial in September 1938, and was called up in July 1939 into the 77th Welsh RA. He took part in the defence of Cardiff during the Blitz. He married Franny (Frances) Blasky in 1941 just before going overseas. She lived at a number of addresses in Cardiff but by the end of the war was at 8/10 Melrose Court, Penhill.

Moss was taken POW in Java on 8 March 1942 after his crew covered the evacuation of the last Allied ship to leave Java. His first camp was at Batavia, where they did all kinds of menial work, but then he was sent to Gloduk camp for six weeks, living in conditions of indescribable filth (it had previously been a camp housing local natives, and consequently, as they lacked any discipline or leadership, the camp had become infested with lice and fleas and disease). It was here that he carved into his aluminium food bowl his name and his wife's name, with the Star of David and the dates he was in the camp.[2]

Among Moss's papers is a chit that was signed by his CO, confirming that he and all other British POWs had been compelled, on 6 July 1942, to sign a declaration of obedience to Nippon, rather like those in Singapore who had been forced to sign the notorious 'no escape' declaration (known as 'the Selarang Incident'). While at Gloduk, each Friday evening and Saturday the Jewish POWs organized a service in an open field, using their memories to recall the prayers – but of course there was no wine for the sanctification (kiddush). In his written statement, Moss says very clearly that all though his incarceration, the communal spirit of the Jewish POWs he was with was a key factor in his survival, as they all helped each other as well as their other comrades.

In January 1943, he and a large number of POWs were packed like sardines into one of the notorious Hell Ships and taken to Changi in Singapore, where he spent five months. Again he helped to organize the Friday and Saturday services for the Jewish POWs, who numbered anything up to 100, from Britain, Australia and the Netherlands. By now their clothes were largely reduced to the loincloth covering – everything else had perished. Here he met the great Rabbi Chaim Nussbaum, whose sole reason for existence was the welfare of his Jewish flock; Moss wrote that 'he suffered with us but his spirit never flagged. When later we went up-country to slave in the jungle (The Railway), he was with us, our spokesman and representative to our Japanese gaolers.'[3]

In May 1943 he was transported to work on the Death Railway in Thailand and Burma. Here many died and he helped bury one Jewish soldier, L—, whose name it has not been possible to obtain, who had been a medical student before the war; he had been beaten unconscious by guards and never recovered. The two Jewish POWs in Moss's party (himself and F—) volunteered for the burial party at 11 p.m. that night. They went into the jungle by the light of a bamboo torch and dug the grave and placed a rough wooden Star of David on the mound, and said Kaddish (the Jewish prayer for dead). They got back to the camp at 4 a.m, and at 6 a.m. they stared work again.

In December 1943, Moss was back in Singapore, this time at Sime Road camp. Again, Nussbaum organized Jewish services. In May 1944, Moss was sent back to Changi, this time witnessing the terrible conditions of the women and children in the nearby civilian part of the camp. He helped in the construction of the synagogue and remembered clearly the dedication ceremony and the invitations sent out to many non-Jews, who attended as well. In April 1945 he was sent to yet another camp on the island, from where he was liberated. Leaflets were dropped to the men on 28 August 1945, telling them to stay in the camps and to list all personnel on nominal rolls, and not to eat large quantities of food if they had been starved, and explaining why. Interestingly, among his papers is a printed warning given to him and all other POWs, from GHQ, to the effect that under no circumstances were they to talk to the press or any welfare organizations or friends or family, on release, about any issues relating to attempted escapes or intelligence matters.

He would rarely, if ever, speak of his experiences. His niece's husband remembers Moss once squashing a wasp on his arm and he asked him whether such insects bothered him. He laughed and replied that when you had loaded and unloaded sacks of sugar covered in wasps onto a Japanese ship, and had been threatened with a beating if you did not, one wasp was never a bother. In conversation with his nephew, David Simon, Moss always ascribed his survival partly to his ability to catnap for short intervals while working, propped up against a tree. On a starvation diet and doing hard labour, this conserved life-saving energy. He always spoke badly of the brutal Korean guards, who were worse, he said, than the Japanese, but he had learnt to understand and speak quite a lot of Korean.

Moss always had a habit of banging his shoes upside down before putting them on, a leftover of his days as a POW when no man ever put on his shoes without first checking for scorpions or nasty insects. After he had seen the premier of the film *The Bridge on the River Kwai*, one of the actors (it may have been Jack Hawkins) asked his view about it. Moss said that compared to what he went through, the film showed 'a bloody holiday camp'. A beautiful pen-and-ink drawing of Moss is at the National Army Museum archives in London, drawn by the Dutch POW war artist Kees van Willigen. Only thirty-seven of the 650 party whom he was with survived the war.

After the war Moss ran a successful jewellery shop in Barry, near Cardiff, but he would never deal in Japanese or German products. He was president of the local Jewish Board of Guardians and took a special interest in the Penylan Jewish home for elderly people. He was also active in the FEPOW Association and local AJEX, never missing a Remembrance Day parade. He did not have any children. He died in January 1990. The archive of Moss Simon can be seen at the National Army Museum in Chelsea, London.

Notes

1. Thanks to Naomi King (niece) and David Simon (nephew) and Doreen Jacobs (AJEX), in Cardiff. His TNA debrief sheet says he was born on 31 May 1912.
2. This bowl, about ten inches in diameter, is on display with other items of his memorabilia, at the National Army Museum (NAM) in Chelsea, London.
3. From an article written under the pseudonym 'Emess' (Hebrew for truth) in a Jewish cultural magazine after the war, title and date unknown, seen at the NAM.

10630110 Pte Mark Simons, RA

Mark, a baker by profession, was born on 15 August 1915 at 373 Roman Road, Bow, London. He was called up on 8 May 1941 and was posted to the Army Catering Corps attached to 85th Anti-Tank Regiment. After six weeks' infantry training he was sent to Aldershot, then Clacton, and on 11 November to Scotland. From here he embarked on the troopship SS *Narkunda*, destination Basra in Iraq, but it was diverted to Singapore after Pearl Harbor was attacked, arriving in the last week of January. As they entered the harbour, Japanese planes were bombing them but there were no casualties.

When the surrender came, Mark was taken to Changi, but in May or October 1942 he was sent to Singapore railway station and packed with hundreds of other men in appalling sealed cattle trucks to be shipped north to work on the Burma Railway. The journey, lasting three days and nights, was vile, with the usual lack of any food or water or sanitation.

At their stop they were marched for miles in the alternate blazing sun and pouring rain to a camp at Wangpo, and put into bamboo huts. Most of their clothes and boots had rotted and were full of lice. Mark slaved for two years on the railway, from dawn until dusk. As his shirt fell apart, his back became blistered and burnt in the intense heat, and as exhaustion built up he also developed malaria. He remembers that hundreds of men died.

One Jewish POW he met was Cyril (Gerald) Gotlieb who was from the same street as Mark; they had been at school together. Cyril was well until he scratched his leg and developed a terrible tropical ulcer which became gangrenous. He was in terrible pain for many days and nights and the doctors could do nothing for him; he died soon afterwards.[1] Mark himself was ill with several other diseases but somehow managed to keep going. Eventually the railway work was finished in that area and they were marched through the jungle to Tarkamon camp. With their few possessions of ragged clothes and mess tins, it was a terrible hardship, marching with no boots, and Mark saw many POWs die in the appalling heat. After working there, they moved yet again, this time to Tarso camp. This was regarded by many as a death camp, as much of the work involved dreadful hard labour with rocks, and blasting them to make way for the track. The camp was tented and so the men were constantly soaked by the monsoon rain during the day and could not sleep in the continual rain at night. Soon cholera broke out and hundreds died. Mark will never forget the cremation of hundreds of bodies of his comrades, carried out to control the killer disease; work stopped for the first time and the men were wired in, to control infection, the Japanese being just as fearful of catching the illness as the prisoners were. Eventually the cholera disappeared and work restarted until the railway was completed.

Some time later Mark was sent, when very ill, to a base camp further south, which he does not name, and he recovered to some extent, being given medical help and better food. But soon afterwards the newly recovered men were ordered to parade before the Japanese and were told they wanted volunteers to go to work in Japan, where they would get better clothes and food. Later regretting it, Mark volunteered, being taken in by the lie.

In August 1944 he was taken by cattle truck to Singapore and then boarded a dilapidated former Scottish coal ship built in 1900, part of a convoy of several vessels. About 1,000 prisoners were herded onto the open decks (the Japanese had cabins below) but there were only about 100 lifejackets. Mark was lucky enough to get one. A large wooden frame with a hole in it was suspended precariously over the side of the ship to act as a prisoners' toilet. They were aboard for two months and conditions became worse and worse on the decks. The rice they were fed was filtered through seawater and had the consistency of slime. Mark met another Jewish POW on board, David Clayton; he had been married just before leaving England. He became ill on board and within a few days had died. Mark witnessed the body being wrapped in a sack and thrown overboard.[2]

Their first stop was Manila, where they saw lots of sunken Japanese vessels in the harbour, bombed by the Americans. In the second week in September they left Manila, but after they had left port, American submarines began sinking vessels in their convoy. The weather was also very poor and for three days and nights they endured a typhoon, with the waves lashing them on the decks. Mark now had dysentery and was also suffering from a bout of malaria, but at least there was plenty of clean rainwater to drink. Then a Japanese tanker sailing alongside was torpedoed and was burning from stern to bow. The heat was intense, and miraculously Mark stopped shivering and felt well again. Soon the storm died away and they approached Formosa, but the damage was done and the ship began to sink – fortunately in shallow waters; the Japanese abandoned them in their lifeboats and the prisoners waited hours until they were picked up by a Japanese warship which took them to Formosa and transferred them to another prison ship, reaching Japan in October 1944.

Mark was taken to a prison hospital with two infections on his arms; an American, Dr Gordon, operated on him without anaesthetic while medical orderlies held Mark down. In November he was discharged and sent to Osaka to work in the shipyards; it was bitterly cold and they had few clothes and very poor food. In April 1945 the shipyards were destroyed by American bombers, and Mark and his comrades were moved yet again, this time to do back-breaking work in the iron mill at Omi, Japan. In August 1945, after Mark had worked a twenty-four-hour shift at the mill, and on his thirtieth birthday, the first bomb was dropped on Japan and Mark was free.

Notes

1. No. 963481, 135th Field Regiment, Royal Artillery, died 3 May 1943 and was buried at Kanchanaburi.
2. No. 5829611 (aka Cohen), Private, 4th Suffolk Regiment, died 18 August 1944, aged 26, and is remembered on the Singapore memorial, column 54.

159938 Capt. Emanuel 'Manny' Snell, RAMC

My thanks to Manny's son Victor for the story.

64a. 159938 Capt. Emanuel Snell, RAMC (standing, centre), before posting overseas with his brothers Monty (left) and Bernard (right).

Manny was born in Giffnock, Glasgow, on 19 December 1915, but at the time of the war he was living in Manchester with his wife Rene at 714 Windsor Rd, Prestwich (another address given was 10 Chorlton Road, Manchester). He was one of three sons of Asher and Rebecca Snell of 49 Merrylee Park Avenue, Glasgow. Asher was a veteran of both the Boer War and the First World War, and was then serving as a lieutenant in an anti-aircraft battery in the Second World War. Manny's brother Bernard was a ship's surgeon in the Merchant Navy and brother Monty a sub-lieutenant in the FAA (Fleet Air Arm).

As Manny arrived in Singapore on the *Empress of Asia*, it was sunk by Japanese aircraft and he barely escaped with his life, managing to get overboard into the sea through a porthole. He was with 197 Field Ambulance when captured in Singapore. In Changi he suffered the commonly described hardships of illness and starvation (ten ounces of rice per day) in cramped conditions. There was little water so he shaved his head to save on washing. He was eventually sent, in June 1943, on an awful train journey into Thailand to work on the Burma Railway. He was also at Thompson Road camp from May to August 1942. At Tonchin camp, Manny was ordered to set up a camp hospital, but there were no drugs. He said that a third of the men died. He was forced to watch as sick men were carried on stretchers to the railway and forced to work. Until September 1943 he was at Hintok River camp and then at Kanchanaburi until December 1943.

Manny himself had severe dysentery and many ulcers. When his brother Bernard saw him after liberation (this is described later in the testimony), he still had many scars from old ulcers, as well as scars from septic scabies, and beri-beri and pellagra. He survived by buying and begging for drugs from the local Chinese who somehow smuggled them into the camp hospitals, and by buying eggs from the local Malays for the patients. This earned him the Australian nickname 'Egg-man', or 'Doover'. He described the Korean guards as especially cruel to the men. In January 1944 he was sent back to Sime Road camp in Singapore and then finally to Changi, where he was liberated in September 1945.

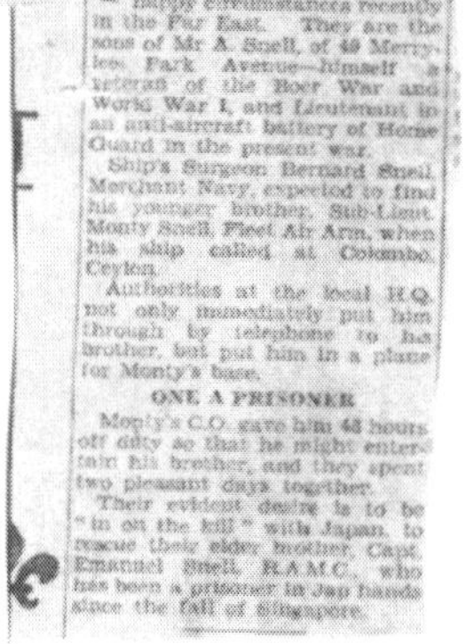

Lt. A. Snell. Emanuel. Monty. Bernard.

Brothers Met In Far East

TWO Giffnock brothers met in happy circumstances recently in the Far East. They are the sons of Mr A. Snell, of 49 Merrylee Park Avenue—himself a veteran of the Boer War and World War I, and Lieutenant in an anti-aircraft battery of Home Guard in the present war.

Ship's Surgeon Bernard Snell, Merchant Navy, expected to find his younger brother, Sub-Lieut. Monty Snell, Fleet Air Arm, when his ship called at Colombo, Ceylon.

Authorities at the local H.Q. not only immediately put him through by telephone to his brother, but put him in a plane for Monty's base.

ONE A PRISONER

Monty's C.O. gave him 48 hours off duty so that he might entertain his brother, and they spent two pleasant days together.

Their evident desire is to be "in on the kill" with Japan, to rescue their elder brother, Capt. Emanuel Snell, R.A.M.C., who has been a prisoner in Jap hands since the fall of Singapore.

64b. A local 1945 Glasgow newspaper cutting showing the Snells, a typical Jewish family at war.

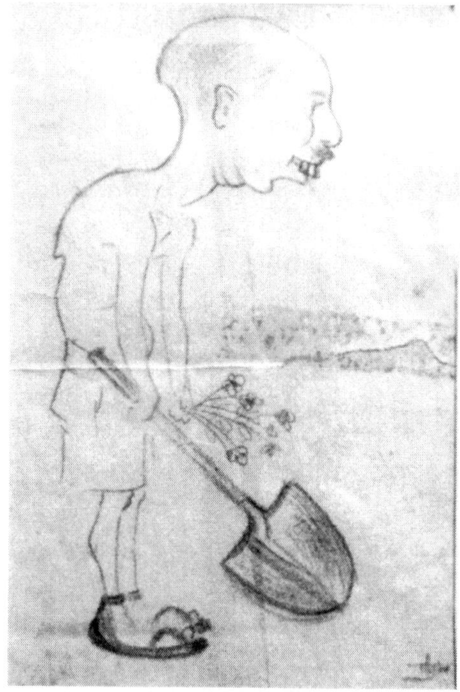

64c. A cartoon of Emmanuel Snell while a POW, by an unknown artist in an unknown camp.

Manny said he was beaten only once but witnessed far worse, especially the bashings meted out to Indians who refused to join the Indian National Army to help the Japanese. He told his son that once a Japanese guard pointed at him and, because he had hair on his chest, said 'monkey'. Manny pointed back at the hairless guard and said 'baby' – whereupon the guard jammed his lighted cigarette into Manny's leg, giving him a permanent scar.

Having threatened to kill all the POWs if the Allies invaded, when the surrender was actually announced the Japanese suddenly became very friendly. One day each man was given a pound of butter and a box of cheese for breakfast, and they all gorged themselves and were sick. A few days later, doctors from the Parachute Regiment were dropped in by British aircraft and the POWs could hardly believe their eyes; the Japanese kowtowed and, soon after, ground occupation troops arrived; the Japanese were forced to scrub the streets under the supervision of Indian troops, but still were being let off lightly, according to Manny.

His brothers Monty and Bernard had met purely by chance in Colombo and had visions of going together to liberate Manny from a Japanese POW camp. Bernard did, in fact, meet Manny on liberation in Colombo, to where he had been sent from Singapore on the Dutch ship SS *Tegelburg* in September 1945. Bernard wrote a long letter home, describing the condition he had found Manny in, and what Manny had told him about all that he had been through. He asked his brother Monty to destroy the letter, so that their mother would not read it, but it was kept in the family and hidden in the family 'archive' until recently. Manny's frame seemed to have shrunk, his teeth protruded and he was stooped and appeared to lack spirit – and this was after he had gained two stone in the previous few weeks (he gained even more weight on the journey home via Port Said – where they were issued with winter clothes – and he looked much better when he arrived home in mid-October). His clothes, of course, were very baggy on him. But the ex-POWs were all put on a special diet and began to slowly recover.

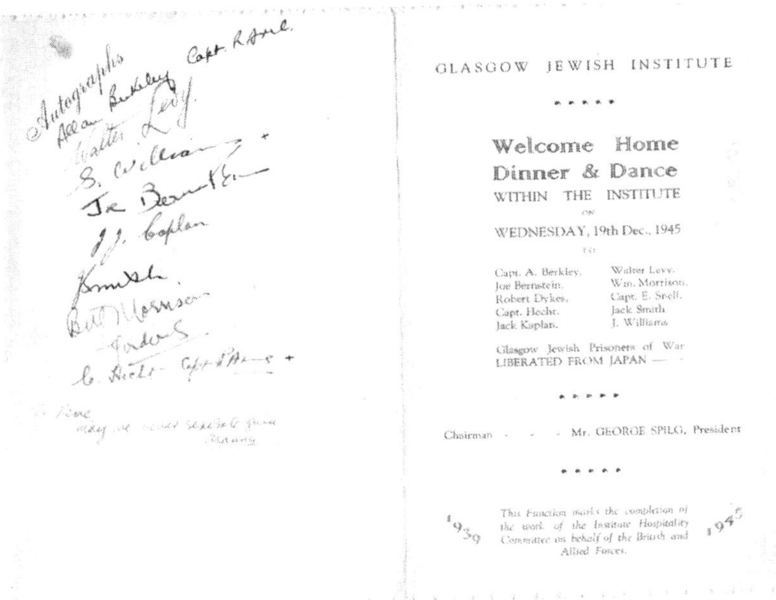

64d. The signed menu card from the Glasgow Jewish Institute welcome dinner for returned POWs from Japanese camps in December 1945.

Despite all this, Bernard said that Manny wanted to stay and work in the Malaya medical service. However, he stayed in the Manchester area as a doctor until 1956, when he emigrated to Newfoundland in Canada. His family joined him in Portage la Prairie, Manitoba, in 1957, moving to Winnipeg eighteen months later, where they joined the Shaarey Zedek synagogue. In Winnipeg, Manny worked in the Provincial Department of Health, rising to become chief epidemiologist. His two sons were bar mitzvah there (sadly, his son Jonathan died in his 20s) and Victor also married in the same synagogue. Manny died in Winnipeg on 21 February 1982.

Sgt Loet/Lou Velmans, Dutch Army

The information here is taken from his book, Long Way Back to the River Kwai *(Arcadia Books, 2003).*

'I was 17-years-old when the Germans invaded my native Holland. I became a fugitive ... but the enemy I came to know intimately had a Japanese face. I spent my POW years trying to figure out what made ... this enemy tick. It has only been in recent years that ... I realised that much of my life since the war has been spent trying to understand the Japanese.'

Loet was born in Amsterdam in 1923, son of Joseph and Anna, an assimilated Jewish family. They fled from Scheveningen in Holland with other Dutch refugees to England in May 1940 as the Nazis invaded, on a local Dutch coastguard vessel, *Seaman's Hope.*[1] Avoiding mines and German aircraft, they made it to safety when they were picked up on the second day by HMS *Venomous* of the Royal Navy, which took them to Dover, and from there they went to London. They lived for some months in a hotel off Russell Square, with other Dutch refugees. Business contacts of Loet's parents helped them and Loet obtain work; he had learnt good English while living in Holland and had visited England before.

Refugee Dutch civil servants suggested that the family should go to the Dutch East Indies, where there were great opportunities far from the war, so they decided to leave. Loet sailed from Liverpool in September 1940, on the Cunard liner *Viceroy of India,* for Batavia/Jakarta via Singapore, with his parents and his Aunt Ro, Uncle John and cousin Dick. They sailed via the Cape and the journey took seven weeks. Off the West African coast they stopped and picked up survivors from a British merchant ship sunk by the German battleship *Graf Spee.* They also stopped at Durban, Mombasa and Bombay. In all these ports Loet excitedly met confident British and Commonwealth troops on their way either to or from the Far East. From Singapore they transferred to a Dutch steamer and soon arrived in Batavia, where they rented a Dutch-style house and found work quickly in the retail trade, settling into colonial life. But they soon discovered that both the local Indonesians and the established Dutch settlers treated them as outsiders. School for Loet was different, however; the pupils were integrated and friendly – Eurasians, Indonesians and Chinese – although they never mixed in each others homes.

On reaching the age of 18 in the summer of 1941, Loet was drafted into the army and sent to the hill-town barracks of Bandung near Batavia. He did well during the harsh training and was soon promoted to sergeant. They were given a lot of faulty Canadian anti-tank guns and radios that did not work properly; even their boots did not fit and caused huge problems. When the Japanese conquered Malaya, their soldiering suddenly became very serious, with lots of practice on the rifle ranges with their British Lee Enfield rifles. Then the Japanese landed in Java on 1 March and they were all confined to barracks. An enemy aircraft bombed their barracks and Loet felt that the totally unprepared and unprofessional army commanders simply panicked, sending troops in all directions to no obvious purpose. Then his company of sixty-plus young men – virtually all Eurasians – were suddenly ordered to go in three trucks along a road on which the Japanese were allegedly advancing. They had one old machine gun per truck, each commanded by a sergeant, with a sergeant major in overall command. They also carried some spare bullets, and one motorcyclist scouted out ahead. Nobody spoke; it was their first time in action.

Loet was in the third truck but fortunately they met no Japanese. They did meet some bored Australian machine gunners who gave them some beer, but by evening the mission was over and they all returned thankfully to base. Suddenly and without warning, on 8 March they were told that the Dutch had capitulated and all the soldiers were confined to barracks. Virtually no Dutch soldiers had fired a shot and many native soldiers deserted and simply went home; some began looting in the Chinese and Dutch neighbourhoods of the towns. Fear gripped all the young men, as they did not know what to expect of the Japanese. On 9 March, at 5 a.m., Loet and his friend Hans decided to escape. They took a staff car and drove out of town but soon were stopped by a Dutch roadblock patrol and forcibly returned to the barracks.

Prison

As elderly reservist civilians arrived in the camp, some tensions developed between them and the regular soldiers, for these older men were wealthier planters, merchants, teachers, officials, and so on. The Japanese appeared only after several days, some on bikes and smartly dressed, others in rags and clearly front line troops. All had one thing in common: rifles with menacing-looking fixed bayonets. Soon they began collecting all the Dutch weapons and trucks. The Dutch officers behaved in a very subservient manner and carried out all the Japanese orders, telling their own men to move and act quickly, barking orders just like the Japanese themselves.

For some minor infringement, Loet soon received his first beating from a Japanese guard, and was seething with anger. During this period, the men kept busy with sports and bridge tournaments and also listening to a secret radio at night; the news was not good but contact kept up morale. Each day the men queued by a window and were allowed to communicate with their families outside the wire, though at a distance. One day Loet's mother arrived with a small basket with some clothes, cigarettes and canned food and he took it eagerly. This was the last time he saw her for four years, and the last gift he received for three years.

After two weeks, new guards arrived and 8 a.m. roll calls were started; Japanese began patrolling the camp and were far more menacing and bullying than the previous guards, who had left. One morning a large group of smartly dressed officers with high boots, white gloves and samurai swords arrived. They escorted a sleek, shiny staff car with a large Japanese flag on it, from which emerged a senior Japanese officer, their new commandant. He mounted a small platform. There was total silence, then he bellowed in Japanese. The interpreter explained: no radios; prisoners were to bow to the north to salute the emperor every morning; they were to salute every Japanese soldier they passed; they would work; there would be no contact with the outside world; any infraction would be punished; attempting to escape meant a death penalty. The commandant then gave a demonstration; he took a soldier's rifle and charged, screaming, with bayonet forward, at a man and just stopped before reaching his target. The message was clear, and Loet remembers that the prisoners, including him, trembled with fear. The officer returned to the platform and gave a long speech about the new Japanese regime in the Far East. After he left, the camp officers began rationing food and medicines, as there was no sign of the Japanese replenishing any supplies. Next day, three Eurasian soldiers who attempted escape were caught and bayoneted to death before the eyes of the whole camp and the laughing guards. This was Loet's first encounter with death. But the men carried on; sports and entertainment were organized to keep up morale; the camp was kept clean; the Japanese got them to build both a bamboo and wire fence around the camp – so they could not see out or get out. Loet was also delegated to clean the Japanese latrines, often getting a beating into the bargain; stick, boot, fist and rifle were freely used by officers and men against the POWs for either the slightest reason or, more often, for no reason. The rationale? Because the Dutch had surrendered and were thus worthless – this was the Japanese code.

On Java, Loet saw that most Indonesians welcomed the humiliation of their former white masters and applauded the Japanese for what they were doing. Once a week the POWs were harangued at roll call by the commandant for half an hour – it was a mixture of insults, threats and the virtues of the emperor. The slightest movement meant a beating by guards who patrolled the rigid, standing ranks of prisoners.

One day, two American merchant seamen arrived and were put in Loet's hut. He befriended them and showed them the ropes, and they lent him their old American magazines which he avidly read over and over again. He is sure that this is why he decided to go to the USA after the war.

After eight months they were given a day's notice to move and were taken from Bandung camp to the railway station on foot, then sent on an eight-hour rail journey to a camp at Tandjong Priok, the outport for Batavia. After three weeks they were taken on foot at night to the port and all 2,000 men were crammed into the holds of an old Japanese freighter. They could not even lie down and for four days they endured suffocating heat, surrounded by faeces, vomit and sweat; watery soup was distributed twice a day. Twice a day they were allowed on deck for one hour while a group of POWs hosed out the stinking holds, and they were allowed to use the toilets on deck if there was time. By the time the ship reached Singapore, several men had died.

Changi

On the quayside they were hosed down by fellow POWs and felt again refreshed, and relieved to be off the ship. They were taken by truck to Changi barracks camp where there were already thousands of Allied prisoners. This camp was well established and had organized education, sport and entertainment infrastructures, not to mention gambling fraternities and a black market. Japanese guards were almost absent and the British ran the camp in an orderly fashion. But beatings were still common; Loet saw one Australian sergeant beaten by guards and sent to gaol, and then die after he was released to the hospital.

The bravest prisoners were the operators of the secret radio, who once a week in the dead of night put on their headsets and memorized the news, and then passed it on to others as discretely as possible. Afterwards, the set would be disassembled and hidden in water bottles, sick bays, hats, underground – and also taken to other camps when men moved on. An Australian would pass on news to Loet, who then gave it to a Dutch officer to pass on to the Dutch POWs. The Japanese knew that there must have been a radio but never found it. Loet, like others, would engage in bartering valuables for food if he could, in the black market – both within and outside the camp. By the time he got to Singapore, Loet had little left, but it included his old puttees and his Omega watch.

He managed to sell the watch, and with his acquired stock of canned goods and cigarettes he became a middleman in trading watches, for which he took a 5 per cent or 10 per cent cut on each sale; he dealt mainly with a Dutch entrepreneur called Max, who himself was working for a Scots sergeant major who closed deals directly with the Japanese guards.

Later, Loet was moved to the main Changi gaol, where his Java group was delegated to camp cleaning each day whilst others were sent to work in the docks. In the evenings he took up acting and often the Japanese guards and officers would attend the shows. They always left before the end, when the anti-Japanese sketches took place, and the men could express their pent-up resentment of the Japanese, to wild cheers of approval from the audience. Loet witnessed how a man's looks and build could determine his fate; his good friend Frans was a large, handsome and outgoing man who attracted the attention of the guards and was frequently beaten. When he was picked to go on the railway, his large body never received the food that it needed, and he died early on – the first of many friends of Loet to die.

To keep active, Loet asked the British if he could start a small café and use the profits to obtain much-needed medicines. Obtaining some benches, tables and chefs, dishes made from bananas, coconut, rice, sugar and salt were concocted and sold after normal meals, and some entertainment was provided. It worked well, as did the black market trade in pens, flashlights, lighters – anything the outside work parties could steal and bring, secreted, back to camp for sale to the local Chinese, who provided food and medicines in return. But by May 1943, the venture came to an end as the men were sent away to work on the railway. Loet was lined up with his group and sent by train on a four-day journey up through Malaya, in the usual crowded and filthy conditions, with infrequent toilet stops and two food stops per day, when they could also buy some extra food from local traders if they had any money. Many became ill from the human waste that littered the inside of the iron wagons, and the discomfort, hunger, thirst and stench in the heat of the day was appalling. There were few guards – but the men were so weak that even if they could have escaped from the wagons, they were too exhausted to get away. Finally they arrived in Bampong.

Thailand

Bampong was a sprawling and filthy mess – both the village and the camp – and gloom descended on the men as they were told of even worse to come, further up the railway. But speaking to a local Dutch POW, Loet was told that any railway supplies they managed to steal were eagerly bought for food by the Japanese themselves, who then sold them on to the locals for profit – especially tools. He also told him of the death of Frans. After a few days, Loet's group were moved – about 300 of them – on a forced march for twenty miles, harried by the guards with fixed bayonets, and carrying their sick and exhausted comrades where they could. Every hour-and-a-half they were given a ten-minute rest. The trail became narrower and more and more muddy and difficult. Starting three abreast they now walked snake-style, and any man who stopped got a beating from infuriated guards. On the third day the first man died of heatstroke. For five interminable nights, in heat and pouring rain, they continued, following a railway bed. At the end of the fourth night, many of the men simply could not go on and the guards began screaming at and kicking the prisoners; but the men stuck to their guns and a stand-off developed whereby the men stood their ground and refused to move. Suddenly the noise died away and the two sides stood glaring at each other. The Japanese leader then ordered the prisoners to sit and rest. After they had rested, the order was given to resume the march and the weaker prisoners were helped slowly along the way until they all finally arrived at their destination;[2] they had walked eighty-six miles. The guards made no further effort to force the pace, but still twelve men died of exhaustion. Finally they arrived at Spring Camp, and a Dutch Army Christian chaplain held a short service of thanksgiving.

Death Camp

The prisoners were immediately put to work, first to finish their poorly built huts and then on a long service road which linked them to the railway, for moving supplies by truck. Tree cutting was the most dangerous work as the Japanese themselves shouted contradictory orders about which direction to pull the ropes in, to bring down a half-cut tree, and many men had close calls as trees fell where they were not meant to. Loet worked on the road and it was a never-ending battle against torrential rain and fast-growing vegetation. Then he was moved to rock cutting on the railway itself, using chisel and hammer, nine hours a day, every day, with just three short breaks and little food. When the odd Sunday half-day was given for rest, the men still had to work in the camp to keep it clean – dig latrines, help the sick, maintain their atap hut roofs, and so on. Most slept badly as the bugs kept men awake at night; Loet said that they were so tired and ill that the men barely spoke to each other after work and just struggled to stay alive. Food was three small portions of rotting rice daily with barely any vegetables; Loet had bad dysentery like many others, and many men died, and of course the sick were forced out to work by the guards. When he caught malaria, he was finally left in bed. Others had ber-beri and appalling tropical ulcers as well, which gave off vile odours, and Loet watched many friends die around him, some going mad with cerebral malaria.

There were virtually no medicines and nothing to do in the sick hut but lie about getting weaker and weaker; Loet tried to focus on killing bed bugs. Some men gave up and died, but others fought back mentally, told feeble jokes, spoke sparingly, and came through. Loet saw it all and became hardened to the sight of death. He was also disgusted at the way the Dutch and British officers sometimes seemed to acquire better food through trading with locals, and although they would eat it in their more isolated hut, the food smell often pervaded the camp and enraged the ordinary soldiers. But there were differences. When the Japanese ordered the officers to work, the Dutch soldiers cheered, but the British soldiers went on strike for a day to support their officers!

Some of the Eurasians knew about local medical remedies obtained from the forest plants and, using these, some of the men – including Loet – recovered from their illnesses. Other plants were edible and the men learnt what to pick and eat. In fact, statistics showed that the Eurasians adapted better to conditions and so survived at a higher rate than the Europeans. The Australians were next because they were more rugged and fit to begin with and more willing to help each other. The British rate of survival appears to have been lowest – but Loet admits that these are only his own experiences and might not hold up to further scrutiny.

Their most urgent need after food and cigarettes was good footwear, as this reduced considerably the chances of infection through cuts. While carrying rocks in baskets for the railway bedding, Loet was beaten by a guard in the small of his back, with a hammer, and lost consciousness. He never worked again on the railway. After he recovered, the Dutch doctor gave him a job in the operating theatre, which was a small tent, as his medical assistant. Most operations were amputations and Loet said that most of these men died soon afterwards. Otherwise he was in the sick hut helping bedridden patients, washing bodies and scraping maggots from wounds. When a man died, he had to remove the bandages and wash and reuse them. The Japanese allowed only a few minutes for a burial service and, to save time, they permitted them only after work, at dusk. In this way, Loet saw many of his friends die in agony and with little time for any compassion, as the sick hut acted as both morgue and funeral chamber. For those trying to care for these men, such as Loet, the drug of exhausted sleep was the only escape. Another strategy for Loet was to take each week as it came and live from Sunday to Sunday; yet another was to imagine the starvation rations as large meals and he thought especially of the delicious white challah bread he had at home.[3]

On occasion, Loet saw the Japanese also in extremis; exhausted infantry, with huge backpacks, trudging the muddy jungle passing their camp, going up to the front in silence; others man-hauling huge flat cars carrying artillery, or pushing huge carts of equipment, as NCOs with whips urged them all on. Meanwhile, Korean guards – themselves brutalized by the Japanese – took over from Japanese guards. On one occasion they ordered the prisoners to witness their execution of a pig by cruel, slow, screaming death from bayoneting, in a wooden cage. When the pig was cooked, the men did not get a single scrap, but could smell the delicious aroma.

On one occasion a Japanese field officer visited them in the sick hut; speaking perfect English he bragged for an hour about his time in the USA, never once asking them about their obvious misery. Then he left, but later returned with a large box of medicines for the dispensary. He said nothing and just departed, shaking hands with them all. Was there a more humane type of Japanese, the men wondered?

Other visitors came at night – Japanese soldiers with VD, fearful of telling their superiors as it was a serious offence. By giving injections of a fluid, alleged to be a cure, that they brought with them, the doctors were able to bargain with them for extra food for the patients.

In September 1944, there was much excitement among the guards as they announced the arrival of the first train to make the journey on the railway that the prisoners had been building – but it came off the rails near the camp. All POWs and every available Japanese in the area, and elephants, were called to the incident. On arrival, Loet and his men found many at work with long poles trying to jack the train out of the ditch back onto the rails. Prisoners deliberately dropped poles and the engine kept sliding back; guards rained down blows on everyone. Eventually, however, the job was done, although an elephant had accidentally killed a POW; from then on, fully loaded trains passed the camp each day. But soon after this, Loet and his men were taken back south by truck to Kanchanaburi transit camp.

Here they rested from work; food was a little better and there were organized entertainments. Food smuggling restarted with the locals and men began to recover a little from their illnesses. Loet counted himself lucky to be alive, even though he had malaria. Beatings by guards continued, however, and the POWs' hatred of their persecutors knew no bounds.

Sometime in November they were sent down to Singapore, but first they had to take part in a bizarre ceremony to mark the inauguration of the railway. The Japanese paraded, shouting, in spotless uniforms and the prisoners received some extra food while being harangued by a general extolling the emperor. Suddenly Loet walked straight into the guard who had used his hammer on him at Spring camp. Loet froze as the Japanese barked at him, demanding a salute. Loet saluted at once; the guard smiled and crisply gave a return salute, then disappeared. Loet hoped they would never meet again.

Sime Road Camp and Changi

This was a clean, wooden-hutted gaol, and food improved – it now included soya beans. The men received their first and last Red Cross parcels, one between two. They were able to wash regularly and there were latrines, entertainments, and electric lights to read by. But in May 1944 the men were moved again, to Changi. By this time, Loet had only one plate, a mug, fork and spoon, a ragged mosquito net, one pair of socks and one pair of pants, one ragged shirt, wooden clogs, one small cake of soap.

Changi was bursting at the seams with tents and huts, but it was also well organized and had small gardens, wood-burning stoves for boiling water, rudimentary showers, and a theatre. Soap made from palm oil and potash was available, workshops made artificial limbs and clothing, and one machine even stamped 'Rolex waterproof' on old tin watches so they could fetch a high price on the black market and thus be used to obtain food. Any stray cat or dog became a nice stew. Whilst some mail came to POWs from abroad, nothing came from the Dutch internee families in Java to the Dutch soldiers. Loet therefore worried deeply about his family. Relations between officers and men were better because the men had noticed that officers had shared the horrors with them on the railway.

Religious services took place regularly and Loet joined the Jewish services run by Lieutenant Rabbi Chaim Nussbaum, the Dutch Army chaplain, with whom he shared a cell in the main building. For Loet the services gave a strong sense of community. Many non-Jews came to Nussbaum's lessons on Judaism and Loet became editor of the Jewish magazine, *Ha Beemah: The Changi Jewish Forum*; non-Jewish artists contributed illustrations, including the great Ronald Searle and George Sprod. Only one edition was produced but it was very popular among all the POWs.

In the Jewish group was a tall, thin, stooped bachelor Dutch soldier called Moishe, who had been trapped in the Indies while on business when the Nazis invaded Holland. He was Rabbi Nussbaum's loyal assistant and not only fetched and carried for him, but also tried to obtain small amounts of extra food where he could; he was also very protective of Loet. As the months passed, however, he became increasingly reclusive and would be seen alone, scribbling furiously on used notepaper. Meanwhile, Loet noted the resurfacing of tensions among the different nationalities of prisoners, and racial and class divisions being reasserted within those groups.

Then the Japanese reintroduced work, and Loet was put on camp maintenance duties. The theft of Japanese goods increased and so all rations were reduced, beatings increased and the morale-boosting entertainment activities were heavily reduced, especially as the Japanese officers objected to patriotic songs featured in the shows.

In May 1945 they heard of the surrender of Germany, and then on 7 August came news about Hiroshima. Celebrations broke out and Loet has a clear memory of Chaim

Nussbaum, short and shrunken, facing the wall of his cell towards the west and Jerusalem, chanting Hebrew prayers and swaying, covered with his prayer shawl. The next day, the Japanese disappeared.

Liberation

The following day, the guards reappeared but only with sidearms, and Allied planes dropped leaflets telling all POWs to stay put until the relief troops arrived. After 15 August, when the surrender was announced, Chinese locals came to the fence and threw food to the men, who were also allowed to send one telegram to loved ones. Soon troops arrived to help with the welfare of the POWs, and a British officer took Loet and a friend to the port to see the British battleships that had arrived there. A naval officer invited them on board to eat, and they had the luxury of gorging on tea, sugar and milk, white bread, butter and marmalade – and then threw up immediately. Back at camp they were given eggs, flour and milk and made a cake, and could not keep that down either. The Japanese guards stood at attention at the gates and saluted when the POWs passed, but were ignored. Then one day they disappeared too.

Admiral Mountbatten arrived and gave a stirring speech on a soapbox about the courage of the prisoners. But this was deeply marred by the arrival later on of a Jewish officer from Australia, who told the men about the Holocaust at a gathering of the Nussbaum congregation. Hardened veterans of the Japanese camps, who thought they had seen and lived through the worst, were filled with rage and tears. That Friday evening, the Jewish service was one of mourning; Loet knew that many of his family in Holland would have perished and, had he been there, he would have died too.

Soon Loet heard that his parents were alive in a camp in Jakarta; Chaim Nussbaum was given a short leave to visit his wife and children there, and took a letter from Loet to his parents. He returned with the grim news that 15 per cent of Dutch civilians had died in the camps and many had suffered just as the soldiers had.

Loet was anxious to move on; he was well fed and recovering his health (except for the bouts of malaria that plagued him for years afterwards) but after one month he just walked out of the camp, found an empty office by the docks and moved in with some other soldiers. He met a Dutch Jewish liberation soldier who asked Loet to help him produce a local newspaper for the Dutch called *Oranje*, which translated into Dutch all the local and international news. Loet also did Dutch-language broadcasts on local Singapore radio. News then came that they may be recruited to fight the Indonesians in Java. Whilst the Dutch planters and other colonists were keen, Loet was Dutch born and bred and had no intention of fighting Indonesians who wanted their freedom.

In the autumn of 1945, Loet was reunited with his parents, who arrived on a Dutch ship in Singapore. They looked very ill and had aged a great deal; for the rest of their lives they never spoke of what had happened to them, nor did Loet ever speak to them of what he had been through. They returned to Holland but Loet said he would stay for a while in Singapore. He attended Chaim Nussbaum's congregation, now swollen by newly arrived Jewish troops from the Commonwealth and Holland. He had learnt that the commandant of Spring Camp had been tried for war crimes and executed, but few others were ever caught.

Then finally he left in February 1946 on the Dutch troop carrier *Alcantara* and returned to Holland.

Epilogue

Post-war Holland was miserable and cold after the warmth and bright lights of Singapore. Everything was rationed and Loet lived in crowded conditions with his surviving family. He enrolled at Amsterdam University and met Edith van Hessen; they married and emigrated to the USA (New York) and had two children (twin daughters, Hester and Marianne), and Loet went into public relations.

He returned to Indonesia in the early 1950s to find his fellow POW Willem who had slaved with him on the railway and who later had married a local Singapore widow and returned to live in Java. He could not find him, but a bizarre event occurred some years later. Loet and Edith were on the New York subway at Times Square one day during the rush-hour, and as they entered the car, Loet stared into the face of Willem, who was sitting down. Loet shouted to him and Willem saluted him back, but he and Edith were pushed out of the car by the crush and the doors closed; they never saw Willem again.

Loet kept some contact with POW friends, especially Chaim Nussbaum whom he met in Toronto, and George Sprod, the artist, in Sydney. In 1985 he attended a reunion of Dutch veterans in Utrecht, Holland; each group of ex-prisoners stood by a banner with the name of the camp where they had been incarcerated. Loet stood by the one for Spring Camp but nobody turned up. For some years he suffered terrible nightmares.

For twenty-five years, Loet had to visit Japan many times on business. He discovered that the Japanese never wanted to discuss the war; it was a closed book for which they never apologized, nor even fully acknowledged. Loet had very mixed feelings towards his Japanese business hosts, however kind they were to him. He often imagined that he saw the faces of his guards among the audiences he was addressing, or heard a commandant's barking order when he heard a Japanese shout in the street.

He returned to the Far East in 2000, and, as a result, wrote his book, for his family and for posterity.

Notes

1. It later became a Dunkirk 'Little Ship'.
2. This is the only account I have ever read whereby Japanese guards actually gave way to the demands of a group of prisoners in this manner.
3. The traditional Jewish plaited bread eaten on Friday evenings to welcome the Sabbath.

6022643 Pte David Leonard Wince, 2nd Battalion, Cambridgeshire Regt, 53rd Brigade

Between 9 March and 20 April 2007, David related his experiences to me in a series of emails. This is his testimony.

65. 6022643 Pte David Wince, Cambridge Regt.

David was born in Stepney on 26 January 1920, son of Samuel Wince, and when war broke out he was living with his parents at 31a Essex Road, Islington, London. Called up in March 1940, he was posted to the Far East in the autumn of 1941.

In Singapore, David's battalion was based on the racecourse. When the Japanese invaded Malaya, David was in constant action on the peninsula with them. They had been trained for desert warfare and their uniforms were khaki, rather than the jungle green they should have been. Their first skirmish was at a bridge at Batu Pahat, but the Japanese moved in behind them and after ten days they had to retreat. A Japanese spotter plane kept strafing them and after three passes David's men were given permission to fire back, and they shot it down.

At Sengarang, their retreat was opposed by a strong Japanese roadblock, so they had to take to the jungle and go round; here David was wounded in his heel by a splinter from a mortar shell. After a night's forced march they stopped to rest and when David woke he was covered in (harmless) land crabs; they hid in a nearby fishing village for two days and then crossed a coastal mangrove swamp where a Royal Navy gunboat spotted them and picked them up. On the journey to Singapore they were again strafed; as the fighter came in the men would move to the lee side to shelter from the gunfire while the navy fired back with an Oerlikon; after three passes the plane flew off.

Back in Singapore they were issued with new uniforms and rifles, the latter having rusted from the wading through the swamps. However, David's wound had turned septic and so he was sent to St Andrew's hospital, a converted school. When the surrender came on 15 February he was in hospital.

David and another soldier decided to try to escape, and a local Eurasian (mixed race) man gave then some civilian clothes. As they wandered Singapore they were stopped by Japanese soldiers but were able to bluff their way through. They did not know what to do and did not want to be shot as spies, so when they heard that civilians were being marshalled on the Padang in the city centre (a large open area used as a cricket pitch), they joined them. They were then taken to Changi civilian prison, but the Japanese ordered any military men to move to the army barracks prison. They decided to do this and rejoined their units as POWs.

POW

Among the Jewish prisoners David met around this time were Len Goldfarb,[1] Bob Dykes (Glasgow) and Alfred 'Ginger' Kaplan.

In March 1942, David was sent upcountry on cattle trucks, in appalling conditions, to Bam Pong, Thailand, to work on the Death Railway. The food and work conditions were horrendous as they slaved to build the notorious railway embankments in the thick jungle, and men began to fall ill and die. The guards were vicious. David witnessed men being punished for the slightest alleged infringement; sometimes they were left standing in the boiling sun until they collapsed. After one month they were sent on a forced march of about twenty miles to another camp at Kanchanaburi; owing to the number of sick men, this took a full two days. In June 1942 they crossed a shaky footbridge over the Kwai river and hiked to Chungkai camp; very few huts were built and so they had to set to work to finish the construction of the camp.

It was possible to walk around the area, and one day David was passing a village when a dog ran out from underneath a hut and bit him. This was a blessing in disguise, because the bite became an ulcer and anyone with such a wound could not be sent upcountry. But he still had to work on the railway in the Chungkai area. Here he witnessed a POW being forced to kneel on a log until he passed out; and another POW, who had one leg, was hung up by his arms. Another POW had lost his mind and was led around the camp on a dog leash in a catatonic state. On many occasions David was slapped, punched and kicked, but by his own admission he has erased many of the atrocities from his memory.

Usually an ulcer would grow rapidly and result in either death or amputation, but David was told to bathe in the river water and allow the fish to nibble the dead flesh; it later healed up. David often helped in the ulcer ward in the camp, where the smell and sights were atrocious; he described it as being like an abattoir, and the smell was unbearable.

On one occasion, when walking to his hut at Chungkai, he passed a small working party some twenty metres away. A guard yelled at him to stand still and then came over, still yelling, in Japanese, this time into his face. David did not understand a word. Screaming to a crescendo, the guard whacked David across the face with all his strength, then walked away. David never understood why. On one working party unloading a barge, the Japanese guard, with no explanation, lined all the POWs up and went down the line smacking their faces. Such was the bizarre behaviour of the guards.

David did remember holding a small Yom Kippur service one September with several other Jewish POWs, but he did not remember the place. He left Chungkai in October 1942 for Tamarkan, where he slaved until December 1943.

David remembered the Korean guards as being particularly cruel; he recalled particularly two guards nicknamed 'Harold Lloyd' and 'Moonface', who were notorious in various camps and named as war criminals by many at the end of the war. At Nakom

Paton camp (described below), Moonface threw a shovel at David, which cut his back and hand. Personally, David never saw any guard exhibit the slightest kindness, but he recalls local Thai people trying to give them help in various ways – by throwing bananas to them, for example, as they marched to and from work.

Burma

After David's ulcer healed he was ordered onto a train with many others, to go to Burma. He gathered his possessions – a canvas sleeping mat, a loincloth (fondly known by the men as a 'Jap-happy' or 'bollock bag'), a pair of shorts, wooden clog flip-flops and a prized tin mug. They were put on open railway cars and the embers from the old wood-burning locomotive flew out on to their bare skin and shirts, causing burns. After about 70 kilometres they came to the Wan Po Viaduct, a huge trestle bridge, 150 metres long, curving around a mountain about 100 metres above the valley below. They had to get off the train and very carefully walk across the unfinished bridge to board another train; this was a special torment, as the men had to step from sleeper to sleeper, with no support if they slipped, and there was a sheer drop to the valley beneath.

The next train was made up of cattle trucks and so they were able to sleep that night on the very slow journey north, eating the meagre rations they were given as they went. They passed many camps, each with their cemetery of Allied dead – Hintok, Takenun, Konkoita – until they reached Nikke-Nikke, from where they could see the hills of the Three Pagodas Pass, the border between Thailand and Burma. As they crossed the border and met some local people, David noticed a difference between the cheerful Thais and the more sullen Burmese. Eventually they arrived at their Burmese camp destination, but David says he has no idea what it was called – he had no memory of ever hearing its name, but in his debriefing notes he names it as 62 Kilo. Camp, where he stayed until April 1944.

The camp appeared to have been previously inhabited by coolies and the sanitation was dreadful; the work here was to maintain the railway (this was therefore possibly 1944, after it had been finished). David was sent to work in the blacksmith shop, where four POWs swung the hammer while the Japanese blacksmith held the white hot metal on the anvil. One day a yell went up when a huge green snake – a poisonous mamba – was seen in the shop, but they chased it away.

Another day a Japanese officer ordered David and a comrade to go and help lift a large piece of machinery; they could not manage it and David got a heavy kick on his body for this. The camp had no fence because it was in dense jungle; one day David was walking during his spare time and found a skull. He took it back to the camp, put it on a stick and stuck it in view of the Japanese guards' huts to cause a stir. The next morning a Japanese officer approached, took his sword and smashed the skull to pieces.

Soon after this, David began to feel shivery and he developed malaria, with alternate attacks of sweating and shivering; this went on every ten days, each bout lasting four days, and was accompanied by dysentery. David was dangerously ill. Soon afterwards, in April 1944, he was sent back by train with some other sick prisoners to the 'hospital' camp at Chungkai.

Chungkai Again

As David struggled to get off the train into the camp, he was met by an astonished Len Goldfarb, who was shocked at his old friend's condition. He got David to a bed and, as Len worked in the cookhouse, he brought him a plate of rice and chicken, which David promptly exuded because of his dysentery. Later David realized that he must have looked similar to the inmates of concentration camps in Germany.

(At this point, David confessed he had never told anyone about these experiences before, not even his family.)

The dysentery hut was large and airy but the smell was horrendous. About seventy men were in various stages of the illness, most in pain. Epsom salts to flush the guts was the only medicine; some men were dying. The doctor was a Canadian (David could not remember his name, but this was almost certainly Jacob Markovitch) and he came round about twice per week to ask how many poops the men had done; some said eight or nine, but one man said sixty. 'Sixty?' said Markovitch, 'Good God, man, there isn't that much s—t in the whole of Thailand!' David laughed and said he had never forgotten that remark, typical of Markovitch's humour. Soon David got better and was returned to his hut, but was still too weak to work. Len Goldfarb helped by bringing him food from the Japanese cookhouse, which was better than the POWs received. For his malaria, David was given a small amount of quinine which helped keep it under control.

One evening, David was woken by the sound of the bass drum of the camp orchestra and roars of laughter; two prisoners had been caught outside the camp trying to smuggle in local alcohol. Parading into the camp were the Japanese guards, followed by the camp band and then the two miscreants wearing sandwich boards on which the Japanese had written 'I bad mans. I bringing liquors in camps', to humiliate them. The reason for the huge amount of laughter from the POWs was that the band was playing 'Roll out the Barrel' – a joke which, of course, went over the heads of the Japanese.

In July 1944 David was moved back to Tamarkan.

Tamarkan

David's health improved slightly. At Tamarkan they were building a bridge; the camp itself was just like most of the camps on the railway, with several atap and bamboo huts, a parade ground and the Japanese huts nearby. The worksite was just 200 metres or so from the camp entrance. They were unloading steel and timber for the bridge and also building a cenotaph-style memorial in concrete for the Japanese and POWs who had died building the railway, which struck the POWs as amazingly bizarre, as men were still dying in the camps upcountry. The memorial can be seen to this day.

On one occasion the men were pile-driving a timber pole into the mud using a frame with a pulley on top, which was tied to a concrete slab; the slab was pulled up by about five POWs on a rope and dropped onto the pole to drive it in. The Japanese guard was perched on a rock and singing a rhythm in Japanese: 'One and two and three and Boomp'. After a while the men began singing to the same rhythm: 'Tojo's a bastard, a bastard, Boomp'. The guard slowly joined in this little ditty, mimicking the men in English, while the men roared with laughter; then the Japanese interpreter turned up and, realizing what they were singing, raged at the guard. He was taken away, but surprisingly nothing happened to David and his friends.

As usual, the camp canteen was controlled by the Japanese, who charged extortionate prices for items available more cheaply from local Thais. The money given to POWs for work was known as 'scrip', printed by the Japanese themselves, and was only valid in Thailand and pretty much worthless. So POWs confined to the camp owing to work fatigues or illness gave items to those who worked outside, to trade pens, rings, watches and so on with the Thais, who hid in the bushes waiting to carry out these transactions. As such trade was severely punished, the dealers charged the other POWs 10 per cent for the risk. Some guards turned a blind eye but others did not.

One day a large group of Australian sick arrived in the camp from Burma with Burmese scrip money, useless in the (Thai) camp. One day a Japanese guard pulled David aside and quietly said that he wanted to buy Burmese scrip using Thai scrip – could David arrange it at seventy cents Thai for one dollar Burmese? So David told the Aussies, who were overjoyed to get rid of Burmese scrip. David gave them sixty-five cents and kept five cents as profit; the Japanese guard got his Burmese scrip, the Aussies got their Thai scrip to spend, and David got his cut!

After two weeks of this the guard became nervous, as he heard that the Kempeitai (Japanese Gestapo-type police) would search the camp soon and find he had a lot of Burmese money. He asked whether the Australians would hide it for him, as it would be expected that they would have Burmese money; he also got David a light job in the camp away from the bridge. But the police never came, the guard got his money back, and everyone was happy: the Aussies bought food and stayed alive, David bought food and stayed alive – and the Japanese guard? He became a rich man because his angle was that he was selling the Burmese money at inflated rates to Japanese soldiers going into Burma!

David's job in the camp was now filling hot water tubs for the Japanese officers. But one night all the men were turned out of their beds by the guards, who went along each man shining a torch on their feet. Two men seen outside the camp had run back in through the mud, so the guards were looking for the muddy feet. They did not find them and so the whole camp was threatened with being made to stand up all night unless the two owned up. They did so; David's friend Bill Palfrey was one of them and he later explained that they had washed their feet on entering the camp, knowing what the Japanese would look for. The two men were taken away and beaten up, but returned a week later relatively unharmed. The Japanese commandant, however, was seen later wearing the watch that Bill had gone to sell in the village.

Nakom Paton

This camp was quite well run but the work was hard. David was here from November 1944 to January 1945. They had to build a high bank of earth around the camp (this was where the incident with Moonface and the shovel took place), but the worst aspect was the outbreak of the killer disease of cholera. Despite the infectious nature of this illness, David did not catch it and does not know why he was somehow saved from its clutches. Many times he had to build the funeral pyres to burn the bodies of dead comrades: the smell and the moving bodies as they burned – it was horror upon horror.

In early 1945, David put his hand on a scorpion he had not seen sitting on the wood pile used for heating the hot tubs, and this put him out of action for the day, as he was in intense pain. Some days later, waiting at evening roll call, the men heard a sound not heard for years – Allied aircraft. They began strafing the bridge and its ack-ack guns, and all the POWs dived for the anti-flood ditches around the huts. Bombs hit the bridge nearby but two fell short and hit some POW huts; several men were killed and Len Goldfarb was injured. Len was amused; the first time he had been wounded was by the Japanese in Singapore. Now the Allies had hit him!

Bangkok

In January 1945, David and comrades were moved to a godown (dockside warehouse) in Bangkok – Harbour camp. They were loading ammunition barges and digging gun emplacements; at least the concrete floors of the godowns were easy to keep clean. Local Thais told them that if anything 'happened' they were to flee to the Thai Naval Academy along the waterfront; this remark puzzled the men because the POWs did not realize at the time that the Thais had caught wind of the Japanese plan to massacre the POWs if the Allies invaded.

Then one day there was no work, nor again the next day, or the next. Then a notice was put up: the war was over; a huge bomb had been dropped on Japan. Everyone was stunned. After a few days, David was put in charge of a party to go and dig latrines at the Race Course for huge numbers of POWs being brought from upcountry to Bangkok, on their way home via Rangoon. The food and accommodation of David's group were hugely improved and they even went into town in the evenings and had tea at a café run by two White Russian émigrés. Soon David was on a Dakota to Rangoon as well, and went straight into hospital to recuperate.

David emigrated to Hamilton, Ontario in 1954.

(*At this point, David wrote that the telling of his story was beginning to give him nightmares and he wanted to stop.*)

Note

1. From Norfolk Regiment. In March 2007, when David was relating his story to me by email from Canada where he lived (at St Catharine's, near Niagara Falls), I was able to put him in touch with Len, for the first time in sixty-two years. They spoke for ages on the telephone. Len's testimony is included in this book.

2092875 Cpl/Dvr Nathan Emmanuel Robert 'Bobby' Woolf, RE/RCOS, 11th Indian Division

66. 2092875 Cpl Nathan Emmanuel 'Bobby' Woolf, RE/RCOS.

Bobby was born in Streatham, London, on 16 September 1916 and was in the radio retail business. Later he lived at 2 Ridgemount, Ridge Road, Hampstead. He volunteered on 5 July 1939 into the Territorial Army, RE, and on 10 July 1940 was transferred to the 4th Air Formation Signals, RCOS, at Bletchley. On 22 October he was promoted to corporal and on 3 January 1941 he embarked for Malaya arriving on 11 March 1941.

Just before the surrender, Bobby was engaged in the fighting at Oxley Rise in Singapore, and was sent to Changi gaol, where like many others he suffered all the horrors of starvation diet and disease. On one occasion he was punished by Japanese guards for some minor misdemeanour by being suspended by his thumbs for several hours. From 13 March 1942 he was sent to River Valley camp and then to Tamjonruh camp on 3 October, and was there until 26 October. In August 1942 he was transported on one of the Hell Ships to Formosa, and kept in battened-down, crowded holds with hundreds of others for weeks on end.

In the horror camp of Heito (Formosa), Bobby was tortured again, this time in the notorious 'sweat boxes' – wooden boxes with mesh wire on one side and exposed to the intense sun. For six weeks he endured this, living on small amounts of rice and tea flung at him by his guards. He alleged that he was the only POW to survive this torture in his camp, and also that one Japanese guard said he would give him extra food because he too was Jewish. However, it has been impossible to verify this.[1]

Later (14 March 1945) he was used as slave labour in a copper mine at Taihoku, beginning at 5 a.m. with a five-mile walk and then going down 2,000 steps to the mine itself, then heavy manual work on starvation rations before climbing up the steps and walking the five miles back to camp, arriving at 7 p.m.

He was liberated from Taihoku by US Marines on 6 September 1945. He was one of the early arrivals in the UK in September 1945, travelling via Manila. After the war he married Pamela Chetham at the Golders Green synagogue, and they had a family. Bobby was awarded the 1939–45 and Pacific stars and the Defence and War medals.

Note

1. For a full picture of the awful conditions in these mines, see John McEwan, *Out of the Depths of Hell* (Barnsley: Pen and Sword, 2005).

Short Testimonies

Lt Alfred Gustave Herbert Bachrach, Dutch Forces Officer

67. Lt Alfred Bachrach, Dutch Forces, from his obituary in *The Times*, 2010.

Alfred was born in Frankfurt on 9 December 1914; his mother was German and his father was French (a professor of French at Frankfurt University). The family moved to Amsterdam soon after Alfred's birth and he took his degree there; he spoke German, Dutch and English. He married Winifred MacManus in 1939, and in 1941 he became English master at Samarang Grammar school in Sumatra. A keen sailor, he explored the waters of the Dutch East Indies, and then became a liaison officer for the Dutch Army Special Forces.

He was interned by the Japanese, and so were his wife and baby son, in a separate camp; his son died in captivity. Enduring six camps, he was finally sent to Changi in Singapore. Allowed one 'holy' book by the Japanese, he chose a complete Shakespeare, which he used for teaching in the camp. His students so valued the book that when it was confiscated by the Japanese as a punishment, one man risked his life and stole it back.

When he was liberated, Alfred weighed just six stone, but he remained in the Dutch spy service in Batavia/Jakarta, Singapore and Saigon until 1947, when he went to work in The Hague Overseas Ministry. After the war he was an English academic in different universities worldwide and was both a British CBE and holder of the Dutch Knight of the Order of the Netherlands Lion. In 1988 he moved to live in Twickenham. He married three times, had three children and died in 2010.

3534885 Pte Jack Berner, Loyal Regt

Jack was from Manchester and belonged to the JLB (Jewish Lads Brigade) and Jewish Working Men's Club; married to Anne née Goldstone, he lived at 193, Great Cheetham Street West, Salford. He was called up on 30 May 1940 in Preston and after eight months of marriage he was posted to the Far East in 1941 and was captured in Singapore. He testified to the regular Jewish services held at Kanchanaburi by David Arkush, while he was working on the Burma Railway. When being transported by Japanese ship to labour in

68. 3534885 Pte Jack Berner, Loyal Regt. From a wartime issue of the *Jewish Chronicle*.

Japan, he fell ill on the stopover at Manila and was left in a camp there. From that camp he was eventually liberated by the US Army in August 1945. He weighed just six stone and six pounds.

On an American troopship home, he celebrated the Passover Seder meal with 300 American soldiers, a moment he never forgot.

6845454 Jack Bloom (aka John Baker), KRRC/RCOS

69. 6845454 Sgmn Jack Bloom (aka John Baker), KRRC/RCOS.

Jack was born in London and enlisted in April 1936 and served on the North-West Frontier of India; he was captured in Singapore and died aged 46 in 1964, from the effects of his captivity. His son Avi, who lives in Israel, sent us a photo of his medals: the 1939–45 and Pacific Stars, War and Defence Medals and the rare NW Frontier India Medal. His family know nothing about his time as a POW. He was captured in Singapore and sent to Changi and then by train to work on the Burma Railway. His main camp was Tarsao but he was also a POW at Wampo, Hintok, Tamarkan and finally Nakhon Nayok, from where he was liberated on 14 August 1945. This is published as a tribute to him.

Yahya Cohen, Singapore

70. Yahya Cohen, Singapore.

[*This account is taken from a synopsis published by the Singapore National Archives, and from information available on the Internet.*]

Yahya was born on 1 April 1920 in Singapore; his father was from Yemen and mother from Iraq. From a very young age he wanted to be a doctor, and he aimed for this during his time at St Andrew's School.

Yahya's parents moved in the 1930s from Singapore to Sandakan (Borneo/Sabah) to start up a business as general merchants. They were arrested there by the Japanese and their ship was bombed while they were en route to Kuching to be interned. They were both killed. The aircraft might have been American, but it is not known for sure.

In Singapore, Yahya studied medicine, but this was interrupted by the war. He witnessed the display of decapitated heads on spikes after the Japanese conquest;

many of his Chinese friends were murdered, though he himself did not suffer any cruelty at the hands of the Japanese. He was one of the 100 or so Jewish men arrested first by the Japanese, at Changi, where the Jewish quarters in an old rice store were known as 'Aldgate'; he actually welcomed internment, as it brought recognition that they were on the side of the British. Eventually he was moved to Sime Road camp. When the Japanese surrendered, the internees maintained order and did not express their joy openly, in case the Japanese were goaded into taking revenge on them.

After the war, Yahya studied in the UK and Singapore and became a famous surgeon, hugely admired by his peers. He married Nina (an Israeli engineer) in 1965, after meeting her when he was a visiting professor at the University of Tel-Aviv in Israel, a place he often travelled to. After his death in February 2003, a series of memorial lectures was established at the University of Singapore in recognition of his work.

7603532 Pte Claude Maurice Dicks, REME, TD

71. 7603532 Pte Claude Dicks, REME.

Claude was born on 28 July 1920. Prior to volunteering on 30 August 1939, aged 19 years, when living at 33 Keyes Road, Cricklewood, he worked as an instrument maker at Avro Aircraft on Lancaster bombers. Posted to the Far East with REME, he sailed on the SS *Empress of India* to Bombay and was then sent to Singapore, where the ship, as it arrived, was dive-bombed by Japanese aircraft and many men were killed and drowned; he was captured on the island.

Captured in February, he was soon sent to Changi and then Thompson Road camp in May 1942, and finally transported in December 1942 to the Death Railway in Thailand, enduring horrific conditions as a slave labourer. Among his acts of sabotage was disabling many vehicles that the Japanese had captured from the British. He was then sent back to Changi, where his essential skills were used to make surgical instruments and artificial limbs; although he was using the most primitive of materials, they worked well and many lives were saved or improved. Claude particularly remembered the lectures, debates and courses run by numerous gifted prisoners from all walks of life, to keep minds active; he described it as a kind of university.

When he finally returned home, he was so ill that he had to spend months recovering at Queen Mary's Hospital, Roehampton. He died in 1987.

T/198549 Dvr Eli Fine, RASC, att. 35th, 77th LAA

Eli Fine was son of Daniel and Lena, and later the husband of Annie Cecilia Fine of 21 Mostyn Avenue, Llandudno. He enlisted on 29 April 1940 and was posted to Singapore, arriving on 9 November 1941. He was captured at the fall of Singapore and imprisoned at Changi gaol.

72. T198549 Dvr Eli Fine RASC.

It is believed that on 23 March 1943 he was sent from Changi with 'D' party, consisting of about 5,000 men, in nine train loads, to Thailand to work on the Death Railway. He returned to Singapore on 20 June 1944 and was put into River Valley camp. It appears that from here he was sent on 4 September, probably to Japan with about 2,000 Australian and British POWs on two ships. His ship was sunk on 12 September and he was killed, aged 32. His name is engraved on the Kranji memorial, column 99, in Singapore.

However, another report states that he was captured in Java and then died of severe malnutrition and illness aboard the Japanese prison ship *Haruyoshi Maru* while being sent from the island of Ambon (where they were building an airstrip) back to Java. He was allegedly buried at sea.

Eli was awarded the Pacific and 1939–45 Stars and War Medal. His daughter Sylvia Kramrisch, who lived in Manchester, never knew her father because she was a baby when he was sent overseas. In 1988 she met a group of Welsh Jewish expatriates in Jerusalem, and she discovered that she had a large family in Cardiff, who told her about her father. Her uncle, Nat Fine, who was Eli's brother and in the army, was also captured by the Japanese and was the last to see his brother alive. His testimony follows, and the information he gives could mean that Eli was back at Changi, and not at River Valley camp as was thought, when he and Nat met, or that Eli was temporarily working in the cookhouse at Changi. As these testimonies show, there was sometimes confusion in the 'fog of war' as to exactly what happened, and when, to those POWs who never came home.

1450601 Nathan/Moss Fine

Nathan was the brother of Eli (above). He wrote:

> I was captured in Java in March 1942. The first six months were bearable as we had a Japanese Commandant who had lived in Britain, but later when we were sent to work on the Death railway in Thailand, it was two years of hell! Then we were brought to Changi, and there on a notice board I saw a list of prisoners and my brother's name Eli Fine was there. I asked and was told he was in the cookhouse and there I found him making rice rissoles. In tears we embraced and he and his friends gave me some food. We were receiving the BBC news on a clandestine radio and I told him the war was being won and he was not to volunteer for any drafts. We met every day for two weeks and then one day he disappeared, shipped away.

Before he died, Nat told his close friend Norman Berg how he used to crawl under the fence of his camp at night and trade or beg for food from locals, until he was caught; he then shared this among his friends and the sick. Tied to a post, he was flogged by the Japanese in front of the parade. After recovering, he continued his night time forays to help others.

1077842 Gnr/Dvr Daniel 'Danny' Goldberg (aka Morgan), 5th Field Regt, RA

73a. 1077842 Gnr/Dvr Daniel Goldberg (aka Morgan), RA.

Danny lived at 24 Darenth Road, Stamford Hill, and 24 Pentland House, Lynmouth Road, London N16. He was the husband of Sadie and father of Brian. He was a tailor by trade and called up on 12 September 1940. His archive at the AJEX Jewish Military Museum shows that he sailed to the Far East via Cape Town in November 1941, arriving in Malaya in early 1942. By March 1943 he was a POW in Bamboo Camp, Taiwan, and was there for much of the war. A surviving item in his archive shows a programme for a play put on by the POWs dated 12 March 1943 – John Galsworthy's *The Skin Game* – in which Danny played the agent Hornblower. His many letters and cards home stated that he was the camp tailor. He apparently received at least twenty-four letters from home during his captivity – he mentioned this in one card – and wrote to his wife that he reread them all every Sunday and they were 'as sweet as sugar can be' for him.

On liberation, in September 1945, he was taken to Melbourne (18 September) and then to Victoria (Gordonhead Camp), British Columbia, on the SS *Gosper*, then across the USA and Canada by train to Nova Scotia (19 October), and he sailed to the UK on the USS *Santee*.

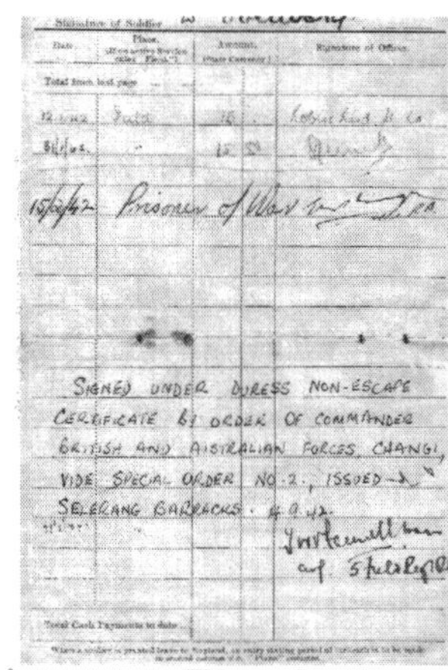

73b. His archive includes rare POW cards sent to his family and the very rare 'Selarang non-escape chit' which survived in his records.

1497683 L/Bdr Henry Irwin Greenwood (aka Greenburgh), 5th Searchlight, 13th Battery

Henry was born on 4 August 1918, son of David and Jennie of 98, Brick Lane, London E1, and later of 42 Mitre House, Western Road, Brighton. Henry enlisted on 18 July 1939, one of the first to join up.

He was captured in Singapore and was in Changi until 9 April 1942, from where he was sent to Saigon on the Hell Ships, until 26 June 1943, to labour in the docks. While here, he participated in sabotage including putting sand in aviation fuel, removing nuts and bolts from aircraft, and damaging rice stocks meant for the Japanese troops by flooding the carriages taking the rice to the front.

From here he was sent to Kinsaiyok on 1 July to work on the Death Railway until 10 September, and then to the hospital at Chungkai until 24 April 1944. He was marched to Tamuang where he stayed until 3 June 1945, and then had a short stint at the 'Rice Field' camp until 31 July 1945. Finally he was sent to Nakom Nai, where he remained until liberated on 2 September 1945.

Frederick Jacob Isaacs

[*This account (like Yahya Cohen's) is taken from a synopsis published by the Singapore National Archives.*]

Frederick was an optician in Singapore, and although at first he was not harassed, he was eventually arrested by the Japanese Kempeitai and kept at the Cockpit Hotel; after two days he was released and sent to Changi prison. One of the guards was a Japanese who once made spectacle frames for Frederick and he saw that he was treated well.

At Changi he took part in Jewish religious services. In early 1942 he was sent to Sime Road camp. Food was poor here but Jews were not harassed in any way and were allowed to hold their services as usual.

LAC Abraham 'Kiki' Kissin/Kissen, RAF, 211 Squadron

Abraham was the last surviving Israeli POW of the Japanese when he died. Married to Zelda, he was posted to Java in January 1942, and was captured there at Tasik Malaya the following month. His first prison was in a converted school at Surabaya, from where he was sent to labour on road and airfield construction at Samarang.

After about ten months, Abraham was one of about 1,500 men herded onto a ship and taken to Haruku Island in the Ambon archipelago; here he suffered appalling torture, illness and forced labour, and saw many of his comrades die; some were killed by American friendly fire from bombers. From here they spent weeks on a ship sailing between islands until they reached Cycle Camp in Batavia. In early 1945 they were taken to Bandung with the intention of finally being shipped to Japan, but the military situation was such that this did not occur and it was here in August 1945 that he was liberated.

1766478 L/Bdr Walter Levy, RA

Walter was born on 12 May 1916, son of Abram and Tillie (née Tytz) of 24 Braemar Street, Greenock, Scotland, and husband of Lavinia Levy of 114 Kirkaldy Road, Glasgow. He was imprisoned at Fukuoka 4 camp, Japan, with Captain Alan Berkeley. As Alan testified, Walter was one of his medical orderlies, responsible for saving many lives by stealing, at risk of execution, valuable medicines from the Japanese clinic in the camp. His story is thus similar to that of Captain Berkeley, but he never spoke of these things, and so his family know little more. He was captured in Singapore in February 1942 and reached home via Manila after liberation. He died in 1993.

1757667 Gnr Nathan Lipman, 48th LAA, RA

74. 1757667 Gnr Nathan Lipman, RA.

Nathan was born on 17 April 1910, probably in the East End of London, son of Solomon and Fanny (Russian Jewish immigrants), and later the husband of Doris of Highbury, London. Nathan was one of four children; his father was killed in a German bombing raid when the family's fishmonger business at 23 York Way, Kings Cross, was hit on 12 January 1941. Nathan was a black-cab driver until he was called up.

Nathan was posted to Singapore, where he was captured on board a ship as it was evacuating troops. Family history says he was on the Burma Railway; it is known that he was sent to Japan at some stage, and his daughter Ros says her research shows that he was at Motoyama camp, working in a coal mine, when he died on 10 December 1942. He is buried at Yokohama in Japan.

3321222 Pte William Morrison, 2nd Battalion Gordon Highlanders

75. 3321222 Pte William Morrison, Gordon Highlanders.

William was born on 13 December 1916, son of Benjamin and Bessie Morrison of 76 Abbotsford Road, Glasgow. He was called up on 15 March 1940 in Glasgow, where he still lived, and captured in Singapore in February 1942. He was first sent to Changi, but in November 1942 was transported to Chungkai in Thailand to work on the notorious Burma Railway. He stayed there until June 1944 and then was sent back to Singapore to be transported on the Hell Ship *Hofuku Maru* to Japan. However, this ship was sunk by American planes (operating off a carrier) off the Philippines on 28 September. William was rescued and taken by the Japanese to Taiwan on 9 November 1944 and held in Heito camp. He was sent by ship to Japan on 28 January 1945, to Sendai 3B, and was finally liberated there in August 1945. After the war he married Janet/June Cohen on 10 September 1947 and became a buyer for Levy brothers in Glasgow. His children were Valerie and Colin (sadly, Colin died aged 6 years).

[*The author thanks to Stewart Mitchell for his research at TNA and the Gordon Highlanders Museum, Aberdeen, Scotland.*]

1286023 LAC2 Joseph Simmonds, RAF

Joseph was born in Bow, London, on 30 May 1914, and later lived at 23 Shore Road, Hackney, London; he was called up in August 1941. Captured in Singapore, he was sent to Changi and helped build the synagogue there in 1944. On one occasion he was incarcerated in a roofless, tiny cell under the intense sun for days for some misdemeanour, and was threatened with execution, but he was saved at the last minute by his officer paying homage to the emperor.

At risk to his life, Joseph kept a secret diary at Changi, with the names of all the POWs, crossing them out as they died of illness, hunger and beatings. He was also a friend of the famous war artist, Ronald Searle. Later he was sent upcountry to work on the Death Railway.

On arriving home after liberation, he discovered that his parents and young brother had been killed in the bombing of London, in Ilford in 1943, but he successfully continued the family scrap metal business from Westcliff after the war. He died in 1993, survived by his wife Winnie and his son.

5573577 Cpl Louis Saunders, 4th Royal Norfolk Regiment

76. 5573577 Cpl Louis Saunders, 4th Norfolks.

Louis was born in London on 3 December 1919, son of Simon and Annie Saunders, and was a prominent member of the Oxford and St George Jewish youth club in Stepney, where he played for the soccer team. At his club he several times won the Mocatta prize for best apprentice. He was quickly promoted in the army to corporal.

He was captured in Singapore and was a POW with Syd Shulman (Shulman's testimony is below). He was transported on a Hell Ship in September 1944 and killed at sea when the ship was sunk on 21 September. In an obituary in the club magazine *Fratres* (vol. 32, July 1946) Syd Shulman wrote that Louis's conduct was always superb as a prisoner; he bore himself with courage and fortitude, rarely complaining, always helpful. Syd had first met him on a forced jungle march in Siam; he arrived at the destination two hours late, staggering in after dark with his kit and carrying that of a weaker comrade as well. Louis had refused to leave his comrade when he fell, and 'brought him in'. Louis is remembered on the Singapore memorial, column 48, as he has no known grave.

Louis had been a reliable and hard working soldier, much respected by his men.

Zaida Short

[*This information is taken from the book* Women beyond the Wire, *by L. Warner and J. Sandilands (London: Michael Joseph, 1982).*]

Zaida and her sister Hilda were born into the great Jewish community of Baghdad, but were taken to live in Singapore, where they were interned after the Japanese invasion. Zaida's husband was a British Army staff sergeant who died on the Death Railway; her brother was tortured to death. She was a solicitor, and in the camps she was head of the rationing committee. She was entrusted with the very dangerous and secretive task of selling the valuables of other internees to Japanese guards and using the money to buy food and medicines which kept people alive; discovery would have meant execution. After the war she again became a successful solicitor.

2889889 Pte Sydney Solomon Shulman, 2nd Battalion, Gordon Highlanders

Sydney was born in London on 27 July 1908, and was later, husband of Mrs S. Shulman of 68 Stepney Green, London E1. He was called up on 31 March 1941, arriving in Singapore on 28 November. He was captured and sent to Changi in February 1942, but from November 1942 to June 1944 he was in various camps on the Burma Railway. Returned to Singapore, he was sent on the Hell Ships to Amagasaki in Japan in September 1944 and was there until May 1945, when he was transferred to Nagoya 6 camp. Here he was liberated in September 1945.

At his debriefing, Sydney said that he and his comrades continually put large planing machines out of action at Otami Industries, Amagasaki, purposely broke tools that could not be replaced and put water in the oil. At Nagoya he took part in extensive sabotage of Japanese planning machines, tools, putting water in oil drums, and mixing metals in a furnace so pourings were all spoilt. These were very dangerous acts, punishable by death if he had been caught.

[The author thanks to Stewart Mitchell for his research at TNA and the Gordon Highlanders Museum, Aberdeen, Scotland.]

2359837 Dvr Lewis Singer, RCOS

Lewis was born on 20 November 1907 at 153 Backchurch Lane, Whitechapel, London, son of Harry and Fay Singer, and later the husband of Sadie of Hackney, and the father of Leon. He was a taxi driver and as a result of this he served as his colonel's personal driver. After capture at Singapore, and imprisonment in Changi, he was taken upcountry in May 1943 to work on the Death Railway as part of the 3,000 strong 'H' Force. He worked at the notorious Hell Fire Pass. When the railway was finished, he was evacuated south to Kanchanaburi Camp. Lewis died of beri-beri on 18 October 1943, aged 36 years. He is buried at Kanchanaburi.

His family only received from him two of the standard Japanese postcards that were allowed, saying that he 'was fine' and 'he was OK'.

7612479 Jack Smith, RAOC

See entry below for Alfred Tropp.

2194505 Alfred Tropp, RE, and 7612479 Jack Smith, RAOC

Alf was from the East End, and Jack was son of Mrs I. Smith of 10 Abbotsford Place, Glasgow. They were serving in Hong Kong in 1940; they met when Jack walked into a barracks and saw some *Jewish Chronicle* newspapers on Alf's bunk, and they became friends. Before the Japanese attack, life in Hong Kong was good, with swimming, parties, outings and the hospitality of the local Jewish Community. But by March 1942 they had both been captured and interned at Sham Shui Po Camp.

At first conditions were bearable, and Jewish services were held, led by Jack with his small siddur (prayer book) for the small group of Jewish POWs. But conditions soon worsened: the starvation diet, long hours of heavy labour, disease, and beatings for the slightest or no reason by the Japanese and Korean guards soon took their toll and many died, mainly of diphtheria and dysentery.

In about late 1942, despite being ill with pellagra, Alf was ordered to prepare to leave for Japan, and as a parting memento he gave Jack an old twenty-first birthday card he had received. He was then sent by one of the notorious Hell Ships – in his case it was a former coal transporter – to Japan. Covered in coal dust, the sweating bodies crammed into the boat reminded Alf of an old slave ship. For three long years he toiled in Japanese Mainland camps, ridden with lice and suffering appalling conditions, until liberation in August 1945. He was taken to Manila by the US Navy and remarkably, after hearing about some British POWs in another tent, he met Jack again. By November they were both home but lost touch for many years. In 1985, however, they met again in London, where they had both been living, unbeknown to each other.

They both took part in a commemoration return to Hong Kong soon after this, with other veteran POWs; the site of the camp was now a park and had a plaque describing what once had stood there; they took part in radio programmes and receptions. They always attended the AJEX parades in Whitehall as POW veterans.

[*This information is taken from an article in the* Jewish Chronicle, *10 November 1989.*]

Mischa Warmen

Mischa was a 4-year-old boy, son of Russian Jewish refugees living in Shanghai who had made their way to Singapore when the Japanese invaded. They were aboard the doomed

SS *Vyner Brooke* when it was sunk trying to leave Singapore in February 1942. When the bombs hit the boat, the father was drowned and his mother collapsed. Another passenger, Eric Germann, calmly told Mischa they were going for a swim, so Mischa gripped Eric around the neck and they both jumped into the sea. They found a lifeboat and were saved. The survivors were interned in Irenelaan camp, Palembang, Sumatra, and other camps later on. Mischa's mother died of pneumonia, leaving him orphaned. Mischa was adopted by several British women prisoners and lovingly cared for. At the end of the war, Mischa's relatives in Shanghia were found and he went to live with them. When driving around, he kept asking where the wire fence was, as he had no memory of life outside the wire.

[*This information is taken from Warner and Sandilands,* Women beyond the Wire.]

1080758 AC1 Arthur Reginald Warshawsky, RAF aka Shaw

Arthur was born on 7 July 1920 in Great Titchfield Street, London, W1, son of Rachel née Abrahams and Adolf (Polish/Russian Jewish immigrants). Adolf was a master tailor. Arthur attended Upper Marylebone Elementary School in New Cavendish Street. Then they moved to Cricklewood (185 Fordwych Road) in 1928 and he attended Broomsleigh Street School in West Hampstead and then he won a scholarship to St Marylebone Grammar School. After leaving school in 1936, he worked at the estate agent, William Whitely's.

Arthur was called up into the RAF in 1940 and did his initial training in Blackpool, then No. 1 Wireless School in ? Cransworth. His first posting was to Croydon where he did some flying as a wireless operator, but his proper role was ground-based wireless. In early December 1941 Arthur was posted to West Kirby (Wirral) to prepare to go overseas to North Africa. On the day they set sail from the Clyde on the SS *Esperance Bay*, Pearl Harbor was attacked and, on changing ships in Durban, they were told they were going to the Far East. He landed in Singapore just as the Japanese were on the causeway and landing on Singapore Island. As a result, his ship was diverted straight on to Java, landing in Batavia. Aircraft on board in packing cases were thrown into the sea as there was no chance of assembling them.

On Batavia he was one of hundreds of Allied personnel fleeing east to Sumatra, but the Japanese had got there already so they went back to Batavia and then to a nearby aerodrome at ? Andea to set up a listening post to receive signals from London to pass on to senior RAF officers. After a couple of weeks, they were ordered to destroy all fuel and radio equipment and Arthur and a few others grabbed a jeep with rifles and ammunition and made for Tilapjap port in the south east, to try to escape on an Allied ship. They saw the last one leave and then go down as Japanese aircraft sank it.

Arthur has no memory of exactly how they were captured, except that they were marched to a camp at Semplak and in August to Makasura. In January 1943 they were sent by ship to Singapore in appalling conditions with no food, drink or sanitary facilities. Eventually they were taken to Sime Road camp and began working in the docks for the Japanese, mostly loading tin ingots, on a starvation diet. He met the Dutch Army rabbi

Chaim Nussbaum in Singapore, and attended Friday evening services. He also met David Arkush there but did not see him again until after the war.

The guards were very brutal and if they saw any trading between prisoners and locals of cigarettes or watches going on through the wire, in order to get extra food, Arthur witnessed how prisoners, locals and even the odd lax Japanese guard, would be subjected to beatings with rifle butts until the victim could not stand up.

In April 1943, Arthur was sent to Bampong with H Force by cattle truck train in terrible conditions to work on the Death Railway. With him was a fellow Jewish RAF Medical Orderly, who Arthur would not name, but who died in the first camp and whom he helped to bury; Arthur clearly remembers saying the 'Kaddish' prayer for him; they had been together since West Kirby. He said his worst experience was when he came home and had to visit the man's parents to tell them that he had died.

He was soon sent to Hintok and work on the railway began in earnest. Like most prisoners Arthur had malaria most of the time, but he saw many military and native workers die of cholera, something he thankfully avoided. For such victims, he saw the huge pits into which bodies were thrown en masse. He also remembers the nightmare of seeing sick men with huge leg ulcers being dragged to work from the sick bay when they were obviously incapable, due to the Japanese demand for labour, especially when the notorious 'Speedo' was on – the last, urgent weeks when the Japanese were determined to get the Railway connected up and working at any cost. As best they could, the fitter men like Arthur tried to help the sick men with their tasks, but often to no avail, and beatings by the guards were frequent and often murderous.

On one occasion, whilst on a break on a railway cutting, a Japanese guard started up a conversation with the men in Arthur's group, about the future of the world. In the sand he drew a circle and said this part of the world would be controlled by Japan; then another circle, which would be controlled by Hitler.

Arthur spoke highly of the superb work done by the POW doctors to save lives, often with virtually no drugs and equipment, but just using ingenuity and fortitude. For him, the worst thing was the endless slavery, with no knowing when it was going to end. Only towards the end did he hear news from the secret radios kept in some of the camps.

Some time in December 1943 (Arthur recalls Japanese troops were being moved north in great numbers to Burma to join the great battle there with the British and Indian forces), he was sent south by train back to Changi gaol; he also recalls a rehabilitation camp on the way (this may have been Chungkai) where they stayed for a short time and did not have to work. At Changi he was actually in a cell, but outside they were set to work building the airport near the camp. At some point he was lucky to be given a job in the kitchen where it was always possible to acquire a bit of extra food while working. But if you were caught – and the temptation to steal was very great – it was a thorough beating by either the Japanese or even your fellow prisoners. Although not forced to work outside, shifts in the kitchen were very long and he remembers the huge amount of work to prepare a Christmas meal in December 1945 for thousands of men. It consisted mostly of rice presented in a hundred different ways!

At no time does Arthur recall ever receiving any post or being able to send post home. It was only in 1944 that his family found out he was still alive when his aunt heard a list of prisoners read out on a radio programme in London, broadcast from Australia.

By May 1945, the prisoners knew the war in Europe had ended in victory, as secret radios were in use. Suddenly one day in August some large Allied aircraft dropped huge containers of food and clothes over Changi airport, followed soon after by paratroopers – and then next day Lord Mountbatten appeared with his wife in a jeep. The Japanese guards had meanwhile disappeared. Soon after many of the men left the camp, walked to the port and went aboard British ships. Arthur remembers going aboard a submarine and having his first bread and jam for four and a half years. They also saw the Japanese being rounded up by Allied troops.

Soon Arthur was on his way home on the SS *Tegelberg*, with new shorts and shirt and good food; in Egypt they were re-equipped with warmer clothing for a British winter and at Gibraltar received their first letters from home. One lasting legacy was a recurrence of malaria for several years after, which was treated at the Hospital for Tropical Diseases in London. Arthur shut out the memory of his terrible experiences for many years and it was only in the late 1990s that he started to talk about it to his wife, son and daughter. This interview in September 2004 was the first he had ever done. Asked if he ever wanted to visit the places he had been in, he said categorically not. Whilst out of respect he belonged to the FEPOW Association, he never wanted to attend reunions; to him it was over and done. Whilst he was a member of AJEX, he did not attend the annual Parade in Whitehall for many years. His medals are the 1939–45 and Pacific Stars, and the War Medal, but he thinks his late wife threw them away in disgust at the visit to the UK of the Japanese Emperor.

1263122 LAC Louis R. Wiener/Weiner (aka Wilton), RAF

Louis was born on 1 September 1914 and lived at 1 Wargrave Avenue, Stamford Hill, London N15. Louis volunteered for the RAF in December 1939 and served in Radio Location, being stationed at Filby. He was posted to Singapore in August 1941. As a specialist, he was evacuated on the *Empire Star* from Singapore as the island surrendered, and despite bombing by Japanese aircraft he reached Bandoeng in Java. Within weeks he was captured at Semplak as the Dutch forces capitulated, and was held in the camp there until September 1942 when he was moved to Makasura until January 1943 and then on to Tanjong Priok Camp. Because he had dysentery he was not sent to work in Burma on the railway; he then developed beri-beri, which made him blind for three months. In April 1943 he was sent to and treated by a doctor at Batavia St Vincetus Hospital, where he worked as an orderly on his recovery. In September 1943 he was sent to Cycle Camp where he stayed until November of that year, and was then moved to Penkalan Camp in Bali.

He was sent by ship to Changi gaol in June 1945; the conditions onboard were horrendous, as the men were all locked in the hold for three days. At Changi, synagogue services were held at in the camp by Jewish POWs – he remembered them clearly. He said that just before the Japanese surrender, the Japanese threatened to gather all the POWs and force them inside tunnels that they had been forced to dig, and told them that they would be blown up inside if the Allies invaded. As it happened, on the day of the surrender, Louis was in the hospital with malaria and was liberated there and sent to India to recover. He then went via Tel-Aviv to England, arriving home in October 1945.

Australian Short Stories

Sydney historian Russell Stern compiled research in 2007 on Australian POWs, some of which was published in the Australian Jewish News, *20 April 2007. Other information on Australian Jewish POWs can be found on the National Archives of Australia website, where digitalized service records are on view for no charge, and include photographs of the men.*

NX38958 L/Sgt Evan/Evsick Altshuler

77. NX38958 L/Sgt Evan/Evsick Altshuler, RAAOC.

Evan/Evsick was son of Morris and Deborah, of 698 Old South Head Road, Bondi, Sydney, and was in the RAAOC, attached to the AIF 8th Division. When captured at Singapore he was sent to Changi gaol and then later sent to slave on the Burma Railway. Sometime in 1944 he was shipped to Japan on the Hell Ships, but his ship was attacked and sunk en route by a US submarine.

His niece, Mrs Sunny Gold of Sydney, says that two witnesses were with him in the water, at first clinging to some wreckage, and they saw Evan swimming off to rescue another man who was in difficulties; they then lost sight of him. The two men visited Evan's mother after being rescued, to tell her the story, but Evan never came home and is presumed to have died at sea on 14 September 1944, aged 26 years. He is remembered on the Labuan memorial on panel 27.

VX65015 Pte Leon Alfred Berliner

Leon was born on 24 June 1920, son of Alfred and Francis of Brighton, Victoria. He was with 2/21 Battalion and captured at Ambon. He was executed on 15 February 1942 at Laha aerodrome. He is remembered on the Ambon memorial.

0329 Wing Comm. Reginald Henry Saville Davis, OBE (Mil.), MiD, RAAF

Reginald was born on 8 April 1912 in Sydney. He joined up in 1933 and was a squadron leader all over Southeast Asia during the start of the war against Japan. As CO of No. 1 Squadron in Malaya, he took off on 8 December 1941, commanding seven Hudsons, to attack approaching Japanese warships shelling the coastal defences. They destroyed one and damaged two others for the loss of two aircraft. In March 1942 he was POW, and was awarded an OBE in 1947 for his courage and leadership as a POW. He died in 1988.

NX56235 Pte Cyril Montefiore Harris

Cyril was born on 16 February 1908, in Johannesburg, South Africa, son of Ernest and Jennie, and later the husband of Theresa of Mosman, NSW. He was in 2/18th Battalion and was in Force E. He died on 20 April 1945 at Labuan and is remembered on the Labuan memorial.

78. VX63884 Sgmn Syme Newell Harris, RACOS.

VX63884 Cpl Sgmn Syme Newell Harris

Syme was born on 22 May 1921, son of Syme and Esther of Middle Park, St Kilda, and became a tailor. He was a member of the 8th Armoured Division Signals and was sent as part of Force E from Singapore on 28 March 1943 on SS *De Klerk* to Sandakan with 1,000 Allied POWs, arriving on 15 April. On 6 June they were sent to Eight Mile camp. Syme died of malaria at Ranau on 5 June 1945, just before liberation, and is buried at Labuan.

VX63270 Pte Alfred Ernest Langford

Alfred was born on 17 August 1922, son of Louis, and lived at 11 Bourke Street, Melbourne. He was an apprentice engineer and posted to 2/21st Pioneer Battalion, AAOC, and attached to a Bren gun unit. He was captured in Java on 27 March 1942 and sent to Hakodate camp in Japan, from where he was liberated in 1945.

VX19346 Lt Samuel Alexander Lazarus

Samuel was born on 17 January 1915, son of Louis and Ruby Elizabeth, brother of Pauline, of St Kilda, and became a barrister. He was in the 2/21 Pioneer Battalion, AAOC, and he was executed on 6 February 1942, allegedly for loss of life when a Japanese minesweeper hit a Dutch mine. He is buried at Ambon.

NX47023 Pte Robert Edwin Levey, 2/10 Field Ambulance

Robert was a salesman, son of Gerald and Emma of Hamilton, NSW, and was held at Sandakan camp in Borneo after being transported there with 'B' Force from Singapore (Changi) with 1,494 others on the Hell Ship *Ubi Maru*, arriving on 18 July 1942. He died at Ranau on 14 June 1945, aged 31, and is remembered on the Labuan memorial.

VX52820 Sgmn Max Nagelberg/Nagel

Max was born on 16 February 1924 in Bercoze, Poland, son of Freida and Isador, and came as a child refugee to Australia on a *Kindertransport* from Hamburg in 1939. His parents perished in the Holocaust. He enlisted in the AIF in April 1941, saying that he was 18, but in fact he was just 16 years old. He was POW in Singapore and was sent on the Hell Ships to Osaska as a slave labourer, and later to a coal mine near Nagasaki, where he survived the atomic bomb which was dropped on that city in August 1945. In the autumn of 1945 he boarded a B-17 bomber bound for home, but it crashed and he was the only survivor. He went to live in Worcester, Massachusetts, where he met and married Nita Goff. In 1983, aged 59, he went to live in Herzliya in Israel. He died in Florida in 2006, leaving his wife Nita and his daughter Joanne.

VX55757 Pte Leon Silman, 2/9 Field Ambulance, AIF, MiD (Malaya)

79. VX55757 Pte Leon Silman, MiD, AIF.

Leon was born on 25 June 1917 in Lodz, Poland, where he worked in a textile factory and was active in the trade union movement. He was often involved in scuffles with pro-Nazi officials. He joined the Polish Army and was in the artillery. After leaving Poland in the late 1930s, it is believed that he fought against Franco in the Spanish Civil War. He came to Australia as a young man in late 1937, to live with his uncle, Samuel Wynn, at 155 Wells Street, Melbourne. He enlisted on 13 May 1941 after harassing the authorities and telling them he wanted to fight, despite not yet being naturalized. He gave his occupation as truck driver and his address as 61 Park Street.

Eventually he was posted to the 2/9th Field Ambulance at Kota Tinggi, where his foreign accent and swarthy looks made him a well-known character. Posted to Malaya, Leon was on the front line at Tanglin as an orderly at an Advanced Dressing Station, taking casualties by ambulance back to St Andrew's Cathedral Field Hospital in Singapore City. Waiting in a slit trench during a bombardment, to take some wounded away, a shell killed the two drivers near him. Although he had never driven that type of vehicle before, he got it started and took the wounded to Singapore. For this he was Mentioned in Despatches for 'gallant and distinguished service in Malaya' (*London Gazette*, 1 August 1946).

After imprisonment at Changi, he worked at the Roberts Prison Hospital and was later sent to work, with 'L' Force, on the notorious Burma Railway. Little is known of his war experiences. Family memories of his stories include his testifying that he stole food from Japanese stores to give to the sick, and stole chickens from the Japanese guards' coops to make soup for the sick men.

On liberation he found out that all his family except one sister had perished in the Holocaust. He married Minnie and had two daughters, Doreen and Suzanne, and worked in the wine and restaurant business. After the war he stayed in regular touch with a group of men with whom he had suffered as a POW. They were all very close and died within six months of each other in 1992. Leon died on 2 June 1992.

British and Commonwealth Short Stories

All these men, save three, have AJEX cards; the surnames of the rest indicate that they are Jewish. Their notes are from the TNA debriefing sheets of POWs.

2357021 Sgmn Joseph Coleman (aka Cohen)

Joseph was born on 10 July 1915 and lived at 142 Bellot Street, Manchester. His debrief testimony sheet states that he was a POW in Changi, Thailand and Indo-China. He assisted in damaging Japanese army rations and military equipment in Singapore, and also witnessed an incident when Lieutenant Primrose and Major Roberts of the Gordons were ordered to watch the shooting of a POW who was dying of cholera. The Japanese sergeant was shaking with fear and could not do the job, so the two British officers took rifles and shot the man together. They were tried after the war for murder, but acquitted.

104686 Capt. Harold Wolfe Cowen, RAMC, 9th Indian Division

Harold was born on 7 April 1906 and lived at 85 Hornsey Lane, Highgate, London. Captured at Singapore, he was at Changi, Kynsyok, Takelin, Shimon-niki, Tamarkan and Chungkai. At Shimon-niki he made contact with local Thai resistance groups and planned to escape by elephant to a pre-prepared hideout in September–November 1944, with Sergeant G. Chanden (?) from the Cambridge Regt, but they were moved to another camp before the escape could take place. Forced to be an MO for the Japanese, he deliberately infected their wounds, but they became suspicious of him and so this had to stop. He also put sand in Japanese petrol tanks whenever he could.

2035400 Spr Alfred Davidovitch

Alfred was born on 27 December 1916 at 47 Brady Street Mansions, London E1. He was captured in Hong Kong and imprisoned at Sham Shui Po camp; he was shipped to Tokyo on 16 September 1942. He mentions helping to destroy Japanese army food stores and stealing flour and beans for the prisoners.

7686971 Cpl Louis Dorfman

Louis was from 47 Hanson Street, London W1, and was with 15th Field Security Section in Malaya. Imprisoned at Changi, Sonkrai, Tembaya (Burma) and Kranji, he helped to set Korean and Formosan guards against the Japanese guards whilst acting as official interpreter.

225882 Lt Edward Felix Elkan, General Service/SOE/Guerrillas, 'Stay Behind Party'

Edward was born on 25 February 1911 and lived at 2 Old Manor Court, Abbey Road, London NW8. He was captured nine miles north of Temerloh on the Pahang River in Malaya on 12 March 1942. He was imprisoned in Kuala Lumpur (April–September 1942); Curran camp, a punishment centre run by Sikh traitors (17–30 September 1942); Changi (September–May 1943); and five camps in Japan (from June 1943 until liberation in August 1945). In Japan he was at Hokkaido and Mitsushima.

He assisted in two escapes from Kuala Lumpur and was put in the punishment cells as a result. He heard the stories of the recaptured escapers at Curran camp, where he met them, and he later learnt they had been shot. He also helped to store hand grenades in case the Japanese attempted a massacre of the POWs at the end of the war. At Yakuma aerodrome in Hokkaido he helped to cause delays by damaging survey equipment and moving survey pegs, between August and October 1943. At Hakkodate, between February and August 1944, he helped make and hide a radio set.

1076994 Gnr Nathan Ernstone, 135 Field Regt, RA

Nathan was born on 24 January 1920 and lived at 132 Wingrove Gardens, Newcastle. Taken at Singapore, he was also at Tamarkan, Kinsayok, Conquita, Nonpladuk and River Valley Singapore. He took part in much pilfering of Japanese food for the POWs.

Pte Moshe Gareh, 2nd Loyals

Moshe was born on 5 July 1918. He was captured in Malaya and sent to Pudu prison in Kuala Lumpur, Changi, Wunlun, Takelin, Nambrudai, Takenun, Chungkai, Arohero and Tamawang. While working in ammunition dumps for the Japanese, he and other prisoners damaged many explosives and weapons, and also stole some to assist in an escape from Pudu gaol. Several men got out but all were eventually recaptured, and disappeared. He wrote especially of the kindness of the Chinese Malays and Tamil prisoners, who always tried to help them with food, even though their own conditions were pitiful.

851820 Gnr Leonard Glickman, 7th Coast Regt, RA

Leonard was born on 26 August 1918. He joined as a regular on 20 January 1936, and lived at 31 Wicklow House, Stamford Hill, London N16. Captured at Singapore, he was at Havelock Road camp, and in October 1942 was sent on the railway to Wampo; he was also in Chungkai, Tamuan, Tamaton and Nacom Nai, ending at Tikuri where he was liberated. In May 1945 he disabled Japanese lorries by putting sand in the fuel tanks. He wrote that at 7 p.m. at 191 kilo mark on the railway, on 8 December 1944, their train was bombed and strafed by Allied four-engine aircraft, and medical orderly Lance Corporal Tomms was killed by shrapnel when lying over a sick man to protect him.

T231908 Dvr Sidney Gordon (aka Goldberg), RASC

Sidney was born on 19 January 1906 and lived at 31 Hendon Way, London NW2. He enlisted on 15 November 1940. He was captured at Singapore and imprisoned first at Changi and then at Non-Pladuk, from June 1942 until 1944, and then at Lampo (?), Chang Mai. He testified in his debriefing notes to a group of fifty Allied POWs, including himself, being helped by an underground group led by local Chinese at Chang Mai and a local Englishman, W.A.R Wood, who had been the local British consul in Chang Mai before the war. At great risk to themselves, they supplied news, medicines, money, cigarettes and food. The POWs were also told of plans by loyalist Thai troops to rescue them in the event of Japanese reprisals if the Allies landed to liberate the area. Plans for an escape were made and five men did get away before liberation, from Lampo camp. He gave the names and addresses of the six most prominent members of the underground group and suggested that they should be given honours for their dangerous assistance. Sidney was a cousin of the historian Sir Martin Gilbert.

2412 NZD Sgmn Nathan/Naphtali Jaffee, HMS *Exeter*

Nathan was born on 22 December 1918 and lived at 5 Belle Vista Road, Herne Bay, Auckland, New Zealand. He was captured at sea on 1 March 1942 and taken to Macasser in the Celebes. He was moved to Maros camp until 12 September 1944 and then back again to Macasser until liberation. He took part in the sabotage of machinery at Macasser docks as well as pouring sand into Japanese vehicle petrol tanks.

918076 Gnr Myer Lerman, 122 Field Regt, RA

Myer was born on 30 March 1918 and lived at 31 Leeds Road, Bradford. Captured at Singapore, he was at Changi until August 1942 and then sent by Hell Ship to Jinsen and Konan in Korea until the end of the war. He helped to sabotage Japanese dynamos and barges in the docks in Korea.

6092322 S/Sgt Leslie Levin, IACC

Leslie was born on 28 September 1918 and lived at 47 Hannan Road, Liverpool. He enlisted on 15 September 1939 and was captured on Banka Island, where he was kept until March 1942; he was sent to Palembang until May and then the airport camp until September 1942. From here he was sent to Palembang 'B' camp until April 1944, and finally Sung Ron until liberation. Among his exploits was stealing Japanese medical supplies for sick POWs.

6019465 Pte Sidney Levinson, 1st Cambridge Regt

Sidney was born in 1918 and lived at 9 Elsie House, Philchurch Street, Commercial Road, London E1. Captured in Singapore, he was at Changi until 23 December 1942, from where he was transported by cattle truck to slave on the Burma Railway at Non Pladuk camp until 6 May 1943. From here he was sent to Tamarkan camp until 4 December 1943 and then down to Kanchanaburi until 5 January 1945. From here he was sent on one of the Hell Ships to Saigon, where he was liberated in August 1945. [He was the author's uncle by marriage.]

Fl. Off. George Melville Louisson, RNZAF Signals

George was born on 8 February 1902 at 30 Alan Street, Palmerston North, New Zealand. He was captured on 8 March 1942 at Tasik Malaja, Java. He was at Maspati camp until 30 April 1942, at the aerodrome, where he took part in sabotage such as breaking open fuel drums. He was also at Jaarmaarkt, Bandoeng and Cycle camps.

PJX 182275 AB Henry Mandelman, HMS *Stronghold*

Henry was born on 23 October 1919. He lived in Hove and also at 121 Long Acre, London WC2. He was captured on 3 March 1942 in the Indian Ocean and was in hospital at Macassar. He was later involved in the sabotage of machinery and equipment (such as refrigerators and engines) being shipped to Japan, and in putting sand in fuel drums and explosives, spiking fuel drums, breaking open packing cases and breaking tools.

277090 Lt Joseph Morris, RASC

Joseph was born on 10 July 1913 and lived at 8 Merryhills Drive, Enfield, London. He was captured at Singapore but made an escape with five NCOs by boarding a sampan loaded with food and water. They rowed hard with makeshift oars, but Morris was sick and asked to be left on an island so that the others could get away to Java. All the men were eventually recaptured but unharmed. Joseph was rescued by a Chinese fisherman and taken back to Singapore prison. After Changi, he was sent to Chungkai in October 1942 and then Nakon Platon in May 1944; finally he was sent to Kanchanaburi in January 1945, from where he was liberated in August 1945.

235005 Sgmn Andrew Pinto, 3rd Indian Division

Andrew was born on 27 November 1911 and lived at 117 Haggerston Road, Hackney, London E2. Captured in Singapore, he was at Changi until November 1942 and in Thailand until June 1944. He was shipped to Manila until September 1944, then went to Formosa until November 1944, and to Japan, where he was finally liberated.

6915371 Cpl Barnett Popper, 18th Recee

Barnett was born on 13 June 1919 and lived at 95 Highfield Avenue, Brent. He was captured at Singapore and was in River Valley camp until July 1942; he was sent to Tan Jan Ru in Singapore until September 1942 and Changi until January 1943. After this he was transported to camps in Thailand. While in the docks at Singapore, he and comrades spiked many Japanese fuel drums; he named his friend Jack Knifes as his accomplice.

1461091 L/Bdr Maurice Albert Rose, 77th HAA, 9th Division

Maurice was born on 23 October 1908 and lived at 61 Westville Road, Cardiff. He enlisted in April 1939. He was captured in Java at Garoet on 8 March 1942 and imprisoned at Garoet/Bandoeng, then sent to Lansdopvoedingestiat until April 1942, Tjimahi until May 1942, Bandoeng until June 1942 and Cycle Camp (Batavia), until October 1943. He was transported to Pangkalan Balia, Sumatra, and was there until November 1943, Changi until May 1945, and liberated at Kranji in August 1945.

1101535 Gnr Morris Rosen, 118 Field Regiment, RA

Morris was born on 2 January 1911 and lived at 7 Cumbrian Gardens, London NW2. Captured at Singapore, he was in Changi until May 1942, Towner Road until November

1942, Tonchan on the railway until September 1943, Tarsao until January 1944, Chungkai until April 1944, Tamawan until September 1944, 152 kilo. camp until February 1945, Petsuri until June 1945, and liberated at Nakun Nai in August 1945. While in Singapore he took part in the spiking of petrol drums and damaging Japanese vehicles.

1391963 WO Albert/Alfred Rosenberg, 203 Squadron, RAF

Albert was born on 20 June 1922 and lived at 221 Navarino Mansions, Hackney, London E8. He was captured on Sebeen Island, Sumatra, after his plane crashed on 14 August 1945, and imprisoned at Palembang until liberation on 20 September 1945.

921478 Gnr Alexander Rosenthal, 77 HAA, RA

Alexander was born on 2 January 1922 and lived at 160 Burnhouse Street, Maryhill, Glasgow. He was captured at Garoet, Java, and moved to Tanjong Priok on 8 March 1942 and Batavia in February 1943. He was sent to Changi and liberated there in August 1945. At Batavia he helped to sabotage oil drums while on working parties.

273555 Cpl Alfred Rosenthall, 42 Coy RASC

Alfred was born on 4 February 1910 and lived at 2 Bristol Road, Leeds. He was captured at Singapore and was at Changi until October 1942, then Kanu until June 1943, Chungkai until October 1943, Tamarang until April 1945, Nakon Platon until May 1945 and Mergoi (?) Road, where he was liberated on 30 August 1945.

155297 Lt Alexander Humphrey Samuelson, 21st LAA

Alexander was born on 1 April 1909 and lived at Old White House, Woking, Surrey. He was captured at Timor on 23 February 1942. He was sent to Oesapa Bisar until September 1942, Changi until May 1943, Hintok Valley and River camps until August 1943, Krian Kri until November 1943, Sime Road until May 1945 and Changi until September 1945.

T210688 L/Cpl Reuben John Segal, 196 Field Ambulance, RAMC

Reuben was born on 7 February and lived at 77 Southway, Hampstead, London. Captured at Singapore, he was at Tarsao until June 1944 and at Non Pladuk; here he helped sabotage Japanese dynamos, machinery and welding plant.

T170482 Dvr Nathan Silverstone, RASC

Nathan was born on 13 December 1916 and lived at 57 Grange Street, St Albans. Captured at Singapore, he was at Changi until April 1942, Thompson Road until December 1942, Kanchanburi until March 1943, Kinsayo until October 1943, Chungkai until May 1944 and Nakon Nai until liberation.

1580710 Gnr Edward Kitchner Silverton

Edward was born on 18 December 1914 and lived at 99 Bromhall Road, Dagenham. Captured at Singapore, he was sent to Saigon in April 1942 and back to Kinsayo in Thailand until June 1943. He was sent to Chungkai until August 1943, Tamuan until May 1944 and Taqiri until July 1944. No other information was available about what happened to him after 1944. He carried out acts of sabotage in Saigon docks, puncturing petrol cans and damaging weapons being transported to the Japanese army.

175(4)741 LAC Albert Bernard Simmonds, 250 Squadron, RAF

Albert was born on 19 September 1915 and enlisted in June 1939. He lived at 44 Menelik Road, London NW2. He was captured at Padang, Sumatra, on 17 March 1942, and in June was taken to Belawen until July 1942, then Gloegoer until March 1944 and Kenijaren until October 1944. He was sent to Sungei-Sunggal until November 1944 and liberated at Pakan Barae in August 1945.

1286023 LAC Joseph Simmonds, MT Section, RAF

Joseph was born on 30 May 1914 and lived at 23 Shore Road, Hackney, London. Captured in Java on 8 March 1942, he was at Boa Glodak until January 1943 and Tanjongpriok until February 1943. He was shipped to Changi until May 1943, and sent to the railway at Wy Hai, Kinsayo, back to Sime Road, Singapore, and Changi until Augst 1945.

1019590 AC1 Abraham Sidney Solity, 100 Squadron, RAF

Abraham was born on 26 April 1912 and lived at 25 Francis Street, Chapeltown, Leeds. Captured on 8 March 1942 at Neglasari, Java, he was at Bandung until November 1943, Cycle Camp (Batavia), until May 1944, and Pakan Baru, Sumatra, until August 1945.

775325 Cpl Baruch Spektor, 84 Squadron, RAF

Baruch was born on14 August 1908, and lived with the Spero family, 47 Lordship Park, London N16. Captured in Java on 8 March 1942, he was at Gloduk Batan until 11 January 1943 and Priok until May 1943. He returned to Gloduk and was there until December 1943, then sent to Macassar until June 1944. Transported to Singapore until July 1944, he was back at Pakan Baroe until August 1945.

904680 Cpl David 'Doddy' Spero, RAF

Doddy was born on 17 December 1915 and had the same address as Baruch Spektor, above – 47 Lordship Park, London N16. He enlisted on 6 October 1939. He was captured in Java and sent to Boei Gladuk until March 1942, Tanjon Priok until January 1943, Cycle Camp (Batavia) until April 1943, back to Gloduk until May 1943, Cycle Camp again until January 1944, then Camps 1, 2 and 3 (?) in Sumatra until August 1945.

7637234 Pte Meyer Stern(e), RAOC

Meyer was born on 5 May 1920 and lived at Plevna Crescent, St Anne's Road, Tottenham, London. Captured at Singapore, he was at Changi until April 1943, on the railway until December 1943 and back at Changi and Kranji until August 1945.

83670 Flt Lt/Fl. Off. Dr Basil Arnold Stoll, 605 Squadron, RAF

80. 83670 Flt Lt/Fl. Off. Dr Basil Stoll, RAF.

Basil was born in Keidan, Lithuania, on 21 March 1917, son of Rev. Solomon and Mrs Golda Stoll of 49 Southborough Rd, Hackney, London E8. He attended Parmiter's School and Westminster Medical College and lived at 22 Heathland Rd, Stamford Hill, London N16. Captured at Garoet, Java, he was at Gladok until January 1943, then Priok until May 1943, Batavia Hospital until March 1945, Cycle Camp (Batavia) until June 1945; and Bandoeng until August 1945. As son of a *Chazan* (Jewish liturgical leader) Basil hand-wrote, from memory, prayers and songs for services held at Cycle Camp for the Jewish POWs in August 1945. These original documents survive with the family. One of them has ten signatures of the participants, four of which are signed in Hebrew by Israeli/Palestinian Jews captured with Basil with the RAF.

111792 Maj. Leslie David Stone, RAMC

Leslie was born on 3 September 1911 and enlisted on 1 December 1939. He was living at Sycamores, Upper Park Road, Salford. He was at Changi until May 1942, Havelock Road until October 1942, Kinsayok/Hindato until January 1944, Nonpladuk 1 and 2 until February 1945 and Ubon until August 1945.

6021032 Pte Louis Stone, 5th Suffolks

Louis was born on 28 September 1916 and lived at 53 Coronation Avenue, Stoke Newington, London N16. He was at Changi until February 1943, Chungkai until August 1944, Conquita until October 1944, Kinsayo until December 1944 and Tamatao/Tamuan until September 1945. At Conquita he sabotaged Japanese fuel and clothing stocks and often broke out of camp to smuggle in food, despite severe punishments if caught by the Japanese.

10908941 Gnr Clyde Sugar, 135 Field Regiment

Clyde was born on 28 January 1907 and lived at 69 Silversea Drive, Westcliff, Essex. He was at Changi until May 1942, Bukit Timah until October 1942, Tamarkan until December 1943, Nonpladuk until June 1944, Nakonplaton until April 1945 and liberated at Mergui Road camp.

1827124 Frederick Velinsky, 35th LAA

Frederick was born on 14 June 1921 and lived at 37 Folgate Street, Spitalfields, London. He was captured in Java and was at Tanjon Priok until 29 March 1942, Cycle Camp until April 1943, Panklan Bali until November 1943 and Changi until June 1945. He was liberated at Kranji.

109341 Maj. Everard Woods, 17th Indian Division, 45 Brigade

This officer appears in the testimony of Captain Arkush. Born on 6 January 1917, Everard enlisted on 12 January 1935 as a regular and gave his address as 4 Fearnshaw Close, Chelsea, London. He was at Changi until June 1942, Bampong until November 1942, Chungkai until March 1943, 203 kilo. until March 1944, and finally at Kanchanaburi. He testified in his notes that there was a scandal at one camp where British NCO quartermasters curried favour with the Japanese to get extra food for themselves.

Canadian Short Stories

B68230 Rif. Louis Brown, RROC

Louis was born on 5 April 1905 and lived at 1 Beatrice Street, Toronto. He was taken POW in Hong Kong on 25 December 1941. He was deported to Fukuoka in Japan and liberated on 18 September 1945.

Capt. David Aaron Golden, Winnipeg Grenadiers of Canada

81. (a) Capt. David Aaron Golden, Winnipeg Grenadiers, as a soldier; (b) David in later life.

David was born in Sinclair, Manitoba, on 22 February 1920, son of S.W. Golden of 548 Burrows Avenue. He attended the Sholem Aleichem school in Vancouver and the J.L. Peretz school in Winnipeg. He was a member of the University of Manitoba OTC and lectured there in military law; although he won a Rhodes Scholarship to Oxford University, he turned this down and instead volunteered and was commissioned in the autumn of 1941. He was first posted to the West Indies as intelligence officer and adjutant to the 1st battalion, and then to Hong Kong; he was taken POW on 25 December 1941 in Hong Kong and held there until 7 September 1945. He was at Sham Shui Po, North Point, Bowen Road, and sent back to Sham Shui Po

David adamantly refused to talk about his captivity and never joined or attended reunions of the Canadian Hong Kong veterans group. In 1946 he successfully defended four soldiers who allegedly collaborated with the Japanese. The trial transcripts have never been found. (Canadian researcher Dr Nathan Greenfield told me this in a telephone conversation in November 2011). In 1954 he became the first Canadian Jewish Deputy Federal Minister and later was prominent in the Canadian satellite technology industry, and awarded the Order of Canada.

H6375 Cpl George Robert Harrison, Manitoba Regt

George was from Winnipeg, son of Rachel, of 78 Hart Avenue. He was taken POW in Hong Kong on 25 December 1941 and was in camps in Hong Kong and Osaka, Japan. He was liberated on 30 August 1945.

B40578 Rif. Frederick 'Zeke' Zaidman, RROC

Zeke was born on 6 December 1917, son of Dora, of 9 St John's Block, Winnipeg, and taken POW on 25 December 1941 in Hong Kong. He was held at St Stephen's Hospital until 26 December, then at Bowen Road until 30 June 1942 and North Point until September 1942. He returned to Bowen Rd until October 1942 and then was at Sham Shui Po until September 1943. While in Hong Kong, he testified that plans were formulated for a mass escape as they were in touch on a secret radio with British Intelligence in Chungking (China). But the Japanese got wind of the plan and it was abandoned.

In September 1943 he was transported on the Hell Ships to Nigata in Japan until the end of the war. His story appeared in the *Toronto Daily Star*, 7 February 1946:

> During the battle for Hong Kong, Zeke was badly wounded in the leg and made his way to St Stephen's Hospital on Dec 15th. There he witnessed the notorious horrific massacre of patients in the hospital by Japanese troops: 'The Japs rushed in and bayoneted my wounded buddies, took many VAD's (Voluntary Aid Detachment nurses) and regular nurses and assaulted (raped) them. The rest of us crowded into a small room. Every hour the Japs would come and in and pick out some poor wretch. His dead and mutilated body would be found later. I was beaten down with the flat of an officer's sword.
>
> On Feb 6th 1944, I was in Camp 5 in Tokyo. It was a day that I wished I was dead. I worked on the coal docks. Beatings, meagre rations and utter fatigue was our daily lot. I was down to 97 pounds and in a fever, and it was my 27th birthday. I was starving and it was then I vowed that if I survived, and got back home, I'd make sure that at least on every birthday, I would have enough to eat.'

On his twenty-eighth birthday, Zeke was still a POW; he saved a can of coffee from a rare Red Cross parcel and with some old cakes and bread made a cake and served it with the coffee to some buddies. Zeke was hospitalized at Chorley Park Hospital in Toronto after liberation.

Epilogue

Written by Jack Caplan in 1985, the poem below is, I think, a fitting end to this book, encompassing as it does his ambivalent thoughts about his time in Asia.

> I've been out East, in war-time years,
> Sampled rice, curry and foreign beers,
> I suffered nostalgia, longed for an English lane,
> Especially when drenched by monsoon rain.
>
> But this year of our Lord '85,
> Twelve wintry months of wind, rain and thunder,
> I long to return to tropical shores ...
> Is it any bloody wonder?

Afterword

In 1995, to mark the 50th anniversary of the end of the Second World War, AJEX gave the veterans of the war against Japan, and especially the ex-POWs, the honour of leading the Annual Parade in Whitehall. The youngest were in their 70s but they marched in a manner that would grace any military parade ground, the batallion led by Capt. Dr David Arkush. My father had made his last AJEX Parade six years before, and passed away the following year in 1990. I was moved to write this poem soon after to honour him and all the Jewish veterans.

The AJEX March

Fallen leaves and distant guns,
Lone, bare trees and muffled drums,
shadows fall across Whitehall,
November sun so weak and cool,
They march in the Autumn of their years.

Israel's children, stooped but proud,
Still march steady, heads held high,
Swaying medals clinking,
Polished silver glinting,
Still in step to soldiers songs,
Gentle slap of shoes on tarmac.

Not so tall now, slow of pace,
Rheumy eyes and hair of grey,
As the years pass, their ranks are streams of memories,
Bittersweet,
They march in the Autumn of their years.

But such ardour in their youth,
But such strength and fire and courage,
Gallant men and women both.

Snap of orders, caps are doffed,
For long lost comrades,
And the Six Million
And those fallen for our Land of Israel,
Together passed the Cenotaph.

A minute's silence, coughs and sobs,
Bugles sound, and flags of sky blue lap in the breeze.

Blood red Star of poppies crimson,
Gently placed,
Lovingly placed,
We weep.
Comfort Ye!
Comfort Ye!

Children search for grandpa, grandma.
Words are not enough.
Words are not needed.

Martin Sugarman, 1991

The Record and Roll of Honour

Where death is not specified in the list of names, the POW is presumed to have survived the war. Any awards named are only for those given for gallantry whilst a POW, unless otherwise stated.

At the National Archive at Kew, in series WO344 361–410, there are hundreds of debriefing sheets, some very detailed, which were compulsorily completed by surviving POWs on liberation.

I appeal to all readers everywhere to send me any further information they may have, via AJEX, so that amendments can be made.

Readers should note that those who died in Taiwan/Formosa were re-interred/commemorated in Hong Kong and elsewhere

Navy

Biderman, WO George
RNVR. Died.

Cohen, Rad/O Joseph
MN. MS *American Leader*.

Drake, Ldg/Tel E.A.
RN. POW sinking of *Prince of Wales*.

Goldberg, AB Solomon
Glasgow.

Grossman, Lt Edward
HMS *Cornflower*. Hong Kong POW.

Herman, AB Isaac
RAN. HMAS *Perth*. Buried Kanchanaburi.

2412 Jaffee, Nathan
RNZN, HMS *Exeter*. POW Sulawesa Island, Indonesia.

Kaye, 1st Rad/O Eric
MN. Buried Kranji after dying as a POW (?).

PJX/182275 Mandelman, AB Henry
HMS *Stronghold*.

PJX182275 Mason, H.
HMS *Stronghold*. POW *Celebes*.

Nissim, Sub Lt Archibald
RNVR. First World War veteran. HMS *Cornflower*. POW Hong Kong.

Webber, Petty Officer Bertram
POW Hong Kong.

Wimbusch
MN. Sime Road camp.

UK and Dutch Army

My thanks to Peter Dunstan for his help in compiling this list.

Aaron, Tpr Abraham
The Loyals, RAC. Buried Chungkai.

Abas, S.
Dutch Infantry. Buried Chungkai.

Abel, Dvr Albert
RASC.

Abett, Pte I.B.
1st SSVF. Singapore. Died Burma Railway, 2 September 1943. Buried Thambyuzayat; not on CWGC.

Abraham, Pte Ezekiel
SVC.

3773453 Abrahams, Alfred
RA/Norfolks. Java.

1634974 Abrahams, Alfred Alex
RA.

Abramson, Barnett/Ben
Duke of Wellington's. Died 3 June 1943.

Amsel, L. (?)
Dutch Army. Batavia (Basil Stoll papers).

Angel, Bdr Eric
RA.

Arkush, Capt. David
RADC. Changi/Thailand.

Auerbach, Gnr Harry Aaron
RA. Buried Labuan.

Bachrach, Alfred G.H.
Liaison to Dutch Special Forces, Far East. Interned with wife and baby son (who died).

Baker, Pte James Morris
Recce Corps. Changi. Buried Kranji (SVC).

165210 Barcellon/Barcilon, Major Hector
RA.

Barkin, Jack
RA. Java.

Barnett, Pte Bernard
RAMC. Alexandria Hospital, Singapore. From Leicester.

Baron, William
Norfolk Regt. Died 22 April 1942.

Basco, Gnr Lionel
RA. Buried Chungkai.

Bass, Pte Joseph
5th Beds and Herts.

Baum, Gnr Henry
RA. Buried Kranji.

Baume, Lt Louis

Beck, Capt. Clifford
RASC.

Bell, C
Java.

Benjamin, Aaron
Died as a POW, 5 August 1943.

Benjamin, Nathan
RAMC.

Benjamin, Pte Vivian
HKVDC. Died 21 September 1945.

Berkley/Berkeley (aka Berkovitch), Capt. Allan
RAMC.

Berkovitch, Pte Frank/Len
13th Kings.

Berler, Bruno
Dutch Resistance. Captured and executed near Jakarta, 1944.

Berner, Pte Jack/Isaac
2nd Loyals.

Bernstein, Dvr Joseph
RASC. Cantor in Synagogue, Changi. Glasgow.

Besbrode, Cpl Mark
Middlesex.

Besser, Pte Barnett
Recce Coy/Rifle Bde. Changi.

Bialestock/Bialstock, Pte Charles
Norfolk Regt. Killed with Saunders.

Birnbaum (aka Goldstein), David
Toronto serving UK forces.

Blitz, Cpl
Dutch Army. Died (?).

Bloch, Cpl M.
RASC.

217704 Bloom, Jack

6845454 Bloom (aka Baker), Rfmn John
KRRC/RCOS. POW Singapore. Died 1964, aged 46.

Bloom, Major Dr Phillip M.
RAMC. From South Africa. Changi .

Bluestone, Cpl W.
RCOS.

Blum, Capt.

Bolter, Gnr Max
RA, SSVF. Buried Chungkai.

Braham, John
RASC.

Braham, Capt. Dr Mark Gordon
RAMC. Died with Saunders.

Brandon, David
Dutch Army.

Brisk, Pte L.V.
1st SSVF.

Brodie, Cpl J./Leo

Brody, Louis
Singapore.

89344 Brosky, R.M.
Dutch Army. Died 16 October 1943, Thanbyuzayat. [possible]

Brown, Rev. Mendel
SVC Chaplain. Shanghai

Brownhood, Gnr Alec
RA. Died as a POW.

Burton, Anthony Lewis
RA.

Burton, Sgmn Joseph
RCOS.

Caggan, Gnr Leon
RA Beds Regt. Buried Kranji.

Caplan, Gnr Harold
RA.

Caplan, Jack
RCOS. Glasgow.

Caplin, Pte Ellis
13th Kings.

Cashman, Daniel
RA. Timor, Java, Japan.

Chernin, Pte Lionel
Singapore/Japan.

Child (aka Burleigh), Kenneth Joseph
RCOS. MiD.

Citroen, J.
Changi.

Clayton (aka Cohen), Pte David/Sid
Suffolks. Buried Formosa/Hong Kong, commemorated at Kranji.

Cohen, Capt.
Doctor, *Dutch Army.*

5932342 Cohen, CSM Arthur R.L.
Regular 1930s.

1099755 Cohen, Gnr Harry
RA. Buried Kuala Lumpur.

Cohen (aka Coleman), Joseph
From Manchester. Changi

T181551 Cohen, L/Cpl Marcus Leonard
RASC.

Cohen, Gen Morris 'Two Gun' (Retd)
Interned Hong Kong.

Cohen, Pte Norman
Middlesex. Buried Yokohama.

1696896 Cohen, Pte Sydney
RA.

Copel, Pte Woolfe
1st Manchesters. Changi/Japan.

Cowen, Capt. Harold Wolfe
RAMC.

Cyfer, Adj. J.
Dutch Artillery. Java.

Dabelstein, Pte Lionel A
HKVDC Hong Kong.

Da Costa, W/OB
Dutch Army. Java.

Dak
Dutch Army.

David, L/Cpl Soloman
FMSVF. Died Sime Road, 27 August 1945.

Davids, Pte
Dutch Army.

Davidson, Sgmn A.
SSVF. Ilford.

Davidovitch, Spr Alfred
RE.

Davis, Bdr David
RA. Buried Kanchanaburi.

De Louw, Pte
Dutch Army.

De Vries, Louis
Dutch Army.

Diamond, Gnr Barnett
RA. Singapore.

Dicks, Claude Maurice
REME.

Donn, Gnr Harry
RA.

Dorfman, Cpl Louis
Field Sec. Intel. Corps.

Dykes, Bdr Robert
RA.

Egalnick, Abraham/Alf Joseph
18th Recce/Rifle Bde. Changi

Elkan, Lt Edward Felix
SOE/General Service. Malaya/Japan.

Ellis, Nurse Leontine
HKVDF. Buried Stanley CWGC Cemetery, Hong Kong.

Elze, Lt
Dutch Army. Died Changi.

Ernstone, Gnr Nathan
RA.

Fay, Capt. Dennis Henry
Gurkhas.

Feldman, Gnr Louis
RA. Lost at sea as POW.

Fine, Dvr Eli
RASC. Died at sea as POW. Brother of Nathan (reported in the *JC*).

Fine, Nathan
RA. Changi, brother of Eli (reported in the *JC*).

Finegold, Lt Martin
RIASC.

Finn, Maurice
RA. Java.

Frank, Pte Leon
Chindit. POW.

Franken, John Jacob Herman
Dutch Navy Air Force.

Franken, Joob/Joseph
Franken, Louis
Franken, Paul
Dutch Army. Brothers of John Jacob Herman Franken, above (a fifth brother, Albert, was murdered at Mauthausen Camp, Austria).

Franken, Rosete
Mother of the brothers named above. Interned by Japanese.

Franklin, Dvr Jacob
RCOS. Buried Kanchnaburi.

Freedman, Bdr Morris S.
RA. Died Kuching, Sarawak.

Freeman, Sgt Sidney
RA. Died Formosa. Buried Hong Kong.

Frenkel, Pte Jacob
HKVDC. Hong Kong.

Gaffin, Cpl Merton Eugene
RCOS. Buried Thanbyuzayat.

Gaisman, Lt Peter
Manchester Regt.

Gareh, Pte Moshe
2nd Loyals.

Gellman, Pte Boris Abraham
HKVDC. Hong Kong/Japan.

Gendleman, A.
Sherwood Foresters. From Shoreditch. POW Nong Pladuk.

Gershonblatt, David
Norfolk Regt.

Gerstein, Pte Lewis
5th Beds and Herts.

Getzels, Gnr A.
RA.

Glaser, Lt/Col J.
Dutch Army.

Glickman, Gnr Leonard
RA.

Gold, Sgmn Benjamin
RCOS. Borneo.

Goldberg (aka Morgan), D.
5th Field Regt, RA.

5245 Goldenberg, Pte Alan Ezra
HKVDC.

Goldenberg, Lt William
SVC.

Goldfarb, Len
Royal Norfolks. Glasgow/London. An alleged 'King Rat' figure,

Goldman, Maj. Alex Percival
FMSVF. MiD. POW Thailand .

Goldman, Lt Laurence

Goldman, Major S.
Singapore. Thailand.

Goldrich, Pte Samuel
Sherwood Foresters. Group 5 Nakon Ni/Taiwan.

Goldstein, David
4th Norfolks. Changi.

Goldstone, Pte Bernard
Manchester Regt. Died at Song Krai. Buried Thanbyuzayat.

Goldstone, Phineas
Royal Fusiliers.

T231908 Gordon (aka Goldberg), Simmy/Sidney
RASC (information from Sir Martin Gilbert).

Gotlieb, Gnr Gerald
RA. Buried Kanchanaburi.

Goudsmit, Cpl
Dutch Army.

Goulding, L/Bdr Harold B.
RA. Changi.

Green, Pte Myer
Beds and Herts.

T198268 Greenberg, Jack Joseph/'Aussie Pom'
RAMC. BEM for courage as a POW.

Greenberg, S.

4865 Greenburgh, Pte Walter William
HKVDC (?).

Greenstein, Gnr Harry
RA. Buried Kanchanaburi.

Greenwood (aka Greenburgh), Bdr Irvin Henry
RA.

Grosberg, Tpr David
18th Recce.

Gubby (aka Gabbay), Nurse Susan
HKVDC. Died, buried Hong Kong 17 May 1942.

1085133 Hart, Pte Sydney

2333737 Haffner, Sgmn Arthur
RCOS.

Hecht, Capt. Calman
RAMC.

2354193 Halpern, Leopold
RCOS.

Helpern, Len
Manchester.

Herman, William
Dutch Army. Died.

Herzberg (aka Hart), Harry
Leeds. Burma Railway.

Hewitt (aka Reich), Jack
Oxford & St George Club. Died as a POW, Malaya.

Hirsch, Lt Derek
Java.

Hirsch, Pte Joseph
RCOS. Changi and Siam. War artist.

1870026 Homberg, Sgt Harry

Hornstein, S.
Glasgow

Hyman, Pte Maurice
Notts & Derby. Died Changi, buried Kranji. From Hendon

Isaacs, Capt. L.
FMSVF. Java.

Jacoby, M.
Dutch (?). Changi.

Jacobs, Harry
RA. Java.

Jacobs, Capt. Noel
SVC.

3534657 Jacobs, Sydney
15th Manchesters.

Joffee, Spr Eugene
HKVDC.

Janis, Capt. Morris
RASC.

Joseph, J.H.
RA. Java.

Kandler, Reuben
RA.

Kant, Jack
RAOC (?).

Kaplan, L/Cpl Alf
1st Cambs Regt.

Katz, Tpr Nathan
RAC.

Kaufman, Dvr Harry
RASC. Buried Chungkai.

Kelmenson, Dvr W.
RASC.

Kersh, Gnr Leslie
RA. Changi.

Kitchner, G.L.
Glasgow.

Kramsky/Kransky, L/Cpl
Swansea.

Krieger, Pte Alfred
Notts & Derby. Buried Kanchanaburi. From Barnet.

Lamkin, Dvr S.
RASC.

Lander, Ronald Sidney
N. Lancs Regt. Died 8 October 1942. Possibly Jewish.

Lapsky, Jack
Middx Regt. Hong Kong.

Lavender, Sgmn Leopold
RCOS.

Leftwich, Cpl R.W.
Java.

Leigh, Capt. Mark
YMCA. Died as POW. Buried Kranji.

918076 Lerman, Myer
RA.

Leslie (aka Zilesnick), Pte Harry
Beds and Herts. Taiwan.

6092322 Levin, S/Sgt Leslie
IACC.

Levie, Lt E.L.
Dutch Army. Java.

427-29 Levinson, CQMS Basil
9th RNF.

Levinson, Sid/Solomon
Cambridge Regt./Suffolks.

Levene
2nd FMSVF. POW.

EC/2723 Levine, Lt Harry Gerald
6/1 Punjabs. From 3 Callendar Road, London NW/40 Catford Hill, London, SE.

Levison, M.
Changi.

Levitton, L/Cpl Henry Louis
RASC. Buried Chungkai.

Levy, Lt

Levy, Cpl D.
RASC.

V4029 Levy, Cpl Judah Felix
HKVDC.

Levy, Michael Meyer George
Escaped from Japanese in Shanghai. Trekked 2,000 miles to British lines and became *SOE* officer, *Malaya Force 126*, MiD.

Levy, Ronald
W. Surrey's Regt. Died 20 October 1945 after liberation as POW.

Levy, Pte Walter
RA.

Lichman, Sgmn Aubrey
RCOS.

Lipman, Gnr Nathan
RA. Died in Fukuoka, buried Yokohama.

Lisser, Yits
Dutch Army. Died.

Litenstone, Capt. Jack
Singapore.

Lithauer, L/Cpl Ronald Louis
RCOS. Buried Yokohama.

Loskey, Pte Henry Samuel
REME.

Lucas, Lt Henry Arthur
RA. Killed as POW aboard a ship.

Lyons, Pte Charles
RAMC.

Lyons, Henry, RAOC
Changi.

Macklin, Sgt Robert
RAEC. Buried Hong Kong. Wife survived.

Magrill, L/Sgt Jack Harold
RA. Died on POW ship.

Mandell, L/Cpl Abraham
RASC. Hong Kong/Japan.

Margolis, D.
1st Battalion Manchester Regt.

Marguleff, Sgt M.
SVC. Formerly *Palestine Police.*

Marks, Pte R.
RASC. Java.

Marks, Samuel
The Loyals. MiD. Changi/Singapore Seawalls/Burma Railway.

Markson, Pte Norman Gilbert
RAOC. Lost at sea as POW. Glasgow.

Marshall, Pte David
SSVF. Singapore community.

Melek, Meyer
RE. Died.

Mendelson, Pte Arthur Leonard
Middx Regt. Died at sea as POW.

Mendelsohn, Arthur
5th Beds and Herts. Singapore and Thailand.

Mendelson, Pte John
Brit. Att. Dutch Army.

Mendes Da Costa, Sgt N.
Dutch Army. Buried Kanchanaburi.

Mendoza, M.
RCOS. Tottenham.

Meninsky, Cpl Philip
War artist.

Mincovitch (aka Minchcom), Sgt Maurice Hillel
RA. MiD.

4204 Morgernstern, Nicolas
HKVDC.

13912 Morris, Pte Eric V.
1st FMSVF. Singapore. POW. Born 1907. Husband of Anna Estel Silvermann.

Morris, Pte G.
RAMC. Newcastle.

Morris, L/Cpl G.L.
FMSVF. Java.

Morris, Lt J.J.
Middlesex Regt.

Morris, 2nd Lt Joseph
RASC.

Morrison, Pte William
Glasgow.

Moser, Lt Carl David

1324 Moses, Gnr Ernest Sternberg
HKVDC.

Moss, Gnr Alfred Aaron
RA. Died.

Mossel, Pte J. (?)
Dutch Army. Batavia.

Mutsemaker, M.
Dutch Infantry. Buried Chungkai.

Nairnsey, Capt. Coleman
RAMC.

6142298 Nall (aka Ivanosky), Cpl James Frances
Born Walthamstow. Captured Singapore (? – info from daughter).

Nardell, Maj. Sidney
RIAMC.

Nathan, Pte
Harrogate.

Nemko, Sam
RCOS. Bethnal Green.

Nickolay, Gnr George Arthur William
RA. Buried Kanchanaburi.

Novick, L/Cpl Harris David
Norfolk Regt. Died at sea with Saunders. Surrey.

Nunes, Pte Morris
Sumatra.

Nussbaum, Rev. Chaim
Dutch Army. Jewish Chaplain.

Onderwyzer, Sgt Sam/L. (?)
Dutch Army. Buried Kanchanaburi.

Oppenheim, Pte Aaron Leonard
Australian 1st SSVF. Died Borneo, 15 February 1945. Kranji memorial.

Oppenheim, Pte Prof. Sir Alex
SSVF. Later Chancellor, University of Malaya.

Oppenheim, Pte L.R.

Orchant, Morris

Orenstein/Ornstien, Edwin Joseph
RA. Changi/Burma.

Paradise, Jack
Escaped to fight with Philippino guerrillas. Reached Australia (see *JC*, 5 June 1942).

Paulden, Gnr R.F.
RA. Prestwich.

Paulden, Pte S.
RAEC (?). Trafalgar St., Salford.

Pearlman (aka Sunderland), L/Bdr Edward Jacob John
RA. MIA/POW (?).

Pearls/Perls, Lt Ormond Israel Rodney
RA. Died Malaya.

Pepper, Pte Frank Robert
Norfolk Regt. Buried Chungkai.

Phillips, Capt. (later Sir) Henry Ellis Isidore
Beds and Herts. MBE. Oxford.

Pinto, Sgmn Andrew
Indian Division. Hackney.

Polak, M.
Dutch.

Popper, Cpl Barnett
Recce Coy/Rifle Bde. Hackney.

Posener, Alf
Aldgate.

Purdy, L/Bdr
RA. Died as POW.

Rabbie, Michael/Myer

Rabin, M
Java.

Rabinovitch, Gnr Simon
RA. Buried Yokohama. From Clapton.

Rabinowitz (aka Rabinoff), Pte David
RAC. Java.

Rakusen, Manasseh ('Nat')
HKDVC. Sham Shui Po camp and Japan.

Redhill (aka Roitenberg), WO Jack
FMSVF Bandmaster.

Reubins/Reubens, Gnr Samuel
RA.

Roit, Pte
Dutch Army.

Rose, Dvr A. Jack (aka Harry Rosinsky?)
AASC. Buried Kranji. From London.

Rose, L/Bdr C.
Cardiff.

Rose, Dr J.
RASC AIF. Died 29 September 1945 of illness on liberation.

Rose, Lt Lionel
Sherwood Foresters. POW 1943, crossing of Chindwin/*Chindits*.

1461091 Rose, Gnr Morris Albert
RA, Cardiff. Returned to London.

1101535 Rosen, Gnr Morris
RA.

Rosen, Gnr Solomon
RA. Buried Labuan.

1075156 Rosenberg, Gnr J.
RA. Buried Kanchanaburi.

Rosenberg, Col R.L.M.
RCOS. Died 3 March 1942.

Rosenthal, Capt. Willi
RASC. Singapore/Thailand. Aged 50+. WWI Croix de guerre (Paris).

Rosenthall, Gnr Alexander
RA. Glasgow.

Rosenthall, Cpl Alfred
RASC. Leeds.

Rosenwald, Capt. Pieter
Dutch. Taiwan POW website. Probably Jewish.

Roumania, Pte Abraham
1st Cambs. Lost at sea as POW.

Rubin, Lt Kenneth W.R.
RCOS.

Russell, M.L.
Intel Corps. Died of illness after liberation, 1946. Buried Kranji.

Sammes, George
Dutch.

Samuel, The Hon. Phillip
HKVDC.

Samsonovitch, Cpl P.
SVC.

Samuel, Herbert
HKVDC. Died 25 December 1941.

5775466 Samuels, N.D.
Royal Norfolk Regt. MIA/POW. Hackney.

Samuelson, Lt Alexander Humphrey
RA.

Saunders, Cpl Louis
Norfolk Regt. Lost at sea while being transferred as a POW.

Schulman, Major Harry Edward
5th Norfolks. Singapore.

Schwenk, H.
RASC. Died 10 April 1942.

76465 Seebohm, Major E./G.
RA. Singapore and Thailand.

Sefton, Capt. Lou
RAMC. Singapore.

Segal, Gnr Mark/Michael
RA.

Segal, Reuben 'Johnny'
RAMC.

Shapiro, Cpl
Lordship Park, Stoke Newington.

Shapiro, Sgmn Israel
RCOS. Buried Kanchanaburi.

Shoop/Shloop, Pte Judah
Beds and Herts.

6472825 Shore, Pte Theodore
Royal Fusiliers. Southgate.

Shulman, Pte Sydney Solomon
Gordon Highlanders.

Silman, Capt. Harry
RAMC. Changi and Burma.

Silver, L/Cpl A.
Norfolk Regt.

5775523 Silver, Abraham
RCOS.

1807866 Silver, Gnr Samuel
Died as POW in Singapore.

Silverstone, Pte Nathan
RASC.

Silverton, E. Kitchener
RA. Saigon.

5950961 Simmons, Michael (aka Morris Sconalinksky)
Beds and Herts. Willesden, London.

Simon, Pte Charles
SSVF.

1447697 Simon, Pte Moss
RA. Cardiff.

Simons/Simon, Mark
Singapore. Burma Railway.

Singer, Dvr Lewis
RCOS. Buried Kanchanaburi.

Sluizer, Pte I.S.
Dutch Army. Died.

Smith, Pte Jack
RAOC.

Snell, Capt. E.
RAMC.

South African Jewish Doctor
Not named by Nussbaum.

Speijer, Simon 'Dik'
Dutch Army.

1829108 Steel, Pte Joseph
Java. *RA*. Hendon.

Steingart, Sydney
RA. Died 24 August 1945.

Stern, Capt. Eugene
WWI *Austrian Army in UK/AIF forces.*

Stern, Pte Nathaniel
Recce Coy/Rifle Bde.

Sterne, Pte Meyer
RAOC, Singapore.

Sternfield, Sol
Dutch Medical Corps.

Stockfisch
Dutch Army.

Stone, Gnr Leslie
RA. Buried Kranji.

Stone, Maj. Leslie David
RAMC.

Stone, Gnr Louis
RA. Buried Kuala Lumpur.

Storch, Pte Adolph G.
1st FMSVF Labuan. Died 22 January 1945. Changi/Borneo.

Strauss, Capt. Richard L.
HQ 11th Indian Division. MIA/POW Singapore/Thailand.

1098941 Sugar, Clyde
From Westcliff.

Sugar, Cpl Jack
Norfolks. From Luton.

Sunshine, Edward
RA. Died 5 March 1943.

Swift, Pte Joseph
Rangoon Memorial.

Symonds
Java.

Talan/Jalan, Lt Emmanuel
HKVDC.

Taylor, Pte H.

Tiller, Gnr Morris
RA.

Tobias, Pte E.
Sherwood Foresters.

Tropp, Spr Alfred
RE.

Tunkle, Benjamin RA
Died at sea as POW.

Turchin, John

Unknown
Two indecipherable signatures on Changi Synagogue leaflet in the Arkush personal archive at his home, August 1945.

Unterman, Frances
RA. Died 1 March 1942.

Van Gelderen, M.J.
Dutch (?). Changi

Vas-Nunes, Sgt Maurice
RASC.

Velinsky, F.
RA. Java.

Velmans, Sgt Loet/Louis
Dutch Army.

Weill, Gnr Leon
HKDVC.

Weinblatt, Pte Raphael Leslie
Suffolk Regt. Buried Thanbyuzayat, Burma.

Weinfass, Pte Leslie
Royal Norfolk. Taiwan.

Weinman, L/Bdr W.
RA. Died 13 February 1942.

Weinstock, Wally

Weiss, Pte Harry
Luton. Mentioned in MacArthur, *Surviving the Sword.*

Weissmann, Geoffrey
Java.

Wenzerul, Gnr Edward
Lost at sea as POW.

Wernick, L/Cpl Cyril
Norfolk Regt. Died after liberation as POW in Singapore.

Willenski, Pte Michael
Cambridge Regt.

Williams (aka Williamovsky), J.
Glasgow.

Wince, L/Cpl David Leonard
2nd Cambs.

Winter (aka Ginsburg), L/Cpl Leonard E.
Middlesex Regt. Buried Hong Kong.

Wolf, Pte
Dutch Army.

Woods, Maj. Everard
RIASC.

Woolf, Robert 'Bobby' (aka Emmanuel Nathan Woolf)
RCOS. Taiwan.

S1826763 Yegliss, Pte Isaac
RASC. Liverpool.

Yudkin, Abraham
RA. Lost at sea as POW.

Zeehandelaar, Pte L.H.
Dutch Army. Java.

Zimmer, Gnr Morris
RA. Lost at sea as a POW.

Zimmern, Bdr A.
HKVDC. Died 18 December 1941. Of Jewish origin.

Zimmern, Sgt E.
HKVDC. Died 19 December 1941. Of Jewish origin.

RAF

Altman, Cpl Lewis
Attached *RAAMC.* Secretary to Col 'Weary' Dunlop.

Baker, Sgt Leonard Alfred
Java. Brother of James. *Army.* Died.

Bloom, Pte Jack

Brogsall, A.
Glasgow.

Camberg, David
Died 28 April 1945.

Circurel, P/O Richard Leon
Java.

Coleman, Samuel Gerald
POW Batavia.

Courts, AC2 Harry A.

De Lange, LAC George Ivor
POW Java. Buried Bandoeng.

Fell, A.
Glasgow.

Fisher, LAC John
BEM for courage giving medical services and interpreter for POWs.

Goldberg, AC2 Charles
Died 29 November 1943.

Grunis, Cpl Abraham
Lost at sea as POW.

Harris, LAC C. Hyman

Hart, LAC Sidney

Kaplan, J.
Glasgow.

Lawrence, LAC Sydney

Leigh/Lee, LAC Harold
Sandakan/Sarawak.

Levinson, LAC Sydney Henry
Buried Kanchanaburi.

Levy, Harold
Java.

Levy, LAC Alf/Sonny
Buried Kanchanaburi.

Louisson , Fl. Officer George
RNZAF, Java.

Marks, AC1 Robert George
Kranji memorial.

Merton, R.M.

Nathan, LAC Gerald M.
Died on Sandekan Death March.

Prechner, AC1 Harold Joseph
Changi.

Rosenberg, WO Alfred/Albert
203 Squadron, RAF. Hackney.

Ruben, LAC S.A.
Leeds.

Shenken, L.
Glasgow.

Simmonds, AC2 Albert Bernard

Simmonds, LAC Joseph

Solity, AC2 Abraham Sidney

Spektor, Cpl Baruch
84 Squadron. Java.

Spero, Cpl David 'Doddy'
Java.

Stoll, Dr F/O Basil Arnold

Warshawsky (aka Shaw), AC1 Arthur Reginald
Changi/Java.

Weiner (aka Wilton?), AC2 Louis
Changi.

Welzer, LAC David
Java.

Australians

Some of this information comes from G. Pynt, *Australian Jewry Book of Honour* (*AJBH*) (Australian Federation of Jewish Ex-Servicemen and Women, 1974). There are a dozen indecipherable names from the Barmitzvah Certificate written at Tamarkan camp, 27 September 1944 (now in the Museum of Australian Jewish History), mentioned in the *AJBH*, which has Australian, British, Dutch, USA and even some Japanese names on it, of those attending the mock bar mitzvah ceremony for the son (Leon) of Mark 'Gil' Hayman, one of the Australian POWs.

The abbreviation 'San' = Sandakan Death March, killed as POWs.

Abrahams, Gnr
Java.

Abramovitch, Capt. Gershom S.
Malaya.

Altshuler, Sgt Evan
Died at sea.

Asher, Pte Keith R
Died at sea.

Bard, Dr Solomon
HKDVC. Kowloon.

Berliner, Pte Leon A.
Executed.

Berman, Izzi

Bernstein, Pte Harry
Died at sea.

Bernstein, Pte Philip
Malaya.

Bines, James
From USA. *AIF* and *FMSVDF.*

Blashki, Sgt Victor
Kranji memorial.

Bliner, Sgt Sam

Boan, Sgt Clive H.
2/29 Infantry. BEM. Changi.

Brand, Capt. Victor
2/29th Battalion 8th Division. MC.

Cavanagh, Pte C
22nd Pioneers, AIF. Orbust, Victoria.

Cohen, Bdr Arthur
Java.

Cohen, Pte Lewis S.
Changi.

Cohen, Pte Norman M.
Changi. Died of illness after liberation.

Cohen, Sgmn Ronald L.
Tokyo.

Danny (?)

Davis, Gp Capt. Reginald H.S., MiD,
OBE, *RAAF*, Manchuria.

De Groen, Sgt Geoffrey
Fukuoka.

Deane, Dvr Nathan
Timor.

Diamond, DFC, Lt Oscar
Changi, Taiwan, Pusan, Mukden.

Edelstein, Dvr Ivor
Died Rabaul.

Falk, Gnr Harry
Timor.

Falk, Sgt Norman
Changi.

Goldburg, Arthur Leslie
AAOC. Died at Kanchanaburi, 7 September 1943.

Goldman, Pte Samuel
Changi.

Griff, Sgt B.G.
Java.

Harris, Lt
Son of Maj. Sam Harris, below.

Harris, Sgmn Syme Newell
San – died 5 June 1945. Second son of Maj. Sam Harris, below.

Harris, Cyril Montefiore
Died 20 April 1945.

Harris, Maj. Samuel
Victoria.

Hayman, Mark/Marcus 'Gill'
Tamarkan.

Jacobs, Sgt Peter A.
Died at Ambon.

Jacobs, Pte David T.
Died Malaya.

Jacobs, L/Cpl Harold
Malaya.

Jonas, Cpl Lewis
Changi.

Joseph, Pte Leonard
Singapore.

Kempler, L/Cpl Edward
Java.

Krantz, Maj. Dr Sydney
RAAMC. Burma, Thailand, Changi.

Krasnostein/Krass, Pte Leslie/Lazar
Malaya.

Langford, Pte Alfred E.
Java.

Lazarus, Lt Samuel A.
Ambon.

Letwin, Pte Gerald
Died Malaya.

Levey, Pte Robert E
Borneo.

Levinsohn, Sgt Harold A.
Changi.

Levy, WO2 Patrick R.
Malaya.

Mansfield, Lt George J.
Changi.

Marks, Pte F.J.
AIF. Died 20 July 1943, Saigon.

Milston, Sgmn Neville
8th Division. Thailand.

Montefiore, Sgt
(Named by Col E. 'Weary' Dunlop.)

Moritz, WO Abraham
RAAF, Java.

Morris, Cpl Alan
Died Ambon.

Nagelberg, Sgmn
Polish serving Australian Army.

Nagle, Max
German refugee.

Noble, Cpl C.H.
Bondi.

Norman, Sgt G. Solomon
Changi.

Orloff, Bert

Pearlman, Leslie
Executed Rabaul, January 1942.

Peck, Flt Lt
RAAF. Melbourne.

Pierce, Pte J.
Alias, as not in *AJBH.* Bondi.

Pyke, Lt R. Elias
Changi.

Rabinov/Rapinow, Alan

Raphael, Lt Geoffrey A.
Died Malaya.

Relik, Sgt Schmuel
(Written in Hebrew on Bar Mitzvah certificate.)

Rosebery, Sgt Norman J.
Osaka.

Rosenberg, Alf

NX23803 Rosenberg, Gnr Allan
RAA. Buried Kanchanaburi.

Rosenberg, Gerald

Rosenbloom, Alex

Rostiker, Pte Samuel
Changi.

Sachs, Lt John
AIF 'Z' Force Commando. Executed as POW.

Samuelson

Sanders, Pte Harry
AIF. Buried Kranji.

Schwartz, Capt
Java.

Segal, Sgmn Norman E.
Timor.

Shadbolt, R.
Sydney.

Sharp, W.
NSW. San – died.

Shaw, R.
NSW. San – died.

Silman, Leon
MiD, Melbourne

Silverman, Capt. H. Nathan
Executed Rabaul.

Soffer, Pte S.
Melbourne.

Solomon, A.
Melbourne.

Solomon, Gnr Neville L.
Died Malaya.

Solonsch, Samuel (aka John Douglas, aka Schmidt)
Executed by Japanese as an Australian Intelligence Officer, while a POW, Java, 11 April 1943 (see *The War Diaries of Weary Dunlop* [Penguin, 2005], p.80).

Spielvogel Sgt L.H./Speilvogal, F.
Balarat, Victoria.

Stolarski, C/J
Victoria. San.

Stone, Pte Joseph E.
Died Malaya.

Swartz/Schwartz, Pte J.
Lawley, Western Australia.

Taylor, Joe
Rumanian refugee in *AIF.* Changi.

Van Der Sluys, Pte Aaron
Java.

Victorsen, E.L.M.
New South Wales/Queensland. San.

Weinberg, Sgmn Coleman

White, Bernard
Victoria. San.

Wood, Sgt Francis E.
Changi.

Woolf, Dvr E.B.
AIF.

Zinn, Spr Aubrey
Born in South Africa. Sent from Sandakan to Labuan.

Ziporkin, Maurice
AIF

Israeli/Palestinian Jews

Alster, Reuben/Rudolf
RAF 211 Squadron. KIA Israel War of Independence, 1948.

775400 Berliner, LAC Alexander
From Nes Ziona. Java, Changi, Fukuoka 2, Japan. Often beaten for deliberately damaging metal shipping plates in shipyard camp.

Cohen, Mrs Moselle Y.
Murdered by Japanese for assisting Sandakan marchers in 1944.

Gompertz, P.
Batavia (Basil Stoll papers).

Kissen, Abraham 'Kiki'
RAF 211 Squadron. Samarang, Haraku, Batavia.

Sikral (?)
Batavia (Basil Stoll papers).

Unknown Third *RAF*, probably Ehrmann, Leonard/Ludwig
RAF. Tel-Aviv (*JC*).

Canadians

From *Canadian Jewry at War* (Montreal: Canadian Jewish Congress, 1947–48).

Allister, Sgmn William
RCCOS. Hong Kong/Tokyo. War artist.

Brown, Rfmn Louis
Canadian Rifles. Hong Kong/Fukuoka.

Golden, Capt. David Aaron
Winnipeg Grenadiers. Hong Kong.

Harrison, Capt. Robert George
Manitoba Regt. Hong Kong/Osaka.

Markowitz, Capt. Dr Jacob
Served with RAMC. MBE for courage as a doctor POW. Malaya.

Rose, Sgmn Jacob
RCCOS. Hong Kong/Tokyo.

Zaidman, Rfmn Frederick 'Zeke'
Canadian Rifles. Hong Kong/Tokyo.

South African

Hefferman, AB Nelson Joseph
HMS *Encounter.* San. Lost at sea as POW.

Totals

Navy	13
Army (including Dutch)	375
RAF	38
Palestinian Jews	7
Canadian	7
Australian	101
South African	1
Total	**542**

Addendum to the Roll

Among the debriefing sheets of liberated POWs of the Japanese at the National Archives at Kew (series WO344 361–410) were the names of men who did not have AJEX Jewish Chaplain cards but who were probably Jewish; they have *not* been included in the final figures. It should be noted that file WO344/405/2 is missing (Thompson, K. – Totterdell) and so the author was unable to examine these sheets. (b. = born)

DSX1188588 Asher, Cecil, *RN*, HMS *Exeter*, b. 27 October 1924, of 41 Ridley Rd, Dalston, London. Captured in the Java Sea.

Barnet, Maj. Kenneth Myer Arthur, *4th Artillery, HKVDF and 8th Coast Regt*, b. 19 March 1911. Assisted in many escapes from Sham Shui Po Camp, Hong Kong, by drawing maps, advising on routes, providing letters in Chinese to escapers for use with locals, distracting sentries and translating Japanese and Chinese newspapers with war news to boost morale.

1577679 Baume, Leonard, *RA*, b. 10 January 1912, from Camberwell. Changi and Thailand.

888791 Berger, Francis Maurice, *RA*, b. 21 January 1916, of 247 Queens Park Rd, Brighton. Changi and Thailand; sabotaged Japanese vehicles and railway rolling stock when on work parties.

7372859 Brand, Leslie G., *RAMC*, b. 31 March 1917, of 21 Sarrie Rd, West Hampstead. Changi and Thailand.

1638025 John Caplan, *RA*, of Totton, near Southampton. Saigon camps.

128---- Davidoff, Nicol, ***SREV*** (?), b. 17 October 1906, from Woolloomooloo, Dowling St, Sydney, Australia.

7360095 Davidson, Leslie Samuel, *197 Field Ambulance, RAMC*, b. 19 April 1919, from 57 Kempster St, Broughton, Manchester.

7532 Deitz, Pte Bertram John, *RAMC*, b. 3 February 1918, from 110 Sackville Rd, Hove. Tarso and Nakonpladuk camps.

Fl. Lt de Vogel, Dick, *Royal Netherlands Air Force*, b. 13 April 1903, with an address in Portman Square, London W1. Captured at Bandoeng; carried out many acts of sabotage, detailed in TNA file WO344/372/2, and attempted several escapes from Burma/Thailand camps.

1111004 Diamond, Gnr Joseph Henry, *135 Field Regt, RA*, b. 9 December 1910.

5775819 Dunkelman, L/Cpl Henry C., *6th Norfolks*, from 97 Tollingon Park, London N4. Imprisoned at Changi, in Thailand, Nagasaki (Japan) and Toyuma (Japan). Helped sabotage the iron works at Nagasaki.

No. 31, Ellis, Asst. Quarter Master Ellis, Sophie, *HKVDC*, b. 28 October 1897, address c/o Mrs E.A. Richardson, 32a Adelaide Rd, Surbiton. She was imprisoned at Bowen Road and Stanley camps, Hong Kong.

4619244 Fairman, Pte Douglas, *RAOC*, b. 29 September 1919, from Holcombe Rd, Rochester.

133939 Falk, Capt. John A., *RAMC*, from 81a Portway, Warminster.

5828939 Gaster, Sydney George, *1st Cambridge Regt*, b. 29 October 1919. Nonpladuk and Chungkai.

PLYX1062 Glass, Sgt David Reuben, *RM*, b. 20 August 1915, from 264 Boothberry Rd, St Anne's, Heesle, Yorks. A regular who enlisted on 27 August 1929. Captured at Muntok; in prison from 15 February 1942 until 2 March 1942; moved to China School gaol, then on 16 March to Chung Wa School gaol. Sent to Airport Camp on 15 May and back to Chung Wa on 20 August. Moved to Songai Ron on 22 February 1944 and there until liberation. Loading ships at Palembang, he assisted in making holes in oil drums bound for Japan, with RM sergeants Railey, Hall and Crampton, and cutting various electric cables.

178401 Goetz, Dvr Charles Alfred, *RASC*, b. 25 December 1913, of 26 Mardon Rd, Acton. Changi, Bukit Timah, Tamarkan, Tamarang, Sanat Su and Takili camps.

H17419 Gold, Cpl Stanley, *Winnipeg Grenadiers*, b. 19 September 1916, of Pine Falls, Manitoba. Captured in Hong Kong, at Sham Shui Po, North Point, Bowen Road, and on 19 January 1943 was shipped to Japan to work in the mines at Ormine; he took part in damaging mining equipment.

929873 Goldring, Gnr Arthur, *88th Field Regt, RA*, b. 7 January 1919, from 49 Greyhound Rd, Tottenham, London.

555610 Gollop, Cpl Frederick, *3rd Hussars*, b. 29 May 1913, from Andover.

H6480 Gurski, Joseph, *Winnipeg Grenadiers*, b. 20 October 1911.

1267085 Hankin, LAC Stanley, *RAF*, b. 27 April 1920, Portsmouth.

182712 Hayman, Dvr Joseph, *35th LAA, RA*, b. 18 November 1919 of 84 Goodwin Ave, Birkenhead. At Changi until February 1942, then Havelock Road until May 1942. On the Railway until November 1943 (he mentions Tanabaya) and back at Changi until May 1945, then liberated at Kranji.

823512 Hillberg/Hilberg, Pte Frederick, *RA*, Bristol.

2360104 Hoffman, Cpl Roland, *Malaya Sigs*, Huddersfield.

1808968 Gnr Jacobson, Samuel, Nottingham.

3186 Kaploon, Pte Boris Lazar, *HKVDC*.

6019225 Koss, Pte Jack/John, *Suffolk Regt*.

Koster, Steward George, *MN*, Liverpool.

1558213 Kramer, Gnr Edward, London E6.

1456110 Lazarus, Gnr Alfred, Gosport.

10533710 Lever, Pte Lionel, *RA*, Clapham.

6142604 Levey, Albert, *East Surreys*, Wandsworth.

1548265 Levey, Harry, *RA*, Clacton.

13215 Levien, Bdr Norman Max, *SVC, New Zealand Territorials*, Auckland.

148693 Louis, Sgt Albert Leonard, *RASC*, b. 8 April 1913, of 34 Hazebury Green, Edmonton. At Changi until October 1942, Chungkai until March 1943, Takamun until October 1943, Tamarkan until March 1945 and Petchburi until liberation.

1920 Meyer, Act. CSM Joseph, *HKVDC*, WWI veteran, b. 8 September 1894. Argyll Street and Sham Shui Po camps.

1289362 Michaelis, Cpl Maurice George, *RAF Air HQ*, b. 4 March 1911, from Old Rectory, Cuddington, Bucks. Captured at Tasik Malaj, Java, on 8 March 1942; there until May 1942, then at Soerabaja until December 1942, Darmo until January 1943, Jaarmarkt until April 1943, Haroekoe Island until July 1944, Liang (Ambon) until August 1944, Ambon Town until end August 1944, Batavia (Makasura) until January 1945, River valley (Singapore) until April 1945, Tanjon Pagan (Singapore) until July 1945, Morse Road (Singapore) until August 1945 and liberated at Tanjon Pagan on 15 September 1945.

973133 Mier, Gnr Arthur, Shore Place, Hackney, London E9.

3959165 Mordecai, Pte Cyril, *2nd Welch*.

845015 Moser, Gnr Arnold Henry, from Manchester.

891055 Moser, Bdr Ernest John, from Blackpool.

3856173 Moser, Gnr Frederick Francis, from Blackpool.

184333 Moser, Lt Peter Howard Moser, from Reigate.

13342 Naktiel, Pte Daniel, *2/10/Baluch* (India), b. 19 August 1907, from 60 Ellerby St, Fulham, London, enlisted January 1938. Captured at Singapore; at Changi until 4 November 1942, Tarsao until July 1943, Tonchan until December 1943, back to Tarsao until November 1944, then Tamuang until December 1944, Wampo until February 1945, Kanchanaburi until July 1945 and liberated at Nakon Nyak in August 1945.

4126515 Orman, Fus. Leonard, *9th RNF*, b. 7 March 1919 from 578 Didsbury Rd, Manchester. Captured in Singapore; in Thailand until May 1943 and at Changi for the rest of the war.

1427423 Pam, Gnr Jack, *RA, 9th Coast Regt*, b. 23 June 1920, enlisted 24 November 1938 in TA; from 104 Glenfell Ave, Romford. Captured in Singapore; at Nonpladuk until January 1943, then in various railway camps until July 1943, returning to Nonpladuk until February 1945. Liberated at Ubon in August 1945.

Paskin, Gnr Stanley, *RA, 79/35th LAA* (no army number visible), b. 4 November 1920, from 33 Leech Ave, Chesterton, Stoke-on-Trent. Captured at Gardet, Java, 9 March 1942; at Tanjong Priok, Batavia, until 9 September 1942; Changi until October 1942; Kuching, Borneo, until September 1945.

134557 Pearl, Dennis F., *RASC*, b. 22 December 1914, of Holland Park, London.

67514/69514 Perez, Capt. Julian Loudon Low, *1st Manchesters*, b. 27 November 1913 and enlisted in TA on 27 January 1938. Captured at Singapore; at Changi until October 1942, Chungkai until January 1943, Bankow until April 1943, Takanum North (Kilo. 211) until October 1943, back at Chungkai until February 1945 and at Kanchanaburi until September 1945.

1709605 Phillipson, George Frances, *5th Searchlight RA*, b. 1 June 1911 and lived at 117 Kingsway Rd, Liverpool. Captured at Singapore; at Nakon Platon, Tarsao, Kanyu and Tamuang.

4857663 Pizer, Pte William, *1st Leic. Regt*, b. 26 September 1915.

5933238 Reuben, Pte Walter Philip, *1st Cambridge.*

95580 Rickman, Capt. Gerald, *RNF*, from Twyford.

3193 Rootstein, Pte Anatol, *HKVDF.* Sham Shui Po and Kowloon Island camps.

4275458 Salkeld, Fus. Richard, *9th RNF*, b. 20 November 1917, enlisted 15 December 1939, 11 Fourth St., Gateshead. Captured Singapore; sent to the Railway at Kinsayo until August 1943, Hindarko until December 1943, Nakom Platon until December 1944, Nonpladuk until February 1945 and Ubon until August 1945. His acts of sabotage included spiking oil drums and throwing metal ingots and other war material into the sea at Singapore Docks.

1543363 Salkeld, Gnr William, *69/71st LAA, RA*, b. 20 May 1908, enlisted 18 November 1939, of 17 Beaufort St, Liverpool. Captured Java 9 March 1942; at Batavia until September 1942, Changi until February 1943; Kuching, Borneo, until September 1945.

1745257 Salmons, Gnr Cyril E., *12/6th HAA*, b. 21 August 1914, from 33 Edith Rd, Stratford, London E15. Captured Garoet, Java; sent to Tanjon Priok until 1 October 1942 and transported to Changi; then to Kuching, Borneo, until September 1945.

273 Samuel, Dvr Phillip Ellis, *HKVDF*, b. 23 December 1900, of 32 Porchester Terrace, London W2. At North Point, Hong Kong, until January 1942, Sham Shui Po until December 1943, then shipped to Narumi (Japan) until May 1945 and Toyama, Japan, until liberation.

1609789 Samuel, Gnr Samuel, *77th HAA*, b. 4 November 1920, of 64 Acton St, Birches Head, near Stoke. Captured 8 March 1942, Java; at Tanjong Priok until May 1942, Changi (Roberts Hospital) and Kanchanaburi until October 1942, Tonchan South until May 1943, Hintok/Conquita until June 1943 and Pasir Padang until late 1944. No more is known.

2125163 Samuels, Spr Alfred T., *RE*, b. 22 July 1920, from 34, Jubilee Rd, Newbury, Berks. At Changi until June 1942, then on the Railway until May 1944 and sent by ship to Mitsushima, Japan, until September 1945.

7390882 Schaefer, Pte Leonard Raymond, *32 Coy, RAMC*, b. 4 May 1920, from 11 Wellgarth Walk, Knowle Park, Bristol. Captured and kept in Singapore, at Changi until May 1943 and Kranji until September 1945.

7228 Schuler, Frederick Alfred, *FMSVF*, b. 5 May 1903, from Perth, Australia. At Changi until May 1942, Kinsayok until October 1942, Tamarkan until May 1943, Non Pladuk until November 1943 and Ubon until September 1945.

948116 Schulz, Gnr Frederick, *134th RA*, b. 8 November 1918, from 261 Warley Rd, Blackpool. At Changi until May 1943, on the Railway until November 1943, back to Changi until June 1945 and Kranji until September 1945.

7687721 Shieff, Cpl Stanley Alfred, *CMP*, b. 17 November 1921, of 72 Welsh Row, Nantwich, Cheshire. Captured Banka Straits, 22 May 1942; at Changi until September 1945.

Sichel, Lt Eduard Charles Sylvester, *SSVF*, b. 1 August 1900, volunteered 1937, from Shanklin, Isle of Wight. At Changi until June 1942, Bampong until October 1942, Chungkai until February 1945, Tamarkan/Kanburi until July 1945 and Nyakon Nye until September 1945.

832059 Silver, Mechanic Joseph, *88th RA*, b. 8 September 15, from 58 Harrington St, Preston. At Changi gaol until April 1942, Towner Road until November 1942, Serangor Road until December 1942 and Changi until September 1945.

D/KX 112322 Simons, Stkr Albert E.R., HMS *Encounter*, *RN*, b. 31 January1919, from 6 Tasman Rd, Stockwell. Captured Java Sea, 2 March 1942; at Macassar until September 1945.

2353043 Simons, Dvr Herbert A., *RCOS*, b. 10 August 1912, of 126 Moorhouse Rd, Hull. Captured Singapore; at Changi until November 1942, Tarsao until August 1943, Rin Tin/Tamuang until April 1944 and Takanun until April 1945.

1531683 Solomon, AC Raymond Stanley, *RAF, 13th MU*, b. 12 November 1910, of 125 Horninghow St, Burton-on-Trent. Captured Garoet, 8 March 1942; at Boei Gloduk until January 1943, Tanjon Priok until April 1943, Ambon until May 1944 and Cycle Camp, Batavia, until August 1945.

5773707 Solomon, Pte Jack, *6th Norfolks*, b. 27 January 1919, of 3 Clarkes Lane, Oulton, Lowestoft. At Singapore until January 1943, the Railway until May 1944 and Japan until September 1945.

6144853 Somner, Cpl Arthur Harry, *2nd E Surreys*, b. 17 June 1919 and enlisted 15 November 1939, of 12 Owen Mansions, Queens Club Gdns, London W14. At Changi until November 1942, Chungkai until February 1943, Bankow until March 1943, Km. 110 until April 1943, Chungkai until May 1945 and liberated at Tamuan.

1180881 Spiers, AC2 Henry John, *RAF*, b. 21 September 1920, of 26, Chestnut Ave, Headington, Oxford. Captured Java at Batavia; at Struiswijk until September 1942 and Soeka Miskin, Batavia, until February 1943, then Tjimihi (Bandoeng) until February 1944.

6202--93 Sprenger, Pte Charles, *1st Middlesex Regt*, b. 4 December 1913, of 28 Halse St, Holloway, enlisted 9 June 1936. Captured Hong Kong; at Shingowa until 4 September 1942, Omory until 2 April 1944 and Camp 24D until liberation.

2338349 Stirzacker, L/Cpl Frederick Arnold, *RAF Signals*, b. 29 April 1919, enlisted 31 January 1939, of 22 Victoria Terrace, Cleckheaton, Yorks. Captured Java; at Bodei Gladok until May 1942, Tanjong Priok until April 1943, 12th Infantry Barracks until May 1943, back to Boei Gloduk until September 1943, Barracks camp until November 1943 and sent to Fukuoka (Yawata steelworks, building locomotives) until liberation. Whilst working in the docks in Java he helped sabotage oil drums and machinery. He also witnessed the giving of an order that any attempted escapes would be dealt with by execution of every tenth man in the camp.

873316 Stirzacker, Bdr Thomas, *137 Field Art., RA*, b. 24 June 1920, enlisted 7 September 1937, from 15 Jeffrey Sq., Blackpool. At Changi until June 1942, Nonpladuk until September 1943, Wun Yai (Wan Yi?) until December 1943, Yosohan until September 1944 and liberated at Kaorin.

620075 Symes, Lawrence

V4309 Tausz, Sgt Joseph, *HKVDC*, b. 24 April 1900. Captured Hong Kong; at North Point until January 1942, Sham Shui Po until September 1944 and Sendai, Japan, until liberation. Took part in sabotaging oil drums at the docks in Japan.

5829648 Weistock, Sgt Kenneth, *4th Suffolks*, b. 27 March 1917, of 26 Allendale Ave, Finchley. At Changi and various jungle camps.

7684586 Werninck, Edward, *CMP*, b. 7 September 1913, enlisted 5 May 1939, of 131 Langham Ave, Tottenham. Captured Java; at Gloduk until 4 April 1942, Tanjong Priok until 4 May 1942 and in Ube, Japan, until liberation.

4756250 Weyman, George

7368795 Whyman, Pte Leslie, *RAMC 197 Field Ambulance*, b. 1 July 1917, of 26 Brereton Rd, Tottenham. At Roberts Hospital, Changi, until March 1942, Salerang Hospital until August 1943 and Kranji Hospital until late 1944. No more is known.

2875509 Woolfson, Pte Cecil, *2nd Gordons*, b. 2 November 1916, enlisted 8 November 1934, from Shotesham, Norfolk. At Adam Park Singapore then Tamarkan, Kinsayok, Chungkai, Nonpladuk and Ubon.

1286396 Zenker, LAC Oscar Anton, *RAF*, of 21 Arlington Rd, Woodford Green. Captured Java; at Bourabaya until April 1943, Haroke Island/Ambon until September 1944 and Batavia until September 1945.

Civilian POWs

There are over 100 other Jewish-sounding names on the list from Jonathan Moffatt (see the introduction to this book), but we cannot yet verify religion. By 7 April 1943, eyewitness Thomas Kitching (surveyor of Singapore) stated that there were 110 Jewish civilian POWs at Changi gaol: see C. Bayly and T. Harper, *Forgotten Armies* (Penguin, 2005), p.340.

Unless specified, the dates refer to when they died.

Aaron, J.E.E. Salimon
Singapore.

Aaron, H.R.
Estate agent, Singapore.

Abbott, Bill

Abbott, Shookee

Abdullah, Abraham Ezekiel
Died.

Abett, Stanley

Abraham, David Ezekiel Joshua
Died Shanghai camp, 27 May 1945.

Abraham, Rabbi, name unknown, and his family
Shanghai camp.

Adair, Randall

Adler, E.S.

Albert, Elias
Died.

Attias, Clara Adelina
Died Sime Road camp, Singapore, 2 August 1944.

Ballas, Jacob

Baruch, Haskel

Bekhon/Bekhore, Yahya
Singapore.

Bennett, Mrs Margaret
Aged 26 in 1942. Changi and Sime Road.

Bloom, Mrs Freddy
US wife of Dr Phillip Bloom (see Roll and Record, UK and Dutch Army list). 'Double Tenth' victim. Singapore.

Bolter, Norman
Died at sea on transport.

Bolter, Max

Bridges, Grace

Bridges, Ralph
Husband of Grace, above.

Brisk, Albert Wolfe
Died aged 62, Changi, 19 December 1942

Brisk, Buyeenish
Died Changi, 18 November 1943; buried Jewish cemetery, Lim; Chu Kang.

Brisk, Mrs E.R.
Singapore.

Brisk, Mr I.A.

Brisk, Max
Changi and Sime Road.

Brisk, Mrs Rebecca
Wife of Max, above. Aged 47 in 1942. Changi and Sime Road.

Citrin, Walter
Also his wife and two sons. Chapel camp, Hong Kong.

Cohen, Mrs A.
Died at sea, April 1942. Singapore.

Cohen, Mr A.
Husband of Mrs A. Cohen, above. April 1942.

Cohen, Flori Enid
Interned China. Born Shanghai.

Cohen, Isaiah Meyer
Husband of Flori Enid, above. Interned China. Born Shanghai.

Cohen, Moses M.

Cohen, Silman
Died.

Cohen, Mr Yahya
Changi and Sime Road. A professor of surgery after the war.

Dabelstein, Winifred E.
Died Hong Kong camp, 21 September 1944.

Daniel, Moses

Daniel, Sion Charlie

David, Jacob Moses
Singapore.

David, Jack M.
Changi. Survived.

David, Joe Brooke
Tin broker.

Davis, S.
Changi. Singapore Jewish Community.

Deborah, Albert
Died.

De Leon, S.
(With the Dutch at Tamarkan Camp) Dutch War Office civilian.

Elias, A. Ezra
Changi and Sime Road.

Elias, Diana (née Hardoon)
Interned Stanley Camp, Hong Kong. Survived.

Elias, E.A.
JP. Interned Stanley Camp, Hong Kong. Survived.

Elias, Helen

Elias, Jack M.
Interned Stanley Camp, Hong Kong. Survived.

Elias, Jacob
Changi and Sime Road.

Elias, Jonah
Changi and Sime Road.

Elias, Joseph
Changi and Sime Road. Died.

Elias, S. Moses
Changi and Sime Road.

Elias, Ezekiel Moses
Died.

Elias, Zeki

Elisha, Joseph

Elisha, Meyer J.
Changi and Sime Road.

Elkins, Mrs K. Lena (née Cohen)
Changi and Sime Road.

Elkins, Philip H.
Husband of K. Lena Elkins, above. JP. Customs and Excise Officer.

Engels, Dr

Ephraim, E

Ephrain, Jacob

Etlin, B

Ezekiel, Abraham
Changi Road, 10 May 1943.

Ezekiel, Elias
Died.

Ezekiel, Maurice
Died.

Frankel, David
JP. Singapore.

Gareh, A.

Goldberg, Dr Mrs
Sumatra. See Betty Jeffrey, *White Coolies* (Angus & Robertson, 2001), pp.82, 113, 114.

Goldenberg, Harry
Died aged 57. Malaya.

Goldman, B.
Changi and Sime Road.

Gozel, Jacob Y.

Grand, Meyer

Green, Mr
In charge of Singapore Electric Company.

Greenburg, Essie/Elsie
Wife of Walter. Stanley, Hong Kong.

Grimberg, S.

Guston, Mrs Thelma

Hardoon, Abraham Isaac

Hardoon, Charles Isaac
Hong Kong Police.

Hardoon, Elizabeth

Hardoon, Isaac Silas
Died 1945 as a result of internment.

Hardoon, Jack/Jacob

Hardoon, John

Hardoon, Loulou/Lily
Wife of Isaac, above.

Hardoon, Mattie/Mathilda
(Martin Sugarman)

Hardoon, Sophie

Hawa, Nissim

Hayasad, Simon

Hayim, R.E.J.
Changi and Sime Road.

Herman, Leo
Czech. Died Omata camp, Japan, 4 February 1945.

Hornin, S.

Ingold, Felix

Isaacs, Frederick

Jonah, Elias

Jonah, Solomon
Died.

Joseph, Albert
Changi and Sime Road.

Joseph, Dolly
Died Singapore camp, 11 August 1945.

Joseph, Elias
Died.

Joseph, Sarah and two children
Died.

Kanturkova, Anna
Shanghai.

Kasir, Mr

Kauffman, Stella
28 February 1942. Husband Bernard Maurice KIA at sea, MV *Rooseboom.*

Kelly, D.

Kelly, J

Khatena, S

Kleinmann, D.H.

Klemperer, Dr

Komaroff, Charles David

Komaroff, Elsa
Wife of Charles David, above. Son fought in Battle of Britain, *RAF.*

Kossick, David
From Newcastle. Died Hong Kong. Son Arnold KIA *RAF.*

Laband, Dr J.F.
German dentist at Sandakan.

Landor, Dr Joseph Victor
Registrar, Singapore General Hospital. Changi and Sime Road.

Lazarus, Mr C.
Died 29 June 1942, Toungoo, Burma.

Lelah, Albert
Born 1913. Changi and Sime Road.

Lelah, M.

Levien, John Lawson
Lindenhafen, New Britain. Died aged 57.

Levy, Phillip Montague
Husband of Mabel. Kavieng, Papua New Guinea.

Lewis, Max
Changi and Sime Road.

Macklin, Mrs R.
Hong Kong.

Macmorine, Levers
Changi and Sime Road.

Macmorine, Mrs L.
Changi and Sime Road.

Marudi, Isaac

Manasseh, Alec

Manasseh, Elsie
Wife of Ezekiel. Sime Road.

Manasseh, Ezekiel S.
JP. 17 May 1944. Changi and Sime Road.

Manasseh, Jacob Isaac
Changi and Sime Road.

Manasseh, J.H.
Changi and Sime Road.

Manasseh, Maurice S.
Changi and Sime Road.

Manasseh, Sassoon
Changi and Sime Road.

Mandelson, Jack
Rubber planter. Glasgow/Sumatra.

Messiah, Joe

Meyer, Ken
Changi

Meyer, L.D.
Superintendant of Surveys. Malaya.

Mordecau, K.J.
Photographer. Changi.

Moser, G.A.
Kuala Lumpur.

Moses, E.
Singapore.

Moses, Hannah

Moses, Hayim

Moses, M.H.
Sime Road. 2 September 1945.

Moses, P.D.
Penang

Moses, Robert

Moses, Simon

Murad, W.K.
Changi and Sime Road.

Nabi, David Abdul (aka Davies, David)
Java.

Naftaly family – Harry, Lilly, Emil, Albert and Nancy
Shanghai.

Nassim, Moshe, and family

Nathan, Eze

Nathan, Ivor
Died.

Nathan, Isaac Elias
Died.

Nathan R.

Nathan, Meyer

Nathans, Esther

Nathans, Maximillian
Husband of Esther, above. 9 May 1944. Malaya, Sime Road.

Neufeld, Rebecca
Lunguha camp.

Neufeld
Brother of Rebecca, above.

Panjarg, David
Died at Sime Road.

Pinhas, Abraham
Died.

Raymond/Rayman, Lazarus
Changi.

Robin, Moses Nissim
Died.

Ruben, Sol

Sadka, Sassoon

Saltoon, R.E.
Changi and Sime Road.

Sharbanee, David
Died.

Sharbanee, E.M.
Changi and Sime Road.

Short
Brother of Zaida. Murdered by Japanese.

Short, Hilda
POW Sumatra.

Short, Zaida
POW Sumatra. Sister of Hilda, above. See Lavinia Warner and John Sandilands, *Women beyond the Wire* (Michael Joseph, 1982).

Silvermann/Silberman, Ann Estel
Survivor Giang Bee, rescued HMS *Tapah.* POW Sumatra. Secretary to Sir A. Young during the 1950s Malaya Emergency. Later Mrs Powell of Shrewsbury.

Silvermann, J.C.
Singapore Municipality. Changi and Sime Road.

Silvermann, S.B.
Changi and Sime Road.

Simon, Edward
Died.

Simon, Leon
Singapore Auxiliary Police.

Soloman, Edgar Samuel
Died at sea, *Montevideo Maru,* 1 July 1942.

Soloman, Moses David
Lost at sea, February 1942.

Solomans/Solomon, S.D.
Sime Road. 27 August 1945, Changi, aged 28.

Sternfeld, Dr
Austrian refugee. Sandakan.

Storch, David

Thomas, Silas

Tuck, Sharon Anne
Died aged 10 months. Daughter of T.W. Tuck and Anna Raphael Tuck. Died Shanghai camp.

Tweg, David

Unidentified family of four
Killed in bombing in Joo Chiat Rd, Singapore (evidence of Eze Nathan).

Waltz, Max
Died.

Warmen, Mischa
Boy POW from HMT *Vyner Brooke.* Later lived in Shanghai. See Warner and Sandilands, *Women Beyond the Wire.*

Weinberg, M.
Changi.

Weinberg, Simone
French. Changi.

Wiseberg/Weisberg, Hyman
Changi and Sime Road.

Wuertzburger, Walter
Singapore camp.

Zimmerman, Herbert
Husband of Sonia. Sumatra.

Commonwealth Internees

The following list of Commonwealth Jewish internees of the Japanese, all over South and East Asia, was drawn up by Brian Gander (himself a child internee from 14–17 years old) of ABCIFER (Association of British Civilian Internees in the Far East) from their memorial book which is kept at St Michael's Church in Cornhill, City of London.

For ease, in some cases the family name has been listed and is then followed by the first names. Where an initial only was known, that has been given. Every entry represents a separate individual.

AARON
Abraham
Abraham
Benjamin
David
Eileen Dorothy
Helen Jena Margaret
John David
Jonah
Katherine
Kelly
Margaret Elizabeth
Meda
Meyer G.
Moses
Pauline

ABRAHAM
Abraham
Abe
Abraham S.
Absalom
Aziza
Celia
Diana
Edgar S.
Elias Ezekiel
Ellis Michael
Emma Moses
Ezekiel
Ezekiel
Ezekiel R.D.
Ezra Gertie
Geulah
Hannah
Harry Ezekiel
Isaac S.
Isaac Reuben
Joseph Hayim
Julian
Katie
Katie Rahma
Leah
Louisa
Masie
Marjorie Aziza
Meda
Mildred Patricia
Moshe Hai
Mozelle
Mozelle
Mozelle Ruby
Nachama
Naomi
Patricia
Rachel
Rachel
Rahma Katie
Rebecca
Regina
Reuben David
Reuben
Rosie
Sarah
Sassoon Hai
Seema
Shulamith
Victoria
Yahya

ACKRILL, Reuben

AKERIB
Ezekiel Manasseh
Saul Manasseh

ALBERT
Ezekiel
Moses

BEKHORE
Dinah Diana
Flora
Hawa Hannah
Joseph Elias
Judah Freddy
Lily B.
Sally Sudah
Toba

BENGHIAT
Abraham Jacob
Fortuna
Lillian
Victoria

BENJAMIN
A.
B.
Benjamin David
Benjamin F.
David
Eileen Grace
Elsa Beate
Estelle Rosemarie
Eva Rahma
Isaac
Judah Ephraim
Liza Matilda
Said Ezekiel
Simon H.
Sophie Isaac
Sylvia Simha

BERAHA
Denise
Michael

COHEN
A.A.
Abraham Meyer
Aziza Meyer
E.P.
Elias Meyer
Florence Frances
Hilda
Hilda Katherine
Hubert
Jack Major
Joseph J.
L.
Mordecai
R.
Rachel Grace

DANIEL
Elias Sion
Frederick
Jacob
Katie
Manasseh Sion
Menehem
Meyer Ezekiel
Mozelle Sion
Rachel Sion
Rebecca Remaah
Sion

DAVID
David B.
Dina
Evelyn
Ezekiel Haim
Flossie Flora
Jacob
Joseph Salem
Maurice
Samuel
Samuel Yahya

ELIAS
Abraham
Abraham
Benjamin
Clara Harry
Clara Reuben
Dinah Dorothy
Dinah Nassim
Dinah Ruben
E.
E.
Edward Leopold
Eleanor
Elias
Elias Jacob
Elias nassim
Esther Reuben
Ezekiel
Ezekiel Ezra
Flora
Flora
Flossie Reuben
Grace Nassim
Hannah Katie
Haron
Harry
Harry Moses
Isaac
Isaac Brooke
Isaac Ezekiel
Jacob
Jacob John
Jacob Moses
Jacob Sasoon
Jonah
Jonah Jacob
Joseph Nassim
Joseph Reuben
Lily
Louise
Louise
Manasseh Nasim
Marina
Meda E.
Meda Nassim
Mollie
Moses Joseph
Moses Reuben
Natty
Olivia
Rachel
Rachel
Ramon
Raymond

Reuben Sassoon
Reuben Solomon
Rochelle
Ronald Jack R.
Rosalind
Rosie
Sarah Blanche
Sarah Catherine
Sarah Joseph
Sassoon Jacob
Saul
Sion
Solomon M.
Sophie
Victor
Victoria
Victoria
Violet

EMANUEL
A.
B.
C.
D.
D.A.
F.
G.
J.
J.F.
M.
M.
M.A.
P.

EZEKIEL
Aba Mannesh
Abraham Saul
Alfred
David
David Shalom
Elias Nissim
Ellis
Ellison S.
Elsie N.
Ezekiel Harry
Flossy Abraham
Hannah
Isaac Saul
Isadore Saul
Jacob Harry
Joseph
Julian Saul
Katie Abraham
Manasseh Nassim
Marguerite A.
Maurice
Meyer
Morris Saul
Moselle
Noel Elias
Rachel
Rachel
Rachel Sassoon
Raymond Aaron
Regina
Remah Abraham
Ritchie
Sassoon
Sasoon George
Saul
Sophie Abraham

EZMANIEL, Ari

EZRA
Densil Marcus
Edward
Ellis Isaac
Frederick
Grace K.
Hannah
Irene M.
Joseph
Judah Isaac
Lydia
Messiah
Rachel
Simon

ISAAC
A.
Aaron Jacob
Albert Solomon
Ezekiel
Freddie J.
George J.
Isaac N.
Jacob N.
Joseph N.
Kitty
Leah
Maurice S.
Mooda
Nahom N.
Ose Moses
R.
Rosie
S.

JACOB
Abraham
Amelia Cambell B.
Edward Fred Meyer
Ephraim
Ezekiel
Hil
Isaac E.
Jack
Jean Anne
Maisie
Majia
Matilda Meyer
Meda
Meyer Manasseh
Reuben
Solomon
Sophie
Sophie

JACOBS
Archibald Gordon
Aurelius Reuben
Beatrice Daisy
Bertram William
Ernest
Harry William Frank
Noel Stanley

JOSEPH
A.
A.
A.V.
Abraham
David George O.
E.
Ellis Menashih
Ezekiel
Ezekiel Sassoon
Flora
Flossie
Freddy H.
H.W.
Hannah
Hannah
John
Jonathan David
Joseph
Joseph Edgar
Judah
Manasseh
Mozelle
Nassim
Nassim Edward
Raphael
Raphael Judah
Rebecca
Sassoon
Solomon Saul
T.

JOSHUA
Elias
Ezekiel
Hannah
John
Katie
Moses
Moses
Nellie
Roy Moses
Sarah
Saul
Theresa

JUDAH
Albert Joseph
Cyril Ralph
David Bernard
Diana Evelyn
Elsie Rosemary
Ezekiel Joseph
Flossie
Gloria Jeanette
Joseph Edwin
Judith Jocelyn
Katie
Katieh
Manya
Moses Joseph
Rachel
Rameh Rosamond
Rebecca
Reva
Samson Albert
Valerie Marian
Valerie Nicholevna

KADOORIE
Elly Horace
Lawrence
Michael David
Muriel
Rita Laura

KLIENE
Bert
Dennis
Ron

LEHEM
Fortuna S.
Harry
Mazaltob S.
Morris
Solomon Marco

LELAH
Aaron Sion
Charles Aaron

LEVENSPIEL, Octave

LEVI
Raphael Zabulon
Renee Zabulon

LEVIS
Bonnie
Constance H.
Esther
Evelyn M.
Hazedella Louisa
Hebe
Isaac Abraham
John Hazedel
Lilian
Luela Iris Suzanne
Rachel
Violet H.

LEVISON
A.H.
Corina M.
Harold
Jacob

LEVY
Bertie Saul
David Ezra
Peter Cecil
Raphael Hyeem
Salam H. Rachel
Seemah

MANASSEH
Albert
Daisy
Ezekiel Hai
Ezekiel
Ezekiel
Flora Dolly
Gertrude
Hilda Kitty
Hone
Jackie
Lily

Liza
Louisa
Louisa Jacob
Mary
Mary Jacob
Meda
Moses Maurice
Sally
Sophie Poppy

MASRY
Esther
Joseph Eskel

MENAHEM
David
Elias Saul
Esther Elias
Esther Reuben
Grace Elias
Katie Elias
Katy Reuben
Max
Sarah

MENEHEM
Elia
Manahem
Margillet
Mary
Mesumer

MEYER
A.
E.
Elias David
Esther
Ezekiel
Ezra
Katie Elias
Kiaassen
Kitty
Morris
Mozelle
Rachel
Raymond Elias
Rebecca
Reuben
Sarah
Saul Snr
Sophie Poppy
Toba

MOALEM
Daniel
Emma M.
Girgee
Grace K.E.
Hannah
Joseph
Joseph R.
Louise R.
Reuben Joseph
Simon Joseph
Sophia Shaker
Sophie Irene
William M.

MOSES
Aaron Ezra
Abraham Elias
Abraham Sassoon
Branita Edith
Cyril David
David Hya Ezra M.
Flora
Flossie
Helen Riddel
Hilda
Isaac Hayoo
Lionel
Lucy A.
Moses Simon
Nissim
Polly
Rachel Simon
Rebecca
Rosa Christina
Sarah
Sion Abraham
Solomon Simon
Sybil Simon
Victoria Simon
Violet

NASSIM
Abraham
Freddy Isaac
Menahem Aaron
Moses J.
Nassim A.
Sarah
Sarah Katie

NATHAN
Edward Jonah
Ezekiel E.
Ezekiel J.
Henry Bernard
Joseph A.

NISSIM
Dorothy
Flora
Matthew Albert
Matook Raymond
Nazira
Nissim Marcus M.
Nooriel Frederick
Rachelle S.
Rahma Nissima
Ramona Mamona
Salem Raphael
Saul Gabriel

RAPHAEL
Eleazar Freddy
Ezekiel
Flora
Flossie
Helen Marcus
Louise
Maurice
Meda
Tobs

REUBEN
Elias Y.

Ezekiel
Ezekiel Elias
Florence
Grace
Helen
Helen Ezekiel Hilda
Ezekiel Maisie
Ezekiel Morris
Nina Sassoon
Naim George
R. Saul
Rachel Yahya
Reuben N.
Reuben Saul
Rosalind
Rose Ezekiel
S.M.S.
Saul Felix Carrady
Shirley Solomon Ezekiel
Sophie Ezekiel
Violet Y.
Yahya

SASSOON
Abraham Albert
Abraham Ezra
Adee Lee
David E.
Elias Albert
Ezekiel Albert
Ezekiel Sassoon
Harry Meyer
Isaac Albert
Jacob
Matilda Albert
Sally Albert
Sassoon Albert
Solomon Albert

SAUL
David
Esther
Flora
Flora Ezekiel
Hannah Meyer
Hilda
Hilda Meyer
Joseph Joseph
Joseph Meyer
Katy Meyer
Lily Meyer
Meda Meyer
Naima Meyer
R.V.
Raymon Meyer
Sarah Ezekiel
Schimea Meyer
Sofia Ezekiel
Walter Emil Medhurst

SCHUKER
Hannah Joseph
Jack Meyer J.
Joseph Saul
Sadie Joseph
Saul Ray Joseph
Sydney H.
Thelma Joseph

SHALOME
Ezekiel I.
Gracie Ezekiel
Halka
Joseph Ezekiel
Menahem I.
Mozelle Ezekiel
Rosie Ezekiel
Victoria Ezekiel

SHEMTOB
Diana
Ella
Ezekiel Jacob
Flora
Gladys
Jacob Silas
Maisie
Rachel Jacob
Silas Jacob

Solomon
Aaron David
Abraham
Abraham George
Abraham S.
Betty Said
Daisy Sassoon
David, died Sime Road camp, 27 August 1945, aged 28 years.
David Sassoon
Dinah Said
Eleanor Sassoon
Eleazar B.
Eleazar Hai
Ellis David
Emma Jacob
Ezekiel
Ezekiel Elias
Flora
Flora
Florence
Florence Sassoon
Freddy
Grace Sassoon
Helen
Helen Said
Isaac Jnr
Isaac Snr
Sophie Sassoon
Jacob
Jacob D.
Jacob H.
Jamila
Judah
Juliett
Lulu Moses
Manasseh
Manasseh David
Mary Said
Menahem
Moses E.
Mozelle
Nassim E.
Nissim
Rachel
Rachel
Rachel, Mrs

Rahama
Rebecca
Rosie
Said
Salim
Sally
Sarah
Sassoon Eleazar
Sophie
Sophie
Sophie
Sophie Sassoon
Thelma
Victoria Jacob
Yahya
Yayish

ZAAB, Jacob Hyam

Zacharias
Hans
R.A.
V.D.

Zelensky, Miriam

Zellensky
George Abraham
Jacob

Zimmerman
E.R.
Ernest
Kathleen
Vlademar

Dutch Civilian Internees

Armand (?)

Benda, Harry

Berler, Bruno
Civilian member of a jungle Resistance group, captured and executed near Jakarta in 1944 (Inside Indonesia website, constructed by his granddaughter, Kathryn Pentecost).

Blitz, John
Died in captivity.

Cassuto, Ernest

Cassuto, Rob

Cassuto
(Parents of Rob, above)

Chagoll, Lydia

Cohen, Mrs Ro

De Vries, Vera and parents (father Michael)

Elburg, Rabbi and family
Changi.

Goldberg, Szyman
Famous Polish violinist.

Goldberg, wife of Szyman née Kraus
Pianist.

Grunwald, Moses

Hollaender, Hans
Died.

Hymans, Henri and family

Hymans (father of Henri – died)

Hymans (mother)

Hymans, two siblings of Henri

Joel, Henri F.
Journalist. Taiwan. Probably Jewish.

Joshua-Sasson, Mrs Regina

Lehrer, Shoshana, and parents

Mussry, David, and family

Nussbaum, Mrs Rachel (wife of Chaim) and children Motke, Brenda and Shifra

Rapaport, Eddie

Roos, Betty Struch

Rotenberg, Frank

Silberberg, Mr Hanokh, and parents

Van Zuiden family

Velmans (later Cohen), Anna

Velmans, Joseph and Anna (parents of Lou/Loet, *Dutch Army*)

Wertheim, William

Appendix: From Israel, with Nostalgia

The following article appeared in the The Jakarta Post, *27 September 2010:*

Indonesia and Israel became modern nations at about the same time, but have yet to establish diplomatic relations. The assumption then would be that Israel is the last place one would hear gamelan music. But, four times a year, a group of people gather to reminisce about Indonesia over a meal of Indonesian food. These are members of Yayasan Tempo Dulu, founded in 1996, Jewish men and women from the former Dutch East Indies who now reside in Israel. Currently, the youngest member is 69 while the oldest is 104 (although no longer active due to age), and they come from all over Israel.

Childhood Memories

Shoshana Lehrer's family fled to Indonesia in 1938 from Vienna when she was only four years old. They first settled in Bandung and then moved to Jakarta, sharing a home with other refugees. She does not recollect any particular Jewish community in Bandung or Jakarta, as her family was secular, but what she does remember is how normal life was back then. 'We went to school, bought secondhand books in Pasar Senen, visited the zoo and swam in Cikini [swimming pool].'

Betty Struch Roos, a 73-year-old writer, was born in Padang, West Sumatra, where her father worked as an agricultural engineer. The family moved to Samarinda, East Kalimantan; Struch Roos waxes nostalgic of her idyllic childhood in the jungle: the exotic animals, birds and flowers in the garden, and going to the market with her cook, where the locals sat with their baskets filled with all kinds of food. They had Javanese household staff of different beliefs (traditional spirituality, Christians and Muslims) that lived in a village inside the jungle, and 'during some nights I could hear their gamelan music'.

All around, relations with local Indonesians were excellent. She recalls her father coming home from his long journeys deep inside the jungles of Borneo with beautiful presents from the Dayaks. When her baby brother was born, her parents threw a selamatan [communal feast] for the servants and their families. A big meal was prepared, and the servants bathed the baby in rose water so he would lead a happy life.

Regina Joshua Sassoon, 94, also experienced good relationships. An Iraqi Jewess who was born in Penang, Malaysia, but whose family moved to Indonesia when she was still a baby, she spent much of the first half of her life in Surabaya. 'Life was wonderful. We got along well with the Arab community, the Chinese and our Indonesian neighbours. We were like one big family. We worked together, went to weddings, and we visited each other's homes. The locals loved my [Indonesian] home cooking. Everybody in Surabaya knew the Sassoon family.'

Surabaya had the largest Jewish community in the country, mostly merchants, and it was a close knit and orthodox society. They observed the rites and festivals and had their own synagogue, and a rabbi came from Singapore once a year to perform Kosher slaughtering of cows and chickens. The meat would then be kept in the freezer for six months. The rest of the time they mostly ate a vegetarian diet to keep kosher. Sassoon's father had a fruit shop in front of the famous Toko Nam, selling, among other things, imported apples, cherries and grapes. As the eldest child, Regina ran it after her parents migrated to Israel.

Not everyone lived a privileged life, however. Hanokh Silberberg, born in Surabaya in 1931 to a German travelling salesman, had to juggle school and work when his father died prematurely. Silberberg, called Sastro by his local friends from nearby neighbourhoods, shared living quarters with another family as his mother worked hard to raise two small children. In 1938, his mother had the extra responsibility of trying to get her own mother and sister out of Nazi Germany. At first, the Dutch authorities didn't approve of her attempts, and they threatened to imprison her if she didn't stop insulting Adolf Hitler, as the head of a then-friendly neighbouring country. It took a long time but eventually she succeeded, and the two women joined the Silberbergs in Bandung. The extended family moved to a bigger house on Naripan Road, where they ran a small catering business and little Hanokh was responsible for delivering food containers all over town on his bicycle.

When the War Began

When Germany invaded the Netherlands during World War II, lives among the Jewish community in Indonesia began to unravel. Families were separated. Germans, including German Jews, were locked up on Onrust, one of the islands off mainland Jakarta.

Struch Roos, who was starting kindergarten at the time and knew only the sounds of gamelan and animals, now heard an alien sound: the wailing war siren. She gave away her metallic lunch box to be melted down (she later recalled seeing a fighter plane and was convinced that the plane was partly made from her lunch box). Her father was drafted into the army and was sent to protect the Dutch oil fields in Tarakan.

For the first time, Hanokh Silberberg was confronted with anti-Semitism. In class at his Christian school, one of his teachers asked students of German origin to identify themselves. He kept quiet, and was sent to the headmaster's office – although another classmate who was German and did not identify himself (but wasn't a Jew) was allowed to stay in class. The next day, Silberberg's mother complained about the incident and enrolled her son in a different school, a decision for which the young boy was grateful.

Then, in late 1941, the Hague declared war on Japan. When Japanese forces arrived, they freed the prisoners at Onrust. These released prisoners worked with the Japanese to turn in others to the Kempeitai (Japanese military police). Sassoon recalls a Japanese business acquaintance who went missing for a while, then came back as a high-ranking official in the Japanese army.

In the Camps Foreigner citizens were now being called to register. The women and children were sent to different camps to their menfolk, and some of the camps were actually working prisons. Lehrer recalled that, in her first camp in Tanah Tinggi, the criminals already in prison were responsible for preparing meals for the new internees.

Over the following several years, the internees were moved around a number of camps.

Transport sometimes involved long train rides in shuttered carriages without food or drink. Conditions in the camps kept deteriorating. Internees had their heads shaved because of lice. Cells and barracks overflowed with people, who slept with no more than 30–55 cm of space per head. Food was near inedible: there were stones in the rice and vegetables still had their roots on; portions were meager and dwindling.

Death was no stranger. Unsanitary conditions, malnourishment, hunger and manual labor led to diseases such as malaria and jaundice, but there was no medicine available. At 12, Lehrer weighed only 23 kg. Naturally, smuggling became popular. Struch Roos, one of the smallest and youngest children in her camp, was chosen by a group of resourceful children to crawl outside the camp to barter stolen items for bananas. This was dangerous: if caught, smugglers and their counterparts outside the camps were heavily punished, even killed.

Sassoon once pleaded with a Heiho (member of the Indonesian auxiliary troops) for an egg that she might share with her small children; he gave it to her. But they were caught. The Indonesian man was beaten to death, while she and 13 other women were kept in the so-called 'death room', a small room with a barrel as a toilet, where they lived on water and tapioca bread for a month.

Life in camps revolved around manual labor and roll calls that would last for hours. At first, the Jews were not separated from the other internees. But as time progressed, they were put in separate blocks. In one camp, the Iraqi Jews were separated from the 'white' Jews. The Iraqi Jews got the worst accommodation, with zinc roofs that made the barracks swelteringly hot. Lehrer believes children coped better than adults. Struch Roos also notes that the adults died first. Lehrer kept herself sane by teaching the younger kids to read and write.

Some women tried to survive by other means. Struch Roos recalled a beautiful fellow internee whose mother lived in the room next door. The young woman used to bring her mother extra food every now and then, which the mother vehemently refused because the daughter was a jugun ianfu (comfort woman). The mother used to scream at her daughter, 'I don't want your food, you are a hooker.'

One day during roll call, Struch Roos mentioned this, and the Japanese heard of it. The mother was made to stand in the sun; she was sent back to her room the next morning, bloodied. Struch Roos was only seven at the time, and everybody was reluctant to tell her what a 'hooker' was.

Hanokh Silberberg and his family managed to avoid the camps. Because his father did not have Dutch citizenship, his mother had difficulties with the occupying forces because of the confusion over her status. After an argument between the Japanese officer and his German adviser, Mrs Silberberg was granted a red, white and red badge to indicate her stateless status.

Contact with the camps was impossible. One day, Silberberg was caught smuggling milk and food into the women's camp, and he was taken to the Kempeitai headquarters where he was beaten badly. He was sent home, but had to return the next day for manual labour with a group of people. They all had to do 12-hour workdays for several weeks with frequent beatings, with only one break for food: a meager bowl of rice with broth.

After the War

Chaos followed the Japanese capitulation as Indonesians fought to keep their independence and keep Europeans out (known as the Bersiap period). The internees, when they could finally leave the camps, needed help to find their relatives or housing; young Europeans and Eurasians who had not been interned, including Silberberg, helped the internees. When Sassoon finally returned home, she found her house occupied by other people; her six brothers died young after the war.

When the opportunity arose, most Europeans decided to leave Indonesia, although their reasons varied. Lehrer and her mother left for Holland because her father felt she was better off continuing her studies there. Silberberg's mother remarried with a Dutch Jew who wanted to recuperate in his homeland, so he took his new family with him. Struch Roos' mother, after discovering that her husband was killed in action, decided to take the family back to the Netherlands. She felt they were only in Indonesia for her husband's work.

They were in for a culture shock once they arrived 'home'. They felt unwelcome in a society still recovering from the trauma of World War II. They learned of the fate of Jews who perished in the European camps. Nobody wanted to hear their stories and some, like Lehrer, were forbidden from talking about their experiences in the Japanese camps.

Meanwhile, the Baghdadi Jews stayed on in East Java. All Sassoon's children were born in Surabaya. Mida and Isadora, her daughters, grew up in the 1950s and 1960s when hardly any Europeans remained in the city. Both girls were completely integrated in the local community and they went to Indonesian schools. They studied wayang, were active in sports and traveled extensively in Java with school, family and for tournaments. When Mida – who was scared of sundel bolong (a traditional ghost) as a child – fell ill, [and] she was cured by a local doctor, Sassoon invited 50 beggars for a selamatan gathering.

Again, the good times didn't last.

Promised Land

In 1966, because of the coup in Indonesia in the previous year, Mida and Isadora were sent to live with their grandparents in Tel Aviv. Sassoon and her husband soon followed. None of them spoke Hebrew, and they only had summer clothes. It took Regina Sassoon three years to feel comfortable in her new surroundings.

The European Jews of Indonesia also relocated to Israel for different reasons. Hanokh Silberberg, who was involved with Jewish youth groups after his mother died not long after they settled in the Netherlands, wanted to help build the then fledgling country. Struch Roos, after she was married, wanted to raise her children in Israel. For them, they were finally home, or as Lehrer put it, 'I saw people with faces like my parents.' That isn't to say others understood them, though. Mida couldn't relate to the Iraqi Jews. 'I am an Indonesian Jew', she asserted.

But the Israeli government doesn't consider people from countries not occupied by Germany as victims of the war. Thus, the story of the war in the Pacific was not generally known in Israel. Struch Roos, who had written four books about child war victims, offered the Hebrew translation of her World War II book to the Holocaust Museum, but the director felt her story had nothing to do with the Holocaust.

She left the translation in his office anyway. 'We're all frustrated, because no one cared', says Lehrer. Thus they found refuge in each other at the Tempo Dulu gatherings. Six people act as board members and manage the foundation's finances, administration and activities. The activities include a newsletter, published two or three times a year, and gatherings. The foundation holds four gatherings a year in different Israeli cities. These gatherings, lovingly called kumpulan, consist of meals, lectures about Indonesia and sometimes music or dance performances. The gatherings, transport for members to attend them, and the newsletter are paid through a small membership fee of 85 shekels per year (about Rp 200,000) and money from the Dutch government taken from the bank accounts of Dutch Jews who perished in the Holocaust. The foundation members also held a demonstration outside the Embassy of Japan in Tel Aviv, asking for acknowledgment and compensation from Japan. Nothing came out of it.

Despite the memories of Japanese occupation and Bersiap, all insisted they treasured their time in Indonesia. Everyone still remembers random Indonesian words. Lehrer and Struch Roos were delighted when they overheard Indonesian spoken by tourists in Jerusalem. Lehrer still considers Indonesia as her other home, and her short time in the archipelago helped her appreciate diversity as she was in daily contact with different cultures.

Mida Sassoon, who follows the news in the Surabaya Post, visited her old home in Surabaya three years ago and wants to visit more often. She misses the people, the rice fields, kretek (clove cigarettes) and banyan trees. Regina Sassoon, who was making nasi pecel and emping crackers when interviewed, listed the food she misses (from the East Javanese staple rawon to Balinese spices). She also wanted us to know that she made the best lupis, rujak (spicy fruit salad) and semur (stew) for the gatherings. Most get their food fix imported from the Netherlands. Struch Roos has replicated her childhood garden with its birds, plants and exotic animals in her current home in Jerusalem.

All think Israel–Indonesia diplomatic relations are long overdue. Some mentioned they would visit Indonesia, with other members of Tempo Dulu or with their family, if it happened.

'I would try to be the first ambassador', Shoshanna Lehrer joked. 'Even with a walking stick!'

Index